Green Solvents in Organic Synthesis

Green Solvents in Organic Synthesis

Edited by

Xiao-Feng Wu, Zhiping Yin, Liang-nian He, and Feng Wang

WILEY-VCH

Editors

Prof. Xiao-Feng Wu
Dalian Nat. Laboratory for Clean Energy
Dalian Institute of Chemical Physics
Chinese Academy of Science
Dalian 116023
China

and

Leibniz-Institut Für Katalyse e.V.
Albert-Einstein-Straβe 29a
18059 Rostock
Germany

Prof. Dr. Zhiping Yin
Jiangsu University
School of Pharmacy
Zhenjiang 212013
China

Prof. Dr. Liang-Nian He
Nankai University
College of Chemistry
Institute of Elemento-organic Chemistry
Tianjin 300071
China

Prof. Feng Wang
Dalian Nat. Laboratory for Clean Energy
Dalian Institute of Chemical Physics
Chinese Academy of Science
Dalian 116023
China

Cover Image: © Iana Kunitsa/Getty Images

Library of Congress Card No.: applied for

British Library Cataloguing-in-Publication Data
A catalogue record for this book is available from the British Library.

Bibliographic information published by the Deutsche Nationalbibliothek
The Deutsche Nationalbibliothek lists this publication in the Deutsche Nationalbibliografie; detailed bibliographic data are available on the Internet at <http://dnb.d-nb.de>.

Print ISBN: 978-3-527-35200-5
ePDF ISBN: 978-3-527-84192-9
ePub ISBN: 978-3-527-84193-6
oBook ISBN: 978-3-527-84194-3

Typesetting Straive, Chennai, India

Printed and bound by CPI Group (UK) Ltd, Croydon, CR0 4YY

C9783527352005_270224

Contents

1 Recent Achievements in Organic Reactions in Alcohols *1*
Lan Zhao, Man Zhao, Meng-Ge Wei, Hong-Ru Li, and Liang-Nian He
1.1 Introduction *1*
1.2 Alcohols as Green Solvents *6*
1.2.1 Hydrogenation/Reduction Reaction *6*
1.2.2 Oxidation Reaction *8*
1.2.3 Substitution Reaction *10*
1.2.4 Addition Reaction *11*
1.2.5 Cyclization Reaction *13*
1.2.6 Coupling Reaction *18*
1.2.7 Condensation/Ring Condensation Reaction *21*
1.3 Alcohols as Green Solvents and Catalysts *28*
1.3.1 Addition Reaction *28*
1.3.2 Cyclization Reaction *28*
1.3.3 Coupling Reaction *30*
1.3.4 Condensation Reaction *30*
1.3.5 Metathesis Reaction *35*
1.4 Alcohols as Green Solvents and Hydrogen Donors *35*
1.5 Miscellaneous *39*
1.5.1 Polyethylene Glycol as a Solvent for CO_2 Capture and Conversion *39*
1.5.2 Polyethylene Glycol Radical-Initiated Oxidation Reactions in Compressed Carbon Dioxide *41*
1.5.3 Ring-Opening Reaction *43*
1.6 Summary and Concluding Remarks *45*
Acknowledgments *46*
References *46*

2 Recent Achievements in Organic Reactions in MeCN *51*
Tongtong Xing, Guizhi Zhai, Linna Wu, Xiaofen Wang, and Zechao Wang
2.1 Introduction *51*
2.2 MeCN in Transition Metal-catalyzed Reactions Without Radicals Involved *52*

2.2.1 Transition Metal-catalyzed Addition Reactions in MeCN *52*
2.2.2 Transition Metal-catalyzed Oxidation Reactions in MeCN *56*
2.2.3 Transition Metal-catalyzed Reduction Reactions in MeCN *64*
2.2.4 Transition Metal-catalyzed Substitution Reactions in MeCN *66*
2.2.5 Transition Metal-catalyzed Cyclization Reactions in MeCN *74*
2.3 MeCN in Transition Metal-free Catalyzed Reactions Without Radicals Involved *80*
2.3.1 Transition Metal-free Catalyzed Cyclization Reactions in MeCN *80*
2.3.2 Transition Metal-free Catalyzed Multicomponent Reactions in MeCN *84*
2.3.3 Transition Metal-free Catalyzed C—X Bond Formation in MeCN *87*
2.4 MeCN in C—X Bonds Formation With Radicals Involved *90*
2.4.1 C—C, C—Si Bond Formation in MeCN *90*
2.4.2 C—N, C—P Bond Formation in MeCN *93*
2.4.3 C—O, C—S Bond Formation in MeCN *96*
2.4.4 C-Halogen Bond Formation in MeCN *98*
2.5 Conclusion *102*
References *102*

3 Recent Achievements in Organic Reactions in Bio-based Solvents *107*
Shaomin Chen, Noman Haider Tariq, and Yanlong Gu
3.1 Introduction *107*
3.2 Glycerol *108*
3.3 Polyethylene Glycols (PEGs) *112*
3.4 2-Methyltetrahydrofuran (2-MeTHF) *114*
3.5 Cyclopentyl Methyl Ether (CPME) *117*
3.6 Organic Carbonates *120*
3.7 γ-Valerolactone (GVL) *125*
3.8 Ethyl Lactate (EL) *128*
3.9 Miscellaneous *130*
3.10 Conclusions and Outlook *131*
References *131*

4 Recent Achievements in Organic Reactions in DMSO *137*
Peng Yuan, Jia-Chen Xiang, and An-Xin Wu
4.1 Pummerer-type Activation of DMSO *138*
4.2 Selectfluor-enabled Activation of DMSO *148*
4.3 Activation of DMSO Enabled by Single-electron Transformation *151*
4.4 Electrocatalytic Synthesis Enabled Activation of DMSO *163*
4.5 Photocatalytic Reaction Enabled Activation of DMSO *164*
4.6 DMSO Acts as the Metal Ligand *171*
4.7 Some Special Activation or Usage of DMSO *174*
4.8 Summary and Outlook *181*
References *181*

5 The Use of DMC as Green Solvent in Organic Synthesis *185*
Xinxin Qi and Xiao-Feng Wu
5.1 Introduction *185*
5.2 Organic Reactions in DMC *185*
References *197*

6 Applications of Green Deep Eutectic Solvents (DESs) in Synthetic Transformations *199*
Zhuan Zhang and Taoyuan Liang
6.1 Introduction *199*
6.2 Cross-coupling Reactions in Deep Eutectic Solvents *201*
6.2.1 C—C Bond Couplings *201*
6.2.2 C—N Bond Couplings *210*
6.2.3 C—O Bond Couplings *211*
6.2.4 C—S Bond Couplings *212*
6.3 Oxidation Reactions in Deep Eutectic Solvents *213*
6.3.1 Metal-catalyzed Oxidation *213*
6.3.2 Other Oxidative Processes *214*
6.4 Reduction Reactions in Deep Eutectic Solvents *217*
6.4.1 Metal-catalyzed Reduction *217*
6.4.2 Other Catalytic Reduction *218*
6.5 Cyclization Reactions in Deep Eutectic Solvents *219*
6.5.1 Synthesis of Five-membered Ring *219*
6.5.2 Synthesis of Six-membered Ring *220*
6.6 Condensation Reactions in Deep Eutectic Solvents *221*
6.6.1 DES as the Catalyst/Solvent System for Condensation *221*
6.6.2 Other Catalytic System for Condensation *223*
6.7 Multicomponent Reactions in Deep Eutectic Solvents *224*
6.7.1 One-pot Three-component Reaction *224*
6.7.2 One-pot Four-component Reaction *227*
6.8 Other Organic Reactions in Deep Eutectic Solvents *228*
6.8.1 Isomerization Reaction *228*
6.8.2 Ring-opening Reaction *230*
6.8.3 Esterification Reaction *230*
6.9 Polymerization in DSEs *231*
6.9.1 Anionic Polymerization of Alkenes *231*
6.9.2 Glycolysis and Polyesterification *231*
6.9.3 Oxidative Polymerization *232*
6.9.4 Visible-light-driven RAFT Polymerization *232*
6.10 Conclusion *233*
References *233*

7 Recent Achievements in Organic Reactions in Ionic Liquids *237*
Jianxiao Li and Huanfeng Jiang
7.1 Introduction *237*

7.2 Transition Metal-catalyzed Reactions *238*
7.2.1 Palladium-catalyzed Cascade Cyclization Reaction *239*
7.2.2 Carbonylation Reactions *248*
7.2.3 Sonogashira Coupling Reactions *252*
7.2.4 Suzuki Coupling Reactions *255*
7.2.5 Copper-catalyzed Coupling Reactions *257*
7.3 Outlook *259*
List of Abbreviations *259*
References *260*

8 Recent Achievements in Organic Reactions in Ketones and Esters *263*
Fan-Lin Zeng and Bing Yu
8.1 Introduction *263*
8.2 Organic Reactions in Ketones *263*
8.2.1 Organic Reactions in Cyrene *263*
8.2.2 Organic Reactions in NBP *266*
8.3 Organic Reactions in Esters *268*
8.3.1 Organic Reactions in Organic Carbonates *268*
8.3.2 Organic Reactions in γ-Valerolactone *270*
8.3.3 Organic Reactions in Ethyl Lactate *273*
8.4 Conclusion *275*
References *275*

9 Recent Achievements in Organic Reactions in Polyethylene Glycol *279*
Zhiping Yin
9.1 Introduction *279*
9.2 PEG in Pd-catalyzed Coupling Reactions *280*
9.2.1 Pd-catalyzed C—C, C—Si Bonds Formation in PEG *280*
9.2.2 Pd-catalyzed C—N, C—P Bond Formation in PEG *290*
9.2.3 Pd-catalyzed C—O Bond Formation in PEG *291*
9.2.4 Pd-catalyzed C—B Bond Formation in PEG *291*
9.3 PEG in Cu-catalyzed Reactions *292*
9.3.1 Cu-catalyzed C—C Bond Formation in PEG *292*
9.3.2 Cu-catalyzed C—N Bond Formation in PEG *293*
9.3.3 Cu-catalyzed C—O, C—S, and C—Se Bond Formation in PEG *296*
9.4 PEG in Ni, Ru, and Pt-catalyzed Reactions *299*
9.5 PEG in Organocatalysis Reactions *302*
9.6 PEG in Multicomponent Reactions *304*
9.7 PEG in Cyclization Reactions *306*
9.7.1 Synthesis of Five-membered Ring Systems *306*
9.7.2 Synthesis of Six and Seven-membered Ring Systems *308*
9.8 Conclusion *309*
Acknowledgments *310*
References *310*

10 Recent Advances in Organic Reactions Using Water as Solvent *317*
Chang-Sheng Wang, Qiao Sun, Guowei Wang, Wei He, Zheng Fang, and Kai Guo
10.1 Introduction *317*
10.2 Cross-Coupling Reactions *318*
10.2.1 C–C Cross-Coupling *318*
10.2.2 C–N Cross-Coupling *336*
10.2.3 C–S Cross-Coupling *342*
10.2.4 C–P Cross-Coupling *346*
10.3 C–H Functionalization *347*
10.3.1 C–C Bond Formation *347*
10.3.2 C–N Bond Formation *364*
10.3.3 C–O Bond Formation *367*
10.3.4 C–X Bond Formation *369*
10.3.5 C–H Annulation/Cyclization *370*
10.4 C–C Activation *374*
10.5 C–O Cleavage Reactions *376*
10.6 Oxidative and Reductive Reactions *377*
10.6.1 Electrochemical Oxidation *377*
10.6.2 Reduction and Related Reactions *379*
10.7 Substitution Reactions *381*
10.7.1 Nucleophilic Substitution *381*
10.7.2 Electrophilic Substitution *383*
10.7.3 Radical Substitution *384*
10.8 Addition Reactions *386*
10.8.1 Nucleophilic Addition *386*
10.8.2 Alkene/Alkyne Functionalization via Radical Addition *392*
10.8.3 Alkene or Alkyne Functionalization via Radical-Free Addition *396*
10.8.4 Cycloaddition Reactions *399*
10.9 Cyclization or Annulation Reactions *403*
10.9.1 Radical-Free Cyclization/Annulation *403*
10.9.2 Radical Cyclization *406*
10.10 Multicomponent Reaction (MCR) *410*
10.11 Domino/Tandem/Cascade Reactions *417*
10.11.1 Chemo-Domino/Tandem/Cascade Reactions *417*
10.11.2 Chemoenzymatic Reactions *422*
10.12 Rearrangement or Insertion Reactions *425*
10.12.1 Rearrangement Reactions *425*
10.12.2 Carbene Insertion/Transfer Reactions *429*
10.13 Amide Condensation Reactions *431*
10.14 Summary and Conclusions *435*
Acknowledgments *435*
References *435*

Index *443*

1

Recent Achievements in Organic Reactions in Alcohols

Lan Zhao[1], *Man Zhao*[1], *Meng-Ge Wei*[2], *Hong-Ru Li*[2], *and Liang-Nian He*[1]

[1] *Nankai University, College of Chemistry, State Key Laboratory and Institute of Elemento-Organic Chemistry, 94 Weijin Road, Nankai District, Tianjin, 300071, China*
[2] *Nankai University, College of Pharmacy, No 38 Tongyan Road, Jinnan District, Tianjin, 300353, China*

1.1 Introduction

Almost all manufacturing and processing industries now rely on the extensive use of solvents, and conventional synthesis depends heavily on environmentally incompatible organic solvents [1]. Most often, the solvents used in organic synthesis are volatile, so they can be removed from the reaction mixture by simple evaporation. The widespread use of these volatile organic solvents raises environmental concerns due to their intrinsic properties. With the growing awareness of the impact of solvents on environmental pollution, energy use, on-air quality, and climate change, sustainable solvents are a topic of increasing interest to the science community and the chemical industry [2].

Alcohols are a more desirable class of green solvents used in a wide range of organic reactions. In the last 15 years, alcohols have started to attract attention as alternatives to traditional organic solvents. These alcohols are able to work as solvents in general and include monohydric alcohols (methanol, ethanol, isopropanol, etc.), dihydric alcohols (ethylene glycol), tertiary alcohols (glycerol), fluorinated alcohols (2,2,2-trifluoroethanol [TFE], hexafluoroisopropanol [HFIP]), perfluoro *tert*-butyl alcohol [PFTB]), and polymers (polyethylene glycol [PEG], polypropylene glycol [PPG]), as shown in Scheme 1.1. Alcohols are typical proton-polar solvents and are widely used as inexpensive general solvents. The physical and chemical properties of some alcohols are depicted in Table 1.1 [3].

Nowadays, numerous solvent selection guides have emerged from the pharmaceutical industry, such as those from Pfizer [4], GlaxoSmithKline (GSK) [5], and Sanofi [6]. In this regard, Table 1.2 lists several commonly used alcohols evaluated by selected guidelines in Table 1.2 [2]. As a consequence, most of the several short-chain aliphatic alcohols appear to be the preferred solvents from a green chemistry perspective. High-boiling alcohols, glycerol, PEG, and PPG are also considered green solvents for various reactions. Furthermore, fluorinated alcohols are frequently used as solvents in chemical reactions due to their unique properties.

Green Solvents in Organic Synthesis, First Edition.
Edited by Xiao-Feng Wu, Zhiping Yin, Liang-nian He, and Feng Wang.

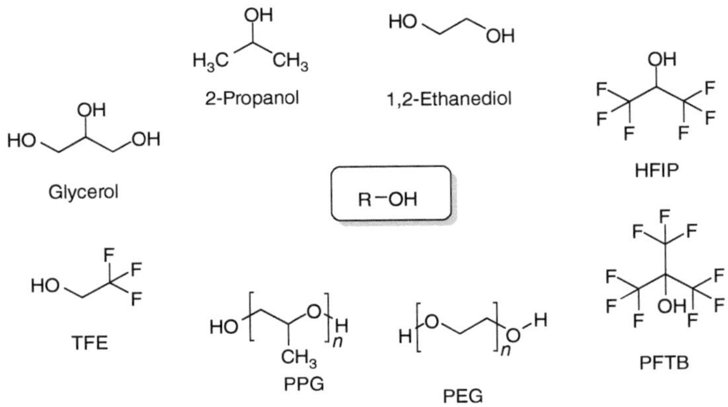

Scheme 1.1 Structure of alcohols.

Table 1.1 Physical and chemical properties of alcohols [3].

Alcohol properties	Methanol	Ethanol	2-propanol	Ethylene glycol	PEG-400	Glycerol	TFE	HFIP
Melting point (°C)	−98	−114	−89.5	−13	64–66	18	−44	−4
boiling point (°C)	65.4	78	82.5	197.6	>250	290	73	58
Relative density (g/mL)	0.791	0.789	0.786	1.113	1.27	1.25	1.391	1.596
Dielectric constant	32.7	24.6	19.9	37.7	—	42.5	24.5	16.7
Dipole moment (D)	1.69	1.7	1.65	2.31	—	2.56	0.61	0.61
Viscosity (mpa·s)	0.59 (20 °C)	1.07 (20 °C)	2.43 (20 °C)	25.66 (16 °C)	—	$1.41*10^3$ (20 °C)	—	—
Flash point (°C)	11.1	12.0	11.7	111.1	270.0	177.0	29.0	4.4
Surface tension (mN/m)	22.7	22.3	22.6	46.5	—	63.4	16.1	14.1
pKa (25 °C)	15.2	15.9	17.1	14.2	—	14.2	12.4	9.3
Polarizability (π*)	0.63	0.54	0.48	0.92	0.91	1.04	0.73	0.65
Hydrogen-bond donor (α)	0.93	0.83	0.76	0.90	0.31	0.93	1.51	1.96
Hydrogen-bond acceptor (β)	0.62	0.77	0.95	0.52	0.75	0.67	0.18	0.03
Polarity	12.3	8.8	6.1	11.0	—	12.1	10.2	11.8
Hydrogen bonding forces	22.3	19.4	16.4	26.0	—	29.3	—	—

Table 1.2 Guidelines for solvents for common alcohols.

Solvent	Pfizer[a)]	GSK[b)]	Sanofi[c)]
Methanol	Preferred	Some issues	Recommended
Ethanol	Preferred	Some issues	Recommended
1-Propanol	Preferred	Some issues	Recommended
2-propanol	Preferred	Some issues	Recommended
1-Butanol	Preferred	Few issues	Recommended
2-Butanol	–	Few issues	Recommended
tert-Butanol	Preferred	Some issues	Substitution advisable
Ethylene glycol	Usable	–	Substitution advisable
2-Methoxyethanol	–	Major issues	Substitution requested
Benzyl alcohol	–	–	Substitution requested

a) Selection guide by Pfizer.
b) GSK's solvent selection guide.
c) Sanofi's solvent selection guide.

Monohydric alcohols are organic compounds containing one hydroxyl group, including methanol, ethanol, and isopropanol. They are the most commonly used solvents in the thermochemical processing of lignocellulose due to their easy recovery and low cost. These alcohols have similar solvent properties, such as solvent strength, dielectric constant, critical point, and hydrogen supply capacity. They are usually soluble in polar solvents such as water, ether, and acetone but not in non-polar solvents such as petroleum ether. The hydroxyl groups in monohydric alcohol molecules give them polarity and can form hydrogen bonds with other polar molecules, resulting in high boiling points, melting points, and surface tension. These alcohols can be obtained from various sources such as fossil fuels, biomass, and waste. Methanol can be used in the production of other chemicals such as formaldehyde, methyl methacrylate, and dicarboxylic acid. Ethanol has wide applications in pharmaceuticals, beverages, coatings, cosmetics, and spices. Ethanol can be used to produce chemicals such as ethyl acetate, vinyl acetate, and acetaldehyde. Isopropanol can be used to produce plastics, paints, coatings, and solvents. It can also be used to prepare other chemicals, such as acetone and methyl isopropanol. They have low toxicity and are important clean fuels that can replace traditional fuels [1].

Ethylene glycol, a colorless, odorless, relatively non-volatile, low hygroscopic, and low-viscosity liquid, is the simplest representative of 1,2-diols. Owing to its unique structure, that is, two hydroxyl groups (OH) at adjacent positions along the hydrocarbon chain, allows it to engage in reactions such as esterification, dehydration, oxidation, and halogenation. In terms of solubility, ethylene glycol is completely miscible with numerous polar solvents (such as water, alcohols, glycol ethers, and acetone), but only slightly soluble in nonpolar solvents such as benzene, toluene, dichloroethane, and chloroform. Other than that, it is difficult to crystallize, but becomes highly viscous supercooled mass that solidifies to deliver a glassy substance upon gradual cooling. With regard to toxicity, ethylene glycol is inherently low in toxicity, but can yield toxic metabolites (Oral-rat LD_{50}: 4700 mg/kg; Oral-mouse LD_{50}: 5500 mg/kg) [7].

Glycerol is a ternary alcohol derived from biomass, which can be obtained from biodiesel production as a coproduct at a low price [8]. It has the same solubility as polar solvents such as water, DMSO, and DMF. In addition, it is immiscible with some common organic solvents such as hydrocarbons, ethers, and esters, which allows the reaction products to be separated by simple liquid–liquid phase extraction when using glycerol as a solvent. Glycerol is nonvolatile at atmospheric pressure and has a high boiling point (290 °C), so the reaction products can be separated using distillation techniques. Compared to most organic solvents, glycerol is a nontoxic (LD_{50} [rat oral] = 12 600 mg/kg), biodegradable, and nonflammable solvent that does not require special handling precautions or storage [9]. Glycerol has a greater polarity value ($\pi^* = 1.04$) than other alcohols, as listed in Table 1. Glycerol molecules contain three hydroxyl groups capable of forming hydrogen bonds with reactants or intermediates. The hydrogen bonding force (29.3) is relatively high compared to other alcohols, as shown in Table 1.1. This strong hydrogen bonding network presumably plays a significant role in facilitating organic synthesis. These characteristics of glycerol render it widely used as an environmentally benign solvent in organic chemistry. In this context, glycerol mainly has four applications in organic reactions: (i) working as a promoting medium for organic synthesis, (ii) serving as a solvent to improve the product selectivity, (iii) enhancing the separation of products, and (iv) facilitating catalyst design and the recovery of catalytic systems. However, there also are some limitations of using glycerol as a solvent. First, its high viscosity can cause poor mass diffusion of the reactants in the medium, eventually leading to reduced reactivity. Second, the three hydroxyl groups of glycerol may provide coordination sites that may cause deactivation of specific organometallic complexes [10].

PEGs are also considered suitable alternatives to conventional organic solvents due to their attractive physicochemical properties (e.g. chemically inert, thermally stable, and nontoxic). In particular, PEGs are stable even at high temperatures up to 150–200 °C. [11]. Low-molecular-weight PEGs, such as PEG-300, have a greater polarity value ($\pi^* = 0.91$) than methanol (0.63) as listed in Table 1.1. The hydrogen bond accepting ability ($\beta = 0.75$) is similar to that of methanol (0.62), while the hydrogen bond donating ability ($\alpha = 0.31$) is less than that of methanol (0.93) [11]. As a hydrophilic oligomer, PEG is readily soluble in water and most organic solvents, such as toluene, methylene chloride, alcohols, and acetone; however, it is insoluble in aliphatic hydrocarbons such as hexane, cyclohexane, or ether. Liquid PEG is proportionally miscible with water, and solid PEG with a higher molecular weight is also extremely soluble in water [12]. For example, PEG-2000 has a solubility of about 20% in water at 60 °C. Water-soluble PEG is considered a cosolvent for organics in water, because it can contribute to a significant reduction in the polarity of aqueous solutions, which can be used as an alternative to other solvents such as ionic liquids, supercritical carbon dioxide, and micellar systems for reactions such as substitution, reduction, oxidation, and conventional or free-phase transfer catalysts (PTCs). PEGs as linear counterparts of crown ethers, are considered "host" solvents with regard to their ability to coordinate with metal cations to form complexes that transfer such cations from the aqueous phase to the organic phase. In order to remain electrically neutral, the PEG–metal complex has to carry the equivalent anion into the organic phase, enabling the anion to react with the organic reactants, so that PEGs can act as PTCs. Several factors influence the catalytic activity such as PEG molecular weight, chain end effects, and the nature of the relevant cations and anions [13].

In coordination with metals, PEGs have a more flexible structure compared to crown ether. Crown ether metal complexation depends on the size of the crown center cavity (fixed size). Whereas PEGs show a more flexible selectivity in the binding of metal cations of different sizes, as a result of the variation in the PEG helical conformation. Hence, PEG can coordinate metal cations with different sizes. Overall, PEG offers a more flexible structure for metal cation complexation. Besides, PEG complexation can also bring significant cost savings in extraction compared to crown ether complexation [14].

PPG is a viscous liquid with negligible vapor pressure, which is stable under normal reaction conditions and is easily recoverable. The majority of commercially available PPG, in contrast to PEG, has a molecular weight ranging from 250 to 4000 Da and is a viscous liquid. While PPG has an inverse relationship with temperature and solubility, low-molecular-weight PPG-250 Da to PPG-425 Da are soluble in water, and such solubility rapidly decreases as the molecular weight increases. These physical characteristics will restrict the use of PPG as a reaction medium because it is considerably more hydrophobic than PEG [15].

The use of fluorinated solvents as chemical reaction media in place of traditional volatile organic solvents has increased significantly. Fluorinated alcohols have unique physicochemical properties, including lower boiling point, higher melting point, high polarity, high hydrogen bond donor capacity, high ionizing power, and solvation capacity [16]. Among several fluorinated alcohols, the most commonly used and cheapest ones are TFE and HFIP, which have reached commercial scale and are often used as cosolvents, catalysts, or additives in addition reactions, condensation reactions, and ring-opening reactions [17]. Reactions in fluorinated solvents are usually highly selective, with no other effluents, so the product is easily separated and the solvent can be recovered by distillation for reuse. Moreover, due to the weak acidity ($pK_a = 12.4$ for TFE, $pK_a = 9.3$ for HFIP) and strong ionization capacity of fluorinated alcohols, the use of Brønsted or Lewis acids is avoided in the reactions. Due to the low boiling point and low viscosity, reactions complete faster. In addition, fluorinated alcohols have a significant stabilizing effect on cationic substances because of their high polarizability and low nucleophilicity. This highly ionized solvent can stabilize the polarity transition state of the reaction well. Despite these advantages, fluorinated solvents are currently not widely used commercially, probably due to disadvantages such as high cost, environmental persistence, and biological half-life time [3c, 18].

In this chapter, alcohols applied to various organic reactions are introduced in detail according to their roles in organic reactions, mainly including methanol, ethanol, ethylene glycol, glycerol, TFE, HFIP, PFTB, PEG, and PPG. Some alcohols are used not only as solvents but also as catalysts or hydrogen donors, mostly for reduction reactions, coupling reactions, condensation reactions, and some other reactions. In particular, PEG plays an essential role in CO_2 capture and conversion due to its unique properties. In addition, fluorinated alcohols (TFE and PFTB) also have significant applications in ring-opening reactions.

1.2 Alcohols as Green Solvents

Based on Green Chemistry principles, alcohols are less hazardous, and more environmentally benign green solvents, mainly due to their low toxicity, non-flammability, non-mutagenicity, and wider availability. In addition, they are cost-effective, easy to handle, and recyclable [19]. As versatile green solvents, alcohols have applications in many types of reactions, including hydrogenation/reduction reactions, oxidation reactions, substitution reactions, addition reactions, cyclization reactions, coupling reactions, and condensation/ring condensation reactions. The properties of alcohols enable the catalytic system or solvent to be recovered after reaction and reused several times.

1.2.1 Hydrogenation/Reduction Reaction

Methanol plays an important role in improving enantioselectivity. The structure derivatives are synthesized through asymmetric hydrogenation in an atomically economical and clean way. Wu et al. successfully achieved the first general and efficient enantioselective Ru-Synphos-catalyzed asymmetric hydrogenation of readily available trisubstituted enamides derived from 3-chromanones (Scheme 1.2) [20]. Polar solvents are found to give higher enantioselectivity than nonpolar solvents, and methanol is the most effective one. This solution is atomically economical and clean, with high chemical yield and enantiomeric excesses up to 96%.

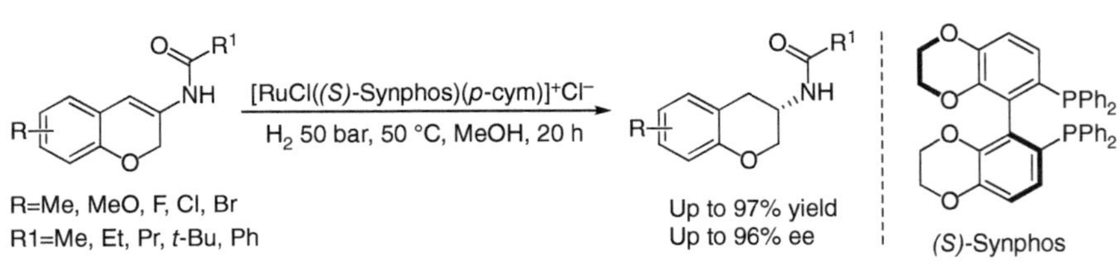

Scheme 1.2 Ruthenium-catalyzed asymmetric hydrogenation of chromanone.

PEG has shown the ability to microregulate fine-tuned reactions via supramolecules, which has been reported in asymmetric hydrogenation reactions and the recycling of chiral metal catalysts, especially as a host molecule participating in the modification of the reaction outcome. For example, the reaction of quinoline with chiral cationic ruthenium diamine complexes in PEG has been employed in asymmetric hydrogenation reactions (Scheme 1.3) [21]. In oligoethylene glycols (OEGs) and PEGs with different molecular weights, the Ru catalyst loses reactivity giving a wide variation in the reactivity of the hydrogenation reaction, which is mainly attributed to the encapsulation of quinoline salts by PEG or long-chain OEG molecules. The hydrogenation of quinoline is "turned on" by the addition of a small quantity of MeOH.

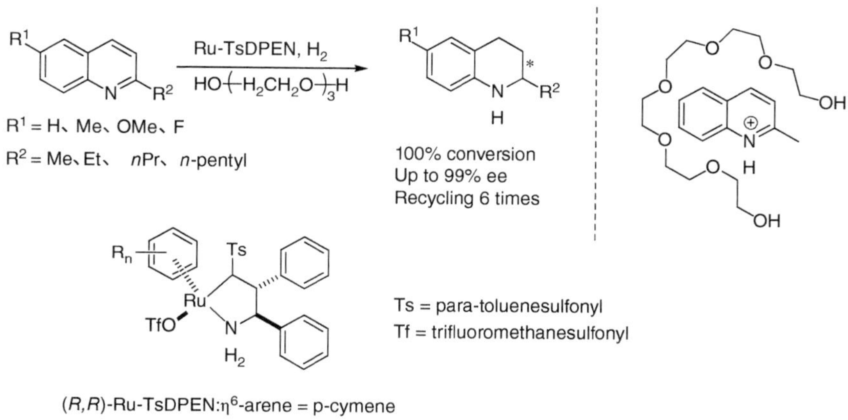

Scheme 1.3 Asymmetric hydrogenation of quinoline derivatives catalyzed by a chiral cationic Ru catalyst in 3-OEG.

Evidence from mass spectrometry and control experiments indicates that long-chain OEG molecules encapsulating quinoline salts through supramolecular interactions are the main reason for this switchable hydrogenation reaction. In addition, an asymmetric hydrogenation of 2-substituted quinoline derivatives performs in triethylene glycol (3-OEG), resulting in excellent reactivity and enantioselectivity (up to 99% ee) for 1,2,3,4-tetrahydroquinoline. In addition, the Ru catalyst is easily recovered and reused in pure 3-OEG and biphasic 3-OEG/hexane systems without any significant loss of reactivity and enantioselectivity.

Similarly, polystyrene-stabilized iridium (Ir@PS) nanoparticles (NPs) have been developed as a heterogeneous catalyst and characterized by infrared spectroscopy (IR), UV–Vis ultraviolet–visible (UV–Vis), energy dispersive X-ray (EDX), X-ray diffraction (XRD), scanning electron microscope (SEM), and transmission electron microscope (TEM) studies. The developed Ir@PS catalyst exhibits outstanding reactivity for the chemoselective and regioselective controlled hydrogenation of functionalized nitroarenes to the corresponding *N*-arylhydroxylamine, utilizing hydrazine hydrate as the reducing source and PEG-400 as the green solvent (Scheme 1.4) [22]. In this method, a wide range of substrates can be adapted to a wide range of reduction-compatible functional groups. The reactions are conducted at 85 °C or ambient temperature and completed in 5–80 minutes with yields as high as 87–99%. The catalyst can be easily filtered out of the reaction mixture and reused.

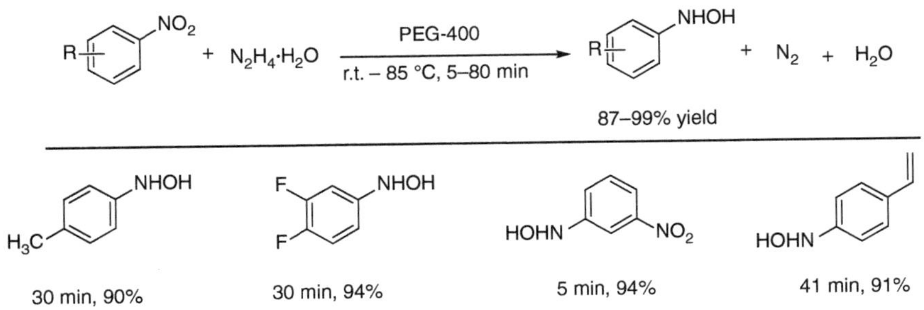

Scheme 1.4 *N*-arylhydroxylamine synthesis via transfer hydrogenation reaction in PEG-400.

Based on these results, a rational mechanistic pathway for the semi-hydrogenation of nitroaromatics to *N*-arylhydroxylamines is mapped. Initially, hydrazine hydrate is chemisorbed on the catalyst surface I, which in turn participates in the H-bonding with the nitro group of the substrate via intermediate II. The N—H bond of the hydrazine hydrate might also be weakened and its disintegration into H_2 and N_2 facilitated by the non-covalent interaction between the hydrazine hydrate and the nitroarenes across the catalytic surface. Together with the temporary nitrosoarene intermediate III, the free hydrogen is chemisorbed onto the heterogeneous surface, enabling the transfer hydrogenation process via intermediate IV to produce *N*-arylhydroxylamine as the final product (Scheme 1.5).

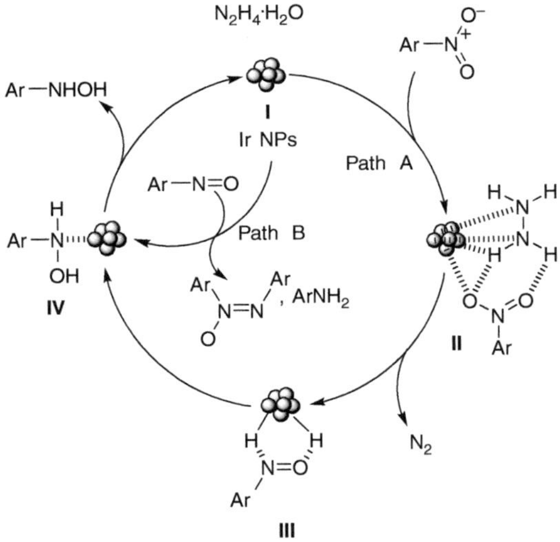

Scheme 1.5 Proposed mechanism for semi-hydrogenation of nitroarenes-catalyzed by Ir@PS in PEG.

1.2.2 Oxidation Reaction

Alcohols have certain applications as green solvents in oxidation reactions. In this context, Jacob et al. developed a method for the dehydrogenation and oxidation of thiols in glycerol by microwave radiation to obtain the corresponding disulfides (Scheme 1.6) [23]. The reaction is fast with simple work-up procedure. Once the reaction is complete, the product can be obtained by simple extraction; thus, the glycerol can be easily recovered and used for next-run reactions. This method is able to extend to the application of green solvents in oxidation reactions.

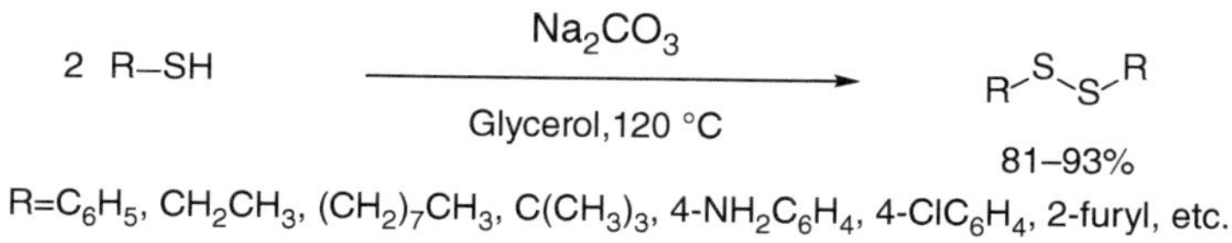

Scheme 1.6 Glycerol as a green solvent in the oxidation of thiols to disulfides.

PEG possesses an acyclic polyether with flexible polyethylene oxide chains, which allows it to have an ionic dipole interaction with metal salts that can effectively immobilize and stabilize metal catalysts that can provide a more active oxidant by binding oxygen to the peroxide species. In the highly selective conversion of various aromatic and aliphatic sulfides to the corresponding sulfoxides, an oxidation reaction regime consisting of an inexpensive and easily available $Fe(acac)_2$ as the catalyst and molecular oxygen as an oxidant in PEG is regarded as an effective method for the highly selective oxidation of sulfides to sulfoxides (Scheme 1.7) [24]. High conversions (>99%) and excellent chemoselectivity of up to 94% at 100 °C can be achieved. This protocol represents an environmentally friendly access to sulfonation oxidation with potential economic performance. It is worth noting that PEG-1000 may provide an electron-rich environment, thereby playing a crucial role in stabilizing the *in situ* formation of the Fe (IV)-oxo species, which are believed to be responsible for sulfide oxidation. Therefore, PEG facilitates the sulfide oxidation to sulfoxide.

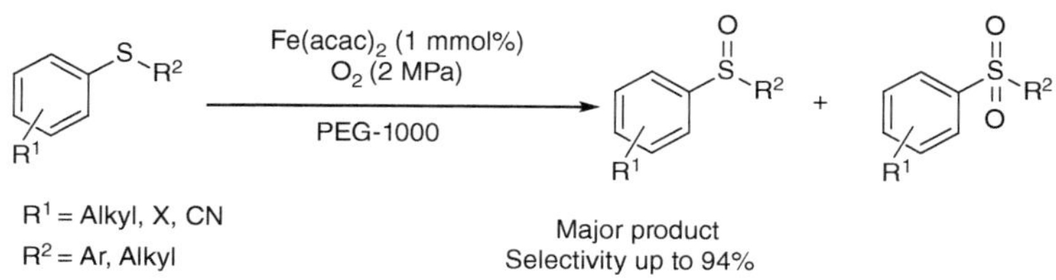

Scheme 1.7 Selective oxidation of sulfides to sulfoxides in PEG-1000.

Furthermore, the oxidation of alcohols to carbonyl compounds also is of great importance in synthetic organic chemistry owing to their practical utility in various products such as pharmaceuticals, agrochemicals, and fragrances. The green catalytic regime using heterogeneous nanocrystalline TiO_2(TiO_2-np) as a multiphase catalyst, H_2O_2 as a reagent, and PEG as a green solvent to perform the oxidation of *para*-alcohols to ketones would be a very economical and simple approach to convert *para*-alcohols into the corresponding ketones (Scheme 1.8) [25]. As a result, this protocol is eco-friendly with simple posttreatment and mild reaction conditions. Over and above this, the reactions are completely clean, consisting of green solvents, green reagents, and catalysts, as well as only water as a by-product.

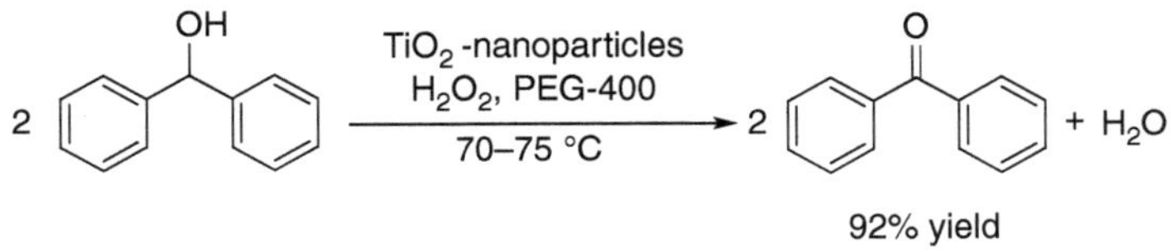

Scheme 1.8 Model reaction for aromatic alcohol oxidation to ketone using TiO_2 nanoparticles in PEG-400.

Besides, iodoaromatics are recognized as essential synthetic intermediates and precursors in organic transformations that can undergo oxidative addition and are also a widespread component of biologically active molecules, including pesticides and drugs. Given the outstanding and versatile properties of iodoaromatics, efforts have been made to optimize their preparation in the field of scientific research and industrial manufacturing. The high-efficiency iron-catalyzed oxidative iodination of electron-rich aromatics in the presence of oxygen with the aid of an environmentally benign and reversible acidic system of *in situ* acidic carbon dioxide/glycol is a very typical instance (Scheme 1.9) [26]. Notably, in the absence of any conventional acid additives or organic solvents, moderate-to-high isolation yields of aryl iodides (up to 97%) with considerable regioselectivity are obtained when $Fe(NO_3)_3{\cdot}9H_2O$ is used as a catalyst with molecular iodine under 1 MPa CO_2 conditions.

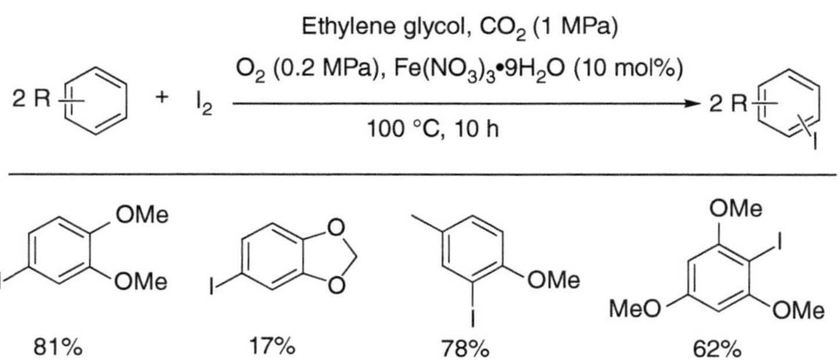

Scheme 1.9 Aerobic oxidative iodination of electron-rich aromatics in ethylene glycol.

1.2.3 Substitution Reaction

Alcohols are also commonly used as green solvents in substitution reactions. For example, the use of glycerol as a solvent and H_3PO_2 as a reducing agent can effectively promote the reaction of 2-chloropyridine with organylselenols generated *in situ* by diorganyl diselenides to afford the corresponding 2-organoselenopyridine, as depicted in Scheme 1.10 [27]. Significantly, the product is easily extracted with hexane/ethyl acetate = 95/5. The glycerol/H_3PO_2 system can be directly recycled more than four times after drying under vacuum, and no additional reducing agent is needed for the reaction.

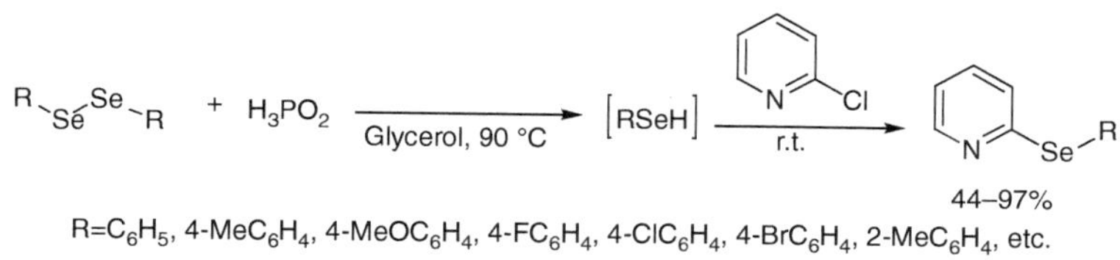

Scheme 1.10 An efficient system for the synthesis of 2-organylselanyl pyridines using glycerol/hypophosphoric acid as solvent-reducing agent.

1.2.4 Addition Reaction

Methanol can significantly increase the reaction rate in the hydrogenation of olefins. A gold(I) carbene complex as a catalyst was first used in the intramolecular *exo*-hydroamination of unactivated alkenes (Scheme 1.11) [28]. The catalyst system has high activity in the external hydroamination of *N*-4-pentenyl or *N*-5-hexenyl urea, forming the corresponding nitrogen-containing heterocyclic ring with excellent yield, good regioselectivity, and 5.5 : 1 non-steroidal selectivity. As for the selection of solvents, experiments show that the hydroamination rate can be increased by two times when dioxane is substituted by methanol.

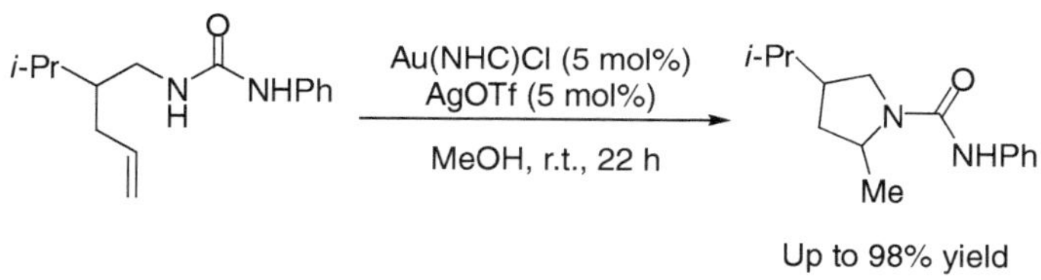

Scheme 1.11 Room-temperature intramolecular *exo*-hydroamination of *N*-alkenyl urea catalyzed by a gold(I) *N*-heterocyclic carbene complex.

Organic solvents are widely used in various chemical processes and gave rise to several serious environmental challenges. Hence, social and economic needs for sustainable development have prompted the scientific community to explore alternative reaction media for replacing solvents that are volatile, spontaneously combustible, toxic, and difficult to recover. For solving the problems caused by volatile organic solvents, PEG has been subjected to an increasing number of scientific studies in view of the nontoxic and nonhazardous properties, in addition to the fact that the chemical reactions in PEG have different thermodynamic and kinetic behavior than those in conventional solvents. More frequently, PEG is typically employed as a green reaction medium due to its low cost, thermal stability, and biodegradability. For example, in the research and development of antitumor-active drugs, 7-[4-alkyl- and (hetero)aryl-1*H*-1,2,3-triazol-1-yl]thio[3,2-*b*]pyridines **3a-f** are afforded in high yield through the formation of the azide **2** via S_NAr of chlorinated heteroaromatic compound **1** with NaN_3, followed by CuAAC reaction with different alkynes in the eco-friendly solvent PEG-400 (Scheme 1.12) [29].

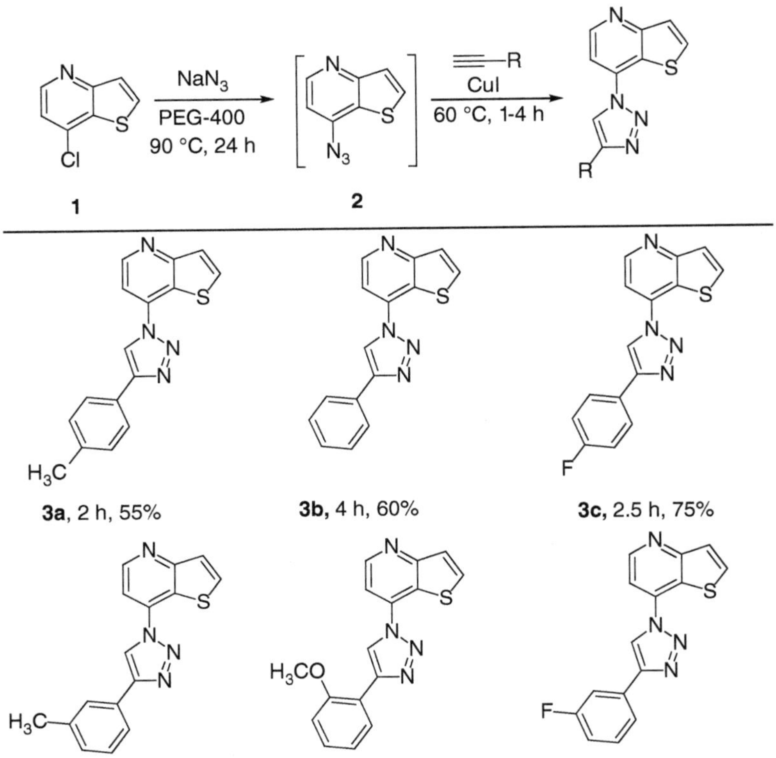

Scheme 1.12 Synthesis of azides from chlorinated heteroaromatic compounds in PEG-400 under the optimal conditions.

Among all the developed methods, the application of PEG-400 as a recoverable reaction medium in asymmetric organocatalytic Michael addition to trans-β-nitrosostyrene is a good model (Scheme 1.13) [30]. A highly stereoselective organocatalytic Michael addition of aldehydes to trans-β-nitrosostyrene using PEG as a recoverable solvent medium has been proposed, as depicted in Scheme 13. The scope of this organocatalytic system is demonstrated by the formation of several Michael adducts with good yields and stereoselectivity. Furthermore, application of this new scheme to acetaldehyde, simple form syntheses of (*R*)-primethalin, (*R*)-phenibut, and (*R*)-baclofluorene have been disclosed with good yields and excellent enantioselectivity. Faster and higher stereoselective reactions could be achieved compared to other environmentally friendly solvents. This innovative and promising result might be attributed to the host–guest complex, PEG–nitrosostyrene, facilitating the nucleophilic addition of enamines formed from the aldehyde and the organocatalyst. Moreover, it is noteworthy that the PEG is efficiently and completely recovered and reused after the extraction of the Michael addition product with ether.

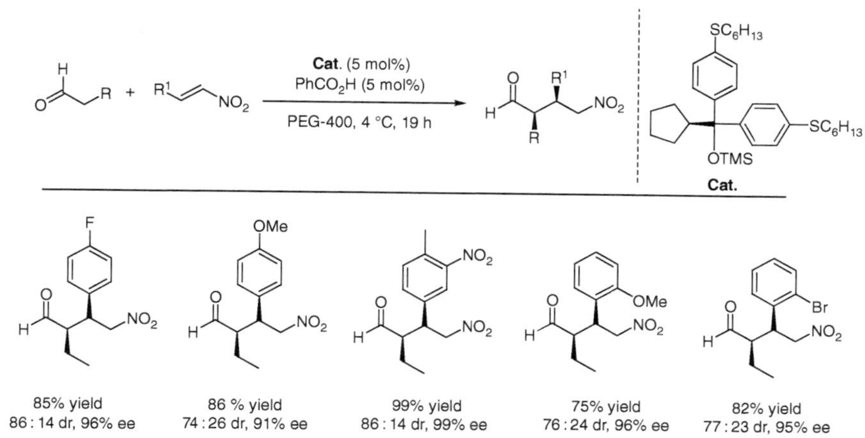

Scheme 1.13 Application of PEG-400 in asymmetric organocatalytic Michael addition to trans-β-nitrosostyrene.

In addition to mono- and diols, glycerol can also be utilized as an environmentally friendly reaction medium in addition reaction for the fast and chemically selective addition of aryllithium reagents to nitriles, to access a range of bis(aryl) ketones at room temperature without an inert atmosphere (Scheme 1.14) [31]. The addition reaction occurs under nonhomogeneous conditions due to the insufficient solubility of the nitrile in glycerol. The nonhomogeneous conditions and the intermolecular hydrogen bonding formed by glycerol may be responsible for favoring the successful nucleophilic addition of organolithium reagents to the nitrile rather than competitive hydrolysis. The application of glycerol in organic lithium chemistry has opened a prospect for the development of more sustainable, air- and moisture-resistant, metal-mediated organic synthesis.

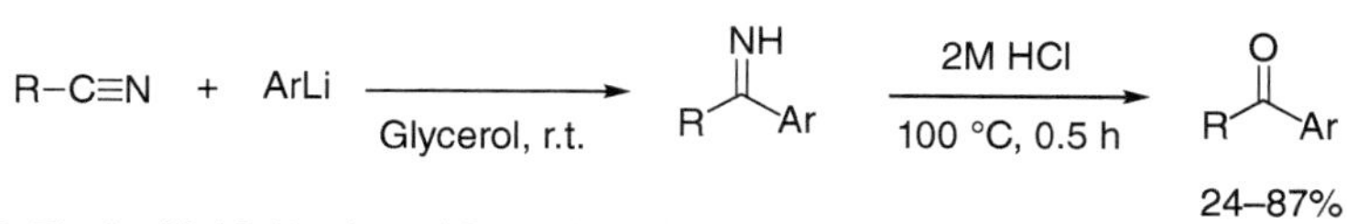

R=Ph, (*m*-Me)C_6H_4, (*o*-Me)C_6H_4, (*o*-F)C_6H_4, (*p*-OMe)C_6H_4, (*p*-Br)C_6H_4, *n*-hexyl, etc.
Ar=(*m*-Me)C_6H_4, (*m*-F)C_6H_4, (*o*-OMe)C_6H_4, [3,5-$(CF_3)_2$]C_6H_3

Scheme 1.14 Synthesis of aromatic ketones by chemically selective addition of aryl lithium reagents to nitriles.

1.2.5 Cyclization Reaction

Alcohols are widely used as green solvents in cyclization reactions, including ethylene glycol, PEG, glycerol, and TFE.

These reactions are grouped into numerous biologically essential compounds such as anticancer, anti-HIV, antibacterial, antimicrobial, and antiallergic drug molecules. For instance, 4-aryl-NH-1,2,3-triazoles, which belong to the group of 1,2,3-triazoles, have gained considerable attention in the field of pharmaceutical industry. Taking the synthesis of polypyrrole (PPy)/Cu(II) nanocomposite as an efficient multiphase catalyst for the synthesis of 4-aryl-NH-1,2,3-triazole as an example, the catalyst together with the solvent PEG-400 can be easily recovered and reused for four consecutive cycles, thus maintaining the consistency of catalytic activity (Scheme 1.15) [32].

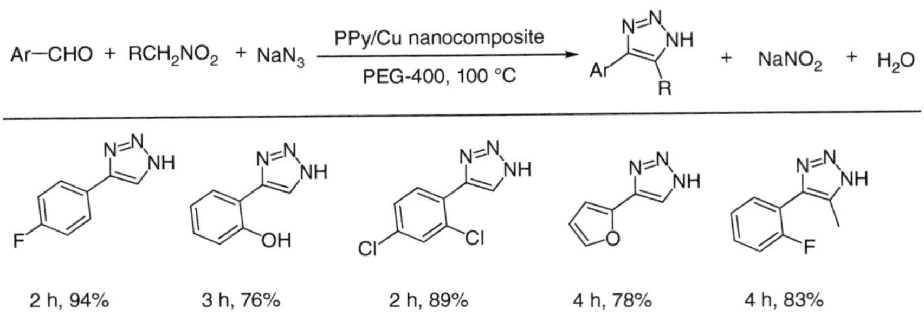

Scheme 1.15 Application of polypyrrole/Cu(II) in the synthesis of 4-aryl-NH-2,3,4-triazoles in PEG-400.

Similarly, the combination of CuI and glycerol for the 1,3-dipole cycloaddition reaction of azides with terminal and 1-iodoalkynes also exhibits versatile and high catalytic activity (Scheme 1.16) [33]. This catalytic system is the first example of the copper-catalyzed 1,3-dipolar cycloaddition reaction carried out in pure glycerol as a solvent. The catalytic system CuI–glycerol can be recycled up to six consecutive runs, and the product can be directly separated through simple filtration.

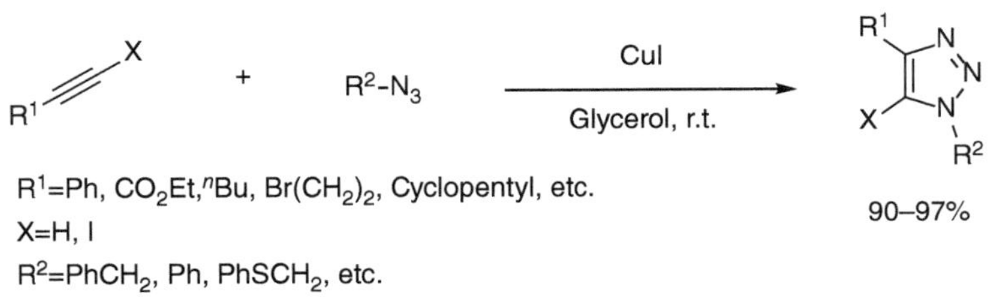

Scheme 1.16 Base-free Cu(I)-catalyzed 1, 3-dipolar cycloaddition of azides with terminal alkyne or 1-iodoalkynes.

The construction of furan rings has gained much attention as it is widely found in natural products. Among them, the ring isomerization of (*Z*)-alkynol has been considered an effective method to synthesize furan rings. In glycerol, hydrophilic palladium(II) *cis*-[$PdCl_2$(DAPTA)$_2$] can effectively promote the metal-catalyzed cycloisomerization of (*Z*) -2-en-4-yn-1-ol derivatives to form corresponding furan compounds (Scheme 1.17) [34]. Although the yield in water is higher, glycerol as solvent is more conducive to the recovery and reuse of the catalyst. With water as the solvent, the activity of the catalyst decreases rapidly with successive reactions and can be recycled up to five times. With glycerol as the solvent, the system can be recycled up to 17 times, and its cumulative conversion number (TON) reached up to 8190. Notably, the high boiling point of glycerol makes product separation easier by Kügelrohr distillation of the crude reaction mixtures, thus avoiding the use of organic solvents.

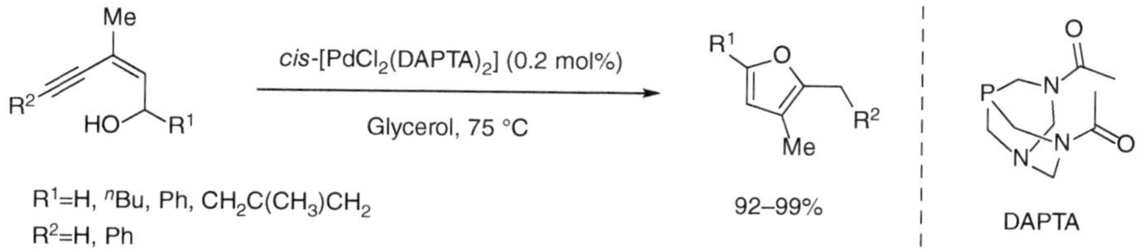

Scheme 1.17 Palladium-catalyzed cycloisomerization of (*Z*)-enynols into furans in glycerol.

In addition to this, K_2CO_3-mediated cyclization and [1,3] rearrangement reaction of γ,δ-alkynyl oximes for the synthesis of 3-hydroxylpyridines was described by Guan and coworkers (Scheme 1.18) [35]. The reaction proceeds through nucleophilic addition of oximes onto alkynes, followed by [1,3] rearrangements of *o*-vinyl oxime intermediates to form pyridols under mild conditions. The reaction works best when the high boiling point proton solvent such as glycerol is used. This process does not require the use of either a transition metal or a stoichiometric amount of oxidant, which makes it an environmentally friendly complement to the existing strategies.

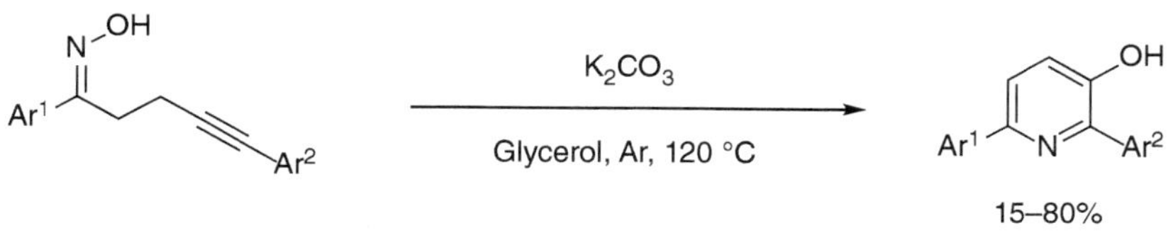

Scheme 1.18 K_2CO_3-mediated cyclization and rearrangement to form pyridols.

Sulfonyl-containing compounds play an overwhelming role in pharmaceuticals, agrochemicals, and materials science. In this context, cascade sulfonation/cyclization reactions triggered by sulfonyl radicals are one of the most promising strategies for constructing valuable heterocyclic scaffolds and introducing sulfonyl groups simultaneously. For this purpose, a versatile and practical photocatalytic system has been developed to conduct various cascade sulfonation/cyclization reactions using ethylene glycol as a unique green medium (Scheme 1.19) [36]. Through this versatile transition metal-free process, sulfone-containing heterocycles, including thioflavones, oxindoles, and quinoline-2,4(1*H*, 3*H*)-diones, are successfully constructed by irradiation with blue light in room temperature. It is intriguing to note that ethylene glycol might be critical for the stability of the ionic intermediates in the reaction.

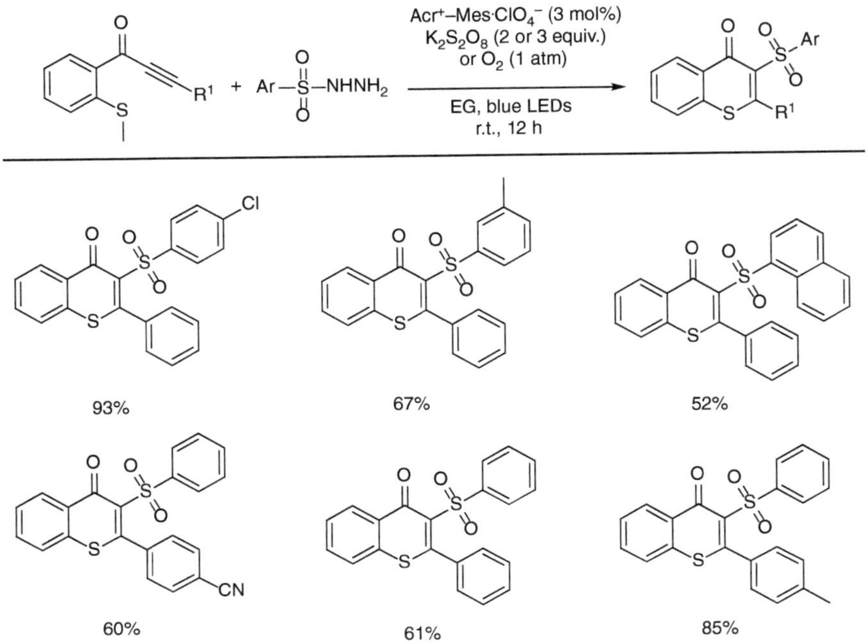

Scheme 1.19 Application of ethylene glycol in visible-light-promoted aerobic transition metal-free cascade sulfonation/cyclization reaction.

Recently, Gu and coworkers developed the $Sc(OTf)_3$-catalyzed [4 + 2] annulation reaction of α-hydroxyacetophenone, the oxidative depolymerization products of lignin, with pyrrole or indole for the synthesis of indole and carbazole derivatives (Scheme 1.20) [37]. The reaction was carried out in the green solvent glycerol with moderate-to-excellent yields, and the use of glycerol is conducive to the separation of the products and the recycling of the $Sc(OTf)_3$ catalyst. Intriguingly, both the solvent and catalyst can be recovered and reused, and high catalytic activity is maintained at least eight times.

Taking advantage of the solubility of glycerol, Lenardao et al. prepared a variety of 1-sulfanylindolizines and 1-selanylindolizines by means of readily prepared chalcogen-containing pyridinium salts (Scheme 1.21) [38]. The intramolecular cyclization of pyridinium salt uses glycerol as a cost-effective and green solvent, which can effectively promote the reaction medium by dissolving the whole substrate.

TFE is commonly utilized as solvent in cycloaddition reactions. TFE shows a booster effect as reaction media and promotes various reactions through the pure solvent effect. An expedient and clean protocol for the synthesis of the easily separable *trans*/*cis*-octahydroacridines, using substituted aniline and citronellal

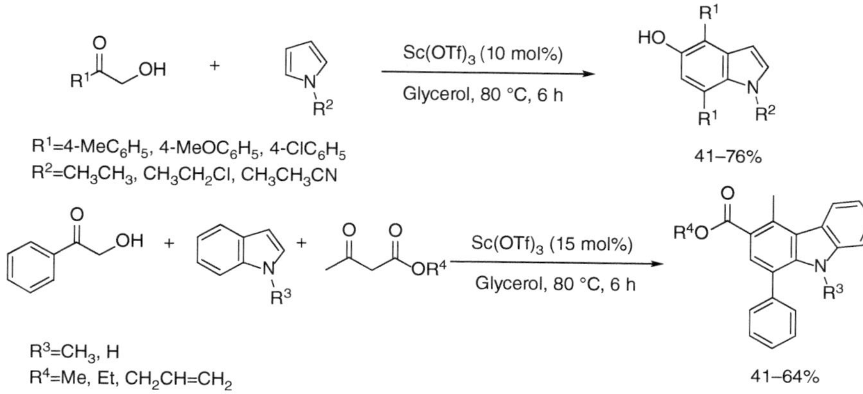

Scheme 1.20 $Sc(OTf)_3$-catalyzed [4 + 2] annulation reaction of α-hydroxyacetophenones with pyrroles or indoles.

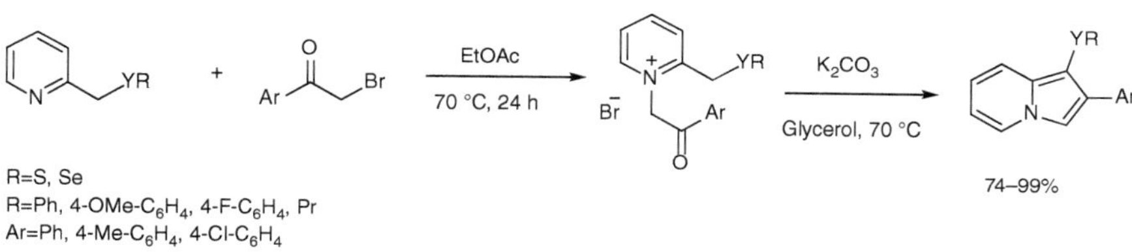

Scheme 1.21 Synthesis of 1-sulfanyl- and 1-selanylindolizines.

catalyzed by $TiCl_3$ in TFE at room temperature, has been reported by Mayekar et al. (Scheme 1.22) [39]. The reaction proceeds via intramolecular Aza–Diels–Alder reaction.

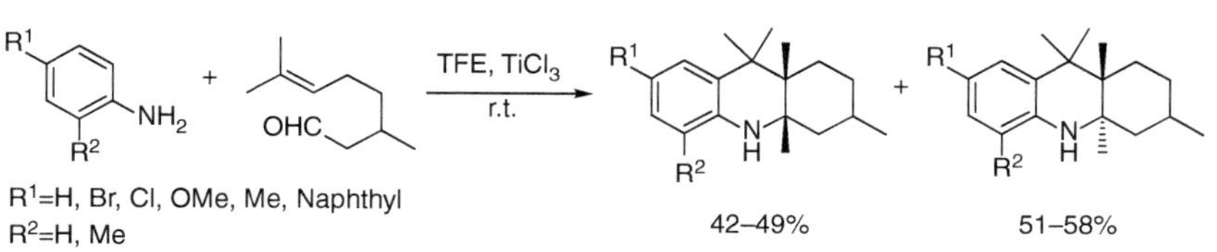

Scheme 1.22 Synthesis of octahydroacridine catalyzed by $TiCl_3$ from substituted aniline and citronellal in TFE.

TFE and HFIP have booster functions as reaction media, and their hydrogen bonding ability has been widely used in the activation of carbonyl functional groups. TFE and HFIP promote various reactions through the pure solvent effect. Propyl quinoline drugs against leishmaniasis have been successfully synthesized. In this study, the Povarov reaction of butyl aldehyde, aromatic amine, and ethyl vinyl ether in TFE, followed by oxidation, provides a convenient pathway for 2-propylquinoline with various substituents (Scheme 1.23) [40].

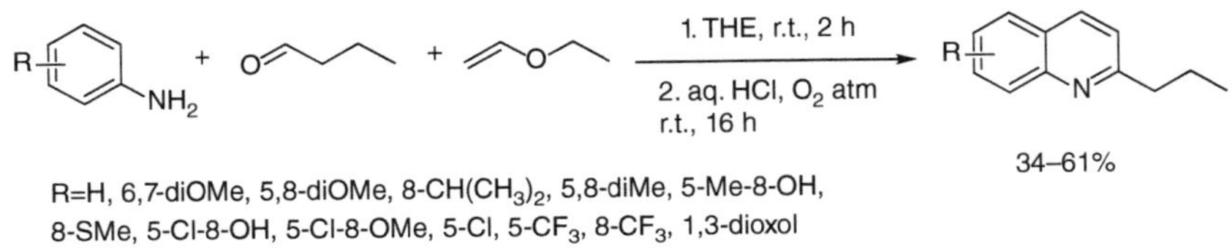

Scheme 1.23 One-pot synthesis of 2-propylquinolines from anilines through the Povarov reaction.

1.2.6 Coupling Reaction

In 2006, Wolfson and Dlugy successfully performed palladium-catalyzed Heck and Suzuki C–C coupling reactions in glycerol as the reaction solvent. The target product can be isolated by simple extraction with glycerol-immiscible solvents in high yields, and the catalyst can be recovered [41]. This work clearly demonstrates the feasibility of using glycerol as a solvent, thus paving a new avenue to find greener organic solvents.

An NCN–pincer–Pd complex has been applied to the Hiyama-type cross-coupling reaction between aryl halides and several organosilanes in a biomass-derived eutectic mixture such as 1 choline chloride (ChCl)/2 glycerol or glycerol (Scheme 1.24). This work is the first report on the use of glycerol as a reaction medium to recover the catalyst in the Hiyama reaction. In addition, the catalytic system can be efficiently used at least three times without the addition of additional catalysts. The corresponding biaryl products can be obtained by simple liquid–liquid extraction with no need for aqueous treatment or chromatographic steps [42].

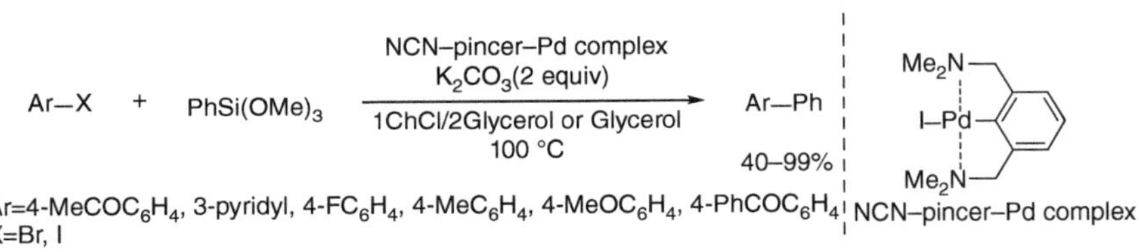

Scheme 1.24 Hiyama-type reaction in neoteric biomass-derived solvents involving glycerol.

Jérôme and coworkers investigated the palladium NP-catalyzed regioselective β, β-diarylation of acrylate derivatives in glycerol (Scheme 1.25) [43]. The palladium NPs are readily prepared by using biomass-derived sugar-based surfactants. Diarylated olefins can be selectively extracted from the glycerol–palladium-catalyzed phase using supercritical carbon dioxide, thus making post-processing more convenient. Interestingly, by adjusting the reaction temperature, the monoarylation and diarylation steps of alkenes can be controlled, providing a convenient route for the synthesis of asymmetric diarylation alkenes. Furthermore, PEG as a solvent in the Mizoroki–Heck coupling reaction has been reported. The use of PEG enhances the activity of the NPs and enables the catalytic system and solvent to be recycled [44].

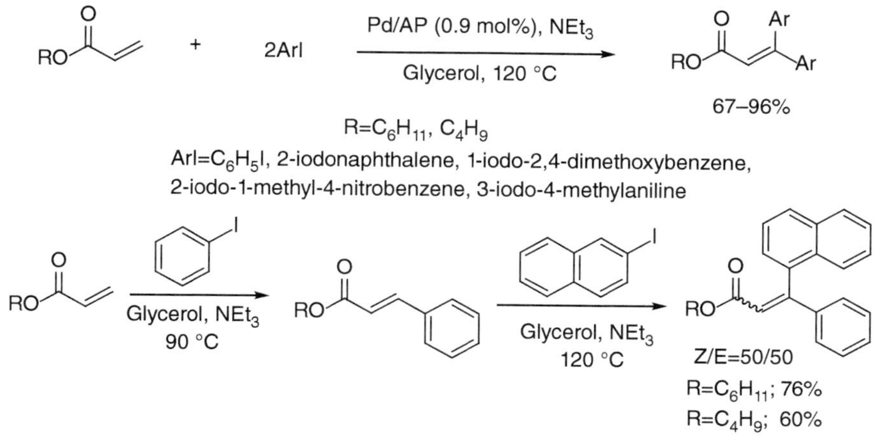

Scheme 1.25 Symmetrical and unsymmetrical β, β-diarylation of acrylates derivatives.

Furthermore, the use of PEG as a green solvent for coupling reactions has also been demonstrated in the synthesis of co-arylated compounds. In this aspect, a solid palladium NP (PdNPs@CA) is prepared from palladium chloride by a novel ultrasonically driven bioreduction method using wasabi extract as a bioreducing agent. In about 30 minutes, aryl halides and aryl boronic acids combine to produce bioaryl compounds with excellent reaction yields (Scheme 1.26) [45]. The results show that the PdNPs@CA could be effectively recovered and reused for seven cycles without loss of catalytic performance using PEG as a green solvent.

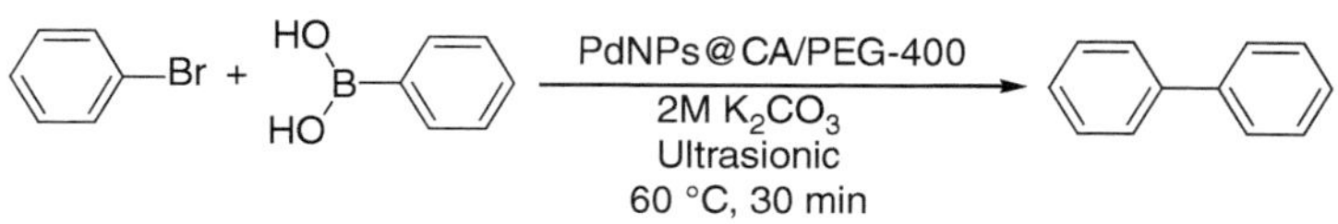

Scheme 1.26 Schematic presentation of Suzuki cross-coupling reaction between bromobenzene and phenyl boronic acid catalyzed by PdNPs@CA in PEG-400.

In addition, the synthesis of aromatic esters has been realized by carbonylation using di-μ-chlorobis[5-hydroxy-2-[1-(hydroxyimino-κN)ethyl]phenyl-κC] palladium(II) dimer, a palladacycle as depicted in Scheme 1.27, as a catalyst in PEG-400 with carbon monoxide as the carbonylation reagent [46]. Comparative study of palladacycle with conventional Pd precursors such as $PdCl_2$, $PdCl_2(PPh_3)_2$, $Pd(OAc)_2$, and $Pd(PPh_3)_4$ shows that palladacycle is superior in activity and selectivity, and brings about higher conversion at very low catalyst loading (0.6 mol%) with high TON and TOF. It is feasible to reuse the Pd/PEG-400 system up to a fifth time without loss of activity and selectivity.

Alcohol is not only used in the palladium-catalyzed coupling reactions, but also in copper-catalyzed coupling reactions. In this regard, glycerol has been proved to be a good medium for the CuI-catalyzed reaction of diaryl diselenides with arylboronic acids (Scheme 1.28) [47]. The catalytic system consisting of CuI/glycerol could

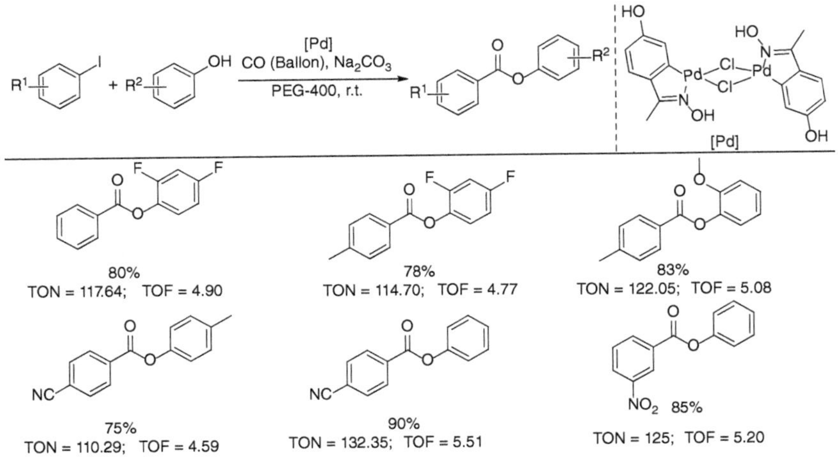

Scheme 1.27 Phenoxycarbonylation of aryl iodides with phenols in PEG.

be easily recovered and used for further cross-coupling reactions, maintaining good efficiency levels of 88–86% after three repeated runs. After four runs, the glycerol efficiency is reduced to 71%.

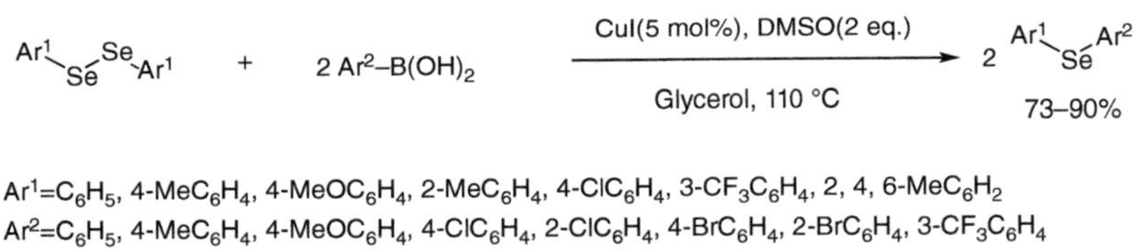

Scheme 1.28 Copper-catalyzed cross-coupling reactions of diaryl diselenides with aryl boronic acids.

Glycerol is a green solvent to the synthesis of metal NPs and their application in catalysis. For instance, Cu_2O NPs (Cu_2ONP) well coated by poly(vinylpyrrolidone) were prepared in glycerol and applied to C-heteroatom cross-coupling reactions by Gómez et al. (Scheme 1.29). The catalyst system can be reused for 10 cycles without loss of activity and selectivity [48].

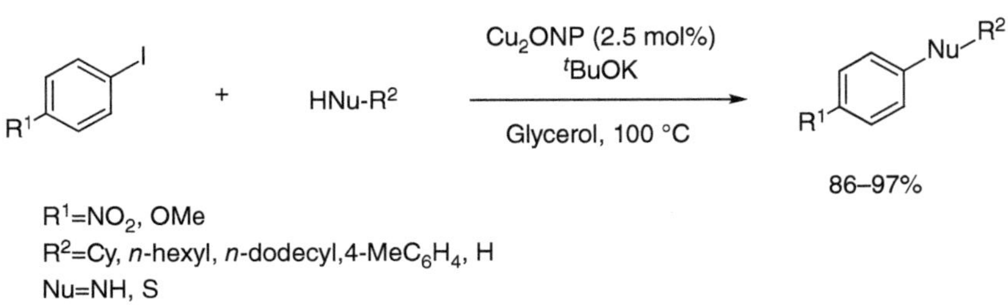

Scheme 1.29 C—N and C—S bond formation reactions catalyzed by Cu_2ONP in glycerol.

In the same way, alcohols can also be used in ruthenium-catalyzed coupling reactions. Organoboronic compounds, especially unsaturated molecules, such as streptavidinyl borates, are one of the most frequently applied intermediates in modern organic synthesis. Alkenyl boronic esters have been introduced using PEG as a green medium and synthesized via a coupling reaction between vinyl boronate and olefin in the presence of a Ru-based catalyst (Scheme 1.30) [49]. Intriguingly, the hydrogenated ruthenium catalyst [Ru (CO)Cl(H)(PCy_3)$_2$] (2 mol%) is immobilized in a series of PEGs with different molecular weights and end groups, allowing the catalyst and PEG to be recycled 8–10 times with high product yields, catalyst activity, and stability. When the catalyst dosage is doubled, it was possible to successfully acquire 16 cycles. Furthermore, this protocol is suitable for the synthesis of various functionalized alkenyl borates, including bis(boryl)ethylene, with favorable selectivity for electronic isomers. The application of the biphasic catalytic system PEG/scCO_2 is found to eliminate aliphatic solvents, especially hexane, which is neurotoxic, from the extraction process and significantly reduce the Ru content in the final product (<0.5 ppm). The high TON values in the PEG-mediated catalytic process hint at the high activity and stability of the catalytic system based on hydrogenated ruthenium complexes. In the single-phase system, the catalyst stability of dimethyl-terminated-PEG-2000 is the best, while in the biphasic boryl coupling reaction, the biphasic PEG-600/scCO_2 system and the biphasic PEG-2000/scCO_2 system exhibit the best catalyst stability. Thus, the advantages of alkenyl borate synthesis in green solvent PEG are (i) effective catalyst immobilization, (ii) recyclability in subsequent use, (iii) reduced amount of transition metals in the final product mixture, (iv) simplified product isolation procedures, and (v) minimized application of environmentally harmful volatile organic solvents.

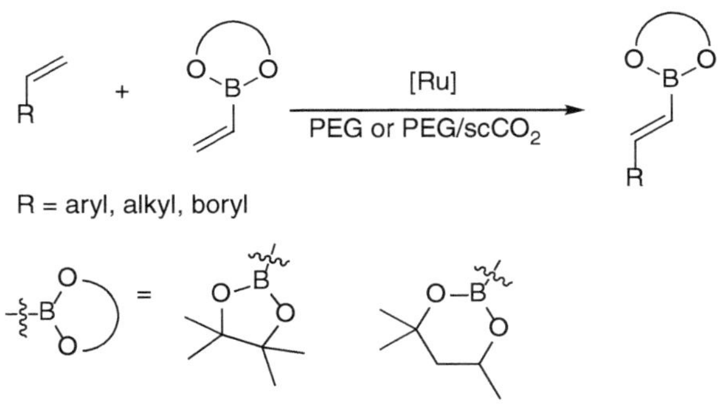

Scheme 1.30 PEG-mediated recyclable borylative coupling of vinyl boronates with olefins.

1.2.7 Condensation/Ring Condensation Reaction

PEG-400 as a solvent for organic transformation has been applied to the reaction of acetophenone, *n*-bromosuccinimide with thiourea or arylthiourea in room temperature to obtain the corresponding 2-aminothiazole derivatives with high yields (Scheme 1.31). The formation of 2-aminothiazole by using PEG-400 might be owing to the hydrogen bonding between the PEG ether-based oxygen and the sulfur of thiourea (enol form) weakening the S–H bond and enhancing the nucleophilicity of sulfur to the electron-deficient α-carbon in acetophenone. The electropositivity of the acetophenone carbonyl carbon can be strengthened by the formation of hydrogen bonds with the –OH group at the end of PEG-400 [50].

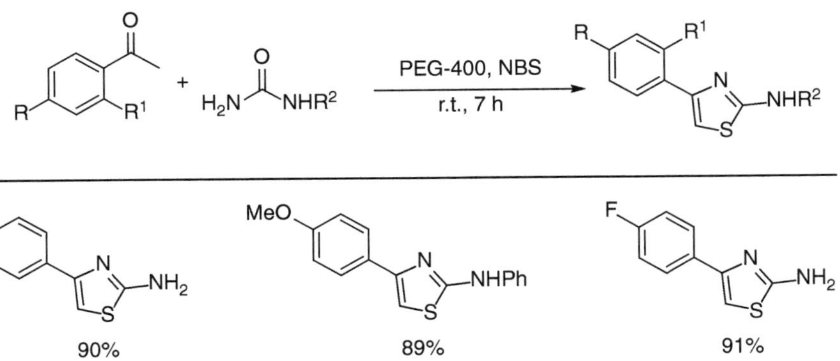

Scheme 1.31 Synthesis of 2-aminothiazoles in PEG-400.

It is no coincidence that heterocyclic pyrrole-containing compounds with pharmacological properties, such as *N*-aryl pyrrole derivatives, are an important class of compounds that have been well prepared using alcohols as reaction media. The synthesis of *N*-substituted pyrroles has been achieved in good-to-excellent yields by the reaction of 2,5-dimethoxytetrahydrofuran and 2,5-hexanedione with arylamines under ultrasonic irradiation or thermal conditions using a PEG green reaction medium (Scheme 1.32) [51].

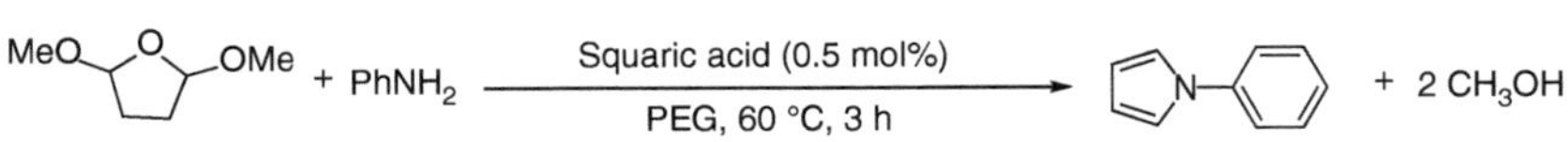

Scheme 1.32 Synthesis of *N*-arylpyrroles from 2,5-dimethoxytetrahydrofuran in PEG.

Pyrrolo[3,2-*c*]pyridines, which are important azo-containing heterocyclic compounds with diverse bioactivities, have been constructed through a one-pot process in the green solvent PEG-400 in an eco-friendly and economically feasible manner with a wide range of substrates (Scheme 1.33) [52]. For instance, 2,3,6,7-tetrahydro-1*H*-pyrrolo[3,2-*c*] pyridin-4(5*H*)-ones have been synthesized by a one-pot domino condensation reaction in the green solvent PEG-400, with 2-methyl-3-carbamoylpyrroles and aldehydes in the presence of ammonium acetate (3.0 equiv.) as a promoter. This high atomic economy reaction, using readily available starting materials, gives good-to-excellent yields with a wide range of substrates.

TFE is also widely used as a solvent in condensation reactions. A simple and convenient procedure is reported for the high-yielding direct reductive amination of

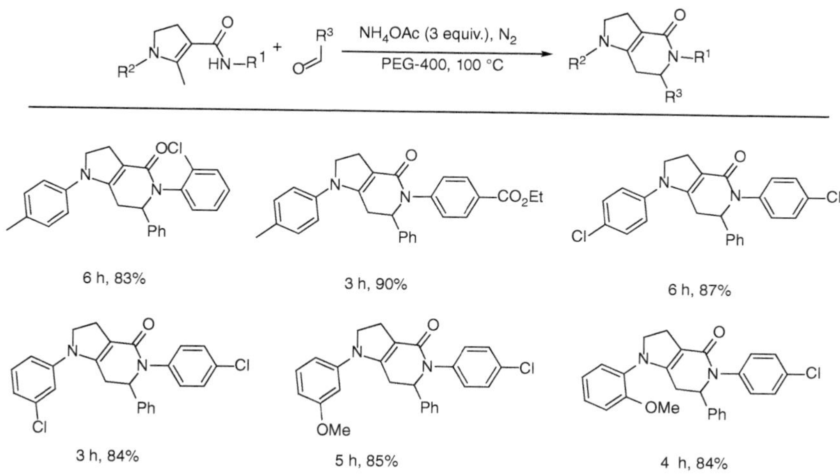

Scheme 1.33 Ammonium acetate-promoted one-pot tandem aldol condensation/aza-addition reactions in PEG-400.

aldehydes and ketones with $NaBH_4$ in TFE (Scheme 1.34). Interestingly, TFE accelerates the amination, which occurs very slowly in ethanol and methanol. Reactions in other solvents such as water, acetonitrile, and tetrahydrofuran (THF) do not yield acceptable product yields even over longer periods of time. Furthermore, TFE can be easily recovered and reused through the distillation process [53].

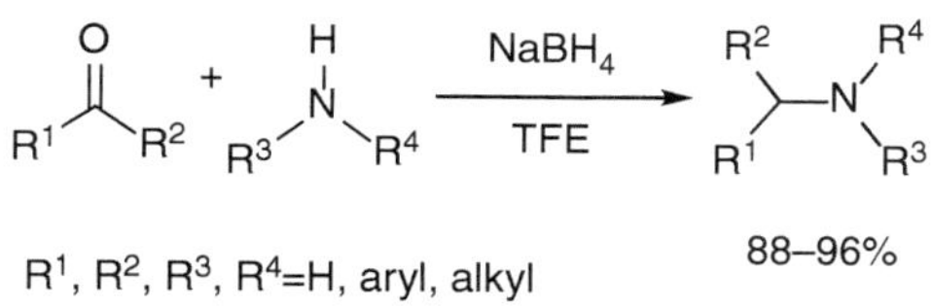

Scheme 1.34 Reductive amination of aldehydes and ketones.

2,3,4-Trisubstituted imidazole derivatives are simply and efficiently synthesized by condensation reaction of aldehyde, benzene, and ammonium acetate with TFE as the solvent (Scheme 1.35). The solution has several attractive features, such as short reaction time, avoidance of catalysts, high chemical selectivity, no side reactions, ease of product separation/purification, and simplicity of process and handling [54].

Similarly, HFIP can also be used as a solvent to synthesize imidazole derivatives by *o*-phenylenediamine and conformal ester (Scheme 1.36). This scheme has the advantages of short reaction time, 85–95% yield, simple operation, direct use of amine, and cost-effectiveness. As strong hydrogen bond donors, HFIP can activate *o*-esters for nucleophilic attacks by the amine group and greatly simplify the separation process. The reusable and nontoxic HFIP media make this method an important alternative to previously reported methods [55].

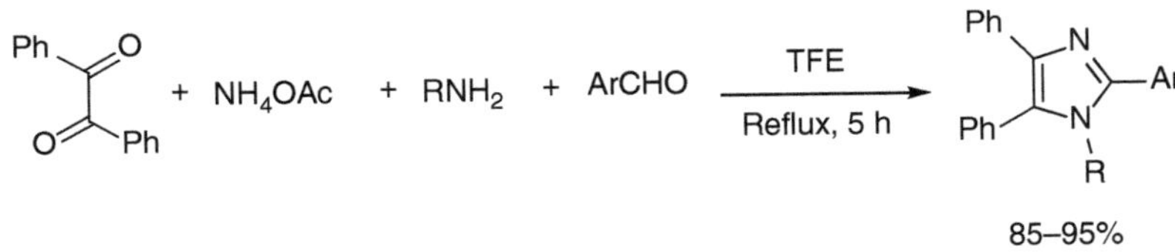

Ar=C_6H_5, 4-ClC_6H_4, 4-MeC_6H_4, 4-$NO_2C_6H_4$, 4-$OMeC_6H_4$, 4-CNC_6H_4
R=C_6H_5, $CH_2C_6H_5$

Scheme 1.35 Condensation reaction of benzil, benzaldehyde, and ammonium acetate.

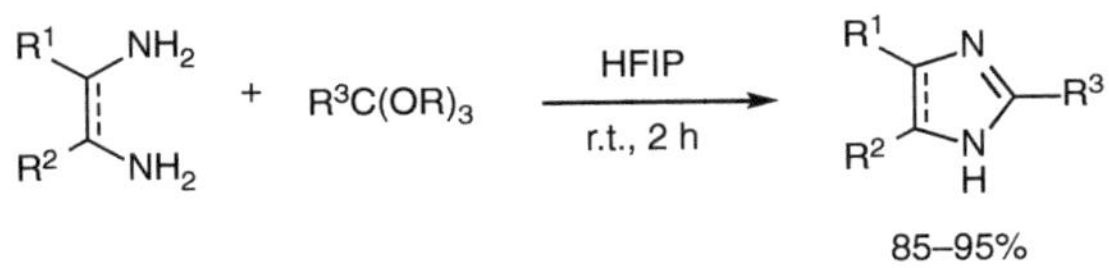

Diamine=*o*-Phenylenediamine, 4-Methyl-*o*-phenylenediamine, 3-Methyl-*o*-phenylenediamine, 4,5-Dimethyl-*o*-phenylenediamine, 4-Chloro-*o*-phenylenediamine
R^3=H, Me, Et

Scheme 1.36 Benzimidazole derivatives were synthesized by reaction of *o*-phenylenediamine with conformal ester in HFIP.

Considering the inherent properties of glycerol, such as high boiling point, unique solubility for polar organic compounds, and the ability to form hydrogen bonds, the direct use of glycerol as a solvent for condensation reactions is consistent with the demand for green chemistry. An efficient and more sustainable catalyst-free method has been developed for the synthesis of benzimidazole, benzothiazole, benzoxazole, quinoxaline, and benzodiazepine heterocyclic compounds in glycerol or glycerol–water system, as depicted in Scheme 1.37 [56]. Aminothiophenol, *o*-aminophenol, or *o*-phenylenediamine can react with aldehydes to produce the corresponding 2-substituted benzothiazole, benzoxazole, and benzimidazole derivatives in good yields, and the reactions are suitable for aromatic, heterocyclic, and unsaturated aldehydes. Under the same reaction conditions, *o*-phenylenediamine can also be condensed with 1,2-dione, or ketones to afford the corresponding quinoxalines and benzodiazepine derivatives in high yields. In particular, glycerol can be reused as a solvent for further condensation reactions, enabling recovery and recycling. The possible reasons for glycerol as a green solvent for this type of condensation reaction are that the high solubility of glycerol for the reactants makes them susceptible to interact with each other, and the carbonyl carbon of the aldehyde is activated due to intermolecular hydrogen bonds formed with the hydroxyl group of glycerol, making them readily reactive. What is more, the intermediate in the reaction can form hydrogen bonds with glycerol to stabilize itself. 2-Arylbenzothiazole can also be synthesized efficiently in PEG-400 as the reaction medium due to the hydrogen bonding between the terminal hydroxyl group of PEG-400 and the substrates [57].

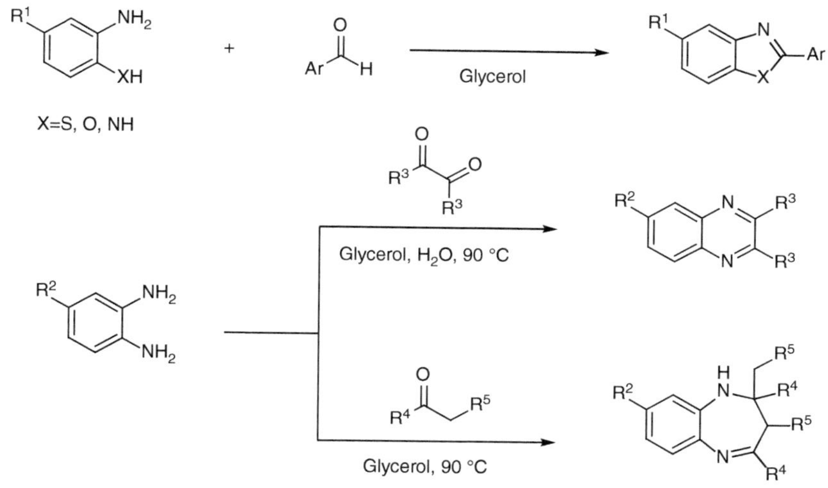

Scheme 1.37 The synthesis of benzimidazole, benzothiazole, benzoxazole, quinoxaline, and benzodiazepine heterocyclic compounds

As shown in Scheme 1.38, Gu and coworkers reported the multicomponent reaction of styrenes with 1,3-cyclohexanediones and paraformaldehyde in glycerol under catalyst-free conditions. This reaction proceeds through a tandem Knoevenagel/hetero-Diels–Alder sequence, where glycerol as a polar proton solvent not only affects the Knoevenagel reaction but also enhances the rate of the Diels–Alder reaction. The recovered glycerol can be reused without significant loss of activity. Furthermore, the method can be extended to other substrates, such as amines, 2-naphthol, 4-hydroxy-6-methyl-2-pyrone, and 4-hydroxy-1-methyl-2-quinolone, which can react with 1,3-cyclohexanediones and paraformaldehyde to yield the corresponding three-component adducts [58].

Subsequently, a stepwise one-pot reaction of phenylhydrazines, β-ketone esters, formaldehyde, and styrenes was further developed, as depicted in Scheme 1.39. This multicomponent reaction is carried out in a glycerol or carboxylic acid-functionalized ionic liquid [MIm-CO_2H]BF_4 through a tandem Knoevenagel/hetero-Diels–Alder sequence. Remarkably, glycerol and [MIM-CO_2H]BF_4 could improve the selectivity of the reaction, and pyrano[2,3-*c*]pyrazole derivatives are obtained with good yields [59].

Methanol is also used as a solvent in condensation reactions for the synthesis of 2,5-dihydropyridine skeleton by β-ketone ester and propylamine catalyzed by gold(III) (Scheme 1.40). Methanol has the best effect according to the screening of solvents, which makes the reaction yield up to 91%. The method is simple, universal, and easy to obtain starting materials [60].

In addition, ethylene glycol has also been reported as a solvent in the synthesis of thiazolopyrimidine derivatives of heterocyclic compounds containing nitrogen and sulfur atoms. The preparation of thiazolopyrimidine–quinoline derivatives by domino reaction in the green solvent ethylene glycol with 5-amino-2,3-dihydro-7*H*-thiazolo[3,2-*a*] pyrimidin-7-one, aromatic aldehydes, and dimeric cyanones as

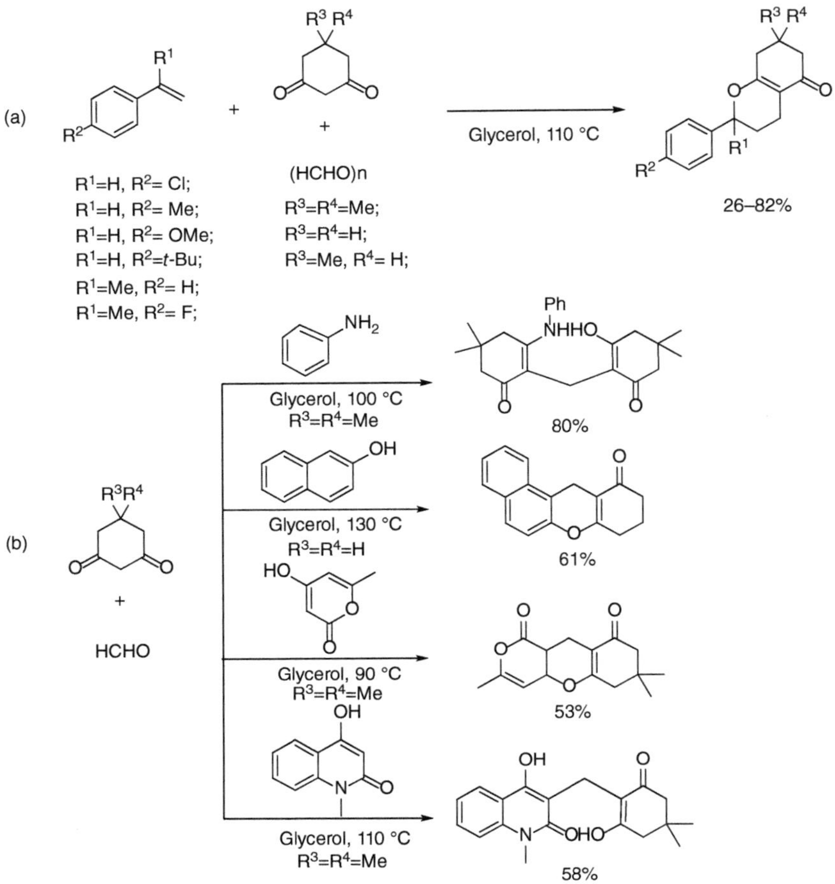

Scheme 1.38 Three-component reactions of 1,3-cyclohexanediones and formaldehyde in glycerol.

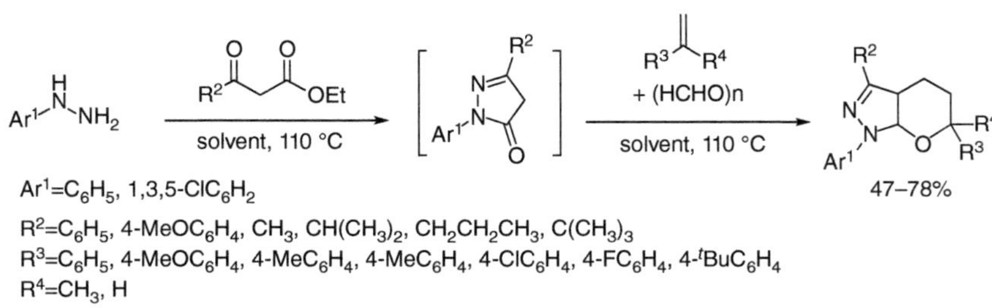

Scheme 1.39 Two-step sequential reactions of arylhydrazines, HCHO, β-ketone esters, and styrenes in glycerol.

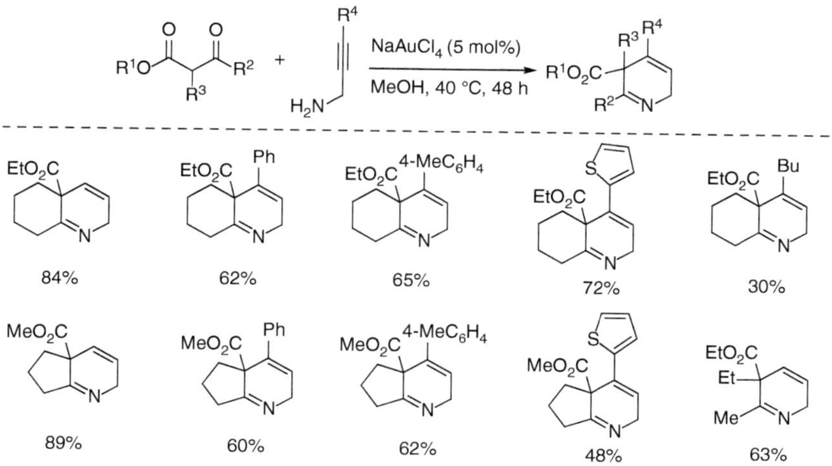

Scheme 1.40 Synthesis of 2,5-dihydropyridine derivatives by gold-catalyzed reactions of β-ketone ester and propylamine.

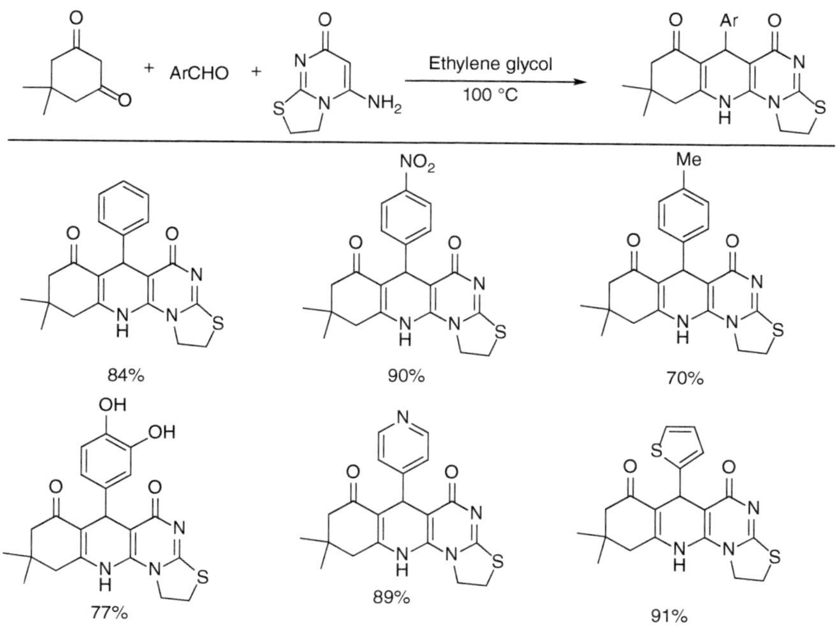

Scheme 1.41 One-pot three-component reaction of 5-amino-2,3-dihydro-7*H*-thiazolo [3,2-*a*] pyrimidin-7-one, aromatic aldehyde, and dimedone in ethylene glycol.

reactants is a good example, which has received considerable attention due to its relatively wide application in bioactive molecules. Thiadiazolo[3,2-*a*]pyrimidine is the main heterocyclic core. This scheme with catalyst-free, green solvent, mild reaction conditions, easy handling procedures, and good-to-excellent yields are the significant merits of this scheme (Scheme 1.41) [61].

1.3 Alcohols as Green Solvents and Catalysts

Alcohols can be used not only as solvents but also as catalysts to accelerate the reaction. Alcohol molecules contain hydroxyl groups, which can form hydrogen bonds with reactants or intermediates and participate in reactions, mainly in addition reactions, cyclization reactions, and condensation reactions. In addition, glycerol as the main green solvent and catalyst in alcohols, can also participate in the catalytic cycle of the system as a ligand in the coupling reaction.

1.3.1 Addition Reaction

Glycerol is able to promote aza-Michael reactions of amines or anilines without the addition of any catalyst, as depicted in Scheme 1.42. Only trace products can be obtained in water or solvent-free conditions, and lots of organic solvents such as toluene, DMF, DMSO, and 1,2-dichloroethane are ineffective for this reaction. This method shows glycerol as a potential catalyst to drive organic conversion, avoiding the use of catalysts, thus simplifying the reaction procedure and improving the greenness of the synthesis method. Similarly, the Michael addition of indoles with nitrite styrene can also be developed in glycerol with a high yield and without a catalyst in Scheme 1.42. In addition, the product separation and recycling of glycerol can be easily realized by simple liquid–liquid phase extraction [62].

Dithiocarbamates and its derivatives are of great importance in medicinal chemistry with respect to their biological activity. In this way, the green solvent, that is, PEG is considered an emerging alternative to traditional hazardous organic solvents. For example, a one-pot three-component reaction of α-amino acid, carbon disulfide, and an electrophilic reagent, such as alkyl halide performs well in a deep eutectic solvent with PEG as both the catalyst and reaction medium to afford the amino acid-based dithiocarbamates, which are regarded as multifunctional synthetic intermediates (Scheme 1.43) [63]. The dithiocarbamate then reacts with epoxides, alkyl halides, and α,β-unsaturated alkenones by addition at room temperature to obtain products in 62–92% yields with short reaction times and without tedious steps. The deep eutectic solvent and PEG can be recovered and reused with no activity and reduced yield.

1.3.2 Cyclization Reaction

In the cyclization reactions, the hydroxyl group in the alcohol molecule can be used to stabilize the reaction substrate and intermediate, thus accelerating the

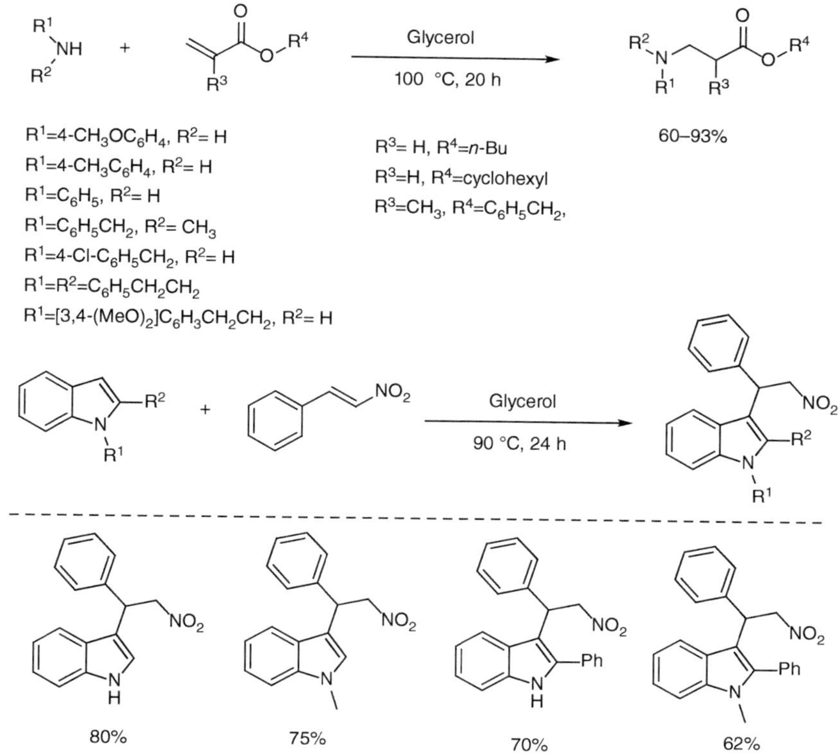

Scheme 1.42 Michael reactions of amines, anilines, and indoles with α,β-unsaturated compounds in glycerol under catalyst-free conditions.

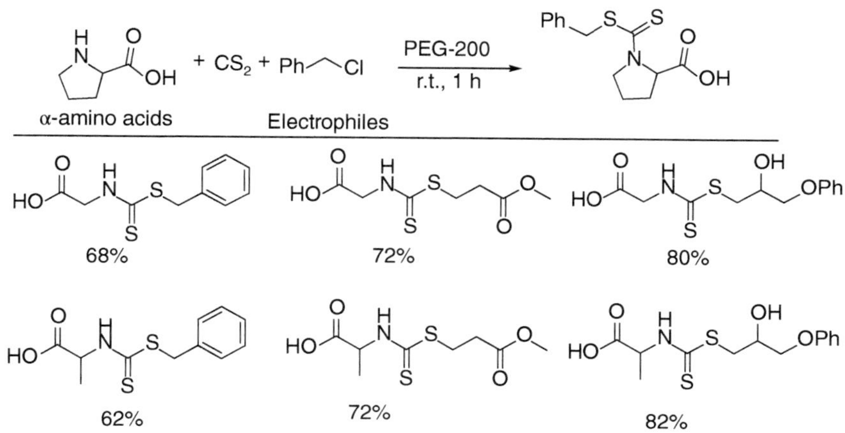

Scheme 1.43 Synthesis of amino acid-based dithiocarbamate by a three-component reaction with PEG as catalyst and reaction medium.

reaction process. Among these alcohols, glycerol is widely used in cyclization reactions because of its high boiling point and nonvolatile properties that allow easy separation of reaction products.

Bhosale et al. has developed for the first time the use of glycerol [2,3] in cycloaddition reaction from organic nitriles and sodium azide under catalyst-free conditions to efficiently furnish 5-substituted 1*H*-tetrazoles (Scheme 1.44) [64]. Interestingly, glycerol enhances the reaction rate by forming intramolecular hydrogen bonds with the nitrile and intermediates, and facilitates subsequent product separation.

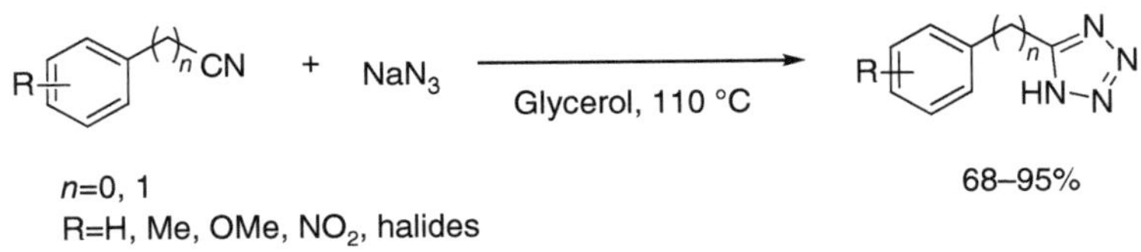

Scheme 1.44 Glycerol-mediated synthesis of 5-substituted 1*H*-tetrazole.

1.3.3 Coupling Reaction

The main application of glycerol as a catalyst and green solvent in coupling reactions is as a ligand or activating metal catalyst. For example, the copper-catalyzed cross-coupling reaction of indole and aryl halides with glycerol as a green recyclable solvent and dimethyl sulfoxide as an additive was reported by Bhanage et al. as depicted in Scheme 1.45 [65]. In this aspect, glycerol acts as a ligand in this coupling reaction and participates in the catalytic cycle. Moreover, glycerol in this reaction might act as a ligand to coordinate with Cu(I) and undergo oxidative addition to aryl halides, and then dehalogenation and reductive elimination of the oxidative addition product to obtain *n*-arylated indoles.

Recently, the "CuO/vitamin C and glycerol"-promoted reaction of aryl halides and imidazoles for the synthesis of *N*-arylation imidazole derivatives has been reported (Scheme 1.46) [66]. Simple and mild conditions render this reaction to be scaled up to the gram scale for the preparation of structurally diverse *N*-arylation products. In this reaction, glycerol serves not only as the solvent but also activates CuO. A possible mechanism is proposed based on controlled experiments, after the coordination of vitamin C with CuO to [Cu^IL] in the presence of glycerol, [Cu^IL] reacts with ArX to form intermediate A by oxidative addition. Subsequently, intermediate A reacts with imidazole anion to obtain intermediate B and then reduces elimination to generate the target product.

1.3.4 Condensation Reaction

The polar hydroxyl groups in the alcohol molecule able to form hydrogen bonds thus promote the condensation reaction. The hydroxyl groups can form hydrogen bonds with the reactants or intermediates involved in the reaction, thus polarizing them to facilitate nucleophilic reagent attack or to stabilize the intermediates in the

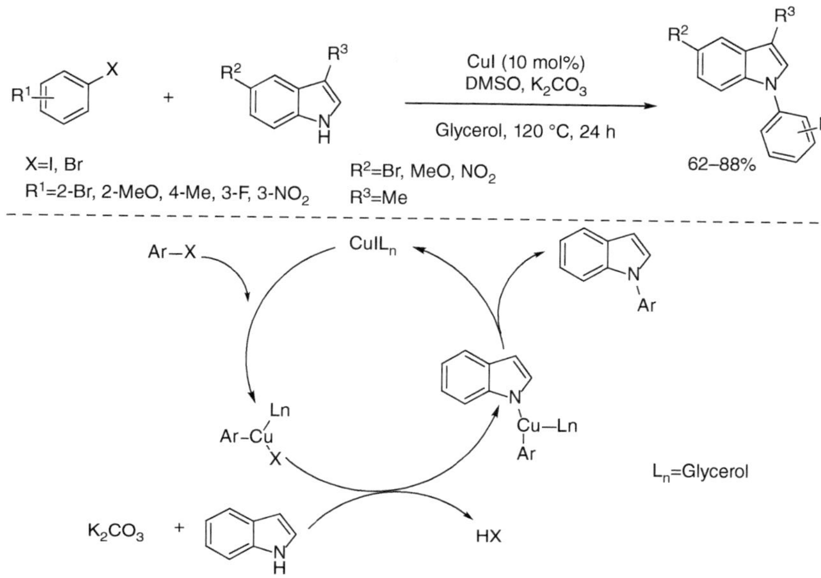

Scheme 1.45 *N*-arylation of indoles with aryl halides using copper/glycerol.

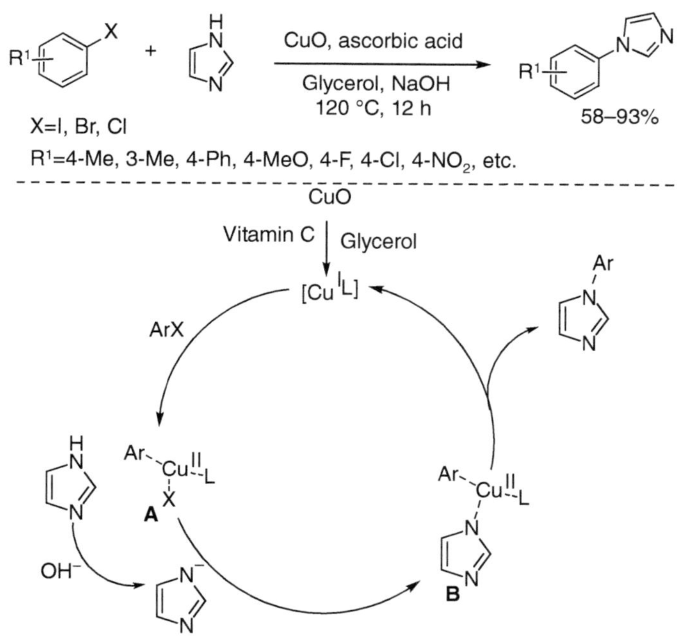

Scheme 1.46 "Vitamin C/glycerol"-promoted copper (II)-catalyzed *N*-arylation of aryl halides and imidazoles.

transition state so that the reaction can proceed under mild conditions. In this kind of condensation reaction, aldehydes are the main substrates.

Glycerol, as a green solvent and catalyst, has important applications in the synthesis of a wide range of heterocyclic compounds. In 2009, glycerol was used as a green and effective medium to promote the electrophilic activity of aldehydes. Some condensation reactions of aromatic aldehydes that are usually carried out using acid catalysts could be performed in the presence of glycerol as the reaction medium without the assistance of any catalyst. Many aromatic aldehydes react with indoles or 1,3-cyclohexanediones in glycerol in excellent yields to afford di(indolyl)methane derivatives (Scheme 1.47a), 3,4,5,6,7,9-hexahydro-9-aryl-1*H*-xanthene-1,8(2*H*)-diones (Scheme 1.47b) and 1-oxo-hexahydroxanthenes (Scheme 1.47c), respectively [67].

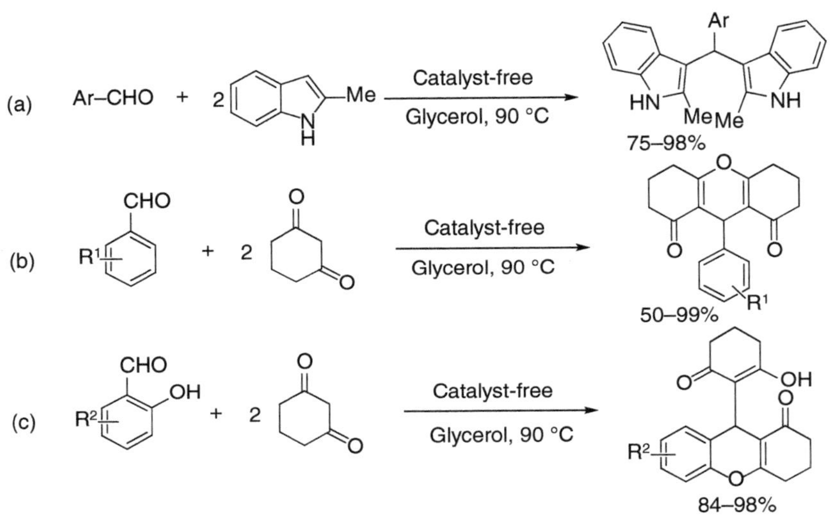

Scheme 1.47 Catalyst-free synthesis of di(indolyl)methanes, xanthene-1,8(2*H*)-diones, and 1-oxo-hexahydroxanthenes with aldehydes and indole derivatives/1,3-cyclohexanedione.

Pyranoid heterocyclic derivatives are an important class of heterocyclic compounds containing oxygen. Moreover, 4*H*-pyrans are useful intermediates for the synthesis of various compounds. The most direct method of synthesizing these heterocyclic rings is multicomponent coupling currently. However, in most of the reported methods, catalysts are not recyclable. In 2012, a highly efficient catalyst-free one-pot, three-component synthesis of carbonyl compounds, specifically those possessing a reactive α-methylene group, and alkylmalonates, using glycerol as a solvent, was developed to synthesize 4*H*-pyrans in high yields and short reaction times, as shown in Scheme 1.48 [68]. The presence of the reactive –OH groups in the structure of glycerol plays a major role in this reaction. It is speculated that it is the polar amphiphilic hydroxyl group of glycerol that facilitates the interaction of the weakly acidic and basic components, due to the stabilization of the corresponding transition states and intermediates through hydrogen bonding.

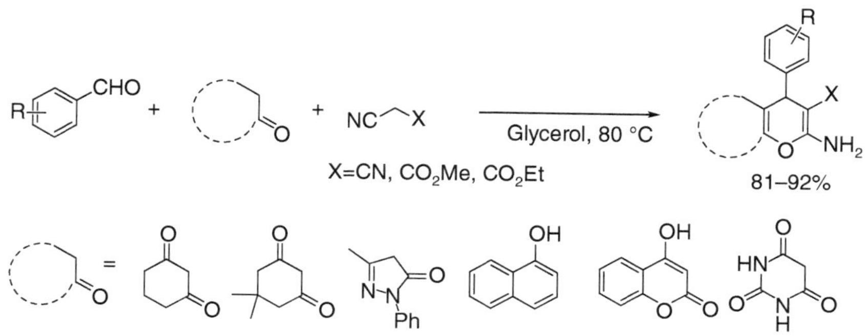

Scheme 1.48 One-pot three-component synthesis of 4*H*-pyrans of carbonyl compounds, carbonyl compounds, and alkylmalonates in glycerol.

In the same way, a one-pot three-component reaction of aldehydes, malononitrile, and resorcinol or dimedone in TFE without the use of catalysts or any other additives is developed as a novel and efficient regioselective synthesis of 4*H*-pyrans derivatives. The weak Brønsted acidity and strong ionizing power may be relevant to its unique role in this transformation. The polarity effect and hydrogen bond donor ability might not be important in this case. Actually, the hydrogen bond donating ability of these solvents drops as temperature rises owing to the fact that hydrogen-bond formation is exothermic. Although fluorinated alcohols are known to have a remarkable stabilizing effect for cationic species due to their high polarizability and low nucleophilicity, the synthesis of pyrans does not go through a discrete carbocation intermediate. The polar transition state of the reaction could be stabilized well by the high ionizing solvent, TFE. After the reaction, TFE is easily separated by distillation and reused without reducing its activity [69].

Glycerol not only enables high product conversion and selectivity, but also is safe to use. It has a low environmental impact and cost, is easy to separate products, and is reusable. 2,4,5-Triaryl and 1,2,4,5-tetraaryl imidazole derivatives in glycerol via one-pot condensation reaction of benzimidazole with various aldehydes, aromatic primary amines, and ammonium acetate are developed, as shown in Scheme 1.49 [70]. In the absence of glycerol, the reaction proceeds slowly, which indicates that glycerol is an important component of the reaction. This may be due to the intermolecular hydrogen bond formed by glycerol with carbonyl compounds and amines, which activates carbonyl compounds and weakens the N–H bond, enhancing the nucleophilicity of nitrogen.

Interestingly, HFIP can be used as both solvent and catalyst to promote Pictet–Spengler reaction between tryptamine derivatives and aldehydes or active ketones to obtain tetrahydro-β-carbolines in high yields (Scheme 1.50) [71]. It is worth mentioning that HFIP-enhanced Pictet–Spengler reaction avoids the use of Brønsted or Lewis acids. In this reaction, the HFIP can be recovered directly by distillation, leaving the product pure enough for most further transformations. HFIP can be recycled several times. This method is simple to operate and harmless to the environment, which has broad application prospects.

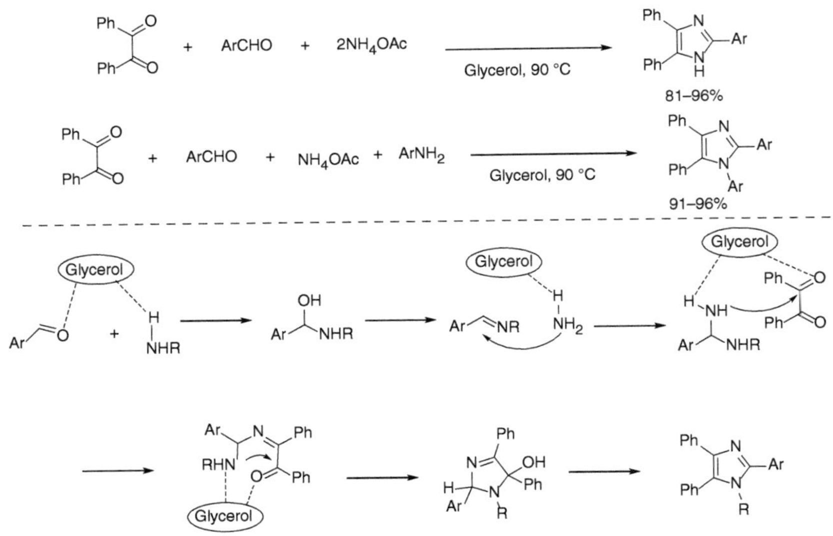

Scheme 1.49 Catalyst-free synthesis of 2,4,5-triaryl and 1,2,4,5-tetraaryl imidazole derivatives in glycerol.

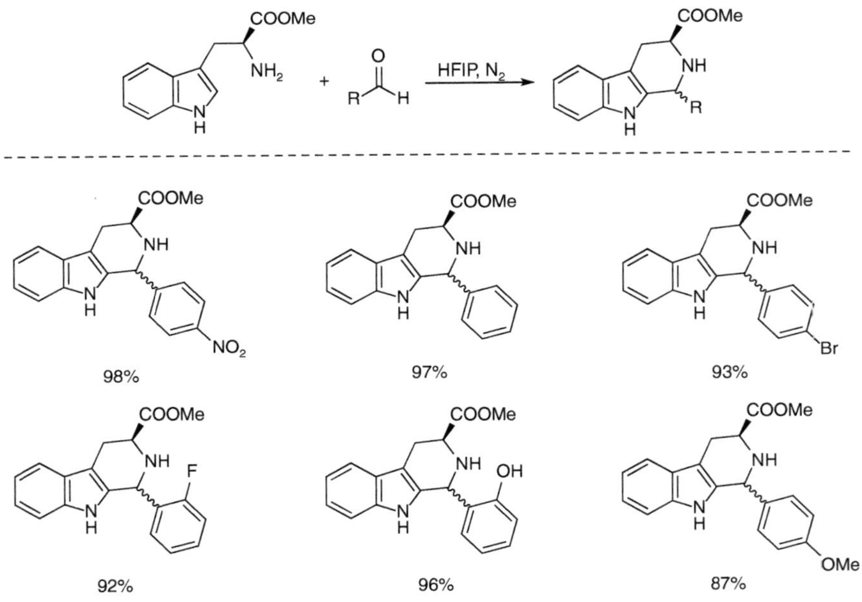

Scheme 1.50 The Pictet–Spengler reactions of L-tryptophan methyl ester and aldehydes.

In addition, the effectiveness of TFE as a reusable catalyst or solvent for the generation of α-amino nitriles has also been explored. Such a scheme uses the one-pot three-component coupling of aldehydes or ketones, amines (primary and secondary), and trimethylsilyl cyanide (TMSCN) using TFE as a reusable solvent (Scheme 1.51) [72]. The main advantage of fluorinated alcohols, such as TFE, is that it can undergo reactions that typically require Lewis acid or catalyst assistance without accelerators. Using TFE as solvent and catalyst can activate $(Boc)_2O$ for tertiary butyl carbonylation of amines and amine derivatives.

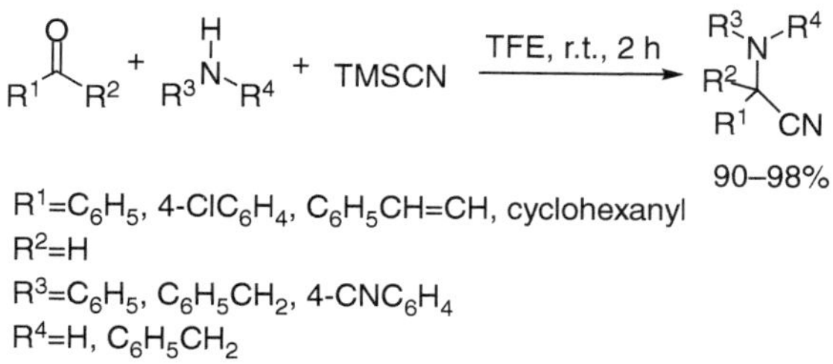

Scheme 1.51 Condensation of trimethyl phosphite with acetaldehyde and aniline.

1.3.5 Metathesis Reaction

The protection and deprotection of amino groups play an important role in organic conversion. *N-tert*-butyloxycarbonyl (*N*-Boc) derivatives are widely used in the synthesis of small organic molecules and natural products as a class of important reagents for protecting amino functional groups. Therefore, the development of a safe, environmentally benign, efficient, and catalyst-free protocol using a cost-effective and recyclable solvent for *N*-Boc protection of amines is attractive. In this context, a catalyst-free, efficient, and green methodology was developed by Ingale and coworkers for the chemically selective *N*-Boc protection of amines by using glycerol as solvent at room temperature (Scheme 1.52) [73]. Glycerol maintains good efficiency after three reuses. A possible mechanism is proposed (Scheme 1.52). The hydrogen bond formation between glycerol and the carbonyl oxygen atoms of Boc_2O makes the carbonyl group more vulnerable to nucleophilic attack. The hydrogen bonding enhances the electrophilicity of carbonyl and the nucleophilicity of nitrogen in the intermediate, making it easier for the nitrogen atom to perform an intramolecular nucleophilic attack on the carbonyl carbon, which subsequently eliminates CO_2 and ^{t}BuOH to form the product *N*-Boc protected amine.

1.4 Alcohols as Green Solvents and Hydrogen Donors

Alcohols have been shown to be reducing agents and are commonly used in transfer hydrogenation reactions under mild conditions in the presence of loaded Pd or other noble metal catalysts.

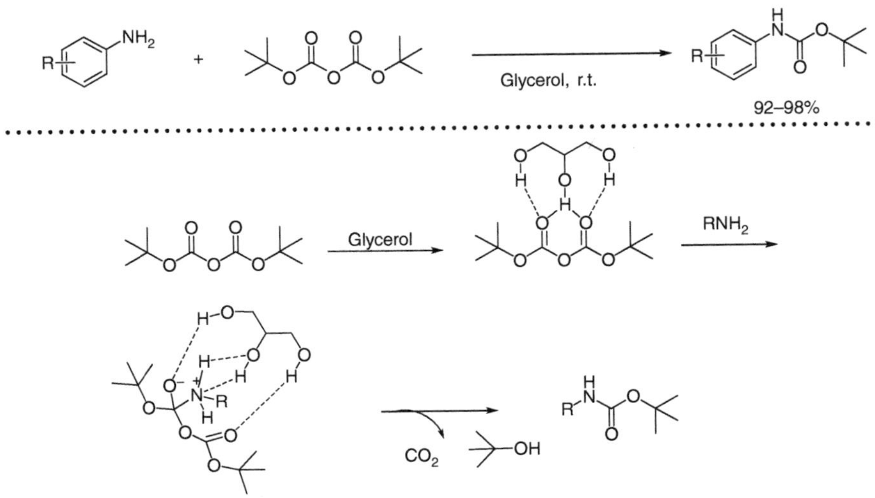

Scheme 1.52 Catalyst-free methodology for chemoselective *N-tert*-butyloxycarbonylation of amines in glycerol.

Glycerol, a biodegradable and virtually nontoxic chemical, can be used as a green solvent and hydrogen donor. Glycerol has been successfully used as an environmentally friendly "donor solvent" in transfer hydrogenation–dehydrogenation reactions for the reduction of alkenes, aldehydes, ketones, and aromatic nitro compounds [74].

In 2009, Wolfson et al. reported the use of glycerol as a green solvent and hydrogen donor in catalytic transfer hydrogenation–dehydrogenation reactions (Scheme 1.53) [75]. In this reaction, glycerol not only donates hydrogen to various unsaturated organic compounds (carbonyl compounds, alkenes, and nitrobenzene), but also acts as a solvent for easy product separation and catalyst recovery. Subsequent studies also demonstrated the design of novel, efficient, recyclable, and inexpensive ferrite-nickel magnetic nanoparticles (Fe_3O_4-Ni-MNPs) and copper nanoparticles (CuNPs) that can also be used for the reduction of aromatic nitro compounds in the presence of glycerol, a sacrificial hydrogen source [76].

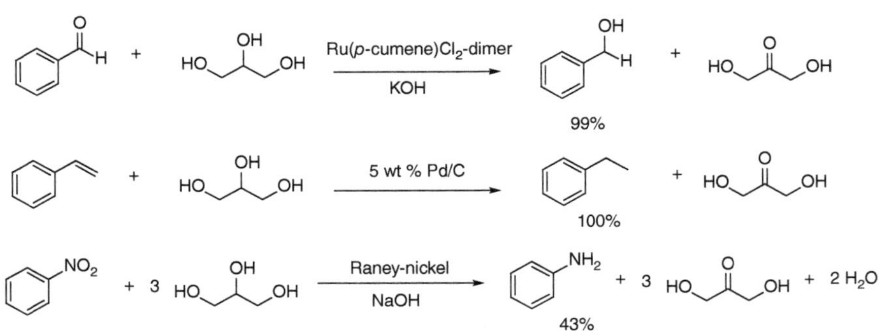

Scheme 1.53 Transfer hydrogenations of representative unsaturated organic compounds in glycerol.

Ethanol is also a renewable resource, a feedstock for the chemical industry, and an environmentally friendly and human-friendly solvent. Similarly, ethanol can be used as a hydrogen donor to reduce ketones to alcohols. Zhang et al. studied a green and efficient iridium-catalyzed asymmetric transfer hydrogenation of acetone to chiral propargyl alcohol (Scheme 1.54) [77]. Using sodium formate and ethanol as hydrogen sources, a series of acetophenones were hydrogenated with the chiral iridium catalyst (*S*)-**1b**, and optically active chiral propargyl alcohol with up to 98% ee was obtained under alkali-free conditions, with high yield and good enantioselectivity. This scheme provides a practical and sustainable method for the preparation of optically active propargyl alcohol.

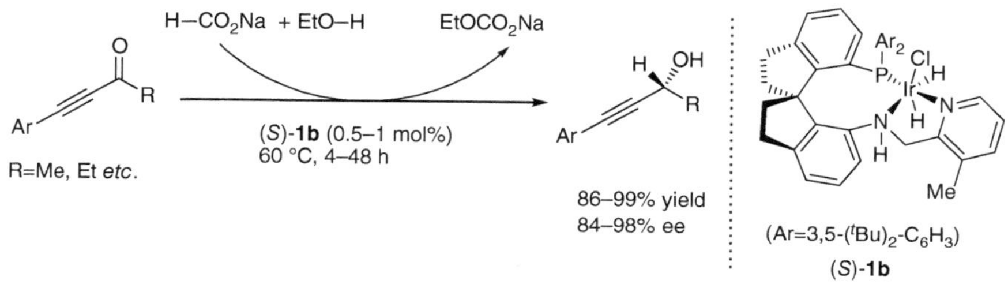

Scheme 1.54 Iridium-catalyzed asymmetric transfer hydrogenation of alkynyl ketones by ethanol as hydrogen donor.

The transfer hydrogenation of nitriles is an important alternative method for producing primary amines or imines. Under alkali-free conditions, using isopropanol as the proton donor and solvent, the optimized Co@NC-900 catalysts can convert nitriles into primary amines or imines with a selectivity of up to 90% (Scheme 1.55) [78]. Compared to traditional hydrogenation processes, transfer hydrogenation utilizes various substitutes, including phosphates, formic acid salts, and inorganic hydrides as proton donors to replace the molecular hydrogen used in traditional schemes. Among them, alcohols (especially isopropanol) are very promising candidates due to their nontoxic and mild reaction conditions. The use of isopropanol as a proton donor for nitrile transfer hydrogenation can avoid handling high-pressure sterilizers and hydrogen gas, making it a green, safe, and economical solution.

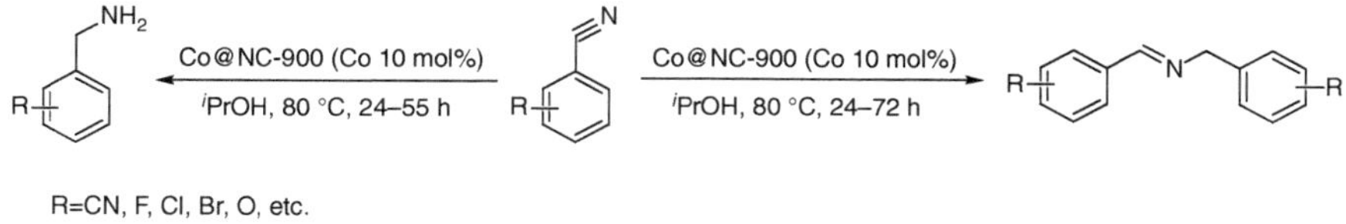

Scheme 1.55 Bifunctional *N*-doped Co@C catalysts for base-free transfer hydrogenations of nitriles.

Glycerol can be directly used as a hydrogen source for reducing amination. The Ru-catalyzed one-pot synthesis of monosubstituted and disubstituted amines is successfully achieved using nitrobenzene and ethanol as raw materials and glycerol as a reducing agent, as shown in Scheme 1.56 [79]. In these catalytic systems, the oxidation products of glycerol occur on the terminal hydroxyl group, and a portion of the alcohol is oxidized to the corresponding aldehyde. Next, nitrobenzene is reduced with the admixture [Ru-H] to produce aniline, which reacts with aldehydes to produce *N*-benzyl-4-methylaniline, and then *N*-benzyl-4-methylaniline is reduced by glycerol to produce *N*-alkylamine.

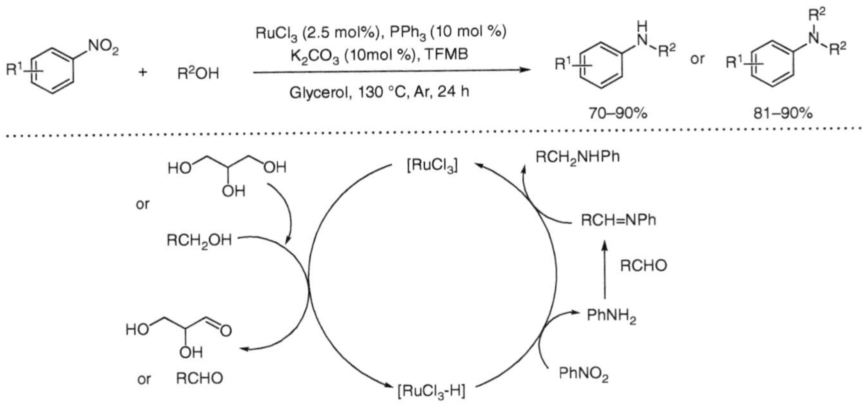

Scheme 1.56 Reductive *N*-alkylation of nitro compounds with an equivalent number of alcohols to *N*-alkyl or *N,N*-dialkyl amines using glycerol as the reducing agent.

Similarly, the catalytic reduction of allyl alcohols is also possible with ruthenium as the catalyst. The hydrophilic arene-Ru(II) [$RuCl_2(\eta_6\text{-}C_6H_6)$(DAPTA)] complex can be associated with KOH to reduce allyl alcohols to the corresponding saturated alcohols in yields up to 90% (Scheme 1.57) [80]. The reaction is a tandem process involving the oxidation of the allyl alcohol and subsequently the transfer hydrogenation of the resulting carbonyl compound in glycerol as the green solvent and hydrogen donor.

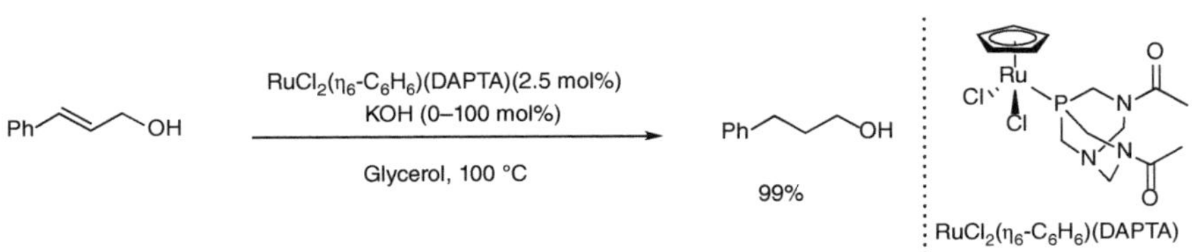

Scheme 1.57 Ruthenium-catalyzed reduction of allylic alcohols.

Recently, it has been reported that the hydrogenation ammomethylation of terminal alkenes in glycerol catalyzed by [Rh(acac)$(CO)_2$]/TPPTS catalyst. The reaction is a multicomponent tandem transformation (Scheme 1.58) [81]. Rh-catalyzed hydroformylation of olefins occurs first, followed by condensation of the resulting aldehyde with an amine to produce an enamine (or imine/imine) intermediate, which then undergoes metal-catalyzed hydrogenation with water as the only by-product. Glycerol acts as a hydrogen transfer agent and facilitates the Rh-catalyzed reduction of the enamine intermediate by ensuring that the reaction proceeds at relatively low pressure.

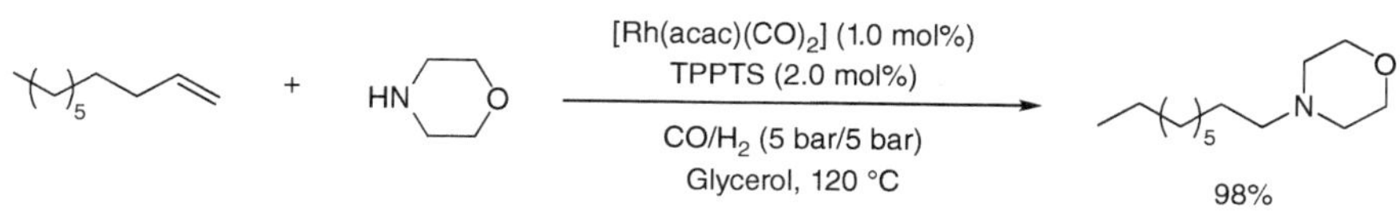

Scheme 1.58 Rh-catalyzed hydroaminomethylation reaction in glycerol.

Furthermore, glycerol has been demonstrated to be used as a solvent and reducing agent for the Mo-catalyzed chemically selective deoxidation of sulfoxides, where it acts only as a reducing agent and not as a hydrogen donor, as depicted in Scheme 1.59 [82].

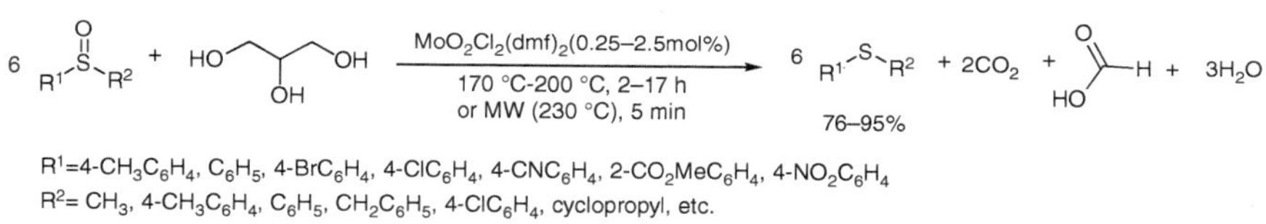

Scheme 1.59 Glycerol as a chemical selective reductant of sulfoxide.

1.5 Miscellaneous

1.5.1 Polyethylene Glycol as a Solvent for CO_2 Capture and Conversion

PEG not only serves as an essential green solvent but also plays an important role in CO_2 capture and conversion. Recently, much attention has been focused on developing cost-effective and powerful absorption agents and CO_2 capture and storage (CCS) technologies. Nevertheless, the large energy input during desorption becomes a key obstacle to achieving practical CCS. Alternatively, reactions involving CO_2 are typically conducted at high pressures, which may be economically undesirable and a concern for safety. The challenge is to develop catalysts that can activate CO_2 at low pressure (preferably 1 atm) and thus catalyze the binding of CO_2 into the structures of organic molecules. In order to settle the problem of energy loss in the CCS procedure, PEG has gained attention as a CO_2-friendly material. A highly efficient binary system consisting of PEG/superbase has been developed to capture and simultaneously activate CO_2, thus successfully converting the captured CO_2 directly into value-added chemicals or fuels and avoiding desorption (Scheme 1.60) [83]. Experimental results have indicated that the superbase/PEG system has proven to be an excellent system for fast and reversible iso-molar CO_2 sorption, and that the amidazolium alkyl carbonate obtained by capturing CO_2 with superbase and PEG may be more reactive than free CO_2. As a consequence, CO_2 uptake may lead to its activation and the desorption process using this strategy requires few external energy inputs. In overview, the CO_2 fixation can be catalyzed into value-added chemicals/fuels under extremely mild reaction conditions (1 atm, 40 °C, metal-free process).

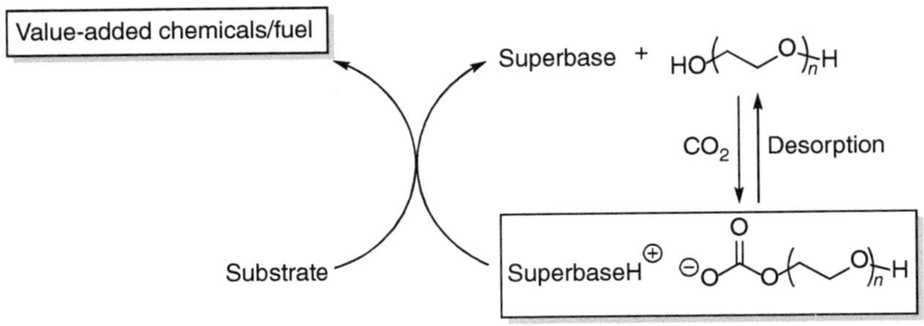

Scheme 1.60 CO_2 capture and activation by superbase/PEG and its subsequent conversion.

Similarly, the flexible PEG chains can coordinate with alkali metal cations, which can increase the capacity of the counterions. Readily available amino acid salts of *N*-substituted groups have a very high CO_2 capacity in PEG solutions, which are nearly equivalent to capture, and have been exploited to some extent in the field of CO_2 capture (Scheme 1.61) [84].

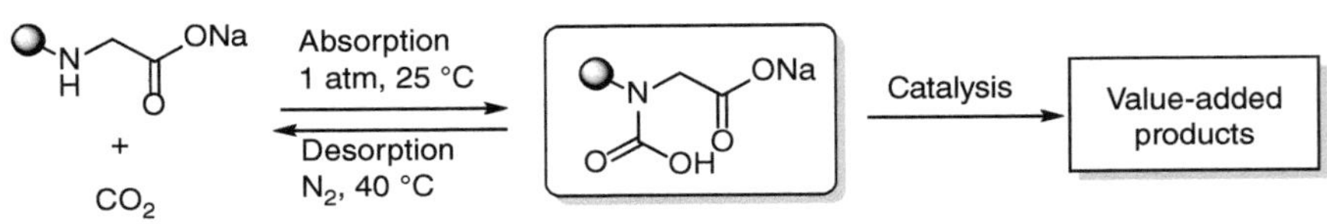

Scheme 1.61 Equimolar CO_2 capture by *N*-substituted amino acid salts and subsequent conversion.

Besides, organo-carbamates have a great variety of applications in the medicinal industry and agriculture and are widely used as protecting groups or key intermediates in synthetic chemistry. Conventional organo-carbamate synthesis involves the use of highly toxic phosgene as a carbonylating agent, which requires organic solvents and can easily cause environmental problems. However, organo-carbamates which are chemically selectively synthesized from the PEG-enhanced amines, carbon dioxide, and alkyl halides have excellent features such as environmental friendliness and high efficiency. In this work, amines, CO_2, and alkyl halides are subjected to a three-component reaction with the help of K_2CO_3 and PEG (MW ¼ 400) to produce organo-carbamates under ambient conditions (Scheme 1.62) [85]. PEG in this reaction is used as a solvent and a PTC. Notably, the presence of PEG can also suppress the alkylation of amines and carbamates, thereby enhancing the selectivity for the targeted carbamates.

From the perspective of environmental protection and resource utilization, the chemical conversion of CO_2 into useful chemicals is of increasing interest. One of the most promising approaches for the chemical fixation of CO_2 is the synthesis of

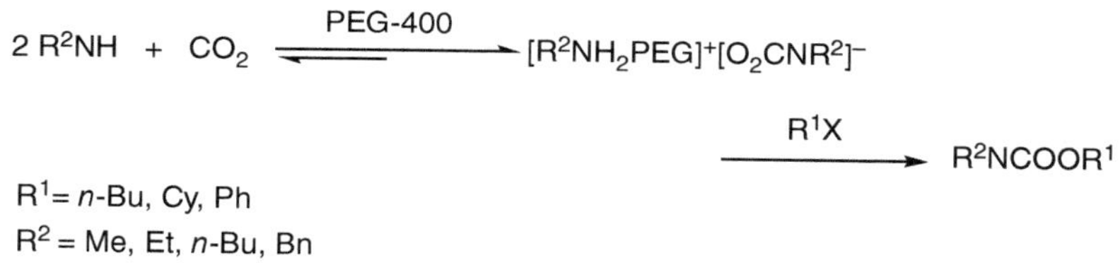

Scheme 1.62 PEG-enhanced chemoselective synthesis of organic carbamates from amines, CO_2, and alkyl halides.

five-membered cyclic carbonates, which have been used in a range of applications, for instance, as polar nonprotonic solvents, intermediates for organic and polymer synthesis, chemical components for pharmaceutical/fine chemicals for biomedical applications, and as lithium secondary battery electrolytic components for lithium secondary batteries (Scheme 1.63) [86]. In this context, PEG has already been used as an efficient reaction medium to synthesize cyclic carbonates by reacting ortho–halo alcohols with carbon dioxide in the presence of a base. Notably, PEG-400 exhibits a unique impact on reactivity as an environmentally friendly solvent compared to conventional organic solvents. First, PEG-400 can come up with complexes via coordinating the potassium cation in the same way that crown ether does, which influences the basicity of K_2CO_3. Second, the "CO_2-expansion of PEG" impact alters the physical characteristics of the reaction mixture, such as viscosity and solubility of the reactants, which enhances the synthesis process. High yields and excellent selectivity of cyclic carbonates are achieved under mild reaction conditions. Furthermore, the product can be isolated easily by extraction, which is considered a cost-effective way to produce cyclic carbonate in an environmentally friendly manner.

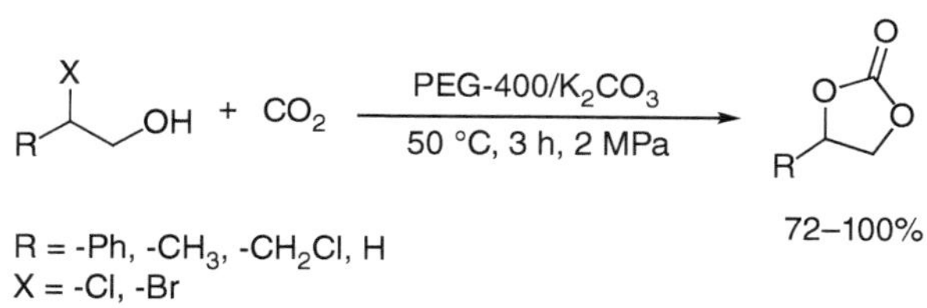

Scheme 1.63 Synthesis of cyclic carbonate from vicinal halohydrins and CO_2 in PEG-400.

1.5.2 Polyethylene Glycol Radical-Initiated Oxidation Reactions in Compressed Carbon Dioxide

PEG and its derivatives are usually recognized as a thermally stable, cheap, toxicologically harmless, and relatively environmentally friendly medium for chemical reactions and as PTCs. On the other hand, PEG, being susceptible to oxidative attack by free radicals in the presence of oxygen at high temperatures (above 70 °C), can afford the formation of numerous complicated low-molecular-weight PEG peroxides through a random chain-breaking process. In this case, when combined with compressed CO_2, the PEG radicals induced by thermal oxidative degradation can initiate a series of radical reactions such as selective formylation of primary and secondary fatty alcohols, oxidation of benzyl alcohols, benzyl C—C bond cleavage and benzyl sp^3 C–H oxidation, as shown in Scheme 1.64 [87], representing great synthetic potential utility in a cost-effective, practical, and environmentally friendly manner. The whole procedure does not require any catalyst or additional free radical initiator. In this work, hexadecanol is chosen as the model compound for the preliminary study. As a consequence, hexadecanol is mixed with PEG-1000, oxygen (2.5 MPa), and carbon dioxide (13.5 MPa) at 100 °C for 12 h, hexadecyl formate in 68% isolated yield, and 15% palmitic acid is obtained along with 15% recovery of the hexadecanol. This method allows for a metal-free, cost-effective synthetic transformation with environmentally friendly characteristics compared to conventional processes.

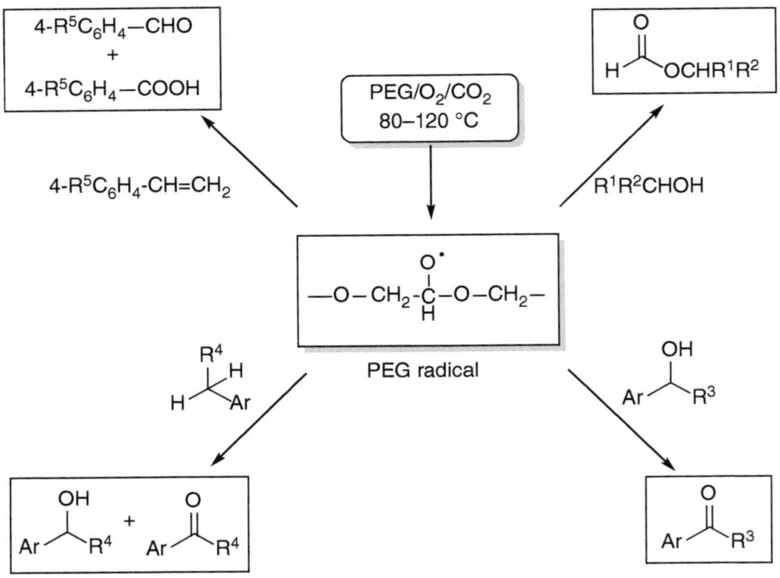

Scheme 1.64 Organic reactions involving PEG radicals in compressed carbon dioxide.

In this similar way, PEG/O_2/CO_2 is also used as an initiator, oxidizer, and solvent to oxidize primary and secondary alcohols to aldehydes or acids and ketones, respectively (Scheme 1.65) [88]. Compressed CO_2 here could offer a relatively safe environment for the oxidation of molecular oxygen as an oxidant in the reaction and could also be used to adjust the selectivity of the target product by altering its pressure to improve reactivity and to render product separation easier.

Based on the abovementioned experimental achievements, the application of PEG radicals in the oxidation of benzyl hydrocarbons is further explored (Scheme 1.66) [89]. In this regard, PEG oxidation/thermal degradation plays a critical role in the oxidation of ethylbenzene to acetophenone.

To investigate the oxidation reaction process during thermal degradation of PEG by *in situ* generation of radicals, the radical species, the trace formylation

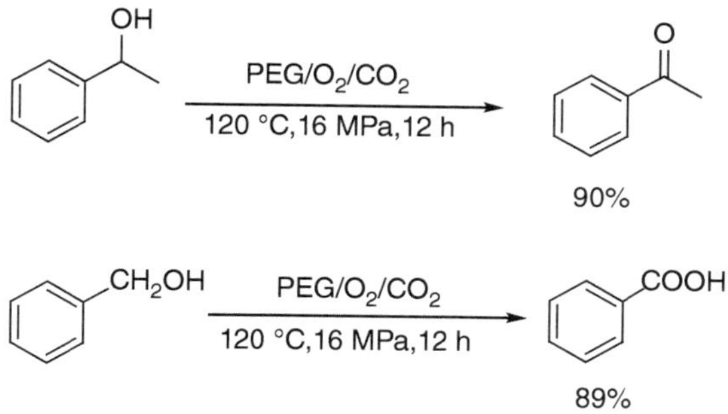

Scheme 1.65 PEG radical-initiated oxidation of benzylic alcohols in compressed carbon dioxide.

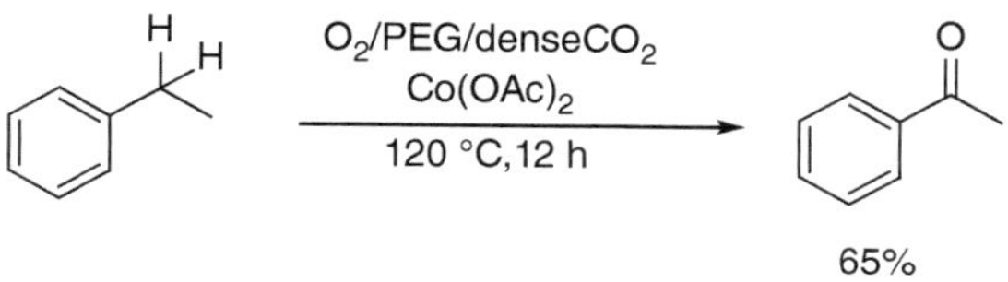

Scheme 1.66 PEG radical-initiated benzylic C—H bond oxygenation in compressed carbon dioxide.

products under the action of PEG radicals, the extended distribution of relative molecular masses of PEG, the generation of peroxide intermediates determined by KI/starch method, and the inhibition of the reaction by TEMPO (2,2,6,6-tetramethylpiperidine-1-yloxy) have been investigated, respectively. Then a rationalized radical mechanism is proposed, as shown in Scheme 1.67 [87]. The PEG radical generated by the reaction of PEG with oxygen initiates the substrate to produce a relatively stable radical, e.g. b–g, depending on the substrate structure. As expected, benzyl or allyl may favor the formation of benzyl radicals or allyl radicals, and thus primarily undergo oxidation, such as benzyl alcohol oxidation, benzyl C–C cleavage reactions, and benzyl/allyl sp_3 C–H oxidation, while aliphatic alcohols may preferentially pass through the formylation pathway.

1.5.3 Ring-Opening Reaction

Fluorinated alcohols are good hydrogen bond donors and have high polarity, high ionizing power, and low nucleophilicity. Therefore, TFE and PFTB can promote the opening of epoxides by heteroatom or carbon nucleophiles. TFE and PFTB both contain highly nucleophilic fluorine atoms, which can act as nucleophilic reagents to react with epoxides, attacking the carbon–oxygen bond in epoxide molecules and opening up the ring structure. The fluorine atoms in fluorinated alcohols can polarize the O—H bond in the alcohol, making it more polar and thus more easily attracting nucleophilic reagents. In addition, fluorine atoms can also increase the electrophilicity of alcohols, thereby reducing the bond energy of the C—O bond and making epoxides easier to open. Therefore, fluorinated alcohols can promote the

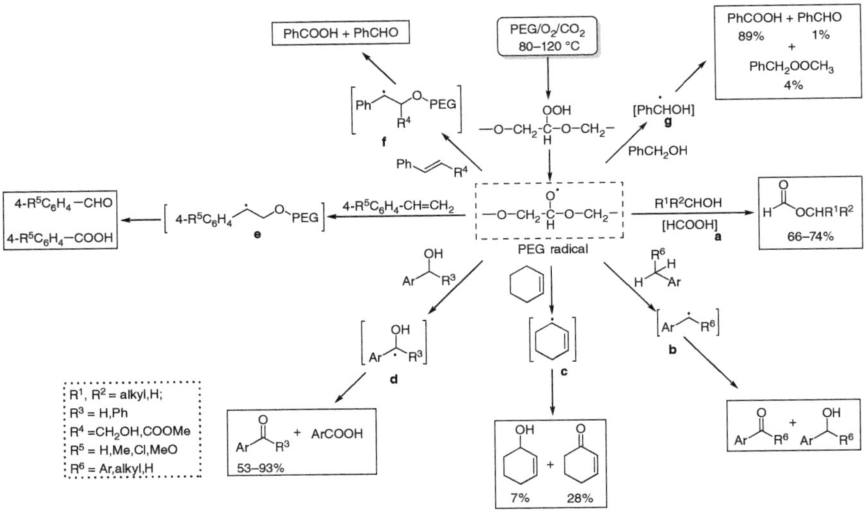

Scheme 1.67 The structural formulae and proposed pathways of organic reactions initiated by PEG radicals generated by thermal oxidative degradation.

ring-opening reaction of epoxides by increasing the electrophilicity of oxygen atoms in epoxides and reducing the bond energy of the C—O bond.

Epoxides are important and useful intermediates in organic synthesis because of their easy formation and high reactivity. In this method, TFE serves as both a reusable catalyst and medium, while aliphatic and aromatic amines act as nucleophiles to efficiently open the epoxide ring, resulting in the formation of β-amino alcohols with high yield and good regioselectivity (Scheme 1.68). It is worth noting that the high polarity and hydrogen bonding interactions between TFE and epoxides may be the reasons for promoting the reaction [90].

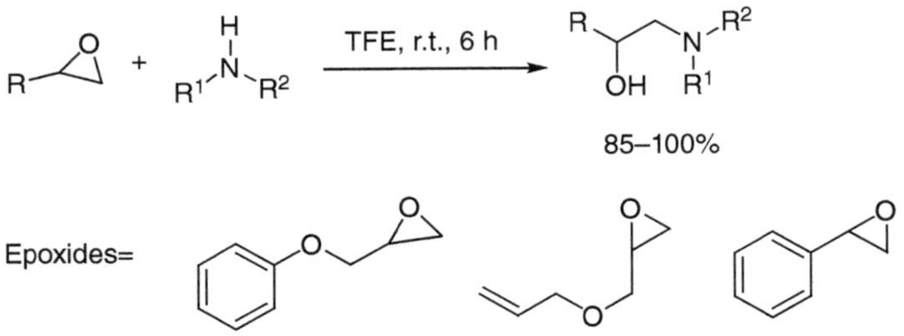

R^1=C_6H_5, 4-ClC_6H_4, 4-BrC_6H_4, $C_6H_5CH_2$,
R^2=H,
R^1, R^2=morpholine, piperidine, tertiary amine

Scheme 1.68 Ring opening of epoxides with amines.

In addition, the ring opening of epoxy alcohols usually occurs through the 5-*exo* selective pathway of producing THF rings, rather than through the desired 6-*endo* selective pathway of producing tetrahydropyran rings. The synergistic effect of PFTB and 1-ethyl-3-methylimidazolium tetrafluoroborate ([EMIM]BF_4) can promote the internal selective ring-opening reaction of trisubstituted epoxides (Scheme 1.69). Starting from readily accessible homochiral polyepoxy alcohols with a methyl group at all the *endo*-cyclization sites, polyethers up to five consecutive fused 6-, 7-, and/or 8-membered rings can be constructed in one step. The partial positive charge on the stable transition state of BF_4 salt is used to further improve the internal selectivity. In the presence of [EMIM]BF_4, the reaction rate increases due to the small amount of HF produced by the dissolution of the BF_4 anion in the PFTB. When using [EMIM]BF_4, which is more easily soluble in PFTB than Ph_4PBF_4, the internal selectivity slightly increases [91]. Therefore, through the combined use of PFTB and [EMIM]BF_4, a selective inversion from *exo*-selectivity to *endo*-selectivity can be achieved, thereby quickly and effectively synthesizing the multi-ring porous core structure of marine stepped polyether molecules.

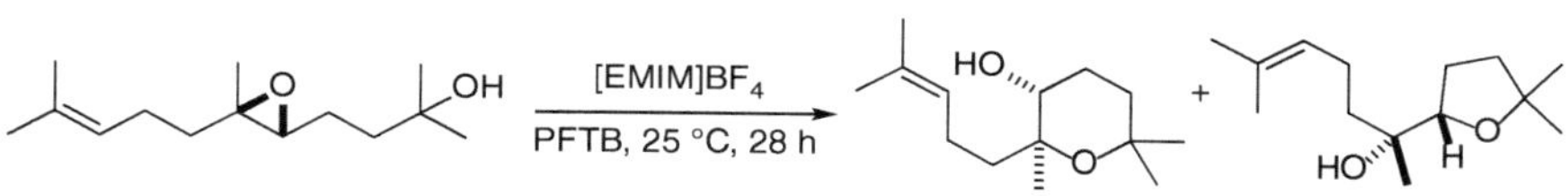

Scheme 1.69 The synergistic action of PFTB and [EMIM]BF_4 promotes the internal selective ring-opening reaction of triple-substituted epoxides.

1.6 Summary and Concluding Remarks

In this chapter, the different reactions involving various alcohol as solvents are summarized, among which the different roles of glycerol, PEG, and fluorinated alcohols in organic reactions are mainly introduced according to the functional characteristics of alcohols.

Alcohols are ideal media for many reaction processes as a cheap, abundantly available, and low-toxic solvent, mainly in reduction reactions, coupling reactions, and condensation reactions, but also in other reactions. Some bio-based alcohols that can be prepared by petrochemical industry procedures or from renewable sources are widely used in laboratories and industries, such as bioethanol, 2-propanol, and glycerol. In particular, glycerol, as an important green solvent combining the advantages of water and ionic liquids, has great advantages in improving the catalytic rate and selectivity in organic synthesis, product separation, and catalyst and solvent recovery, which provides an innovative solution to replace traditional volatile solvents. Interestingly, PEG is used as a green and cheap solvent for various types of organic transformations due to its good compatibility. Secondly, it has remarkable cationic complexing ability as an acyclic analog of crown ethers. In metal-catalyzed reactions, the catalytic system "PEG-metal" can be easily recovered and reused without significant loss of activity. Furthermore, fluorinated alcohols have unique properties and are a very attractive solvent for organic reactions, allowing for many challenging reactions, although it is not very green.

Although great strides have been made by utilizing alcohol as a solvent and a variety of valuable chemicals have been obtained, there is still major room for improvement in the future:

(1) The reason for glycerol as a solvent to facilitate the reaction needs to be further investigated in depth. Second, technical-grade glycerol (80%) is more desirable as a green solvent for direct industrial and laboratory applications, but there are few successful examples. Finally, glycerol as a high-boiling-point polar solvent faces the same problem as many ionic liquids, How to remove the polar compounds from glycerol? This is an important question that needs to be addressed.
(2) The interactions and mechanical details of the multiphase catalytic interface in the presence of PEG as a solvent are yet to be explored in depth. Second, the solvation of solutes in PEG is largely unknown, and the interaction of PEG with metal cations and how it affects the solvation of resistant anions also remains to be addressed. In addition, studies on how the phase behavior of PEG enhances product recovery are yet to be thoroughly investigated.
(3) Fluorinated alcohols are less used in green chemistry due to their corrosiveness, toxicity, and cost, and it is critical to develop methods that reduce the use of fluorinated solvents while continuing to utilize their inherent properties. Moreover, the interaction mechanism between fluorinated alcohols and substrates in the reaction process needs to be further studied.

In short, we hope that the content presented in this chapter will inspire organic chemists to give more consideration to alcohol solvents in their future research to achieve the goal of green chemistry.

Acknowledgments

This work was financially supported by Nankai Cangzhou Bohai New Area Green Chemical Research Co. Ltd. (grant no. 20220130), the Natural Science Foundation of Tianjin Municipal Science and Technology Commission (21JCZDJC00100), and the Fundamental Research Funds for the Central Universities, Nankai University.

References

1 Clarke, C.J., Tu, W.C., Levers, O. et al. (2018). *Chem. Rev.* 118: 747–800.

2 (a) Shanab, K., Neudorfer, C., Schirmer, E. et al. (2013). *Curr. Org. Chem.* 17: 1179–1187. (b) Calvo-Flores, F.G., Monteagudo-Arrebola, M.J., Dobado, J.A. et al. (2018). *Top. Curr. Chem.* 376: 18.

3 (a) Jessop, P.G., Jessop, D.A., Fu, D. et al. (2012). *Green Chem.* 14: 1245–1259. (b) Quispe, C.A.G., Coronado, C.J.R., and Carvalho, J.A. Jr., (2013). *Renewable*

Sustainable Energy Rev. 27: 475–493. (c) Shuklov, I.A., Dubrovina, N.V., and Börner, A. (2007). *Synthesis* 2007: 2925–2943. (d) Freed, B.K., Biesecker, J., and Middleton, W.J. (1990). *J. Fluorine Chem.* 48: 63–75. (e) Kamlet, M.J., Abboud, J.L.M., Abraham, M.H. et al. (2002). *J. Organomet. Chem.* 48: 2877–2887. (f) Hansen, C.M. (2007). *Hansen Solubility Parameters: A User's Handbook*, 2e. Boca Raton, London, New York: CRC Press. (g) Reichardt, C. (2003). *Solvents and Solvent Effects in Organic Chemistry*, 3e. Weinheim: Wiley.

4 Alfonsi, K., Colberg, J., Dunn, P.J. et al. (2008). *Green Chem.* 10: 31–36.

5 (a) Jiménez-González, C., Curzons, A.D., Constable, D.J. et al. (2001). *Clean Prod. Processes* 3: 35–41. (b) Prat, D., Wells, A., Hayler, J. et al. (2016). *Green Chem.* 18: 288–296. (c) Jiménez-González, C., Curzons, A.D., Constable, D.J.C. et al. (2004). *Clean Technol. Environ. Policy* 7: 42–50.

6 Prat, D., Pardigon, O., Flemming, H.-W. et al. (2013). *Org. Process Res. Dev.* 17: 1517–1525.

7 Yue, H., Zhao, Y., Ma, X. et al. (2012). *Chem. Soc. Rev.* 41: 4218–4244.

8 Anitha, M., Kamarudin, S.K., and Kofli, N.T. (2016). *Chem. Eng. J.* 295: 119–130.

9 Gu, Y. and Jérôme, F. (2010). *Green Chem.* 12: 1127–1138.

10 Sonnati, M.O., Amigoni, S., Taffin de Givenchy, E.P. et al. (2013). *Green Chem.* 15: 283–306.

11 Chen, J., Spear, S.K., Huddleston, J.G. et al. (2005). *Green Chem.* 7: 64–82.

12 Hoffmann, M.M. (2022). *Curr. Opin. Colloid. In.* 57: 101537–101553.

13 (a) Santaniello, E., Manzocchi, A., and Sozzani, P. (1979). *Tetrahedron Lett.* 20: 4581–4582. (b) Wang, M.-L. and Chang, K.-R. (1991). *Can. J. Chem. Eng.* 69: 340–346.

14 Totten, G.E. and Clinton, N.A. (1998). *J. Macromol. Sci., Polym. Rev.* 38: 77–142.

15 Vafaeezadeh, M. and Hashemi, M.M. (2015). *J. Mol. Liq.* 207: 73–79.

16 Joule, J.A., Dover, T.L., and McDonald, F.E. (2021). *Arkivoc* 2021: 85–114.

17 Khaksar, S. (2015). *J. Fluorine Chem.* 172: 51–61.

18 Colomer, I., Chamberlain, A.E.R., Haughey, M.B. et al. (2017). *Nat. Rev. Chem.* 1: 0088–0099.

19 Capello, C., Fischer, U., and Hungerbühler, K. (2007). *Green Chem.* 9: 927–934.

20 Wu, Z., Ayad, T., and Ratovelomanana-Vidal, V. (2011). *Org. Lett.* 13: 3782–3785.

21 Wang, T., Chen, Y., Ouyang, G. et al. (2016). *Chem. Asian J.* 11: 2773–2777.

22 Bhattacherjee, D., Shaifali, K. et al. (2021). *Mol. Catal.* 514: 111836–111841.

23 Cabrera, D.M.L., Líbero, F.M., Alves, D. et al. (2011). *Green Chem. Lett. Rev.* 5: 329–336.

24 Li, B., Liu, A.-H., He, L.-N. et al. (2012). *Green Chem.* 14: 130–135.

25 Kidwai, M., Bhardwaj, S., and Jain, A. (2012). *Green Chem. Lett. Rev.* 5: 195–202.

26 Ma, R., Huang, C.-B., Liu, A.-H. et al. (2014). *Catal. Sci. Technol.* 4: 4308–4312.

27 Thurow, S., Webber, R., Perin, G. et al. (2013). *Tetrahedron Lett.* 54: 3215–3218.

28 Bender, C.F. and Widenhoefer, R.A. (2006). *Org. Lett.* 8: 5303–5305.

29 Rodrigues, J.M., Calhelha, R.C., Ferreira, I.C.F.R. et al. (2020). *Tetrahedron Lett.* 61: 151900–151903.

30 Feu, K.S., de la Torre, A.F., Silva, S. et al. (2014). *Green Chem.* 16: 3169–3174.

31 Rodriguez-Alvarez, M.J., Garcia-Alvarez, J., Uzelac, M. et al. (2018). *Chem. Eur. J.* 24: 1720–1725.
32 Phukan, P., Chetia, R., Boruah, R. et al. (2021). *Mater. Adv.* 2: 6996–7006.
33 Vidal, C. and García-Álvarez, J. (2014). *Green Chem.* 16: 3515–3521.
34 Francos, J. and Cadierno, V. (2010). *Green Chem.* 12: 1552–1555.
35 Wang, S., Guo, Y.Q., Ren, Z.H. et al. (2017). *Org. Lett.* 19: 1574–1577.
36 Jiang, Y.-Q., Li, J., Feng, Z.-W. et al. (2020). *Adv. Synth. Catal.* 362: 2609–2614.
37 Chen, Z., Huang, W., Yi, L. et al. (2022). *Green Chem.* 24: 2919–2926.
38 Penteado, F., Gomes, C.S., Perin, G. et al. (2019). *J. Organomet. Chem.* 84: 7189–7198.
39 Mayekar, N.V., Nayak, S.K., and Chattopadhyay, S. (2004). *Synth. Commun.* 34: 3111–3119.
40 Venkateswarlu, C., Balaji, P.V., De, K. et al. (2013). *J. Fluorine Chem.* 152: 94–98.
41 Wolfson, A. and Dlugy, C. (2007). *Chem. Pap.* 61: 228–232.
42 Marset, X., De Gea, S., Guillena, G. et al. (2018). *ACS Sustainable Chem. Eng.* 6: 5743–5748.
43 Delample, M., Villandier, N., Douliez, J.-P. et al. (2010). *Green Chem.* 12: 804–808.
44 (a) Bagherzadeh, M., Hosseini, H., and Salami, R. (2019). *Appl. Organomet. Chem.* 34: e5287–e5296. (b) Hajipour, A.R. and Azizi, G. (2013). *Green Chem.* 15: 1030–1034.
45 Bathula, C., Subalakshmi, K., Ashok Kumar, K. et al. (2020). *Colloids Surf., B* 192: 111026–111032.
46 Gaikwad, V.V. and Bhanage, B.M. (2019). *Appl. Organomet. Chem.* 33: e4741–e4748.
47 Ricordi, V.G., Freitas, C.S., Perin, G. et al. (2012). *Green Chem.* 14: 1030–1034.
48 Chahdoura, F., Pradel, C., and Gómez, M. (2014). *ChemCatChem* 6: 2929–2936.
49 Szyling, J., Walkowiak, J., Sokolnicki, T. et al. (2019). *J. Catal.* 376: 219–227.
50 Jawale, D.V., Lingampalle, D.L., Pratap, U.R. et al. (2010). *Chin. Chem. Lett.* 21: 412–416.
51 Azizi, N., Davoudpour, A., Eskandari, F. et al. (2013). *Monatsh. Chem.* 144: 405–409.
52 Zhang, Z., Gao, X., Wan, Y. et al. (2017). *ACS Omega* 2: 6844–6851.
53 Tajbakhsh, M., Hosseinzadeh, R., Alinezhad, H. et al. (2010). *Synthesis* 2011: 490–496.
54 Khaksar, S. and Alipour, M. (2012). *Monatsh. Chem.* 144: 395–398.
55 Khaksar, S., Heydari, A., Tajbakhsh, M. et al. (2010). *J. Fluorine Chem.* 131: 1377–1381.
56 (a) Radatz, C.S., Silva, R.B., Perin, G. et al. (2011). *Tetrahedron Lett.* 52: 4132–4136. (b) Zhou, W.J., Zhang, X.Z., Sun, X.B. et al. (2013). *Russ. Chem. Bull.* 62: 1244–1247. (c) Bachhav, H.M., Bhagat, S.B., and Telvekar, V.N. (2011). *Tetrahedron Lett.* 52: 5697–5701. (d) Sadek, K.U., Mekheimer, R.A., Hameed, A.M. et al. (2012). *Molecules* 17: 6011–6019.

57 Mali, J.R., Jawale, D.V., Londhe, B.S. et al. (2010). *Green Chem. Lett. Rev.* 3: 209–212.

58 Li, M., Chen, C., He, F. et al. (2010). *Adv. Synth. Catal.* 352: 519–530.

59 Tan, J.-N., Li, M., and Gu, Y. (2010). *Green Chem.* 12: 908–914.

60 Fañanás, F.J., Arto, T., Mendoza, A. et al. (2011). *Org. Lett.* 13: 4184–4187.

61 Tabibi, T. and Esmaeili, A.A. (2023). *Mol. Diversity* 27: 477–486.

62 Gu, Y., Barrault, J., and Jérôme, F. (2008). *Adv. Synth. Catal.* 350: 2007–2012.

63 Azizi, N. and Marimi, M. (2013). *Environ. Chem. Lett.* 11: 371–376.

64 Nandre, K.P., Salunke, J.K., Nandre, J.P. et al. (2012). *Chin. Chem. Lett.* 23: 161–164.

65 Yadav, D.K.T., Rajak, S.S., and Bhanage, B.M. (2014). *Tetrahedron Lett.* 55: 931–935.

66 Wu, F., Yan, F., Wu, L. et al. (2022). *Appl. Organomet. Chem.* 36: e6618–e6624.

67 He, F., Li, P., Gu, Y. et al. (2009). *Green Chem.* 11: 1767–1773.

68 Safaei, H.R., Shekouhy, M., Rahmanpur, S. et al. (2012). *Green Chem.* 14: 1696–1704.

69 Khaksar, S., Rouhollahpour, A., and Talesh, S.M. (2012). *J. Fluorine Chem.* 141: 11–15.

70 Nemati, F., Hosseini, M.M., and Kiani, H. (2016). *J. Saudi Chem. Soc.* 20: S503–S508.

71 Wang, L.-N., Shen, S.-L., and Qu, J. (2014). *RSC Adv.* 4: 30733–30741.

72 Heydari, A., Khaksar, S., and Tajbakhsh, M. (2009). *Tetrahedron Lett.* 50: 77–80.

73 Ingale, A.P., More, V.K., Gangarde, U.S. et al. (2018). *New J. Chem.* 42: 10142–10147.

74 Díaz-Álvarez, A. and Cadierno, V. (2013). *Appl. Sci.* 3: 55–69.

75 Wolfson, A., Dlugy, C., Shotland, Y. et al. (2009). *Tetrahedron Lett.* 50: 5951–5953.

76 (a) Gawande, M.B., Rathi, A.K., Branco, P.S. et al. (2012). *Chem. Eur. J.* 18: 12628–12632. (b) Moran, M.J., Martina, K., Stefanidis, G.D. et al. (2020). *Front. Chem.* 8: 34.

77 Zhang, Y.M., Yuan, M.L., Liu, W.P. et al. (2018). *Org. Lett.* 20: 4486–4489.

78 Long, J., Shen, K., and Li, Y. (2016). *ACS Catal.* 7: 275–284.

79 Cui, X., Deng, Y., and Shi, F. (2013). *ACS Catal.* 3: 808–811.

80 Díaz-Álvarez, A.E., Crochet, P., and Cadierno, V. (2011). *Catal. Commun.* 13: 91–96.

81 Serrano-Maldonado, A., Dang-Bao, T., Favier, I. et al. (2020). *Chem. Eur. J.* 26: 12553–12559.

82 García, N., García-García, P., Fernández-Rodríguez, M.A. et al. (2013). *Green Chem.* 15: 999–1005.

83 Yang, Z.-Z., He, L.-N., Zhao, Y.-N. et al. (2011). *Energy Environ. Sci.* 4: 3971–3975.

84 Liu, A.-H., Ma, R., Song, C. et al. (2012). *Angew. Chem. Int. Ed.* 124: 11468–11472.

85 Kong, D.-L., He, L.-N., and Wang, J.-Q. (2011). *Synth. Commun.* 41: 3298–3307.

86 Wang, J.-L., He, L.-N., Dou, X.-Y. et al. (2009). *Aust. J. Chem.* 62: 917–920.

87 Wang, J.-Q., He, L.-N., Miao, C.-X. et al. (2009). *ChemSusChem* 2: 755–760.
88 Wang, J.-Q., He, L.-N., and Miao, C.-X. (2009). *Green Chem.* 11: 1013–1017.
89 Wang, J.-Q. and He, L.-N. (2009). *New J. Chem.* 33: 1637–1640.
90 Khaksar, S., Heydari, A., Tajbakhsh, M. et al. (2010). *J. Fluorine Chem.* 131: 106–110.
91 Li, F.X., Ren, S.J., Li, P.F. et al. (2020). *Angew. Chem. Int. Ed.* 59: 18473–18478.

2

Recent Achievements in Organic Reactions in MeCN

Tongtong Xing, Guizhi Zhai, Linna Wu, Xiaofen Wang, and Zechao Wang

Zhengzhou University, Henan Institute of Advanced Technology, Division of Molecular Catalysis & Synthesis, No.100 Science Avenue, Zhengzhou, 450001, P. R. China

2.1 Introduction

Acetonitrile, also known as cyanomethane or methane carbonitrile, is a chemical compound with the formula CH_3CN. It is a clear colorless liquid with a sweet ethereal odor [1]. Compared with water, it is less dense than water. Acetonitrile is easily soluble in water and is quite flammable [2]. Usually, it is used as a medium-polarity solvent in the laboratory [3]. Due to its properties and low toxicity, it is predominantly used as a solvent in the pharmaceutical manufacturing processes, for spinning fibers and for casting and molding of plastic materials, in lithium batteries, for the extraction of fatty acids from animal and vegetable oils, and in chemical laboratories for the detection of materials such as pesticide residues [4]. Acetonitrile is also used in dyeing textiles and coating compositions as a stabilizer for chlorinated solvents and in perfume production as a chemical intermediate.

Acetonitrile is classified as a nitrile in terms of its functional group. In organic chemistry, nitrile is defined as a carbon atom that contains a triple bond to a nitrogen atom [5]. It is the simplest organic nitrile, which contains a carbon-nitrogen triple bond. There are different reactivities between the two carbons and the nitrogen atom of acetonitrile. The proton on the methyl group shows slight acidity and can be deprotonated to form a nucleophile. The triple-bonded carbon behaves as an electrophilic center. The lone pair of electrons on the nitrogen can act as a nucleophile on reactive electrophiles and as a coordination site for metals [6]. Owing to the structure of CH_3CN, it can act as a multipurpose building block in organic synthesis [7]. Therefore, acetonitrile can participate and act as a green solvent in many reactions such as addition, oxidation, reduction, and substitution reactions.

Recently, some reviews on the different roles of acetonitrile in reactions have been published [8]. The aim of this chapter is to give an overview of the various reactions that involve acetonitrile.

Green Solvents in Organic Synthesis, First Edition.
Edited by Xiao-Feng Wu, Zhiping Yin, Liang-nian He, and Feng Wang.

2.2 MeCN in Transition Metal-catalyzed Reactions Without Radicals Involved

2.2.1 Transition Metal-catalyzed Addition Reactions in MeCN

In 2005, Ozerov and coworkers explored an efficient nickel-catalyzed coupling of acetonitrile with aldehydes (Scheme 2.1) [9]. The catalyst was robust and easy to handle, and the catalysis was suitable for various aldehydes under mild conditions. *p*-Fluorobenzaldehyde was chosen as the test substrate because of the convenience of the flourine nuclear magnetic resonance (F NMR). The reaction was not only catalyzed by the Brønsted acids or bases but also Ni-catalyst was crucial for selective catalysis. 1,8-Diazabicyclo[5.4.0]undec-7-ene (DBU) was employed as the base and the optimal temperature was 45–50 °C.

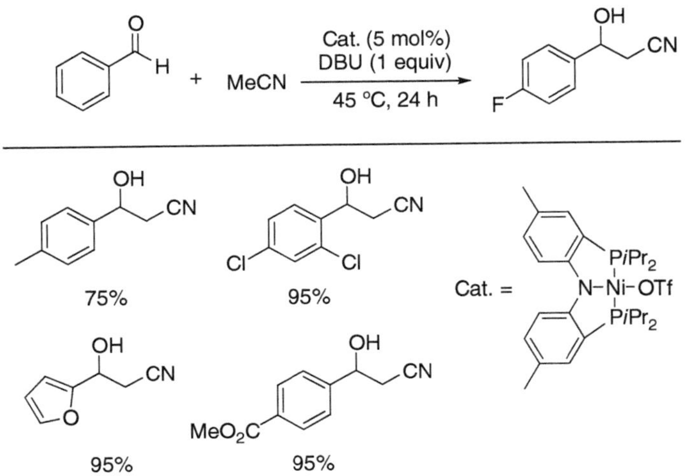

Scheme 2.1 Ni-catalyzed coupling of acetonitrile with aldehydes.

Guan and coworkers showed a cyanomethylation of aldehydes catalyzed by an amazingly robust nickel catalyst (Scheme 2.2) [10]. In this reaction, acetonitrile worked as a reactant and as a solvent at the same time. This reaction yielded good-to-high yields with its corresponding products. The aldehydes were inserted into a C—Ni bond cyanomethyl complex reversibly and then the acetonitrile was activated to form an alkoxide intermediate. The paper ascribed the thermodynamic driving force for the latter step to the relief of dp–pp repulsion present in the nickel alkoxide species.

Santos et al. described the diboration of substituted propargylic alcohols using the bimetallic Pd/Cu catalyst system (Scheme 2.3) [11]. The desired organoboron products were necessary in medicinal chemistry. The authors discovered that pentrafluoroboronic acid intermediate was formed *in situ*, which sufficiently activated the C—O bond to achieve dual catalysis to obtain stereoselective *Z*-allyl and

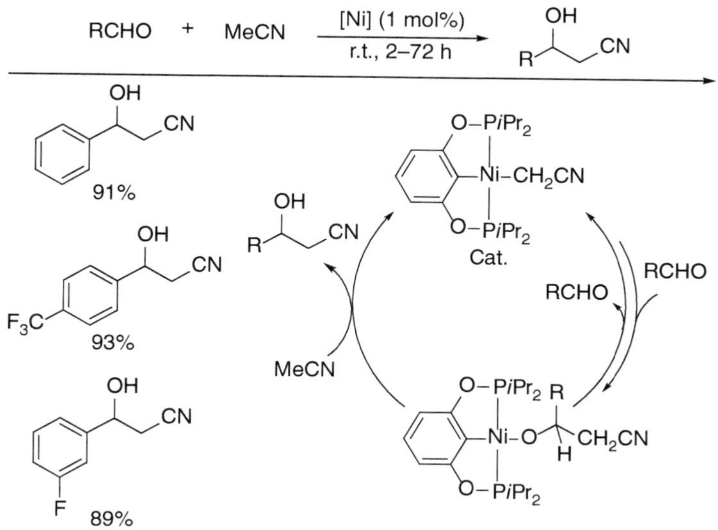

Scheme 2.2 Ni-catalyzed cyanomethylation of aldehydes.

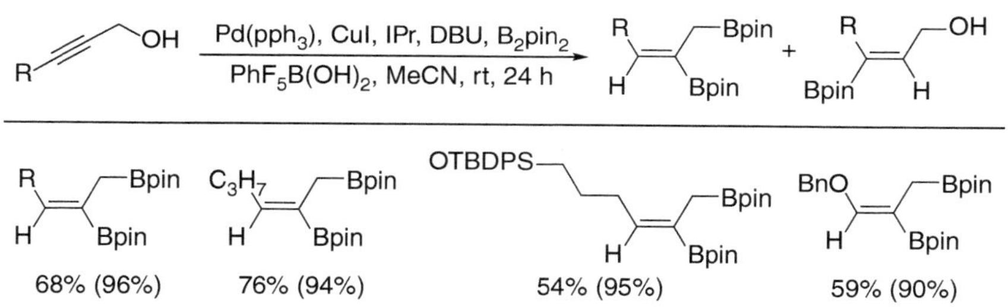

Scheme 2.3 Pd/Cu-catalyzed diboration of propargylic alcohols.

vinyldiboronates. The reaction was carried out at room temperature in acetonitrile with moderate-to-good yields and high stereoselectivity.

Tokuyama and coworkers developed a copper-mediated tandem-catalyzed double functionalization process (Scheme 2.4) [12]. The reaction underwent copper-catalyzed aziridination and nucleophilic ring-opening. These, in turn, were triggered by the addition of $(NH_4)_2S_2O_8$, which worked as an oxidant for the *in situ* conversion of the copper catalyst to form various C—C and C—X bonds. The strategy consisted of three contiguous catalytic cycles and also could be carried out in a one-pot process to obtain functionalized indolines with moderate-to-excellent yields.

Li and coworkers reported a FeI_2-catalyzed reductive Strecker reaction, which was widely used for the synthesis of amino acids from aldehydes and cyanides (Scheme 2.5) [13]. This methodology produced aminoacetonitriles, which were key intermediates to many herbicides and pharmaceuticals through reductive cyanation of formamides. The big challenge of over-reduction of active

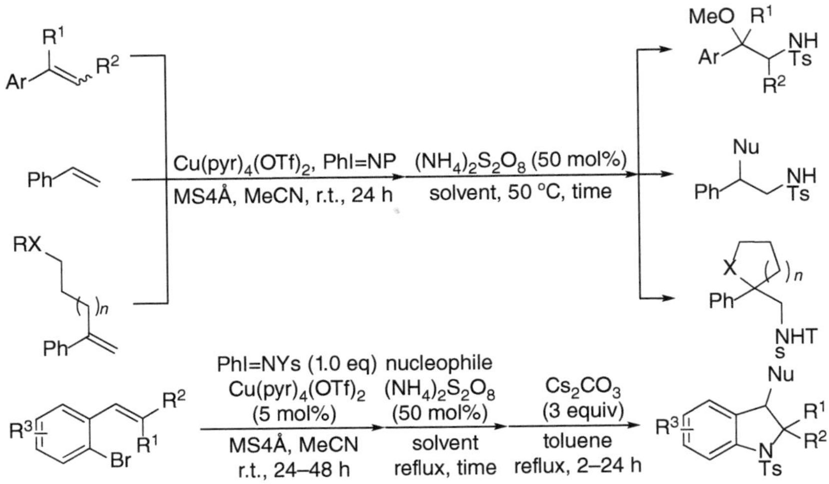

Scheme 2.4 Cu-mediated assisted tandem catalyzed double functionalization.

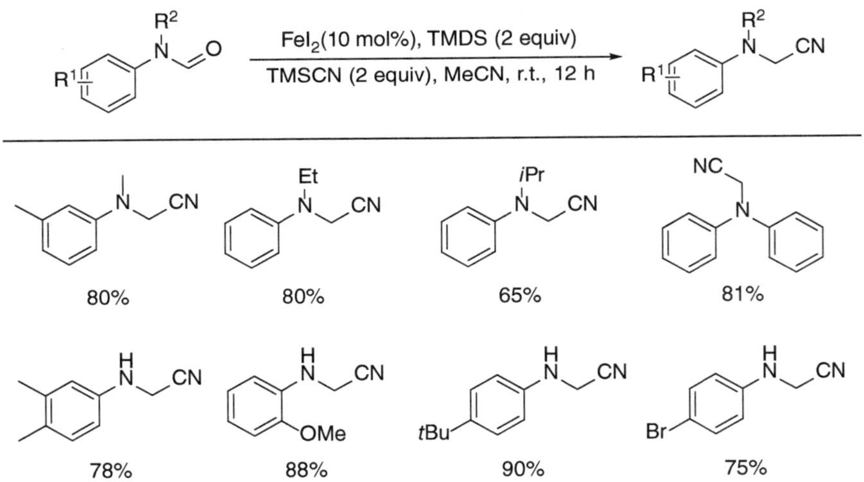

Scheme 2.5 Iron-catalyzed reductive Strecker reaction.

minimum/carbinolamine intermediates was resolved via Fe catalysis. This reaction tolerated various secondary and tertiary formamides under this catalytic system and exhibited moderate-to-excellent yields.

Bao and coworkers developed a method for the copper-catalyzed synthesis of diamines (Scheme 2.6) [14]. These diamines occupied an important position in natural products, drug design, and organic synthesis. This method achieved a regioselective intermolecular three-component vicinal deamination of styrenes with azodicarboxylates and acetonitrile, which was also used in subsequent Ritter reaction. The reaction produced vicinal diamines in moderate-to-excellent yields, featuring great substrate scope and functional group tolerance.

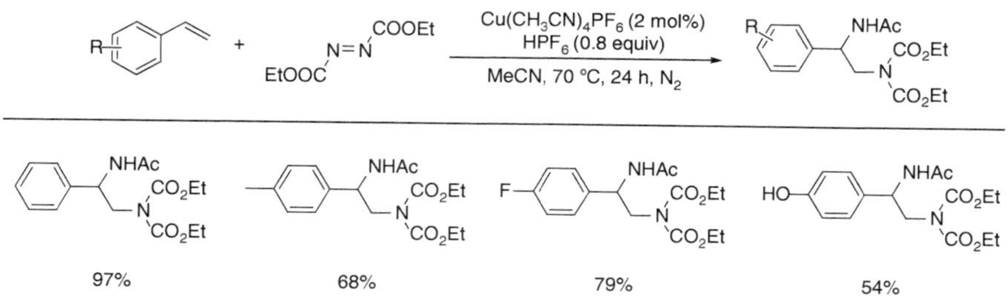

Scheme 2.6 Cu-catalyzed diamination of styrenes.

You and coworkers synthesized tetrahydrofuroindoles through a palladium-catalyzed asymmetric dearomative formal [3+2] cycloaddition of nitroindoles with epoxybutenes (Scheme 2.7) [15]. The polarity of the solvent played a pivotal role in the diastereoselectivity. In toluene and acetonitrile, tetrahydrofuroindoles with different diastereoisomers were produced with good-to-excellent yields and stereoselectivity. When MeCN was used as the solvent, the reactions occurred faster to produce the diastereoisomer with higher enantioselectivity. Substrates bearing both electron-withdrawing groups and electron-donating groups reacted successfully with satisfactory yields, diastereoselectivity, and enantioselectivity.

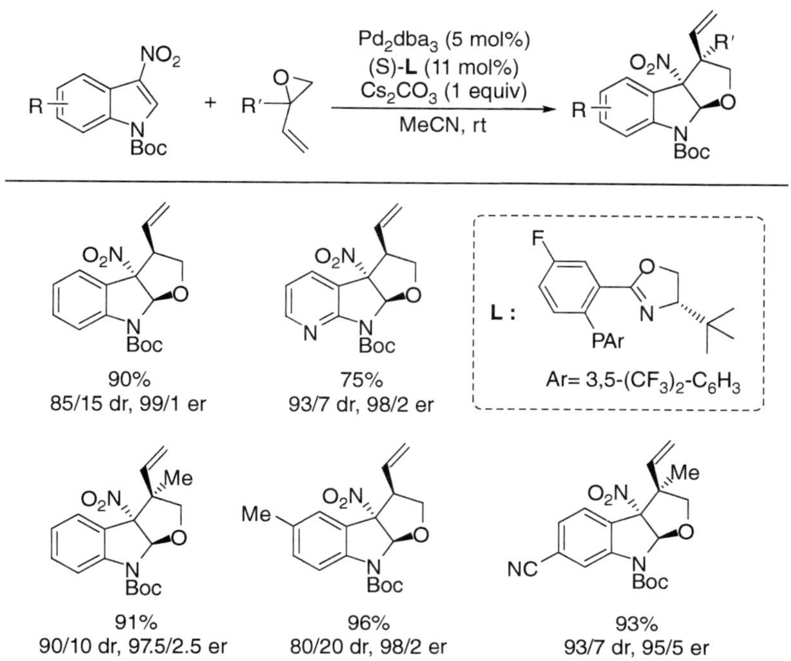

Scheme 2.7 Palladium-catalyzed synthesis of tetrahydrofuroindoles.

2.2.2 Transition Metal-catalyzed Oxidation Reactions in MeCN

In 2002, Andrus and Zhou reported a highly selective process using a panel of eight malonyl-derived bisoxazolines together with *t*-butyl-*p*-nitroperbenzoate to yield enantiomerically enriched ester products with cyclic olefins (Scheme 2.8) [16]. Asymmetric allylic oxidation of cyclohexene and oxidation of cyclopentene/cycloheptene/COD were included. The copper-bisoxazoline-catalyzed allylic oxidation of cyclic olefins with *t*-butyl-*p*-nitroperbenzoate was carried out in acetonitrile at −20 °C with high enantioselectivity.

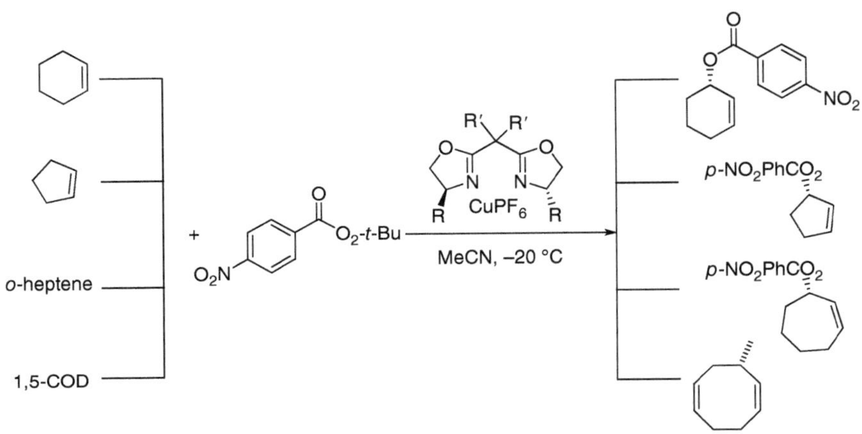

Scheme 2.8 Copper-bisoxazoline-catalyzed allylic oxidation.

In 2006, Wilson and coworkers developed the application of chiral nonracemic and C2-symmetric 2,2′-bipyridyl ligand in copper(I)-catalyzed asymmetric allylic oxidation reactions, which had the highest reported enantioselectivity (up to 91% ee) for a bipyridyl ligand copper(I) complex to date (Scheme 2.9) [17]. The ligand was used in the copper(I)-catalyzed asymmetric allylic oxidation reactions of the cyclic alkenes with tert-butyl peroxybenzoate. The yields and enantioselectivities

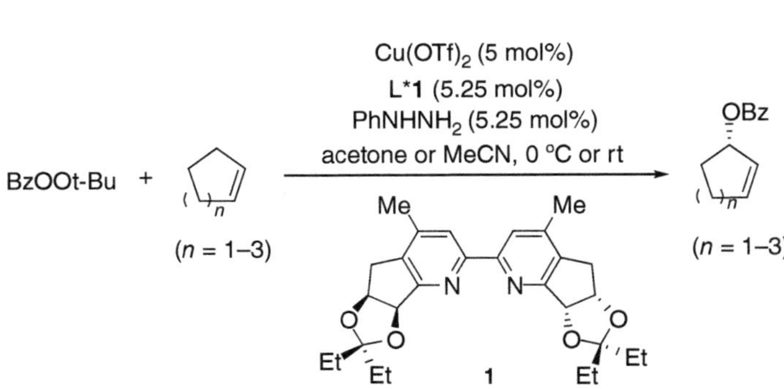

Scheme 2.9 Copper(I)-catalyzed asymmetric allylic oxidation.

were found to be dependent on the substrate, solvent, and reaction temperature. Acetonitrile was an important solvent in this reaction.

In 2010, Kirillov and coworkers demonstrated a homogeneous catalytic system of copper catalyst, *t*-BuOOH (TBHP), water, and acetonitrile solvent (Scheme 2.10) [18]. The system was applied for the mild oxidation of alkanes and alcohols, including glycerol. The oxidation reactions of alkanes and alcohols were carried out in round-bottom flasks with vigorous stirring and using CH_3CN as the solvent under air. The copper catalyst was introduced into the reaction mixture in the form of a stock solution in water. The substrate was then added, and the reaction started when TBHP (70% in H_2O) or hydrogen peroxide (50% in H_2O) was added in one portion. The reactions were stopped by cooling.

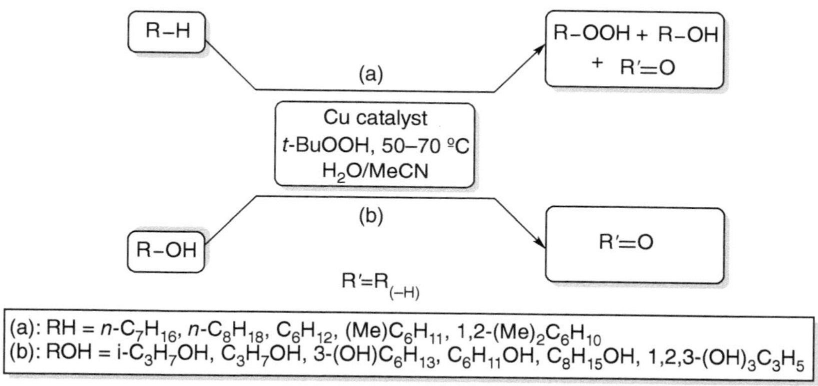

Scheme 2.10 Oxidation of alkanes and alcohols via Cu catalyst/TBHP system.

In 2011, Baiker and coworkers synthesized Cu-BTC and mixed-linker Cu-BTC-PyDC (Scheme 2.11) [19]. Both these copper catalysts could catalyze the demanding direct hydroxylation of toluene, both in acetonitrile and in neat substrate. Acetonitrile was used as the solvent because of the solubility of both reactants (toluene and hydrogen peroxide) in this method. Furthermore, it was also conducive to the reaction. The Cu-BTC and the mixed-linker copper catalyst indicated surprisingly high selectivity in comparison with homogeneously dissolved Cu salts.

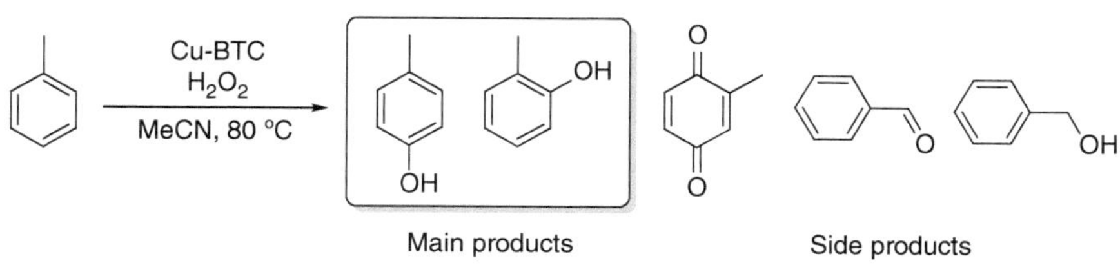

Scheme 2.11 Oxidation of toluene.

In 2016, He and coworkers illustrated the Cu/Fe-catalyzed intermolecular oxidative amination of benzylic C—H bonds to carboxylic groups using acetonitrile as the nitrogen source and solvent (Scheme 2.12) [20]. They used 2-methylbenzoic acid as

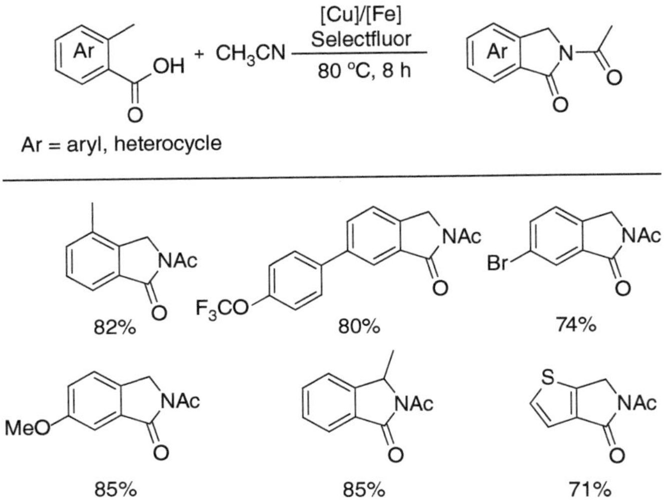

Scheme 2.12 Oxidative amination of benzylic C—H bonds.

the model substrate with Selectfluor as the oxidant. The procedure was carried out at 80 °C for 8 hours. Heterocyclic substrates and primary and secondary benzylic were sufficiently reactive.

In 2017, Liu and coworkers reported the synergetic oxidation of ethylbenzene catalyzed by manganese(II) complexes bearing pendant iodophenyl groups (Scheme 2.13) [21]. Electron paramagnetic resonance (EPR) and Fourier transform infrared spectroscopy (FTIR) data demonstrated that Mn(IV) was in the active oxidation state. The 1:1 mixture of CH_3CN and H_2O was employed as solvent. The reaction was found to give the best result at room temperature for 16 hours with Oxone as an oxidant. The reaction procedure was successfully extended to various hydrocarbons.

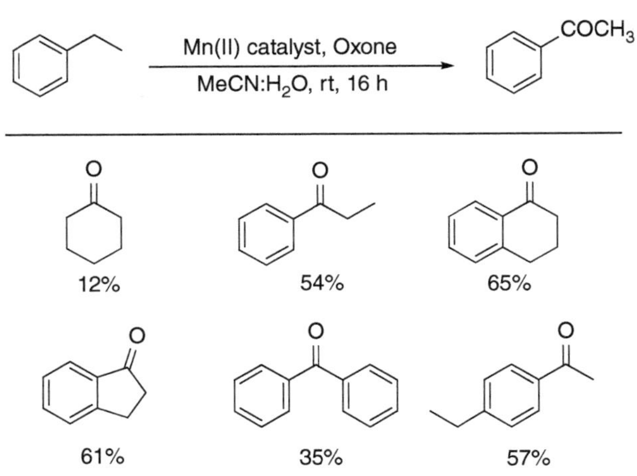

Scheme 2.13 Oxidation of ethylbenzene.

In 2019, Ray and coworkers trapped a highly reactive oxoiron(IV) complex in the catalytic epoxidation of olefins by hydrogen peroxide (Scheme 2.14) [22]. They reported the spectroscopic characterization of the oxoiron(IV) intermediate **D** [(cyclam)Fe^{IV}(O)(CH_3CN)]$^{2+}$ and explained the reaction mechanism. They slowly added H_2O_2 to the acetonitrile in the presence of iron(II) cyclam complex **A** [(cyclam)Fe^{II}(OTf)$_2$] and excess cyclohexene at 25 °C under an inert atmosphere. Mechanism study showed that typical radical reactions were not involved.

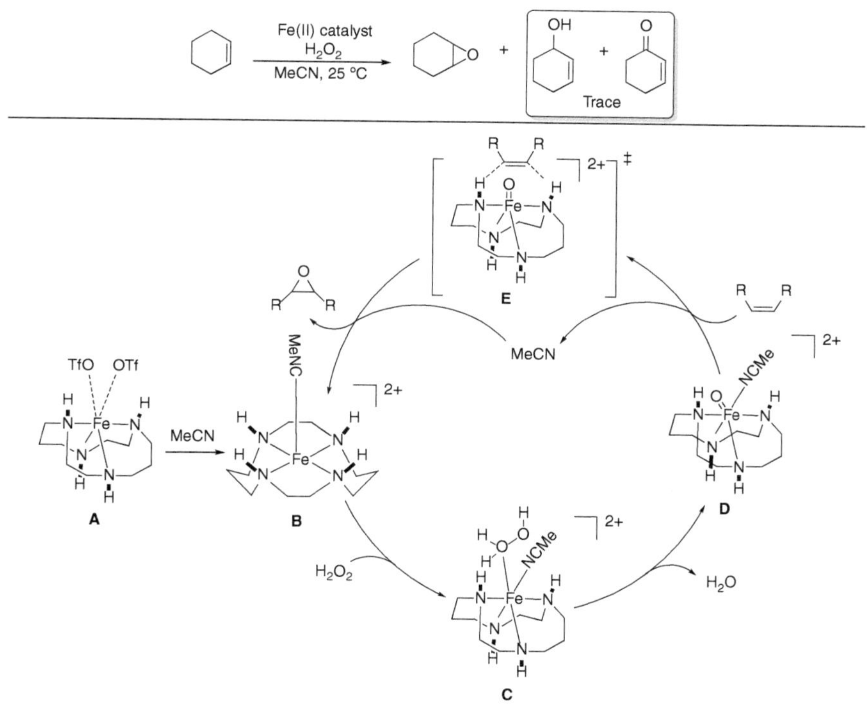

Scheme 2.14 Epoxidation of olefins by hydrogen peroxide.

In 2019, Li and coworkers developed a mild and convenient approach for Pd/Bi-catalyzed direct synthesis of α,β-unsaturated nitriles using aromatic alcohols and acetonitrile (Scheme 2.15) [23]. They employed Pd/C as the catalyst, $BiCl_3$ as the co-catalyst and O_2 as the oxidant. The solvent effect was of vital importance, since the addition of toluene, 1,4-dioxane, dimethyl formamide (DMF), dimethyl sulfoxide (DMSO) to CH_3CN would reduce reaction efficiency. Control experiments of oxidation and Knoevenagel condensation showed that benzaldehyde was formed as an intermediate.

In 2020, Mayilmurugan and coworkers synthesized cobalt(II) complexes of tetradentate ligands as catalysts for the single-step synthesis of phenol (Scheme 2.16) [24]. The 30% H_2O_2 was employed as an oxygen source and acetonitrile was utilized

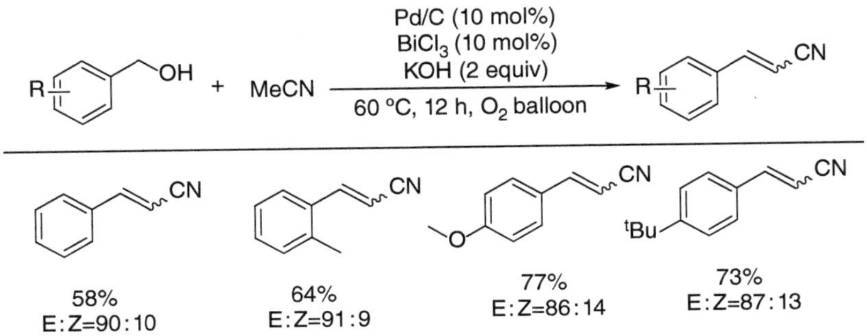

Scheme 2.15 Synthesis of α,β-unsaturated nitriles.

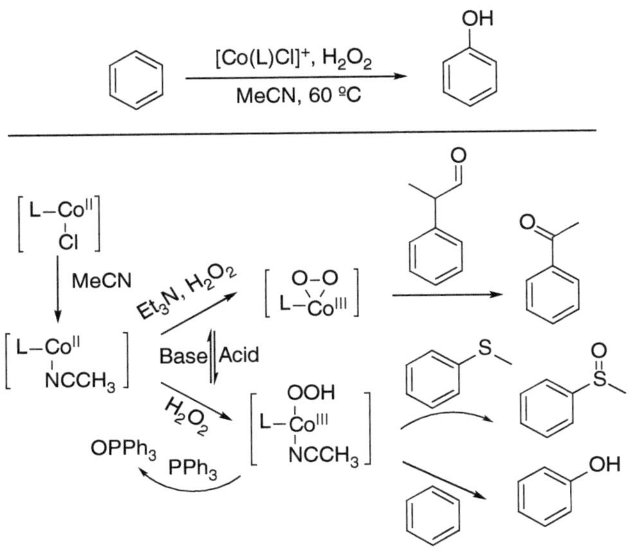

Scheme 2.16 Synthesis of phenol.

as the solvent. The reaction mixture was continuously stirred at 60 °C for 5 hours. This reaction procedure afforded phenol selectively as a major product. It was the highest catalytic performance reported using cobalt(II) complexes as catalysts to date with a maximum yield of phenol up to 29%. The isotope-labeling experiments demonstrated that H_2O_2 was the main oxygen supplier.

In 2022, Iskra and coworkers developed a cheap, green, mild, and environmentally friendly method for the selective oxidative cleavage of the C=C double bond in cinnamic acids with hydrogen peroxide catalyzed by vanadium(V) oxide (Scheme 2.17) [25]. A 30% aqueous solution of hydrogen peroxide was used as the oxidant. The choice of solvent determined the selectivity of products. Cinnamic acid derivatives were selectively converted to benzoic acid derivatives in MeCN.

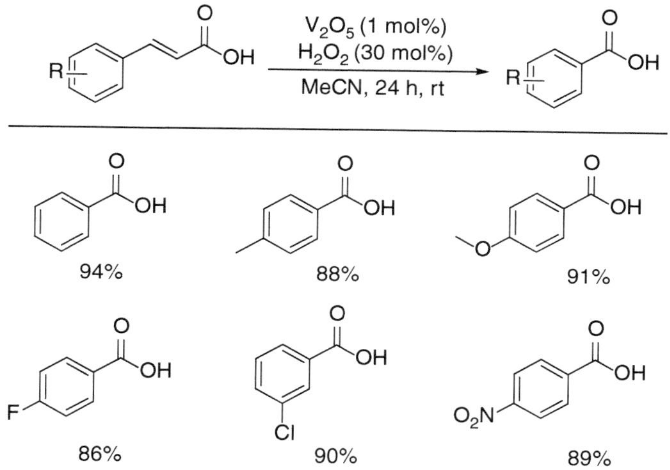

Scheme 2.17 Synthesis of benzoic acids.

In 2022, Ando et al. developed one-pot oxidation procedures for primary alcohols generating *Z*-α,β-unsaturated esters (Scheme 2.18) [26]. The one-pot O_2-oxidation- Horner–Wadsworth–Emmons (HWE) reaction was characterized by high selectivity and high yield. They chose 2,3-isopropylideneglycerol as substrate, TEMPO (2,2,6,6-tetramethylpiper-idine-1-oxy), and Cu salt as the oxidant and catalyst. The procedure was carried out in MeCN at room temperature for 31 hours, yielding the desired product with $Z/E = 96{:}4$ in high yields. Varieties of substrates were extended.

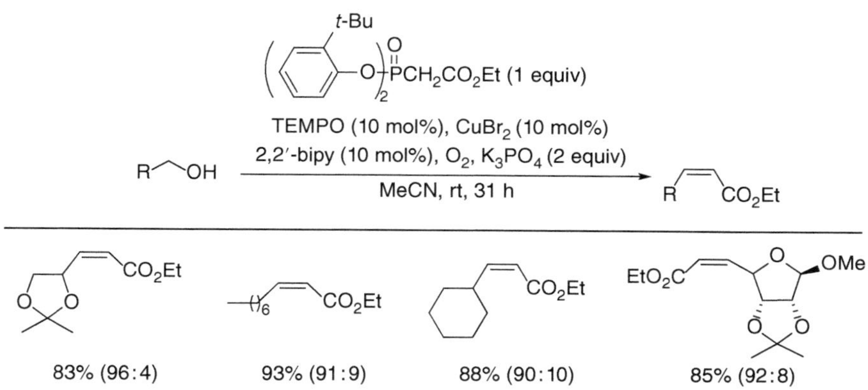

Scheme 2.18 Synthesis of *Z*-α,β-unsaturated esters.

In 2018, Ding and coworkers reported a new efficient Ag(I)-catalyzed aminooxygenation reaction of an azido-Ugi adduct for the preparation of 2-tetrazolyl-substituted 3-acylpyrroles, involving a 5-endo-dig cyclization and aerobic oxidation (Scheme 2.19) [27]. The reaction condition of $AgNO_3$ as the catalyst and DMAP as

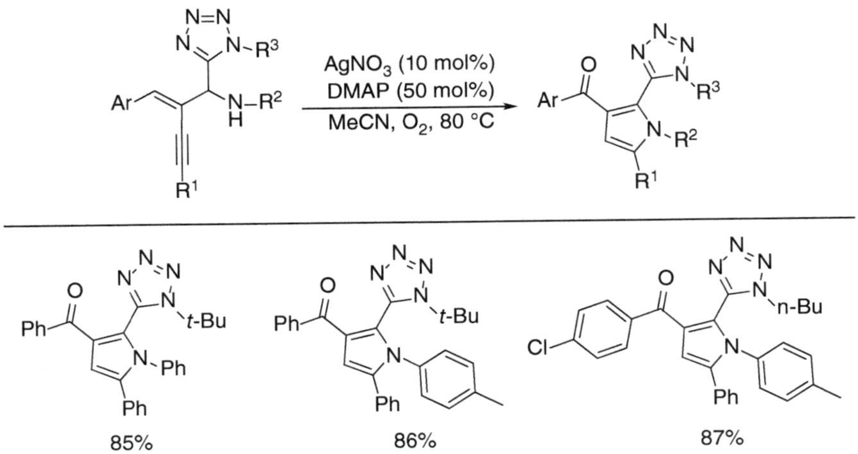

Scheme 2.19 Cascade aminooxygenation reaction.

the base under O_2 atmosphere was optimal for the formation of aminooxygenation product with 85% yield. A variety of Ugi-azide adducts were employed for the oxidative cycloisomerization reaction and gave the desired products in good yields.

Kholdeeva and coworkers developed nucleophilic versus electrophilic activation of hydrogen peroxide over Zr-based metal-organic frameworks (Zr-MOF) (Scheme 2.20) [28]. They carried out catalytic oxidation reactions of methyl phenyl sulfide (MPS) with H_2O_2 in the presence of Zr-MOFs. The oxidation procedure was carried out at a near room temperature of 27 °C and was completed within 15–60 minutes. The corresponding sulfone was produced with surprisingly high selectivity when acetonitrile was employed as the solvent. Moreover, the oxidant utilization efficiency was up to 98–99%.

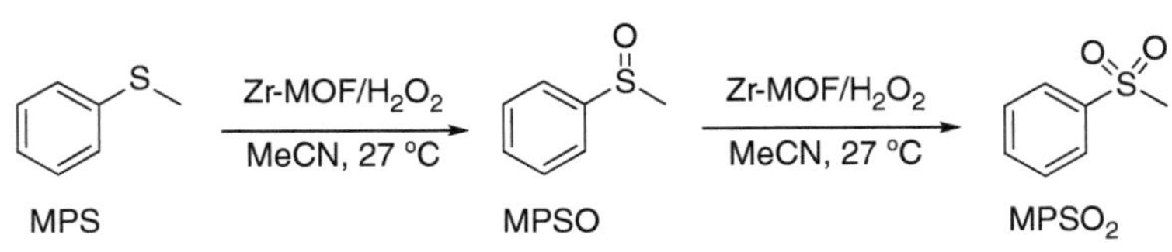

Scheme 2.20 Oxidation of MPS.

Jiang and coworkers disclosed a palladium-catalyzed oxidative cross-trimerization between alkynes and alkenes, yielding 1,3,5-trienes or 1,2,4,5-tetrasubstituted benzenes with dioxygen (Scheme 2.21) [29]. The reaction system was green and atom-economic with high chemoselectivity. Ethyl phenylpropiolate and methyl acrylate were used as model substrates. Molecular oxygen was chosen as the sole oxidant. Various substrates proceeded efficiently, and a series of products were synthesized in good yields by one-pot synthesis.

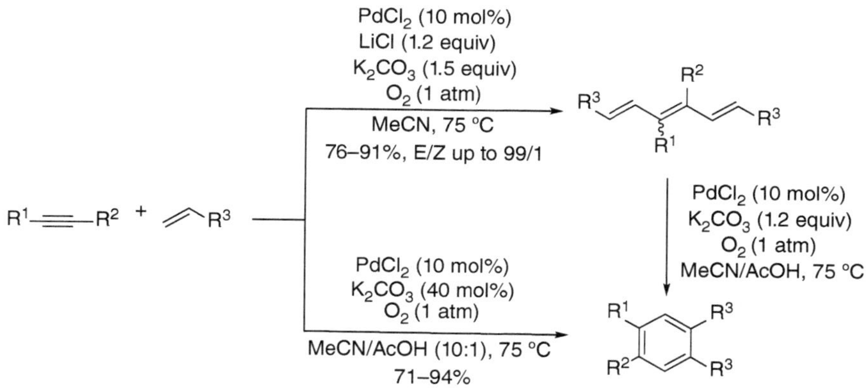

Scheme 2.21 Pd-catalyzed oxidative cross-trimerization.

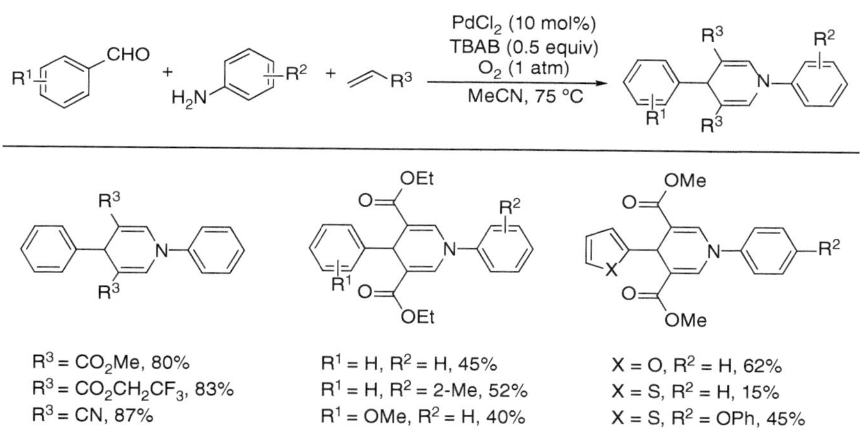

Scheme 2.22 Pd-catalyzed synthesis of 1,4-dihydropyridines.

Jiang and coworkers established a novel and highly efficient protocol for the synthesis of 1,4-dihydropyridines via palladium-assisted multicomponent cyclization of aromatic aldehydes, arylamines, and terminal olefins under an O_2 atmosphere (Scheme 2.22). [30] The additive tetrabutyl ammonium bromide (TBAB) had a vital effect on the reaction. Acetonitrile was the most effective solvent. Various substrates smoothly and successfully reacted in moderate-to-excellent yields. Molecular oxygen was a cost-effective, easily available, and environmentally benign oxidant in this reaction.

Liu and coworkers reported a novel palladium-catalyzed oxidative arylalkylation of activated alkenes, which involved dual C—H bond cleavage of an arene and acetonitrile (Scheme 2.23) [31]. They carried out a series of reactions and discovered that the addition of $PhI(OPiv)_2$ and AgF was crucial for this transformation, and AgF contributed to the C—H bond cleavage in acetonitrile. These reactions afforded various nitrile-bearing indolinones in excellent yields. Furthermore, the presence of a radical scavenger 2,2,6,6-tetramethyl-1-piperidinyloxyl (TEMPO) had no effect on this reaction.

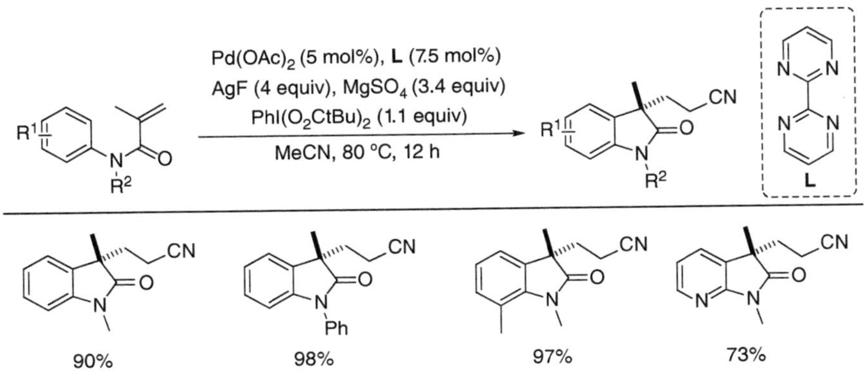

Scheme 2.23 Pd-catalyzed oxidative arylalkylation of alkenes.

2.2.3 Transition Metal-catalyzed Reduction Reactions in MeCN

In 2008, Clarke and coworkers described a convenient and chemoselective method for the deoxygenation of pyridine N-oxides through palladium-catalyzed transfer oxidation of trialkylamines (Scheme 2.24) [32]. Acetamido-pyridine N-oxide was chosen as the model substrate. The results showed that triethylamine was the most appropriate amine reagent, and the presence of a ligand, palladium, and base were all essential for the reaction to proceed. The polar solvent acetonitrile was employed. The generality and functional group tolerance of this palladium-catalyzed transfer oxidation process were then investigated. The pyridine derivatives were obtained in good-to-excellent yields without the reduction or cleavage of other functional

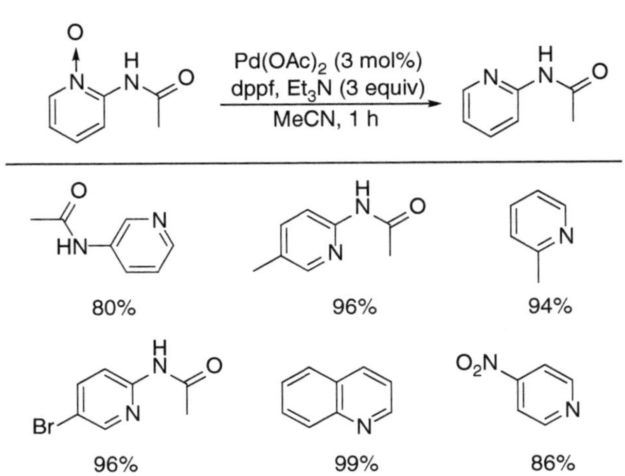

Scheme 2.24 Palladium-catalyzed deoxygenation reaction.

groups. The reactions proceeded smoothly in the presence of amido, nitro, ester, hydroxy, carbonyl, and Br groups, and there was no influence on the position in the pyridine ring of the substituent.

In 2010, Gebbink and coworkers reported Rh-catalyzed hydroformylation reactions (Scheme 2.25) [33]. Several ligands were applied in the Rh-catalyzed hydroformylation of styrene. Acetonitrile was used as a solvent because it was the only solvent in which all five ligands were soluble. Reactions were processed at 50 °C for 24 hours. They also carried out hydrogenation reactions of dimethyl itaconate at 30 °C. In that case, reactions yielded low conversions when CH_2Cl_2 was used as the solvent. The addition of a small amount of acetonitrile could significantly improve the catalytic activity.

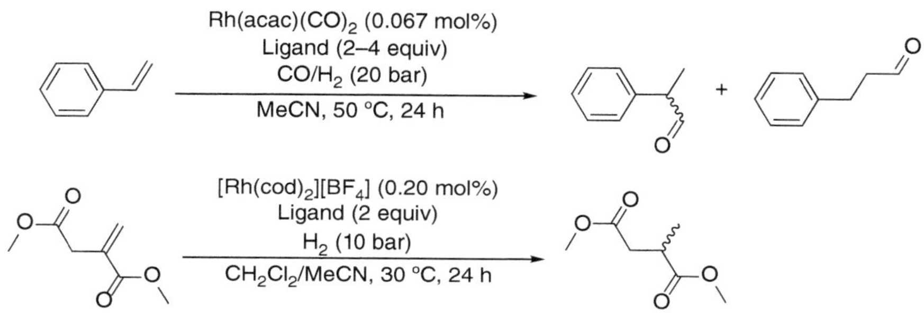

Scheme 2.25 Rh-catalyzed hydroformylation and hydrogenation.

Thomas and coworkers reported nitro reduction and hydroamination using a single iron catalyst under mild conditions (Scheme 2.26) [34]. The reduction was highly chemoselective for nitro groups. Solvents were of great significance to both

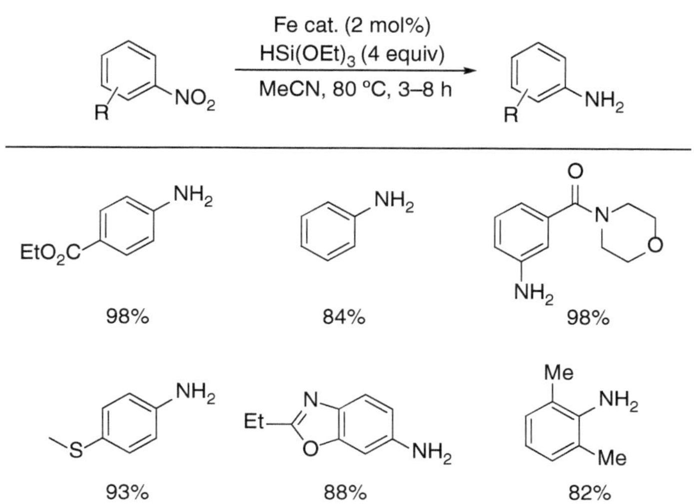

Scheme 2.26 Reduction of nitroarenes.

reactivity and selectivity in the reduction of nitroarenes. MeCN was employed as the best solvent. Various substrates yielded good-to-excellent yields and chemoselectivity in all cases. $HSi(OEt)_3$ was employed as the reducing agent. The reactions for different substrates were required to react at 80 °C for 3–8 hours.

In 2011, MacArthur and coworkers described the hydrodehalogenation of aryl chlorides and aryl bromides (Scheme 2.27) [35]. The reaction system used a microwave-assisted, copper-catalyzed concurrent tandem catalysis methodology. In the first step, halogen exchange, an aryl chloride or aryl bromide underwent halide exchange with a source of iodide under the catalysis of a metal catalyst to produce a more reactive aryl iodide substrate. The newly generated aryl iodide then served as the substrate in the subsequent second reaction to yield the desired product.

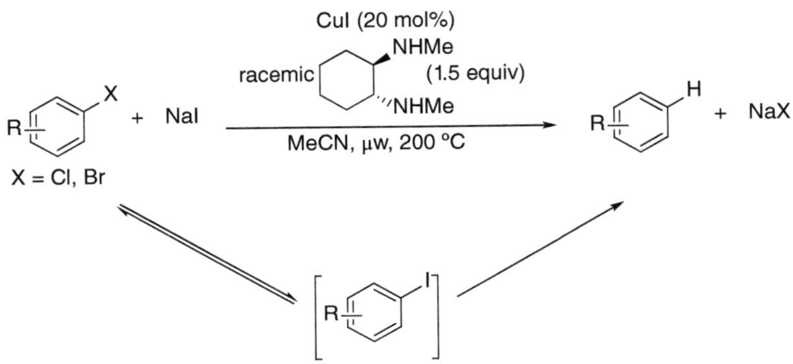

Scheme 2.27 Hydrodehalogenation of aryl chlorides and aryl bromides.

In 2018, He and coworkers found that cupric subcarbonate ($Cu_2(OH)_2CO_3$) was effective for the reductive functionalization of CO_2 to yield formamides and methylamines using phenylsilane as a reductant [36] (Scheme 2.28). Both secondary aromatic amines and aliphatic amines could be methylated and formylated. Nitro- and allyl-group tolerance was observed in this system. They also found the ligand-controlled selective N-methylation or N-formylation of amines with CO_2 and phenylsilane in the presence of copper catalysis. The ligand effect had an important influence on the selectivity of products. A variety of substrates could successfully react under the reaction conditions with moderate-to-excellent yields.

2.2.4 Transition Metal-catalyzed Substitution Reactions in MeCN

Shi et al. disclosed a method for both 1,2,4-trisubstitution and dearomative 1,2,4-trifunctionalization of benzyne. It was accomplished using sulfoxides bearing a penta-2,4-dien-1-yl moiety (Scheme 2.30) [38]. In the presence of acetonitrile, these cascade transformations proceeded through benzyne insertion into the S=O bond and an uncommon regiospecific anionic [4,5]-sigmatropic rearrangement, furnishing a C—O, C—S, and C—C bond on the C1, C2, and C4 positions of a benzene ring, respectively. Notably, all three new chemical bonds originated from the same sulfoxide substrate, demonstrating a highly economical cascade process.

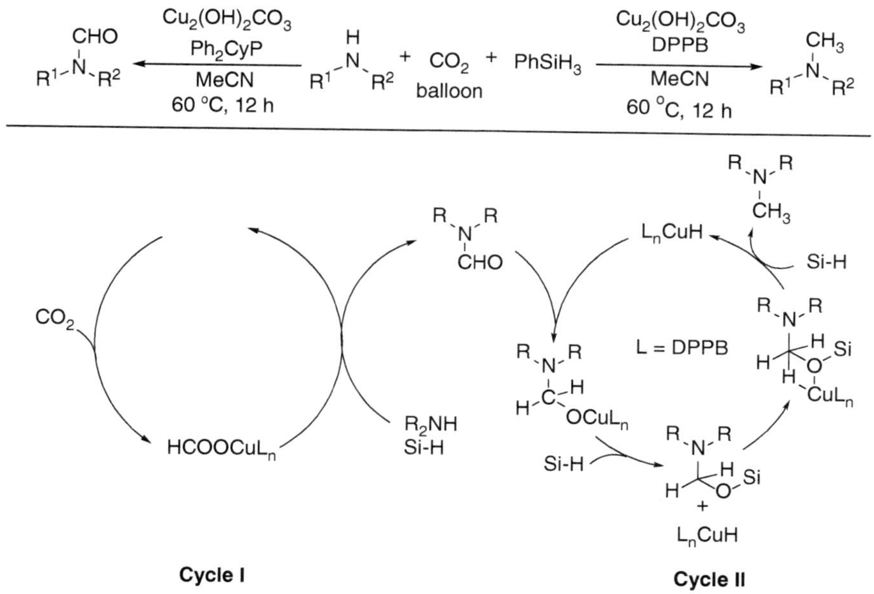

Scheme 2.28 Selective N-methylation or N-formylation. Nakazawa and coworkers developed an unprecedented selective double hydrosilylation of nitriles catalyzed by an iron complex containing indium trihalide using tertiary and secondary silanes (Scheme 2.29) [37]. An iron complex containing indium trihalide $[Fe(MeCN)_6][Fe(CO)_4(InX_3)_2]$ could also catalyze the reaction successfully. In this double hydrosilylation reaction, MeCN was employed as both the substrate and the solvent. All substrates were placed in a sealed glass tube under a dry nitrogen atmosphere, and the solution was heated at 80 °C for 24 hours.

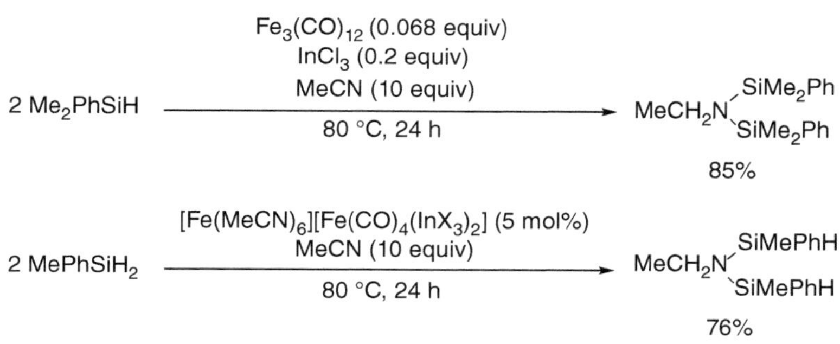

Scheme 2.29 Double hydrosilylation of acetonitrile.

Qian et al. developed a new route for the direct allylation of various alkyl halides catalyzed by simple cobalt(II) bromide (Scheme 2.31) [39]. This method was very straightforward and environmentally friendly. It was efficient for the coupling of a large variety of alkyl halides (primary, secondary, and tertiary) with substituted allylic acetates and carbonates, and provided good-to-excellent yields with good functional group tolerance by using acetonitrile solvent. Moreover, in the case of substituted allyl acetates, the reaction yields the linear product as the major or

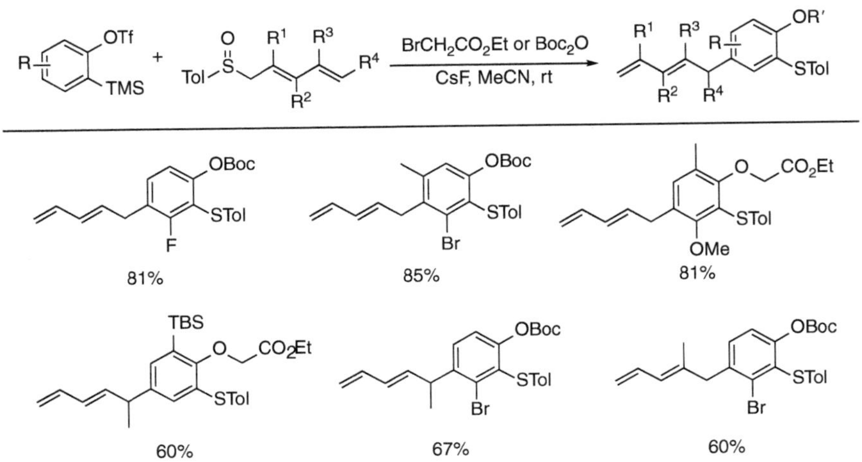

Scheme 2.30 Benzyne 1,2,4-trisubstitution and dearomatization.

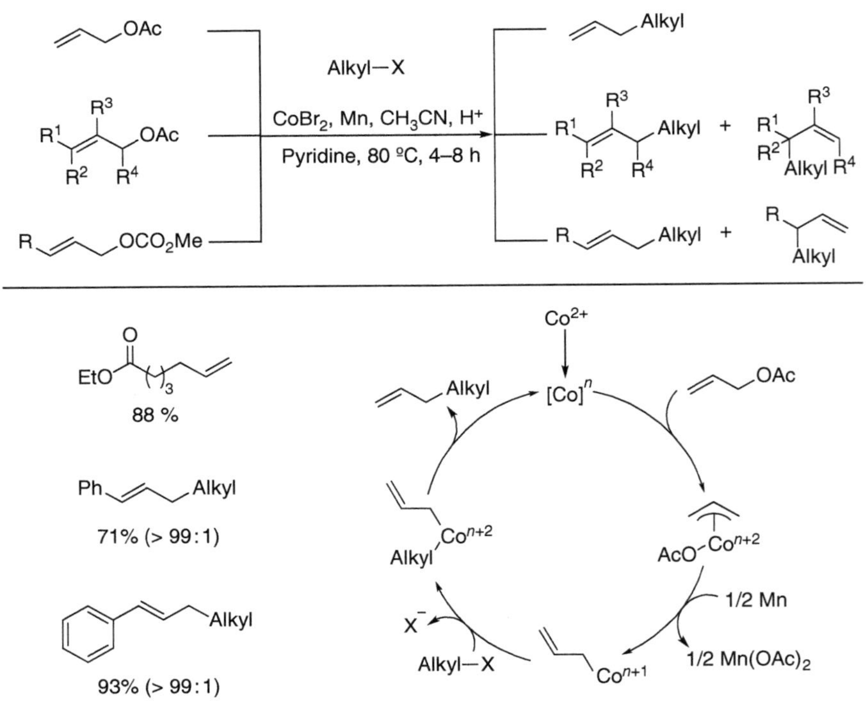

Scheme 2.31 Cobalt(II) bromide-catalyzed allylation of alkyl halides.

sole product. Both sterically hindered secondary allyl acetates, and secondary and tertiary alkyl halides, were acceptable as substrates.

Ma and coworkers reported two easily available *N,N'*-disubstituted oxalamides, which were efficient ligands for Cu-catalyzed coupling of phenols with (hetero)aryl bromides and iodides (Scheme 2.32) [40]. These reactions worked well using MeCN as the solvent at 90 °C. It could be completed even with 0.2–2 mol% catalyst loadings and yielded a broad range of diaryl ethers in good-to-excellent yields. The results further demonstrated the power of *N,N'*-disubstituted oxalamide ligands in promoting Cu-catalyzed arylation of nucleophiles.

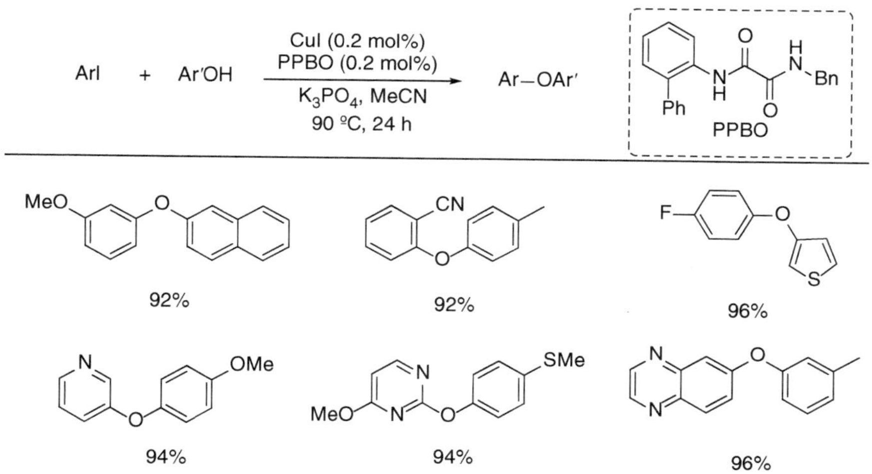

Scheme 2.32 CuI/PPBO-catalyzed coupling of (hetero)aryl iodides with phenols.

Li et al. developed a highly enantioselective cobalt-catalyzed reverse prenylation of *β*-ketoester with tertiary allylic carbonate (Scheme 2.33) [41]. This method constructed vicinal all-carbon quaternary carbon centers using the cobalt/(*S,S*)Ph-BPE complex, which was generated *in situ* with zinc reduction as the catalyst. The reaction completed in MeCN yielded excellent branched-to-linear selectivity (>20 : 1) and up to 92% ee.

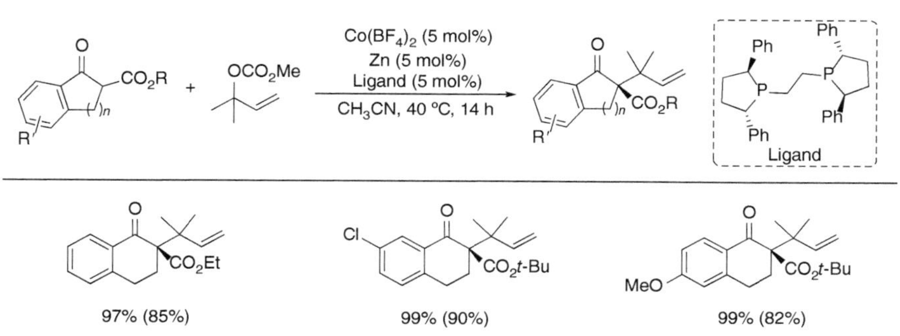

Scheme 2.33 Allylic substitution.

Shen and coworkers developed a highly efficient and selective method for the synthesis of 3-cyanoindoles through the cyanation of indoles (Scheme 2.34) [42]. The reaction was promoted by the Cu/TEMPO/$(Me_3Si)_2$ system and completed using acetonitrile as the cyano source, which also worked as the solvent. It was carried on a sequential iodination/cyanation process in one pot, with moderate-to-good yields and great tolerance for a series of functional groups.

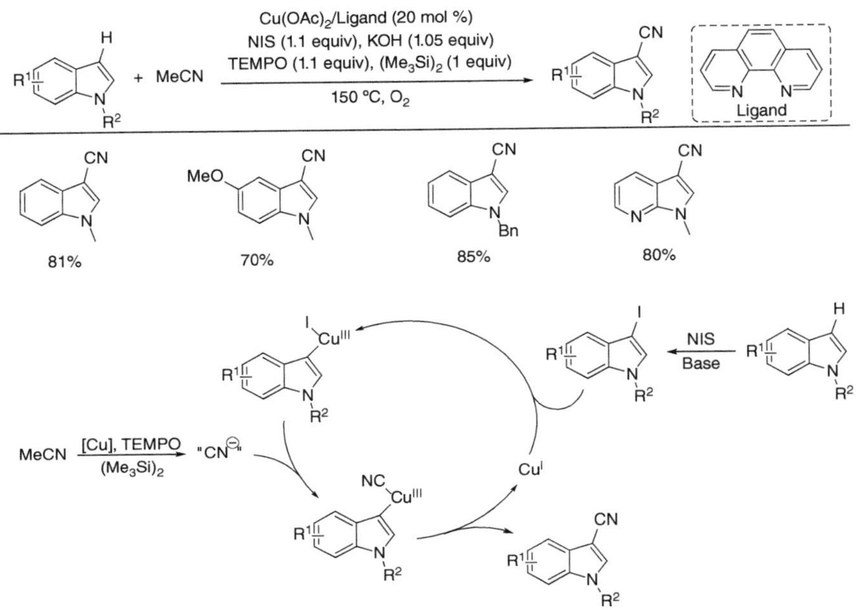

Scheme 2.34 Cu-catalyzed cyanation of indoles with acetonitrile.

Weix et al. disclosed a cross-electrophile coupling of aryl halides with alkyl halides without amide solvents, which usually cooperated with stoichiometric metallic reductants (Scheme 2.35) [43]. The organic reductant tetrakis(dimethylamino) ethylene (TDAE) allowed the reaction to proceed smoothly in acetonitrile with high

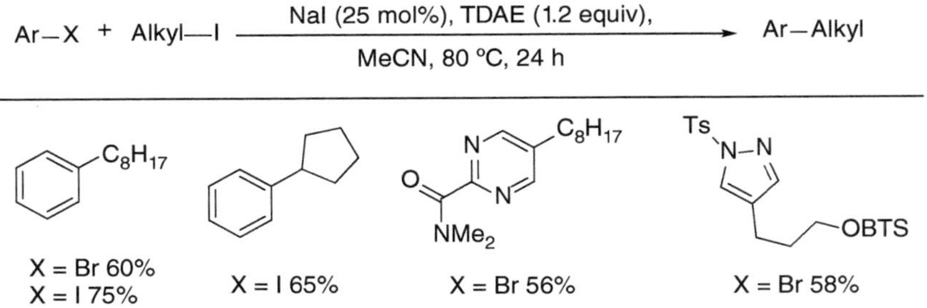

Scheme 2.35 Ni-catalyzed cross-electrophile-coupling.

yields and great efficiency. Furthermore, these conditions could also be applied for several heterocycles, which were easily reduced by manganese, and broadened the versatility of nickel-catalyzed cross-electrophile-coupling.

Liu et al. achieved a ligand-free palladium-catalyzed direct C—H functionalization (Scheme 2.36) [44]. This protocol completed C3-cyanation of indole for 3-cyanoindole, which was a value-added basic unit in organic synthesis. The methodology utilized acetonitrile as a green and readily available cyanide source and solvent, obtaining corresponding products in moderate-to-good yields with good tolerance for different functional groups. In these reactions, the pre-functionalization of indole C—H was unnecessary, while a transition-catalyzed C—CN bond cleavage was included.

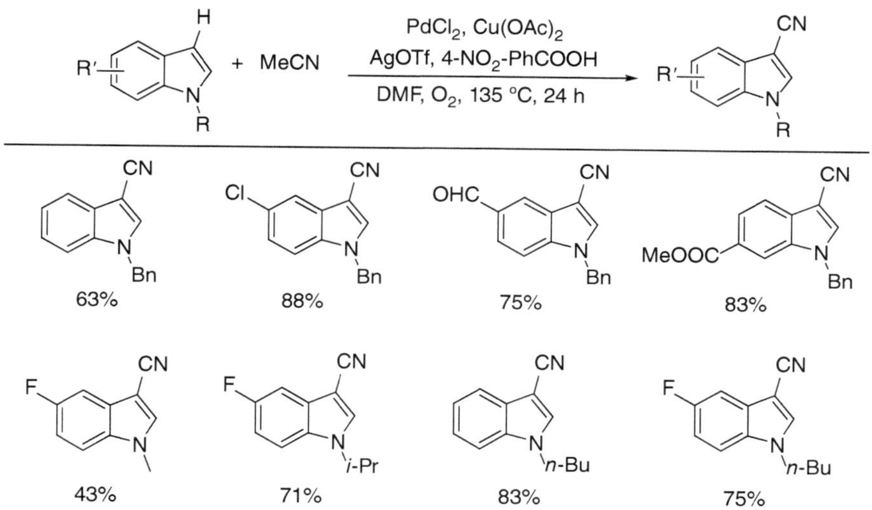

Scheme 2.36 Ni-catalyzed C3-cyanation of *N*-aryl indoles.

Singh and coworkers developed a practical and mild nickel-catalyzed approach for C—S bond formation to obtain unsymmetrical thioethers (Scheme 2.37) [45]. This method was conducted through the cross-coupling of arylsulfonyl hydrazides and aryl boronic acids in acetonitrile while exposed to air. The reaction exhibited good-to-excellent yields and a wide tolerance for functional groups. Electron-donating as well as electron-withdrawing groups were well tolerated during the course of the reaction at different positions.

Li et al. reported a Sonogashira-type, synergistic Rh and Cu-catalyzed alkylation for the synthesis of highly branched and enantioselective 1,4-enynes (Scheme 2.38) [46]. The reaction was conducted through the interaction between readily available terminal alkynes and racemic allylic carbonates in acetonitrile. Aliphatic and aromatic terminal alkynes with various functional groups could be directly applied to this method with generally good yields and high enantioselectivity. Notably, the use of acetonitrile solvent could retard alkyne trimerization caused by Rh(I).

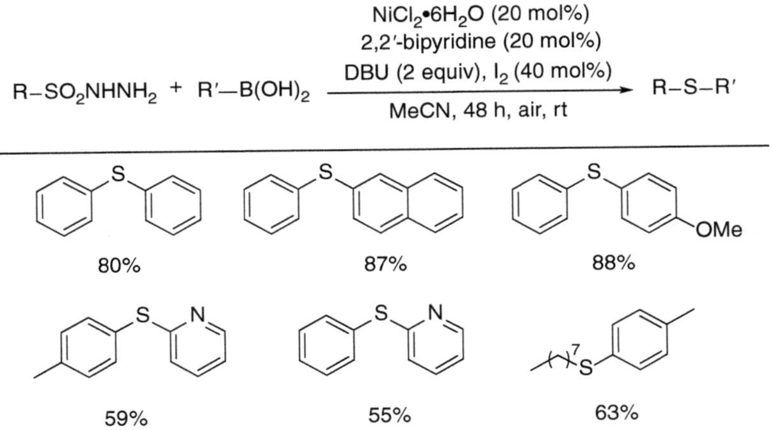

Scheme 2.37 Ni-catalyzed cross-coupling for unsymmetrical thioethers.

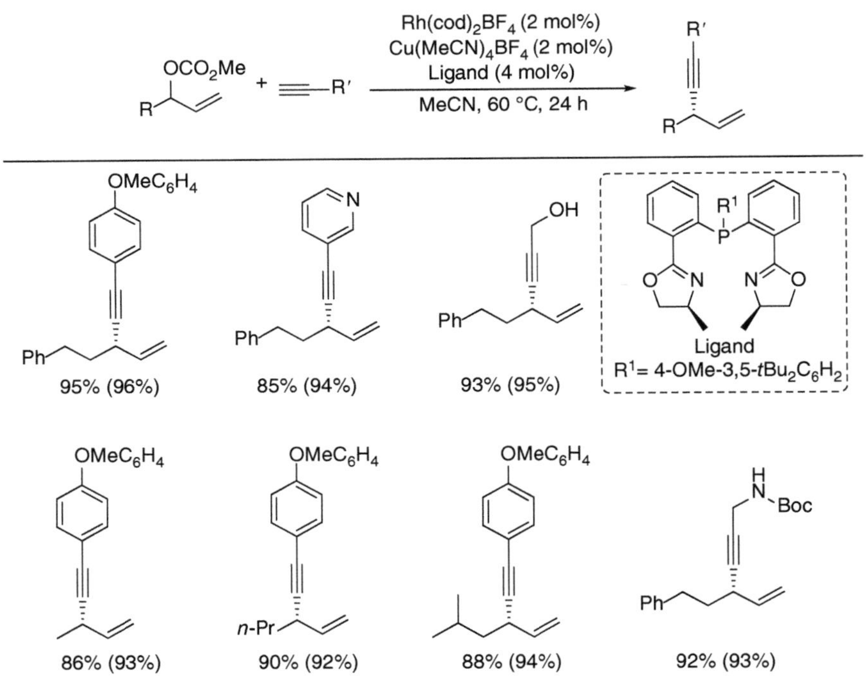

Scheme 2.38 Rh/Cu-catalyzed asymmetric allylic alkylation of terminal alkynes.

Peng and coworkers described the formation of carbon-heteroatom bonds catalyzed by copper(II) acetate monohydrate through a Cham-Lam cross-coupling reaction in acetonitrile at room temperature under air (Scheme 2.39) [47]. They utilized $NaBAr_4$ as the arylation reagent, which was stable, nontoxic, and commercially available. This reaction showed that several aliphatic and aromatic amines were

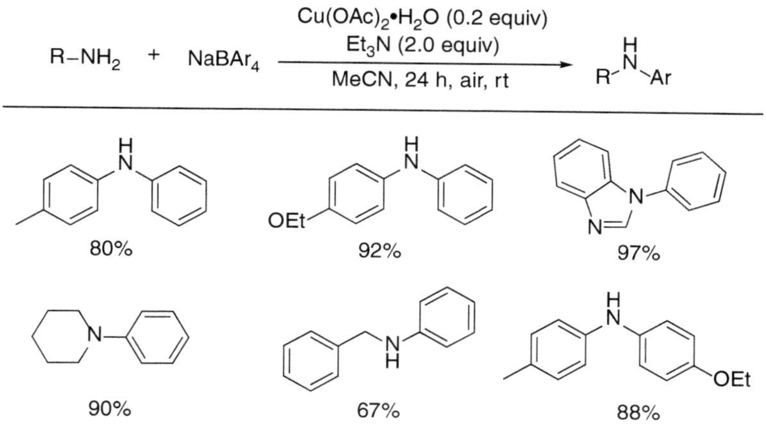

Scheme 2.39 Copper-catalyzed Chan-Lam cross-coupling.

suitable. Especially, alkylamines proceeded smoothly and the corresponding products were generated in good-to-excellent yields.

Fernandes et al. disclosed a cross-coupling of thiophenols with arylboronic acids catalyzed by $NiCl_2$/2,2′-bipyridine system to obtain symmetric and unsymmetric diarylsulfides (Scheme 2.40) [48]. This methodology was applied to synthesize arythioethers mildly and easily at room temperature in the air with a wide substrate scope and good-to-excellent yields. C—S bond construction has usually wide-ranging attention because of its ubiquitous role in various biologically active molecules such as pharmaceutical and agricultural applications. This methodology also avoided the use of expensive transition metals and ligands.

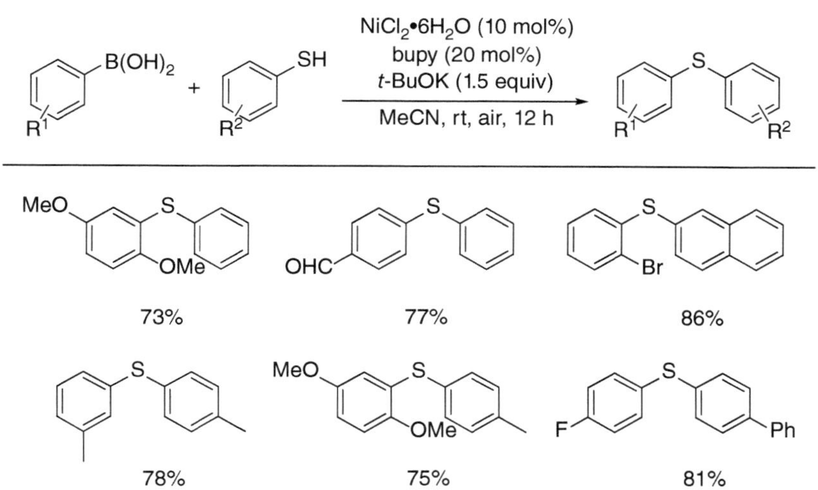

Scheme 2.40 $NiCl_2$/2,2′-bipyridine-catalyzed synthesis of diarylsulfides.

2.2.5 Transition Metal-catalyzed Cyclization Reactions in MeCN

Fox and coworkers reported the rhodium-catalyzed C—H insertion of metal carbenes appended with two electron-donating groups to form benzodihydrofurans (Scheme 2.41) [49]. Unlike other metal carbenes with electron-withdrawing groups, the faded electrophilicity enabled reactions to be carried out in the presence of acetonitrile and water. Another reactant, diazo precursors, could be prepared *in situ* from hydrazone catalyzed by a mild and chemoselective oxidant (MnO_2). The reaction could be completed through two paths while the "two-pot" was more efficient and had high enantio- and diastereoselectivity. This synthetic method might play an important role in the synthesis of natural products.

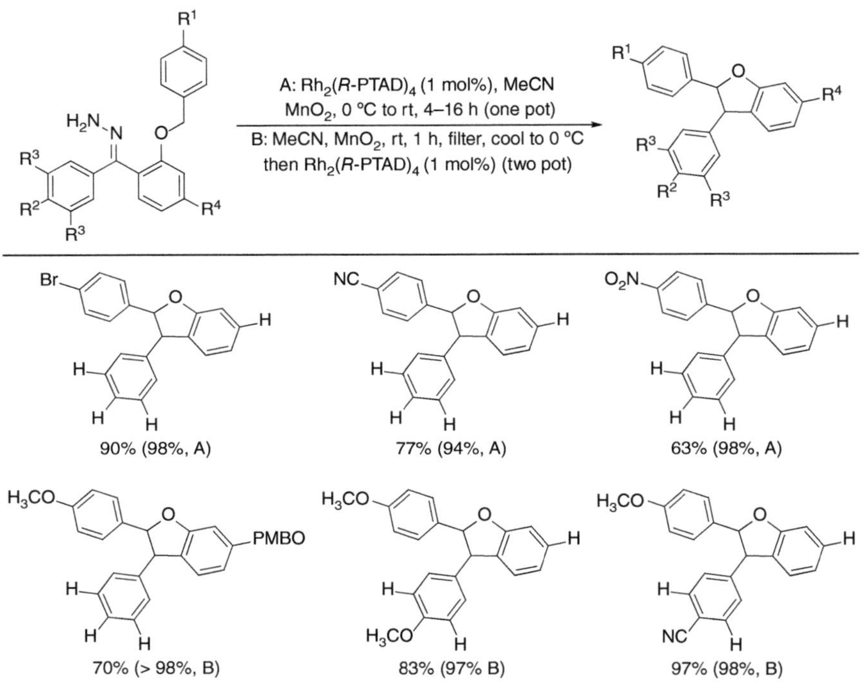

Scheme 2.41 Rh-catalyzed cyclization of benzodihydrofurans.

Sun and coworkers disclosed a [3+2+1] methodology to synthesize pyridine skeleton via a cascade carbopalladation/cyclization of acetonitrile, arylboronic acids, and aldehydes (Scheme 2.42) [50]. The reaction went through six steps, including the carbopalladation of acetonitrile, nucleophilic addition, condensation, intramolecular Michael addition, cyclization, and aromatization. The reaction used Pd as catalyst and showed excellent functional group tolerance via the one-pot domino-style formation of C—C/C—C/C—N bonds. These protocols of diverse products would be used as potentially useful synthetic building blocks for functional materials.

Gabriele et al. developed a novel and multicomponent approach to synthesize 2-(imidazo[1,2-*α*]pyridine-3-yl)-*N*,*N*-dialkylacetamides, an indispensable class of

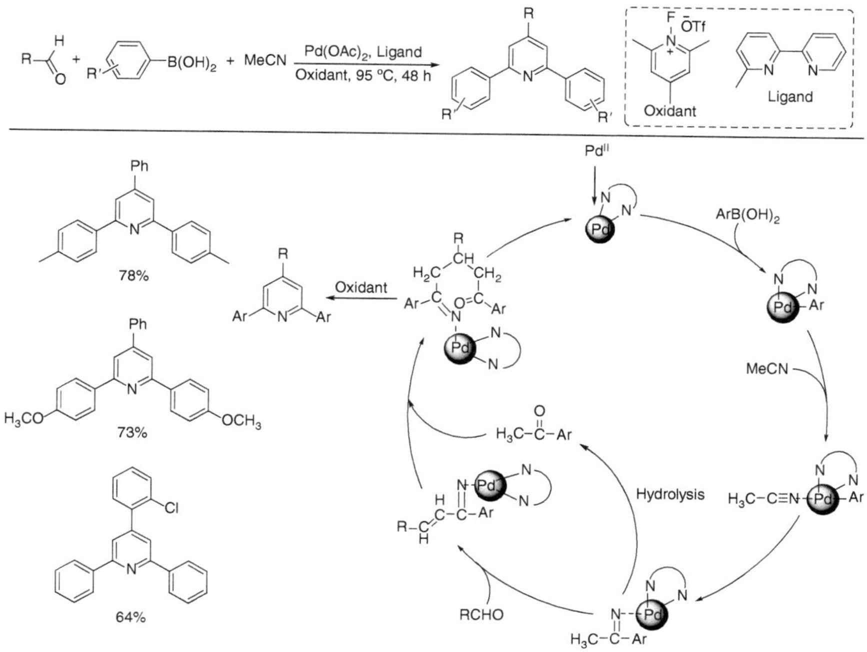

Scheme 2.42 Pd-catalyzed synthesis of pyridines.

fused heterocyclic derivatives acting on the central nervous system by enhancing the activity of the GABA (Scheme 2.43) [51]. This was achieved through the carbonylation of *N*-Boc-(prop-2-yn-1-yl)pyridine-2-amines in the presence of secondary amines and an oxidant. The reaction used oxygen as the oxidant and was catalyzed by the PdI_2/KI system in acetonitrile solvent at 100 °C. The method obtained the corresponding products in good-to-high yields.

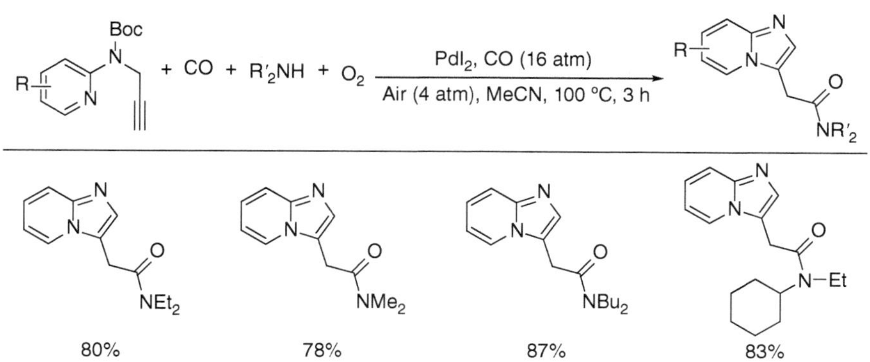

Scheme 2.43 PdI_2-catalyzed carbonylation.

Tabatabaia and coworkers developed an efficient simple method to synthesize substituted benzo[*d*]imidazo[5,1-*b*]thiazol-1-amine derivatives with biological activity, ranging from antineoplastic, antibacterial to labeled molecules (Scheme 2.44) [52]. The reaction was carried out using a one-pot cycloaddition of benzothiazole, trichloroacetonitrile, various amines, and terminal alkynes catalyzed by copper in acetonitrile at room temperature. The method provided mild conditions, high yields, and available raw materials over earlier methods.

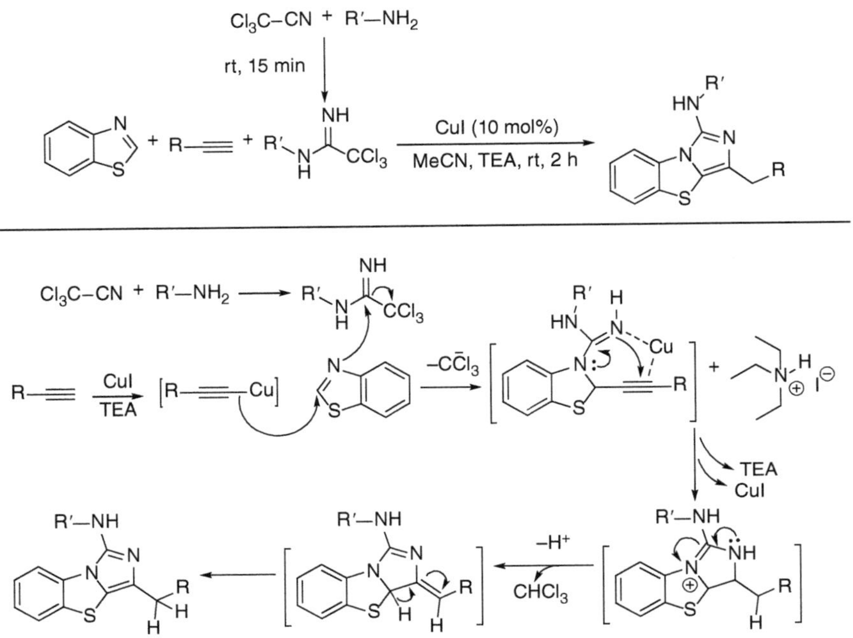

Scheme 2.44 Cu-catalyzed intermolecular cycloaddition.

Sarkar et al. disclosed the synthesis of trisubstituted nicotinonitriles catalyzed by palladium carbon (Scheme 2.45) [53]. These products were synthesized from easily available homopropagylic or homoallylic aromatic alcohols and nitriles in the presence of acetonitrile solvent. The pyridine core was formed through the generation of Pd(II)-C palladacycle followed by oxidative aromatization. The reaction offered a fast and straightforward route to obtain biologically active trisubstituted nicotinonitriles with elite efficiency, moderate yields, and wide reaction scope.

Liang and coworkers disclosed a C—H bond activation reaction catalyzed by a cooperative Pd/Cu system (Scheme 2.46) [54]. Electron-deficient polyfluoroarenes reacted with aryl halides in acetonitrile at 120 °C, providing various polyfluoroarene-substituted benzofuran derivatives. This protocol was unique and strategically distinct to obtain products with moderate yields and a great tolerance for different reaction substrates. The reaction was also an example of introducing electron-deficient arenes to a Catellani-type reaction.

Yoshikai et al. developed a cobalt–diphosphine-catalyzed ring-opening [3+2] annulation reaction that affords multisubstituted cyclopentenol derivatives in

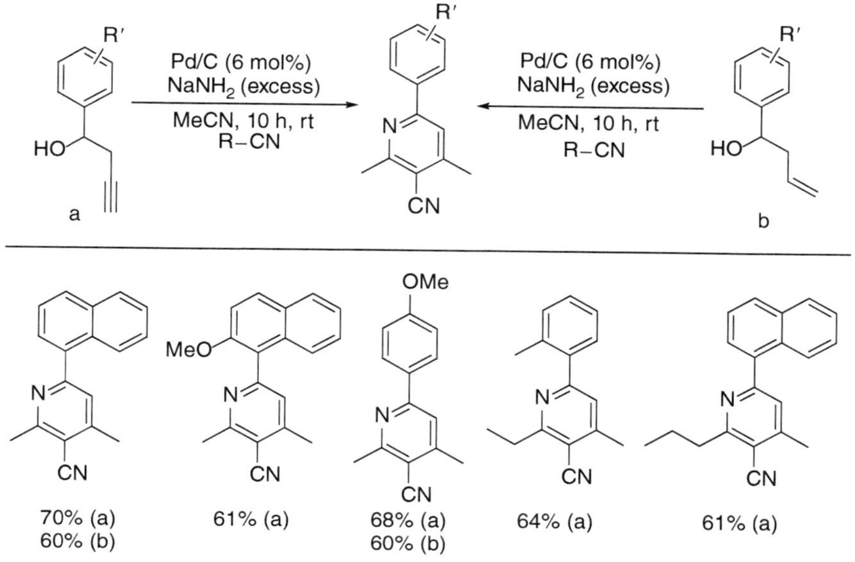

Scheme 2.45 Pd/C-catalyzed synthesis of trisubstituted nicotinonitriles.

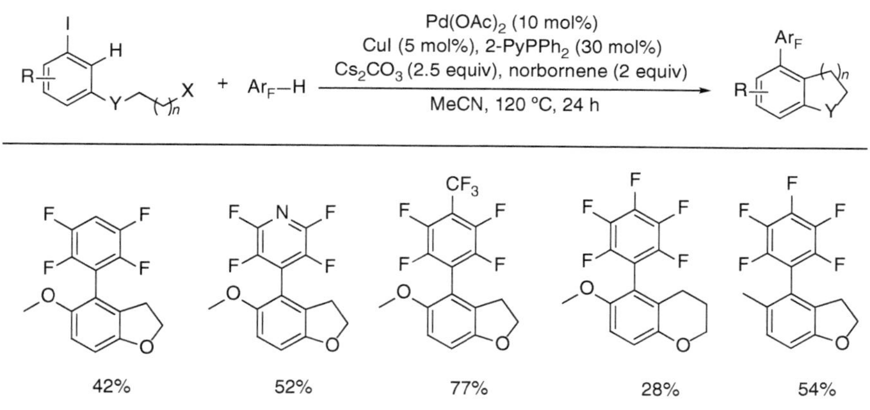

Scheme 2.46 Pd/Cu-catalyzed synthesis of benzofurans.

moderate-to-high yields (Scheme 2.47) [55]. The reaction carried on between cyclopropanols and unactivated internal alkynes in acetonitrile at 80 °C. It also could obtain another β-alkenylation product. The chemoselectivity was exquisitely controlled by the reaction conditions, especially the solvent, which can affect the key catalytic intermediate, a cobalt homoenolate. The authors suggested the reaction involved ring-opening of cobalt cyclopropoxide into homoenolate, migratory insertion of the alkyne into the Co—C bond, and protodemetalation or intramolecular carbonyl addition of the resulting alkenylcobalt species. They also proved the feasibility of these reaction steps by DFT calculations.

Rabbani and coworkers synthesized a novel series of diamino dihydroquinazoline derivatives from isocyanides and guanidines (Scheme 2.48) [56]. This

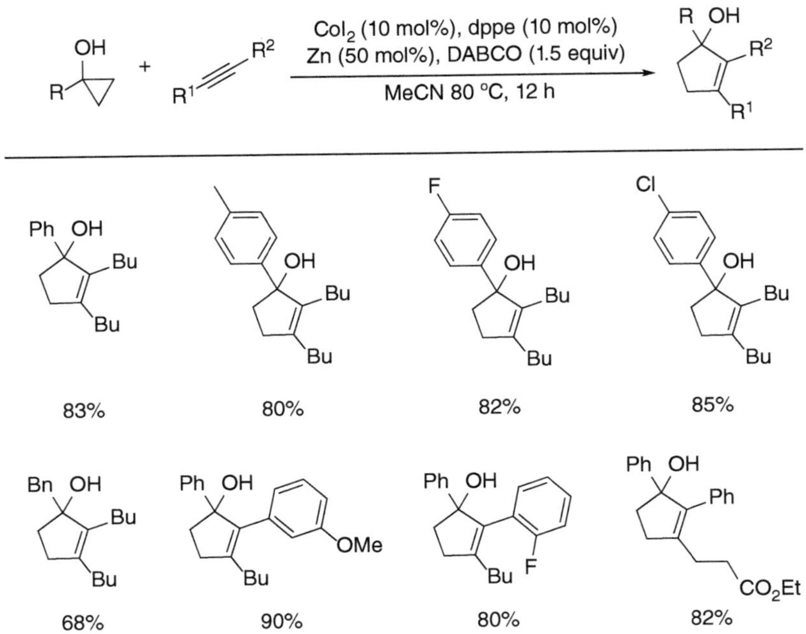

Scheme 2.47 Co-catalyzed ring-opening [3+2] annulation.

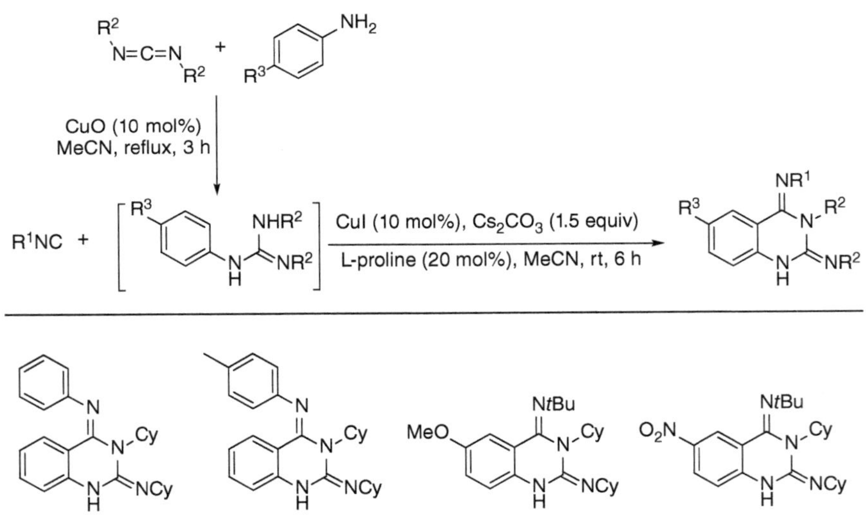

Scheme 2.48 Synthesis of diamino dihydroquinazolines.

transformation applied an efficient, inexpensive, one-pot, copper-catalyzed intramolecular C—H activation reaction under mild conditions and afforded diamino dihydroquinazoline derivatives in quantitative yields. This method was described as a new way to generate heterocyclic compounds. The reaction was carried out under moderate stirring at room temperature for six hours.

Ma et al. disclosed an efficient method to synthesize pyrrolidinedione-fused hexahydropyrrolo[2,1-*a*]isoquinolines catalyzed by Pd through a three-component [3+2] cycloaddition (Scheme 2.49) [57]. The reactions of amino esters or amino acids generated two kinds of [3+2] cycloaddition intermediates followed by a one-pot N-allylation and an intramolecular Heck reaction. The products could be obtained in moderate yields in the presence of acetonitrile.

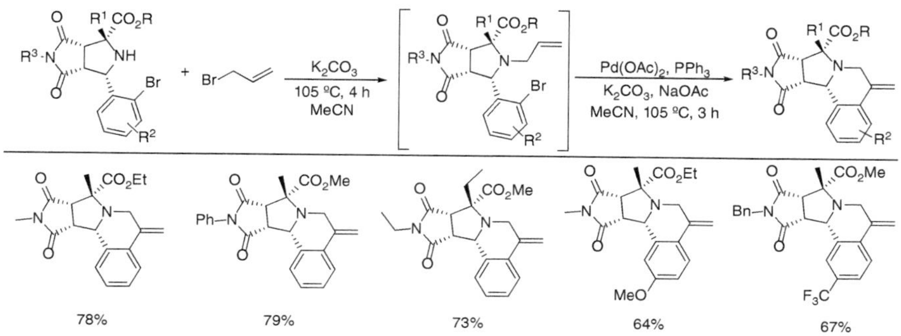

Scheme 2.49 Pd-catalyzed [3+2] cycloaddition of amino esters or amino acids.

Wang and coworkers developed an unexpected Cu-catalyzed heteroaromatization/sulfonyl transfer reaction between propargylic alcohols and isocyanide in the presence of acetonitrile (Scheme 2.50) [58]. The reaction was used to obtain a variety of 3-sulfonyl benzofurans and indoles in good-to-high yields and good tolerance, while these two products were the primary components in numerous natural products and pharmaceuticals. It also provided a controlled, modular, and convenient method for sulfonyl benzoheterocycle scaffolds.

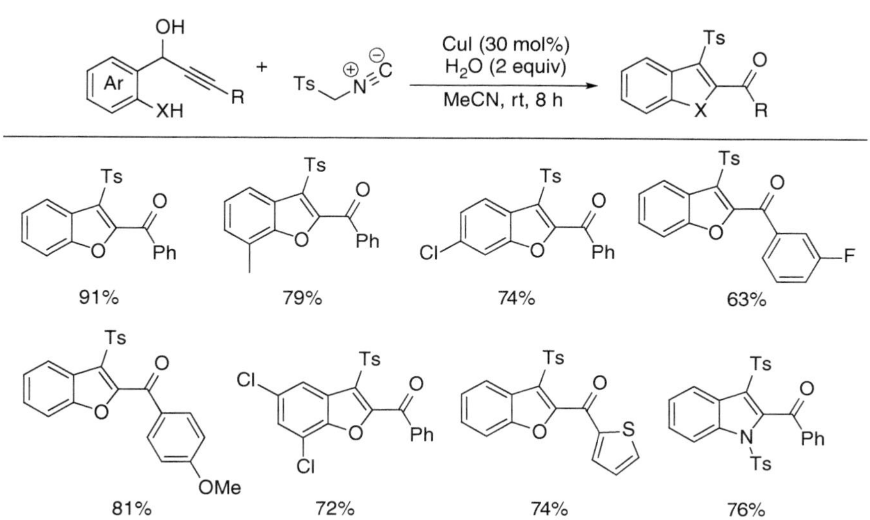

Scheme 2.50 Cu-catalyzed synthesis of 3-sulfonyl benzofurans and indoles.

2.3 MeCN in Transition Metal-free Catalyzed Reactions Without Radicals Involved

2.3.1 Transition Metal-free Catalyzed Cyclization Reactions in MeCN

Ma and coworkers developed a sequential S_N2/formal [3+2] cycloaddition process starting from nonactivated aziridines, which provided an efficient protocol to assemble polysubstituted indolizidines (Scheme 2.51) [59]. Aziridines and their reactions were of great importance because of their synthetic and pharmacological values. Heating a mixture of ethyl 7-iodo-2-heptynoate (or its analogs), 2-aryl aziridines, and K_2CO_3 in dry CH_3CN delivered polysubstituted indolizidines.

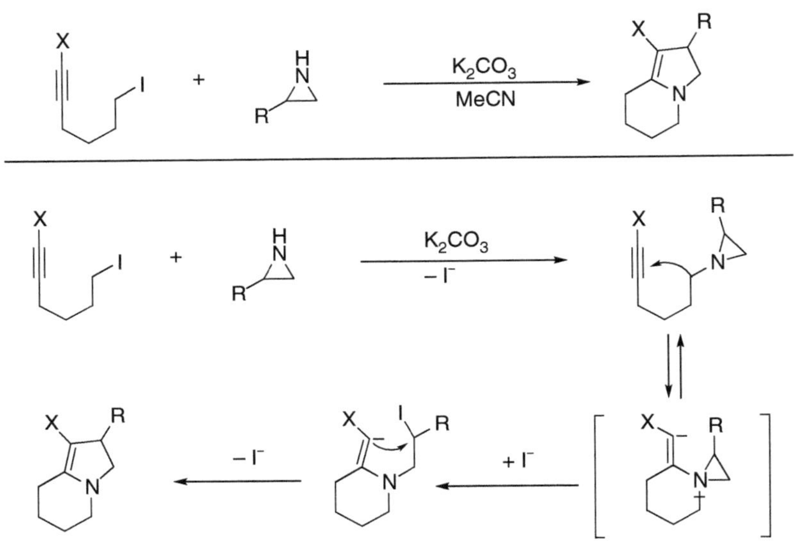

Scheme 2.51 [3+2] Cycloaddition process from nonactivated aziridines.

Ma and Lu reported an efficient route to 3-oxabicyclo[3.1.0]hexan-2-ones via the highly stereoselective iodolactonization of alkylidene-cyclopropyl esters with iodine or N-Iodosuccinimide (NIS) under aqueous conditions (Scheme 2.52) [60]. The study showed that the product was obtained in 79% yield after using 2 equivalents of I_2 (in MeCN aqueous solution, CH_3CN /H_2O, 2 : 1) at 80 °C for 3 hours. The research group developed a highly stereoselective iodolactonization reaction of alkylene cyclopropyl ester with iodine or NIS under aqueous conditions,

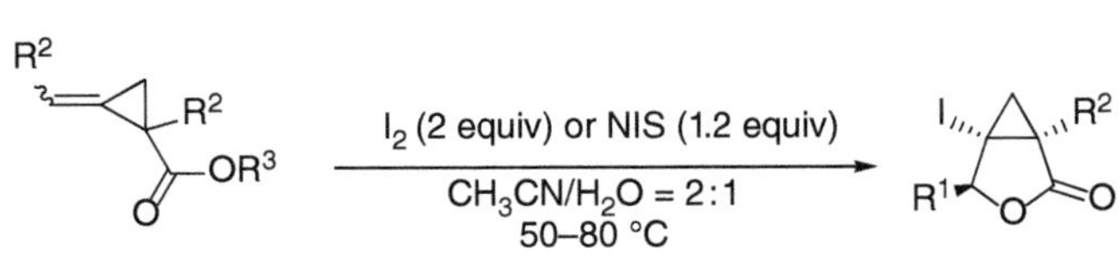

Scheme 2.52 Iodolactonization of alkylidene-cyclopropyl esters.

resulting in the formations of 4,5-trans-1,5-scis-3-oxabicyclo[3.1.0]hex-2-ones in medium-to-good yields.

Li and coworkers reported a new nucleophilic diamination reaction between *α*,*β*-unsaturated ketones, and esters without using any metal catalyst (Scheme 2.53) [61]. Three types of nitriles (CH_3CN, CH_3CH_2CN, and $CH_3CH_2CH_2CN$) were used as nucleophilic nitrogen sources. Studies showed that diamine products were extremely important in pharmaceutical chemistry and drug research. Among them, the electrophilic nitrogen source *N*,*N*-dichloro-*p*-toluenesulfonamide (4-$TsNCl_2$) could be easily synthesized from cheap starting materials, making this method inexpensive.

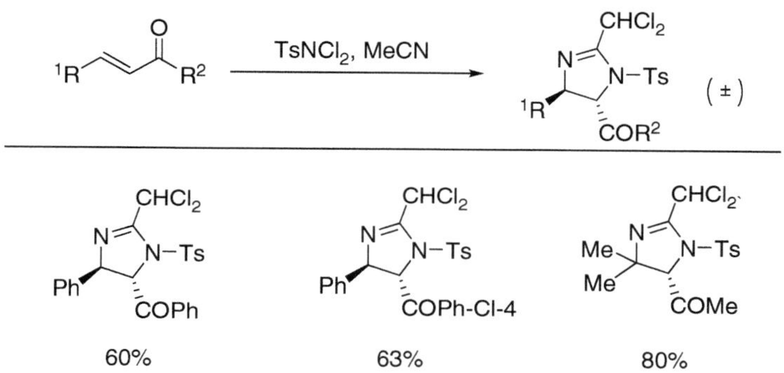

Scheme 2.53 Electrophilic deamination reaction of *α*,*β*-unsaturated ketones.

Johnson and coworkers reported a mild, alkali-free, Lewis acid-catalyzed [3+2] cycloaddition reaction of 2-(1-ethoxyethylideneamino) malonic acid dimethyl ester and negative electron-dipolar molecules (Scheme 2.54) [62]. Azomethine ylides were important intermediates in many [3+2] cycloaddition reactions and have achieved great success in the synthesis of pyrrolidine and other nitrogen-containing heterocycles. Mechanistic studies suggested that the dipolar cycloaddition of imide could be catalyzed by $MgCl_2$, which affected the addition at ambient temperature.

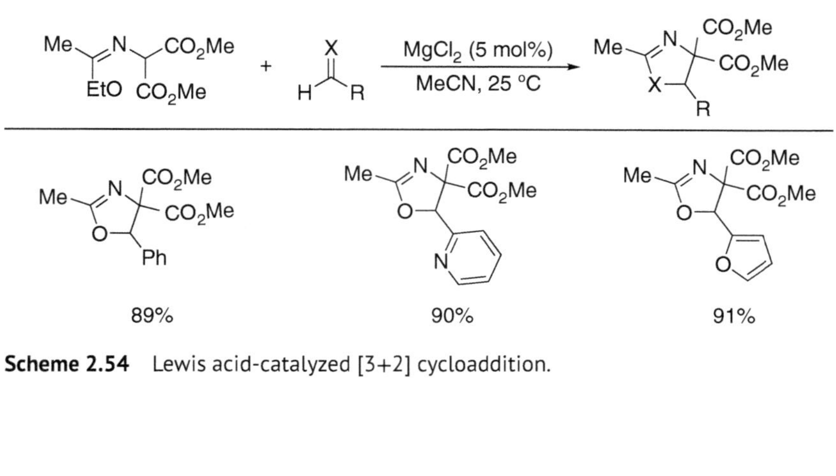

Scheme 2.54 Lewis acid-catalyzed [3+2] cycloaddition.

This process could be applied to several negative electron dipoles to form oxazoline, imidazoline, and pyrroline.

Biju and coworkers developed the synthesis and application of benzotriazoles, using NaN_3 as an azide component in [3+2] cyclization (Scheme 2.55) [63]. The cyclization of aryl groups generated by 2-(trimethylsilyl) aryltriflates led to the transition metal-free synthesis of *N*-H benzotriazole. When the reaction was carried out in CH_3CN, CsF was used as the fluorine source in CH_3CN to form *N*-H benzotriazoles in high selectivity, and medium-to-good yields.

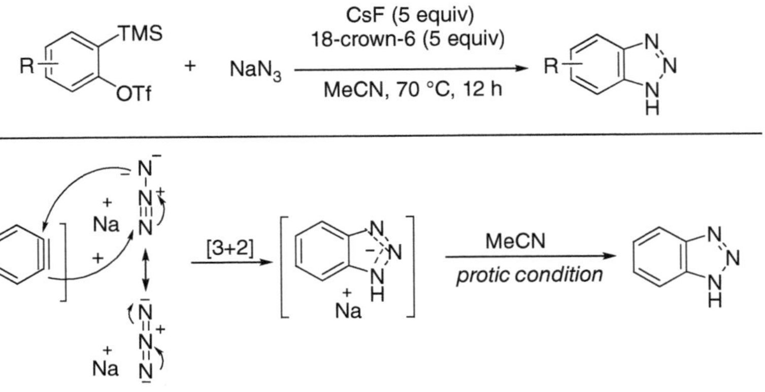

Scheme 2.55 Synthesis of benzotriazoles.

Namboothiri and coworkers developed a new [3+3] cyclization method for the synthesis of 2-aryl terephthalates from nitroallyl acetate and stable sulfur ylides (Scheme 2.56) [64]. The functional aromatic hydrocarbon of terephthalate was an attractive component in organic synthesis, as well as a component of many natural products, bioactive molecules, and functional materials. Among various cyclization methods, [3+3] cyclization was an efficient cascade strategy for rapidly constructing six-membered carbon rings and heterocycles from open-chain compounds in a

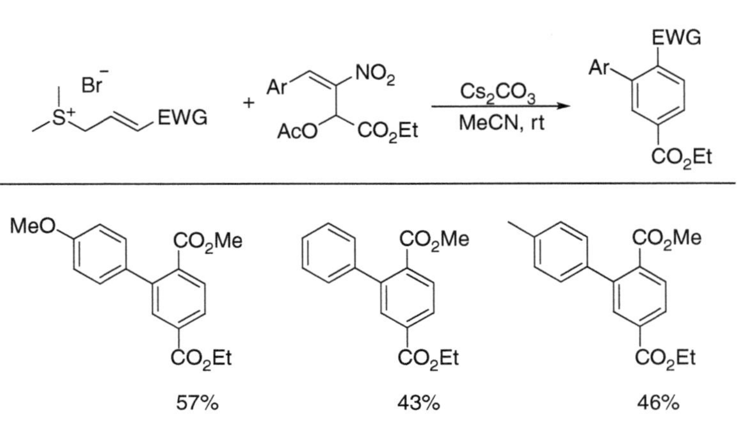

Scheme 2.56 Synthesis of 2-aryl terephthalates via [3+3] cyclization.

highly regioselective manner. This cyclization reaction involved γ-selective S_N2 reaction and γ-selective intramolecular Michael addition with high regioselectivity and moderate-to-good yields.

Tian's group developed a simple method to prepare *N*-acylbenzoxazides using phenols, nitriles, and dimethyl sulfoxide as substrates (Scheme 2.57) [65]. The preliminary mechanism study showed that, as the source of methylene synthon, dimethyl sulfoxide generated HCHO *in situ* by the decomposition of methoxy-dimethylsulfonium salt. The regeneration cycle of methoxydimethylsulfonium salt was proposed, which was initiated by a catalytic amount of $(COCl)_2$. Phenols with either electron-donating groups or electron-withdrawing groups were converted to benzoxazines efficiently.

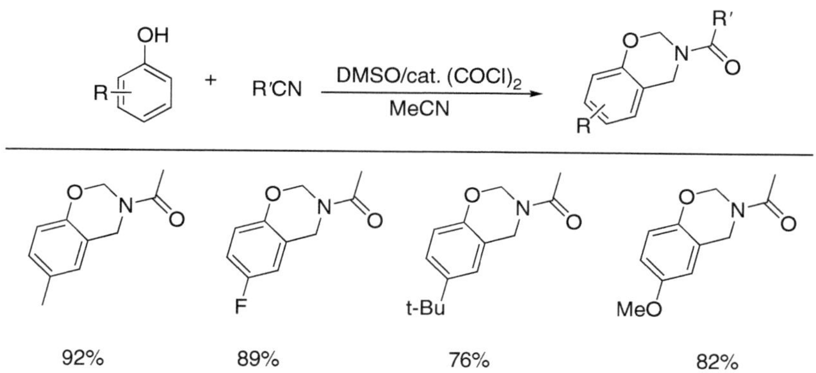

Scheme 2.57 Synthesis of *N*-acetylbenzoxazine.

Zhang and coworkers reported a new method for the synthesis of 2-substituted indoles by C–H amination with 2-alkenylanilines and (phenyliodonio)sulfamate (PISA) in aqueous CH_3CN (Scheme 2.58) [66]. PISA was a new water-soluble hypervalent iodine(III) reagent, (phenyliodonio)sulfamate. In this method, the authors used a ligand-exchange approach for the synthesis of PISA. When synthesizing 2-methylindole, the CH_3CN/H_2O system was found to be the best in solvent

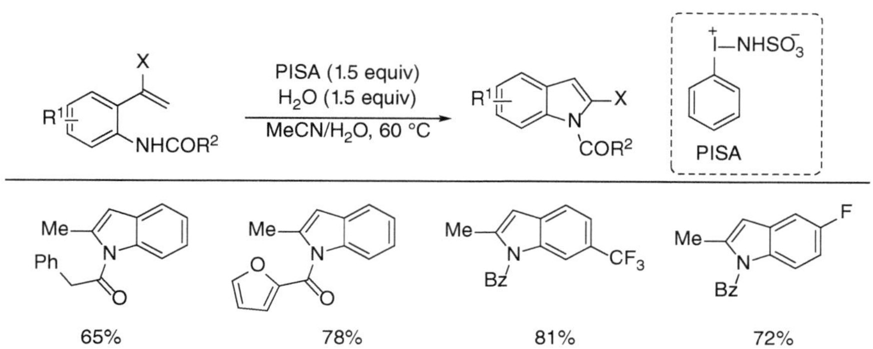

Scheme 2.58 Synthesis of 2-substituted indoles.

screening. It was found that 2-methylindole with different substituents on the benzene ring could be obtained, and various N-protecting groups could also form 2-substituted indoles well. Besides, this method was used to synthesize bioactive molecules rapidly.

2.3.2 Transition Metal-free Catalyzed Multicomponent Reactions in MeCN

Wang's group and Ji's group developed two efficient domino reactions based on isocyanates, which were used for the chemoselective synthesis of new polycyclic spiroindoles and polysubstituted pyrrole from 2-isocyanoethylindole, gem-activated olefins, and secondary amines (Scheme 2.59) [67]. Acetonitrile was used as solvent in the synthesis of indole. First, when secondary amine was used in the reaction system, the intermediate provided by isocyanate undergoes nucleophilic attack by amine to form the next intermediate. Next, the nitrogen atom carries out the nucleophilic attack on the intermediate nitrile to form polysubstituted pyrrole. These reactions proceeded smoothly and yielded the necessary products in moderate-to-good yields without a catalyst.

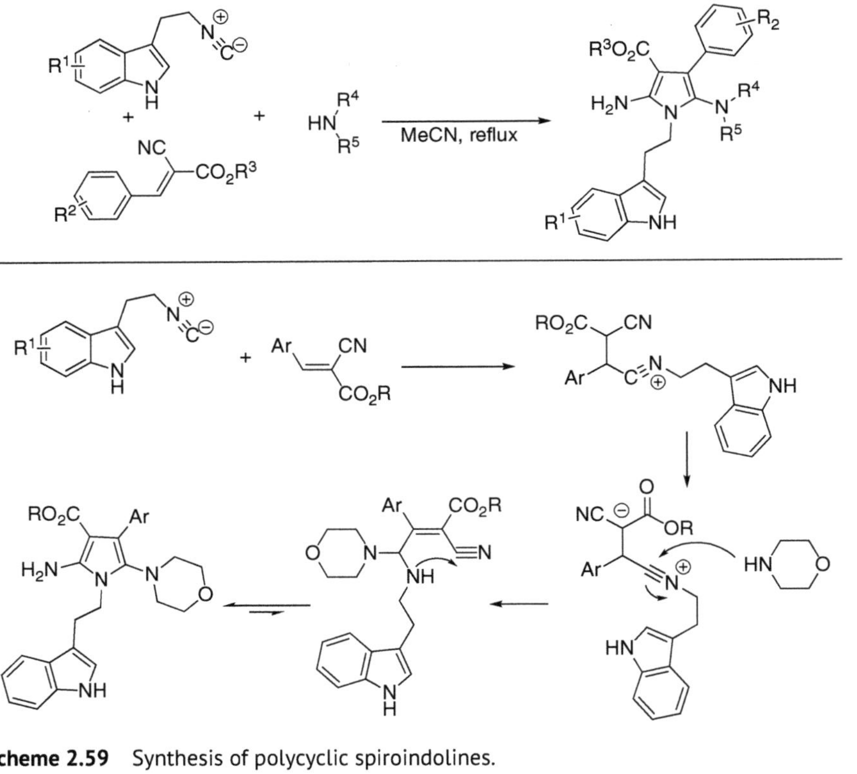

Scheme 2.59 Synthesis of polycyclic spiroindolines.

Tian and coworkers developed an unprecedented three-component arylation for the synthesis of 2,2,2-trichloroethanamines and 2,2-dichloroethenamines from inactive imines, aryl, and carbon nucleophiles (Scheme 2.60) [68]. When aromatic imines with electron-withdrawing groups participated in the reaction, 2,2-dichloroethylene amine was formed instead of 2,2,2-trichloroethylene amine. Mechanistic studies suggested that the elimination of HCl readily occurred in the case of aromatic imines with electron-withdrawing groups.

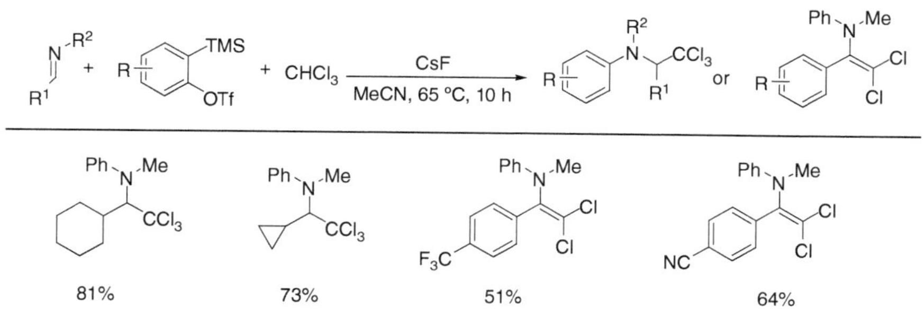

Scheme 2.60 Synthesis of 2,2,2-trichloroethanamines and 2,2-dichloroethenamines.

Larionov's group developed a new three-component reaction of small-ring cyclic amines, arynes, and acetonitrile to synthesize *N*-aryl-γ-amino nitriles with good yields under mild reaction conditions (Scheme 2.61) [69]. In this method, the small cyclic amines were mainly aziridines and azetidines with varied N-substituents. Acetonitrile was not only used as a reaction reagent but also participated in the reaction. *N*-aryl-γ-amino nitriles were found in many natural products and medicines.

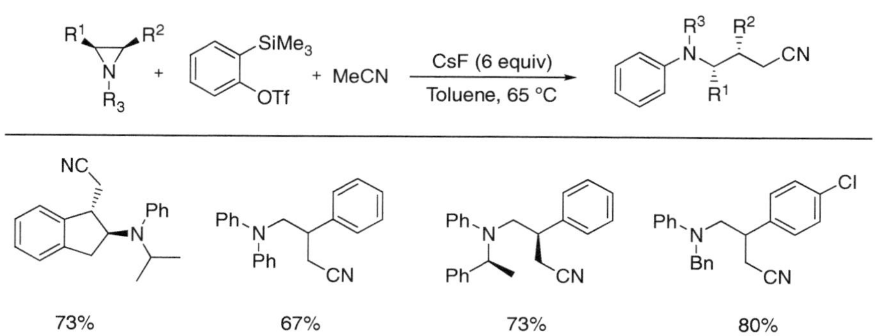

Scheme 2.61 Synthesis of *N*-aryl-γ-aminonitriles.

Shaabani et al. demonstrated a new and efficient method for the synthesis of highly functionalized benzo[g]- and dihydropyrano[2,3-g]chromene derivatives (Scheme 2.62) [70]. This process was carried out in the presence of a catalytic amount of Et_3N in CH_3CN at room temperature. The reaction performed good tolerance to functional groups with easy workup procedures. The products could be obtained with simple filtration in high yields without additional purification. The mechanism of this route involved the addition and cyclization of

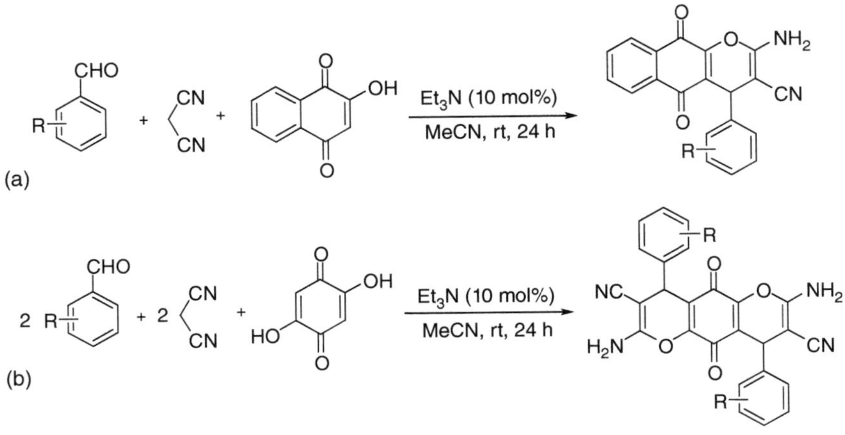

Scheme 2.62 Synthesis of benzo[*g*]- and dihydropyrano[2,3-*g*]chromene derivatives.

2-hydroxynaphthalene-1,4-dione or 2,5-dihydroxycyclohexa-2,5-diene-1,4-dione to the condensation product of an aldehyde with malononitrile.

Subsequently, Shaabani et al. developed an efficient synthetic route for the synthesis of highly functionalized bis(4*H*-chromene)-and 4*H*-benzo[g]chromene-3,4-dicarboxylate derivatives in moderate-to-high yields with good tolerance to functional groups (Scheme 2.63) [71]. The differences between this reaction and the above were the substrates. The mechanism was similar. The reaction was performed under neutral reaction conditions at room temperature without any undesirable by-products.

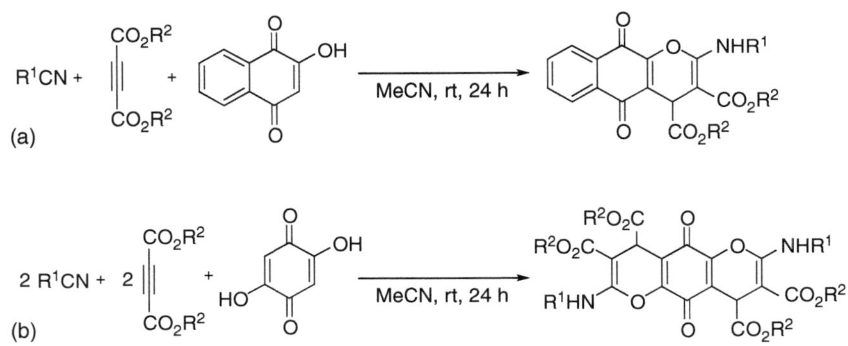

Scheme 2.63 Synthesis of bis(4*H*-chromene)-and 4*H*-benzo[*g*]chromene-3,4-dicarboxylate derivatives.

Lee and coworkers demonstrated a novel metal-free decarboxylative three-component coupling reaction for the synthesis of propargylamines in good yields (Scheme 2.64) [72]. The reaction was performed in CH_3CN at 65 °C for 3 h using alkynyl carboxylic acids, paraformaldehyde, and amines as substrates. This procedure could also work well on a gram-scale, generating the desired propargylamine

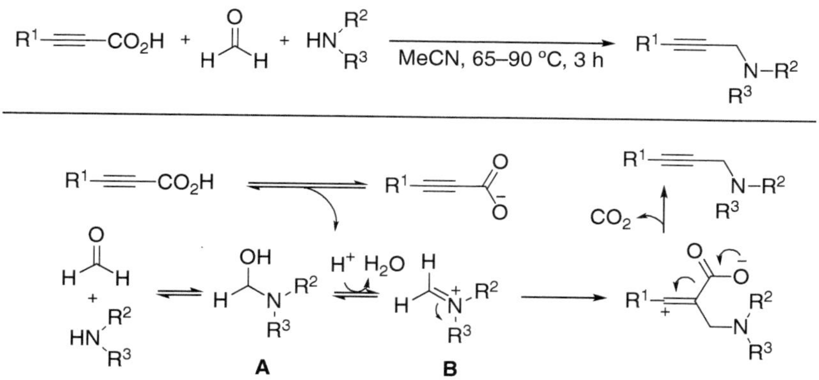

Scheme 2.64 Synthesis of propargylamines.

in 98% yield without column chromatography. This reaction exhibited higher selectivity toward alkynyl carboxylic acids compared to terminal alkynes. The mechanism was similar to Eschweiler–Clarke methylation. The corresponding hemiaminal **A** was formed by the reaction between an aldehyde and an amine. The acid proton of an alkyl carboxylic acid speeded up the generation of iminium salt **B**. The alkyl carbon bonded to the carboxylate attacked iminium salt **B** to generate intermediate **C**, which was then utilized to yield the desired propargylamine via decarboxylation.

2.3.3 Transition Metal-free Catalyzed C—X Bond Formation in MeCN

Oberhauser's group developed a novel methodology for the selective monobromination of phenols with NBS in CH_3CN, promoted by $HBF_4 \cdot Et_2O$ (Scheme 2.65) [73]. However, NBS and $HBF_4 \cdot Et_2O$ in CH_3CN were not suitable for the bromination of hydroxybenzaldehyde or hydroxyacetophenone. ^{13}C NMR measurements were conducted to study the nature of the active brominating species. The authors found protonation of NBS by $HBF_4 \cdot Et_2O$, followed by a slow cleavage of the N–Br bond,

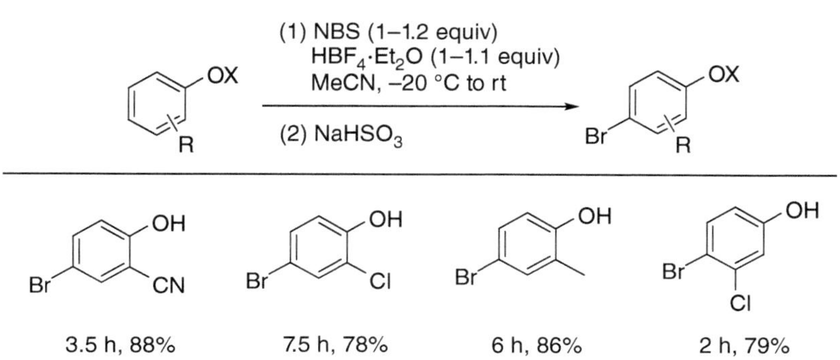

Scheme 2.65 Monobromination of phenols and anisoles.

resulting in the formation of succinimide and possibly bromonium tetrafluoroborate, which might be the actual brominating species.

Katritzky's group reported a strategy for the selective synthesis of *S*-acyl- and *N*-acylcysteines (Scheme 2.66) [74]. In this methodology, the authors utilized *N*-acylbenzotriazoles as substrates under mild reaction conditions to obtain the desired products in good yields. They identified the structures of *S*-acyl and *N*-acylcysteines through X-ray crystallography and 2D NMR. At a low pH value, *S*-acylcysteine was separated, and at a high pH value, *N*-acylcysteine was separated. Cysteine-containing peptides were valuable intermediates for the synthesis of more complex peptides and proteins using natural chemical linkages. Natural chemical linkages are widely used in protein synthesis.

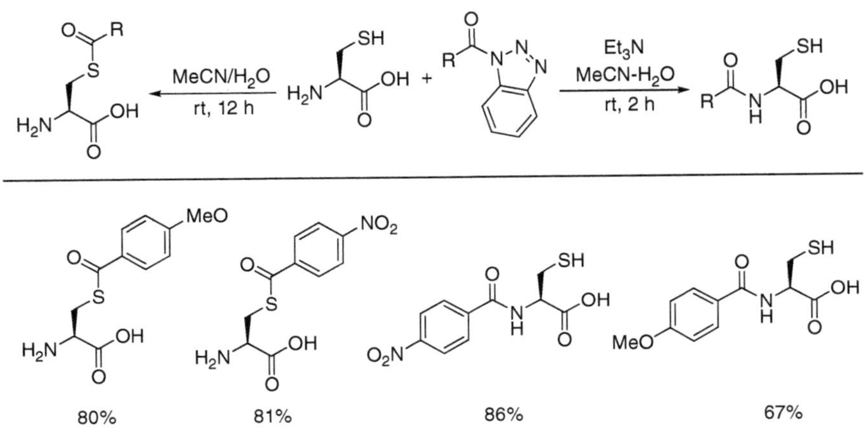

Scheme 2.66 Synthesis of *S*-acyl- and *N*-acylcysteines.

Kurosu's group reported the selective esterification of primary alcohols using oxyma or oxyma derivatives, 1-Ethyl-3-(3-dimethylaminopropyl)carbodiimide (EDCI), and $NaHCO_3$ (Scheme 2.67) [75]. A wide range of carboxylic acids could be esterified with excellent yields without detectable racemization. The strict anhydrous conditions were not necessary for these esterification reactions. The authors pointed out that the separation of oxyma from the product was extremely difficult via silica gel chromatography. However, the coupling additive, glyceroacetonide–oxyma could be easily removed after the reaction via a simple basic or an acidic aqueous workup procedure.

Plumet and coworkers achieved a new method for the stereoselective synthesis of *Z*-thiovinyl sulfones from alkyl and aryl sulfides without using any catalytic reagents (Scheme 2.68) [76]. This developed method was simple and efficient. It also avoided the use of organometallic compounds. A variety of vinyl sulfones was obtained from readily available aliphatic thiols with good yields. When acetonitrile was used as solvent, the protocol involved high diastereoselectivity, a short reaction time, and excellent yield. Vinyl sulfone was an attractive component in organic synthesis, as well as a component of many natural products, bioactive molecules, and functional materials.

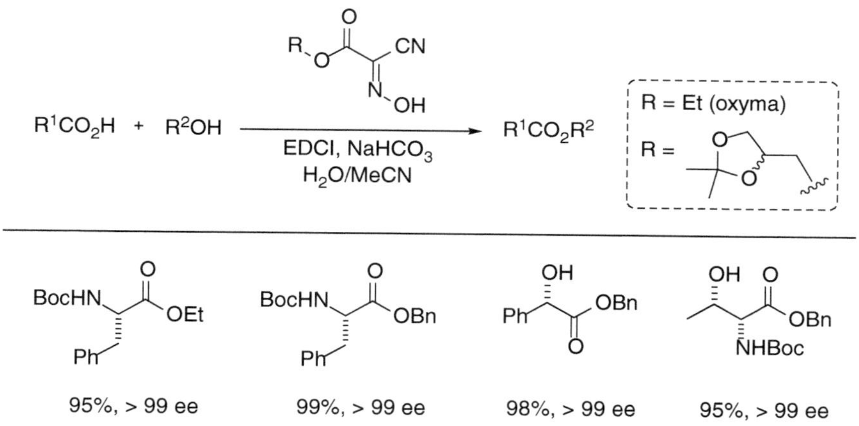

Scheme 2.67 Selective esterification of primary alcohols.

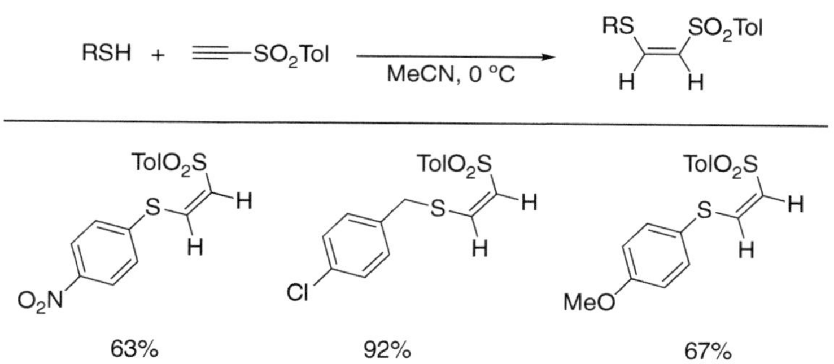

Scheme 2.68 Synthesis of *Z*-thiovinyl sulfone.

De Kimpe and coworkers illustrated a mild and efficient methodology for the synthesis of α-fluoro- and α,α-difluoroimines (Scheme 2.69) [77]. Various *N*-alkylimines derived from acetophenones were successfully monofluorinated using *N*-fluorobenzenesulfonimide (NFSI). Depending on the solvent system used

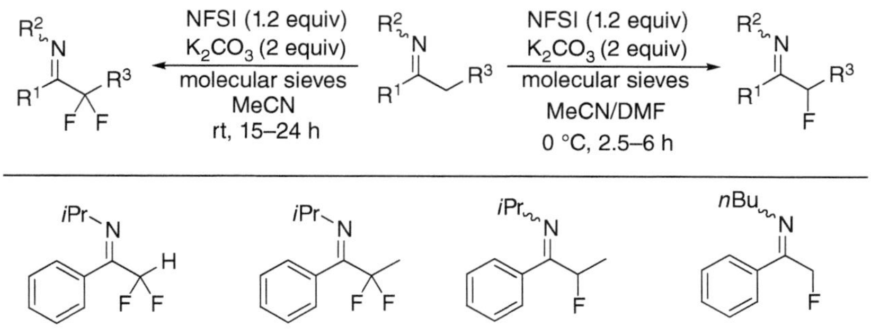

Scheme 2.69 Synthesis of mono- and difluoroimines.

and the reaction temperature, either mono- or difluoroimines were obtained. The authors reduced the desired products, α-fluoro- and α,α-difluoroimines, to obtain the corresponding β-fluoro- and β,β-difluoroamines in good yields.

Ju and coworkers disclosed a practical and scalable one-pot process for the preparation of florfenicol phosphodiester by adopting the phosphorylating system of $POCl_3$/pyridine/CH_3CN (Scheme 2.70) [78]. The authors found that the most common phosphorylating reagent, $POCl_3$, was highly efficient in this synthetic route. To facilitate large-scale production, they developed an improved and practical procedure to avoid column chromatographic purification. Thus, the process could work smoothly on a 500 g and 4 kg scale in the pilot plant.

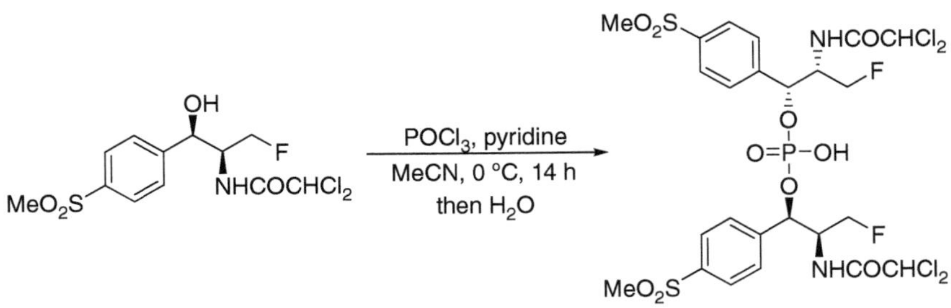

Scheme 2.70 Synthesis of florfenicol phosphodiester.

2.4 MeCN in C–X Bonds Formation With Radicals Involved

2.4.1 C–C, C–Si Bond Formation in MeCN

In 2019, Cheng et al. developed an Ag_2CO_3-mediated coupling reaction between aryl enol acetate and alkyl nitrile to prepare γ-ketonitriles based on direct C(sp^3)-H oxidative functionalization of alkyl nitrile (Scheme 2.71) [79]. This was a low toxic and facile route to synthesize nitrile-containing compounds. The reaction exhibited moderate-to-good yields with good tolerance to the functional group. Acetonitrile served as both the solvent and the source of CN-containing radicals.

Subsequently, the authors proposed a radical mechanism (Scheme 2.72). First, Ag_2CO_3 and MeCN coordinated to form intermediate **A**, which was then deprotonated to form intermediate **B**. Alkyl radical **C** was generated subsequently through ligand-to-metal charge transfer (LMCT). The generated radical **C** reacted with the enol ester to form intermediate **D**, which formed intermediate **E** through single electron transfer (SET). Finally, the desired product, γ-ketonitrile, was obtained via hydrolysis.

In 2020, Liu and coworkers developed a strategy to prepare 1-cyanoethylated 3,4-dihydronaphthalenes (Scheme 2.73) [80]. This reaction involved a radical pathway and proceeded with the formation of cyanomethyl radical, radical addition, and ring expansion. $Cu(OAc)_2$ was used as the catalyst to coordinate with the carbon-carbon double bond in vinylcyclopropanes to stabilize the intermediate,

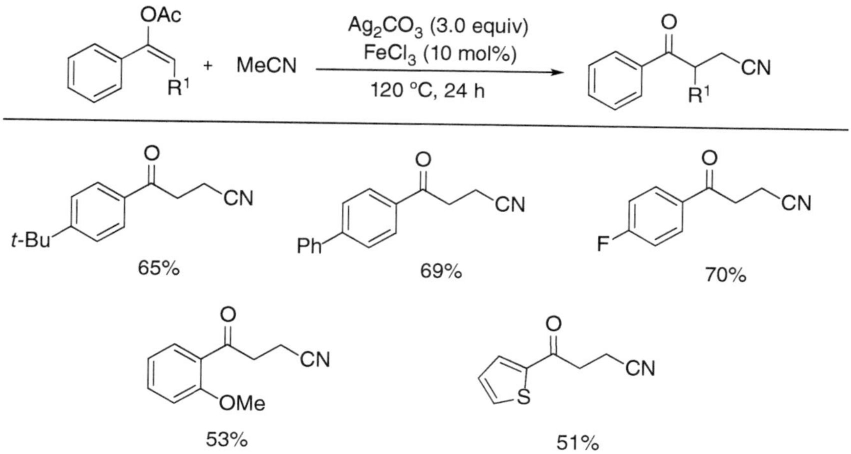

Scheme 2.71 Synthesis of γ-ketonitriles.

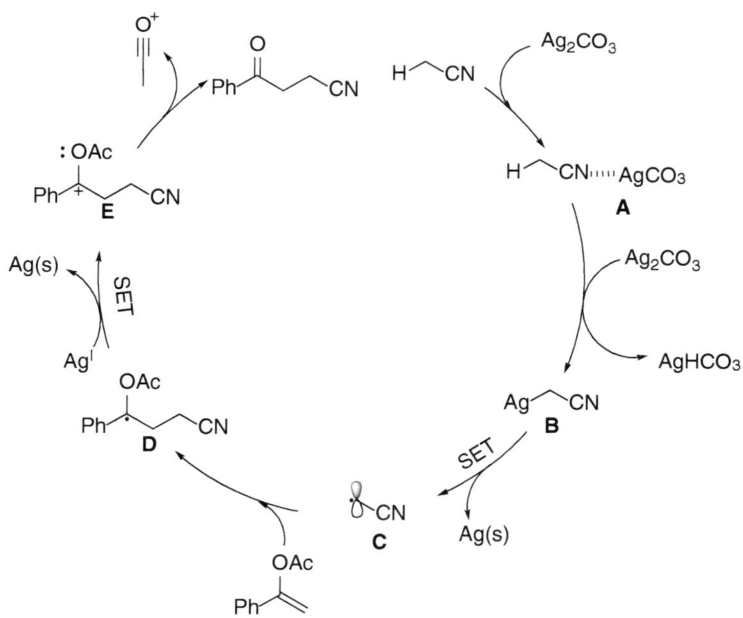

Scheme 2.72 Mechanism of synthesis of γ-ketonitriles.

which could promote the reaction between vinylcyclopropanes and α-C(sp^3)-H bond in acetonitrile. In this highly atomic economic way, acetonitrile served both as the solvent and as the source of CN-containing radicals.

In 2019, Aleman and coworkers achieved intramolecular homolytic substitution through photoredox catalysis and synthesized silicon heterocycles from aryl halides (Scheme 2.74) [81]. This reaction proceeded in an efficient metal-free dehalogenation of aryl halides under mild organo-photoredox conditions. This approach

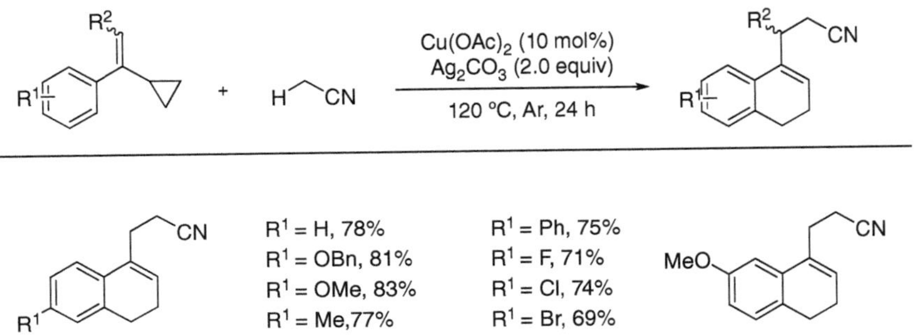

Scheme 2.73 Synthesis of 1-cyanoethylated 3,4-dihydronaphthalenes.

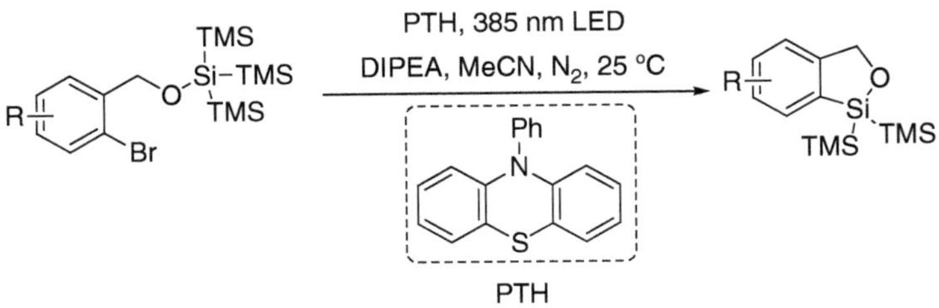

Scheme 2.74 Silicon heterocycle synthesis.

provided a large and diverse application scope, as well as efficient and chemoselective conditions. The authors established the photocatalytic single-electron reduction of aryl halides as a premier activation pathway to reach these intermediates.

Vila and coworkers established the visible-light photocatalytic radical addition reaction between dihydroquinoxaline-2-ones and trifluoromethyl ketones (Scheme 2.75) [82]. Dihydroquinoxaline-2-ones containing trifluoromethyl-substituted tertiary alcohols were synthesized by the addition of dihydroquinoxaline-2-one photocatalytic radicals to trifluoromethyl ketones through the reduction quenching cycle of $Ru(bpy)_3Cl_2$. $Ru(bpy)_3Cl_2$ as the photocatalyst and visible light

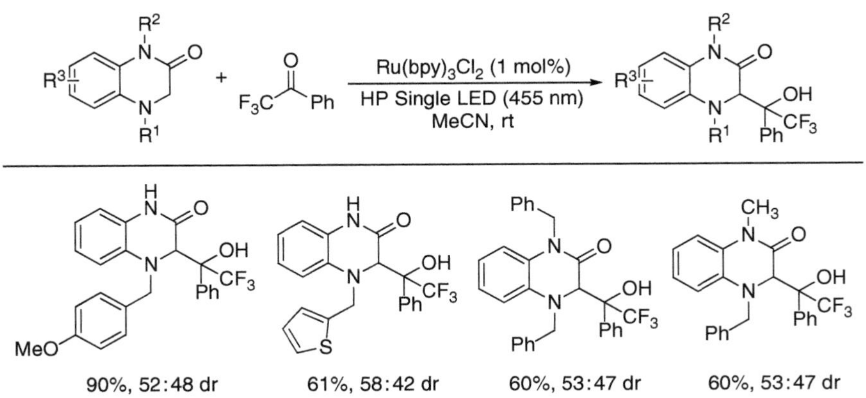

Scheme 2.75 Synthesis of trifluoromethyl tertiary alcohols.

irradiation were two important conditions for the reaction, which could not occur without either one. Besides, this method was effective for the late-stage functionalization of a value-added indomethacin-derived trifluoroacetophenone substrate.

In 2020, Wang and coworkers developed a new strategy to effectively generate silyl radicals through the photo-induced decarboxylation of silacarboxylic acids (Scheme 2.76) [83]. In this process, silyl radicals could be prepared by gram-scale reactions in high yields and stored in the air for a long time. Silyl radicals could further react with various alkenes to yield the corresponding silylation products in good-to-high yields with wide functional group compatibility. The authors proposed the possibility of a cascade reaction that involved both a hydrogen atom transfer (HAT) process and decarboxylation. This work demonstrated the value of photocatalysis for effective synthetic transformations of functional elements by providing a mild, selective, and synthetically useful method for generating reactive hetero-radical intermediates.

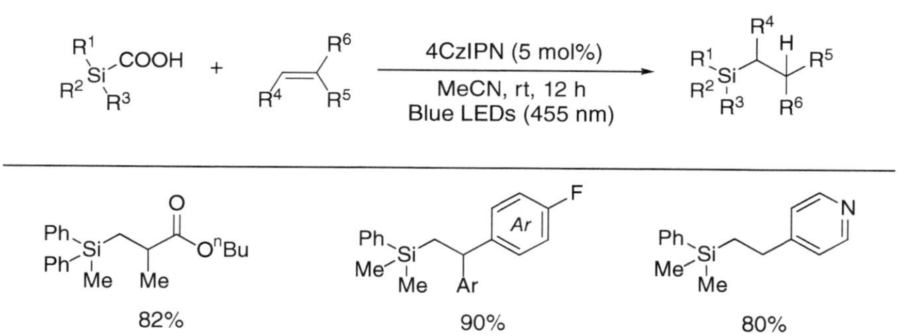

Scheme 2.76 Photo-induced decarboxylation of silacarboxylic acids.

2.4.2 C–N, C–P Bond Formation in MeCN

In 2017, Zhang's group and Bao's group developed a new strategy for synthesizing the core structure of γ-amino butyric acid through the carboamination of olefins catalyzed by copper (Scheme 2.77) [84]. The authors used acetonitrile as the carbon and nitrogen functional group to realize the carboamination of olefins, which had potential value in the synthesis of analogs of γ-aminobutyric acids, substituted pyrrolidines, pyrrolidones, and other drug molecules. Besides, the research group proposed the radical polarity cross mechanism for copper catalysis through density functional theory (DFT) studies.

In 2020, Protti's group and Fagnoni's group developed a novel methodology for the diazenylation of enol silyl ethers under visible light irradiation (Scheme 2.78) [85]. This reaction could bear different substituents on azaderivatives, including cyano groups, different halogens, and even a triple C—C bond. Despite the fact that aryldiazenyl radical intermediates were uncommon due to their poor stability in solution at room temperature, this method could use arylazo sulfones as photoactive precursors to prepare aryldiazenyl radicals while also preventing nitrogen loss by trapping 1,3-dicarbonyls with appropriate enol silyl ethers.

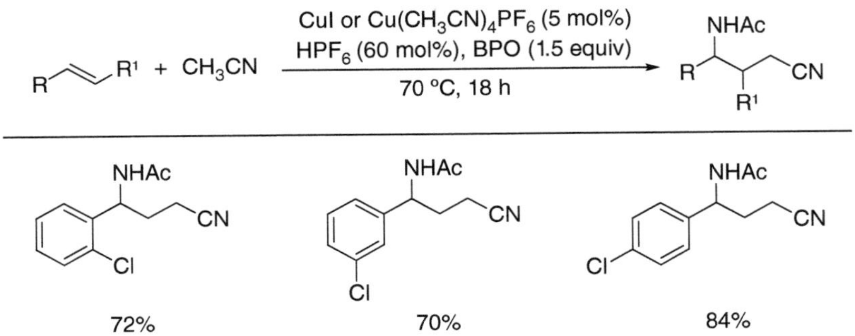

Scheme 2.77 Copper-catalyzed carboamination of alkenes.

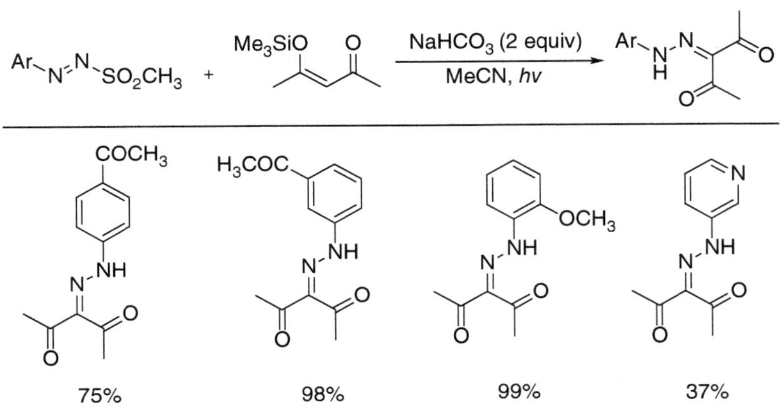

Scheme 2.78 Visible light promoted diazenylation of enol silyl ethers.

In 2016, Tang and colleagues successfully developed a method for producing benzophosphole oxides through the phosphorylation–cyclization of various alkynes with secondary phosphine oxides in the presence of $CuSO_4 \cdot 5H_2O$ as the catalyst (Scheme 2.79) [86]. This methodology provided a quick and easy way to produce benzophosphole oxides in moderate-to-good yields, with a variety of broad scopes.

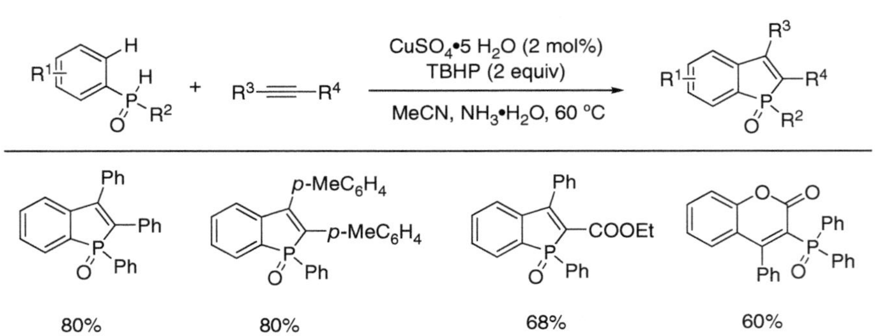

Scheme 2.79 Synthesis of benzophosphole oxides.

In addition, the authors used cost-effective catalysts and easy-to-obtain substrates, which could be used for large-scale production.

In 2019, Chen's group and Zhu's group reported the silver-promoted iminophosphorylation of γ,δ-unsaturated oxime esters through a cascade radical cyclization, which enabled the quick production of a variety of phosphine-containing pyrrolines in good-to-high yields (Scheme 2.80) [87]. This strategy offered a novel method to construct new bulky trivalent phosphine ligands with a pyrroline motif. Besides, this reaction could be easily performed on a gram scale, and the resulting phosphorylated pyrrolines were useful for further transformations.

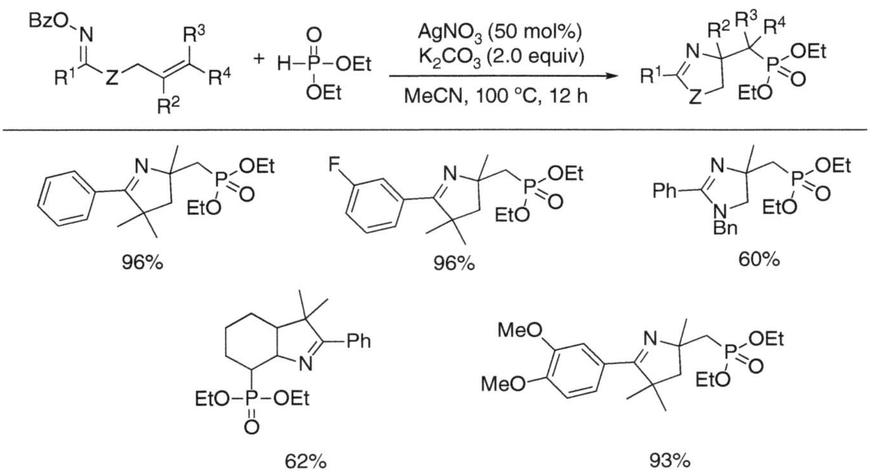

Scheme 2.80 Synthesis of phosphorylated pyrrolines.

In 2021, Chen and coworkers developed a radical cascade reaction between 2-aryloxy phenylacetylenes and phosphine oxides by employing $K_2S_2O_8$ as a radical initiator, providing diphosphonyl xanthenes as products in moderate-to-good yields (Scheme 2.81) [88]. The reaction was conducted without transition metals under

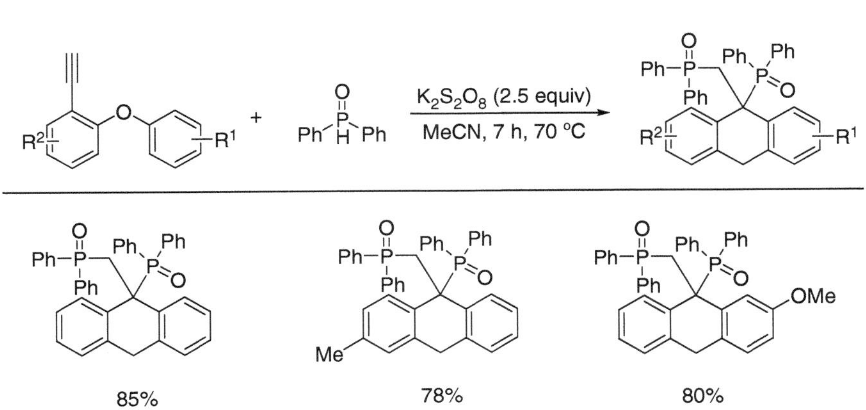

Scheme 2.81 Synthesis of 2-aryloxy phenylacetylenes.

mild conditions. 2-aryloxy phenylacetylene and its various derivatives could be efficiently and environment-friendly produced with this method. According to the mechanistic investigation, phosphine oxide was transformed into a phosphonyl radical, which underwent a double radical addition/cyclization with 2-aryloxy phenylacetylenes and generated bisphosphonyl xanthenes.

2.4.3 C—O, C—S Bond Formation in MeCN

In 2019, Guan and coworkers successfully developed a visible-light-induced cyanomethylation/cyclization tandem reaction of olefinic amides with readily available bromoacetonitrile (Scheme 2.82) [89]. The reaction completed the synthesis of C—O bonds in one pot without using any oxidizing agent. By using this method, several benzoxazines containing cyanomethyl groups could be obtained, which were useful pharmaceutical frameworks. This reaction featured easy operation, mild reaction conditions, a wide range of substrates, and good functional group tolerance.

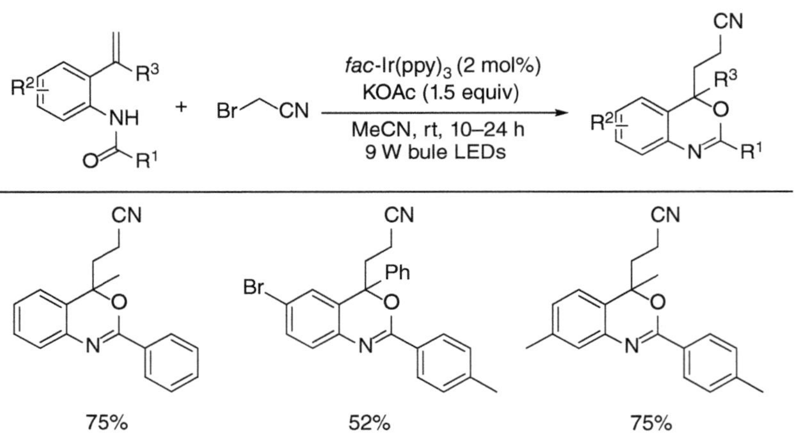

Scheme 2.82 Synthesis of cyanomethylated benzoxazines.

Subsequently, they proposed a mechanism (Scheme 2.83). Under irradiation of visible light, [*fac*-Ir(III)(ppy)$_3$] was excited to [*fac*-Ir(III)(ppy)$_3$*], which was quenched with bromoacetonitrile to produce the cyanomethyl radical and [*fac*-Ir(IV)(ppy)$_3$$^+$] complex. Following this, the radical intermediate **A** was generated through the cyanomethyl radical reaction with olefinic amide. Then, [*fac*-Ir(IV)(ppy)$_3$$^+$] oxidized **A** to produce the intermediate **B** and [*fac*-Ir(III)(ppy)$_3$]. The product **C** was obtained through the cyclization reaction of intermediate **B**.

In 2017, Wu and coworkers described a method for the direct preparation of thiophosphates through a three-component reaction under visible light irradiation (Scheme 2.84) [90]. This reaction worked smoothly, generating the corresponding thiophosphates in moderate-to-good yields, which were functional skeletons existing in agrochemicals and pharmaceuticals. Various diaryliodonium

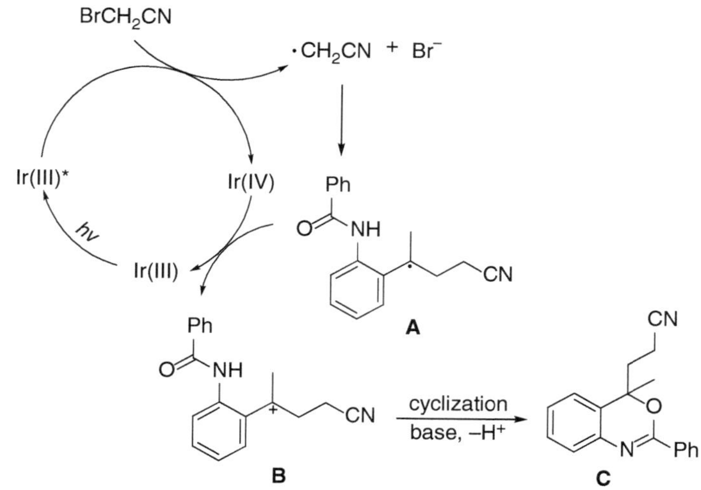

Scheme 2.83 Mechanism for synthesis of benzoxazines.

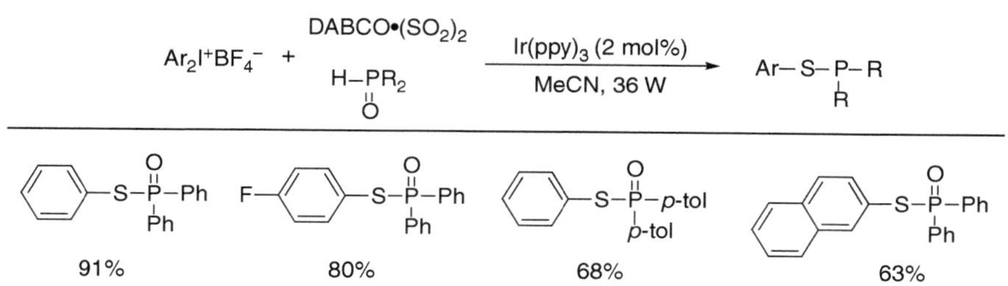

Scheme 2.84 Synthesis of thiophosphates.

tetrafluoroborates, including dinaphthyliodonium tetrafluoroborates were all suitable substrates in this transformation. The sulfur dioxide in 1,4-diazabicyclo[2.2.2] octane-sulfur dioxide (DABCO·$(SO_2)_2$) was the sulfur source. The reaction underwent a radical mechanism, and one of the most important steps in the thiolation of diarylphosphine oxides was the reduction of arylsulfonyl radicals.

In 2019, Tang and coworkers developed a metal-free iodine-promoted radical alkyl sulfuration of imidazopyridines and dialkyl azo compounds using elemental sulfur (Scheme 2.85) [91]. Dialkyl azo compounds were found to be effective alkyl radical sources for the direct alkyl sulfuration with imidazopyridines. The authors suggested that iodine could inhibit the self-coupling of the cyanoalkyl radical, and the *in situ* 2-iodo-2-methylpropanenitrile intermediate acted as a relay cyanoalkyl radical, allowing for efficient transformation. This reaction could successfully synthesize a variety of imidazoheterocycles containing cyanoalkylthio groups without odorous thiols.

In 2020, Wu and coworkers successfully demonstrated an efficient method for synthesizing various tetrasubstituted β-sulfonyl N-unprotected enamines (Scheme 2.86) [92]. This route used copper to catalyze the three-component reaction of *O*-acyl

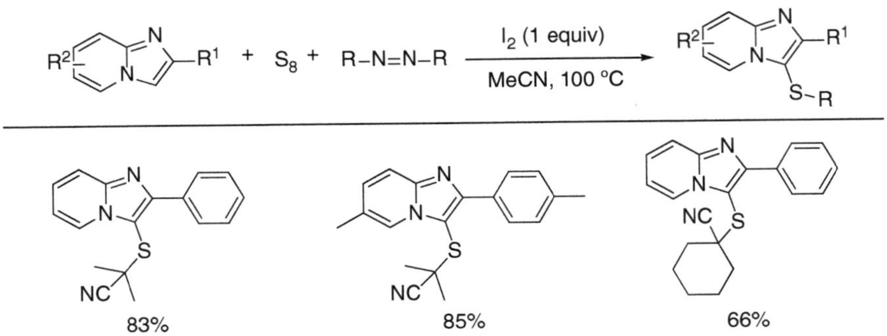

Scheme 2.85 Synthesis of imidazopyridines.

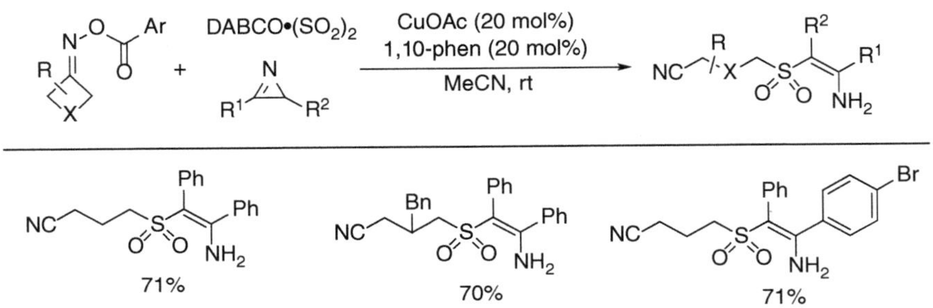

Scheme 2.86 Synthesis of tetrasubstituted β-sulfonyl enamines.

oximes, DABCO · $(SO_2)_2$, and 2H-azirines under mild conditions. This method had good yields with excellent stereoselectivity and regioselectivity. Notably, this was an example of the innovative and useful synthons for β-functionalized N-unprotected enamines that 2H-azirines provided. Preliminary mechanistic studies indicated that the reaction proceeded through the coupling of sulfonyl radical and α-carbon radical via copper-catalyzed ring-opening C–C bond cleavage of *O*-acyl oxime and C—N bond cleavage of 2H-azirine with the insertion of sulfur dioxide.

In 2021, Wu and coworkers reported a practical and efficient method for synthesizing (*E*)-β-sulfonyl enamines under mild conditions using an iron(II)-catalyzed three-component reaction of *O*-acyl oximes, sulfur dioxide, and *N*-vinylacetamides (Scheme 2.87) [93]. This protocol exhibited good stereoselectivity and regioselectivity with a wide range of reactions. In this reaction, various sensitive functional groups were compatible. They suggested that the reaction underwent a radical pathway through an α-carbon radical intermediate generated through iron-catalyzed ring-opening C—C bond cleavage of *O*-acyl oxime.

2.4.4 C-Halogen Bond Formation in MeCN

Li and Yu developed a facile synthetic route for the highly selective fluorination of tertiary alkyl bromides and iodides with Selectfluor (Scheme 2.88) [94]. The practical

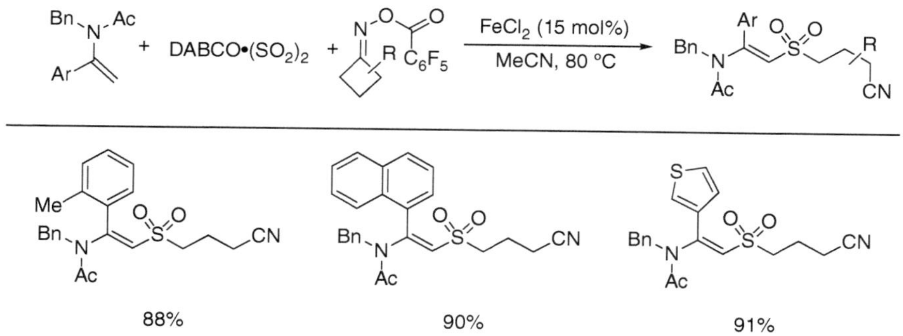

Scheme 2.87 Synthesis of (*E*)-*β*-sulfonyl enamines.

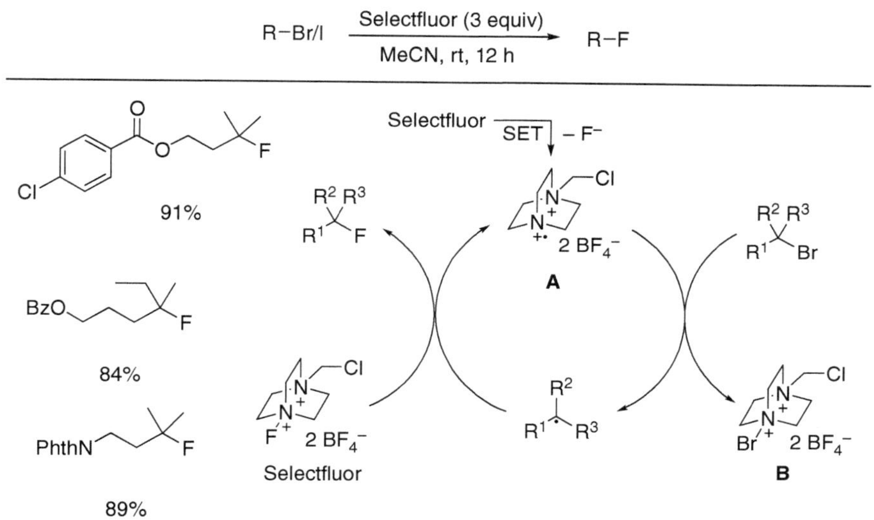

Scheme 2.88 Radical fluorination of tertiary alkyl halides.

protocol was operationally simple, free from metal ions, broad in scope, and tolerant of sensitive functional groups. The halogen-exchange fluorination performed efficiently in acetonitrile at room temperature with good-to-excellent yields. Alkyl chlorides, as well as primary and secondary alkyl bromides, worked smoothly under the reaction conditions. The authors proposed a radical mechanism for this selective fluorination. The aminium radical cation **A** was released via single-electron reduction of Selectfluor. *N*-bromoaminium salt **B** and the corresponding alkyl radical were generated after intermediate **A** removed a bromine atom from the tertiary alkyl bromide. The fluorination product and the aminium radical cation **A** were formed through the subsequent transfer of fluorine from Selectfluor to the alkyl radical.

Jiang's group demonstrated an efficient and cost-effective copper-mediated aerobic oxidative C—H iodination and nitration of indoles via double C—H functionalization (Scheme 2.89) [95]. The domino process performed smoothly under mild

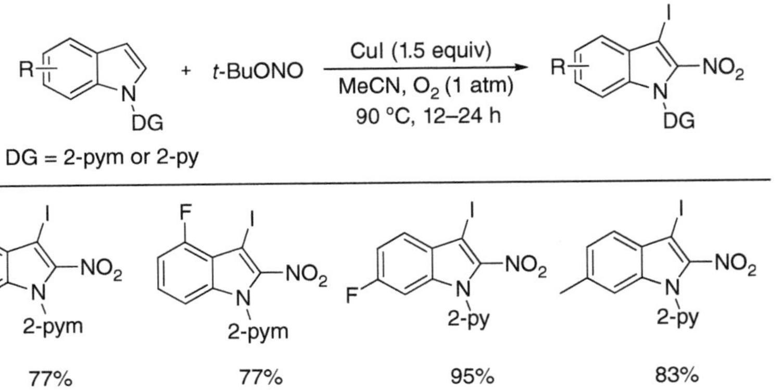

Scheme 2.89 C—H iodination and nitration of indoles.

aerobic conditions to afford 3-iodo-2-nitroindoles in one step with high regioselectivity and a broad substrate scope. The authors suggested that the mechanism of the iodination step was to proceed via a Cu(III)-iodide species that underwent electrophilic addition at the C3 position of the indole motif. A sequence of C—H activation, nitro radical oxidative addition, and reductive elimination were included in the nitration step.

Liu and coworkers developed an alkoxyl radical synthetic route for site-selective functionalization of unactivated methylene and methine C—H bonds enabled by an iron(II)-catalyzed redox process (Scheme 2.90) [96]. This mild methodology allowed for the efficient remote aliphatic fluorination and chlorination of structurally and electronically varied primary, secondary, and tertiary hydroperoxides with high efficiency and selective and predictable fluorine incorporation at the δ-methylene or methine sites of linear alkane substrates. The iron-catalyzed redox process had an excellent functional group tolerance, with common functionalities, including aryl, alkynyl, carbonyl, hydroxy, benzyl ether, acetate, azide, amide, halide, and tetrahydropyran moieties. The generality and modularity of the directed C—H functionalization strategy, its predictable regioselective manner, and the omnipresence and versatility of hydroxy groups indicated that this approach had broad applications in late-stage, diversity-oriented functionalization of complex molecules.

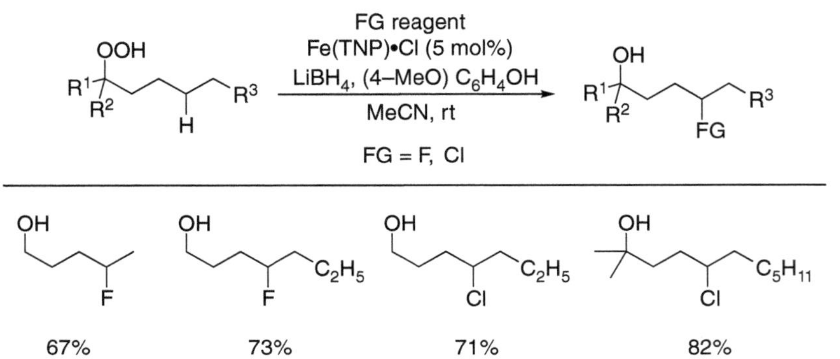

Scheme 2.90 Iron(II)-catalyzed functionalization of C—H bonds.

Su et al. established a metal- and additive-free strategy for the synthesis of oxindoles via a chlorine radical-induced cascade chlorination of *N*-aryl acrylamides with moderate-to-good yields (Scheme 2.91) [97]. Trichloroisocyanuric acid (TCCA) acted as both the radical initiator and chlorine source. A variety of substrates could be applied smoothly in this synthetic route to directly afford chlorinated oxindoles via C—Cl bond formation. A wide range of α-substituted acryl amides such as phenyl, benzyl, alcohol, silyl ether, and ester were used in this cascade chlorination reaction, furnishing the chlorinated oxindoles effectively.

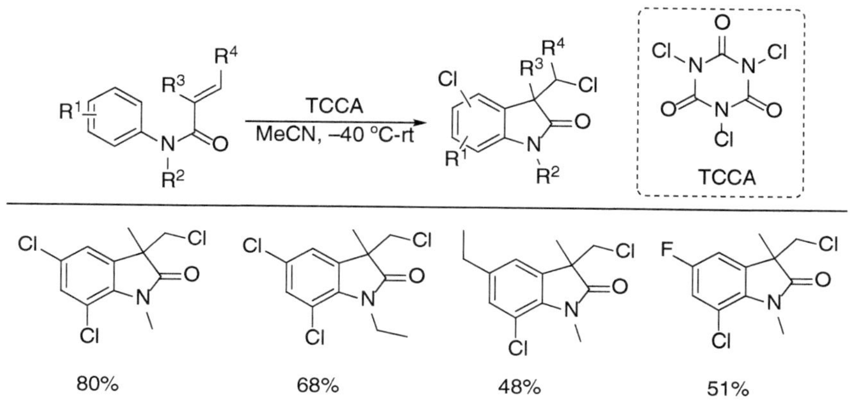

Scheme 2.91 TCCA-induced chlorine radical cascade chlorination.

Wang and coworkers reported ring-opening iodination and bromination of unstrained cycloalkanols with NaI or NaBr and $PhI(OAc)_2$ under visible light irradiation (Scheme 2.92) [98]. In this protocol, the concentration of I_2 was modulated through the generation of triiodide (I_3^-), thus significantly reducing the undesired

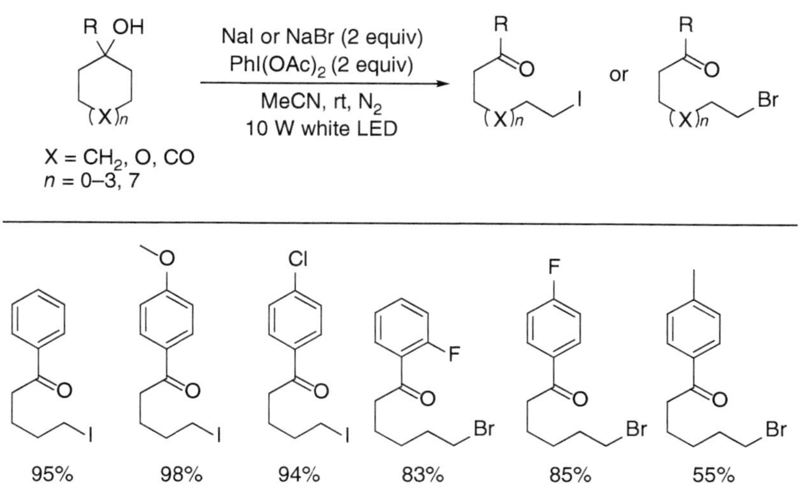

Scheme 2.92 Ring-opening iodination and bromination.

side reactions. The reaction was carried out under mild conditions, thus providing a practically useful method for accessing ω-iodo or ω-bromoketones. The reaction was operationally simple and exhibited wide substrate scope. Particularly in the case of ring-opening iodination, the substrates could include those bearing electron-rich aromatic ring substituents.

2.5 Conclusion

This chapter conveys that acetonitrile serves as a green solvent and a multipurpose reagent in various reactions utilizing its structure for addition, oxidation, reduction, and substitution reactions. Although many significant achievements in the research of acetonitrile have been made, there is still plenty of work to be carried out, especially in the mechanism of the reactions in which acetonitrile is involved. With the development of green and sustainable chemistry, we are looking forward to the utilization of acetonitrile in organic synthesis that will be developed continuously.

References

1 Robles, H. (2014). *Encyclopedia of Toxicology*, 3e, 40–42. Elsevier Inc.
2 Amoore, J.E. and Hautala, E. (1983). *J. Appl. Toxicol.* 3: 272–290.
3 Boschloo, G. and Hagfeldt, A. (2009). *Acc. Chem. Res.* 42: 1819–1826.
4 Pearson, J.D. (2004). *Encyclopedia of Biological Chemistry*, 398–403. Elsevier Inc.
5 Yang, Y. (2016). *Side Reactions in Peptide Synthesis*, 311–322. Elsevier B.V.
6 S. Creager, *Handbook of Electrochemistry*, 2007, 57-72.Elsevier Inc.
7 Fleming, F.F., Yao, L., Ravikumar, P.C. et al. (2010). *J. Med. Chem.* 53: 7902–7917.
8 (a) Hoff, B.H. (2018). *Synthesis* 50: 2824–2852. (b) Bagal, D.B. and Bhanage, B.M. (2015). *Adv. Synth. Catal.* 357: 883–900.
9 Fan, L. and Ozerov, O.V. (2005). *Chem. Commun.* 4450–4452.
10 Chakraborty, S., Patel, Y.J., Krause, J.A., and Guan, H. (2013). *Angew. Chem. Int. Ed.* 52: 7523–7526.
11 Peck, C.L., Nekvinda, J., and Santos, W.L. (2020). *Chem. Commun.* 26: 10313.
12 Kawauchi, D., Ueda, H., and Tokuyama, H. (2019). *Eur. J. Org. Chem.* 10: 2056–2060.
13 Yan, F., Huang, Z., Du, C.-X. et al. (2021). *J. Catal.* 395: 188–194.
14 Cao, J., Lv, D., Yu, F. et al. (2021). *Org. Lett.* 23: 3184–3189.
15 Cheng, Q., Zhang, F., Cai, Y. et al. (2018). *Angew. Chem. Int. Ed.* 57: 2134–2138.
16 Andrus, M.B. and Zhou, Z. (2002). *JACS* 124: 8806–8807.
17 Lyle, M.P.A. and Wilson, P.D. (2006). *Org. Biomol. Chem.* 4: 41–43.
18 Kirillova, M.V., Kirillov, A.M., Mandelli, D. et al. (2010). *J. Catal.* 272: 9–17.
19 Marx, S., Kleist, W., and Baiker, A. (2011). *J. Catal.* 281: 76–87.
20 Liu, C., Zhang, Q., Li, H. et al. (2016). *Chem. Eur. J.* 22: 6208–6212.
21 Yang, Y., Zhong, W., Nie, B. et al. (2017). *J. Organomet. Chem.* 853: 136–142.

22 Engelmann, X., Malik, D.D., Corona, T. et al. (2019). *Angew. Chem. Int. Ed.* 58: 4012–4016.

23 Zhang, Q., Zheng, M., Song, G. et al. (2019). *Asian J. Organomet. Chem.* 8: 1824–1826.

24 Anandababu, K., Muthuramalingam, S., Velusamy, M., and Mayilmurugan, R. (2020). *Catal. Sci. Technol.* 10: 2540–2548.

25 Horvat, M. and Iskra, J. (2022). *Green Chem.* 24: 2073–2081.

26 Ando, K., Takaba, C., and Kodama, M. (2022). *J. Organomet. Chem.* 87: 9723–9728.

27 Kong, H.-H., Pan, H.-L., and Ding, M.-W. (2018). *J. Organomet. Chem.* 83: 12921–12930.

28 Zalomaeva, O.V., Evtushok, V.Y., Ivanchikova, I.D. et al. (2020). *Inorg. Chem.* 59: 10634–10649.

29 Zhou, P., Huang, L., Jiang, H. et al. (2010). *J. Organomet. Chem.* 75: 8279–8282.

30 Jiang, H., Ji, X., Li, Y. et al. (2011). *Org. Biomol. Chem.* 9: 5358–5361.

31 Wu, T., Mu, X., and Liu, G. (2011). *Angew. Chem. Int. Ed.* 50: 12578–12581.

32 Fuentes, J.A. and Clarke, M.L. (2008). *Synletter* 17: 2579–2582.

33 Snelders, D.J.M., Kunna, K., Müller, C. et al. (2010). *Tetrahedron: Asymmetry* 21: 1411–1420.

34 Zhu, K., Shaver, M.P., and Thomas, S.P. (2016). *Chem. Sci.* 7: 3031–3035.

35 Cannon, K.A., Geuther, M.E., Kelly, C.K. et al. (2011). *Organometallics* 30: 4067–4073.

36 Li, X.-D., Xia, S.-M., Chen, K.-H. et al. (2018). *Green Chem.* 20: 4853–4858.

37 Ito, M., Itazaki, M., and Nakazawa, H. (2016). *ChemCatChem* 8: 1–4.

38 Shi, J., Li, L., Shan, C. et al. (2021). *JACS* 143: 10530–10536.

39 Qian, X., Auffrant, A., Felouat, A., and Gosmini, C. (2011). *Angew. Chem. Int. Ed.* 50: 10402–10405.

40 Zhai, Y., Chen, X., Zhou, W. et al. (2017). *J. Organomet. Chem.* 82: 4964–4969.

41 Sun, M., Chen, J.-F., Chen, S., and Li, C. (2019). *Org. Lett.* 21: 1278–1282.

42 Zhao, M., Zhang, W., and Shen, Z. (2015). *J. Organomet. Chem.* 80: 8868–8873.

43 Anka-Lufford, L.L., Huihui, K.M.M., Gower, N.J. et al. (2016). *Chem. Eur. J.* 22: 11564–11567.

44 Liu, B., Liu, M., Li, Q. et al. (2020). *Org. Biomol. Chem.* 18: 6108–6114.

45 Singh, R., Allam, B.K., Singh, N. et al. (2015). *Adv. Synth. Catal.* 357: 1181–1186.

46 Huang, W.-Y., Lu, C.-H., Ghorai, S. et al. (2020). *JACS* 142: 15276–15281.

47 Yang, Q., Lei, X., Yin, Z. et al. (2019). *Synthesis* 51: 538–544.

48 Bhowmik, A., Yadav, M., and Fernandes, R.A. (2020). *Org. Biomol. Chem.* 18: 2447–2458.

49 Lamb, K.N., Squitieri, R.A., Chintala, S.R. et al. (2017). *Chem. Eur. J.* 23: 11843–11855.

50 Bai, C., Guo, H., Liu, X. et al. (2021). *J. Organomet. Chem.* 86: 12664–12675.

51 Veltri, L., Russo, P., Prestia, T. et al. (2020). *J. Catal.* 386: 53–59.

52 Bayanati, M., Parizadeh, N., Nematpour, M. et al. (2020). *J. Sulfur Chem.* 42: 121–130.

53 Sarkar, D., Rout, N., Ghosh, M.K. et al. (2017). *J. Organomet. Chem.* 82: 9012–9022.
54 Shen, Y., Wu, X.-X., Chen, S. et al. (2018). *Chem. Commun.* 54: 2256–2259.
55 Yang, J., Shen, Y., Lim, Y.J., and Yoshikai, N. (2018). *Chem. Sci.* 9: 6928–6934.
56 Nematpour, M., Dastjerdi, H.F., and Rabbani, M.M. (2022). *Lett. Org. Chem.* 19: 34–40.
57 Ma, X., Meng, S., Zhang, X. et al. (2020). *Beilstein J. Org. Chem.* 16: 1225–1233.
58 Chen, X., Shatskiy, A., Liu, J.-Q. et al. (2021). *Chem. Asian J.* 16: 30–33.
59 Zhu, W., Cai, G., and Ma, D. (2005). *Org. Lett.* 7: 5545–5548.
60 Ma, S. and Lu, L. (2005). *J. Organomet. Chem.* 70: 7629–7633.
61 Chen, D., Timmons, C., Wei, H.-X., and Li, G. (2003). *J. Organomet. Chem.* 68: 5742–5745.
62 Bowman, R.K. and Johnson, J.S. (2004). *J. Organomet. Chem.* 69: 8537–8540.
63 Guin, A., Gaykar, R.N., Bhattacharjee, S., and Biju, A.T. (2019). *J. Organomet. Chem.* 84: 12692–12699.
64 Satham, L. and Namboothiri, I.N.N. (2018). *J. Organomet. Chem.* 83: 9471–9477.
65 Wang, H., Xi, Z., Huang, S. et al. (2021). *J. Organomet. Chem.* 86: 4932–4943.
66 Xia, H.-D., Zhang, Y.-D., Wang, Y.-H., and Zhang, C. (2018). *Org. Lett.* 20: 4052–4056.
67 Wang, X., Wang, S.-Y., and Ji, S.-J. (2014). *J. Organomet. Chem.* 79: 8577–8583.
68 Xu, J.-K., Li, S.-J., Wang, H.-Y. et al. (2017). *Chem. Commun.* 53: 1708–1711.
69 Stephens, D., Zhang, Y., Cormier, M. et al. (2013). *Chem. Commun.* 49: 6558–6560.
70 Shaabani, A., Ghadari, R., Ghasemi, S. et al. (2009). *J. Organomet. Chem.* 11: 956–959.
71 Shaabani, A., Ghadari, R., Sarvary, A., and Rezayan, A.H. (2009). *J. Organomet. Chem.* 74: 4372–4374.
72 Park, K., Heo, Y., and Lee, S. (2013). *Org. Lett.* 15: 3322–3325.
73 Oberhauser, T. (1997). *J. Organomet. Chem.* 62: 4504–4506.
74 Katritzky, A.R., Tala, S.R., Abo-Dya, N.E. et al. (2009). *J. Organomet. Chem.* 74: 7165–7167.
75 Wang, Y., Aleiwi, B.A., Wang, Q., and Kurosu, M. (2012). *Org. Lett.* 14: 4910–4913.
76 Medel, R., Monterde, M.I., Plumet, J., and Rojas, J.K. (2005). *J. Organomet. Chem.* 70: 735–738.
77 Verniest, G., Van Hende, E., Surmont, R., and De Kimpe, N. (2006). *Org. Lett.* 8: 4767–4770.
78 Gu, S.-X., Du, J.-W., Ju, X.-L., and Chen, Q.-P. (2014). *Org. Process Res. Dev.* 18: 552–554.
79 Cheng, P., Wang, W., Wang, L. et al. (2019). *Tetrahedron Lett.* 60: 1408–1412.
80 Chen, Z., Zhou, Q., Chen, Q. et al. (2020). *Org. Biomol. Chem.* 18: 8677–8685.
81 Garrido-Castro, A.F., Salaverri, N., Maestro, M.C., and Alemań, J. (2019). *Org. Lett.* 21: 5295–5300.
82 Rostoll-Berenguer, J., Martín-López, M., Blay, G. et al. (2022). *Organomet. Chem.* 87: 9343–9356.

83 Xu, N.-X., Li, B.-X., Wang, C., and Uchiyama, M. (2020). *Angew. Chem. Int. Ed.* 59: 10639–10644.

84 Zhu, N., Wang, T., Ge, L. et al. (2017). *Org. Lett.* 19: 4718–4721.

85 Abdulla, H.O., Scaringi, S., Amin, A.A. et al. (2020). *Adv. Synth. Catal.* 362: 2150–2154.

86 Zhang, P., Gao, Y., Zhang, L. et al. (2016). *Adv. Synth. Catal.* 358: 138–142.

87 Chen, C., Bao, Y., Zhao, J., and Zhu, B. (2019). *Chem. Commun.* 55: 14697–14700.

88 Fan, T., Liu, Y., Jiang, C. et al. (2021). *Org. Biomol. Chem.* 19: 6609–6612.

89 Liu, L., He, Y.-H., and Guan, Z. (2019). *Tetrahedron* 75: 130716.

90 Gong, X., Chen, J., Liu, J., and Wu, J. (2017). *Org. Chem. Front.* 4: 2221–2225.

91 Gao, Y.-C., Huang, Z.-B., Xu, L. et al. (2019). *Org. Biomol. Chem.* 17: 2279–2286.

92 Zhang, J., Yang, M., Liu, J.-B. et al. (2020). *Chem. Commun.* 56: 3225–3228.

93 Tu, X., Huang, J., Xie, W. et al. (2021). *Org. Chem. Front.* 8: 1789–1794.

94 Chen, H., Liu, Z., Lv, Y. et al. (2017). *Angew. Chem. Int. Ed.* 56: 15411–15415.

95 Tu, D., Luo, J., and Jiang, C. (2018). *Chem. Commun.* 54: 2514–2517.

96 Guan, H., Sun, S., Mao, Y. et al. (2018). *Angew. Chem. Int. Ed.* 57: 11413–11417.

97 Su, Y., Cao, L., Shi, Y. et al. (2019). *Synthesis* 51: 2331–2338.

98 Shi, J.-L., Wang, Y., Wang, Z. et al. (2020). *Chem. Commun.* 56: 5002–5005.

3

Recent Achievements in Organic Reactions in Bio-based Solvents

Shaomin Chen, Noman Haider Tariq, and Yanlong Gu

Huazhong University of Science and Technology, School of Chemistry and Chemical Engineering, Key Laboratory for Large-Format Battery Materials and System Ministry of Education, Wuhan, Luoyu Road 1037, 430074, P. R. China

3.1 Introduction

Solvents are vital components in many industries, such as pharmaceuticals, coatings, adhesives, printer inks, cosmetics sectors, and so on. With respect to fine chemical and pharmaceutical industries, large amounts of organic solvents are usually indispensable. As it can typically implement multiple functions, not only serving as the reaction medium to facilitate contact between reactants or change the course of the reaction, but also being utilized for the work-up processes, including extraction, purification, and cleaning steps to obtain the final product with sufficient purity [1]. In addition, solvents can also display prominent performance in improving reactivity and selectivity. Generally, solvents reach around 80% of the entire volume of chemicals used in the process [2]. Consequently, the waste generated from solvents represents the primary source of chemical production waste. [3] Moreover, life cycle assessments have confirmed that the use of solvents accounts for a substantial part of the environmental footprint of industrial processes, especially for complex pharmaceutical molecules [4]. It should be noted that most of the reaction processes mainly rely on the usage of toxic, nonrenewable, and fossil-based solvents. For instance, some dipolar aprotic solvents, such as dimethylformamide (DMF), dimethylacetamide (DMA), dimethyl sulfoxide (DMSO), *N*-methyl-2-pyrrolidinone (NMP), nitromethane (CH_3NO_2), and 1,4-dioxane are frequently employed as solvents in industrial synthetic processes [5], which are apparently associated with several concerns related to human health, safety, environment, and sustainability issues. Besides, the excessive consumption of fossil resources intrinsically leads to the energy crisis [6], subsequently initiating an exorbitant increase in the cost of petroleum-derived solvents, which leads to high manufacturing costs. Meanwhile, to overcome the pressure from the environmental directives and legislations issued by governments and commissions, such as the European Union Solvents Emission Directive and the Clean Air Act [1], the research interest toward developing green,

Green Solvents in Organic Synthesis, First Edition.
Edited by Xiao-Feng Wu, Zhiping Yin, Liang-nian He, and Feng Wang.

environmentally friendly, sustainable, and cost-effective reaction media is continuously growing and has become one of the major topics in the synthetic community nowadays [7].

To clearly define the greenness and sustainability of solvents, a large number of guides have been put forward in the past decade. Gu and Jérôme established a suite of evaluation criteria based on comprehensive considerations that covered the entire process, from production to storage and consumption [8]. In addition, some pharmaceutical companies and collaborative organizations, such as GSK [9], Pfizer [10], Sanofi [11], ACS Green Chemistry Institute Pharmaceutical Roundtable (ACS GCI-PR) [3], and the Innovative Medicines Initiative (IMI)-CHEM21 [12], have also proposed more detailed principles for solvent selection, in which the solvents were ranked after making a broad range of metrics, especially focusing on their environmental, safety, and health characteristics [13]. Taking account of all solvent guides, it is unanimously concluded that bio-based solvents (solvents derived from renewable biomass feedstock) are the preferred choices, which have always been recommended in a myriad of chemical syntheses. Currently, various plant biomass sources are mainly considered the source of renewable feedstocks, including agricultural residues, agroforestry products, energy crops, and aquatic biomass [1, 7e, 14]. Owing to bearing the inherent advantages of renewability and sustainability, bio-based solvents are generally less toxic, safe, and highly biodegradable, and often bring about much reduced waste generation and environmental harm, thus enabling them to serve as greener alternatives to traditional petroleum-based solvents.

Over the past decade, much attention has been devoted to exploring novel biomass-derived solvents, exemplified by glycerol, polyethylene glycols (PEGs), 2-methyltetrahydrofuran (2-MeTHF), cyclopentyl methyl ether (CPME), organic carbonates, and γ-valerolactone (GVL), as well as ethyl lactate (EL), and have found wide application in organic synthesis. In this review, we will summarize and discuss recent progress in the use of bio-based solvents in organic reactions from 2020 to the end of 2022.

3.2 Glycerol

Glycerol is a cost-effective, non-hazardous, renewable, and biocompatible viscous liquid and is usually produced as the primary by-product of the biodiesel industry [15]. It shows excellent and unique solubility with a broad range of compounds, such as acids, bases, inorganic salts, and many transition metal complexes, especially for compounds that are poorly miscible in water. Glycerol has a high boiling point (290 °C) and negligible vapor pressure, making it a safer option [16]. Owing to the hydrogen bonding capacity, glycerol can promote the polarization of the reactants or stabilize the intermediates involved in the reactions, which allows some polar reactions to proceed under mild conditions [17]. Most strikingly, glycerol is immiscible with several organic solvents such as hydrocarbons and ethers, thus allowing the removal of products by simple extraction. These prominent properties also make glycerol feasible to serve as an alternative eco-friendly bio-based solvent in various organic transformations [16, 18].

In the past few years, glycerol has been extensively used as a green medium for the synthesis of diverse heterocycles. For example, Behbahani and coworkers reported a mild and efficient approach to access 1,8-dioxo-decahydroacridines from the three-component reaction of aromatic aldehydes, dimedone, and amines in glycerol (Scheme 3.1) [19]. It should be mentioned that a notable solvent effect was shown in this reaction. Various solvents containing hydroxyl group were investigated, in which it was surprisingly found that the viscous glycerol can lead to an excellent yield of the target product without any catalyst at ambient temperature. Conversely, H_2O and EtOH were not efficient for this transformation, both generating the corresponding product in trace amounts. This significant difference could be ascribed to the existence of polyhydroxyl in glycerol since the high hydrogen bonding capability provided by glycerol as the medium can allow the occurrence of intermolecular polar reactions more easily.

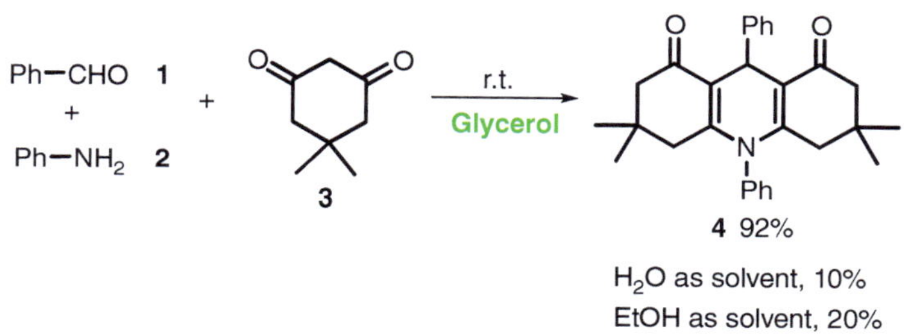

Scheme 3.1 Catalyst-free synthesis of 1,8-dioxo-decahydroacridines in glycerol.

Aryloxy propanolamines are an important class of medicinal molecules, many of which can be regarded as β-adrenergic receptor antagonists [20]. Kumar and Hussain's group described the synthesis of aryloxy propanolamines through the ring-opening aryl glycidyl ethers with amines in bioderived glycerol solvent (Scheme 3.2) [21]. A yield of 87% was obtained in glycerol at room temperature after 4 h in the absence of any catalyst. Control experiments suggested that other eco-friendly protic solvents, such as ethanol, water, PEG, and butanol, were less efficient than glycerol. Probably owing to the high viscosity, a better yield (95%) of the target product was achieved when the temperature was adjusted to 40 °C. The use of glycerol can facilitate the ring opening by the hydrogen bonding of epoxide with the hydroxy group of the glycerol, which weakened the C–O bond. Given that one face of the epoxide might be hindered by the hydrogen-bonded glycerol molecule, the nucleophilic attack on the epoxide ring readily occurred on the less hindered side, resulting in the formation of the product with high regioselectivity. Moreover, the isolation of the product was demonstrated to be simple. After completing the reaction, warm water was added to the reaction mixture, and the product was precipitated out and then separated by filtration. Afterward, the pure glycerol was obtained by the evaporation of water *in vacuo*, which could be reused for the next cycle. The authors also calculated an E-factor of 0.15 and reaction mass efficiency (RME) of 87% for the synthesis of the β-blocker drug propranolol, which well complied with the requirements of fine chemical production.

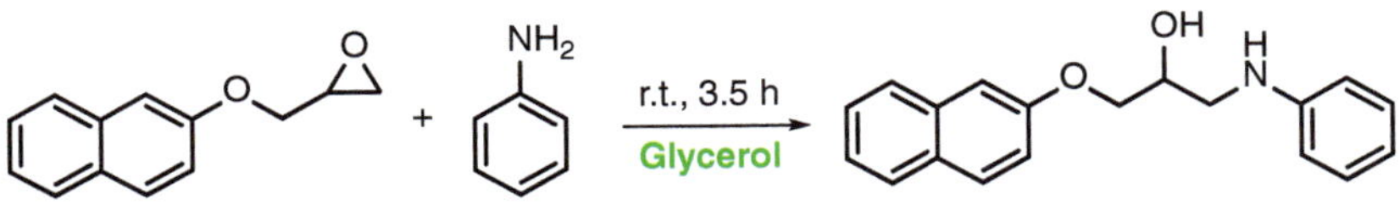

5 + **2** → **6** 87%

Solvent	Time (h)	Yield (%)
Ethanol	6	63
Water	6	40
PEG-200	6	74
Butanol	6	52

Scheme 3.2 Ring-opening of glycidyl ethers with amines in glycerol.

Lewis acid-catalyzed cascade reactions also proceeded well in glycerol. Recently, our group developed $Sc(OTf)_3$-mediated cascade cyclizations of α-hydroxyacetophenone with arene-based nucleophiles pyrroles in glycerol to construct indole scaffolds with good yields (Scheme 3.3) [22]. Other solvents, such as 1,4-dioxane CH_3OH, CH_3CN, and toluene, were found to be less effective than glycerol. Intriguingly, the generated products can be isolated by simple extraction operation, and the residual $Sc(OTf)_3$/glycerol could be recovered and reused for eight consecutive runs without appreciable loss of catalytic activity. It should be mentioned that the catalytic system of $Sc(OTf)_3$/glycerol can be further used for the three-component reaction of α-hydroxyacetophenone, *N*-methylindole, and 1,3-dicarbonyl compounds, resulting in the formation of carbazole products (Scheme 3.4). In contrast to some classic approaches for synthesizing indoles

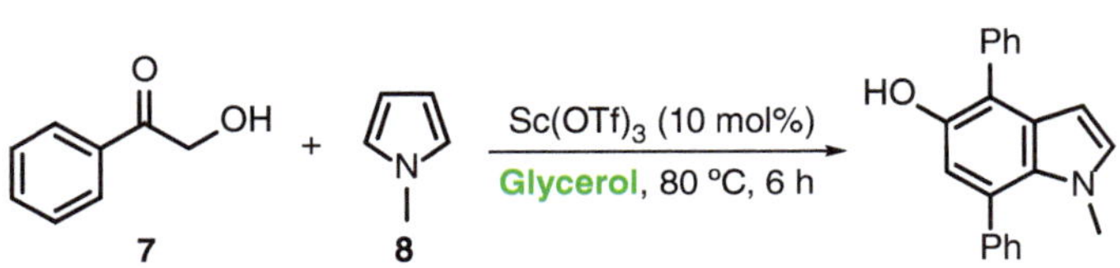

7 + **8** → **9** 74%

Solvent	Tempt. (°C)	Yield (%)
Glycerol	60	54
1,4-dioxane	60	31
CH_3OH	60	44
CH_3CN	60	33
Toluene	60	31

Scheme 3.3 The reaction of α-hydroxyacetophenone and *N*-methylpyrrole in glycerol.

and carbazoles, the use of glycerol obviously strengthened the greenness and sustainability of this protocol.

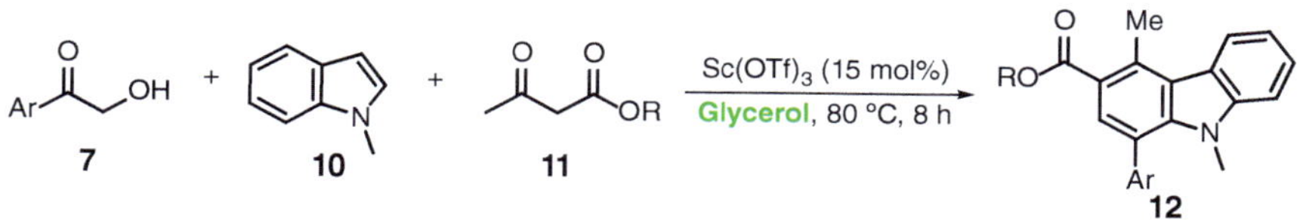

Scheme 3.4 The three-component reaction of α-hydroxyacetophenone, *N*-methylindole, and 1,3-dicarbonyl compound in glycerol.

The common problem encountered in the case of glycerol as a medium is the slow diffusion resulting from its high viscosity. Microwave (MW) heating has served as an effective tool to overcome this problem, which might benefit from the capability of efficient heat transfer and mass transfer [23]. For example, the Andrade group used microwave to facilitate the synthesis of pyrido[2,3-*d*]pyrimidine *via* the three-component reaction of 6-aminouracil, aromatic aldehydes, and malononitrile within 15 minutes (Scheme 3.5) [24]. In the reaction, glycerol played dual roles, not only serving as the green solvent but also working as the catalyst to activate the starting materials, thus avoiding the addition of a catalyst. When the solvent was changed into the PEG-400, MeOH, EtOH, H_2O, or DMF, no yield or traces of products were obtained. Also, the temperature demonstrated a remarkable impact on this reaction. When the temperature was less than 150 °C, it was necessary to add some additives to achieve the desired product. Apparently, these results confirmed that the combination of glycerol with microwave heating provided a highly effective and green platform for the synthesis of complex organic molecules.

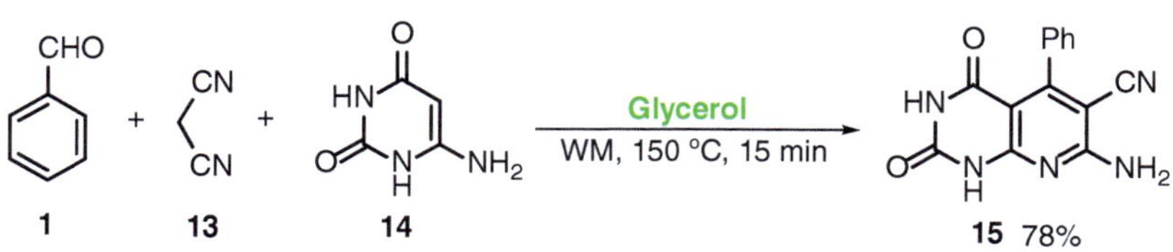

Scheme 3.5 Synthesis of pyrido[2,3-*d*]pyrimidine under microwave heating in glycerol.

Glycerol has been also an ideal solvent for electrochemical transformations because of its high conductivity. The Ackermann group reported a cobalt-electrocatalyzed C–H/N–H functionalization of amides with alkyne in an undivided cell, with aqueous glycerol as the medium (Scheme 3.6) [25]. It is worth noting that the addition of water was essential for ensuring high yields of the target products. In stark contrast, other commonly used solvents, including some alternative renewable solvents, such as 2-MeTHF, furfuryl alcohol, and GVL, were also mixed with water and yielded much less than glycerol. This result can be attributed to its high dielectric constant ($\varepsilon = 42.5$ at 25 °C) to a large extent, meanwhile obviating the use of additional expensive electrolytes. Also, they found that this catalytic system was well applicable to the annulation reaction of amides with allenes, yielding the corresponding product in a slightly reduced yield. Thus, this protocol exemplified the superiority of glycerol as a green medium in organic electrochemistry. Most importantly, it was also feasible to directly employ renewable solar and wind energy for the electrocatalytic C–H activations in aqueous glycerol, which rendered this strategy more environmentally friendly and sustainable.

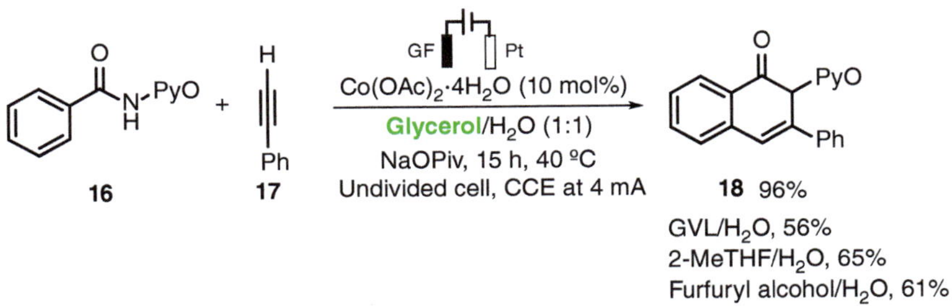

Scheme 3.6 Electrochemical cobalt-catalyzed C–H activation in glycerol.

Glycerol can serve not only as a solvent but also as a hydrogen donor in the catalytic transfer hydrogenation of some unsaturated compounds. In 2020, Cravotto and coworkers developed a simple and reproducible transfer hydrogenation of nitrobenzene to aniline assisted by copper nanoparticles (CuNPs) with glycerol as the solvent and a "sacrificial" hydrogen source (Scheme 3.7) [26]. It is noteworthy that glycerol was also added as a co-solvent in the preparation of CuNPs, forming a protective layer to stabilize NPs so as to enhance the catalytic efficacy. In the reaction, ultrasound and microwave irradiation were also proven to be expedient approaches to boost the reaction, and substantially they can shorten the reaction time.

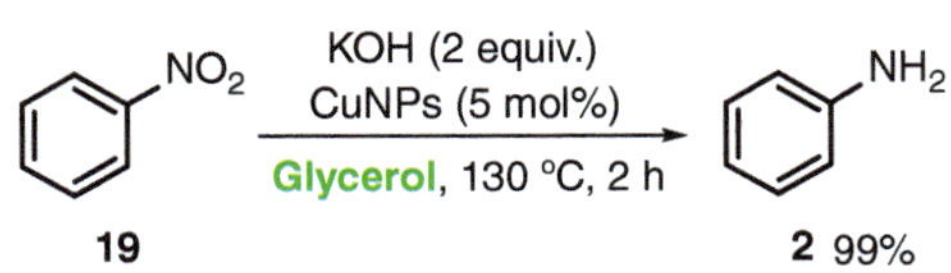

Scheme 3.7 CuNPs-catalyzed reduction of nitrobenzene in glycerol.

3.3 Polyethylene Glycols (PEGs)

PEGs are a sort of low-cost liquid polymers that are characterized by their high thermal stability, negligible toxicity, low volatility, and biodegradability, and they are typically produced from ethylene glycol (EG) that is a by-product of petrochemical industries [27]. Recently, sustainable sources of EG have been of great interest. Bio-based EG or PEGs can be obtained from renewable sugar or lignocellulosic biomasses [28]. PEGs are also readily available from agricultural products such as sugar cane waste. PEGs show good solubility for most organic compounds and metal complexes, which makes them more popular for organic transformations.

Rawat and coworkers described a one-pot tandem coupling of indole, aldehyde, and tetrahydroisoquinoline using biodegradable and low-toxic PEG-600 as a solvent (Scheme 3.8) [29]. It was demonstrated that PEG-600 was superior to other several solvents, including polar organic solvents such as DMSO and DMF, and nonpolar organic solvents such as benzene and toluene, delivering the desired C1-indolyl-*N*-alkyl tetrahydroisoquinoline product with generally good-to-excellent yields in the absence of catalyst within a short reaction time of two hours. Remarkably, the solvent PEG-600 could be recovered through liquid–liquid phase extraction by simply adding ethyl acetate and water, and subsequently, PEG-600 was reused four times with a negligible decrease in the yield. The reaction underwent a successive condensation/tautomerization/nucleophilic addition process, which involved the formation of diverse imine- or enamine-type intermediates. It was considered that the flexibility of the PEG molecule also allowed it to form a helical structure, likely acting as a microvessel to promote this reaction. In addition, the hydrogen bonding interaction between the functional groups in reactants and intermediates with PEG also led to a positive effect on the reaction efficiency.

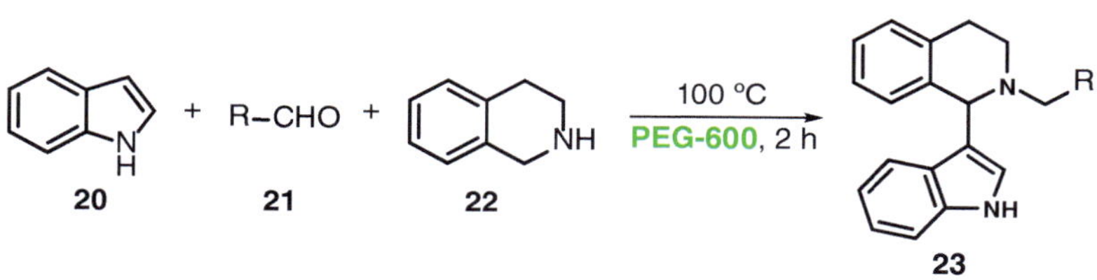

Scheme 3.8 One-pot tandem C1-indolylation and *N*-alkylation of tetrahydroisoquinoline in PEG-600.

Similarly, PEG-400 can also be applied to the three-component reaction of indole, C–H activated acids, and aromatic aldehydes. This reaction was conveniently achieved at 80 °C without any catalyst (Scheme 3.9) [30]. Likewise, the liquid polymer, PEG-400, can generate a helical microvessel accommodating the reactants, so as to promptly enhance the reaction efficiency through hydrogen bonding of its terminal hydroxyl and ethereal oxygen linkage with the substrates, functionalizing as an acidic catalyst. Thus, PEG worked with a dual role as both a solvent and a catalyst in the reaction. On the other hand, the utilization of PEG selectively promoted the formation of heterodimer products and avoided homodimer adducts, such as bis(indolyl)methanes and xanthines.

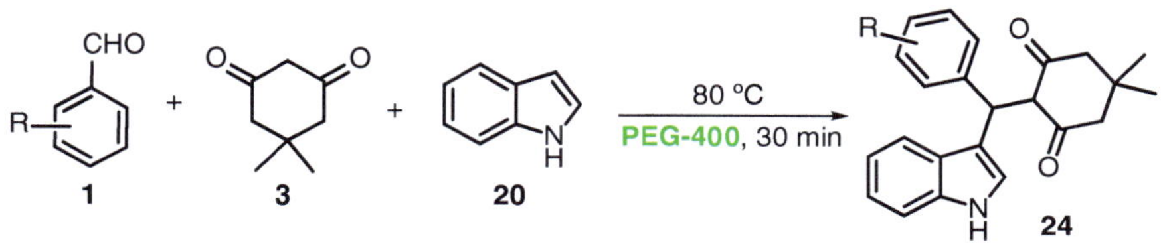

Scheme 3.9 The three-component reaction of indole, dimedone, and aromatic aldehydes in PEG-400.

3.4 2-Methyltetrahydrofuran (2-MeTHF)

2-Methyltetrahydrofuran (2-MeTHF) is a readily available ethereal solvent that can generally be obtained from renewable resources such as levulinic acid or furfural [31], both of which are derived from agricultural wastes, such as corn stover, rice straw, and sugar cane bagasse [32]. Structurally, 2-MeTHF resembles tetrahydrofuran (THF), differing in the addition of a methyl group, which enables it to possess several significantly different physiochemical properties compared to THF, such as high boiling point, good stability toward acids and bases, low miscibility with water, low volatility, and considerable biodegradability. For these reasons, 2-MeTHF has been frequently employed as a green alternative to some toxic solvents in numerous synthetic transformations, especially for organometallic reactions.

Transition metal-catalyzed carbonylation reactions are a powerful synthetic tool to access several privileged scaffolds containing carbonyl functional groups. In 2020, Cai and coworkers reported a novel and green palladium-catalyzed cyclocarbonylation method for the construction of 4*H*-3,1-benzoxazin-4-one skeleton with readily available 2-iodoanilines and acyl chlorides as starting materials in 2-MeTHF as a green reaction medium (Scheme 3.10) [33]. Among a representative set of solvents, such as NMP, DMSO, CH_3CN, THF, and 1,4-dioxane, THF can deliver a comparable yield, whereas other solvents result in much lower yields of products. A broad range of substrates were well tolerant with the catalytic system, resulting in the formation of the desired products with good yields. Moreover, recycling experiments suggested that there was no apparent decrease in catalytic activity even after the catalyst was reused eight times, giving excellent reproducible yields.

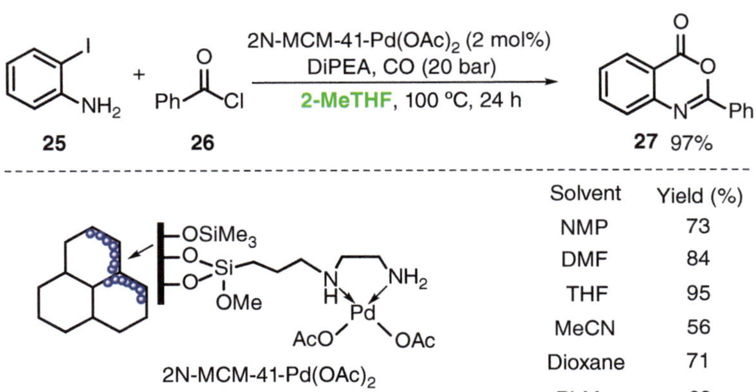

Solvent	Yield (%)
NMP	73
DMF	84
THF	95
MeCN	56
Dioxane	71
PhMe	69
DMSO	62

Scheme 3.10 Palladium-catalyzed cyclocarbonylation in 2-MeTHF.

Palladium-catalyzed allylative dearomatization of naphthyl halides was also feasible in 2-MeTHF (Scheme 3.11) [34]. In this reaction, the use of a Grignard reagent as a nucleophile could lead to the rapid occurrence of dearomatization with the assistance of a low catalyst loading at room temperature. Also, the utilization of 2-MeTHF as the reaction medium was essential to ensure good chemoselectivity under mild reaction conditions and could generate the corresponding product in 95% yield, which was presumably attributed to its peculiar properties. 2-MeTHF was an effective solvating agent, giving rise to a strong interaction with the Grignard reagent. On the other hand, due to the bulkiness of 2-MeTHF, the reactivity of Grignard reagent could be decreased to some extent. In sharp contrast, DCM was used as the solvent under otherwise identical conditions, yielding the dearomatic product with a lower yield of 81%.

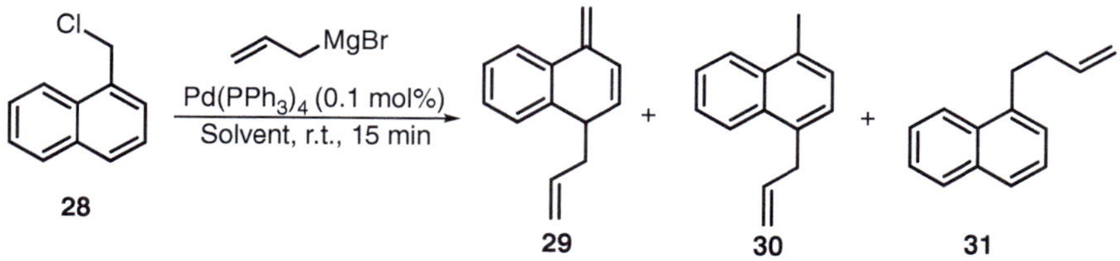

Solvent	Yield (%)	29/30/31
DCM	81	85/15/<1
2-MeTHF	95	>99/<1/<1

Scheme 3.11 Palladium-catalyzed allylative dearomatization in 2-MeTHF.

2-MeTHF can also demonstrate a remarkable effect on metal-mediated remote functionalization. Most recently, the Vasseur group reported a zincate-catalyzed remote functionalization of *p*-iodobenzyl methanesulfonates with aldehydes as coupling partners through metallotropy in the bio-sourced solvent, 2-MeTHF (Table 3.1) [35]. A noticeable solvent effect was shown in the reaction. Upon switching from 2-MeTHF to THF, the dehalogenative compound **33** would serve as

Table 3.1 Zincate-promoted remote functionalization in 2-MeTHF.

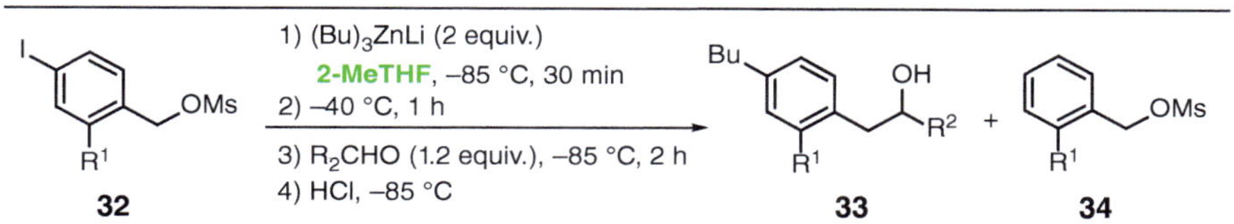

Entry	Solvent	R^1	R^2	Yield (%)	
				33	34
1	THF	Cl	1-naphthyl	33	59
2	2-MeTHF	Cl	1-naphthyl	66	—
3	2-MeTHF	H	phenyl	66	—
4	2-MeTHF	Me	4-chlorophenyl	72	—

the primary product. To investigate the role the solvents played in the reaction, they also performed some control experiments (Scheme 3.12). Treating **32a** with lithium trialkylzincate in THF, only the dehalogenative product **35** was obtained in 54% yield. In contrast, a concomitant formation of **35** and **36** products was observed in 2-MeTHF. Kinetic experiments also confirmed that the solvent indeed had a strong effect on the conversion of **32a** at the later stage. It was inferred that 2-MeTHF could promote some favorable changes in the aggregation state of crucial organometallic species, such as I/Zn exchange and alkyl 1,2-migration.

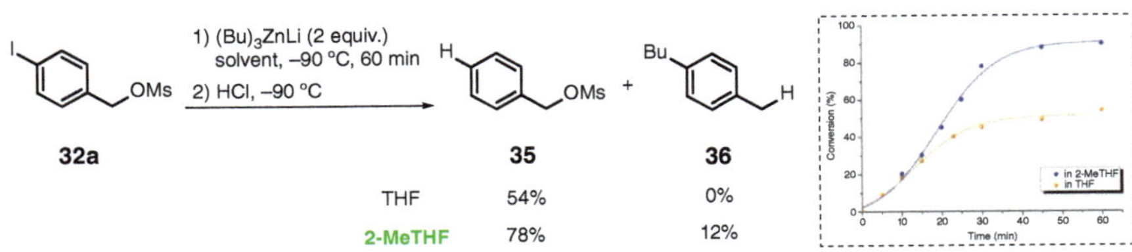

Scheme 3.12 Effect of 2-MeTHF on both the I/Zn exchange step and the butyl 1,2-migration step.

2-MeTHF can perform subtle differences in the solubilizing and stabilizing effects, which has been leveraged by Alcántara and Pace's group [36]. They successfully utilized the bio-renewable 2-MeTHF solvent to boost Kirmse's α-elimination of lithium monohalocarbenoids starting from epoxide (Scheme 3.13). Owing to the special property of 2-MeTHF with a limited stabilizing effect, lithium salts were degraded to carbenoids, which subsequently reacted *in situ* with epoxides to provide the corresponding β-halohydrins. When the coordinative solvents, such as THF or diethyl ether (Et_2O) were employed to stabilize the lithium salts under identical conditions, the transformations occurred with lower efficiency. This result suggested that increasing the medium-coordination effect on the carbenoid had an adverse impact on the yield of the chlorohydrin, resulting from a negative modulating effect on Kirmse's elimination. Later on, the same group reported on $NaBH_4$-assisted reduction of diselenides in the presence of ICH_2F to afford α-fluoromethyl selenoethers (Scheme 3.14) [37].

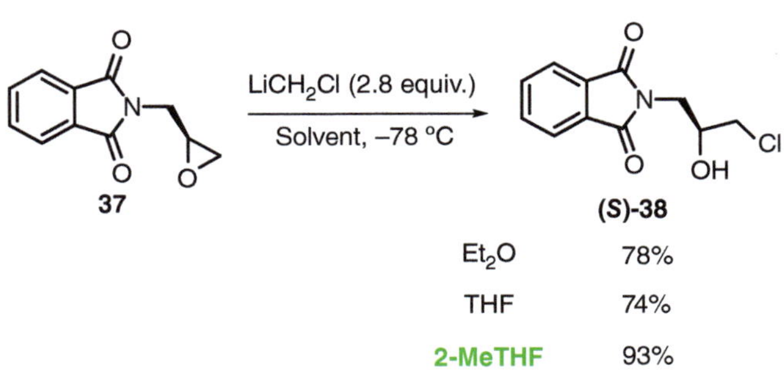

Scheme 3.13 The ring-opening of epoxide in 2-MeTHF.

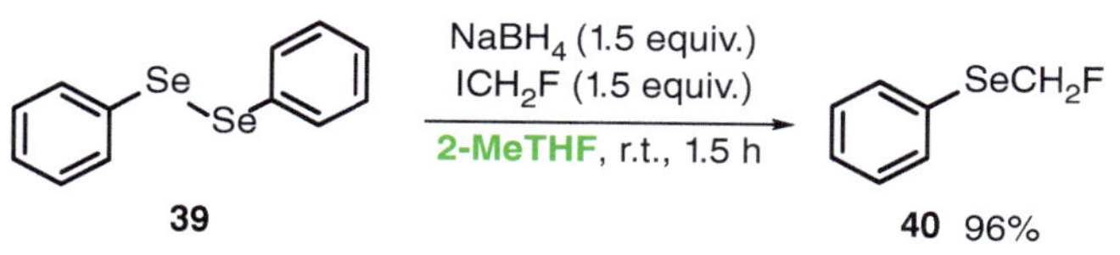

Scheme 3.14 Chemoselective reduction-monofluoromethylation of diselenides in 2-MeTHF.

3.5 Cyclopentyl Methyl Ether (CPME)

CPME is a hydrophobic ethereal solvent and has a low solubility in water (1.1 g CPME/100 g). Additionally, it is characterized by a relatively manageable boiling point (106 °C), low melting point, and low heat of vaporization, as well as the formation of a positive azeotrope with H_2O [38]. Because of these salient properties, CPME can be engaged as a solvent for a variety of reactions, and additionally, can be easily recovered after aqueous workup. Furthermore, it is' highly stable under strong acidic and basic conditions, and toward metal reagents, Grignard reagents, and organolithium derivatives [39], compared to other classical ethereal solvents, such as THF, Et_2O, 1,4-dioxane, and methyl *tert*-butyl ether (MTBE). Most importantly, CPME shows a low tendency to form peroxides that is the largest safety issue and is often encountered in most ethereal solvents, perhaps due to the extraordinarily high dissociation energy of its secondary α-C—H bond [40]. In addition, CPME also shows low toxicity, with moderate irritation and negative mutagenicity [41]. Although in most cases the current production of CPME still relies on the petrochemical industry through the addition reaction of cyclopentene and methanol [39], several bio-based production routes have been developed from the precursors such as cyclopentanol or cyclopentanone, which can be generated from renewable furfural or adipic acid [38b, 42], giving a minimized influence on the environment. Overall, these prominent features confer much promise on CPME to become a sustainable reaction medium.

In light of the ethereal nature of CPME and its unique characteristics, along with the previously well-established employment of analogous solvents in organometallic reactions, CPME has been frequently employed in this field [43]. For instance, the Degannaro and Luisi group reported a straightforward and sustainable mixed flow-batch protocol to access *NH*-aziridines starting from vinyl azides (Scheme 3.15) [44]. Considering that two different reaction steps, including azide cyclization and organolithium addition reactions, were involved in the process, the use of a compatible solvent was of high significance. Intriguingly, CPME worked well for the mixed flow-batch process. It not only rapidly completed the initial cyclization in the flow reactor within 45 minutes but also ensured the occurrence of subsequent organolithium addition under batch conditions. However, a much longer time was necessary to fulfill the complete conversion of azides into 2*H*-azirines when 2-MeTHF and toluene were used as solvents. Evidently, this approach elegantly exemplified the importance and compatibility of CPME as a green and safe reaction medium in the mixed flow-batch technology.

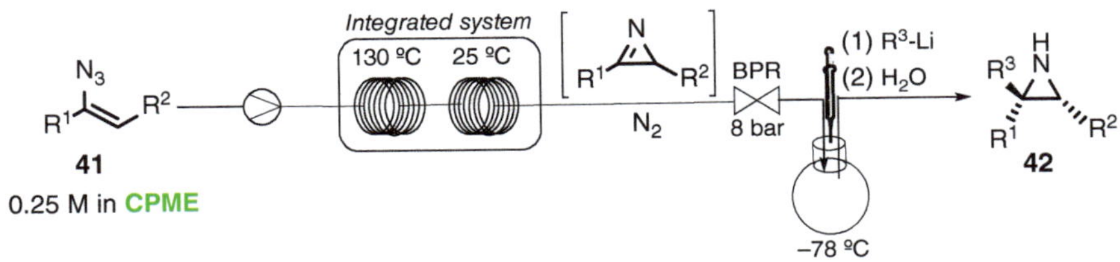

Scheme 3.15 Flow synthesis of 2*H*-azirines from vinyl azides in CPME.

In recent years, the Vaccaro group has made significant progress in the utilization of CPME as a green solvent [45]. They developed an efficient and waste-minimized methodology for the synthesis of aminophenoxazinones *via* the C–H oxidative coupling reaction of 2-aminophenols by a combination of CPME and flow technology (Scheme 3.16) [45b]. Notably, CPME displayed superior efficacy over some other ethereal solvents, such as 2-MeTHF, MTBE, and *tert*-amyl methyl ether (TAME). Also, this catalytic system was applicable to other analogs, such as *o*-phenylenediamines and benzenetriol, providing the corresponding diaminophenazines and purpurogallin products with excellent yields. The combination of flow technology and CPME as the reaction medium not only ensured the high durability of the solid catalyst as a result of negligible manganese leaching but also enabled a minimal E-factor value due to the efficient recovery and reuse of the reaction media. Later on, CPME as a safe and green reaction medium was also proven to be efficient for the preparation of oxindoles based on the heterogeneous palladium-catalyzed intramolecular C(sp^3)–H α-arylation of amides by the same group (Scheme 3.17) [45c]. Although the conversion is comparable for most solvents, CPME can perform a slightly higher selectivity for the target product, oxindoles.

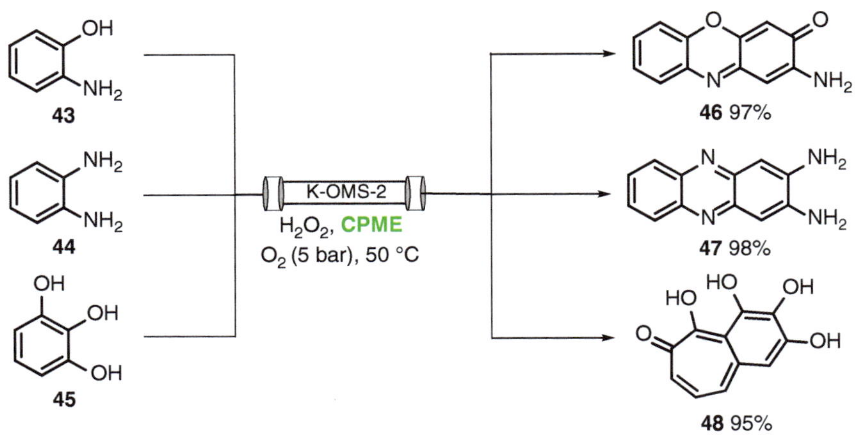

Scheme 3.16 Heterogeneous manganese-catalyzed C–H oxidation under continuous flow conditions in CPME.

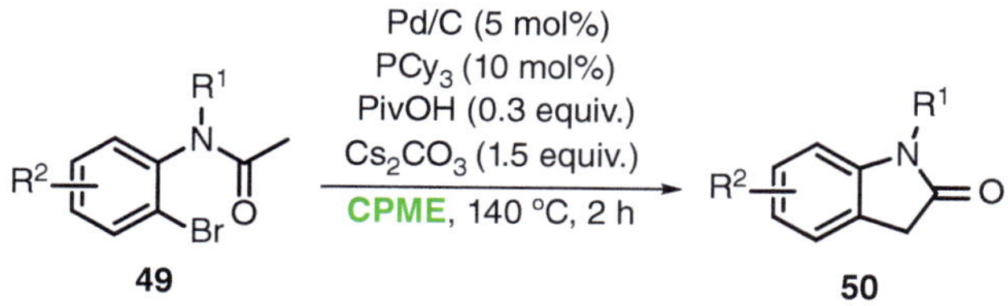

Scheme 3.17 Palladium-catalyzed intramolecular C(sp3)–H α-arylation for the synthesis of oxindoles in CPME.

CPME, as the green medium, can also be applicable for the synthesis of 1,3-dicarbonyl compounds. In 2022, Kandasamy and coworkers established an expedient and efficient LiHMDS-promoted C–C coupling reactions of *N*-Cbz amides with manifold enolizable esters to obtain functionalized β-ketoesters under mild conditions (Scheme 3.18) [46]. With this strategy, ketones and amides were also feasible in combination with *N*-Cbz amides to produce diverse β-keto amides and 1,3-diketones. Solvent screening showed that other ethereal solvents such as THF, diethyl ether, and 1,4-dioxane, which resulted in lower yields, whereas CPME by comparison outperformed these solvents.

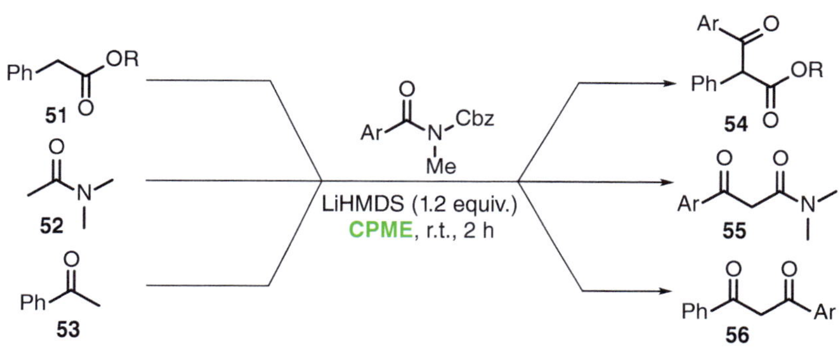

Scheme 3.18 Divergent synthesis of 1,3-dicarbonyl compounds in CPME.

The thioamination of 1,4-naphthoquinone with thiophenols and amines also successfully proceeded in the eco-friendly solvent CPME, as reported by the Yu group (Scheme 3.19) [47]. It was demonstrated that the oxygen from the air could satisfy the need for the cascade reaction. The mechanistic investigation also confirmed that thiolation preceded the amination step in this three-component reaction. Interestingly, further solvent optimization proved that CPME was superior to various green solvents, including H_2O, alcohols such as ethanol and PEG-200, and organic carbonates, and even outperformed ethers such as 2-MeTHF and MBTE in terms of yield. As a result, these protocols demonstrated the unique properties of CPME as a sustainable reaction medium.

CPME can be in conjunction with water to form a heterogeneous aqueous azeotropic mixture (CPME/H_2O = 84/16 wt%), which is also able to serve as

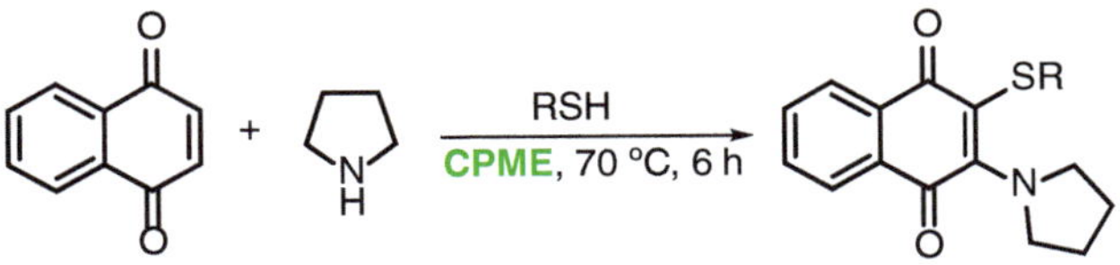

57 58 59

Scheme 3.19 Catalyst-free thioamination of 1,4-naphthoquinone in CPME.

a recoverable medium for organic transformations. In 2021, Vaccaro et al. demonstrated a biomass waste-derived heterogeneous palladium catalyst for the Sonogashira reaction with an azeotropic mixture of CPME and water as the reaction medium in the presence of 1,4-diazabicyclo[2.2.2]octane (DABCO) (Scheme 3.20) [45d]. It was proven that the utilization of CPME/water azeotrope was crucial for promoting the transformation. Upon changing the ratio of CPME and H_2O to 16/84 wt% or 50/50 wt%, the reaction conversion decreased dramatically. Furthermore, the strategy could be exploited in continuous flow technology, in which CPME could be recovered from the water and reused *via* easy liquid–liquid separation, leading to a significant reduction in waste.

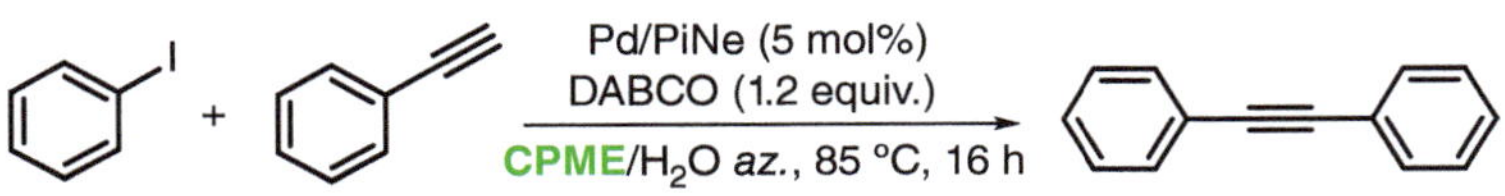

60 17 61 90%

Solvent	Conversion (%)
CPME/H_2O = 84/16 (azeotrope)	99
CPME/H_2O = 16/84	45
CPME/H_2O = 50/50	50

Scheme 3.20 Sonogashira reaction in the CPME–water azeotropic mixture.

3.6 Organic Carbonates

Organic carbonates have received growing interest since they can present several attractive features, such as high biodegradability, negligible vapor pressure, and non-corrosiveness, in the context of green and sustainable chemistry [48]. Organic carbonates are typically produced from CO_2 as the abundant and economical feedstock, thus not only providing opportunities for the utilization of CO_2 in the chemical industry but also contributing to reducing CO_2 pollution in the environment [49]. Among the family of organic carbonates, linear carbonates such as dimethyl carbonate (DMC) and diethyl carbonate (DEC) have become well-established green, bio-based solvents. These compounds are nontoxic and odorless liquids; on the other hand, their bio-accumulations are also low, as they merely degrade into CO_2 and the corresponding alcohols when released into the environment [50]. Therefore, it is not surprising that they have been widely engaged as a significant class of bio-based solvents in organic synthesis [51].

Recently, our group employed DEC as the benign bio-based solvent for the synthesis of pyrrolo[1,2-*a*]quinolines. In this transformation, *N*-arylglycine ester could react with two molecules of versatile reagent atropaldehyde acetals through a cascade oxidative C–C cleavage/multiple cyclizations with the help of trifluoroacetic acid (TFA) (Scheme 3.21) [52]. Atropaldehyde acetals performed a dual role, not only working as a C3-synthon but also converging into the final product as a C2 fragment *via* the C–C cleavage. A noticeable solvent effect was shown in the reaction, in which the best result in terms of product yield was achieved in DEC, rather than in commonly used nonpolar solvents, such as toluene and xylene, and green ethereal solvents, such as anisole and CPME. The analogous solvent, DMC, was also slightly inferior to DEC. In contrast to the classical transition metal-mediated intramolecular cycloisomerization, the present methodology characterized by a cost-effective and sustainable catalytic system was more attractive, from the academic or industrial level.

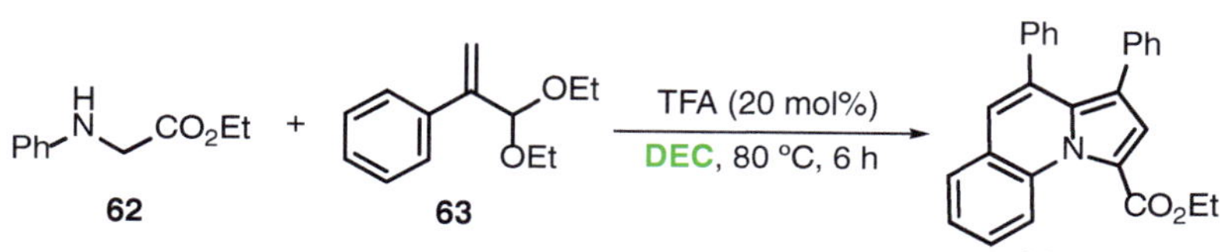

62 63 64 81%

Solvent	Yield (%)
Toluene	73
Xylene	62
Anisole	54
CPME	32
DMC (reflux)	76

Scheme 3.21 Metal-free synthesis of pyrrolo[1,2-*a*]quinoline in DEC.

Photocatalysis is a clean and economical chemical process, and it generally requires inexhaustible sunlight or ubiquitous household lamps for irradiation, which makes photocatalysis a powerful tool for organic synthesis. Organic carbonates can be capable of photocatalytic reactions. Yu and coworkers carried out some research on the use of this class of sustainable solvents in photo-induced reactions [53]. In 2021, this group reported a carbon nitride (g-C_3N_4)-promoted approach for the construction of thiocyanated heterocycles from inexpensive and easily available NH_4SCN as a raw material and irradiation of blue LED (Scheme 3.22) [53a]. Compared to other conventional volatile organic solvents employed in the reaction, the use of a nontoxic and biodegradable DMC solvent allowed the transformation to proceed under mild conditions in the open air and room temperature, yielding a diverse thiocyanated heterocycles with generally good yields. Furthermore, the recyclable photocatalyst g-C_3N_4 could be used at least eight times without an obvious reduction in activity.

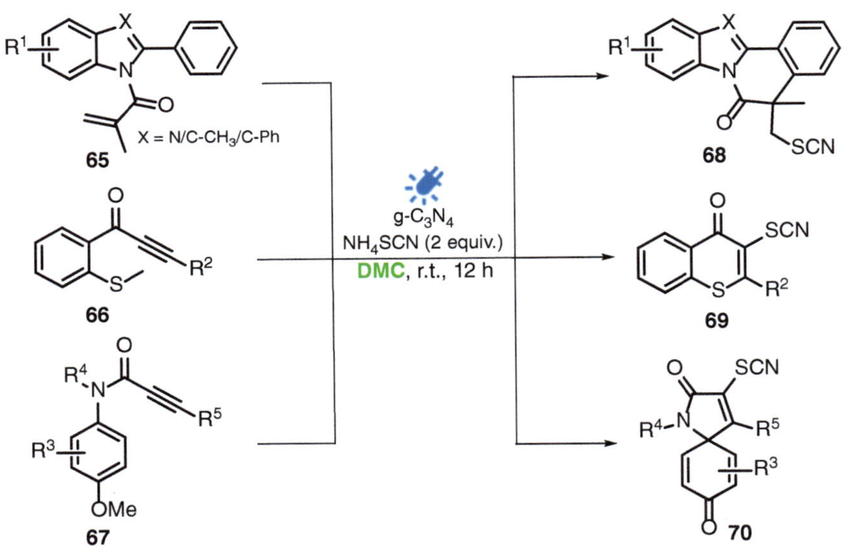

Scheme 3.22 Visible light-induced synthesis of thiocyanated heterocycles in DMC.

In a similar manner, the same group developed an efficient catalyst-free aroylation strategy under the irradiation of blue light (Scheme 3.23) [53b]. The protocol could be simply achieved in DMC using 4-acyl-1,4-dihydropyridines (acyl-DHPs) as an acylating reagent at room temperature within three hours, showing broad

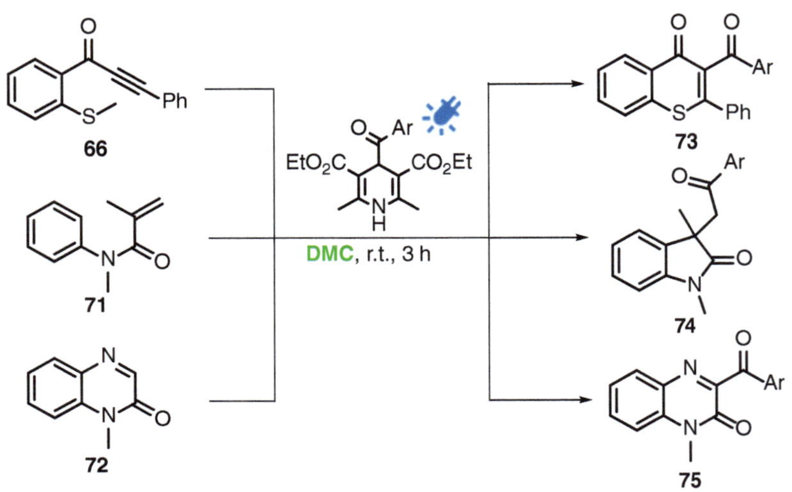

Scheme 3.23 Visible-light-promoted synthesis of aroylated heterocycles in DMC.

compatibility with various heterocyclic molecules, such as thioflavones, 3,3-dialkyl 2-oxindoles, and quinoxalin-2(1*H*)-ones. Notably, conventional organic solvents, including MeCN, THF, DCE, and DMF, resulted in significantly poor yields compared to the reaction conducted in DMC. More surprisingly, the renewable energy source, sunlight, can also be recognized as an efficient light source for the transformation with DMC as the medium.

A combination of organic carbonates and ionic liquids has been considered an effective catalytic system for organic transformations. For instance, Miao et al. designed a novel acidic ionic liquid [BsPy][OTf] for the coupling of 4-hydroxycoumarins with benzyl alcohols in DMC (Scheme 3.24) [54]. Albeit comparable yields obtained in other commonly used solvents, such as CH_3CN, 1,4-dioxane, toluene, and cyclohexane, DMC was still the best option with respect to the greenness and sustainability profile. Intriguingly, excellent catalytic activity could be displayed merely in the presence of pyridine-based ionic liquid. Moreover, the sulfonic acid group in the side chain of ionic liquids was proven to be essential for the reaction, which also suggested the significance of pyridine-based ionic liquids with multi-active sites due to the possible synergistic effect.

Scheme 3.24 C3 substitution of 4-hydroxycoumarins with benzyl alcohols in DMC.

DMC can not only serve as a green and safe medium, and meanwhile also engage in diverse reactive agents in various organic reactions [55]. In 2021, Wu and coworkers reported a facile protocol for the direct synthesis of aryl methyl sulfides based on the utilization of DMC as both the green solvent and methylating reagent (Scheme 3.25) [56]. In this reaction, the formation of CH_3I through the reaction of NaI with DMC plays a crucial role, ensuring the subsequent methylation to produce the desired methyl sulfide products. Afterward, this group developed a

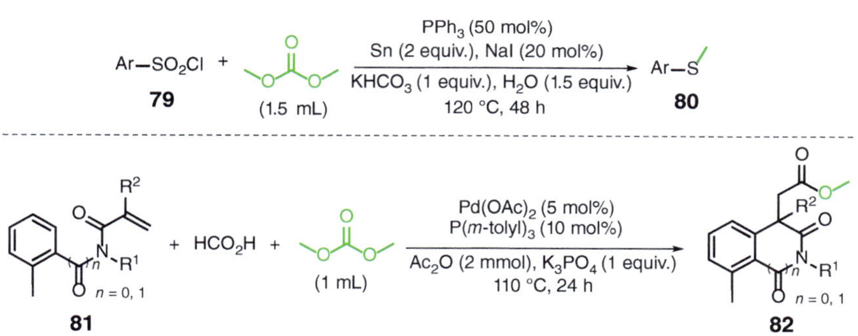

Scheme 3.25 DMC as both solvent and reactant in organic reactions.

palladium-catalyzed domino carbopalladation/carbonylation reaction in DMC for the preparation of functionalized isoquinoline-1,3-diones and oxindoles. Different from the methylation function of DMC in the previous work, DMC acted as a methoxylating reagent *via* an acyl cleavage to incorporate the final products [57]. It is notable that under standard reaction conditions when dibutyl carbonate is used instead of DMC in the reaction, no corresponding product could be detected. This result revealed the uniqueness of DMC as a versatile reagent.

Apart from these, DMC can serve as a sustainable and low-toxicity methoxycarbonylating reagent for the synthesis of non-symmetric carbonate derivatives. The Monbaliu group developed a convenient and selective methoxycarbonylation process to access a library of benzyl methyl carbonates from benzyl alcohol under flow conditions, in which DMC played a dual role as a methoxycarbonyl donor and solvent (Scheme 3.26) [58]. It must be mentioned that the utilization of a large excess of DMC was imperative, since it could facilitate the shift of the equilibrium toward the desired product and led to a marked increase in terms of both conversion and selectivity. Also, the excess DMC can be recycled through distillation, thus maintaining a favorable environmental footprint. Remarkably, the short residence time (10 minutes) in a continuous flow regime also ensured the high efficiency and practicability of the protocol. All these results well demonstrated that dialkyl carbonates warrant further exploration as versatile reagents in a myriad of organic reactions.

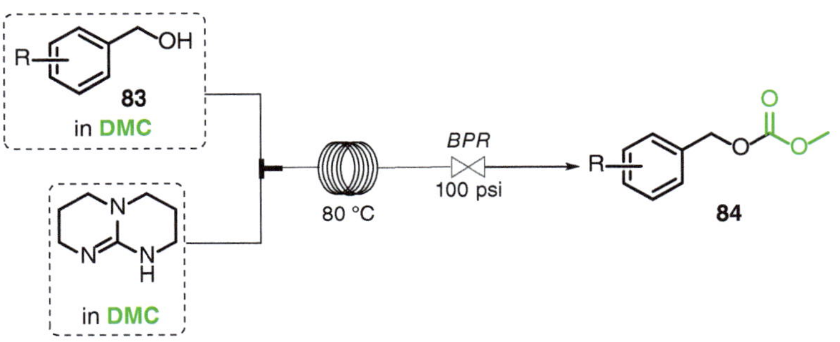

Scheme 3.26 Continuous flow organocatalyzed methoxycarbonylation of benzyl alcohol derivatives in DMC.

In addition to dialkyl carbonates, cyclic carbonate, propylene carbonate (PC), also known for serving as an eco-friendly carbon dioxide-neutral solvent, since it can be generated from propylene oxide and CO_2, or through a microwave-assisted lipase-catalyzed transesterification of propanediol with DMC [59]. PC has higher boiling points than other solvents, thus having lower atmospheric emissions, which makes its utilization and storage simpler and safer [60]. Most recently, the Bhanage group proposed a highly effective Pd/C-catalyzed oxidative aminocarbonylation reaction of arylhydrazines, which was performed in the environmentally benign solvent PC (Scheme 3.27) [61]. In the reaction, the C(sp^2)—N bond of arylhydrazine initially underwent an oxidative C—N bond cleavage with the assistance of KI and O_2, which was accompanied by CO insertion to form an acylpalladium complex.

This complex was subsequently combined with tertiary amines to form the final product. After several screenings of additives and solvents, it was proved that the use of KI as the additive in combination with PC as the medium improved the reaction efficiency, furnishing good yields of the tertiary amide product. After the completion of the reaction, the heterogeneous Pd/C catalyst could be recovered *via* centrifugation and was reused for at least five runs without an appreciable loss in its catalytic activity. Meanwhile, this strategy also circumvented the use of moisture-sensitive ligands, which made this approach simple and appealing from the viewpoint of sustainability.

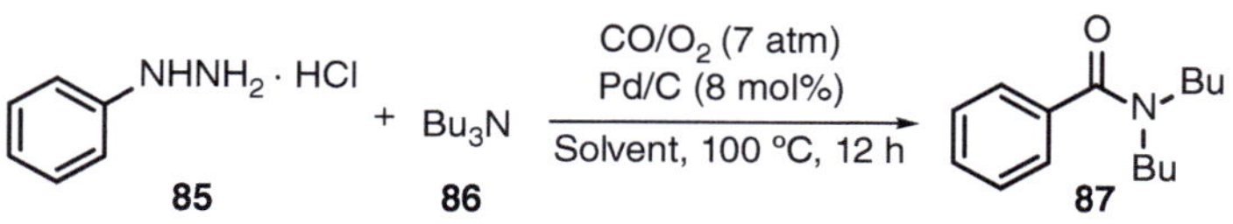

Solvent	Yield (%)	Solvent	Yield (%)
Toluene	72	THF	38
Anisole	68	DMSO	52
CH_3CN	48	Dioxane	40
DMF	22	PC	87

Scheme 3.27 Palladium-catalyzed oxidative aminocarbonylation in PC.

3.7 γ-Valerolactone (GVL)

GVL is a popular bioderived platform chemical and can be produced *via* hydrogenative cyclization from levulinic acid, which is derived from waste lignocellulosic biomass [62]. GVL has demonstrated multiple excellent properties, including a low melting point (31 °C) and a high boiling point (207 °C), as well as low vapor pressure and a high flash point (96 °C), and it shows good chemical stability to acidic and basic conditions [14a]. In particular, GVL has a polarity similar to common dipolar aprotic solvents, such as DMF, DMA, NMP, and MeCN. Together with its environmentally benign nature, such as low toxicity and good biodegradability, these favorable characteristics make it an attractive alternative for widely used dipolar aprotic solvents [63]. Currently, GVL has been frequently used as a sustainable green medium across a wide variety of chemistry applications [64].

In this regard, the Vaccaro group pioneered the application of GVL and has made great contributions to several heterogeneous catalytic cross-coupling reactions [65]. In 2021, the Vaccaro group reported an aerobic Pd/C-catalyzed C–H alkenylation in an environmentally friendly solvent GVL by means of a tailor-made tube-in-tube heterogeneous flow reactor (Scheme 3.28) [66]. As proved by a comparison of GVL and DMF, examined under the same experimental conditions, the employment of the green medium GVL not only provided a higher yield but also resulted in a very limited metal leaching into the solution (0.2–0.02 ppm); on one hand, it contributed significantly in improving the recyclability of the catalyst to a large extent, meanwhile well validating the distinctive effect of GVL. Importantly, the recovered GVL could be reused for further flow reactions after a simple distillation, thus minimizing the generation of waste. The practical utility of this protocol could also be successfully confirmed by a multi-gram scale reaction.

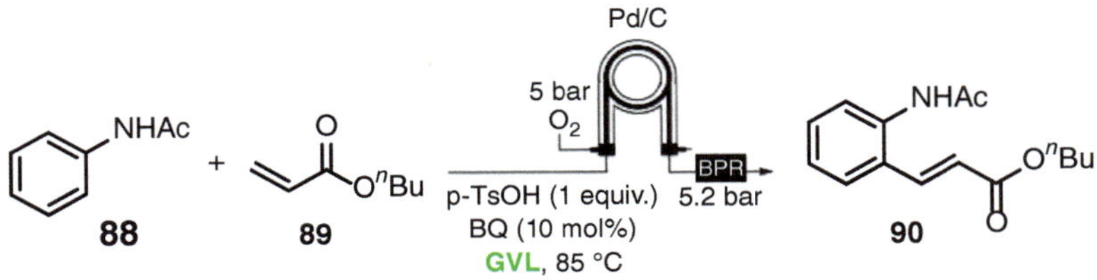

Scheme 3.28 Palladium-catalyzed C–H alkenylation in GVL.

In subsequent studies, this group used GVL as a co-solvent for the Hiyama reaction, which was catalyzed by a palladium-based heterogeneous catalyst supported on pine needle-derived active carbon (Pd/PiNe) (Scheme 3.29) [67]. Likewise, the reaction of 4-bromonitrobenzene and triethoxyphenylsilane proceeded well in the GVL/water mixture, whether in terms of efficiency or metal leaching, and avoiding the need for ligands or additives, which were generally necessary for this process. Once performing the reaction in toluene or DMF, only traces of product **93** could be detected, which further highlighted the important role of GVL.

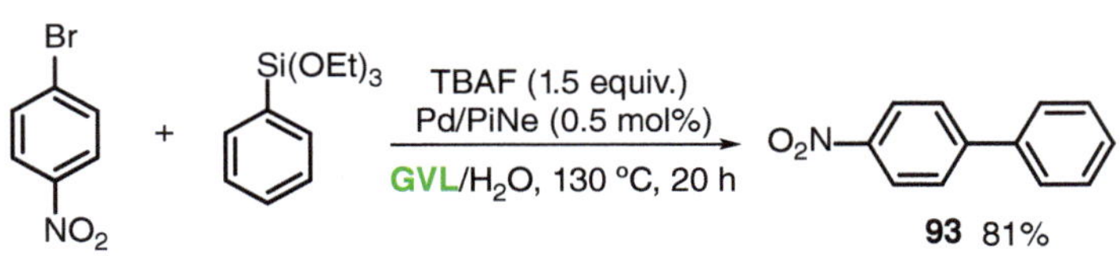

Solvent	Yield (%)
Toluene	18
DMF	12

Scheme 3.29 Hiyama cross-coupling between aryl bromide and triethoxyvinylsilane in GVL/H_2O.

GVL has also been utilized as a reaction solvent for the C–H arylation of indoles assisted by a palladium-containing metal-organic framework (MOF) (Scheme 3.30) [68]. The presence of the MOF increased the activity of $Pd(OAc)_2$, leading to the formation of C2–H arylation products in higher yields and excellent selectivity in

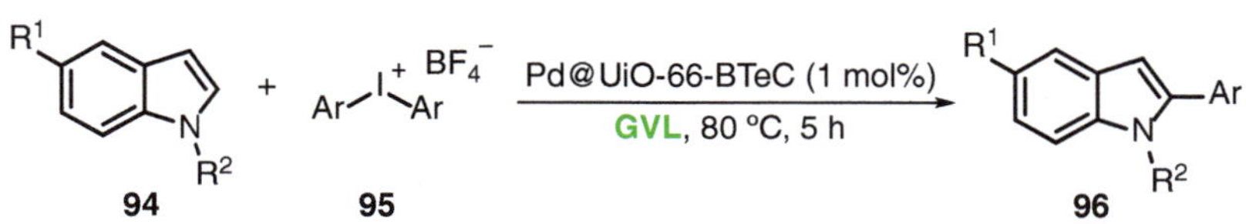

Scheme 3.30 C2–H arylation of indoles catalyzed by palladium-containing metal-organic framework in GVL.

contrast with the reaction performed under homogeneous conditions. It was also found that the amount of palladium in the solution after three runs remained low (approximately 0.5 ppm), whereas the reaction conducted in DMF not only resulted in lower reactivity but also was accompanied by increased metal leaching in the solution (277 ppm). This result verified the ability of GVL to reduce metal leaching from a heterogeneous catalyst.

The microwave-assisted direct arylation of thiophene, catalyzed by Pd/β-cyclodextrin cross-linked catalyst (Pd/CβCAT), was equally applicable in the green medium GVL, which was developed by Cravotto et al. (Scheme 3.31) [69]. This process made full use of the cooperative effect between MW and Pd/CβCAT, because the reaction could not occur using traditional heating, even in the presence of the same catalyst. The reaction proceeded well with as little as 0.2 mol% of the catalyst in the bio-derived GVL solvent, selectively giving the desired 5-arylated products in generally moderate-to-good yields within two hours. In this transformation, the formation of catalytically active palladium nanoparticles was a crucial step, in which the active palladium species could be stabilized by the polymeric network of β-cyclodextrin.

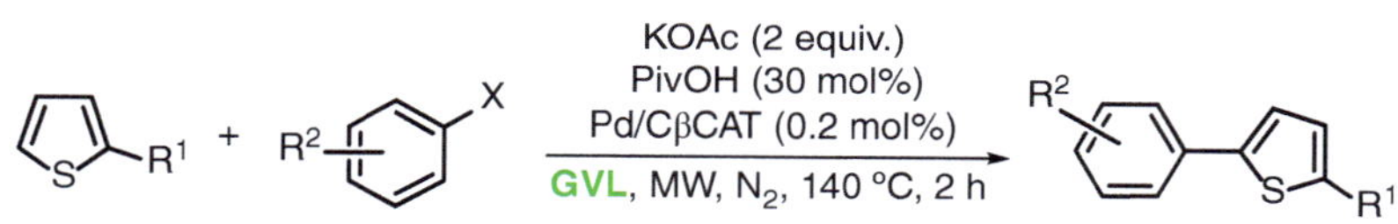

97 98 99

Scheme 3.31 MW-assisted coupling of heteroaromatics and aryl halides in GVL.

Recently, the Zhang group has demonstrated that GVL could be effectively utilized in the synthesis of isoquinolino[1,2-*b*]quinazolines, employing vinylene carbonate as an oxidizing acetylene surrogate (Scheme 3.32) [70]. The reaction was successfully achieved with the co-assistance of $[RhCp^*Cl_2]_2$ and $AgSbF_6$ in the green medium of GVL. Conversely, the use of DMF led to only trace amounts of the product, whereas other conventional solvents, such as DMSO, DCE, 1,4-dioxane, and CH_3CN, resulted

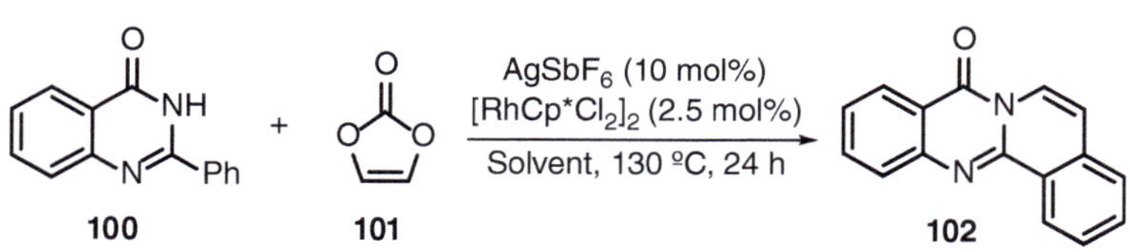

100 101 102

Solvent	Yield (%)
GVL	92
DMF	Trace
DCE	36
DMSO	31
CH_3CN	84
Dioxane	65

Scheme 3.32 Rhodium-catalyzed synthesis of isoquinolino[1,2-*b*]quinazolines in GVL.

in much lower reactivity compared to GVL, indicating the remarkable promotion of GVL on this C–H annulation. Furthermore, this catalytic system was effectively applied to 2 mmol of scale-up synthesis of isoquinolino[1,2-*b*]quinazoline.

3.8 Ethyl Lactate (EL)

EL, which naturally occurs mainly as an optically active L-isomer [71], is industrially produced *via* esterification from the renewable feedstock lactic acid and ethanol, both of which are easily accessible by microbial fermentations of carbohydrates, including starchy and cellulosic agricultural wastes [72]. In recent years, EL has been introduced as a promising environment-friendly bio-based medium for organic transformations due to its intrinsic characteristics, such as low production cost, nontoxicity, and high stability, as well as low environmental impact (completely biodegradable) [73]. Notably, the presence of a hydroxyl group in EL makes it compatible and miscible with both water and organic compounds [74]. Also, its high boiling point (154 °C) allows its use with a broad range of temperatures. Taking the aforementioned advantageous properties into account, EL can be well regarded as a greener replacement for some environmentally detrimental solvents [75].

A three-component benzylation of coumarin, styrene, and KSCN can be performed in EL at room temperature in the presence of peroxodisulfate, leading to the formation of 3-(1-aryl-2-thiocyanatoethyl)-2*H*-chromen-2-one in high yield (Table 3.2) [76]. It was found that the reaction atmosphere had a

Table 3.2 Peroxodisulfate-assisted benzylation of coumarin with styrene and KSCN in EL.

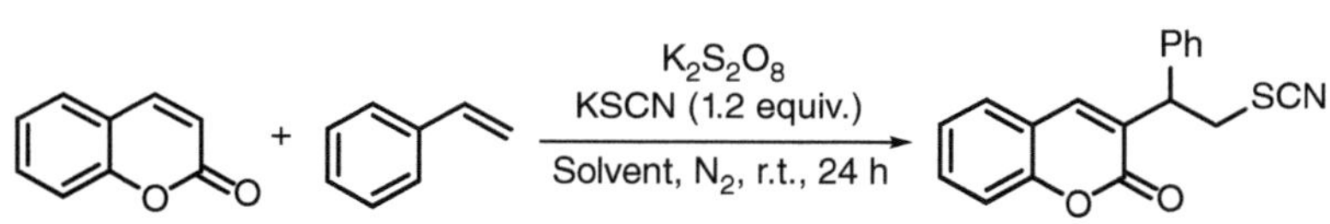

103 104 105

Entry	Solvent (mL)	Oxidant (equiv.)	Yield (%)
1	CH_3OH (5)	$K_2S_2O_8$ (1.5)	21
2	CH_3CN (5)	$K_2S_2O_8$ (1.5)	27
3	DMSO (4)	$K_2S_2O_8$ (1.5)	39
4	H_2O (2)	$K_2S_2O_8$ (1.5)	30
5	CH_3CN/H_2O (3)	$K_2S_2O_8$ (1.5)	45
6	EL (2)	$K_2S_2O_8$ (1.5)	72
7	EL (2)	$K_2S_2O_8$ (2.0)	93

significant influence on the outcome of the reaction, in which the reaction proceeded well under a nitrogen atmosphere, while oxygen or air brought down the yield. Moreover, the conventionally common solvents such as CH_3OH, CH_3CN, DMSO, H_2O, and CH_3CN/H_2O mixture, were much less effective under the same conditions, suggesting the unique advantage of the sustainable reaction medium, EL. More importantly, the pure products could be obtained by a simple water-based workup followed by recrystallization, making this protocol more convenient and practical.

EL, as an efficient green mediator, was also reported by Ghosh and coworkers for the preparation of symmetric and unsymmetric azobenzenes (Scheme 3.33) [77]. For the synthesis of symmetric azobenzenes, a solvent-free condition gave trace amounts of the product. The utilization of a range of other solvents, including CH_3OH, EtOH, PEG- 400, and toluene, all resulted in inferior yields. Intriguingly, significant improvements were observed in EL or aqueous solutions of EL, indicating the remarkable acceleration effect of EL on this transformation. Likewise, a similar promotion effect of EL was also observed in the reaction of nitrosobenzene with aniline.

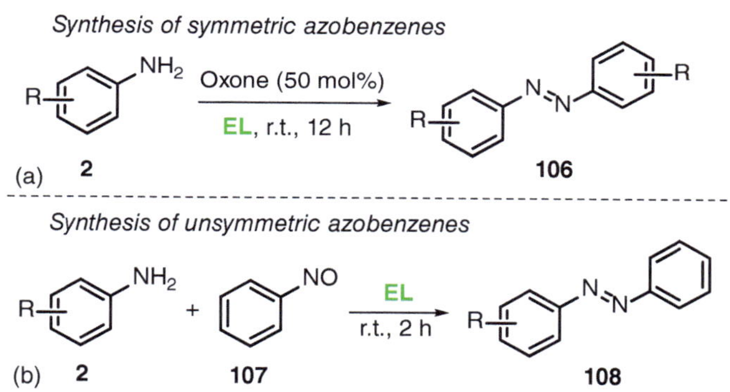

Scheme 3.33 EL-mediated synthesis of azobenzenes.

EL can not only serve as a green reaction medium but also as a green building block for organic synthesis. In this respect, Wan et al. demonstrated a $FeCl_3$-assisted and EL-involved three-component dehydrogenative reaction for the synthesis of quinolines with EL as the green solvent, in which only water and hydrogen were formed as by-products (Scheme 3.34) [78]. In this strategy, EL as the feedstock was transformed into ethyl pyruvate *via* dehydrogenation, which was captured by the *in situ*-generated imine resulting from anilines and aldehyde, and finally incorporating into the product as a C2-donor. This method highly performed the compatibility of biomass-derived EL acting as the material and reaction medium in the clean synthesis of valuable molecules.

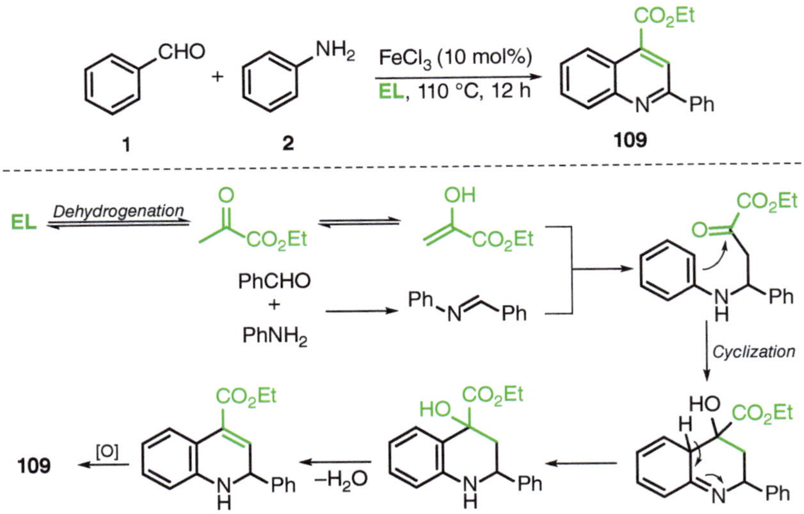

Scheme 3.34 EL-participating three-component dehydrogenative reaction for the synthesis of quinoline.

3.9 Miscellaneous

Eucalyptol (Euc) is a saturated oxygenated terpene (Scheme 3.35), structurally belonging to the bicyclic ether, which is widely distributed in the leaves of numerous renewable eucalyptus plants and their essential oil fractions [79]. Euc is characterized by many favorable properties, such as high polarity, hydrogen bonding nature, low immiscibility with water, and attractive thermal properties. Most importantly, eucalyptol has a more competitive price in the commercial market compared to 2-MeTHF. These distinctive superiorities make it a novel green and safe bio-solvent candidate. So far, it has been applied to some classical reactions, such as Migita–Kosugi–Stille coupling [80], Hiyama coupling [81], Buchwald–Hartwig coupling [82], and Michael addition reaction [83].

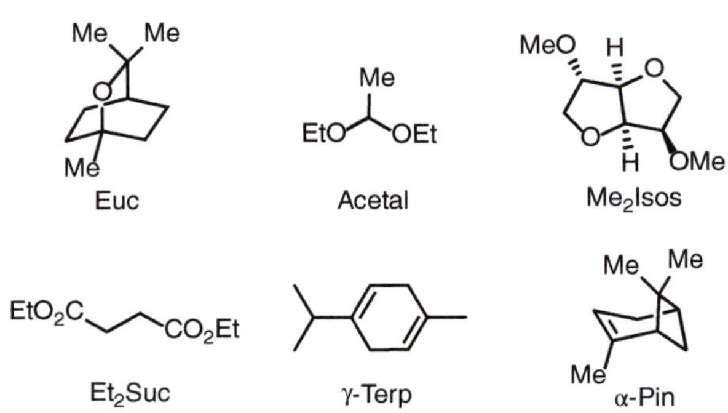

Scheme 3.35 Some novel biomass-derived chemicals proposed as potential solvents.

Apart from the above-mentioned green solvents, some other biomass-derived chemicals, such as acetaldehyde diethyl acetal (Acetal), isosorbide dimethyl ether (Me_2Isos), and diethyl succinate (Et_2Suc), all derived from cellulose (Scheme 3.35) [84], have been proposed as promising candidates for green and safe solvents. Furthermore, the emerging γ-terpinene (γ-Terp) and α-pinene (α-Pin) available from various coniferous plants also deserve to be explored and evaluated serving as novel and versatile biomass-derived solvents in organic synthesis.

3.10 Conclusions and Outlook

Green solvents are increasingly adopted, both at the academic and the industrial level, as alternatives to traditional solvents. In this review, we have reported an up-to-date overview of the typically used sustainable bio-derived reaction medium in organic synthesis, demonstrating their advantageous features in terms of increasing the reaction efficiency, minimizing the generation of waste, and enhancing the recyclability of catalyst, as well as simplifying the work-up procedures, by making a stark contrast to the utilization of conventional petroleum-based solvents. Although replacing all solvents that carry reproductive toxicity warnings with eco-friendly and biodegradable bio-based solvents is not a real possibility across a myriad of reaction types, this review provides a clear and comprehensive depiction of the current state-of-the-art in the realm of potential substitute solvents. Moreover, their sustainable production and favorable physicochemical properties have been compiled, contributing to a more judicious solvent selection in some cases.

Currently, the diversity of bio-based solvents is far from satisfactory, and continuous efforts are needed to develop new types of renewable chemicals that are promising to be engaged as the reaction medium. On the other hand, it is also imperative to explore their novel reactivity and versatility, further broadening their utilization, not limited to the reaction medium, such as working as divergent building blocks for the construction of more valuable functionalized molecules. Meanwhile, seeking industrially simpler and sustainable synthetic pathways of these green solvents should also be taken into consideration.

References

1 Clarke, C.J., Tu, W.C., Levers, O. et al. (2018). *Chem. Rev.* 118: 747–800.

2 Anastas, P.T. and Kirchhoff, M.M. (2002). *Acc. Chem. Res.* 35: 686–694.

3 Diorazio, L.J., Hose, D.R.J., and Adlington, N.K. (2016). *Org. Process Res. Dev.* 20: 760–773.

4 Santoro, S., Ferlin, F., Luciani, L. et al. (2017). *Green Chem.* 19: 1601–1612.

5 Jordan, A., Hall, C.G.J., Thorp, L.R., and Sneddon, H.F. (2022). *Chem. Rev.* 122: 6749–6794.

6 (a) Motagamwala, A.H., Huang, K., Maravelias, C.T., and Dumesic, J.A. (2019). *Energy Environ. Sci.* 12: 2212–2222. (b) Wang, C., Chen, X., Qi, M. et al. (2019).

Green Chem. 21: 6089–6096. (c) He, J., Chen, L., Liu, S. et al. (2020). *Green Chem.* 22: 6714–6747.

7 (a) Albano, G., Punzi, A., Capozzi, M.A.M., and Farinola, G.M. (2022). *Green Chem.* 24: 1809–1894. (b) Bijoy, R., Agarwala, P., Roy, L., and Thorat, B.N. (2021). *Org. Process Res. Dev.* 26: 480–492. (c) de Albuquerque, D.Y., Teixeira, W.K.O., Narayanaperumal, S., and Schwab, R.S. (2022). *J. Braz. Chem. Soc.* 33: 637–663. (d) Miele, M., Pillari, V., Pace, V. et al. (2022). *Molecules* 27: 6701. (e) Gandeepan, P., Kaplaneris, N., Santoro, S. et al. (2019). *ACS Sustainable Chem. Eng.* 7: 8023–8040. (f) Sangon, S., Supanchaiyamat, N., Sherwood, J. et al. (2020). *React. Chem. Eng.* 5: 1798–1804.

8 Gu, Y. and Jérôme, F. (2013). *Chem. Soc. Rev.* 42: 9550–9570.

9 (a) Henderson, R.K., Jimenez-Gonzalez, C., Constable, D.J.C. et al. (2011). *Green Chem.* 13: 854–862. (b) Alder, C.M., Hayler, J.D., Henderson, R.K. et al. (2016). *Green Chem.* 18: 3879–3890.

10 Alfonsi, K., Colberg, J., Dunn, P.J. et al. (2008). *Green Chem.* 10: 31–36.

11 Prat, D., Pardigon, O., Flemming, H.W. et al. (2013). *Org. Process Res. Dev.* 17: 1517–1525.

12 Prat, D., Wells, A., Hayler, J. et al. (2016). *Green Chem.* 18: 288–296.

13 Tobiszewski, M., Namieśnik, J., and Pena-Pereira, F. (2017). *Green Chem.* 19: 1034–1042.

14 (a) Alonso, D.M., Wettstein, S.G., and Dumesic, J.A. (2013). *Green Chem.* 15: 584–595. (b) Fegyverneki, D., Orha, L., Lang, G., and Horvath, I.T. (2010). *Tetrahedron* 66: 1078–1081. (c) Ciriminna, R., Lomeli-Rodriguez, M., Cara, P.D. et al. (2014). *Chem. Commun.* 50: 15288–15296.

15 Abd-Elmonem, M., Mekheimer, R.A., Hayallah, A.M. et al. (2019). *Curr. Org. Chem.* 23: 3226–3246.

16 Diaz-Alvarez, A.E., Francos, J., Lastra-Barreira, B. et al. (2011). *Chem. Commun.* 47: 6208–6227.

17 Wolfson, A., Dlugy, C., and Shotland, Y. (2007). *Environ. Chem. Lett.* 5: 67–71.

18 (a) Dubey, A.V., Gharat, S.B., and Kumar, A.V. (2017). *ChemistrySelect* 2: 4852–4856. (b) Rosholm, T., Gois, P.M.P., Franzen, R., and Candeias, N.R. (2015). *ChemistryOpen* 4: 39–46. (c) Garcia, J.I., Garcia-Marin, H., and Pires, E. (2014). *Green Chem.* 16: 1007–1033.

19 Karimirad, F. and Behbahani, F.K. (2021). *Polycyclic Aromat. Compd.* 41: 2238–2246.

20 Jin, J.Y., Miao, C.X., Wang, Z.L. et al. (2018). *Eur. J. Med. Chem.* 150: 757–770.

21 Saquib, M., Khan, M.F., Singh, J. et al. (2022). *Sustain. Chem. Pharm.* 30: 100860.

22 Chen, Z.Y., Huang, W.B., Yi, L.Q. et al. (2022). *Green Chem.* 24: 2919–2926.

23 Cintas, P., Tagliapietra, S., Gaudino, E.C. et al. (2014). *Green Chem.* 16: 1056–1065.

24 Martinho, L.A. and Andrade, C.K.Z. (2022). *J. Heterocycl. Chem.* 59: 1417–1429.

25 Meyer, T.H., Chesnokov, G.A., and Ackermann, L. (2020). *ChemSusChem* 13: 668–671.

26 Moran, M.J., Martina, K., Stefanidis, G.D. et al. (2020). *Front. Chem.* 8: 34.

27 Zhu, Y.Q., Romain, C., and Williams, C.K. (2016). *Nature* 540: 354–362.

28 (a) Sun, J.Y. and Liu, H.C. (2011). *Green Chem.* 13: 135–142. (b) Colacino, E., Martinez, J., Lamaty, F. et al. (2012). *Coordin. Chem. Rev.* 256: 2893–2920. (c) Chen, J., Spear, S.K., Huddleston, J.G., and Rogers, R.D. (2005). *Green Chem.* 7: 64–82. (d) Pang, J.F., Zheng, M.Y., Sun, R.Y. et al. (2016). *Green Chem.* 18: 342–359. (e) Wang, A.Q. and Zhang, T. (2013). *Acc. Chem. Res.* 46: 1377–1386. (f) Xi, J.X., Ding, D.Q., Shao, Y. et al. (2014). *ACS Sustainable Chem. Eng.* 2: 2355–2362.

29 Gulati, U., Rawat, S., and Rawat, D.S. (2020). *Tetrahedron Lett.* 61: 152304.

30 Kardooni, R. and Kiasat, A.R. (2021). *Polycyclic Aromat. Compd.* 41: 1883–1891.

31 Pace, V., Hoyos, P., Castoldi, L. et al. (2012). *ChemSusChem* 5: 1369–1379.

32 (a) Khoo, H.H., Wong, L.L., Tan, J. et al. (2015). *Resour. Conserv. Recycl.* 95: 174–182. (b) Cai, C.M., Zhang, T.Y., Kumar, R., and Wyman, C.E. (2014). *J. Chem. Technol. Biotechnol.* 89: 2–10. (c) Mariscal, R., Maireles-Torres, P., Ojeda, M. et al. (2016). *Energy Environ. Sci.* 9: 1144–1189.

33 Hao, W.Y., Xu, Z.T., Zhou, Z.B., and Cai, M.Z. (2020). *J. Organomet. Chem.* 85: 8522–8532.

34 Boldrini, C. and Harutyunyan, S.R. (2021). *Chem. Commun.* 57: 11807–11810.

35 Pierret, A., Denhez, C., Gros, P.C., and Vasseur, A. (2022). *Adv. Synth. Catal.* 364: 3805–3816.

36 Ielo, L., Miele, M., Pillari, V. et al. (2021). *Org. Biomol. Chem.* 19: 2038–2043.

37 Senatore, R., Malik, M., Urban, E. et al. (2021). *Tetrahedron* 85.

38 (a) Azzena, U., Carraro, M., Pisano, L. et al. (2019). *ChemSusChem* 12: 40–70. (b) de Gonzalo, G., Alcantara, A.R., and de Maria, P.D. (2019). *ChemSusChem* 12: 2083–2097.

39 Watanabe, K., Yamagiwa, N., and Torisawa, Y. (2007). *Org. Process Res. Dev.* 11: 251–258.

40 Hoshino, H., Sakakibara, K., and Watanabe, K. (2008). *Chem. Lett.* 37: 774–775.

41 (a) Antonucci, V., Coleman, J., Ferry, J.B. et al. (2011). *Org. Process Res. Dev.* 15: 939–941. (b) Watanabe, K. (2013). *Molecules* 18: 3183–3194.

42 (a) Zhang, G.S., Zhu, M.M., Zhang, Q. et al. (2016). *Green Chem.* 18: 2155–2164. (b) Guo, J.H., Xu, G.Y., Han, Z. et al. (2014). *ACS Sustainable Chem. Eng.* 2: 2259–2266. (c) Sudarsanam, P., Katta, L., Thrimurthulu, G., and Reddy, B.M. (2013). *J. Ind. Eng. Chem.* 19: 1517–1524. (d) Piutti, C. and Quartieri, F. (2013). *Molecules* 18: 12290–12312. (e) Corona, A., Biddy, M.J., Vardon, D.R. et al. (2018). *Green Chem.* 20: 3857–3866.

43 Kobayashi, S., Shibukawa, K., Miyaguchi, Y., and Masuyama, A. (2016). *Asian J. Org. Chem.* 5: 636–645.

44 Andresini, M., Degennaro, L., and Luisi, R. (2021). *Beilstein J. Org. Chem.* 17: 203–209.

45 (a) Ferlin, F., van der Hulst, M.K., Santoro, S. et al. (2019). *Green Chem.* 21: 5298–5305. (b) Ferlin, F., Luque Navarro, P.M., Gu, Y. et al. (2020). *Green Chem.* 22: 397–403. (c) Salameh, N., Anastasiou, I., Ferlin, F. et al. (2022). *Mol. Catal.* 522: 112211. (d) Ferlin, F., Valentini, F., Sciosci, D. et al. (2021). *ACS Sustainable Chem. Eng.* 9: 12196–12204.

46 Singh, S. and Kandasamy, J. (2022). *Asian J. Org. Chem.* 11: e202200416.

47 Xiong, L., He, S.Q., Pan, J., and Yu, B. (2022). *New J. Chem.* 46: 4550–4554.

48 Schaffner, B., Schaffner, F., Verevkin, S.P., and Borner, A. (2010). *Chem. Rev.* 110: 4554–4581.

49 (a) Pescarmona, P.P. (2021). *Curr. Opin. Green Sustainable Chem.* 29: 100457. (b) Putro, W.S., Ikeda, A., Shigeyasu, S. et al. (2021). *ChemSusChem* 14: 842–846. (c) Sakakura, T. and Kohno, K. (2009). *Chem. Commun.* 1312–1330. (d) Sakai, T., Tsutsumi, Y., and Ema, T. (2008). *Green Chem.* 10: 337–341. (e) Tan, H.Z., Wang, Z.Q., Xu, Z.N. et al. (2018). *Catal. Today* 316: 2–12.

50 Shukla, K. and Srivastava, V.C. (2016). *RSC Adv.* 6: 32624–32645.

51 (a) Qi, X.X., Lai, M., and Wu, X.F. (2019). *Org. Chem. Front.* 6: 3397–3400. (b) Schaffner, B., Holz, J., Verevkin, S.P., and Borner, A. (2008). *ChemSusChem* 1: 249–253.

52 Chen, S., You, B., Zhang, T. et al. (2022). *ChemSusChem* 15: e202201301.

53 (a) Zeng, F.L., Zhu, H.L., Chen, X.L. et al. (2021). *Green Chem.* 23: 3677–3682. (b) Zeng, F.L., Xie, K.C., Liu, Y.T. et al. (2022). *Green Chem.* 24: 1732–1737.

54 Li, Z.M., Wen, Y.T., Wang, N. et al. (2021). *Eur. J. Org. Chem.* 2021: 3819–3826.

55 (a) Tundo, P. and Selva, M. (2002). *Acc. Chem. Res.* 35: 706–716. (b) Tundo, P., Musolino, M., and Arico, F. (2018). *Green Chem.* 20: 28–85. (c) Tundo, P., Rossi, L., and Loris, A. (2005). *J. Organomet. Chem.* 70: 2219–2224. (d) Pyo, S.H., Park, J.H., Chang, T.S., and Hatti-Kaul, R. (2017). *Curr. Opin. Green Sustainable Chem.* 5: 61–66.

56 Miao, R.G., Qi, X.X., and Wu, X.F. (2021). *Eur. J. Org. Chem.* 2021: 5219–5221.

57 Wen, D., Shen, Z., Qi, X., and Wu, X.F. (2022). *Eur. J. Org. Chem.* 2022: e202200971.

58 Toupy, T., Bovy, L., and Monbaliu, J.C.M. (2022). *J. Flow Chem.* 12: 207–217.

59 (a) Millan, D., Rojas, M., Tapia, R.A., and Pavez, P. (2020). *Org. Biomol. Chem.* 18: 7868–7875. (b) Oncel, N., Kasumov, V.T., Sahin, E., and Ulusoy, M. (2016). *J. Organomet. Chem.* 811: 81–90. (c) Yadav, G.D., Hude, M.P., and Talpade, A.D. (2015). *Chem. Eng. J.* 281: 199–208.

60 (a) Forero, J.S.B., Munoz, J.A.H., Jones, J., and da Silva, F.M. (2016). *Curr. Org. Synth.* 13: 834–846. (b) Yang, G.-P., Zhang, N., Ma, N.-N. et al. (2017). *Adv. Synth. Catal.* 359: 926–932. (c) Zamani, P., Phipps, J., Hu, J.Y. et al. (2019). *ACS Comb. Sci.* 21: 557–561.

61 Kolekar, Y.A. and Bhanage, B.M. (2022). *New J. Chem.* 46: 14421–14426.

62 (a) Yang, Z., Fu, Y., and Guo, Q.X. (2015). *Chin. J. Org. Chem.* 35: 273–283. (b) Ismalaj, E., Strappaveccia, G., Ballerini, E. et al. (2014). *ACS Sustainable Chem. Eng.* 2: 2461–2464. (c) Liguori, F., Moreno-Marrodan, C., and Barbaro, P. (2015). *ACS Catal.* 5: 1882–1894. (d) Zhang, Z.H. (2016). *ChemSusChem* 9: 156–171. (e) Zada, B., Zhu, R., Wang, B. et al. (2020). *Green Chem.* 22: 3427–3432.

63 Gao, F., Bai, R., Ferlin, F. et al. (2020). *Green Chem.* 22: 6240–6257.

64 (a) Song, J.L., Zhou, B.W., Liu, H.Z. et al. (2016). *Green Chem.* 18: 3956–3961. (b) Rasina, D., Kahler-Quesada, A., Ziarelli, S. et al. (2016). *Green Chem.* 18: 5025–5030. (c) Tian, X., Yang, F.Z., Rasina, D. et al. (2016). *Chem. Commun.* 52: 9777–9780.

65 (a) Strappaveccia, G., Luciani, L., Bartollini, E. et al. (2015). *Green Chem.* 17: 1071–1076. (b) Strappaveccia, G., Ismalaj, E., Petrucci, C. et al. (2015). *Green Chem.* 17: 365–372. (c) Ferlin, F., Santoro, S., Ackermann, L., and Vaccaro, L. (2017). *Green Chem.* 19: 2510–2514. (d) Ferlin, F., Luciani, L., Santoro, S. et al. (2018). *Green Chem.* 20: 2888–2893.

66 Ferlin, F., Anastasiou, I., Carpisassi, L., and Vaccaro, L. (2021). *Green Chem.* 23: 6576–6582.

67 Valentini, F., Ferlin, F., Lilli, S. et al. (2021). *Green Chem.* 23: 5887–5895.

68 Anastasiou, I., Van Velthoven, N., Tomarelli, E. et al. (2020). *ChemSusChem* 13: 2786–2791.

69 Tabasso, S., Gaudino, E.C., Acciardo, E. et al. (2020). *Front. Chem.* 8: 253.

70 Wang, L., Jiang, K.C., Zhang, N., and Zhang, Z.H. (2021). *Asian J. Org. Chem.* 10: 1671–1674.

71 Dolzhenko, A.V. (2020). *Sustainable Chem. Pharm.* 18: 100322.

72 Planer, S., Jana, A., and Grela, K. (2019). *ChemSusChem* 12: 4655–4661.

73 Dandia, A., Jain, A.K., and Laxkar, A.K. (2013). *Tetrahedron Lett.* 54: 3929–3932.

74 (a) Pereira, C.S.M., Silva, V., and Rodrigues, A.E. (2011). *Green Chem.* 13: 2658–2671. (b) Gao, G., Han, Y., and Zhang, Z.H. (2017). *ChemistrySelect* 2: 11561–11564.

75 (a) Wei, L., Chen, X.W., Liu, Y.Y., and Wan, J.P. (2016). *Chin. J. Org. Chem.* 36: 954–961. (b) Zhang, M., Fu, Q.Y., Gao, G. et al. (2017). *ACS Sustainable Chem. Eng.* 5: 6175–6182. (c) Shen, G.S., Zhou, H.F., Du, P. et al. (2015). *RSC Adv.* 5: 85646–85651.

76 Natarajan, P., Priya, and Chuskit, D. (2021). *Green Chem.* 23: 4873–4881.

77 Pariyar, G.C., Kundu, T., Mitra, B. et al. (2020). *ChemistrySelect* 5: 9781–9786.

78 Yang, L. and Wan, J.P. (2020). *Green Chem.* 22: 3074–3078.

79 Campos, J.F., Scherrmann, M.M.C., and Berteina-Raboine, S. (2019). *Green Chem.* 21: 1531–1539.

80 Campos, J.F. and Berteina-Raboin, S. (2020). *Catal. Today* 358: 138–142.

81 Campos, J.F., Ferreira, V., and Berteina-Raboin, S. (2021). *Catalysts* 11.

82 (a) Berteina-Raboin, S., Campos, J.F., Pacheco-Benichou, A. et al. (2020). *Synthesis* 52: 3071–3076. (b) Campos, J.F. and Berteina-Raboin, S. (2019). *Catalysts* 9: 840.

83 Dave, A.Y., Mishra, A., Talukdar, M., and Begari, E. (2022). *ChemistrySelect* 7: e202200954.

84 (a) Ismael, A., Gevorgyan, A., Skrydstrup, T., and Bayer, A. (2020). *Org. Process Res. Dev.* 24: 2665–2675. (b) Gevorgyan, A., Hopmann, K.H., and Bayer, A. (2020). *ChemSusChem* 13: 2080–2088.

4

Recent Achievements in Organic Reactions in DMSO

Peng Yuan[2], Jia-Chen Xiang[2], and An-Xin Wu[1]

[1] *Central China Normal University, International Joint Research Center for Intelligent Biosensor Technology and Health, National Key Laboratory of Green Pesticide, 152 Luoyu Road, Wuhan 430079, P.R. China*
[2] *Southeast University, School of Chemistry and Chemical Engineering, 2 Southeast University Road, Nanjing 211189, P.R. China*

Dimethyl sulfoxide (DMSO) has been widely used as an aprotic polar solvent in organic synthesis. Whether in laboratory experiments or industrial production, DMSO is favored by chemists because not only it is a green, low-toxic, and cheap solvent, but also due to its easy-to-handle operation and relatively simple workup [1]. DMSO has some special utilities in reactions as auxiliary reagent, such as forming a super base with the base to improve alkalinity or stabilizing charged intermediates. It is known that beyond being a solvent, an auxiliary reagent, and a well-known oxidant [2], DMSO often acts as a synthetic building block in reactions, providing components for the assembly of the final product through certain activation and fragmentation processes. It provides a green, efficient, and convenient way to deliver molecular fragments such as oxygen sources (–O–, =O), methine (=CH–), formaldehyde group (–CHO), sulfur linkages (–S–), methyl thio group (–SMe), methyl sulfoxide group (–SOMe), methyl thiomethylation donors (–CH_2SMe), and methyl sulfoxide methylation donors (–CH_2SOMe). This important function allows DMSO to be used as a versatile synthon in both designed and beyond-designed synthesis processes, and to some extent, facilitates the development of novel synthetic methodologies and the creation of new molecular backbones.

A review of the application of DMSO as a synthon was reported by Wu's group in 2016 [3]. These were also discussed in Chapter 7 of Wu's book "Solvents as Reagents in Organic Synthesis," the applications of DMSO [4]. In 2020, Mahdavi's group updated a review on the applications of DMSO as a single or dual synthon [1b]. Very recently, Qiu's group reported a comprehensive and updated review on the applications of DMSO in organic synthesis covering literature contributions published from January 2016 to March 2022 [5]. However, DMSO is considered a relatively inert reagent, therefore it is important to highlight the mechanism in different activation patterns of DMSO, which provides its versatile functionalities for synthetic applications. In the chapter herein, we summarize some special and representative transformations using DMSO as a green solvent in organic synthesis

Green Solvents in Organic Synthesis, First Edition.
Edited by Xiao-Feng Wu, Zhiping Yin, Liang-nian He, and Feng Wang.

in the last 5 years. In contrast to the above reviews, we focus more on the *in situ* activation pattern of DMSO by summarizing different fragmentation processes of DMSO to show the reaction initiation pathway. For each activation pattern, several recent representative transformations are listed, along with their selected products to show the generality, while a detailed mechanism is also given after each example to facilitate the reader's reading and to illustrate the chemical principles. We believe that this chapter provides a unique perspective to lead to a better understanding of the DMSO-mediated transformation and may provide integrated information for chemical researchers who are keen to select a method for DMSO activation, thus facilitating the application of DMSO in synthetic design.

4.1 Pummerer-type Activation of DMSO

Being discovered in the first decade of the twentieth century [6], the Pummerer reaction has received continued attention and applications from synthetic chemists due to its extensive use in the design of synthetic methodologies for heterocyclic compounds, as it can provide versatile synthons such as $-CH_2$, $-CH_3$, $-CHO$, and other C1 units [7].

In an acid-based Pummerer reaction, the transformation of DMSO to a thionium ion intermediate could be captured not only by nucleophilic addition to generate an α-substituted sulfide but also by trapping with H_2O to produce the corresponding decomposition product. For example, Sun et al. reported a H_3PO_4-promoted thermal decomposition of DMSO to generate CH_3SH and HCHO, in which the latter could act as a C1 unit [8]. Guided by such strategy, they developed a synthesis of symmetrical methylene-bridged imidazoheterocycles from 2-phenylimidazo[1,2-*a*] pyridine in a toluene solution in the presence of 3 equivalents of H_3PO_4 and 6 equivalents of DMSO (Scheme 4.1).

In 2019, Yi's group reported a synthesis of β-amino ketones from ketones and amides via cobalt(III)-catalyzed cross-coupling reaction [9]. Acting as both a solvent and a carbon doner, DMSO contributes two distinct C1 units: a methylene linker and a terminal olefin, to the final product skeleton via a $K_2S_2O_8$-promoted Pummerer-type reaction (Scheme 4.2).

This strategy was used again by the same group in 2019 [10]. They reported a cobalt-catalyzed allylation of amides employing styrenes as the reaction partner. DMSO was used as an α-methylene source to build the linkage between the amides and styrenes (Scheme 4.3). It undertakes a $K_2S_2O_8$-promoted Pummerer-type reaction to generate thionium ion species, which could be trapped by amides and further participate in the Co(II)-mediated coupling of styrenes.

An important application of the Pummerer-type reaction mode for DMSO activation is the construction of heterocycles where DMSO provides the C1 unit and plays as the oxidant to realize the aromatization for some cases. In 2017, Yi's group developed a Co(III)-catalyzed synthesis of 4-substituted, 3-substituted, or 3,4-disubstituted quinolines from easily available anilines with alkynes [11]. DMSO, as a single solvent and carbon unit donor, is involved in the transformation,

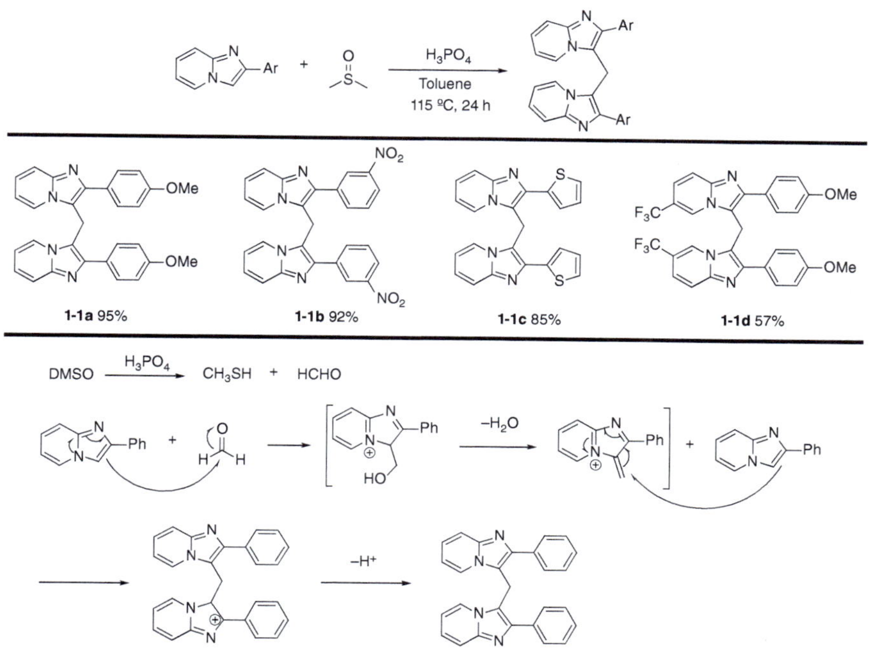

Scheme 4.1

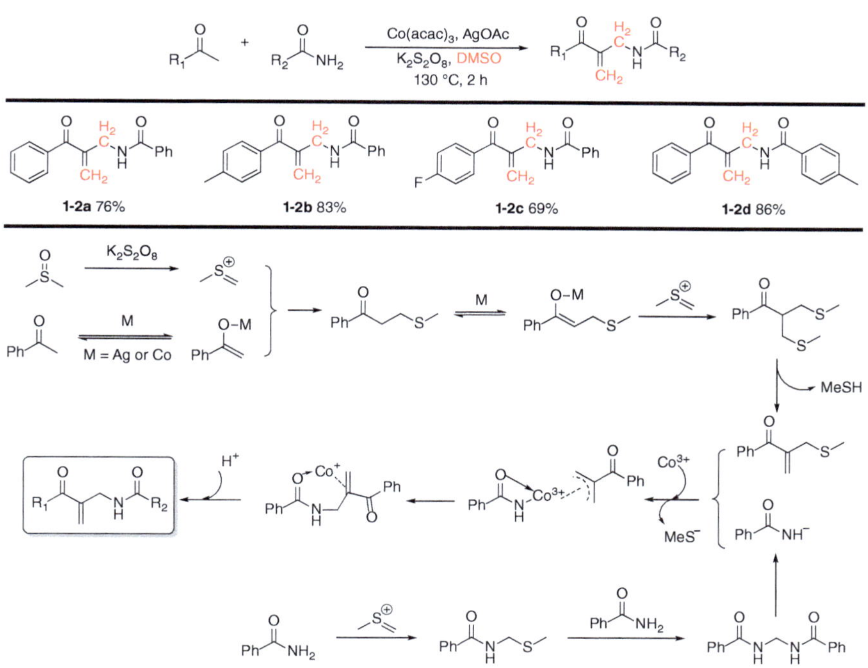

Scheme 4.2

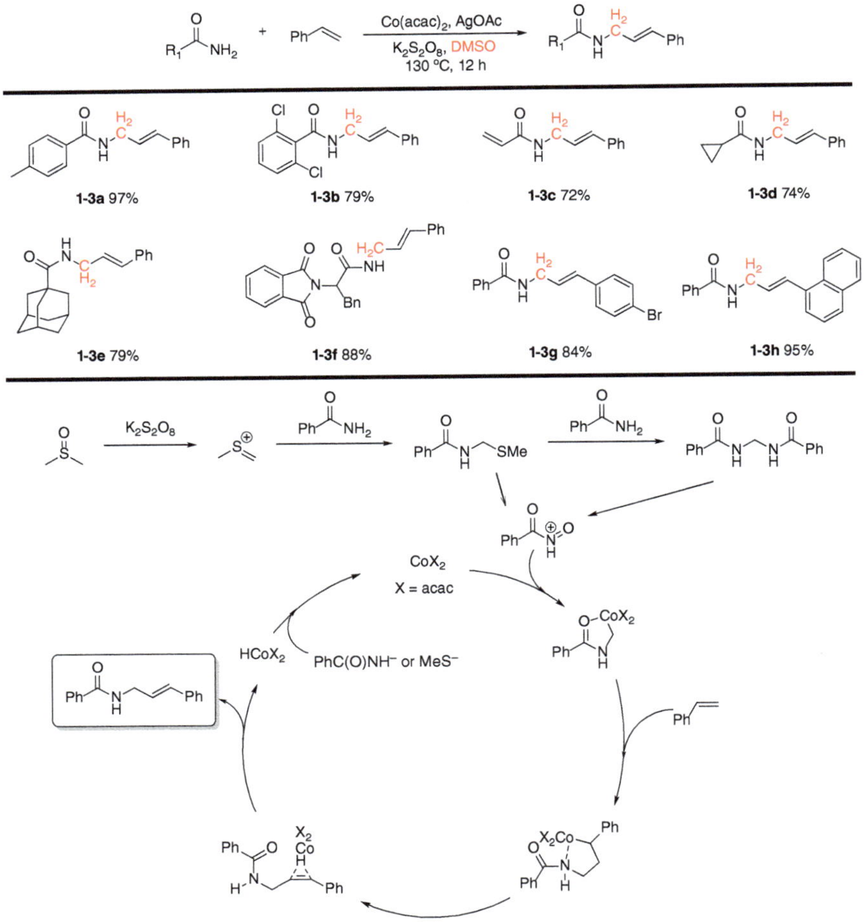

Scheme 4.3

acting as a carbon unit in the 2-position of the quinoline through the activation of $K_2S_2O_8$. Exclusive regioselectivity, broad substrate/functional group tolerance, and good-to-excellent yields were detected in the substrate expansion studies for this reaction (Scheme 4.4).

In the same year, Cheng's group developed a synthesis of 4-aryl quinolines from ortho-vinylanilines and DMSO via a palladium-catalyzed annulation [12]. The key mechanism involves the Pummerer-type activation of DMSO by 1,4-Diazoniabicyclo[2.2.2]octane (DABSO). The whole reaction was inhibited when no DABSO was added. A range of *n*-substituted-4-aryl quinolines (n = 3, 5–8) bearing functional groups could be prepared by this facile method (Scheme 4.5).

In 2022, Guo and Liu et al. developed a [3+1+1+1] annulation process using simple arylaldehydes, arylamines, and DMSO to build a 3-arylquinoline structure [13]. *t*-BuOK was used as the base and $K_2S_2O_8$ as the oxidant. Mechanistic

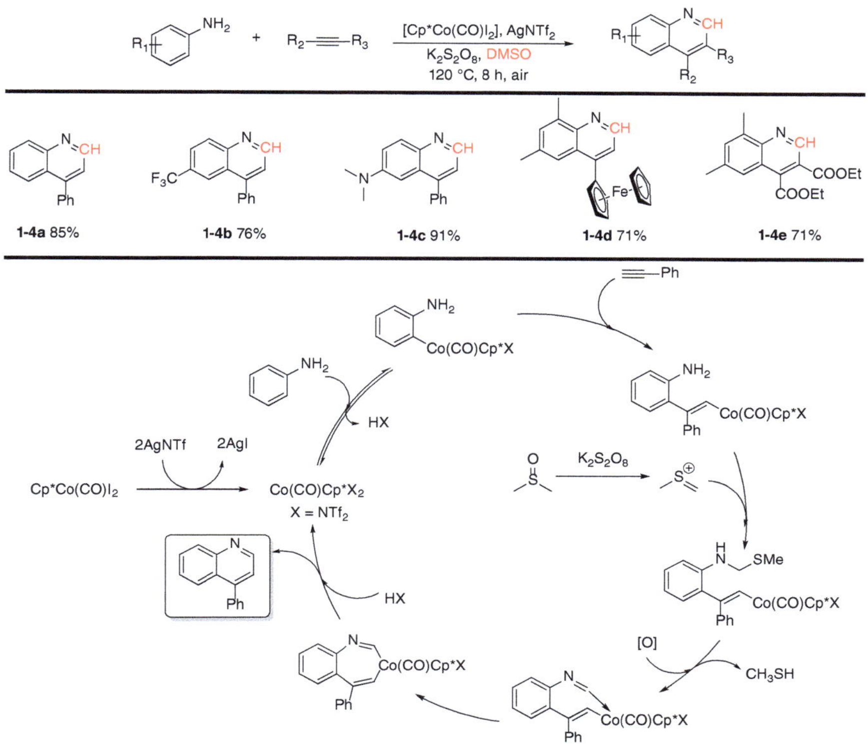

Scheme 4.4

studies show that both the 2 and 4 positions of the quinoline product are derived from DMSO. First, the condensation reaction between benzaldehyde and DMSO in the presence of a base gives intermediate A. This intermediate is then reduced by the *in situ* generation of DMS in the system to form intermediate B. At the same time, $K_2S_2O_8$ activates the DMSO based on the Pummerer reaction, which could be trapped by aniline to give the imine C. The [4+2] cycloaddition of B with C forms the quinoline structure (Scheme 4.6).

Another synthesis of functionalized quinolines was developed by Tiwari et al. from readily available acetophenones, anthranils, and DMSO, in which DMSO acts as the C1 unit to contribute to the C2-position of quinoline [14]. Similarly, DMSO was activated by $K_2S_2O_8$ to generate the electrophilic intermediate thionium ion. α,β-Unsaturated ketone could form from the reaction between $CH_3S^+{=}CH_2$ and acetophenones. With the ring-opening reaction of anthranils and the following domino cyclization process, 3-ketoquinoline could be obtained (Scheme 4.7).

Recently, this group further reported a one-pot synthesis of 3-substituted-1-aryl-1H-pyrazolo-[3,4-*b*]quinoline from anilines and pyrazolones [15]. DMSO acts as the sole solvent and carbon source in this transformation. The activation process of DMSO was achieved by acid (TFA)-promoted Pummerer reaction (Scheme 4.8).

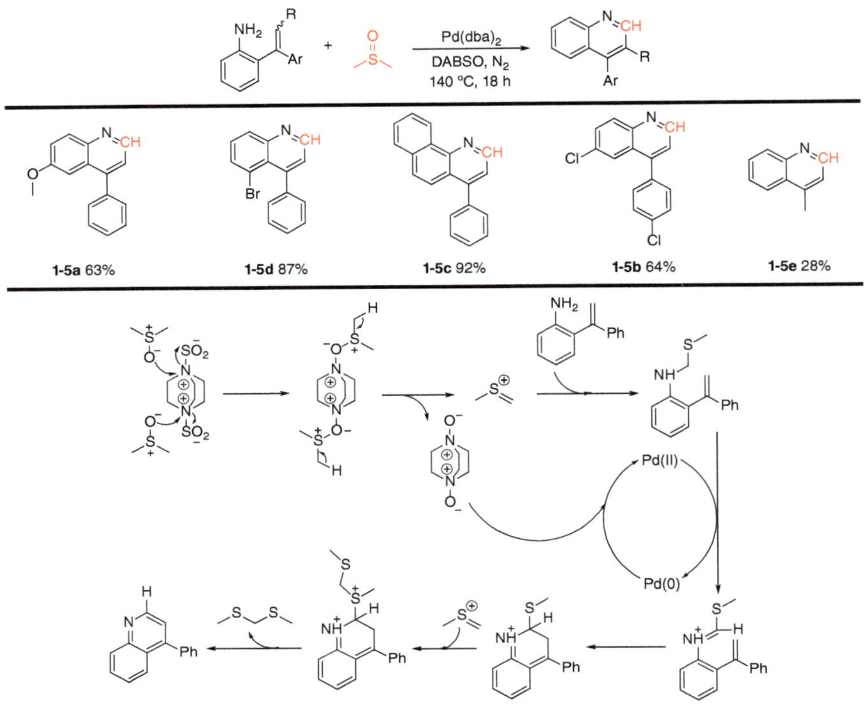

Scheme 4.5

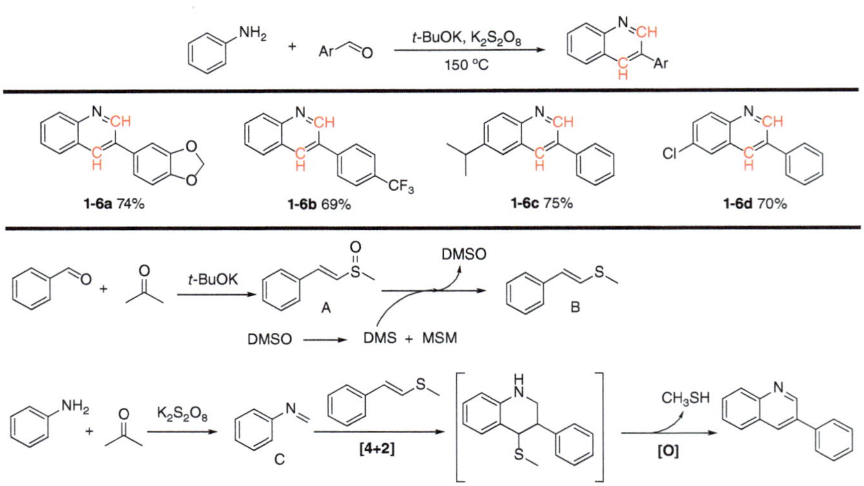

Scheme 4.6

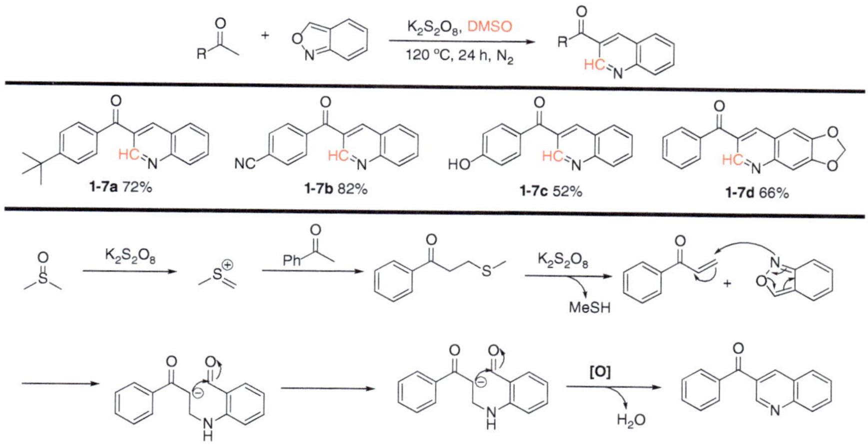

Scheme 4.7

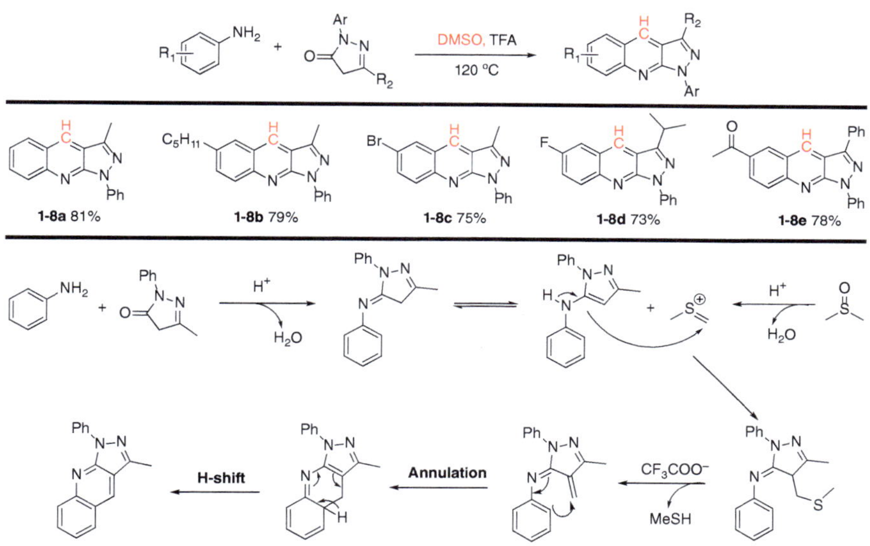

Scheme 4.8

Recently, Wu's group has consecutively reported two cases of quinoline synthesis using amino acids as the starting material and DMSO as the carbon unit. Interestingly, in these two reactions, DMSO was involved in the 2 and 4 positions of the quinoline as the C1 unit through a unified Pummerer reaction activation pattern. In those two reactions, aniline, DMSO, and HI were employed, but phenylglycine (Scheme 4.9) [16] and phenylalanine (Scheme 4.10) [17] (or aspartates) were used, respectively. The DMSO was activated by HI to generate the thionium ion intermediate, while HI could also be converted into I_2 by DMSO. I_2 further triggers the catabolism and reconstruction reaction sequence of amino acid to transform

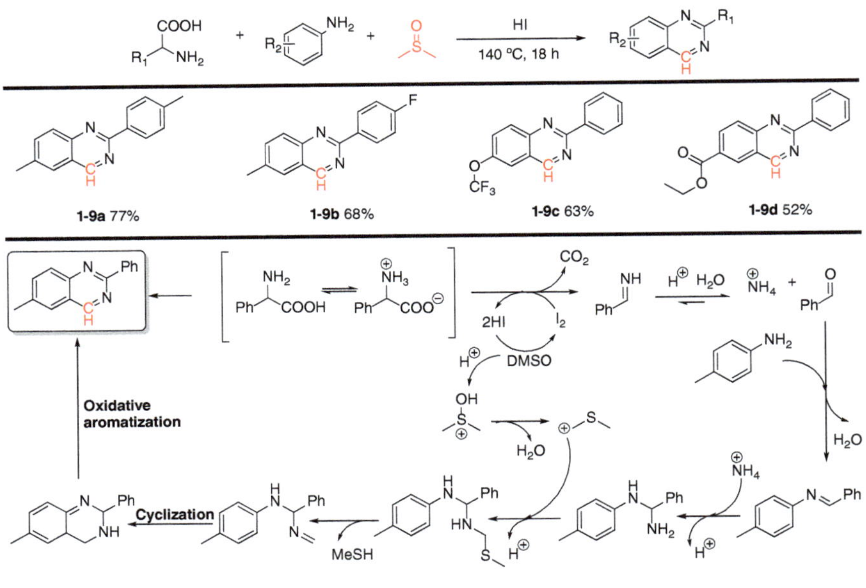

Scheme 4.9

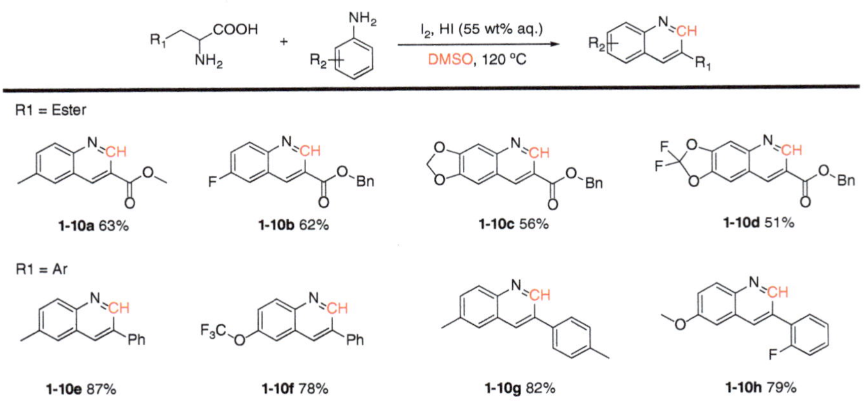

Scheme 4.10

this biomass material into corresponding aldehyde species based on Strecker degradation.

Ma group reported a three-component reaction for the synthesis of 3-aroylimidazo [1,2-*a*]-*N*-heterocycles from aryl ketones and 2-amino-*N*-heterocycles in the presence of I_2, $K_2S_2O_8$, and NaOAc in a DMSO solution [18]. This five-membered ring formation is compatible with a range of heterocyclic substrates, providing access to a diverse 3-arylimidazole backbone. Again, DMSO was activated by $K_2S_2O_8$ to generate a sulfenium ion intermediate to react with a nucleophilic amino group to produce corresponding positive imine waiting for the attack of other nucleophilic reagents to further build the linkage (Scheme 4.11).

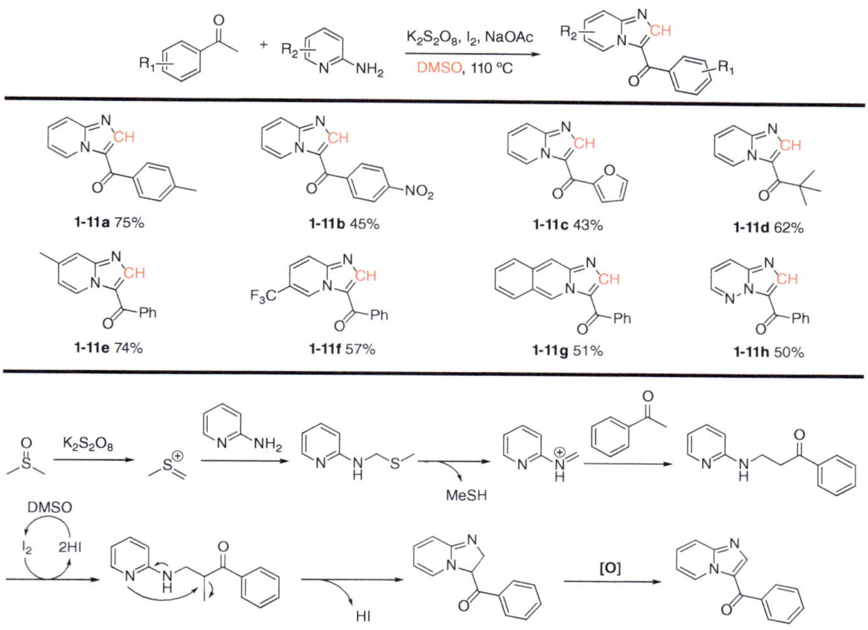

Scheme 4.11

In 2020, Liang's group developed a three-component reaction that provided the facile synthesis of 2-unsubstituted benzothiazoles from *o*-iodoanilines, K_2S, and DMSO [19]. Notably, DMSO plays three vital roles in this transformation, i.e. as a carbon source, solvent, and oxidant. NH_4OAc was chosen as the optimal promoter to activate DMSO, forming a thionium ion based on a Pummerer-type mechanism. Interestingly, when K_2S is not added and under otherwise identical conditions, the product is transformed into 2-unsubstituted benzimidazoles, with the second nitrogen atom in this molecule being derived from ammonium acetate (Scheme 4.12).

In 2021, Wu and Shu et al. developed a method for synthesizing polyfunctional furans using DMSO as the one-carbon source [20]. This approach was similar to the one previously described in that it starts from sulfonium salts and DMSO as the solvent, 2,3,5-trisubstituted furans with good functional group compatibility and moderate-to-good yields that can be easily obtained. Mechanistic studies show that DMSO is involved in the carbon construction at the 4-position of the furan ring system through a Pummerer-type fragmentation (Scheme 4.13).

In addition to the synthesis of heteroaromatic rings, DMSO is also competent for the synthesis of other non-aromatized ring systems through the activation mode of the Pummerer-type reaction. In 2021, Zhang's group reported a cobalt(II) catalytic divergent synthesis of cyclopentenes and cyclohexenes from diketones and alkenes [21]. In this transformation, DMSO was used as the solvent and carbon source. For the construction of cyclopentenes, a single molecule of DMSO was involved as the C1 unit, while in the formation of cyclohexenes, two molecules of DMSO were involved, providing two induvial C1 units. The use of different substrates, such as styrene or

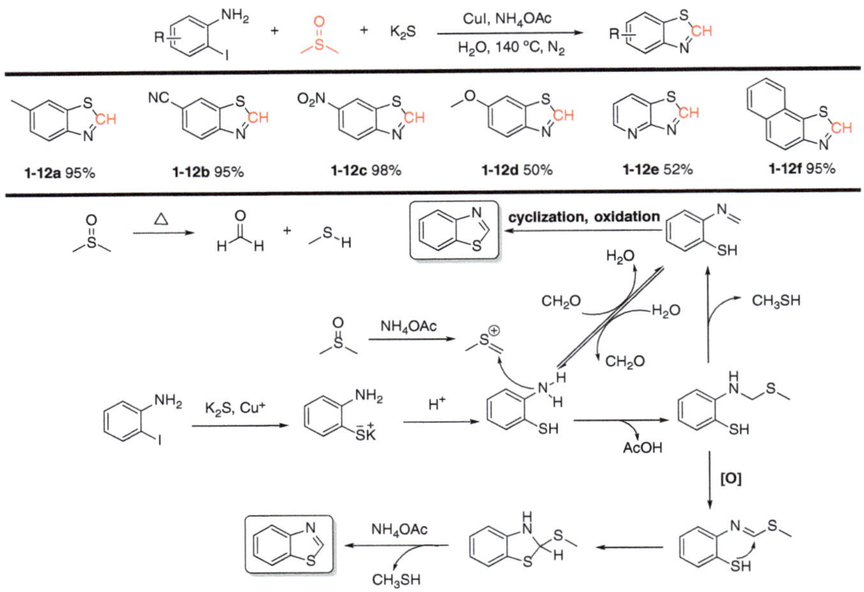

Scheme 4.12

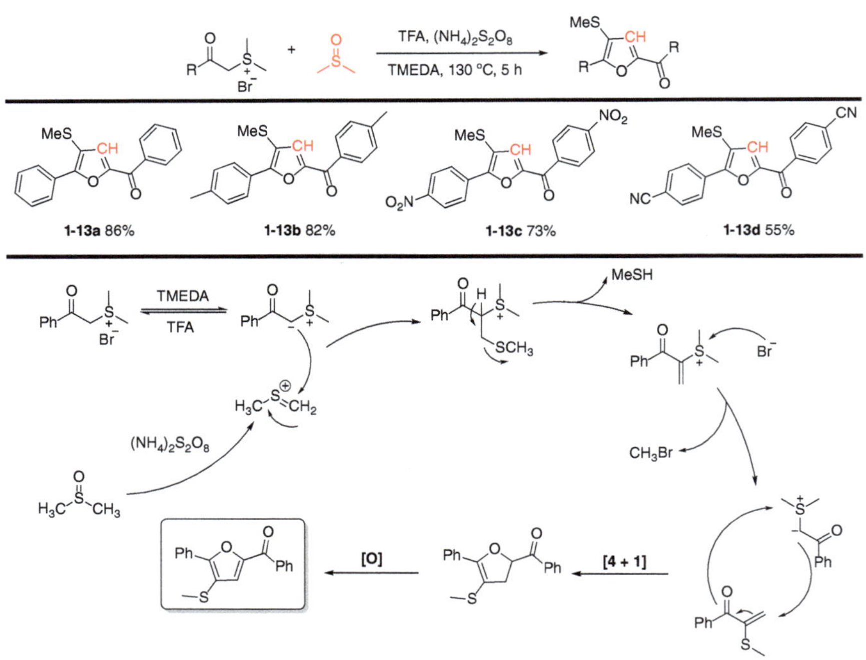

Scheme 4.13

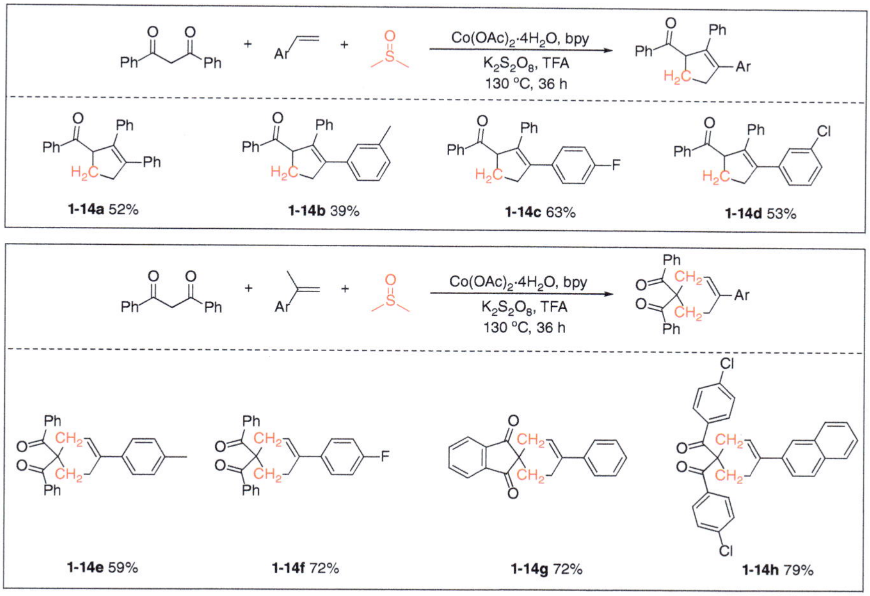

Scheme 4.14

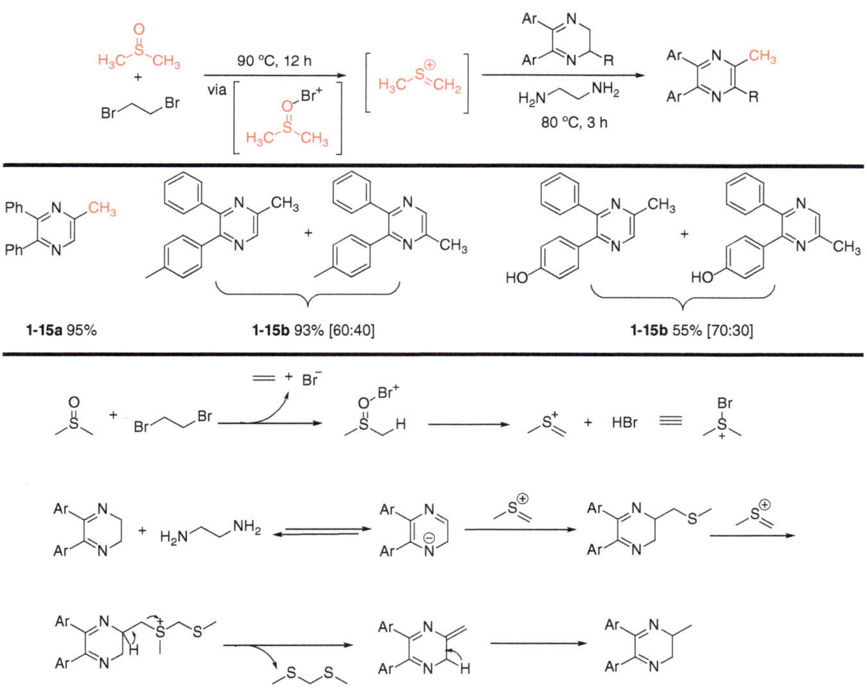

Scheme 4.15

α-methyl styrene, is the key leading to different reaction outcomes. Two different mechanisms have been proposed by the authors to explain the divergent processes of the reactions; however, these uniformly undergo activation of DMSO by $K_2S_2O_8$ in a Pummerer reaction mode, resulting in the formation of an electrophilic thionium ion. This intermediate can be captured by the nucleophilic reagent, a dicarbonyl compound in the reaction, resulting in the formation of the corresponding ketene species (Scheme 4.14).

Panda's group realized the α-C–H methylation of pyrazines and pyrazinones using DMSO as a nonconventional methylating agent. This reaction took place in the DMSO solution with the addition of 1,2-dibromoethane as the initiator [22]. Mechanistically, DMSO reacts with 1,2-dibromoethane by the elimination of ethylene to generate bromine cation pools; this will undergo Pummerer-type fragmentation to yield methyl(methylene)sulfonium ion intermediate. This electrophilic species reacts with the nucleophilic 2 or 3-position of the substrate and results in the formation of the corresponding olefin through a fragmentation process. Finally, the intramolecular isomerization and aromatization process results in the formation of the target methylation product (Scheme 4.15).

4.2 Selectfluor-enabled Activation of DMSO

Discovered in 1980, 1-chloromethyl-4-fluoro1,4-diazoniabicyclo[2.2.2]octane bis (tetrafluoroborate) (Selectfluor, F-TEDA-BF_4) has become a commonly used electrophilic fluorine source and an efficient single-electron oxidant [23]. Recently, it has been frequently reported that the interaction between Selectfluor and DMSO leads to a special activation pattern of DMSO to transform DMSO into an excellent electrophilic reagent. For example, Wan's group disclosed a three-component reaction between enaminones, hydrazines, and DMSO (as solvent) for the direct synthesis of 1,4-disubstituted pyrazoles [24] with the assistance of I_2. This transformation represents the first example of using DMSO as the source of C1 for the formation of the pyrazole ring. Control experiments indicated that the methyl group of DMSO could be fluorinated by Selectfluor to form an electrophilic reagent, which further participates in the domino cyclization. By losing HSOMe, the fluorinated methyl group turns into a C1 unit to join in the formation of a five-membered ring (Scheme 4.16).

Similarly, Qiu and Peng group in 2023 reported a reaction between the aryl methyl ketones with DMSO in the presence of Selectfluor [25]. This divergent transformation provides a convenient synthesis of 2,5-diacylthiophenes and β-acyl allylic methylsulfones, respectively, in which the chemoselectivity was manipulated by changing the solvent from DMSO to toluene. Mechanically, the generation of both 2,5-diacylthiophenes and β-acyl allylic methylsulfones is based on the sulfoxide intermediate, which is a product of the reaction of aryl methyl ketones with Selectfluor-activated DMSO. In addition, during the formation of the coupling product, β-acyl allylic methylsulfones, the sulfoxide intermediate is further converted into a more stable sulfone intermediate by a Selectfluor-mediated (SCE = 0.33 V) single-electron oxidation (Scheme 4.17).

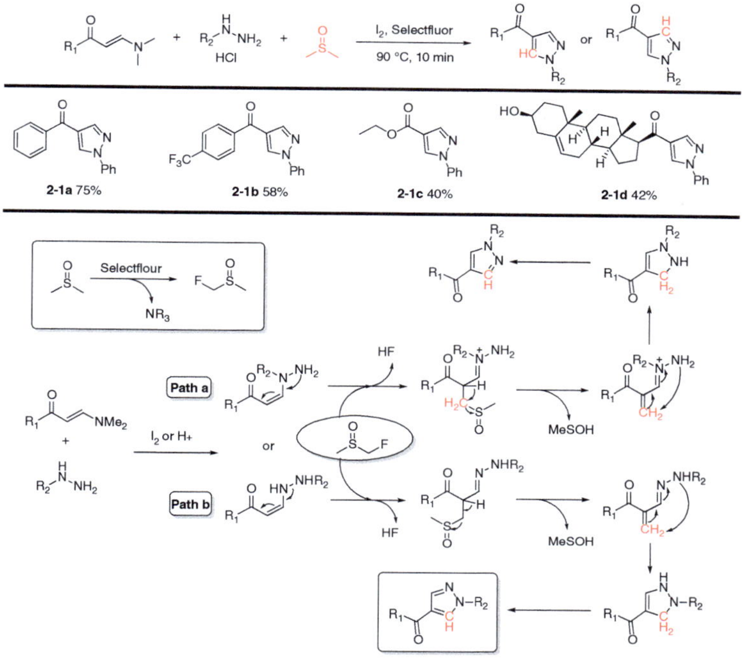

Scheme 4.16

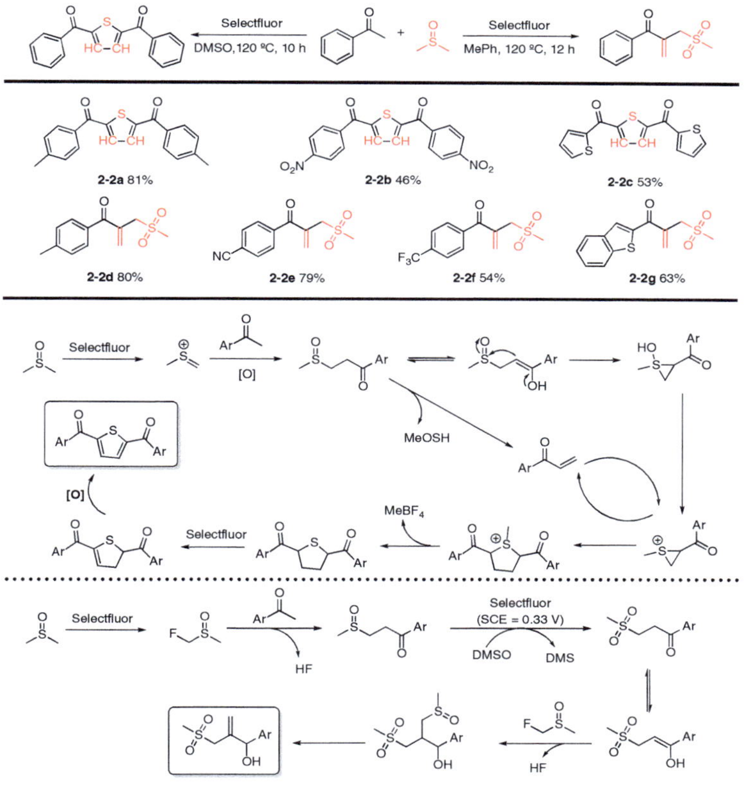

Scheme 4.17

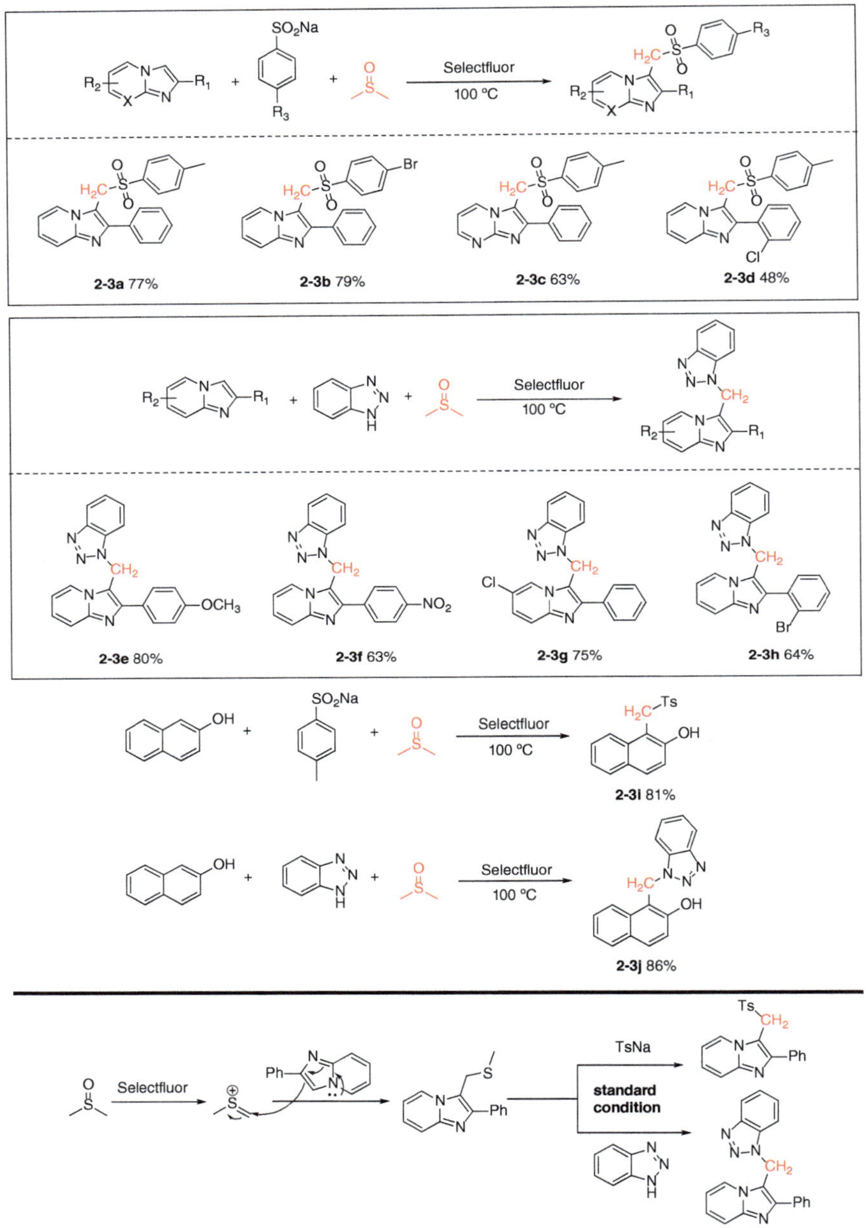

Selectfluor
100 °C
2-3a 77%
2-3b 79%
2-3c 63%
2-3d 48%
2-3e 80%
2-3f 63%
2-3g 75%
2-3h 64%
2-3i 81%
2-3j 86%
TsNa
standard condition

Scheme 4.18

Alternatively, the fluorination of the methyl group in DMSO by Selectfluor was not the only way to the reaction between these two reagents. In a specific case, the F^+ species generated from the fragmentation of Selectfluor could activate the carbonyl group of DMSO, which leads to the formation of methyl(methylene)sulfonium cation. This highly active intermediate could lose an HOF to produce a thionium ion intermediate, which awaits subsequential nucleophilic addition. H. B. Rode employed this Selectfluor-mediated Pummerer-type reaction in their synthesis of methylene-tethered arylsulfonation and benzotriazolation of imidazopyridines from imidazopyridine and benzotriazole, respectively, where DMSO acts as the sole solvent and a one-carbon synthon (Scheme 4.18) [26].

4.3 Activation of DMSO Enabled by Single-electron Transformation

It is frequent to see the fragmentation of DMSO via a two-electron reaction process, as discussed previously. However, single-electron chemistry is also allowed in DMSO-based transformations, and such reaction pattern has rapidly emerged in recent years. In such transformations, these can be broadly classified into two categories according to the role of DMSO: (i) DMSO itself provides the corresponding radical reagent by fragmentation through interaction with other radical initiators. (ii) DMSO acts as a single-electron oxidant to complete the catalytic cycle by oxidizing the metal catalyst in the reaction.

In 2017, Lei's group reported the first example in which DMSO could act as an oxidant to promote the oxidation of Cu(I) to Cu(II) [27]. The oxidation potential for DMSO is between +1.56 and +1.76 V (vs SCE) [28]. The interaction between DMSO and Cu(I) was also detected by electron paramagnetic resonance (EPR). In addition, *in situ* X-ray absorption spectra (XAS) experiments were conducted to monitor the oxidation process. Based on such experimental observations, they developed a copper-catalyzed oxidative cyclization of aryl ketones with styrenes to form furans. Using 15 mol% of $CuCl_2$ and DMSO as the solvent, a wide range of styrene and methyl-substituted aryl ketones are tolerated in this transformation showing good reaction efficiency. Mechanistic studies indicated that the reaction was triggered by Cu(II)-based single-electron transfer (SET) of aryl ketone to form a carbon-centered radical, which would react with styrene via radical addition. The resulting radical species would further attack the carbonyl oxygen, while the final product furans could be generated by hydrogen radical elimination and aromatization sequence (Scheme 4.19).

In 2019, Loh et al. developed a tertiary alkylation reaction of olefins with 1,3-dicarbonyl compounds [29]. Such interesting radical addition of active methylene radicals to the styrenes was also initiated by Cu(II) catalyst, and the catalytic cycles were completed by DMSO. Notably, in the different reaction atmospheres,

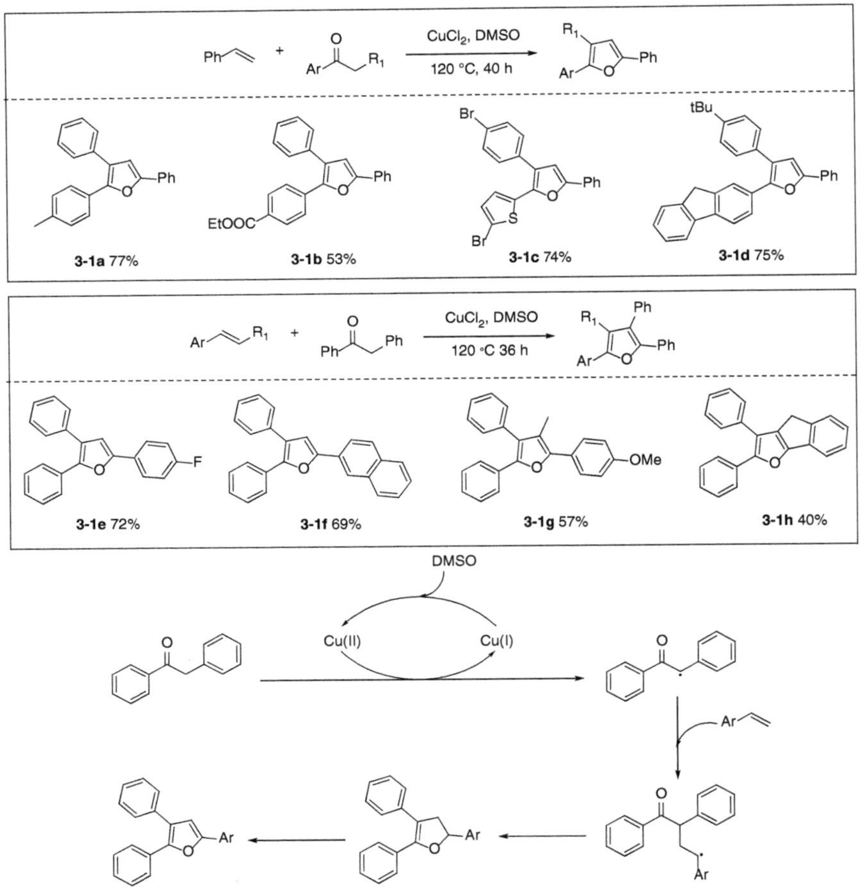

Scheme 4.19

such as nitrogen and air, the reaction outcomes switched from alkylation to alkylation–oxygenation products. A series of Cu(II) salts were capable of promoting the transformation, such as $Cu(OTf)_2$, $CuCl_2$, $Cu(OAc)_2$, and $CuBr_2$, and the last two were selected as the optimal catalysts for the synthesis of alkylation to alkylation–oxygenation products, respectively. These two reactions were triggered by the addition of *in situ* generated α-carbonyl alkyl radical species to styrene, to generate benzyl radical intermediate. This radical could be further oxidized into carbon cations by Cu(II) or directly trapped by O_2 to generate divergent products. Performing as both a single-electron oxidant and solvent, DMSO oxidized Cu(I) to Cu(II) (Scheme 4.20).

In the same year, Bhadra et al. developed a copper-catalyzed thiolation of 2-alkylbenzoxazoles and benzothiazoles using DMSO as both a twofold oxidant

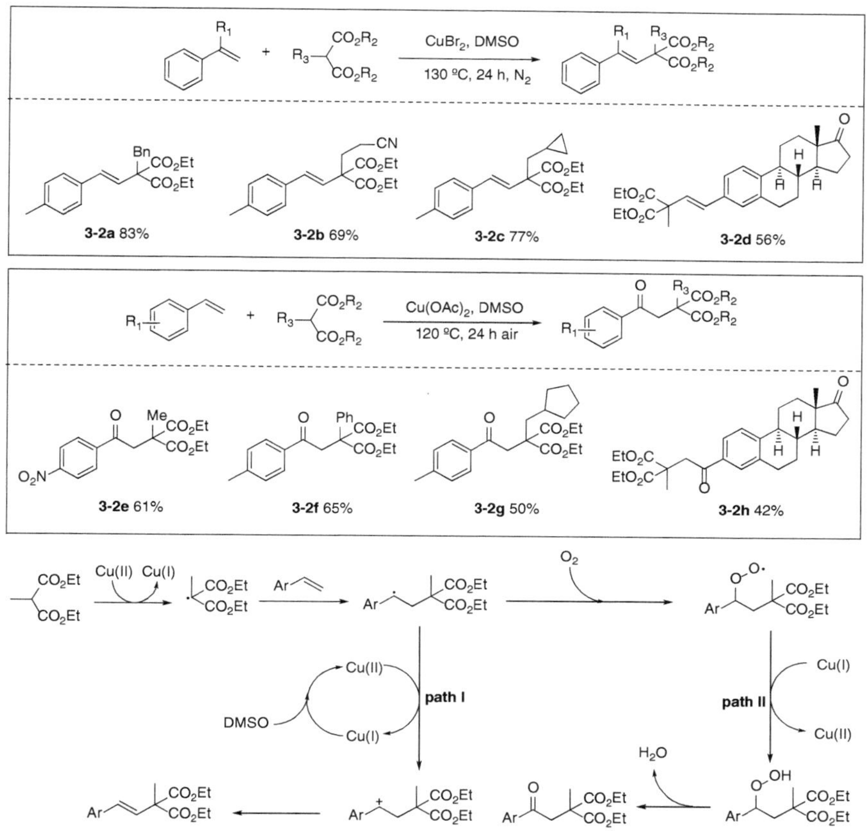

Scheme 4.20

and solvent [30]. In addition, C(sp^3)–H phenylselenylation, which is otherwise challenging to achieve was also realized by such means using Ph_2Se_2 as the donor. These products could act as key precursors for the synthesis of biologically pertinent molecules. Mechanically, DMSO promotes the generation of ArSSAr from ArSH, where the former acts as the radical initiated through fragmentation to form ArS$^{\cdot}$ species. On the other hand, DSMO also enabled the catalytic cycling of Cu(I) to Cu(II), where Cu(II) species could oxidize the benzylidene position in the substrate into the corresponding radical. A radical recombination of ArS$^{\cdot}$ and the benzyl radical leads to the formation of the coupling products (Scheme 4.21).

Besides [Cu] salt, DMSO could also oxidize other metal catalysts through single-electron oxidation, such as [Co] complex. Very recently, Abbasi et al. reported the synthesis of thioamides from benzyl thiols and amines in the presence of cobalt phthalocyanine catalyst and DMSO [31]. Primary, secondary, and heterocyclic

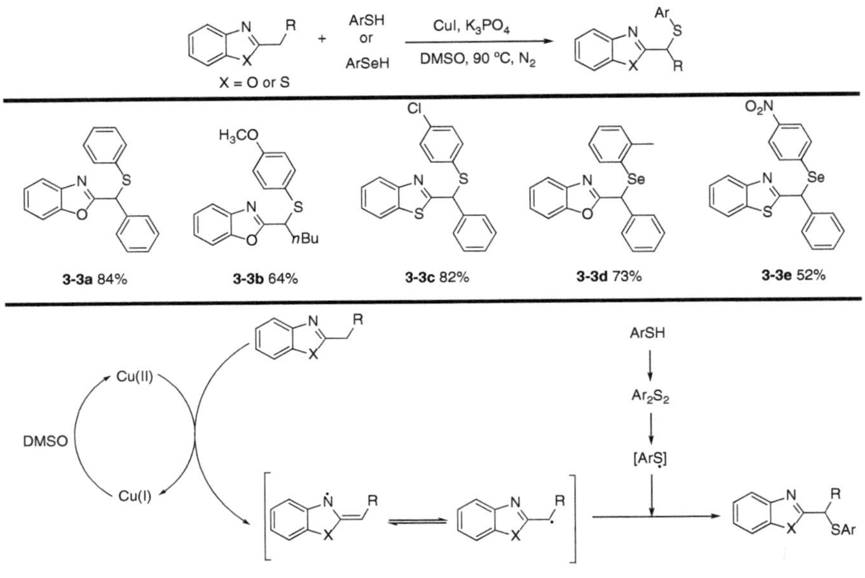

Scheme 4.21

amines are allowed to yield the corresponding benzothioamides in high yields. Notably, reaction solvents apart from DMSO, such as CH_3CN, $CHCl_3$, H_2O, DMF, EtOH, THF, and toluene, have also been tried in the optimizing process, but showed completely neglective results to conversion. The role of DMSO was illustrated by oxidizing Co(II) to Cu(III) to complete the catalytic cycle with the assistance of the amine, where Co(III) is responsible for the oxidative coupling of $ArCH_2S^-$ to $ArCH_2SSCH_2Ar$ (Scheme 4.22).

In 2017, Wen et al. established the cross-coupling reactions between methyl ketones and DMSO to yield β-acyl allylic methylsulfides and sulfones [32]. DMSO was used as the co-solvent together with acetic acid. When TsOH·H_2O and NaOAc were employed as additives, β-acyl allylic methylsulfides could be generated from acetophenone substrates. However, without adding NaOAc, the product switched to keto allylic methylsulfones. The reaction mechanism of the latter transformation was illustrated as follows: The reaction of DMSO with acetic acid in the presence of TsOH yields methylthiosulfonate according to Freeman's impactful work [33]. Then, methyl sulfonyl radical could be generated through the homolysis of the S—S bond in methylthiosulfonate, which would attack the double bond in the *in situ* generated β-acyl allylic methylsulfides. By losing a methyl mercaptan radical, keto allylic methylsulfones could be formed via this radical chain reaction (Scheme 4.23).

The interaction between DMSO and persulfides, such as $K_2S_2O_8$ triggers a single-electron oxidation of DMSO to generate α-sulfinyl radicals, which are electrophilic radical species. Hajra's group developed C3 functionalization of 2H-indazoles [34]. DMSO was used as a co-solvent with acetic acid while acting as

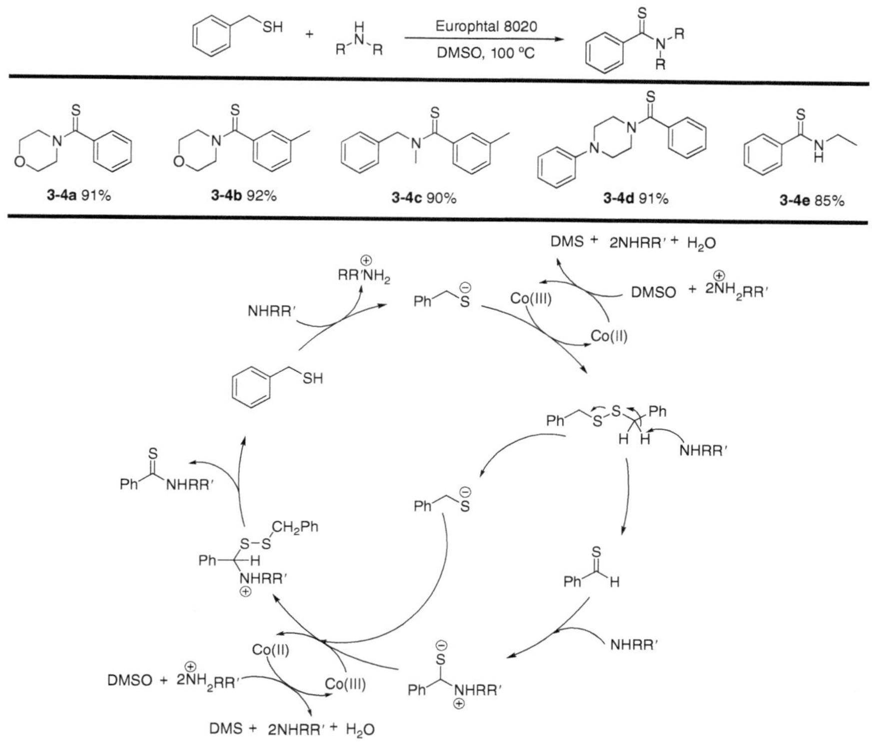

Scheme 4.22

either a formylating agent or methylene units according to the reaction incomes, respectively. A plausible mechanistic pathway was also suggested by the author to illustrate such processes. First, the thermal decomposition of persulfate leads to the formation of an SO_4 radical anion, which could further abstract an H radical from DMSO to form an α-sulfinyl radical intermediate. This would attack the nucleophilic C3 position of *2H*-indazoles, which result in the formation of a methyl radical species through fragmentation. This highly reactive species could either be tapped by O_2 to afford an aldehyde group. Alternatively, the methyl radical species could be trapped by the RCOO radical, which is generated by the reaction of $SO_4^{\cdot}$ radical anion with RCOOH, to yield the desired alkylated products (Scheme 4.24).

In SET reactions, DMSO is often used as a methyl radical precursor in synthetic design. In 2016, Ma's group reported a copper-catalyzed three-component reaction for the synthesis of quinazoline derivatives from 2-aminophenyl, NH_4OAc, and DMSO, in which DMSO acts as both the solvent and the C1 unit. 2-aryl and 2-alkyl quinazolines bearing a range of functional groups such as halogens, thiomers, amines, and so on can be obtained in moderate-to-good yields. The generation of a methyl radical from DMSO in the presence of oxygen and copper

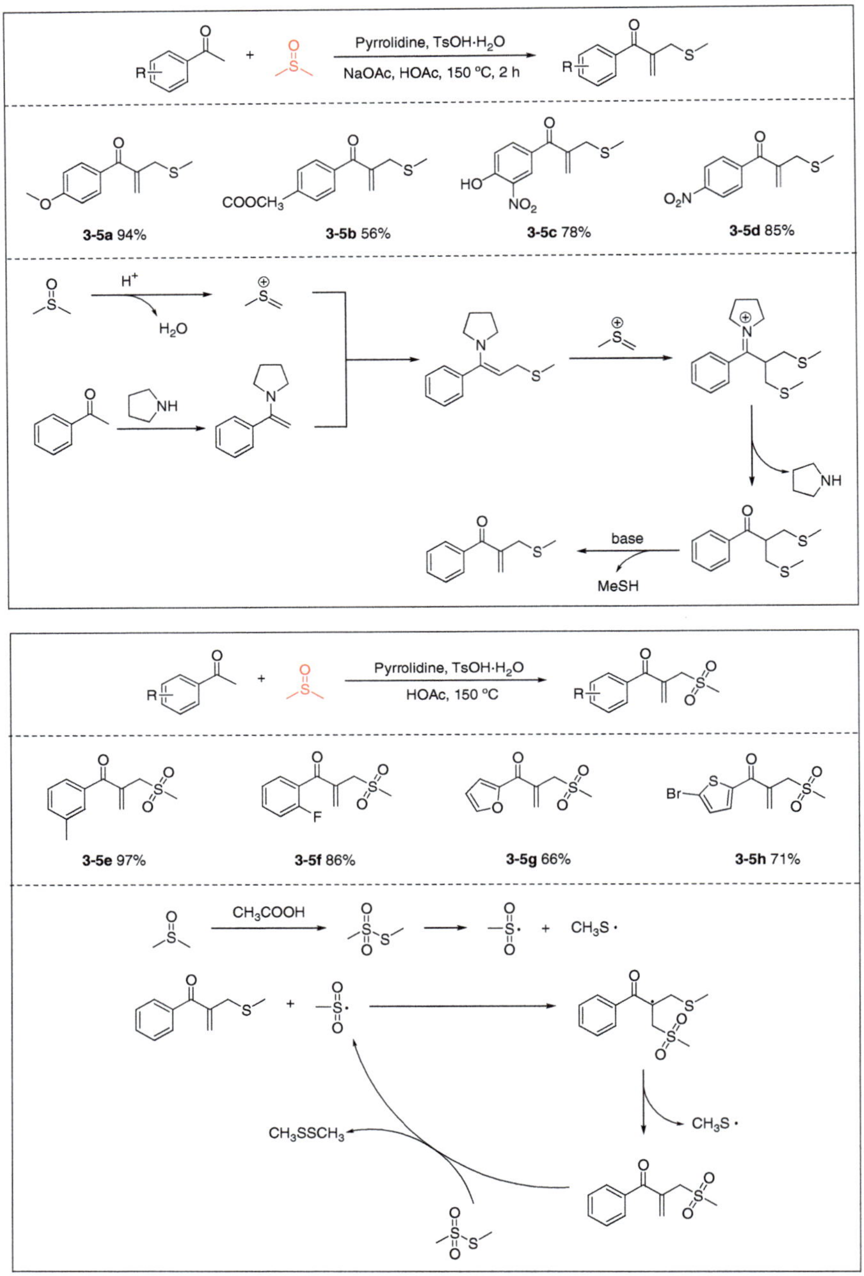
Pyrrolidine, TsOH·H2O
NaOAc, HOAc, 150 °C, 2 h
3-5a 94%
3-5b 56%
3-5c 78%
3-5d 85%
H+
H2O
base
MeSH
Pyrrolidine, TsOH·H2O
HOAc, 150 °C
3-5e 97%
3-5f 86%
3-5g 66%
3-5h 71%
CH3COOH
CH3S·
CH3SSCH3

Scheme 4.23

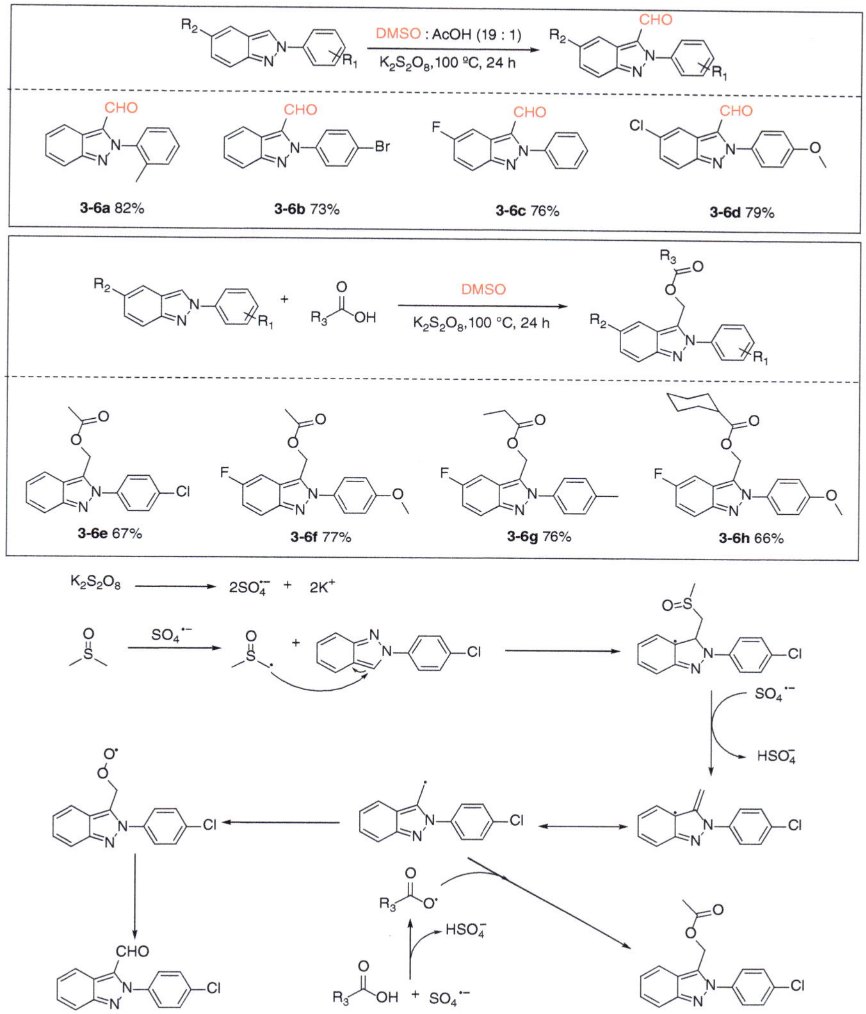

Scheme 4.24

salts via a SET process was illustrated as the initiation step of this transformation. Subsequently, the methyl radical transforms into a peroxy radical by trapping O_2 and was then converted into formaldehyde as the C1 unit. Alternatively, this C1 unit could also be obtained by heating DMSO at high temperatures (Scheme 4.25) [35].

Wang et al. realized a methylation reaction of arylacrylamides to produce an indolin skeleton using DMSO as both the solvent and the methyl donor [36] This transformation took place under Blue LEDs irradiation conditions with the assistance of an iron catalyst and H_2O_2 as the oxidant. Sunlight could also promote the reaction but in relatively decreased yield, while carrying the reaction in the dark

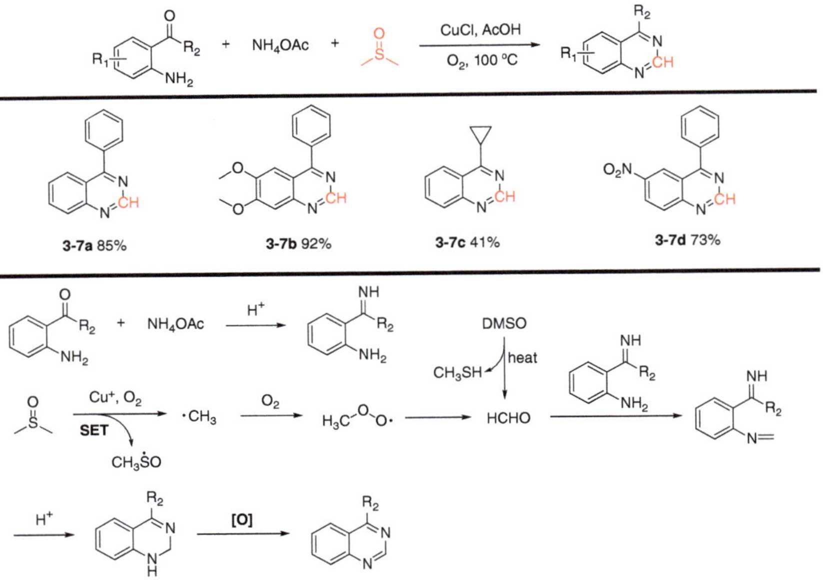

Scheme 4.25

would totally shut down the reaction. Mechanistically, hydroxyl radicals, which are generated through the oxidation of Fe(II) by hydrogen peroxide, can add to the S=O double bond of DMSO, which then provides methyl radicals through fragmentation. The radical addition of the methyl radical to the olefin of arylacrylamide leads to the formation of aryl radicals, which can further yield the target product through a SET transfer process with the superoxide radical (Scheme 4.26).

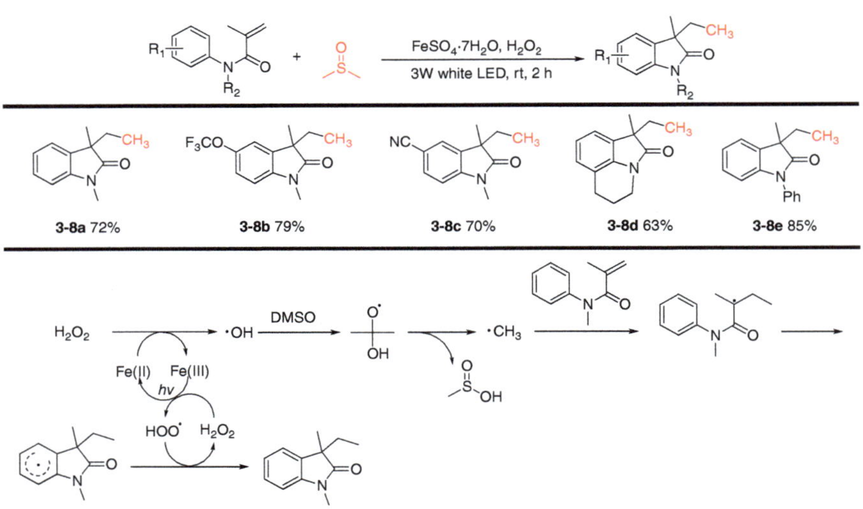

Scheme 4.26

Another methylation reaction using DMSO as the C1 doner was reported in 2019 by Chen et al. in their study on the metal-free α-methylation of 1,8-naphthyridine derivatives [37]. In this transformation, the conversion of the substrate is less affected by the electrical properties or the substituents. Various 1,8-naphthyridines bearing electron-withdrawing or electron-donating groups were tolerated in such methylation. However, other N-heterocycles such as pyridine, isoquinoline, and indole are not suitable substrates in the transformation. Control experiments indicated that the methyl radical was generated from DMSO via a light and base-promoted SET process. A Minisci-type reaction, involving the radical addition of methyl radical to the N-heterocycles forms the target products (Scheme 4.27).

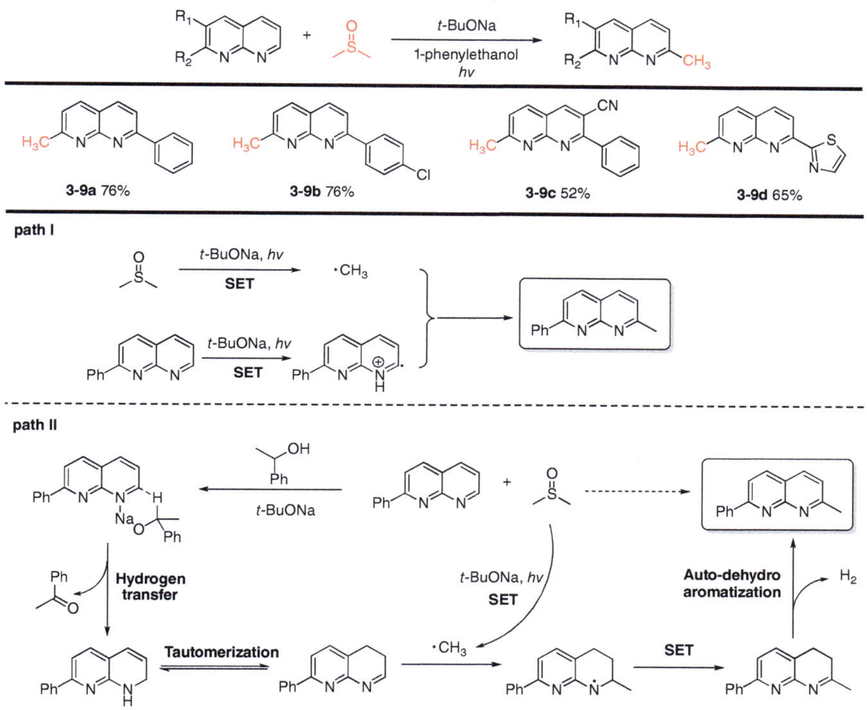

Scheme 4.27

In the presence of copper(II) salts and oxygen, DMSO not only provides the methyl group but also this methylation product can be further oxidized to yield the corresponding aldehyde product in a domino sequence. For example, Patel et al. reported a Pd(II)/Cu(II) catalyzed synthesis of tricyanovinylindoles from *o*-alkynylanilines, malononitrile, and NH_4SCN in a DMSO solution [38] The Pd(II) catalyst is mainly responsible for the conversion of *o*-alkynylanilines into the indole skeleton while Cu(II) is essential for the SET process that activates DMSO into a formylating agent (Scheme 4.28).

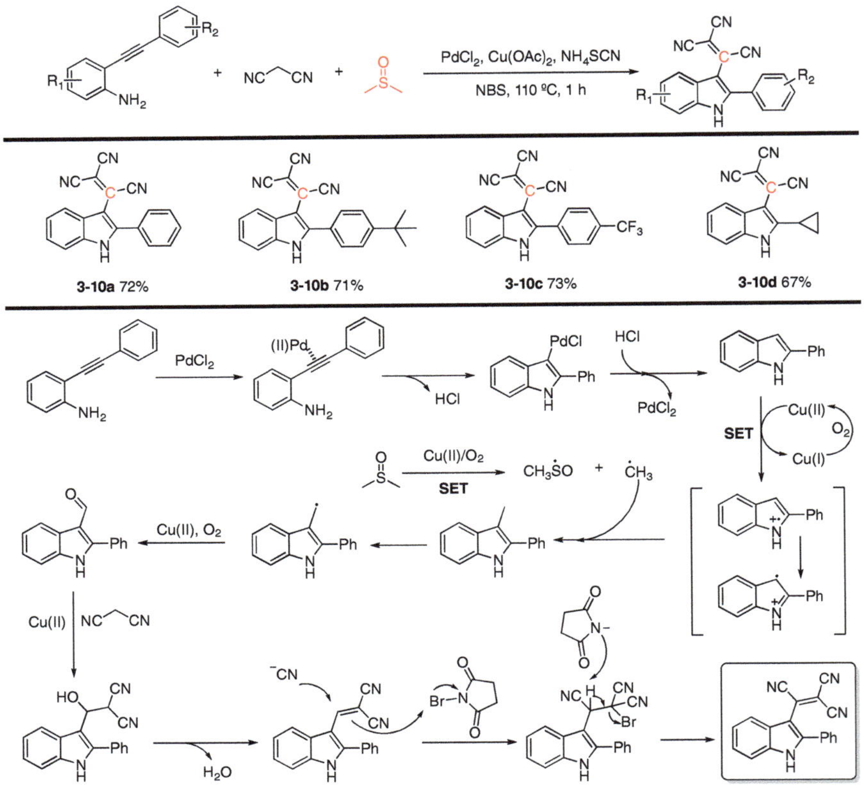

Scheme 4.28

Guo et al have developed a direct alkylation or arylation of alkenes using sulfoxides as alkyl or aryl reagents [39]. This effective transformation allows the reaction between styrene and DMSO to form (*E*)-prop-1-en-1-ylbenzene in DMSO solution with the assistance of $FeCl_3{\cdot}6H_2O$ and H_2O_2. This methylation reaction can be extended to ethylation, benzylation, or even arylation by using different sulfoxide substrates. Mechanistically, hydroxide radicals are obtained by the oxidation of hydrogen peroxide in the presence of Fe(III). Then, the hydroxide radical acts on the DMSO to form the corresponding decomposition product, methyl radical, together with CH_3SO_2H. The methyl radical attacked the terminal alkenes through anti-Markovnikov addition since the resulting benzyl radicals were relatively more stable. Finally, the benzyl-free radical intermediate reacts with hydroxide radical via radical recombination reaction and then undergoes elimination to give the alkene product. This example shows the ability of DMSO to react as a methyl radical to realize the methylation reaction (Scheme 4.29).

In addition to methylation reagents, DMSO can also play other roles in SET-type reactions. In 2020, Li et al. reported a direct hydroxylation of heterobenzylic methylenes to secondary alcohols using DMSO as the –OH donor [40]. Using

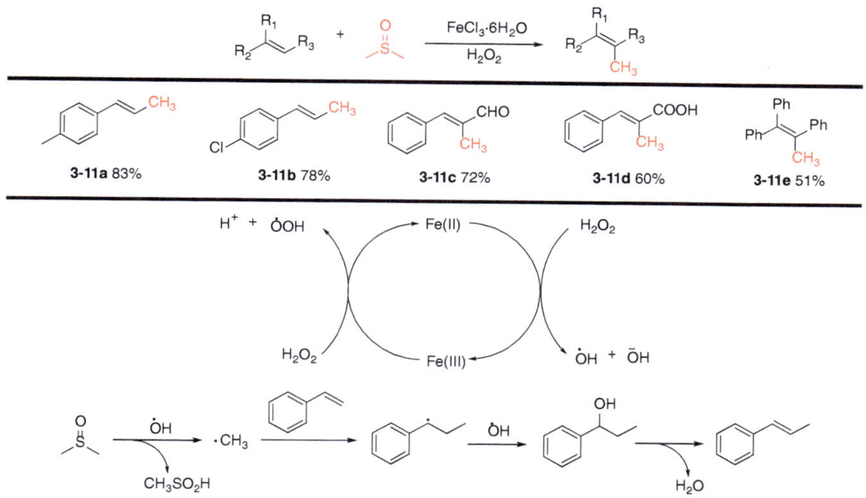

Scheme 4.29

t-BuOK/DMSO/air reaction system, a variety of heterobenzylic compounds bearing isoquinoline, quinoline, pyrazine, and pyrimidine units could convert into corresponding secondary alcohols with high chemoselectivity. No overoxidation to ketone products was observed in each of the experimental cases. In addition, α-deuterated secondary alcohols could be obtained in high incorporation by simply using DMSO-d_6 as the solvent. A mechanism is proposed by the authors. The DMSO-based methyl radical was initially generated by the SET of *t*-BuOK. The hydrogen atom in the benzyl position will be by this DMSO-based methyl radical to form the corresponding benzyl radical intermediate, which could be further trapped by O_2. The resulting hydroperoxide was reduced by DMSO to yield the hydroxylation product together with dimethyl sulfone (Scheme 4.30).

Another way in which DMSO is involved in SET is by generating hydroxyl-attached sulfur radical. In 2022, Liang and coworkers reported the conversion of thioacids to carboxylic acids with DMSO as the oxygen source [41]. Both the aryl and alkyl thioacids bearing functional groups or sensitive groups such as halogens, ternary rings, and tetracycles could be tolerated in this transformation giving carboxylic acids in high yield. This atomic substitution process was enabled by using DMSO as the solvent via an excited electron donor–acceptor (EDA) complex as the key intermediate and was promoted by visible-light (Blue LEDs), involving the noncovalent interaction between thiobenzoic acid and DMSO. Through fragmentation, the disulfide intermediate, and the hydroxyl-attached sulfur radical could be generated from the EDA complex. The hydroxyl-attached sulfur radical could further transform into a hydroxyl radical. The oxygen atom in the carboxylic acid product was derived from this hydroxyl radical through the elimination of elemental sulfur (Scheme 4.31).

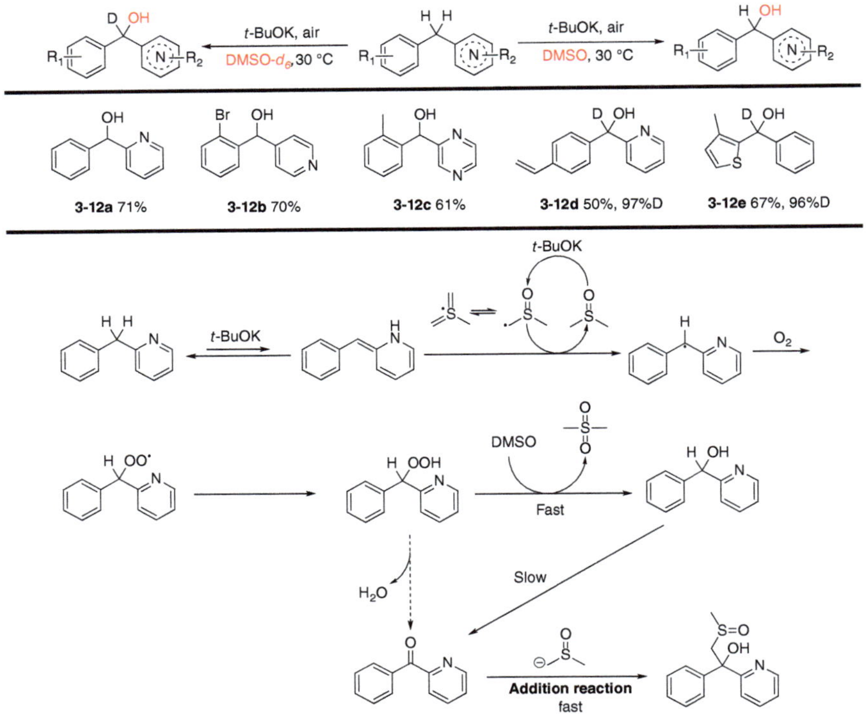

Scheme 4.30

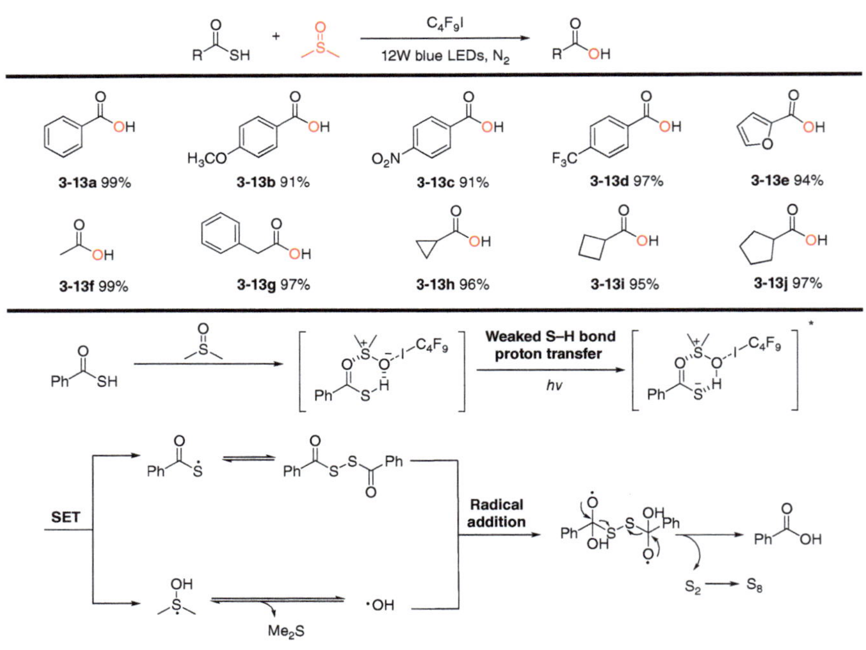

Scheme 4.31

4.4 Electrocatalytic Synthesis Enabled Activation of DMSO

In 2018, Yoshida's group reported an alkenyl C–H/aromatic C–H cross-coupling reaction using terminal or internal olefins and aromatic nucleophiles as substrates [42]. This one-pot, three-step protocol combines the following steps: (i) Electrochemical oxidation of I_2 and sulfoxides (DMSO) first form the iodosulfonium ion intermediates. Then these intermediates react with vinylarenes to generate *β*-iodoalkoxysulfonium ions. (ii) By adding the aromatic nucleophiles after 30 minutes of the reaction, 1,1-diaryl-2-iodoethane was generated as an isolatable intermediate. (iii) After adding DBU in the third stage, 1,1-diaryl-2-iodoethane converts into the final product. Various sulfoxides were screened, and diphenyl sulfoxide was chosen as the optimal one (Scheme 4.32).

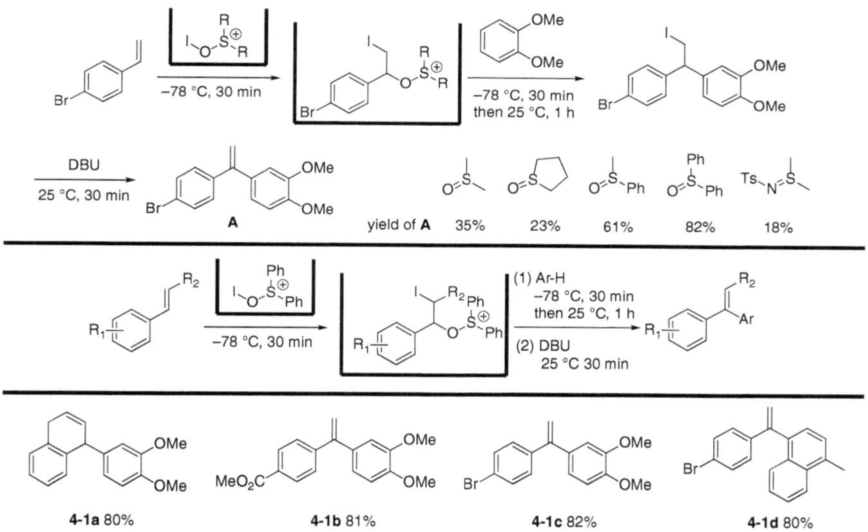

Scheme 4.32

An amination reaction of DMSO with amines was developed for the synthesis of sulfonamides whose motifs hold the potential for a variety of important activities such as anticonvulsants, HIV protease inhibitors, antiviral agents, antibacterials, anti-inflammatory substances, and so on. This electrochemical oxidization reaction used DMSO as the sole solvent and was performed under an undivided cell with Pt and Ni plates in the presence of KI. Different aliphatic and aromatic amines are suitable in this transformation, resulting in the formation of the corresponding sulfonamides in moderate-to-high yields. Through the oxidation at the anode, DMSO is converted into an S-center radical cation, which can react with another molecule of DMSO to yield a dimethyl sulfoxonium cation and a dimethyl sulfoxide radical, respectively. After fragmentation, CH_3SH is formed from dimethyl sulfoxonium cation and reacts with iodine radical to form a corresponding MeS· radical. In the subsequent process, *N*-benzyl and *S*-dimethylthiohydroxylamine intermediates are obtained via the radical recombination of nitrogen radicals with methylthio radicals. DMSO is responsible for the oxidation of these intermediates in a twofold pathway, giving the final product, sulfonamides, by providing two oxygen atoms. In 2021, Yuan et al. reported this work that represents an interesting example of using anodic oxidation to oxidize DMSO and thus activating its participation in the synthetic transformation (Scheme 4.33) [43].

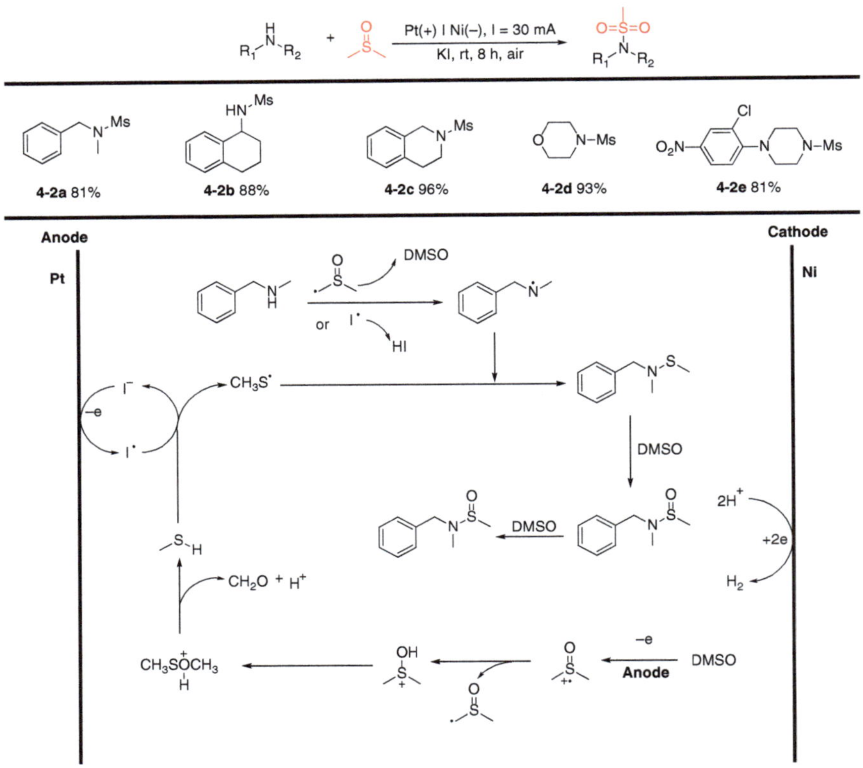

Scheme 4.33

4.5 Photocatalytic Reaction Enabled Activation of DMSO

In the past decade, photocatalytic reactions have developed rapidly in the field of organic synthesis. Photocatalytic reactions have a wide choice of solvents due to their relatively mild reaction conditions that can be carried out effectively at room temperature. DMSO is also a compatible solvent for photocatalytic reactions and often provides some expected or unexpected activation patterns to participate in the reaction.

One of the most common ways in which DMSO is involved in these reactions is by acting as an oxygen donor. Specifically, the oxygen atom in DMSO is fully capable of attacking the carbon cation intermediate that is generated *in situ* in the photocatalytic process, building the C=O bond through a reaction pattern similar to that of Kornblum oxidation by losing SMe_2.

In 2018, Glorius's group reported the synthesis of α-alkyl-acetophenones from styrene and *N*-(acyloxy)phthalimides in a DMSO solution (0.1 M) under Blue LED irradiation conditions [44]. This method has a tolerance for a wide range of commercially available electron-donating and electron-withdrawing groups. For the alkyl radical source, *N*-(acyloxy)phthalimides bearing thioanisole, anisole, veratrol, ketone, fluorine, and Boc-protected piperidine units were efficient substrates in this transformation. A mechanistic proposal indicated *N*-(acyloxy)phthalimides transforming into corresponding radical anions through the reduction of **fac*-Ir(ppy)$_3$ catalyst. By N—O bond cleavage and a decarboxylation domino process, this radical anion delivered the reactive alkyl radical, CO_2, and phthalimide anion. Next, the alkyl radical was added to the styrene via an anti-Markov addition, resulting in a relatively more stable benzyl radical. This species undergoes SET oxidation with [*fac*-Ir(ppy)$_3$]$^+$ to form a carbon cation, which could be tapered by DMSO to generate an α-alkyl-acetophenone product (Scheme 4.34).

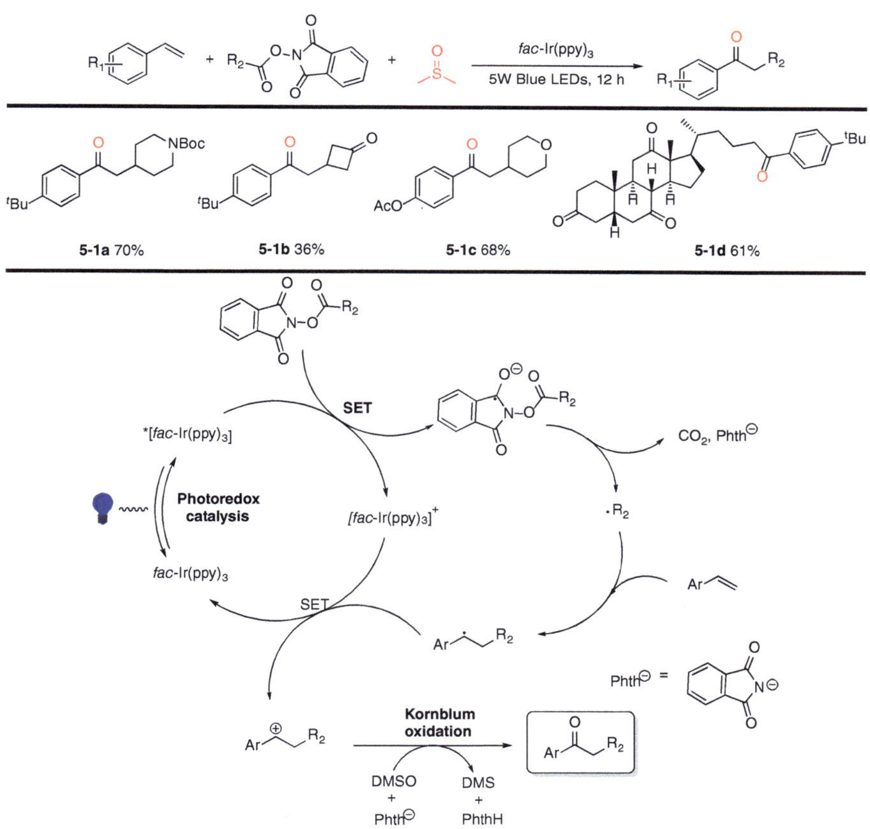

Scheme 4.34

Notably, Ye's group independently applied a similar strategy in their visible-light-promoted oxo-difluoroalkylation of alkenes [45]. In this transformation, $BrCF_2CO_2Et$ acts as the radical donor and styrene as the acceptor, while DMSO is used as the solvent and oxygen source. Kornblum oxidation mechanism is suggested for explaining the formation of the C=O bond by attacking the benzylic cation, which is generated through the single-electron oxidation of benzylic radical (Scheme 4.35).

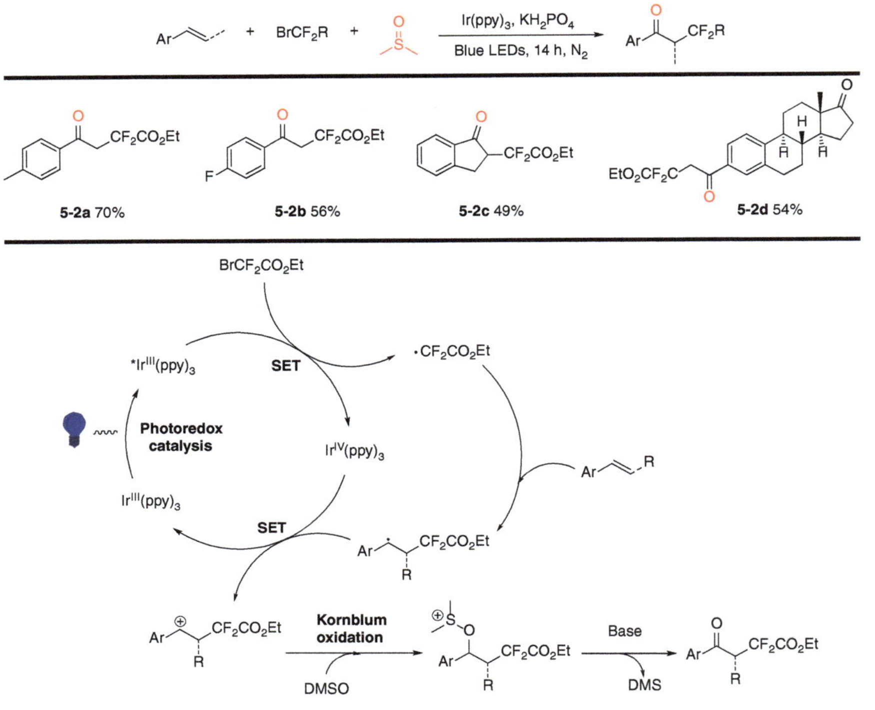

Scheme 4.35

Akita's group reported a photoredox-catalyzed keto-trifluoromethylation, producing α-trifluoromethylated ketones from aromatic alkenes [46]. DMSO was employed as the sole solvent in this transformation. The mechanism involves the excitation of Ir(III) to *Ir(III) by irradiation with visible light, leading to formation the of the CF_3 radical from Togni's reagent. The addition of the $CF_3^{\cdot}$ radical to an alkene produces a radical intermediate, which could take a single-electron oxidation process to give a benzylic cation intermediate. A Kornblum oxidation reaction of this benzylic cation with DMSO yields the desired α-trifluoromethylated ketone (Scheme 4.36).

A radical-relayed cyclization process can also be integrated into this synthetic strategy. For example, Zhu's group, in 2021, reported a new scheme to reach trifluoromethyl group-containing tetrahydrofurans or tetrahydropyrroles. In this

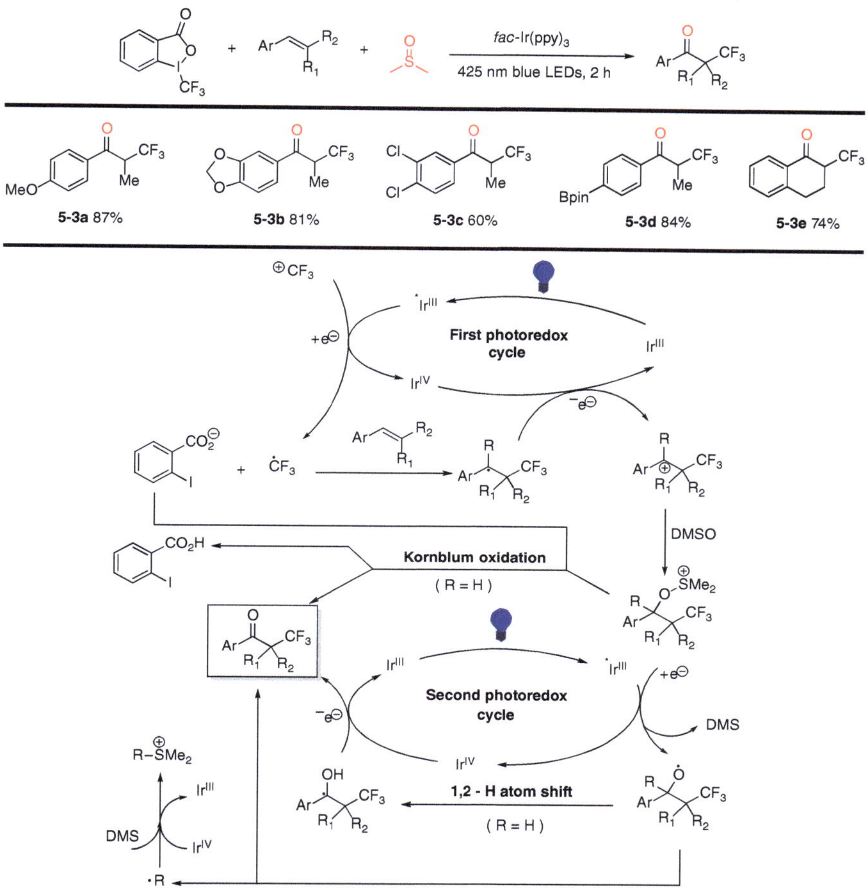

Scheme 4.36

paper, Togni-II was used as the source of $CF_3^{\cdot}$ radical donor, and 1,6-diene substrate was used as the radical acceptor to form the five-membered heterocycles through the radical cascade. Excellent diastereoselectivity was detected in each case. Mechanistically, after the first radical addition, the resulting benzylic radical intermediate undergoes intramolecular cyclization to form another benzylic radical intermediate. The latter could be oxidized into a cation to further accept the attack of DMSO (Scheme 4.37) [47].

1,5-Hydrogen atom transfer process could also be integrated into such synthetic tactics. In 2019, Ma and coworkers developed remote difunctionalization of alkenes to produce fluoroalkyl ketones using DMSO as the solvent and oxygen source [48]. AgOBz was used to promote the generation of $CF_2COOEt^{\cdot}$ radical from $BrCF_2COOEt$ in the presence of $Ir(ppy)_3(III)^*$. The key 1,5-hydrogen atom transfer occurs after the radical addition, which leads to the formation of a more stable benzyl radical intermediate. The final product arises from this *in situ*-generated benzyl cation intermediate through Kornblum oxidation with DMSO (Scheme 4.38).

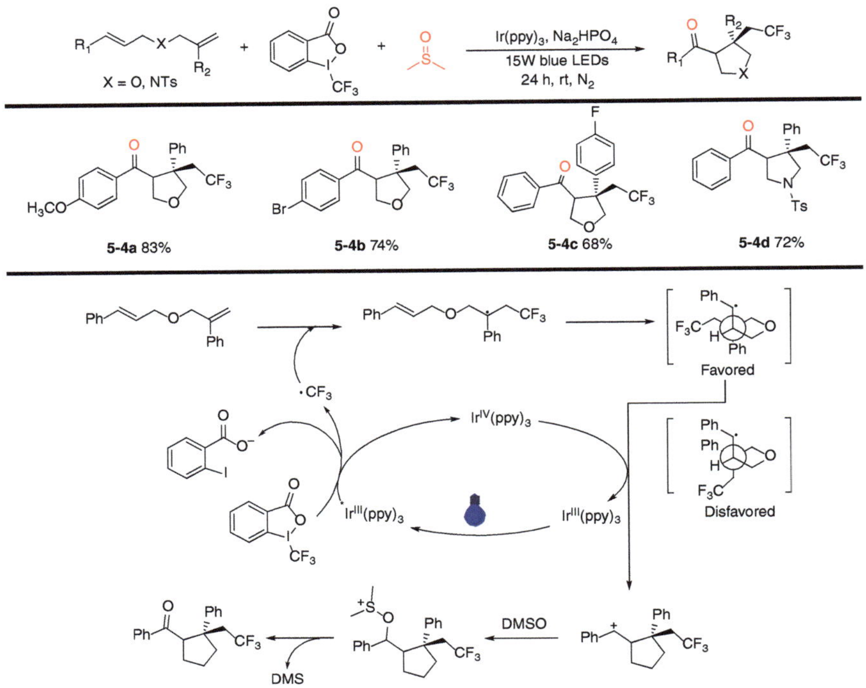

Scheme 4.37

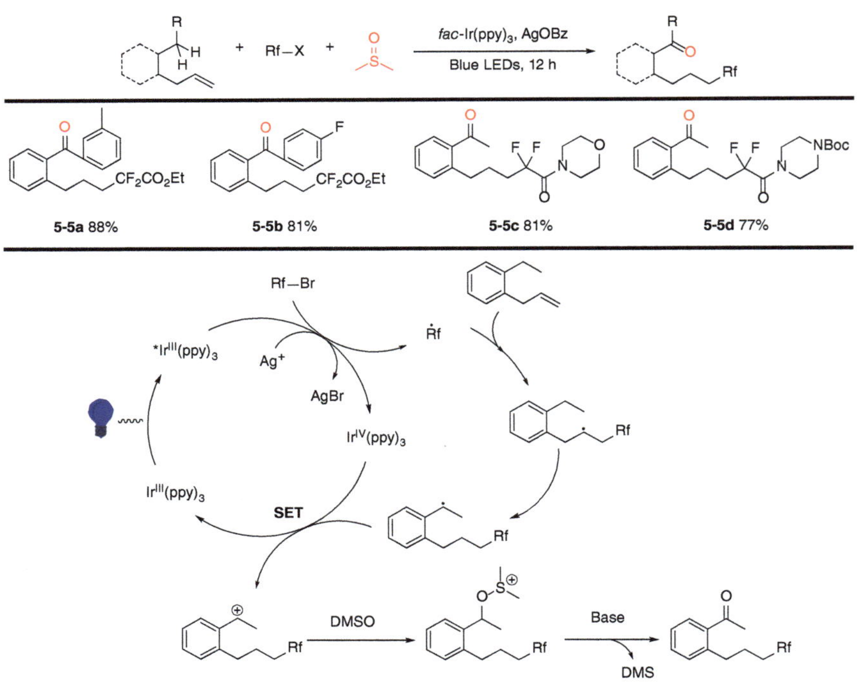

Scheme 4.38

Such a 1,5-hydrogen atom transfer process was also employed by Zhu's group in their remote oxyfluoroalkylation of heteroalkynes enabled by photocatalysis. When DMSO was used as the solvent, ketofluoroalkylation could be realized to furnish the synthesis of fluoroalkylated (*Z*)-alkenes with excellent control over regio-, stereo-, and site-selectivity. The radical intermediate can be produced by $BrCF_2COOEt$, BrC_4F_9, and Umemoto's reagent [49]. Notably, when acetone/H_2O was used to replace DMSO, hydroxytrifluoromethylation products were formed from this transformation through a similar mechanism (Scheme 4.39).

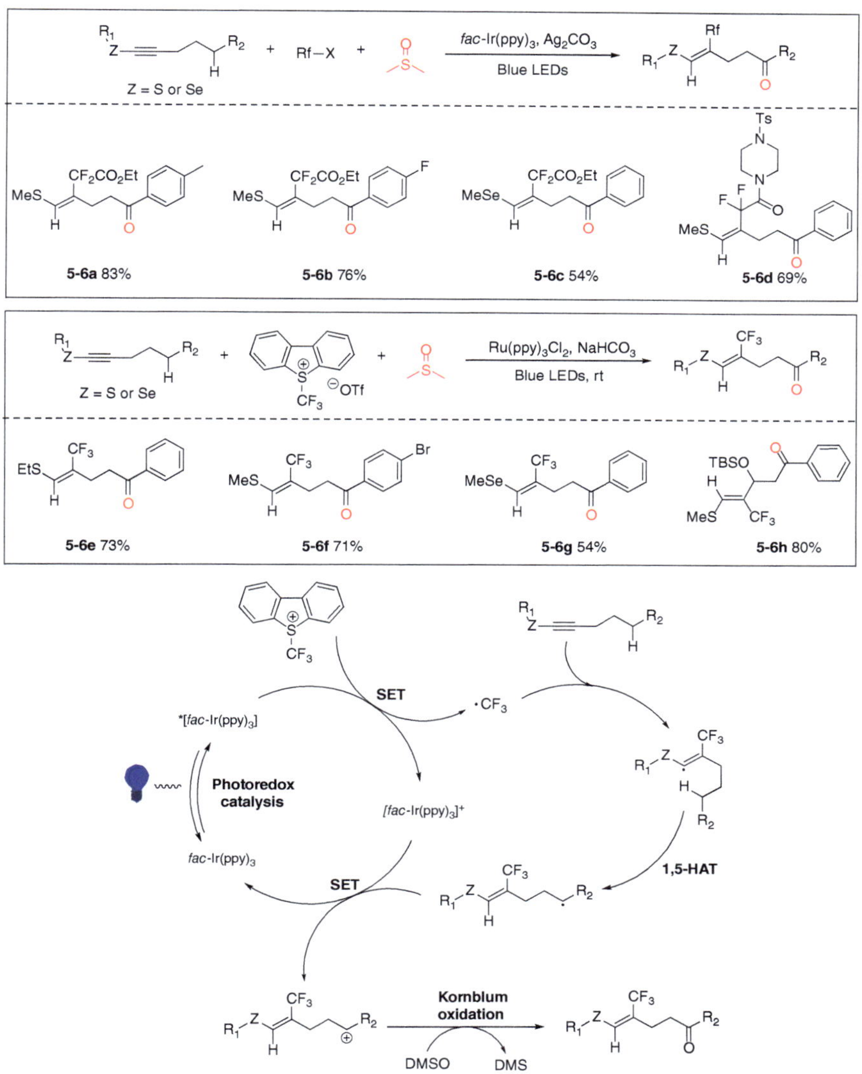

Scheme 4.39

Shi et al. developed the synthesis of cyano-containing ketones by photocatalyst-enabled C—C bond cleavage-oxidation sequence of cycloketoxime esters [50]. *fac*-Ir(ppy)$_3$ was employed as catalyst, while DMSO was used as both the solvent and the oxygen source. This ring-opening reaction was triggered by the SET-reduction of cycloketoxime esters, which leads to the formation of the iminyl radicals that could lead the C—C bond cleavage to generate a benzylic radical species. The oxidation of benzylic radical to benzylic cation furnishes the cycle of photocatalyst and leads to the follow-up of Kornblum oxidation with DMSO. Substrates containing not only the five-membered ring but also the four-, six-, seven-, and eight-membered rings were suitable for this transformation, providing the corresponding ring-opening oxidation products (Scheme 4.40).

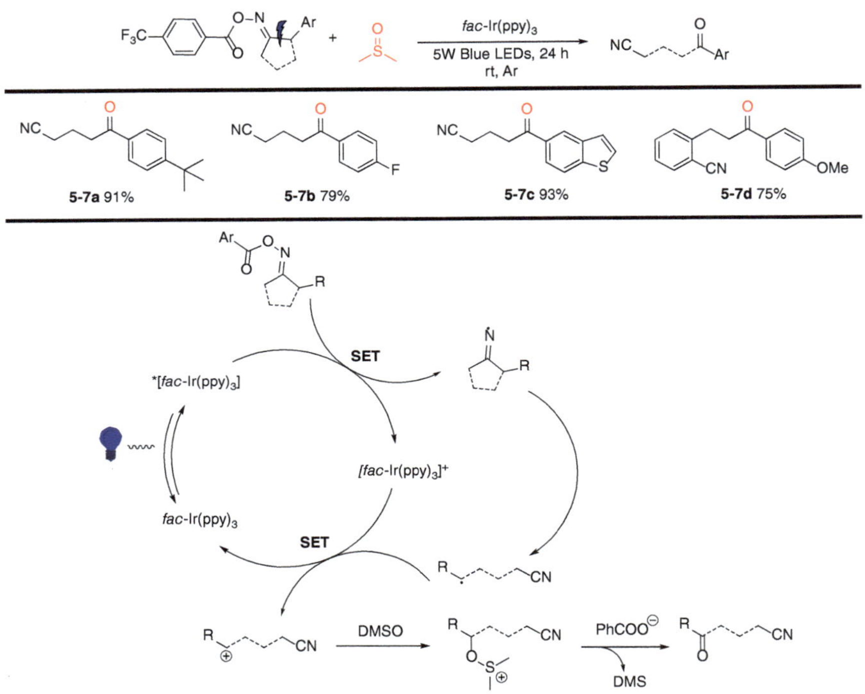

Scheme 4.40

In addition to Kornblum oxidation-type capture for carbon cation intermediates, DMSO can also be involved in photocatalytic conversions in another reaction pattern. In 2018, Polyzos's group reported a visible-light-mediated thiolation of aryl, heteroaryl, and vinyl iodides, and DMSO was used as a crucial addictive to realize this transformation [51]. Using dialkyl disulfides, this thiolation was performed in the presence of visible light photoredox catalysis *fac*-Ir(ppy)$_3$, DABCO, DMSO, and blue LED irradiation. Mechanistic studies showed that DMSO assisted in the S—S bond cleavage in the sulfur radical adduct (Scheme 4.41).

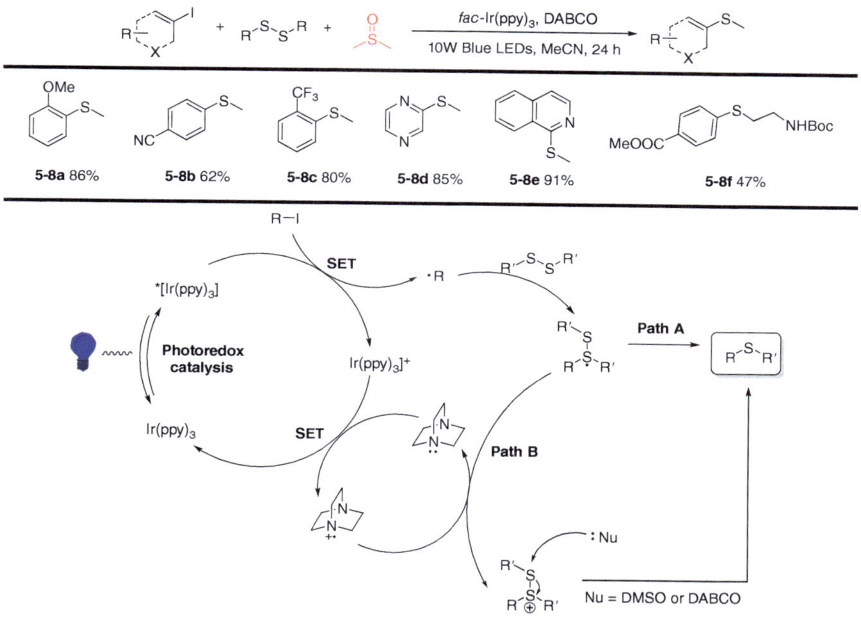

Scheme 4.41

4.6 DMSO Acts as the Metal Ligand

In metal-catalyzed reactions, especially in palladium(II)-catalyzed C–H functionalization, DMSO was widely reported as a co-oxidant to promote reoxidation of Pd(0) with O_2 [52]. However, sulfoxide ligation to palladium is another role of DMSO to promote such transformations. In 2004, M. C. White et al. [53] reported the regioselective synthesis of allylic acetates from monosubstituted olefins by using DMSO as the co-solvent (DMSO/AcOH = 1 : 1) (Scheme 4.42). The high reaction pathway selectivity and regioselectivity of this transformation were rationalized by the effect of sulfoxide ligation of Pd(II) salts. Such a tactic was employed to realize a Pd(II)-catalyzed allylic C–H alkylation by the same group in a few years [54]. This time, DMSO was used as a co-solvent (dioxane/DMSO = 4 : 1) to activate π-allyl-Pd complexes toward alkylation (Scheme 4.43).

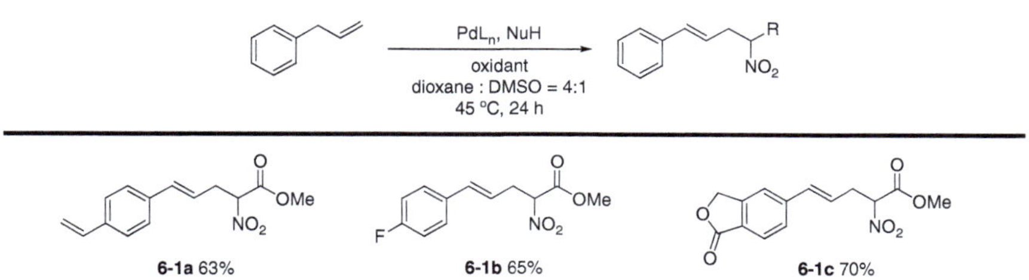

Scheme 4.42

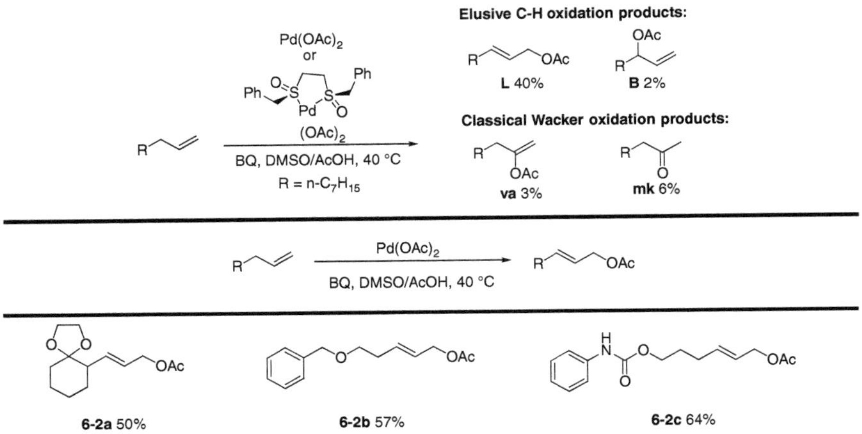

Scheme 4.43

In 2012, Stahl group disclosed the characterization of DMSO coordination to palladium(II) in both the solid state and in solution [55]. At the same time, they also reported using $Pd(DMSO)_2(TFA)_2$ as the pre-prepared catalyst for aerobic oxidative amination of alkenes and five-, six-, and seven-membered nitrogen heterocycles [56].

In addition to these achievements, the strategy that using DMSO as a ligand emerged again only recently. In 2020, Wang et al. reported a palladium-catalyzed tandem dehydrogenative[4+2] annulation of terminal olefins with *N*-sulfonyl amides [57]. This reaction was tolerated by various benzamides, heterocyclic arylamides, alkenyl carboxamides, and commercial olefins leading to the target annulation product, which holds bioactivity and medicinal importance. A significant positive effect to improve the yield was detected by adding 20 mol% DMSO as the ligand. Interestingly, other conventional Pd ligands such as PPh_3, benzoquinone, and binaphthol ligands did not give a higher yield than DMSO. However, only a trace product was detected when directly using DMSO as a solvent. The existence of $Pd(DMSO)_2(TFA)_2$ catalyst, which is formed *in situ* by the combination of DMSO with $Pd(TFA)_2$ accounts for the possible mechanism (Scheme 4.44).

A palladium-catalyzed formal [4+2] cycloaddition of benzyl and allylamines to allenes was developed to access tetrahydroisoquinoline and tetrahydropyridine skeletons in an enantioselective manner [58]. 15 equivalents of DMSO were added as the co-solvent with toluene to improve the yield by more than 10%. In this work, J. L. Mascareñas and M. Gulías et al. demonstrated the ability of DMSO to participate in asymmetric synthesis as a ligand. Though the role of the DMSO was still unclear, the authors believed that it may stabilize Pd(0) intermediates and favor a more effective reoxidation to Pd(II) (Scheme 4.45).

In 2020, J. H. Chu and coworkers reported a palladium-catalyzed cross-dehydrogenative coupling (CDC) of pyrazolo[1,5-*a*]pyridines [59]. A range of 3,3′-bipyrazolo[1,5-*a*]pyridine derivatives were synthesized in high yield and with good functional group tolerance. Different from previous works, DMSO was used as the sole solvent in this palladium-catalyzed reaction. Density functional theory

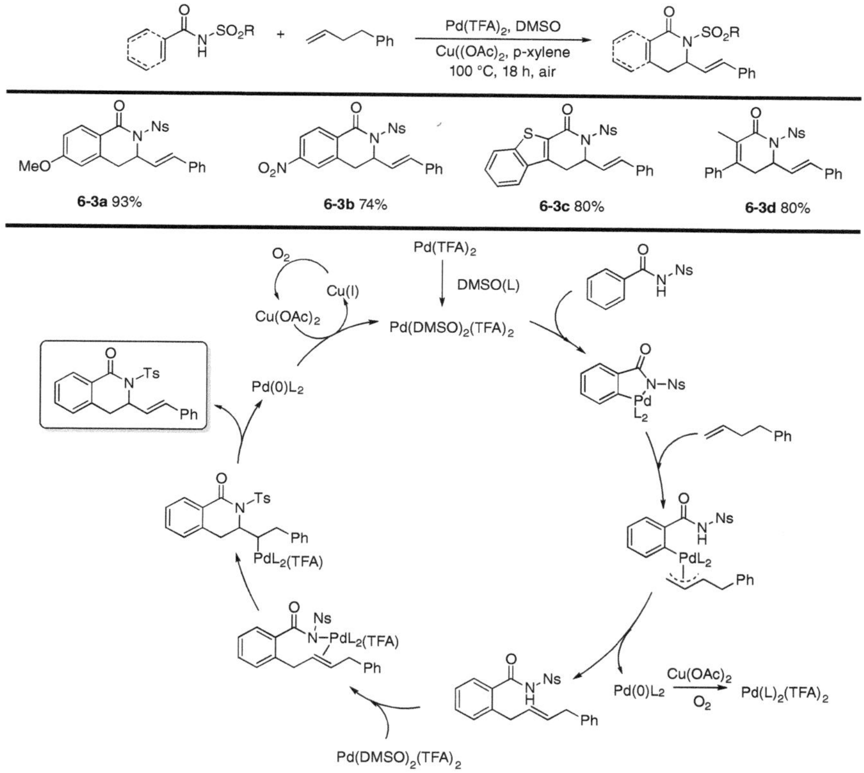

Scheme 4.44

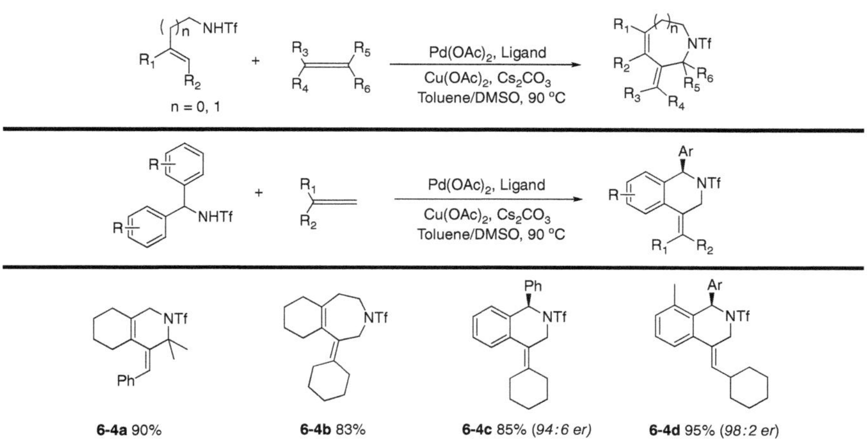

Scheme 4.45

(DFT) calculations of this transformation revealed that the reaction is initiated through the *in situ* formation of $Pd(OTFA)_2(DMSO)_2$. After participating in a concerted metallation-deprotonation (CMD) process, this active catalyst was released as $Pd^0(DMSO)_2$ species during reductive elimination (Scheme 4.46).

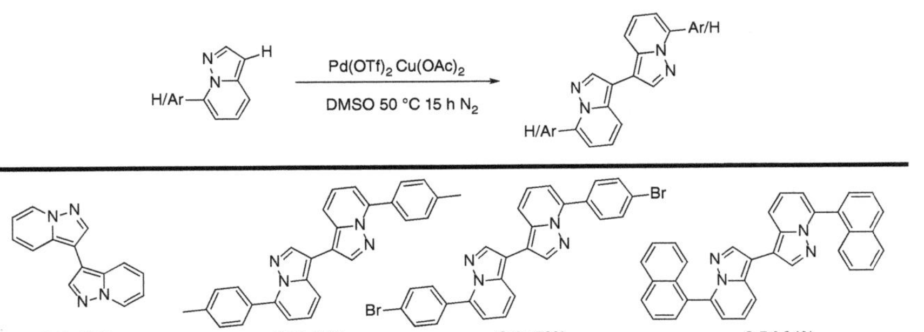

Scheme 4.46

4.7 Some Special Activation or Usage of DMSO

DMSO serves as a promoter and solvent, and enables C—P bond functionalization. This was also realized by the Masarwa group in 2021, through the post-functionalization of benzylic $C(sp^3)$-PPh_3^+ groups to achieve aminations, thiolations, and arylations resulting in a variety of benzhydrylamines, benzhydrylthioethers, and triarylmethanes (Scheme 4.47) [60]. Furthermore, Chu et al. developed a difluoroalkylation reaction of organophosphonium salts with difluoroenol silyl ethers to realize the facile synthesis of β,β-di(hetero)aryl-α,α-difluorinated ketones. In this study, the author demonstrated that DMSO would interact with the PPh_3 moiety in phosphonium salt, resulting in a C—P bond cleavage through the bonding of the phosphorus atom (Scheme 4.48) [61]. The resulting benzyl carbocation

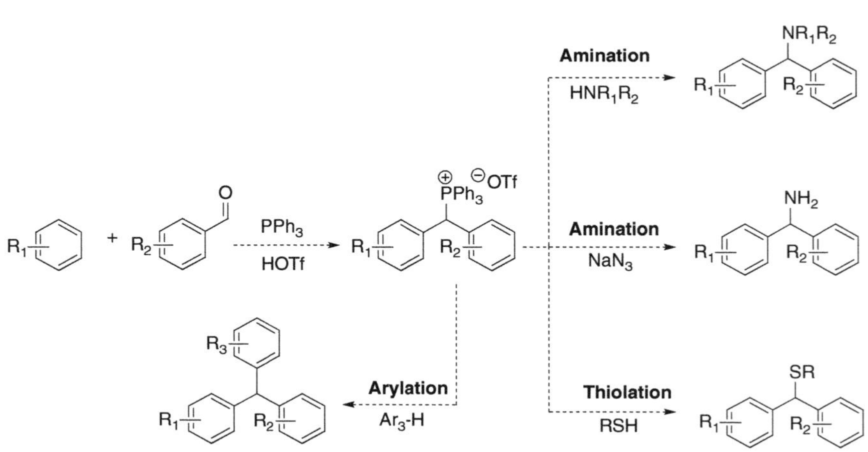

Scheme 4.47

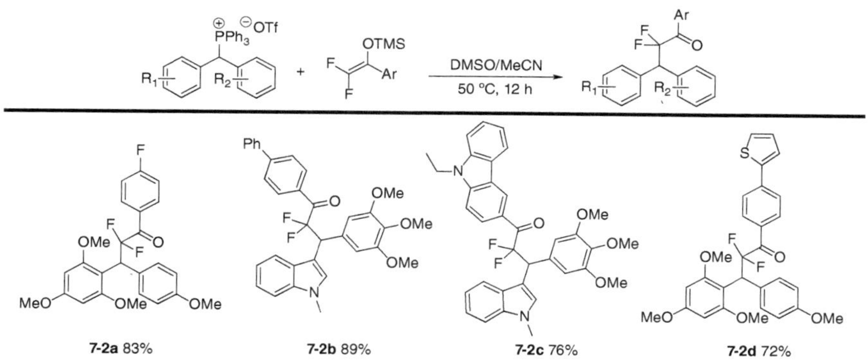

Scheme 4.48

intermediates serve as an efficient nucleophilic reagent to accept the electrophilic attack of fluorinated enolate and produce the difluoroketone product. The whole process is coincident with the classic S_N1 substitution mechanism, but the S_N2 mechanism also cannot be ruled out.

It is widely known that DMSO is capable of oxidating HI to I_2. This strategy is frequently used in I_2–DMSO catalytic systems reported by Wu's group. However, the oxidation of HBr into Br_2 by DMSO is relatively rare to see in the literature, especially in the field of heterocyclic synthesis. Chen et al. demonstrated the application of such a process through their efficient synthesis of (dihydro)thiophenes in which the redox-divergent selectivity could be manipulated by the dosage of DMSO and HBr [62]. From readily available allylic alcohols, dihydrothiophenes, thiophenes, and bromothiophenes could be obtained in one-step synthesis in moderate-to-good yields with high chemoselectivity. Mechanistic studies indicated the sulfur atom in these thiophene-based rings is derived from the fragmentation of DMSO. Specifically, the redox reaction of DMSO and HBr leads to the product, DMS and Br_2, respectively. The reaction between the diene and Br_2 gives a bromonium ion, in which the former one was generated *in situ* from the allylic alcohols by acid-promoted elimination. This bromonium ion then undergoes the nucleophilic attack by DMS, followed by subsequent nucleophilic substitution, yielding cyclic sulfonium salt. Interestingly, as the amount of hydrobromic acid and potassium bromide increases, three types of (dihydro)thiophenes derivatives with different redox levels could be obtained in each case (Scheme 4.49).

Besides the reaction between the DMSO and aryne, another strategy to activate the DMSO through S=O bond cleavage is by [1,3]-rearrangement. Mo and Ma reported a $MnSO_4$-promoted synthesis of functionalized sulfonium ylides from diethyl acetylenedicarboxylate and sulfoxides, including DMSO [63]. The product could be easily obtained on a gram scale by recrystallization without column chromatography. Notably, when DMSO was used as the sole solvent, the reaction was not efficient resulting in the product yield of only 17%. The optimal conditions were determined by using 5 equivalents of DMSO in the toluene solution. Such rearrangement is conceptually different from the previously reported

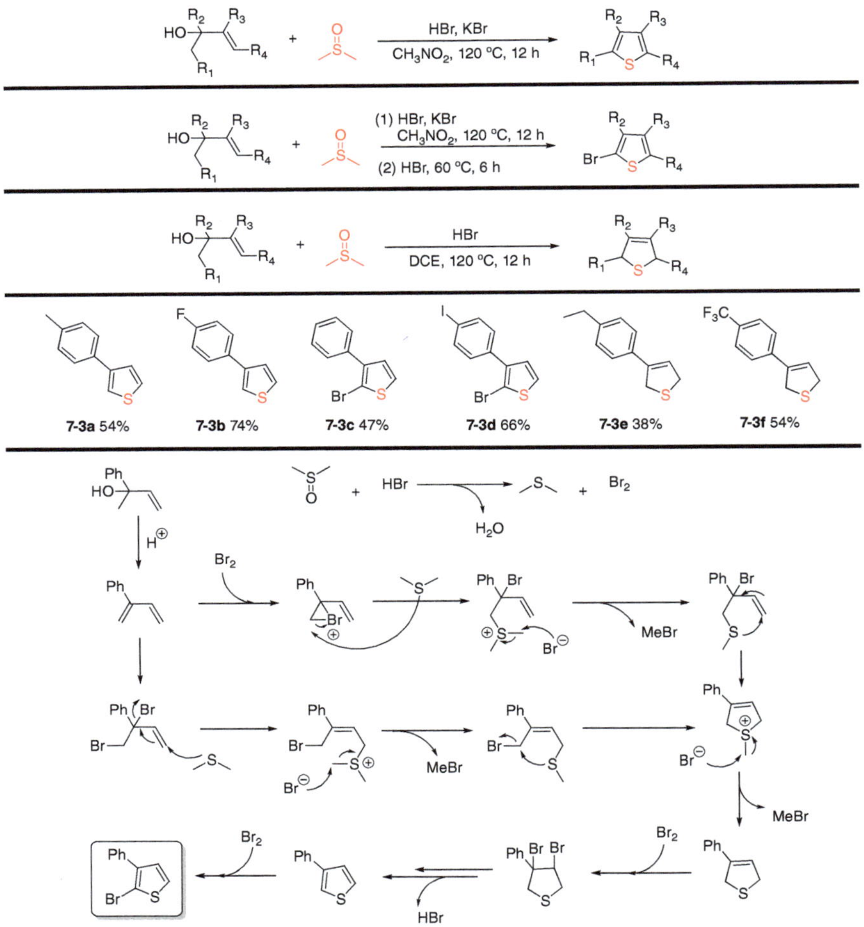

Scheme 4.49

[3,3]-rearrangement of sulfoxides. A possible mechanism was proposed as follows: The Mn(II) catalyst activated the diethyl acetylenedicarboxylate through coordination in the first place. This intermediate is attacked by sulfoxide to undergo a 1,4-addition. Then, subsequent S—O bond cleavage through a [1,3]-rearrangement leads to the formation of the sulfonium ylide. In another way, a ring-opening process of the four-membered ring intermediate, which formed via the addition of an S=O bond to the allene intermediate, could not be excluded as an explanation for the generation of sulfonium ylide (Scheme 4.50).

In 2021, Xu et al. achieved the reaction between DMSO and another important triple bond substrate, ynamides [64]. With the assistance of a microwave, DMSO acted as both an oxidant and a solvent to oxidize the *N*-arylethynylsulfonamides into *N*-sulfonyl-2-aryloxoacetamides rapidly (the reaction lasts only 15 minutes). Various α-keto-imides bearing different arene substituents could be formed in

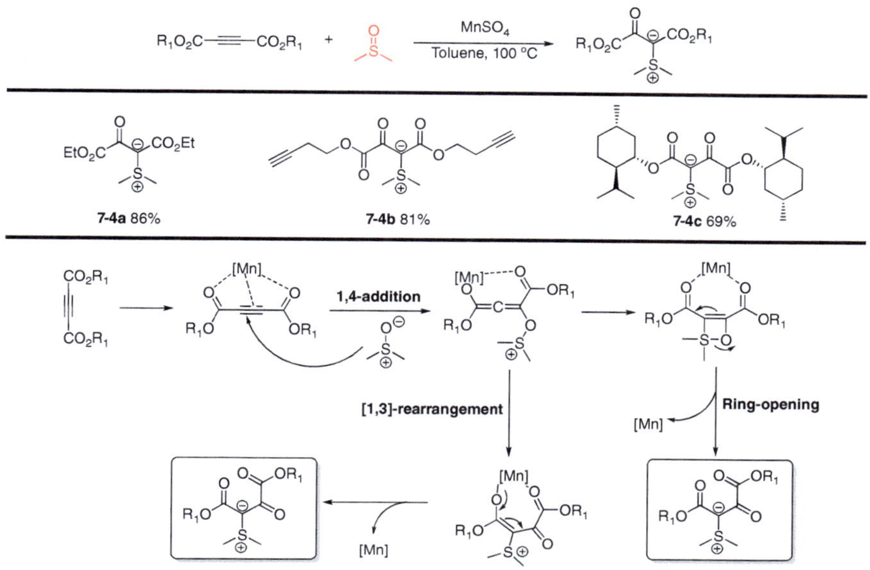

Scheme 4.50

moderate-to-good yields without adding any transition metal catalysts or external oxidants. DFT calculations indicate that DMSO provided both two oxygen atoms in the α-keto-imides by nucleophilic attacks and subsequent S—O bond cleavage (Scheme 4.51).

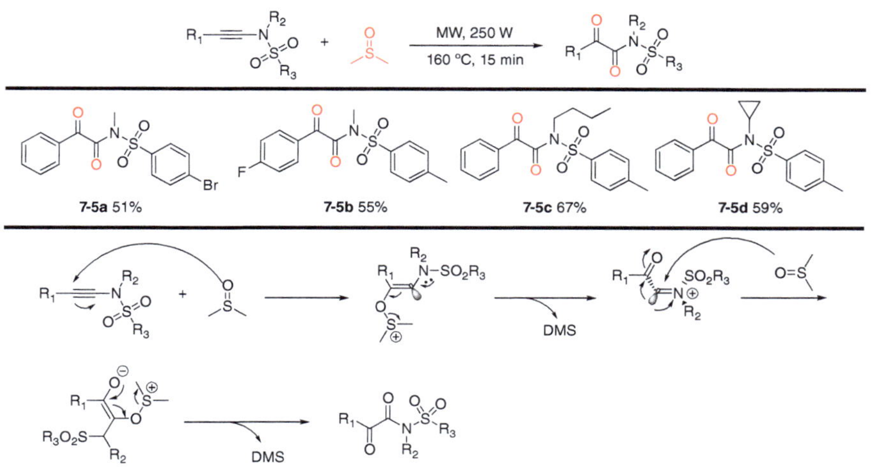

Scheme 4.51

Another example of the special usage of DMSO is to act as a traceless promoter to assist the ring-opening of cyclopropyl ketones by nucleophilic attack of the cyclopropane part. Recently, Ramasastry's group reported an uncatalyzed cascade ring-opening/recyclization reaction from cyclopropyl aryl ketones accessing five-membered ring keto products using DMSO as both a promoter and a solvent [65]. Based on the different functional groups in the *ortho*-position of cyclopropyl aryl ketones, such as aldehyde, ketone, or enone, indenones, and fluorenones could be formed divergently according to a similar mechanism (Scheme 4.52).

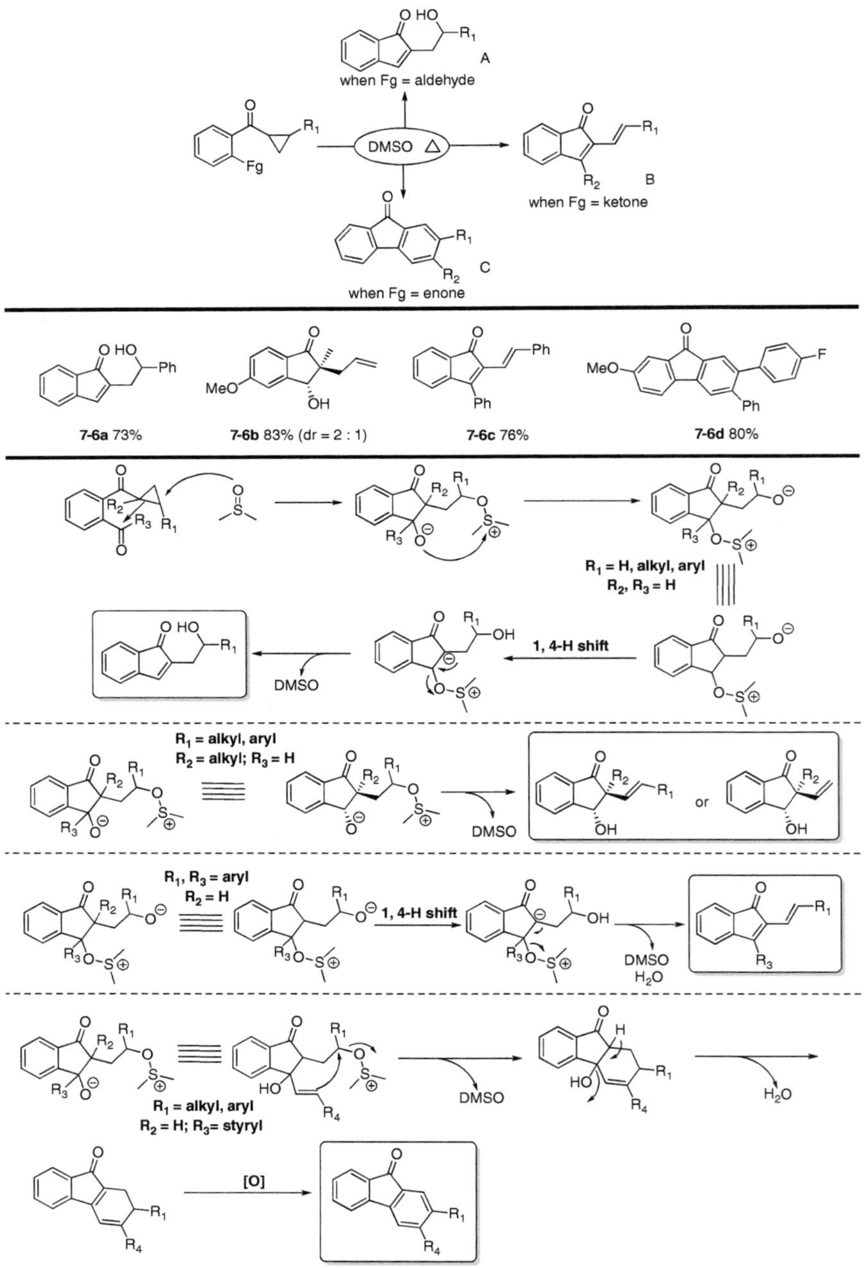

Scheme 4.52

The interaction between DMSO and hydroxyl groups through hydrogen bond accesses the selective C–H activation. In 2020, Jiao et al. disclosed a selective radical O–H activation of two hydroxyls in diols, which have almost identical bond-dissociation energy [66]. Mechanistic studies indicated that DMSO was an essential reagent in this transformation. By adding 3 equivalents of DMSO as the addictive, phenyl migration products could be formed from 1,1-diphenylpropane-1,3-diol using $AgClO_4$ as the single-electron oxidant and $K_2S_2O_8$ as the terminal oxidant in an aqueous solution. It is noteworthy that DMSO was essential to the regioselectivity of such an unprecedented distal C–C cleavage reaction. By the increase of DMSO loading from catalytic among to 5 equivalents, the regioselectivity improved significantly. Mechanistic studies indicated DMSO promoted the proton-coupled electron transfer (PCET) of a certain –OH group to form the primary alkoxyl radical and the regioselectivity was achieved probably due to the steric hindrance (Scheme 4.53).

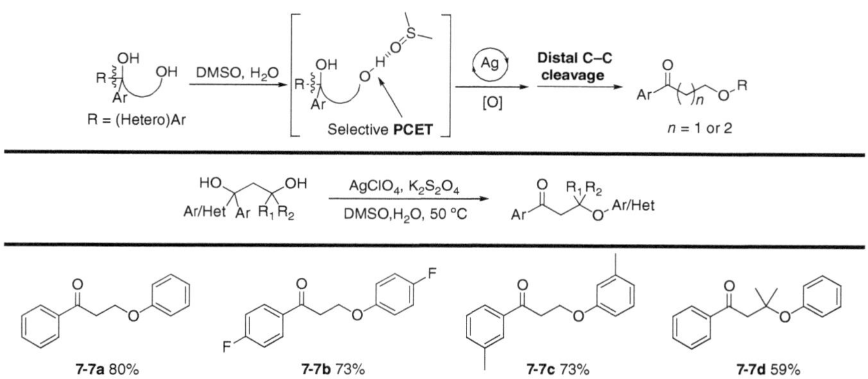

Scheme 4.53

The reaction between aryne and DMSO results in a special pattern of DMSO activation by providing both a methylthiolating agent and an oxygen source into the product (Scheme 4.54) [67]. This synthetic strategy was frequently reported by many research groups such as Xiao, Wang, and Gogoi et al. The key to achieving this reaction mode is the *in situ* generation of the four-membered intermediate through the insertion into the S=O π-bond of DMSO by aryne. This reactive intermediate undergoes ring opening and leads to the formation of *o*-quinone sulfide, which is a highly active nucleophilic reagent for oxygen alkylation reactions. Gogoi, in 2020, reported the direct synthesis of ortho-methylthio allyl and vinyl ethers using an aryne precursor (2-(trimethylsilyl)phenyl trifluoromethanesulfonate) and DMSO as the solvent, providing a new example using this activation strategy.

It is well-known that DMSO could contribute a part of it such as =O, –SMe, –Me, and so on through fragmentation into the final product. However, it is still rare and interesting to see that DMSO could act as a whole substrate and directly assemble into the product. In 2017, Vaitla et al. reported a synthesis of sulfur ylides via rhodium-catalyzed coupling reactions with sulfides or sulfoxides [68].

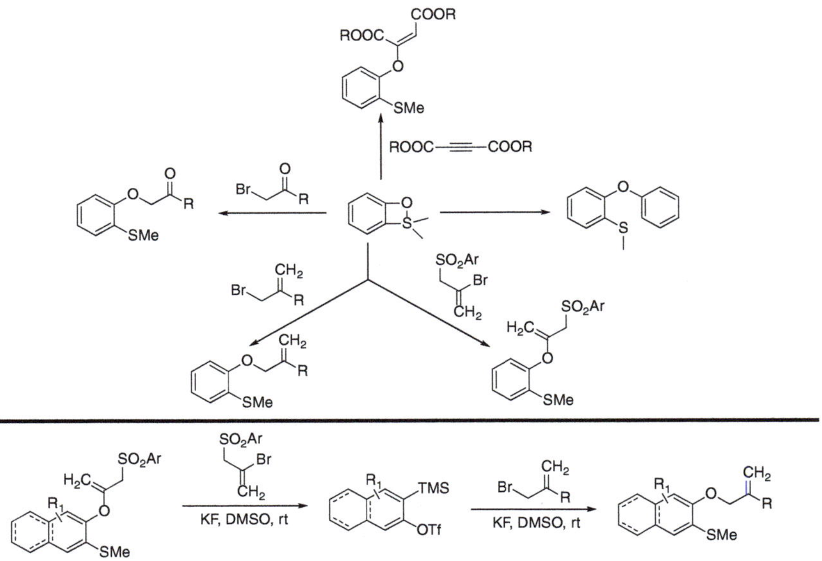

Scheme 4.54

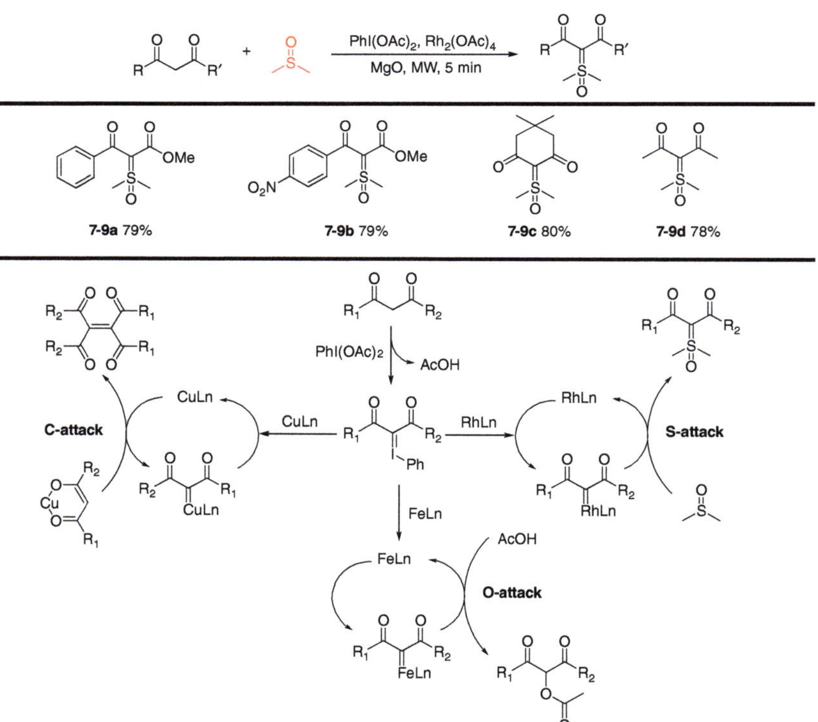

Scheme 4.55

Starting from active methylene compounds and phenyliodonium diacetate, MgO and $Rh(OAc)_2$ as the catalyst in a DMSO solution with microwave-assisted heating, sulfoxonium, and sulfonium ylides could be obtained in high yields. The active methylene compound first reacts with $PhI(OAc)_2$ to yield an iodonium ylide, then it reacts with $Rh(OAc)_2$ to give a rhodium carbene. This carbene could be trapped by sulfide or sulfoxide to generate the sulfur ylide product (Scheme 4.55).

4.8 Summary and Outlook

Recent decades have witnessed the development of using DMSO as a green solvent as well as a versatile synthetic building block in organic synthesis, especially in the research area of synthetic methodologies. The addition of DMSO, as a relatively unreactive small molecule, in some cases does not significantly affect the course of the reaction outcomes, and as a result, although there has been increasing use of DMSO to design synthetic methodologies, a large number of works reported the beyond-designed transformation involving DMSO. All these examples have greatly expanded the application and potential capability of DMSO as a synthon. This chapter classified the ways in which DMSO is activated and summarized the different reactions involving DMSO as a solvent and reactant over the last five years. We believe our review would be helpful to the researchers who want to enter this field. We predict that in the near future, more reaction types for DMSO will be developed to further enrich the synthetic applications of this useful small molecule.

References

1 (a) Soroko, I., Bhole, Y., and Livingston, A.G. (2011). *Green Chem.* 13: 162–168. (b) Tashrifi, Z., Khanaposhtani, M.M., Larijani, B., and Mahdavi, M. (2019). *Adv. Synth. Catal.* 362: 65–86.
2 (a) Omura, K., Sharma, A.K., and Swern, D. (1976). *J. Org. Chem.* 41: 957–962. (b) Pfitzner, K.E. and Moffatt, J.G. (1963). *JACS* 85: 3027–3028. (c) Corey, E.J. and Chaykovsky, M. (1962). *JACS* 84: 867–868. (d) Epstein, W.W. and Sweat, F.W. (1967). *Chem. Rev.* 67: 247–260.
3 Wu, X. and Natte, K. (2016). *Adv. Synth. Catal.* 358: 336–352.
4 Xiang, J.-C., Gao, Q.-H., and Wu, A.-X. (2017). *Solvents as Reagents in Organic Synthesis*, 315–353.
5 Lu, H., Tong, Z., Peng, L. et al. (2022). *Top. Curr. Chem.* 380.
6 Feldman, K.S. (2006). *Tetrahedron* 62: 5003–5034.
7 (a) Smith, L.H.S., Coote, S.C., Sneddon, H.F., and Procter, D.J. (2010). *Angew. Chem. Int. Ed.* 49: 5832–5844. (b) Bur, S.K. and Padwa, A. (2004). *Chem. Rev.* 104: 2401–2432. (c) Padwa, A. and Waterson, A.G. (2000). *Curr. Org. Chem.* 4: 175–203.
8 Liu, P., Shen, Z., Yuan, Y., and Sun, P. (2016). *Org. Biomol. Chem.* 14: 6523–6530.
9 Xu, X., Zhou, Z., Wang, Z. et al. (2019). *Adv. Synth. Catal.* 361: 4278–4285.

10 Zhang, X., Zhou, Z., Xu, H. et al. (2019). *Org. Lett.* 21: 7248–7253.

11 Xu, X., Yang, Y., Zhang, X., and Yi, W. (2018). *Org. Lett.* 20: 566–569.

12 Yuan, J., Yu, J.T., Jiang, Y., and Cheng, J. (2017). *Org. Biomol. Chem.* 15: 1334–1337.

13 Yang, T., Li, H., Nie, Z. et al. (2022). *J. Org. Chem.* 87: 2797–2808.

14 Wakade, S.B., Tiwari, D.K., Ganesh, P.S.K.P. et al. (2017). *Org. Lett.* 19: 4948–4951.

15 Yadav, P., Awasthi, A., Gokulnath, S., and Tiwari, D.K. (2021). *J. Org. Chem.* 86: 2658–2666.

16 Ma, J.T., Wang, L.S., Chai, Z. et al. (2021). *Chem. Commun.* 57: 5414–5417.

17 Ma, J., Chen, T., Tang, B. et al. (2023). *J. Org. Chem.* 88: 3760–3771.

18 Zhang, Y., Chen, R., Wang, Z. et al. (2021). *J. Org. Chem.* 86: 6239–6246.

19 Zhu, X., Zhang, F., Kuang, D. et al. (2020). *Org. Lett.* 22: 3789–3793.

20 Shu, W.M., Wang, S., Zhang, X.F. et al. (2019). *Org. Biomol. Chem.* 17: 4311–4316.

21 Shi, Y., Xu, X., Chen, X. et al. (2021). *J. Org. Chem.* 86: 15335–15344.

22 Palit, K., Sepay, N., and Panda, N. (2023). *J. Org. Chem.* 88: 2931–2941.

23 (a) Banks, R.E. (1998). *J. Fluorine Chem.* 87: 1–17. (b) Singh, R.P. and Shreeve, J.M. (2004). *Acc. Chem. Res.* 37: 31–44.

24 Guo, H., Tian, L., Liu, Y., and Wan, J. (2022). *Org. Lett.* 24: 228–233.

25 Lu, H., Chen, J., Zhou, W. et al. (2023). *Org. Lett.* 25: 389–394.

26 Kalari, S., Shinde, A.U., and Rode, H.B. (2021). *J. Org. Chem.* 86: 17684–17695.

27 Wu, Y., Huang, Z., Luo, Y. et al. (2017). *Org. Lett.* 19: 2330–2333.

28 Kishore, K. and Asmus, K. (1989). *J. Chem. Soc., Perkin Trans.* 2: 2079–2084.

29 Tian, M., Wang, C., Hu, X., and Loh, T. (2019). *Org. Lett.* 21: 1607–1611.

30 Gupta, A., Rahaman, A., and Bhadra, S. (2019). *Org. Lett.* 21: 6164–6168.

31 Zarinderakht, N., Abbasi, M., and Nowrouzi, N. (2022). *Appl. Organomet. Chem.* 36.

32 Wen, Z., Liu, X., Liu, Y., and Chao, J. (2017). *Org. Lett.* 19: 5798–5801.

33 Freeman, F. and Angeletakis, C.N. (1983). *JACS* 105: 4039–4049.

34 Bhattacharjee, S., Laru, S., Ghosh, P., and Hajra, A. (2021). *J. Org. Chem.* 86: 10866–10873.

35 Duan, T., Zhai, T., Liu, H. et al. (2016). *Org. Biomol. Chem.* 14: 6561–6567.

36 Xie, Z., Li, P., Hu, Y. et al. (2017). *Org. Biomol. Chem.* 15: 4205–4211.

37 Jiang, S., Yang, Z., Guo, Z. et al. (2019). *Org. Biomol. Chem.* 17: 7416–7424.

38 Chakraborty, N., Dahiya, A., Rakshit, A. et al. (2021). *Org. Biomol. Chem.* 19: 6847–6857.

39 Su, M., Liu, Y., Nie, Z. et al. (2022). *J. Org. Chem.* 87: 7022–7032.

40 Liu, Y., Yu, Y., Sun, C. et al. (2020). *Org. Lett.* 22: 8127–8131.

41 Wang, R., Xie, K., Fu, Q. et al. (2022). *Org. Lett.* 24: 2020–2024.

42 Hayashi, R., Shimizu, A., Davies, J.A. et al. (2018). *Angew. Chem.* 130: 13073–13077.

43 Lin, Z., Huang, L., and Yuan, G. (2021). *Chem. Commun.* 57: 3579–3582.

44 Tlahuext-Aca, A., Garza-Sanchez, R.A., Schäfer, M., and Glorius, F. (2018). *Org. Lett.* 20: 1546–1549.

45 Xia, Z., Gao, Z., Dai, L., and Ye, S. (2019). *J. Org. Chem.* 84: 7388–7394.
46 Tomita, R., Yasu, Y., Koike, T., and Akita, M. (2014). *Angew. Chem. Int. Ed.* 53: 7144–7148.
47 Chun, J., Zhang, H., Meng, F. et al. (2021). *Adv. Synth. Catal.* 363: 751–756.
48 Li, L., Luo, H., Zhao, Z. et al. (2019). *Org. Lett.* 21: 9228–9231.
49 Shang, T., Zhang, J., Zhang, Y. et al. (2020). *Org. Lett.* 22: 3667–3672.
50 Zhao, B., Tan, H., Chen, C. et al. (2018). *Chin. J. Chem.* 36: 995–999.
51 Czyz, M.L., Weragoda, G.K., Monaghan, R. et al. (2018). *Org. Biomol. Chem.* 16: 1543–1551.
52 (a) Steinhoff, B.A., Fix, S.R., and Stahl, S.S. (2002). *JACS* 124: 766–767. (b) Grennberg, H., Gogoll, A., and Baeckvall, J.E. (1991). *J. Org. Chem.* 56: 5808–5811.
53 Chen, M.S. and White, M.C. (2004). *JACS* 126: 1346–1347.
54 Young, A.J. and White, M.C. (2008). *JACS* 130: 14090–14091.
55 Diao, T., White, P., Guzei, I., and Stahl, S.S. (2012). *Inorg. Chem.* 51: 11898–11909.
56 Lu, Z. and Stahl, S.S. (2012). *Org. Lett.* 14: 1234–1237.
57 Sun, M., Chen, W., Xia, X. et al. (2020). *Org. Lett.* 22: 3229–3233.
58 Vidal, X., Mascarenas, J.L., and Gulias, M. (2019). *JACS* 141: 1862–1866.
59 Hsiao, P., Chang, R., Sue, A.C.H. et al. (2022). *J. Org. Chem.* 87: 9851–9863.
60 Babu, K.N., Massarwe, F., Shioukhi, I., and Masarwa, A. (2021). *Angew. Chem. Int. Ed.* 60: 26199–26209.
61 Yu, Z., Chen, J., Chen, Y. et al. (2022). *Org. Lett.* 24: 5557–5561.
62 Liu, H., He, G.C., Zhao, C.Y. et al. (2021). *Angew. Chem. Int. Ed.* 60: 24284–24291.
63 Chen, W., Liu, Z., Wang, Y. et al. (2022). *Org. Biomol. Chem.* 20: 1656–1661.
64 Dong, J., Fu, D., Sheng, D. et al. (2021). *RSC Adv.* 11: 40243–40252.
65 Mishra, U.K., Patel, K., and Ramasastry, S.S.V. (2020). *Org. Lett.* 22: 3815–3819.
66 Zhu, Y., Zhang, Z., Jin, R. et al. (2020). *Angew. Chem. Int. Ed.* 59: 19851–19856.
67 Hazarika, H. and Gogoi, P. (2020). *Org. Biomol. Chem.* 18: 2727–2738.
68 Vaitla, J., Hopmann, K.H., and Bayer, A. (2017). *Org. Lett.* 19: 6688–6691.

5

The Use of DMC as Green Solvent in Organic Synthesis

Xinxin Qi[1] *and Xiao-Feng Wu*[2,3]

[1]*Zhejiang Sci-Tech University, Department of Chemistry, Key Laboratory of Surface & Interface Science of Polymer Materials of Zhejiang Province, 928 Second Avenue, Hangzhou, 310018, China*
[2]*Chinese Academy of Sciences, Dalian National Laboratory for Clean Energy, Dalian Institute of Chemical Physics, No. 457 Zhongshan Road, Dalian, Liaoning 116023, China*
[3]*Leibniz-Institut für Katalyse e. V., Albert-Einstein-Straße 29a, 18059, Rostock, Germany*

5.1 Introduction

Dimethyl carbonate (DMC) is a biodegradable product widely used as a green solvent in organic chemistry [1]. As an environmentally friendly compound, one of the most attractive and sustainable transformations to prepare DMC mainly relies on CO_2 as the feedstock, by insertion of CO_2 into ethylene oxide to give ethylene carbonate, followed by the reaction with methanol to yield DMC [2]. It does not produce any waste during this process, which makes DMC a nontoxic reagent. Besides, DMC exhibits diverse properties, for example, it could act as a safe and environmentally benign alternative for some methylation and carboxymethylation reactions. It can be handled easily without special precautions, and no disposal issues. Therefore, DMC has been found an appealing reaction medium and a promising building block in organic synthesis. Since many reactions with DMC as a solvent have been reviewed before, this chapter would only focus on the literature during these 10 years.

5.2 Organic Reactions in DMC

A coupling reaction between aryl iodides and arylboronic acid has been explored under metal-free conditions (Scheme 5.1) [3]. DMC, a green solvent in this reaction, played an important role in enhancing the coupling reaction. No transition-metal catalysts were needed in this reaction and with K_3PO_4 as the only reagent, various biaryl compounds have been synthesized in good yields.

Rapid access to aminals has been achieved through an efficient and green method using $FeCl_3$–SiO_2 as the catalyst (Scheme 5.2) [4]. By simple filtration, the silica-supported $FeCl_3$ could be used as a stable and recyclable catalyst in this reaction, and less deactivation was observed after the third cycle. Moreover, DMC

Green Solvents in Organic Synthesis, First Edition.
Edited by Xiao-Feng Wu, Zhiping Yin, Liang-nian He, and Feng Wang.

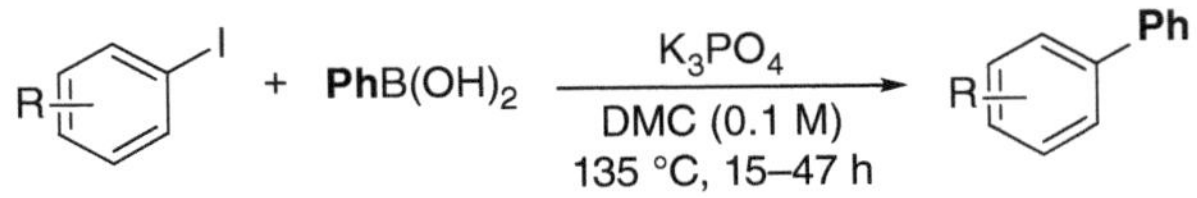

Scheme 5.1 Biaryl coupling of aryl iodides and organoboron reagents.

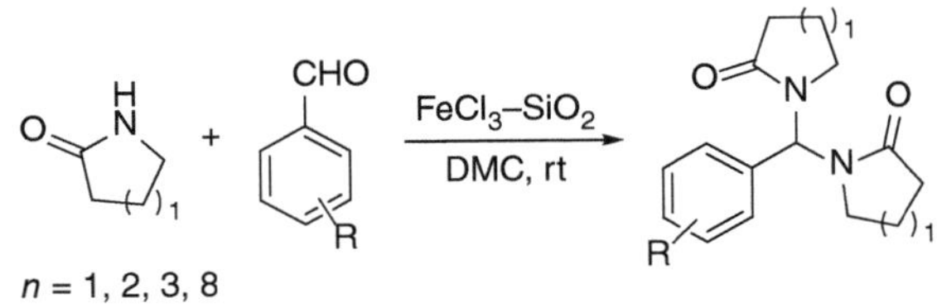

Scheme 5.2 Silica-supported $FeCl_3$-catalyzed aminal synthesis.

was used as a green solvent, and no ligands and additives were used in the current methodology, which would provide an environmentally benign strategy for the synthesis of aminals with potential applications in the pharmaceutical and organic synthesis fields.

The Gorin group developed a copper-catalyzed oxidative coupling reaction of carboxylic acids with methylboronic acid via a Chan–Lam methylation process (Scheme 5.3) [5]. The reaction allows for the preparation of esters with a broad substrate scope, and a broad range of carboxylic acids, including aryl, aliphatic, and alkenyl carboxylic acids could react with methylboronic acid smoothly without any additional oxidant under air conditions with DMC as the solvent. This method presents a good supplement for *O*-alkylation with boronic acid by employing alkylboronic acids as a good alternative to toxic and electrophilic alkylating reagents. Furthermore, an isotope-labeling study supports an oxidative cross-coupling reaction mechanism analogous to that proposed for Chan–Lam arylation.

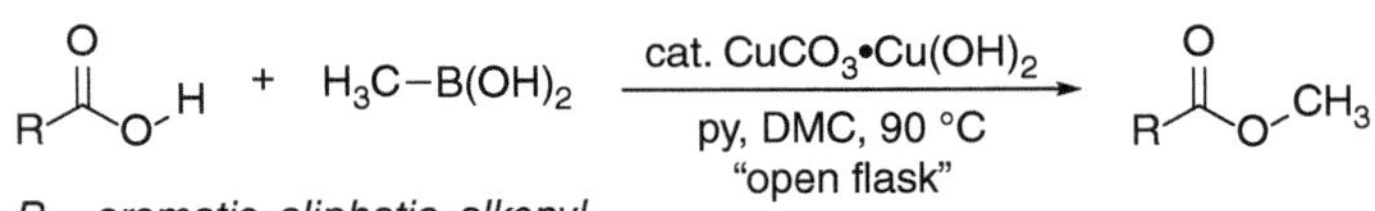

Scheme 5.3 Copper-catalyzed oxidative coupling reaction of carboxylic acids with methylboronic acid.

An efficient strategy for the intramolecular hydroalkoxylation of double bonds has been developed for the synthesis of cyclic ethers via a sustainable catalytic system by using montmorillonite-supported benign metals, iron and bismuth, as catalysts (Scheme 5.4) [6]. As a non-VOC solvent, DMC was employed as the solvent, and a variety of cyclic ethers were formed in moderate-to-excellent yields from the corresponding unsaturated alcohols. Furthermore, a recycling study has been conducted under the standard reaction conditions, and the Fe-MMT still showed good reactivity even after five cycles.

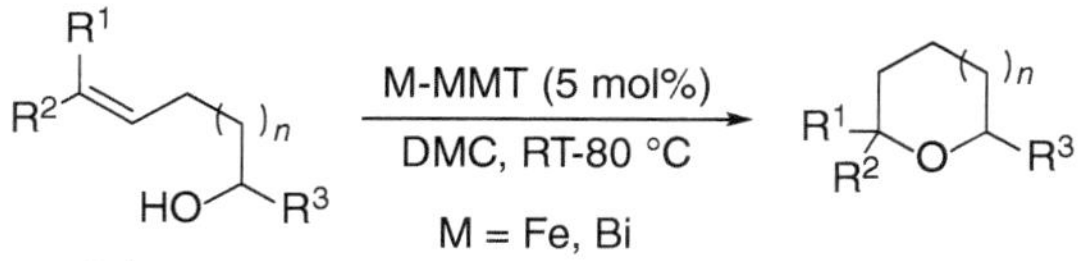

$n = 0,1$
R^1, R^2, R^3 = alkyl, alkenyl, phenyl, benzyl...

Scheme 5.4 Intramolecular hydroalkoxylation of double bonds.

An effective $NaAlO_2$-catalyzed carboxymethylation for the synthesis of mixed carbonate esters has been explored (Scheme 5.5) [7]. Compared to classical methods, this reaction provides a practical and environmentally friendly approach with DMC as a green solvent, and a broad range of alcohols could transform into the corresponding carbonate esters with good yields. $NaAlO_2$, prepared by a simple spray drying method, acts as a highly active heterogeneous and recyclable catalyst with excellent activity.

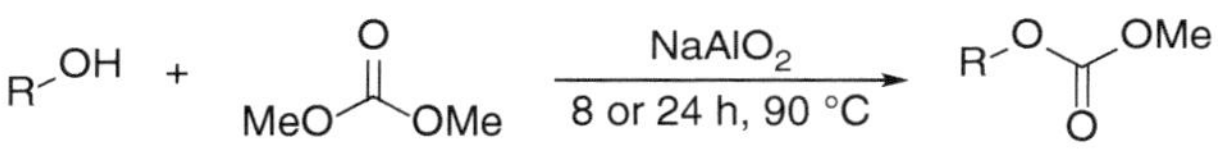

Scheme 5.5 $NaAlO_2$-catalyzed carboxymethylation of DMC.

Wu and coworkers reported a novel palladium-catalyzed strategy for synthesizing methyl 2-arylacetates through a direct carbonylative transformation of benzyl amines (Scheme 5.6) [8]. The reaction showed good tolerance toward various amines, including primary, secondary, and tertiary amines, resulting in a variety of methyl 2-arylacetates in good-to-excellent yields. Notably, the reaction was

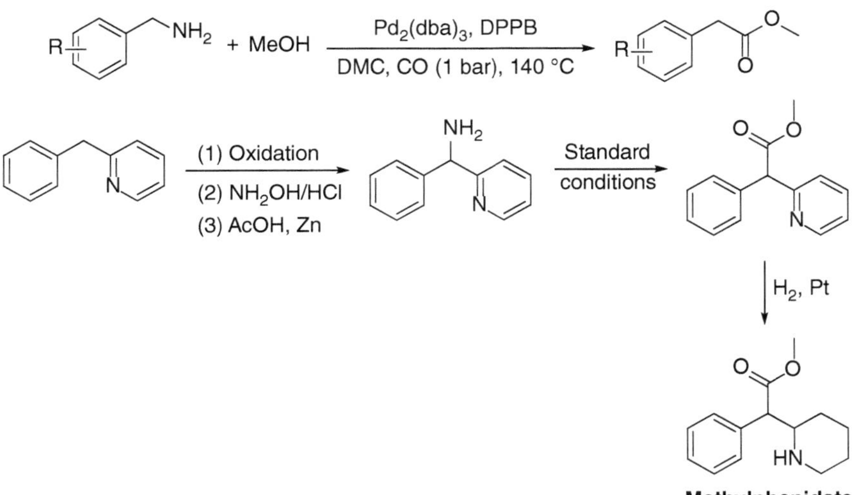

Scheme 5.6 Palladium-catalyzed carbonylative transformation of benzyl amines in DMC.

conducted with DMC as a green solvent, and no base or additional additives were needed. In addition, this approach could be successfully applied in the construction of methylphenidate [9], which is a marketing drug used in the treatment of narcolepsy and hyperactivity disorders.

5-Hydroxymethylfurfural (HMF), a platform molecule that could be used as an important intermediate to access a variety of high-value compounds. Thus, significant efforts have been put into their synthesis. The described method allows for the preparation of HMF using DMC as the main solvent under very mild reaction conditions. Amberlyst-15 or $BF_3O(Et)_2$ could be used as the catalyst to promote the triple dehydration of D-fructose efficiently (Scheme 5.7). Large-scale reaction (up to 20 g) could also be performed successfully to give HMF in good yield. Additionally, the reduction of HMF to BHF was carried out, and an overall 66% yield was obtained from the corresponding starting material, D-fructose [10].

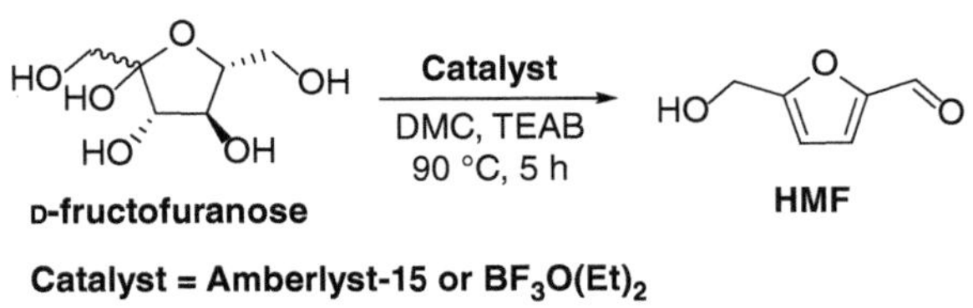

Scheme 5.7 Preparation of HMF from D-fructose in DMC.

Bolm and coworkers have developed a new approach for the synthesis of sulfonimidates and sulfonimidamides under metal-free conditions (Scheme 5.8) [11]. In the presence of *N*-methylpiperidine, diazonium salts reacted with *N*-tritylsulfinylamine and hydroxybenzotriazole hydrate, using DMC as an inexpensive and environmentally benign solvent, resulting in a variety of *N*-trityl-*O*-benzotriazolylsulfonimidates with good yields with notable functional group tolerance. Furthermore, sulfonimidamides were obtained from the corresponding sulfonimidates with the assistance of both aliphatic primary and secondary amines through a radical mechanism.

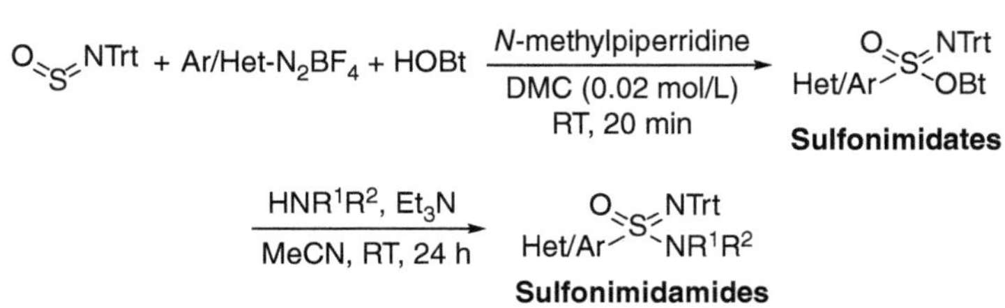

Scheme 5.8 Synthesis of sulfonimidates and sulfonimidamides in DMC.

Ollevier's group disclosed an efficient asymmetric copper(I)-catalyzed method for the insertion of 1-aryl-2,2,2-trifluoro-1-diazoethanes into Si—H bonds (Scheme 5.9) [12]. This method features cost-effective, excellent enantioselectivities, and mild reaction conditions, using $Cu(CH_3CN)_4PF_6$/bis((2,6-dichlorobenzylidene)diimino) cyclohexane catalytic system at room temperature, with DMC as a green solvent, a

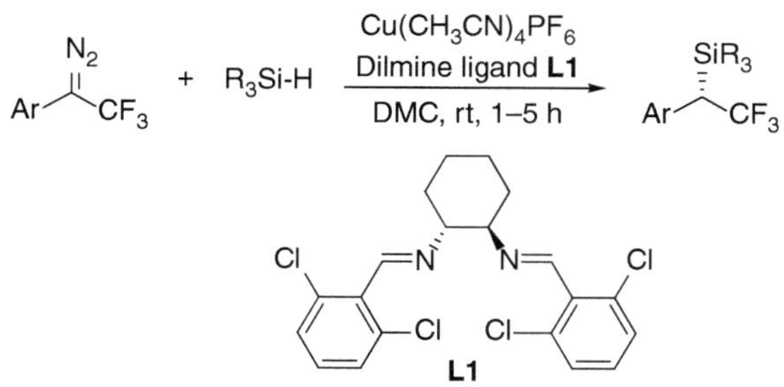

Scheme 5.9 Asymmetric copper-catalyzed Si–H insertion of 1-aryl-2,2,2-trifluoro-1-diazoethanes in DMC.

wide range of (1-aryl-2,2,2-trifluoroethyl)silanes were obtained in excellent yields with very high stereoselectivities (up to 98:2 *er*). Derivatization of these obtained (1-aryl-2,2,2-trifluoroethyl)silanes via a modified Tamao–Fleming oxidation was demonstrated to provide a practical method for the synthesis of the corresponding alcohols while retaining stereochemistry.

An insertion reaction of α-diazo compounds into Si—H bonds has been successfully developed (Scheme 5.10) [13]. The reaction features the application of $Fe(OTf)_2$ as an efficient catalyst and a wide range of α-silylated products were prepared in good-to-excellent yields by using DMC as a suitable green alternative to chlorinated solvents. The mechanistic study showed that the formation of iron carbine intermediate is a rate-limiting step. Moreover, polar X—H bonds, including S—H, N—H, and O—H bonds could react successfully with methyl α-phenyl-α-diazoacetate to form the corresponding products in good yields under standard reaction conditions.

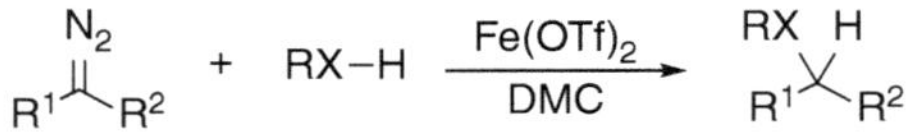

Non-polar insertions: X = Si
Polar insertions: X = O, N, S

Scheme 5.10 Iron-catalyzed insertion reaction of α-diazocarbonyls in DMC.

An iron(III)-catalyzed synthesis of α-substituted homoallylamines has been described via a cationic 2-aza-Cope rearrangement (Scheme 5.11) [14]. With readily

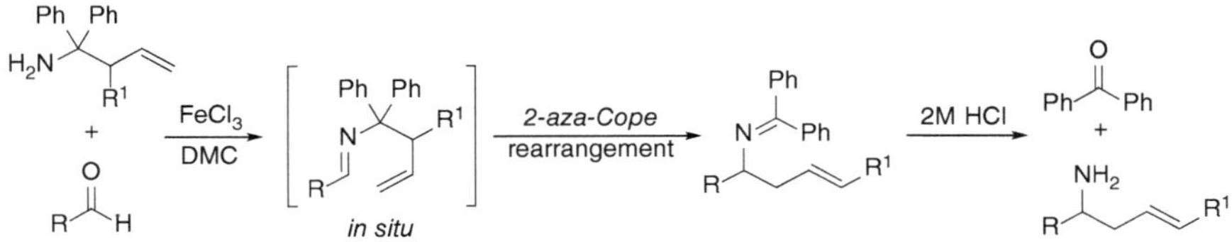

R = H, alkyl & heteroaryl
R^1 = H & Me

Scheme 5.11 Iron(III)-catalyzed synthesis of α-substituted homoallylamines.

available aldehydes as starting materials, a wide range of α-alkyl-, alkenyl-, aryl-, and heteroaryl homoallylamines were obtained in high yields with very broad substrate scope. Notably, DMC was used as an environmentally friendly solvent with water as the sole by-product. Furthermore, *N*-unprotected α-substituted homoallylamines could also be prepared under mild acidic reaction conditions from the corresponding ketamines.

An efficient Diels–Alder reaction of α,β-unsaturated carbonyl/*N*-acyloxazolidione derivatives with cyclopentadienes was achieved through a green catalytic system using ionic salt/iron triflate as recyclable catalysts (Scheme 5.12) [15]. The catalyst was prepared from iron(II) triflate combined with a caffeine-derived xanthinium salt. A variety of cycloadducts were synthesized in high yields with low catalyst loading (1 mmol%) by applying DMC as a green solvent. Finally, in recycling tests for the catalyst, the ionic salt/iron triflate catalyst showed good reactivity and the yield of the product slightly decreased after 5 runs.

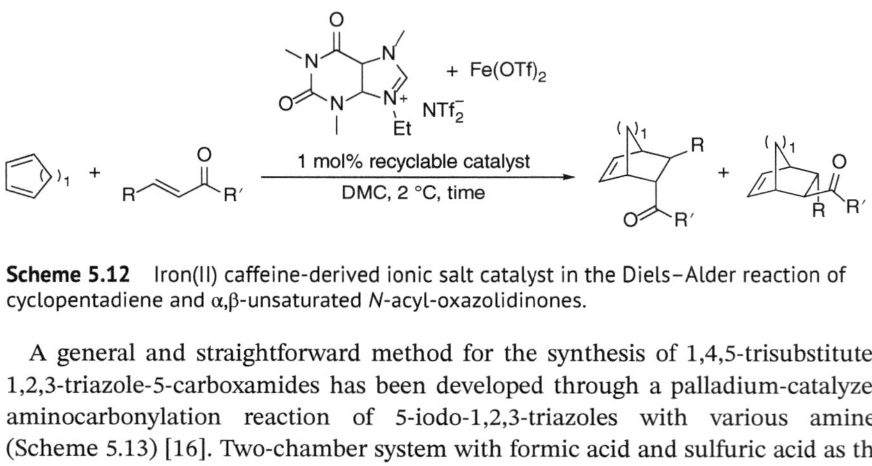

Scheme 5.12 Iron(II) caffeine-derived ionic salt catalyst in the Diels–Alder reaction of cyclopentadiene and α,β-unsaturated *N*-acyl-oxazolidinones.

A general and straightforward method for the synthesis of 1,4,5-trisubstituted 1,2,3-triazole-5-carboxamides has been developed through a palladium-catalyzed aminocarbonylation reaction of 5-iodo-1,2,3-triazoles with various amines (Scheme 5.13) [16]. Two-chamber system with formic acid and sulfuric acid as the CO source was applied. The reaction features phosphine ligands-free conditions, and DMC served as a green solvent, which makes this strategy more sustainable. A variety of 1,4,5-trisubstituted 1,2,3-triazole-5-carboxamides were obtained in good-to-high excellent yields with good substrate scope.

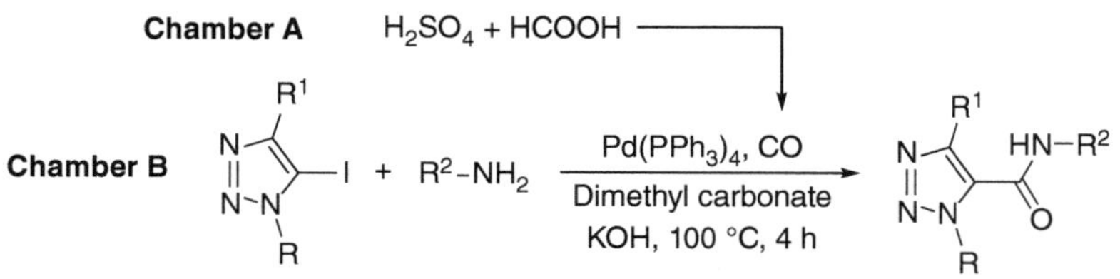

Scheme 5.13 Palladium-catalyzed aminocarbonylation reaction of 5-iodo-1,2,3-triazoles with various amines.

Darcel and coworkers developed a chemoselective approach for the synthesis of N-substituted cyclic amines through iron-catalyzed hydrosilylation of dicarboxylic acids with amines (Scheme 5.14) [17]. By using hydrosilanes as the hydride sources, with DMC as a green solvent, a series of cyclic amines were obtained in good yields with good functional group tolerance. Furthermore, the present strategy allowed for the preparation of drug molecules, Fenpiprane and Prozapine.

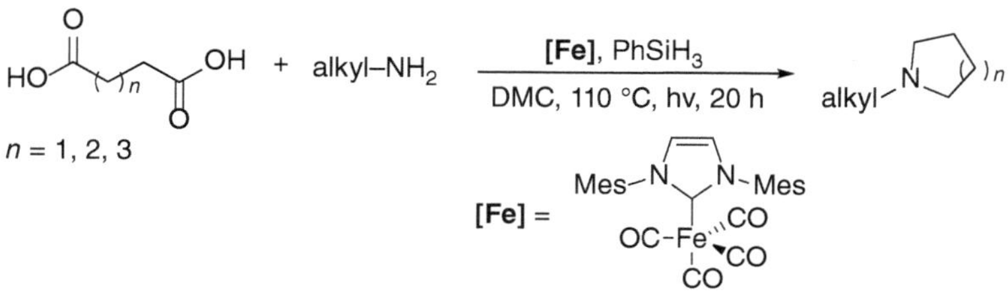

Scheme 5.14 Iron-catalyzed hydrosilylation of dicarboxylic acids with amines.

An interesting fluoride-triggered catalytic system for the preparation of 1-aryl-2,2-difluoroalkenes via 1,2-desilylative defluorination process has been developed (Scheme 5.15) [18]. With DMC as a green solvent at room temperature, a series of gem-difluoroalkenes were obtained in very high yields from the corresponding (1-aryl)-2,2,2-trifluoroethylsilanes in the presence of a catalytic amount of fluoride source. This method allowed for one-pot synthesis of perfluoroalkyl substituted alkenes from perfluoroalkyl-silanes, which was easily generated from copper(I)-catalyzed insertion of perfluoroalkyl diazo substrates into Si—H bond of tertiary organosilanes. When 1,2-desilylative defluoriantion occurred in a concerted manner, (*Z*)-1-benzylideneperflfluoroalkanes were formed as the major (*Z*)-isomer. Furthermore, (*E*)-β-(perfluoroalkyl)styrene derivatives could be obtained with very high yields and complete stereoselectivity upon hydrodefluorination of 1-benzylideneperfluoroalkanes.

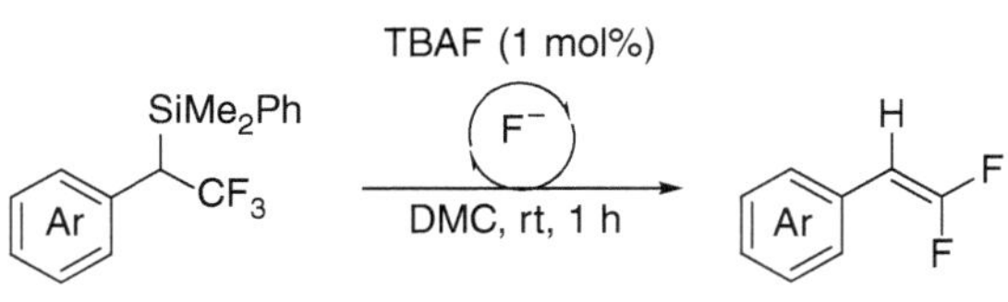

Scheme 5.15 Preparation of 1-aryl-2,2-difluoroalkenes via 1,2-desilylative defluorination.

As an environmentally friendly compound, DMC exhibits diverse reactivities for multiple applications and plays an important role as a versatile building block, such as DMC-mediated methoxycarbonylating and methylating reactions. Due to their intrinsic nature, a series of cases of applying DMC as both green solvent and reagent will also be discussed here.

A metal-free method for synthesizing cyclic carbonates from oxiranes with carbon dioxide has been achieved in the presence of catalytic amount of tetrabutyl ammonium bromide (Scheme 5.16) [19]. It was the first time using DMC as both a recyclable reaction medium and a promoter to accelerate the reaction rate, and

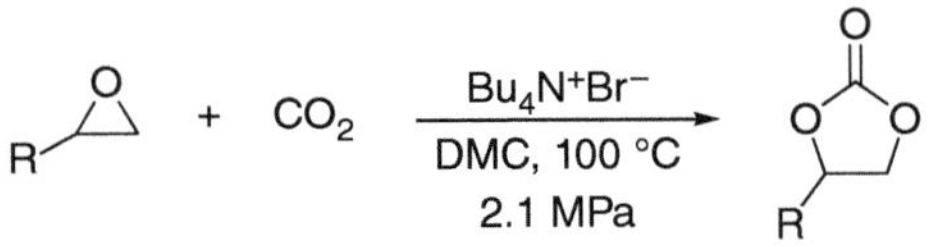

Scheme 5.16 Synthesis of cyclic carbonates from oxiranes with carbon dioxide.

the desired cyclic carbonate products were obtained in very good yields at 100 °C and 2.1 MPa with 1 mol% of TBAB under metal-free conditions. The DMC immobilized TBAB could be easily recycled for at least six runs with almost quantitative conversion.

The synthesis of five-membered *N*-heterocyclic compounds could be easily achieved through a cyclization between aliphatic and aromatic 1,4-bifunctional compounds with DMC as both reagent and solvent (Scheme 5.17) [20]. Under mild reaction conditions, the desired products were obtained in high yields without any chlorine-based chemicals and strong acids. Furthermore, DMC could be recovered and reused, which makes this approach more environmentally friendly.

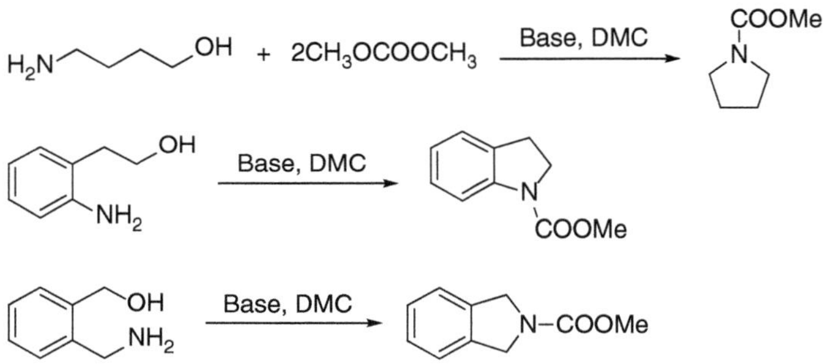

Scheme 5.17 Cyclization between aliphatic and aromatic 1,4-bifunctional compounds with DMC as both reagent and solvent.

Meier and coworkers reported a straightforward and efficient method for the direct conversion of saturated fatty acid methyl esters to malonate derivatives bearing long aliphatic residues (Scheme 5.18) [21]. In the presence of sodium hydride, the desired products were obtained in good yields by using DMC as the solvent. These malonates bearing long aliphatic chains could be further polymerized to form polyesters, which are highly viscous and tacky materials with low melting points. While polyamides, with very high melting points, could also be constructed from the above malonates.

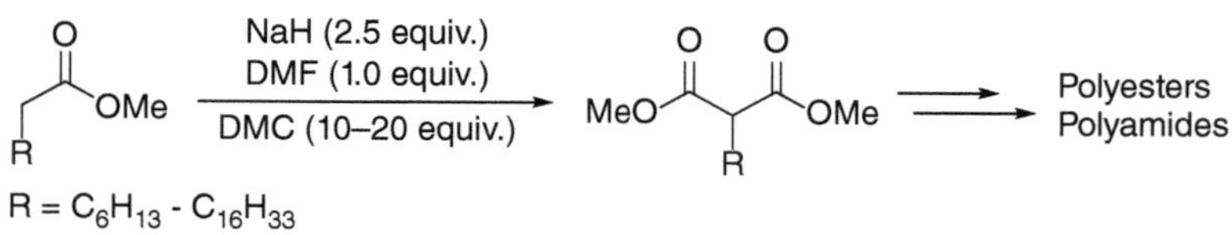

Scheme 5.18 Synthesis of malonate derivatives from saturated fatty acid methyl esters in DMC.

Levulinic acid (LA), a precious C5 building block with wide accessibility, low cost, and renewable source, is used as a useful platform molecule for chemical synthesis. The reaction of LA with DMC as both a reagent and a solvent has been described (Scheme 5.19) [22]. By tuning the reaction conditions, methyl levulinic could be obtained selectively along with dimethyl succinate and the dimethyl ketal of methyl levulinate.

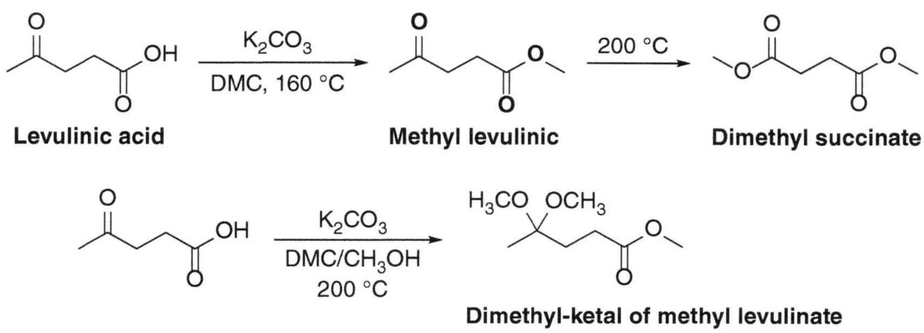

Scheme 5.19 Upgrading of levulinic acid with DMC as solvent/reagent.

A novel and efficient Lossen rearrangement has been explored using catalytic amounts of 1,5,7-triazabicyclo[4.4.0]dec-5-ene (TBD) (Et_3N, DBU, or DABCO) (Scheme 5.20) [23]. Both aliphatic and aromatic hydroxamic acids were well-tolerated in this reaction by employing DMC as *in situ* activation reagent. Methyl carbamate products were produced in moderate-to-good yields from the corresponding aliphatic hydroxamic acids, while using aromatic hydroxamic acids, anilines were obtained from the hydrolysis of methyl carbamates. Notably, the reagent/solvent mixture of DMC/methanol could be recycled several times without observing decreased yields by simple distillation.

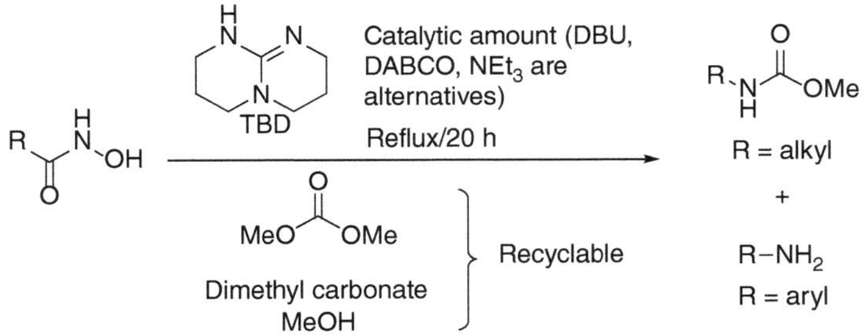

Scheme 5.20 Upgrading of levulinic acid with DMC as solvent/reagent.

The methylation of secondary amines has been achieved using a well-defined iron catalyst bearing an *N*-heterocyclic carbene ligand (Scheme 5.21) [24]. This reaction can be carried out under mild reaction conditions with hydrosilanes as the reductant, with DMC as a C1 source and solvent, resulting in the formation of corresponding amines in good-to-excellent yields. Furthermore, this methodology could

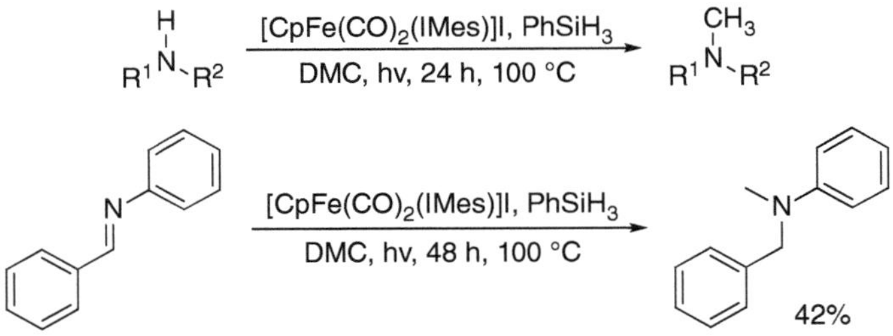

Scheme 5.21 Iron-catalyzed methylation of secondary amines and imines.

be applied in a one-pot reduction–methylation reaction of *N*-benzylideneaniline without isolation of the *N*-benzylaniline intermediate, and the desired product was isolated in a 42% yield.

Bolm's group disclosed an attractive method for the cleavage reaction of lignin β-*O*-4 model compounds with catalytic amounts of base (Scheme 5.22) [25]. By using Cs_2CO_3 as the catalyst, DMC as both a solvent and a reagent, the corresponding methoxy benzene derivatives were obtained as major products in good yield, while 2-aryloxyvinyl benzene derivatives were produced in good-to-excellent yields when the substrates with more hindered group substituted at the aryl ring next to the benzylic hydroxyl group. Moreover, when LiO*t*-Bu was employed as the catalyst, 2-aryloxyvinyl benzene derivatives were generated in high yields from the corresponding non-phenolic model compounds.

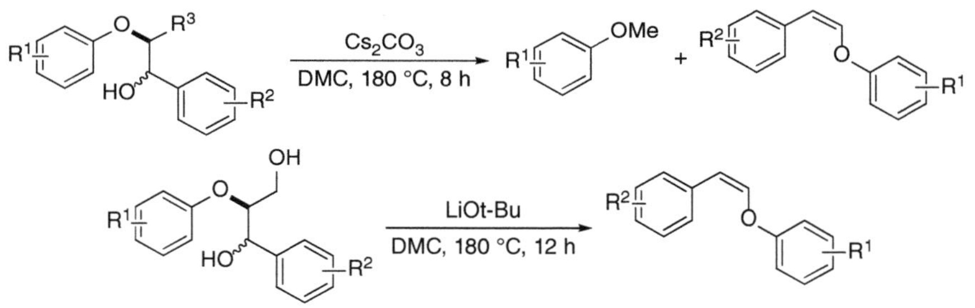

Scheme 5.22 Base-catalyzed cleavage of lignin β-*O*-4 model compounds in DMC.

A general palladium-catalyzed direct coupling of benzyl/allyl alcohols with malonates has been explored via a Tsuji–Trost type reaction (Scheme 5.23) [26].

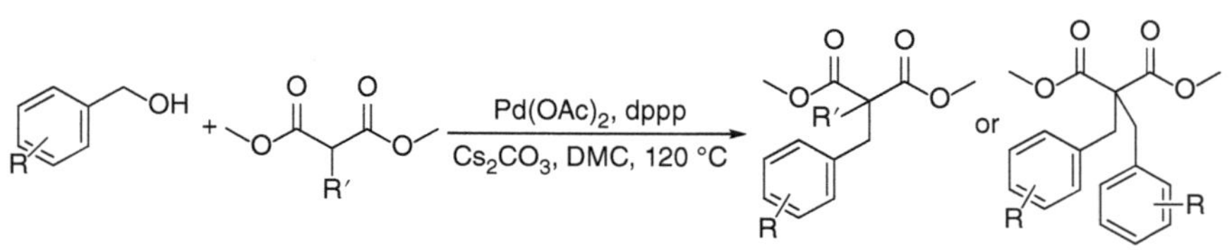

Scheme 5.23 Palladium-catalyzed coupling of benzyl/allyl alcohols with malonates in DMC.

Instead of traditional pre-activated carbonates, DMC was used as both the solvent and the *in situ* activator for benzyl/allyl alcohols. The reaction allowed the selective formation of mono-substituted and bis-substituted malonates in good yields under different reaction conditions. The process demonstrates a step- and atom-economical method for directly using alcohols in coupling reactions.

An environmentally benign protocol for the synthesis of *N*-alkyloxaziridines has been reported for the first time (Scheme 5.24) [27]. Under the DMC/H_2O_2 system, with *N*-alkylamines and aldehydes as the starting materials, a series of *N*-alkyloxaziridines were obtained in good yields. In this reaction, DMC was used as both the green solvent and reagent that would undergo perhydrolysis to *in situ* generate the active oxidant, monoperoxy carbonic acid methyl ester, which leads to the formation of only carbon dioxide and methanol after the epoxidation of the imines. Moreover, *N*-sulfonyloxaziridines could also be prepared in high yields in the presence of 5 mol% of $Zn(OAc)_2 \cdot 2H_2O$ under basic conditions from the corresponding *N*-sulfonylimines.

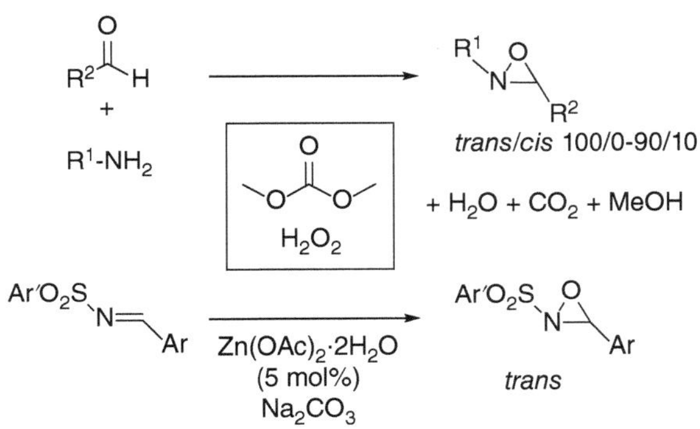

Scheme 5.24 Synthesis of *N*-alkyloxaziridines with alkylamines and aldehydes.

Wu and coworkers reported a sustainable palladium-catalyzed carbonylation reaction for the synthesis of alkyl arylacetates (Scheme 5.25) [28]. By directly using benzyl alcohols as the benzylic substrates without any halogen additives, and with DMC as both the green solvent and *in situ* activator, a series of alkyl arylacetates were formed in moderate-to-excellent yields.

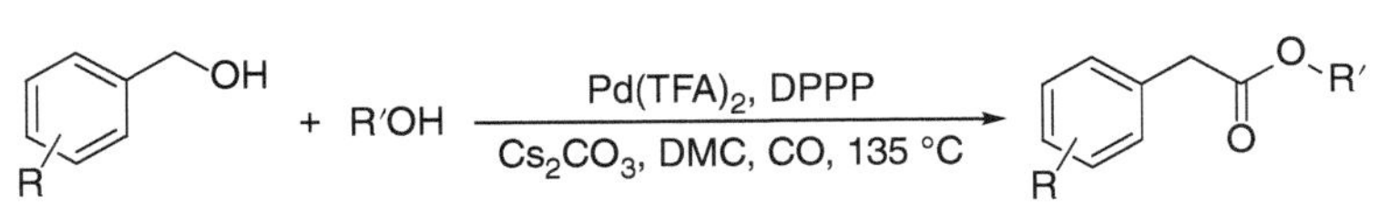

Scheme 5.25 Palladium-catalyzed carbonylation of f benzyl alcohols for the synthesis of alkyl arylacetates.

A novel synthesis of non-symmetrical alkyl carbonates has been developed via a heterogeneous acid (NZSM-5 zeolite)-catalyzed carboxymethylation of alcohols (Scheme 5.26) [29]. To avoid the use of toxic reagents, DMC was applied as both

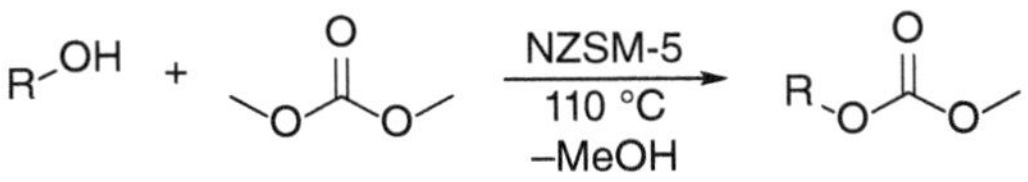

Scheme 5.26 NZSM-5 Zeolite catalyzed synthesis of non-symmetrical alkyl carbonates.

the green solvent and reagent in this reaction. The protocol offers an environmentally benign pathway to access a wide range of non-symmetrical alkyl carbonates in very good yields. Notably, the catalyst could be reused for up to five cycles without obvious activity loss.

An efficient and sustainable catalytic system has been developed using $MnCO_3$ as a heterogeneous catalyst (Scheme 5.27) [30]. The calcined $MnCO_3$ showed good performance for the transesterification of alcohols with DMC, which was selected as both the reagent and green solvent, providing a wide range of final products in high yields and good selectivity. This $MnCO_3$-based catalysis highights the potential applications of DMC in organic synthesis.

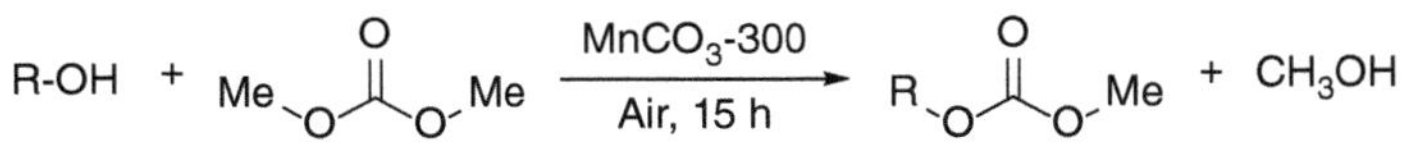

Scheme 5.27 $MnCO_3$-300 catalyzed transesterification of alcohols.

A borrowing carbonate-enabled *N*-allylation has been discovered under additive-free nickel catalysis through an acyl/allyl C—O bond activation (Scheme 5.28) [31]. The reaction is initiated by an acyl C—O bond activation of DMC, resulting in an acyl–Ni complex, which then undergoes nucleophilic trapping with allylic alcohol to afford mixed carbonate. The π-allyl Ni complex, generated from the coordination of Ni to the C—C double bond of the mixed carbonate, and subsequent oxidative addition, followed by allylic cross-amination, leading to the formation of an allylic amination product. The reaction was performed under additive-free conditions with DMC as the solvent, resulting in a wide broad range of *N*-heterocycles with very good chemo-, regio-, and stereoselectivity. Furthermore, scale-up preparation of pharmaceuticals (cinnarizine, flunarizine, and naftifine) could be successfully achieved by using this sequential and selective allylic amination protocol.

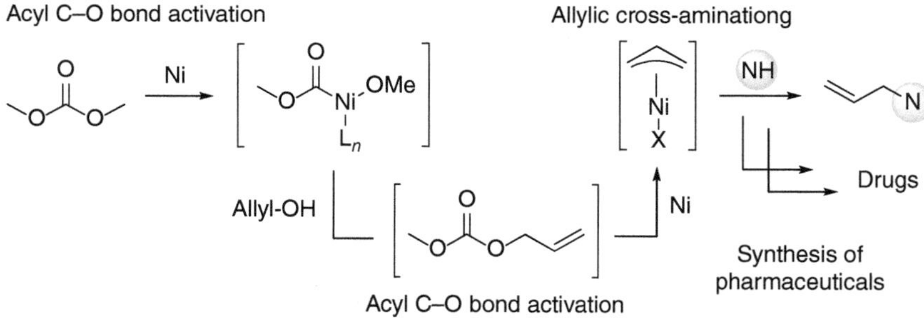

Scheme 5.28 Borrowing carbonate-enabled allylic amination reactions in DMC.

References

1 (a) Shaikh, A.-A.G. and Sivaram, S. (1996). *Chem. Rev.* 96: 951–976. (b) Horvath, I.T. (2008). *Green Chem.* 10: 1024–1028. (c) Schäffner, B., Schäffner, F., Verevkin, S.P., and Börner, A. (2010). *Chem. Rev.* 110: 4554–4581. (d) Huang, S., Yan, B., Wang, S., and Ma, X. (2015). *Chem. Soc. Rev.* 44: 3079–3116.

2 Fukuoka, S., Tojo, M., Hachiya, H. et al. (2007). *Polym. J.* 39: 91–114.

3 Inamoto, K., Hasegawa, C., Hiroya, K. et al. (2012). *Chem. Commun.* 48: 2912–2914.

4 Ramachandran, G., Mandlimath, T.R., Sathesh, V. et al. (2014). *Chem. Lett.* 43: 1631–1633.

5 Jacobson, C.E., Martines-Muñoz, N., and Gorin, D.J. (2015). *J. Organomet. Chem.* 80: 7305–7310.

6 Francesco, I.N., Cacciuttolo, B., Pucheault, M., and Antoniotti, S. (2015). *Green Chem.* 17: 837–841.

7 Ramesh, S., Indukuri, K., Riant, O., and Debecker, D.P. (2018). *Org. Process Res. Dev.* 22: 1846–1851.

8 Li, Y., Wang, Z., and Wu, X.-F. (2018). *ACS Catal.* 8: 738–741.

9 (a) Kimko, H.C., Cross, J.T., and Abernethy, D.R. (1999). *Clin. Pharmacokinet.* 37: 457–470. (b) Markowitz, J.S., Straughn, A.B., and Patrick, K.S. (2003). *Pharmacotherapy* 23: 1281–1299.

10 Musolino, M., Andraos, J., and Aricò, F. (2018). *ChemistrySelect* 3: 2359–2365.

11 Bremerich, M., Conrads, C.M., Langletz, T., and Bolm, C. (2019). *Angew. Chem. Int. Ed.* 58: 19014–19020.

12 Carreras, V., Besnard, C., Gandon, V., and Ollevier, T. (2019). *Org. Lett.* 21: 9094–9098.

13 Tanbouza, N., Keipour, H., and Ollevier, T. (2019). *RSC Adv.* 9: 31241–31246.

14 Gadde, K., Daelemans, J., Maes, B.U.W., and Tehrani, K.A. (2019). *RSC Adv.* 9: 18013–18017.

15 Meng, D., Li, D., and Ollevier, T. (2019). *RSC Adv.* 9: 21956–21963.

16 de Albuquerque, D.Y., de Moraes, J.R., and Schwab, S. (2019). *Eur. J. Org. Chem.* 6673–6681.

17 Wei, D., Netkaew, C., Wu, J., and Darcel, C. (2020). *ChemCatChem* 12: 5449–5455.

18 Carreras, V. and Ollevier, T. (2021). *J. Organomet. Chem.* 86: 13160–13168.

19 Kumar, S., Jain, S.L., and Sain, B. (2011). *Tetrahedron Lett.* 52: 6957–6959.

20 Aricò, F., Toniolo, U., and Tundo, P. (2012). *Green Chem.* 14: 58–61.

21 Kolb, N. and Meier, M.A.R. (2012). *Green Chem.* 14: 2429–2435.

22 Caretto, A. and Perosa, A. (2013). *ACS Sustainable Chem. Eng.* 1: 989–994.

23 Kreye, O., Wald, S., and Meier, M.A.R. (2013). *Adv. Synth. Catal.* 355: 81–86.

24 Zheng, J., Darcel, C., and Sortais, J.-B. (2014). *Chem. Commun.* 50: 14229–14232.

25 Dabral, S., Mottweiler, J., Rinesch, T., and Bolm, C. (2015). *Green Chem.* 17: 4908–4912.

26 Cao, X. and Zhang, Y. (2016). *Green Chem.* 18: 2638–2641.

27 Kraïem, J., Ghedira, D., and Ollevier, T. (2016). *Green Chem.* 18: 4859–4864.

28 Li, Y., Wang, Z., and Wu, X.-F. (2018). *Green Chem.* 20: 969–972.

29 Chevella, D., Macharla, A.K., Banothu, R. et al. (2019). *Green Chem.* 21: 2938–2945.

30 Bi, X., Yao, N., Meng, X. et al. (2021). *Catal. Lett.* 151: 454–462.

31 Vaidya, G.N., Nagpure, M., and Kumar, D. (2021). *ACS Sustainable Chem. Eng.* 9: 1846–1855.

6

Applications of Green Deep Eutectic Solvents (DESs) in Synthetic Transformations

Zhuan Zhang and Taoyuan Liang

Guangxi University, School of Chemistry and Chemical Engineering, Nanning, Guangxi, 530004, P. R. China

6.1 Introduction

The rising concerns about environmental pollution and the energy crisis have resulted in an urgent demand for green chemistry [1]. The aim of green chemistry is to design environmentally benign chemical processes and synthetic methodologies to reduce or eliminate the utilization of hazardous and toxic chemicals at any stage of production in the laboratory or industrial scales. There is no doubt that sustainable synthetic strategies are needed in response to environmental problems [2]. Solvents always play a significant role in the synthesizing process. Conventional solvent systems can be difficult to meet the increasing demand for production and research. The vital reason is that safety and sustainability are as important as productivity [3]. Compared to traditional solvents, deep eutectic solvents (DESs) are regarded as an emerging class of inexpensive, biodegradable, non-toxic, and easier-to-prepare sourced solvents, which have gained great attention after Abbott first mentioned this term in 2003 to elaborate a new generation of sustainable solvents. There are at least two compounds to combine for the formation of DESs. When the compounds were mixed in a suitable molar ratio, they had melting points much lower than those of either individual component to produce a DESs. One of the most widespread components employed for the formation of DESs is choline chloride (ChCl), which is a very cheap, nontoxic, and biodegradable quaternary ammonium salt and can be either extracted from biomass or readily prepared from fossil reserves through a higher atom economic process. For example, when the complexation of ChCl with urea in a 2 : 1 molar ratio (Scheme 6.1), the ChCl/urea mixture which yields a eutectic solvent has a melting point of 12 °C, which is abnormally deep lower than the individual components (the melting point of pure ChCl and urea is 302 °C and 133 °C, respectively) [4].

According to the types of compounds that are mixed to generate the eutectic mixture, DESs can be classified into five different categories. Different categories of DESs are summarized in Scheme 6.2. Among these categories, the types I, II, and IV DESs contain metal salts and are widely applied in metal processing

Green Solvents in Organic Synthesis, First Edition.
Edited by Xiao-Feng Wu, Zhiping Yin, Liang-nian He, and Feng Wang.

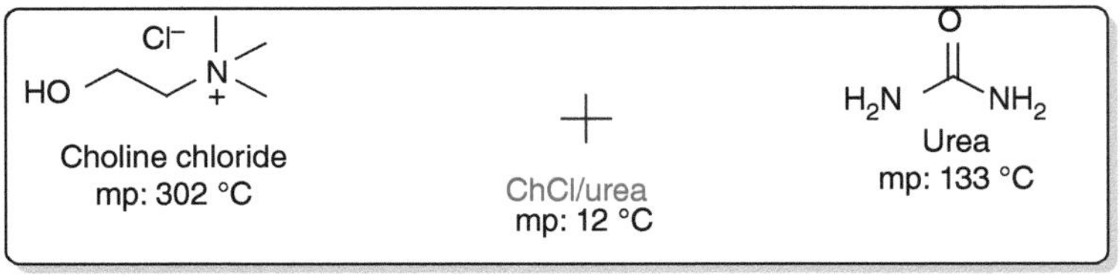

Scheme 6.1 The case of choline chloride + urea DES.

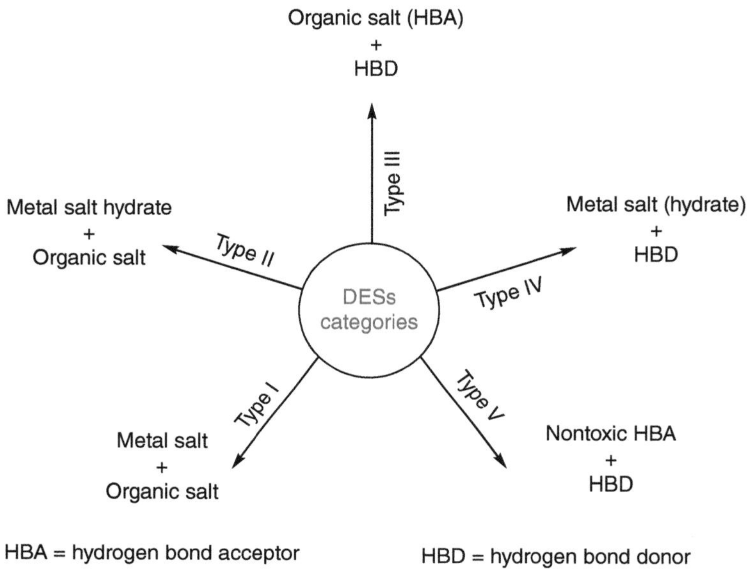

Scheme 6.2 Deep eutectic solvents (DESs) classification.

applications. Type V DESs are a relatively new class composed of only nonionic, molecular hydrogen bond acceptors (HBAs) and hydrogen bond donors (HBDs). Therefore, the overwhelming majority of research, especially for the properties and applications in the energy and fuels area, has focused on Type III DESs. Type III DESs are formed from organic salts HBAs and HBDs (typically organic molecular components such as amides, carboxylic acids, or polyols), which are strongly associated *via* hydrogen bond interactions. This leads to a decrease in the lattice energy of the system and thus, the melting point [5].

Representative structures of HBAs and HBDs commonly used for DESs synthesis are displayed in Scheme 6.3 [6].

Traditional ionic liquids (ILs) have been intensively investigated as an alternative reaction media due to their properties (*e.g.* high thermal stability, negligible vapor pressures, conductivity, nonflammability, and easy recycling), DESs also exhibit similar physicochemical properties to the ILs. As compared to ILs, DESs have other obvious advantages. First, the atom economy of DESs' convenient synthesis is 100%. Second, the price of DESs is very low since most DESs can be prepared from readily accessible chemicals. Thirdly, DESs are low toxic as they are derived from

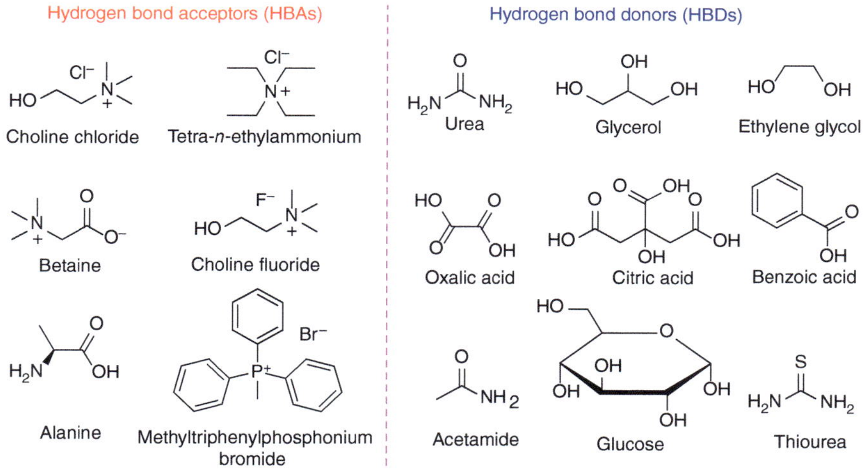

Scheme 6.3 Representative structures of HBAs and HBDs used for DESs synthesis.

ChCl and renewable chemicals. Moreover, DESs do not generate toxic metabolites and are biodegradable. Based on their minimal ecological footprint, tunability of their physicochemical properties, cheapness of their constituents, and ease of preparation, DESs are a better choice to replace often hazardous and volatile organic compounds in many fields of science [7].

Due to the aforementioned features of DESs, it is not surprising that DESs have found a wide variety of applications in different fields of modern chemistry, including organic synthesis, biocatalytic reactions, dissolution of metal oxides, electrodeposition of metals, and material chemistry [8]. This brief chapter aims to cover the progress made in the application of DESs as environment friendly reaction media, catalysts, or even reagents in various synthetic transformations, including cross-coupling reactions, oxidative reactions, hydrogenation reactions, cyclization reactions, condensation reactions, multicomponent reactions, other organic reactions, and polymerization, which are hot research areas in strong expansion, with a focus on the published works in recent years.

6.2 Cross-coupling Reactions in Deep Eutectic Solvents

Cross-coupling reactions conducted in DESs have become a powerful tool for the preparation of valuable organic compounds due to the increase of sustainability of the process, which has attracted the interest of many researchers worldwide since the first work reported by Imperato and coworkers [9].

6.2.1 C—C Bond Couplings

Suzuki-Miyaura reaction establishes a practical route for the direct C—C bond formation, which was widely applied in the synthesis of fine chemicals, functionalized materials, and pharmaceuticals. Suzuki reactions have been reported in organic solvents, in water, in supercritical carbon dioxide, and ILs. König et al. have extended the scope by using DESs as the solvent for the first time [10]. The desired biaryl products were obtained in good yield when phenyl boronic acid was coupled with aryl bromides in the sugar/urea/salt melts using $Pd(OAc)_2$ as a catalyst and Na_2CO_3 as a base (Scheme 6.4). Simple extraction can be used to work up the reaction mixture.

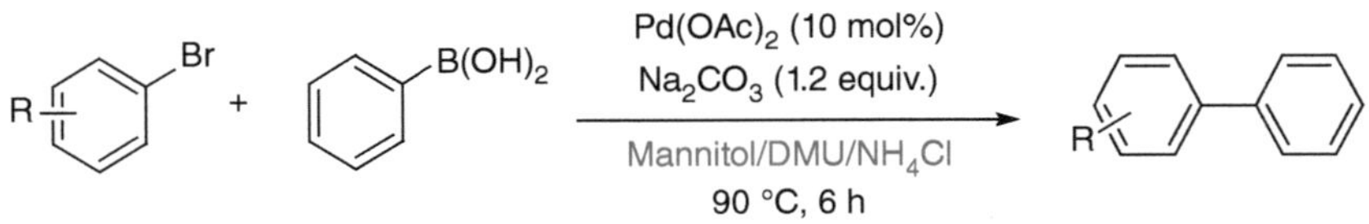

Scheme 6.4 Suzuki–Miyaura reactions performed in melt.

Cyclodextrin is widely used in food, chemical industries, pharmaceutical, agriculture, and environmental domains. An efficient catalytic system in the presence of fresh native-CD-capped Pd^0 nanoparticles has been successfully explored by Zhao's team (Scheme 6.5) [11]. This natural and convenient protocol can make Suzuki–Miyaura reactions perform well in the air to give the corresponding products with a very low Pd loading (0.05 mol%). Moreover, the catalytic system can be recycled and reused without loss of catalytic activity.

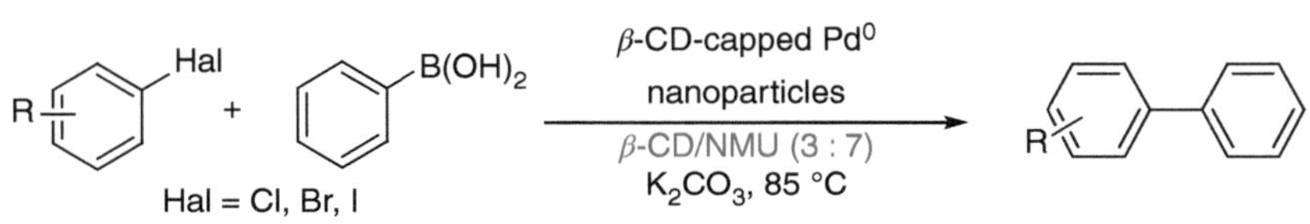

Scheme 6.5 Suzuki–Miyaura reactions performed in melt.

Since the traditional palladium ligands for Suzuki–Miyaura reactions seemed to be ineffective in DESs. Cationic pyridiniophosphine ligands have been prepared by Ramón and coworkers in order to develop DESs compatible catalytic systems [12]. The structurally engineered phosphine ligands with $PdCl_2$ have been successfully employed in Suzuki–Miyaura coupling reaction using both aryl bromides and iodides in a ChCl-based DES (Scheme 6.6). Notably, DES together with the catalyst could be recovered and reused up to five times without an obvious drop in the catalytic activity.

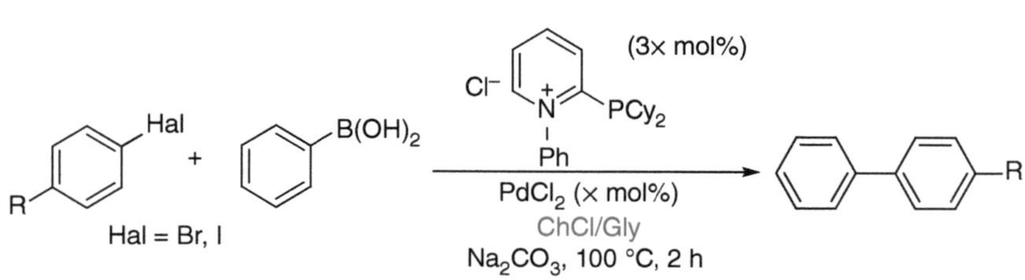

Scheme 6.6 Ligand-based Suzuki–Miyaura coupling with $PhB(OH)_2$ in DES.

To promote the catalytic activities, the design and synthesis of organometallic complexes compatible with DESs is a great challenge for industrial applications. In 2019, a DES-compatible bipyridine palladium complex was newly prepared by the same group to perform the Suzuki–Miyaura reaction using ChCl/ethylene glycol (1 : 2) with a variety of functionalized aryl bromide derivatives. The catalyst and DES could be easily and successfully reused (Scheme 6.7) [13].

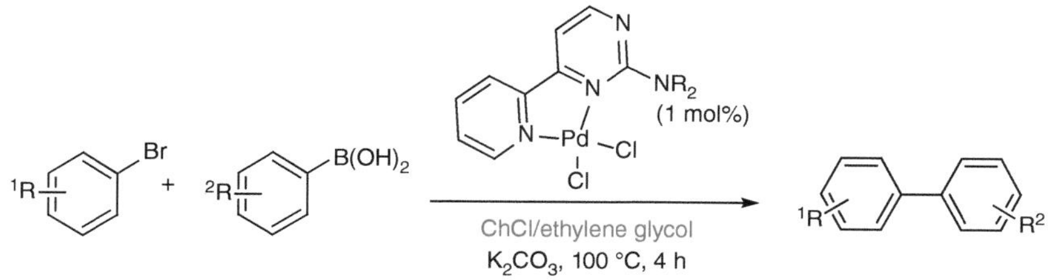

Scheme 6.7 DES-compatible bipyridine palladium for Suzuki–Miyaura reaction.

Due to the less work focus on the synthesis of heterocyclic products using sustainable natural DES, the first Suzuki–Miyaura cross-coupling in DES applied to biologically relevant imidazo-fused scaffolds of imidazo [1,2-*a*]pyridine and imidazo [1,2-*b*] pyridazine have been reported (Scheme 6.8). The reaction can be conducted using readily available $Pd(OAc)_2$ as a catalyst and ChCl/glycerol as natural DES to allow the functionalization of diverse positions on the heterocycles. Remarkably, the catalytic system underwent well with no ligands or additives [14].

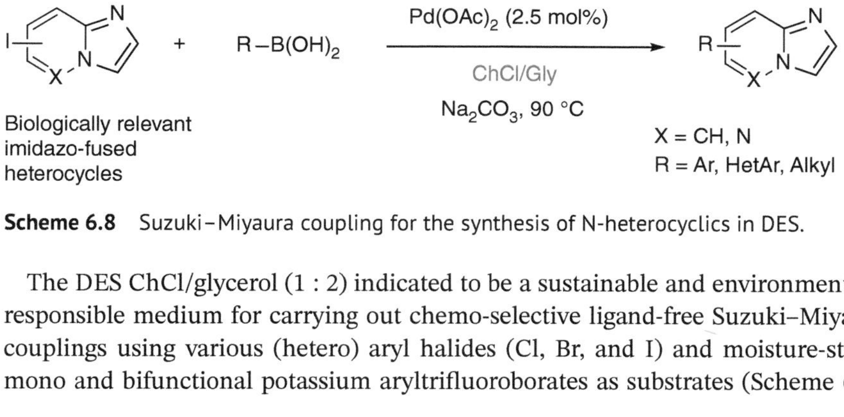

Scheme 6.8 Suzuki–Miyaura coupling for the synthesis of N-heterocyclics in DES.

The DES ChCl/glycerol (1 : 2) indicated to be a sustainable and environmentally responsible medium for carrying out chemo-selective ligand-free Suzuki–Miyaura couplings using various (hetero) aryl halides (Cl, Br, and I) and moisture-stable mono and bifunctional potassium aryltrifluoroborates as substrates (Scheme 6.9).

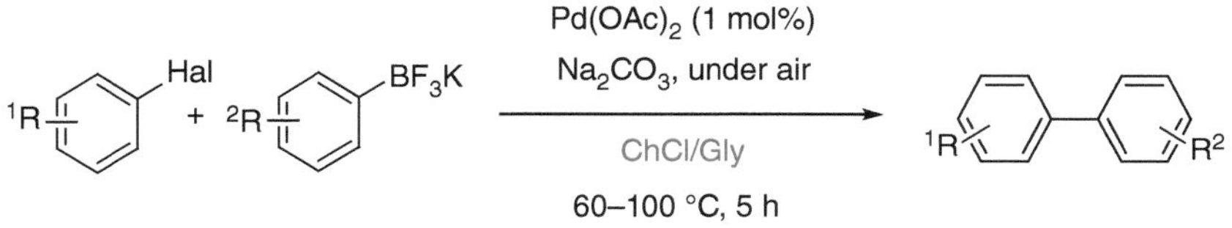

Hal = Cl, Br, I; ^{1}R, ^{2}R = BF_3K, alkyl, aryl, various functional groups

Scheme 6.9 Suzuki–Miyaura couplings of aryltrifluoroborates in DES.

The reaction is performed well under generally mild conditions with an efficient catalyst [1 mol% $Pd(OAc)_2$], Na_2CO_3 as a base. The methodology provided a greener route to prepare valuable biaryl and terphenyl derivatives [15].

An efficient magnetic cobalt nanocatalyst has been applied in the Suzuki reaction using nonhazardous and sustainable DESs as green medium (Scheme 6.10). The DES and catalyst were recovered easily with the help of a magnetic field and reused repeatedly five times with a slight, gradual decrease in the yield of the reactions. Moreover, the low-cost cobalt catalyst could be regarded as an alternative to the expensive Pd catalyst, making the present protocol more promising for the clean synthesis of biaryl products [16].

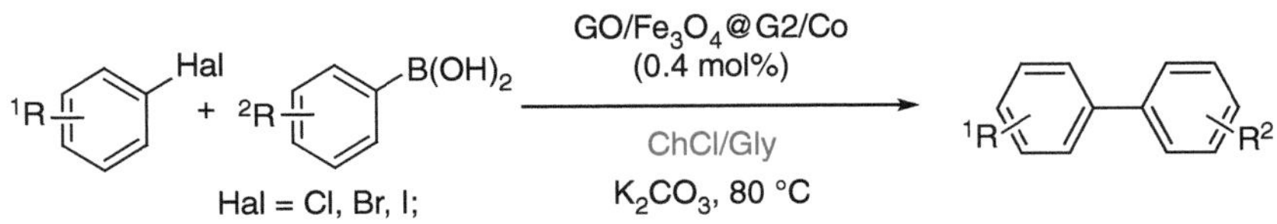

Scheme 6.10 GO/Fe_3O_4@G2/Co for Suzuki coupling reaction in DES.

The Mizoroki–Heck reaction is the key route for the vinylation of aryl/vinyl halides or triflates. There are many reports focusing on Mizoroki–Heck reaction in common organic solvents while very few examples run in DESs. König and coworkers have pioneered the studies on the coupling of iodoarenes with acrylates in two types of DESs indicating that *D*-mannose/DMU (3 : 7) is a better medium than one obtained from *L*-carnitine/urea (2 : 3). However, this sustainable catalytic work suffered from a lack of diversity (Scheme 6.11) [17].

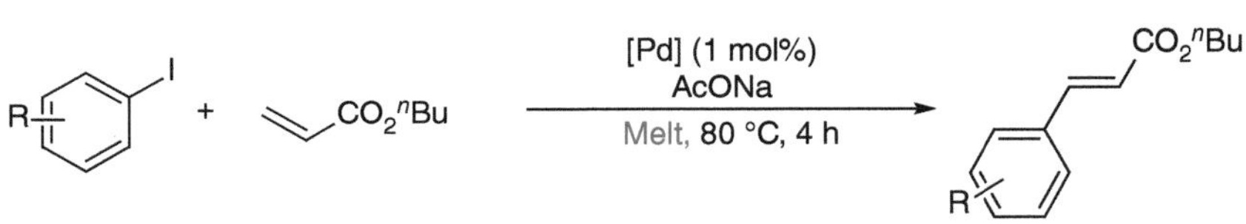

Scheme 6.11 Mizoroki–Heck reaction performed in melt.

Ramón and coworkers later reported the Mizoroki–Heck coupling of aryl iodides with methyl acrylate in ChCl-based eutectic mixtures using $PdCl_2$ with a cationic pyridiniophosphine ligand or a bipyridine–palladium catalyst complex (Scheme 6.12). All those results highlight the broad potential of DES in place of hazardous organic solvents in the cross-coupling reaction, just by choosing the

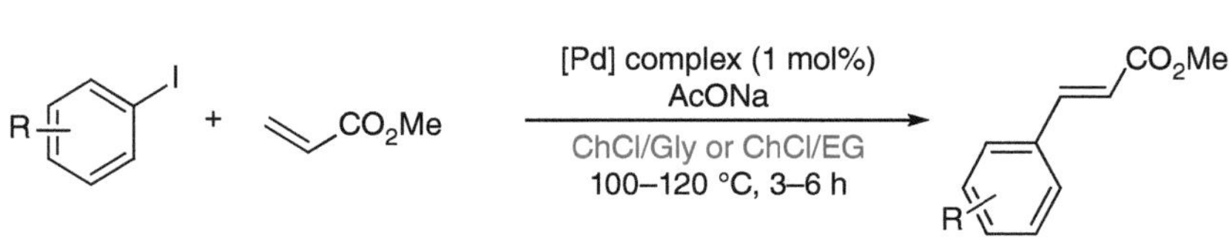

Scheme 6.12 Mizoroki–Heck reaction in DES in the presence of ligand or Pd complex.

suitable pre-catalyst, promoting organic synthesis more sustainable and reducing its negative impact on the environment [12, 13].

Hoang Doan and coworkers disclosed an efficient and recyclable catalyst *via* a palladium-incorporated metal-organic framework (MOF) (Pd@BUT-11) for the Mizoroki–Heck reaction between aryl iodides and styrenes in a ChCl/ethylene glycol (EG) eutectic mixture, in the presence of 1,4-diazabicyclo [2.2.2] octane (DABCO) as a base, at 120 °C, and under a nitrogen atmosphere (Scheme 6.13). Notably, the catalyst and the solvent could be recovered and reused for several trials without loss of catalytic activity. The protocol provides a better choice than the traditional catalysts and solvents, leading to a green and efficient stilbene synthesis process [18].

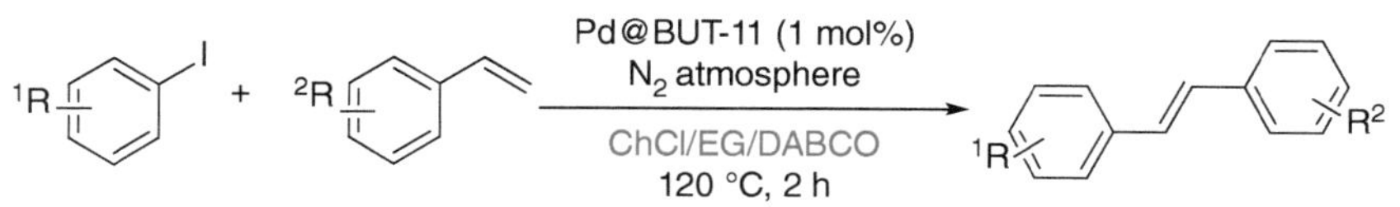

Scheme 6.13 Mizoroki–Heck reaction in DES in the presence of a Pd–MOF complex.

Capriati and coworkers developed one-pot Mizoroki–Heck cross-coupling/reduction processes successively between 2,3-dihydrofuran or 3,4-dihydro-2*H*-pyran and several (hetero) aryl halides to easily access valuable 2-(hetero) aryl tetrahydrofuran (THF) or tetrahydropyran (THP) derivatives in higher yield (Scheme 6.14). The successive two-step reactions were performed under aerobic conditions, in the absence of additional ligands, and with an effective and sustainable reaction medium (ChCl/glycerol). Notably, the practicability of the protocol has been proven to be a more sustainable way of the synthesis of two key THF derivatives [19].

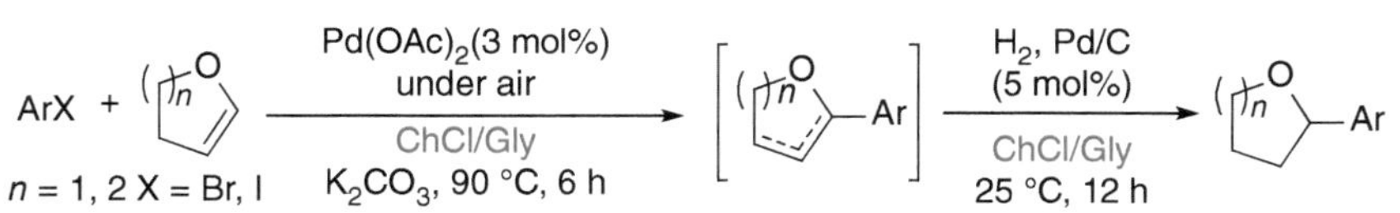

Scheme 6.14 Mizoroki–Heck coupling in DES for the synthesis of 2-(hetero)aryl-THF and THP derivatives.

In modern organometallic chemistry, the Sonogashira–Hagihara cross-coupling has provided a suitable way to construct new carbon–carbon bonds between terminal alkyne motifs and vinyl or aryl halides (pseudohalides) to produce a variety of conjugated enynes, which is of great interest in material sciences, pharmaceutical industry, and the total synthesis of natural products. König and coworkers described Pd-catalyzed Sonogashira coupling of aryl bromide with phenylacetylene in *D*-mannose/DMU (3:7) without copper co-catalyst (Scheme 6.15) [17].

Alonso, Ramón, and coworkers developed the $PdCl_2$ and cationic pyridiniophosphine ligand for the Sonogashira reaction between aryl halides and phenylacetylene in Ph_3PMeBr/gly (1 : 2), leading to the expected products in good yields

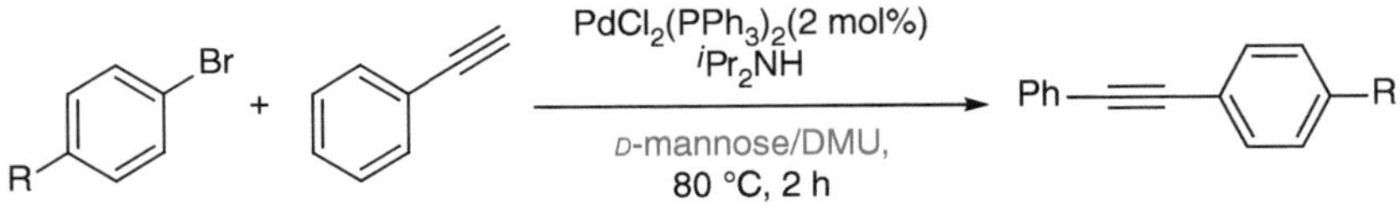

Scheme 6.15 Pd-catalyzed Sonogashira reaction in melt.

(Scheme 6.16). Notably, Ph_3PMeBr/gly, as a green solvent provided higher yields in comparison to ChCl-based DES systems such as ChCl/urea and ChCl/gly [12].

Scheme 6.16 Sonogashira reaction in Ph_3PMeBr/gly.

In 2019, Guillena and coworkers disclosed the Sonogashira reaction between phenylacetylene and several aryl halides in an AcChCl/urea (1 : 2) as a reaction medium by using the strong σ-donor ability of a mesoionic carbene ligand to improve the propensity of Pd to undergo oxidative addition (Scheme 6.17) [20].

Scheme 6.17 Sonogashira reaction in AcChCl/urea.

Salomone, Capriati, and coworkers demonstrated the commercially available and cheap Pd/C for promoting Sonogashira reactions in the green ChCl/glycerol eutectic mixture without external ligands (Scheme 6.18). By working under heterogeneous conditions, (hetero) aryl iodides were successfully coupled with both aromatic and aliphatic alkynes in good yields in short reaction time (3 h) at 60 °C. The aforementioned catalytic methodology is indicated to be effective toward electron-rich iodides as well, which are notoriously known to be poorly reactive in Pd-catalyzed Sonogashira coupling reactions. The DES and the catalyst could easily and successfully be reused up to four times with an E-factor as low as 24.4 [21].

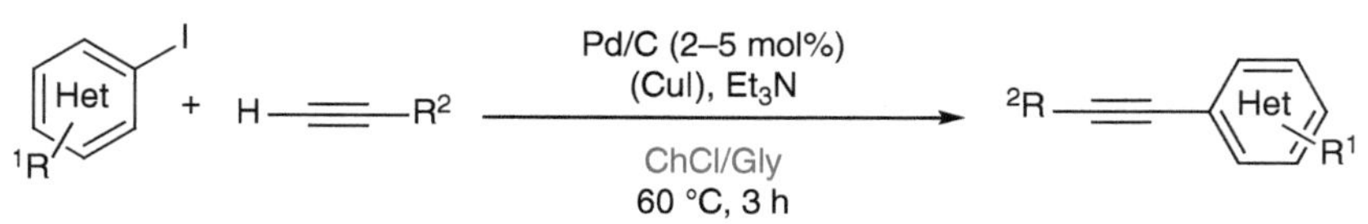

Scheme 6.18 Pd/C-catalyzed ligand-free Sonogashira reaction in DES.

Masteri–Farahani and coworkers prepared a cellulose-modified magnetite-graphene oxide nanocomposite *via* click reaction and utilized it for immobilization of palladium (Pd) nanoparticles in the absence of an additional reducing agent. The as-prepared nanocomposite served as a heterogeneous catalyst for the Sonogashira coupling reactions in a series of hydrophilic and hydrophobic DESs as sustainable reaction media. The hydrophilic DES composed of dimethyl ammonium chloride and glycerol proved to be the optimal medium. Because of the low miscibility of catalyst and DES in organic solvents, the separated aqueous phase, including both the catalyst and DES can be easily and successfully recovered by removing water and retrieved up to eight times with negligible loss of catalytic activity (Scheme 6.19) [22].

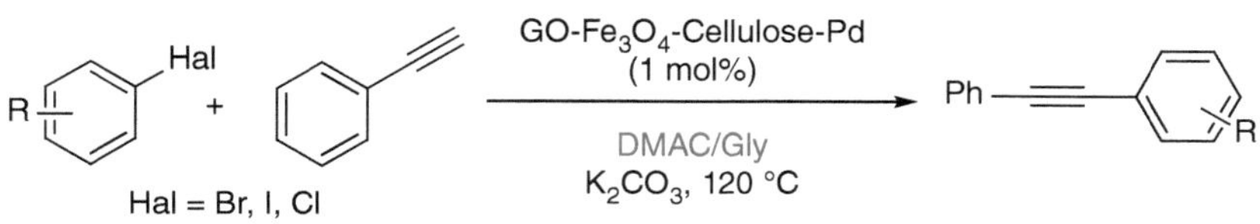

Scheme 6.19 A magnetic Pd-nanocomposite mediated Sonogashira reaction in DES.

The Hiyama reaction represents a Pd-catalyzed C—C bond formation *via* organic halides and organosilanes discovered by Tamejiro Hiyama and Yasuo Hatanaka in 1988. Organosilicon reagents are non-toxic, eco-friendly, commercially available, and stable in air and moisture while they are generally less reactive than the corresponding boron counterparts. Ramón and coworkers disclosed an engineered NCN–pincer–Pd complex catalyzed Hiyama reactions between aryl halides and some organosilanes in a biomass-derived eutectic mixture. The catalytic system could be successfully reused up to three times without adding extra catalyst or solvent. Moreover, the reaction could be conducted on a gram scale to yield biaryl products in high yields (Scheme 6.20) [23].

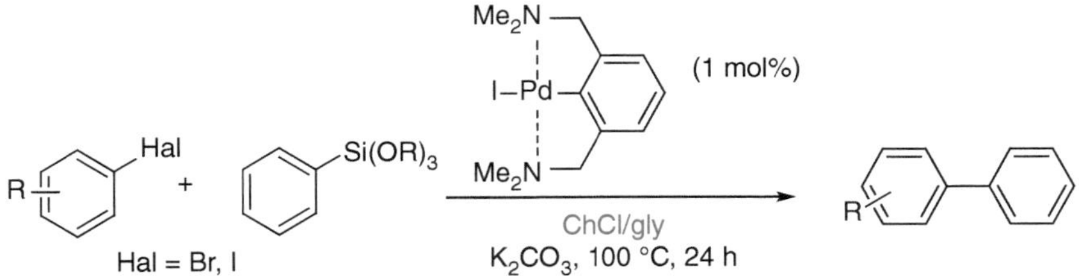

Scheme 6.20 A robust NCN–Pd–pincer catalyst for Hiyama reaction in DES.

In 2019, the same group achieved the Hiyama reaction with mesoionic *N*-heterocyclic carbene (NHC) –Pd complex in ChCl/gly using K_2CO_3 as a base to obtain the desired products in good yields [12]. Furthermore, they also developed a bipyridine–Pd complex as a general pre-catalyst for the Hiyama reaction in ChCl/gly (Scheme 6.21) [13].

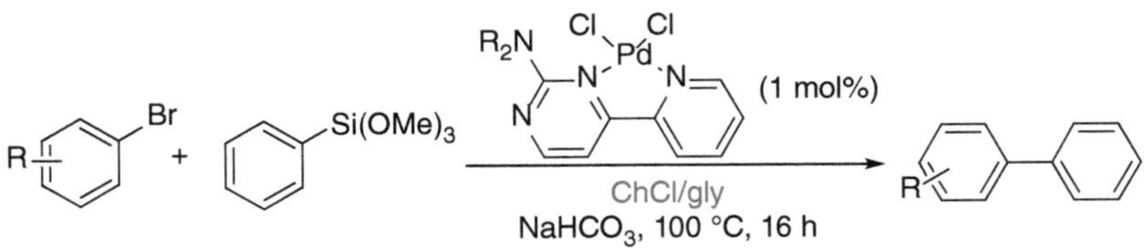

Scheme 6.21 DES-compatible bipyridine palladium for Hiyama reaction.

The Negishi reaction is a Ni- or Pd-mediated cross-coupling reaction of organozinc compounds with a variety of halides or pseudo-halides for the formation of new C—C bonds. Capriati, Perna, and coworkers reported a scalable and air-tolerant catalytic system for Negishi cross-coupling reactions between aryl bromides and organozinc substrates. The utilization of a palladium catalyst in ChCl-based DES proceeds smoothly with high chemo-selectivity under mild conditions (Scheme 6.22) [24].

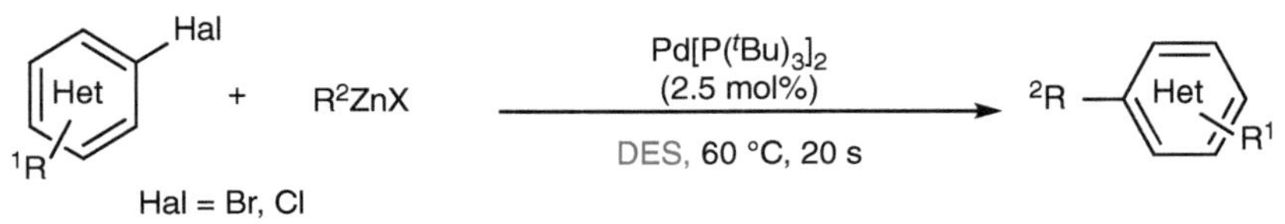

Scheme 6.22 Negishi cross-coupling reactions performed in melt.

Biaryl compounds can be also obtained *via* the Stille reaction between halides and tetraalkyl stannanes or phenyltrialkyl stannanes. Stille reactions generally performed well in aprotic, expensive, and relatively toxic solvents such as DMF, dioxane, and HMPA. König and coworkers disclosed $Pd_2(dba)_3$-catalyzed Stille cross-coupling reaction in the presence of triphenylarsine as a ligand using low-melting mixtures of sugar/urea/salt as reaction medium. The reaction proceeded smoothly due to the high polarity and nucleophilic character of the solvent melt (Scheme 6.23) [9].

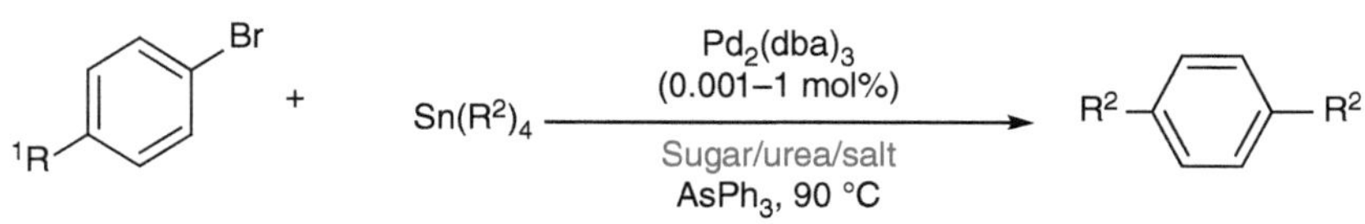

Scheme 6.23 Stille cross-coupling reaction in DES based on sugar/urea/salt.

The C—H bond functionalization reactions have become the extremely atom-efficient protocol for C—C bond formation. Farinola and coworkers demonstrated a thiophene–aryl direct coupling reaction through C—H bond activation with $Pd_2(dba)_3$ as a catalyst and in the existence of $P(o\text{-}MeOPh)_3$ as a ligand in a ChCl/urea mixture. The direct arylation in DESs is conducted in non-anhydrous conditions and without the exclusion of air (Scheme 6.24) [25].

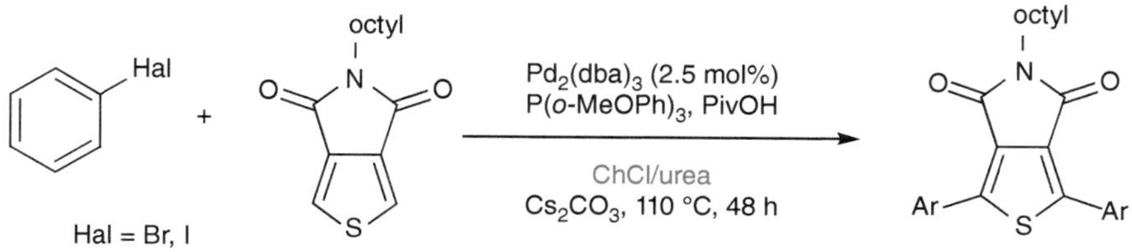

Scheme 6.24 Thiophene–aryl coupling *via* direct arylation in DES.

Khan and coworkers developed the application of palladium-tetrakis (triphenylphosphine) $[Pd(PPh_3)_4]$ in C_{sp3}—H functionalization of acetophenones with benzyl alcohols to the corresponding α,β-saturated ketones using ChCl/malonic acid mixture as solvent. The reaction underwent a domino-dehydrogenation–aldol-condensation process. Hydrogenation borrowing has been successfully achieved with a palladium-tetrakis(triphenylphosphine) catalyst–xantphos ligand combination. Furthermore, this C_{sp3}—H functionalization involved good functional group tolerance, high atom efficiency, and good chemo-selectivity (Scheme 6.25) [26].

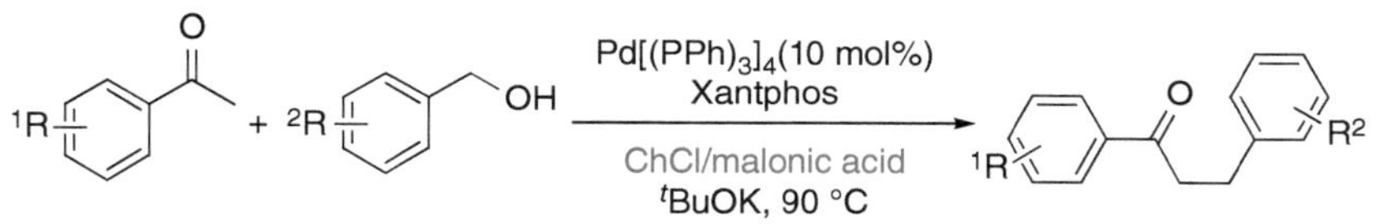

Scheme 6.25 Csp3—H functionalization of ketones with alcohols in DES.

Heydari and coworkers utilized a reusable heterogeneous catalyst, modified magnetic reduced graphene oxide-supported palladium catalyst (MRGO@DAP-Pd), and alkaline DES consisted of potassium carbonate and glycerol systems for the regioselective C—H arylation of imidazoles. Notably, this new route showed high catalytic efficiency, easy recovery of catalyst, good functional group tolerance, and sustainable reaction conditions utilizing green medium (Scheme 6.26) [27].

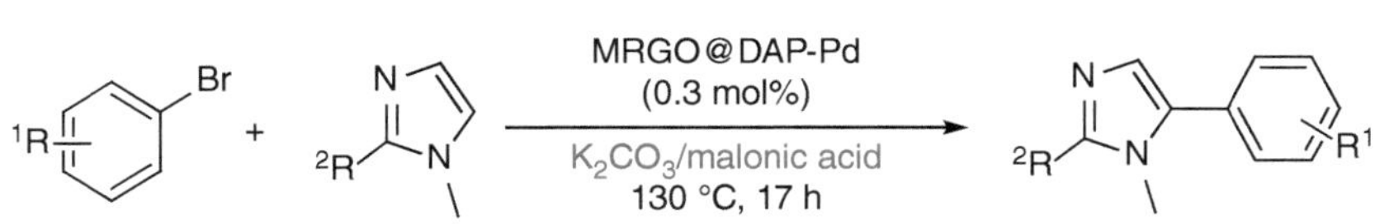

Scheme 6.26 Direct C5 arylation of imidazole derivatives in DES.

In 2022, Marset described the C_{sp3}—H arylation and alkynylation of unactivated 8-aminoquinoline amides mediated by palladium in DESs in the absence of Ag salts. The utilization of DES allowed the catalyst to be reused twice. Furthermore, post-synthetic modifications can also be conducted using DES as a solvent (Scheme 6.27) [28].

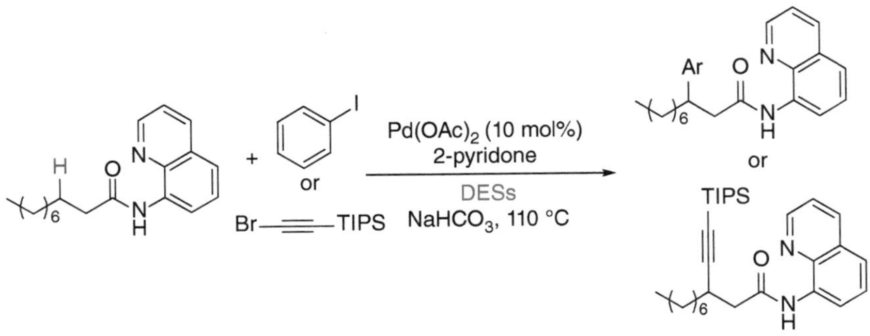

Scheme 6.27 Csp3–H arylation and alkynylation of 8-aminoquinoline amides in DES.

6.2.2 C–N Bond Couplings

Nitrogen- containing compounds are significant due to their interesting and diverse biological activities. The C—N bond formation is very important because it provides avenues for the introduction of nitrogen in organic molecules. Despite great improvements in this field, it is necessary to explore more sustainable methodologies for the construction of C—N bonds. To perform C—N bond coupling reactions in DES is a better choice. Shaabani and coworkers presented a practical and efficient copper nanocatalyst for the Ullmann cross-coupling reaction to yield the corresponding *N*-aryl amines in a DES with high yields. Notably, both the nanocatalyst and DES can be successfully recovered and reused for five turns (Scheme 6.28) [29].

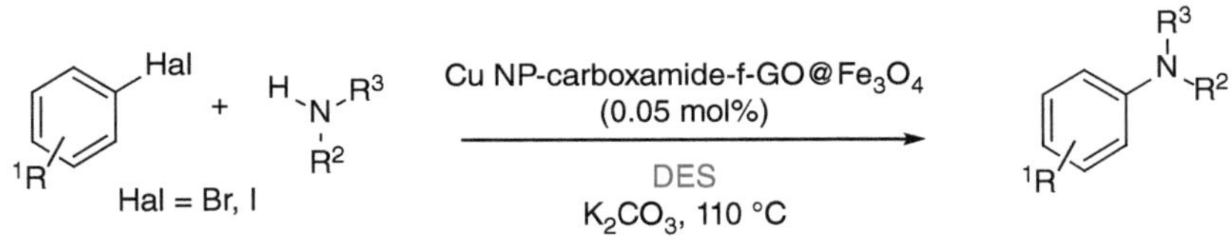

Scheme 6.28 N-arylation of amines mediated by Cu nanocomposite in DES.

In 2018, Capriati and coworkers developed the aminocarbonylation of (hetero) aryl iodides mediated *via* $Pd(OAc)_2$ in DESs as environmentally benign and recyclable solvent for the first time. The reactions performed smoothly with a range of substrates to obtain a series of amides in high yields (Scheme 6.29) [30].

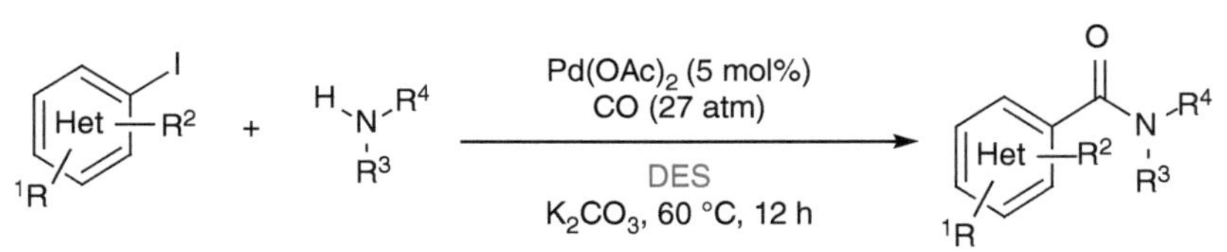

Scheme 6.29 Synthesis of amides *via* aminocarbonylation of (hetero)aryl iodides in DES.

In 2019, the same group reported CuI-catalyzed Ullmann amine cross-coupling between (hetero)aryl halides and amines in DESs as sustainable reaction media. The reaction proceeded well under an air atmosphere without ligands (Scheme 6.30) [31].

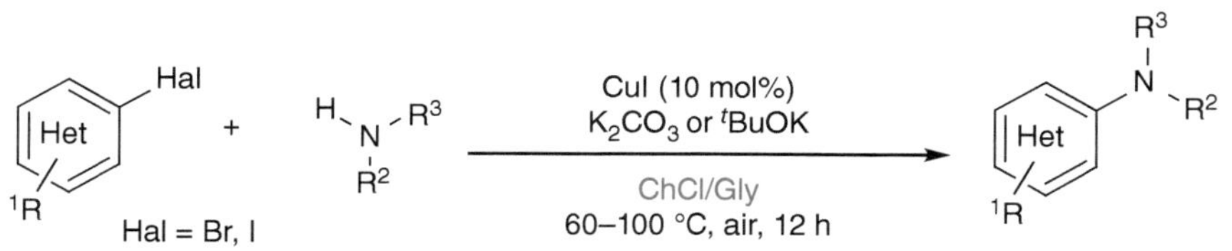

Scheme 6.30 Ullmann amine synthesis in DES.

In 2021, they also demonstrated a CuI-catalyzed Goldberg-type C—N coupling reaction between aryl iodides and amides using ChCl/H_2O DESs as green reaction media in air. The protocol allowed the possibility to scale up the reaction without obvious decrease in the yield, and the recyclability of both, the catalyst and the DES (Scheme 6.31) [32].

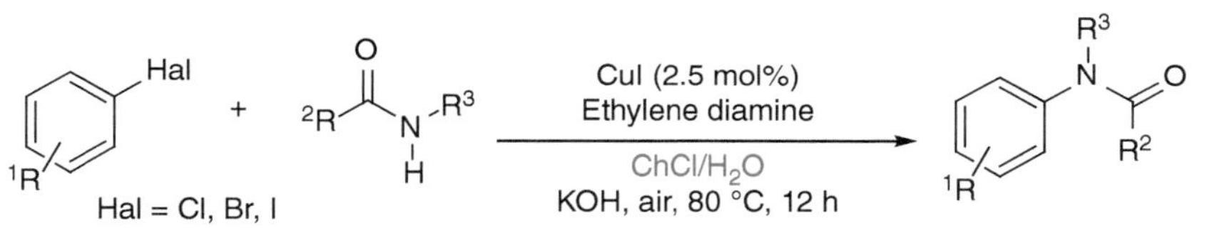

Scheme 6.31 C—N coupling of aryl iodides with primary/secondary amides in DES.

6.2.3 C—O Bond Couplings

Diaryl ether motifs have played a significant role in various medicinally important compounds and natural products. The effective methods for C—O bond formation to synthesize diaryl ethers have attracted great interest from organic chemists. Shankarling and coworkers described an efficient and facile protocol for the preparation of *O*-arylated ethers using DES. In this reaction, DES used to be regarded as both catalyst and reaction medium (Scheme 6.32) [33].

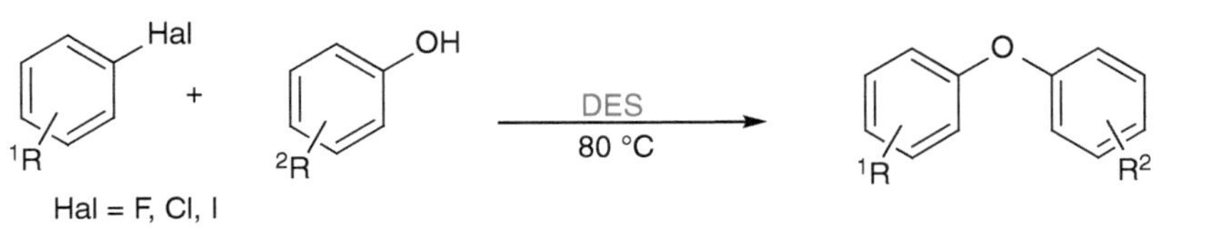

Scheme 6.32 DES promoted coupling of activated aryl halides with substituted phenols.

Capriati, Perna, and coworkers disclosed a Cu-catalyzed route for the Ullmann-type aryl alkyl ether synthesis by using various (hetero)aryl halides and alcohols as substrates. The alcohols can be also considered as active components of environmentally friendly ChCl-based eutectic mixtures at the same time (Scheme 6.33) [34].

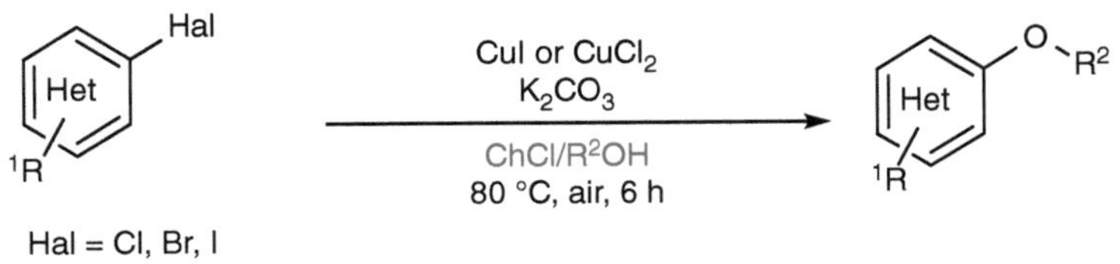

Scheme 6.33 Ligand-free Cu(I) or Cu(II)-catalyzed C—O coupling reactions in DES.

Masteri–Farahani and coworkers presented an efficient magnetic heterogeneous catalyst in promoting *O*-arylation reaction of phenols with aryl halides using DESs based on ChCl as reaction media. The protocol allowed the catalyst and DES to recycle easily up to five runs without an obvious loss on the reaction activity (Scheme 6.34) [35].

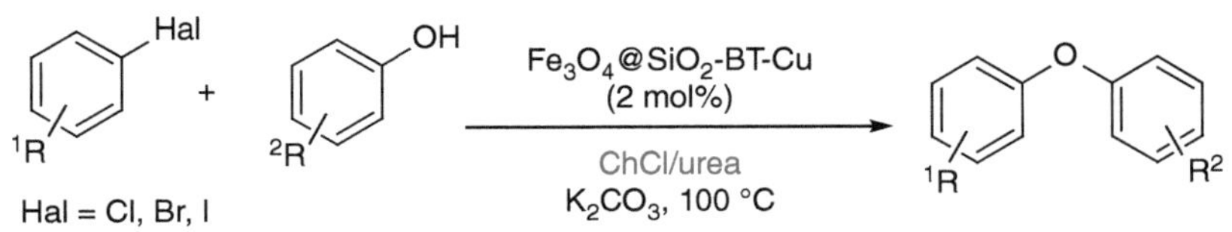

Scheme 6.34 Fe_3O_4@SiO_2-BT-Cu-catalyzed O-arylation of phenols in DES.

6.2.4 C—S Bond Couplings

Thioethers are important building blocks for the synthesis of biologically active molecules. Ramón, Guillena, and coworkers developed a unique jigsaw catalytic system for C—S bond formation using aryl boronic acids and sodium metabisulfite as substrates. A series of thioethers has been obtained applying ChCl/acetamide as reaction medium and $PdCl_2$–phosphine ligand as catalyst. This versatile protocol allowed the synthesis of different types of products in a sustainable medium by selecting suitable reaction components (Scheme 6.35) [36].

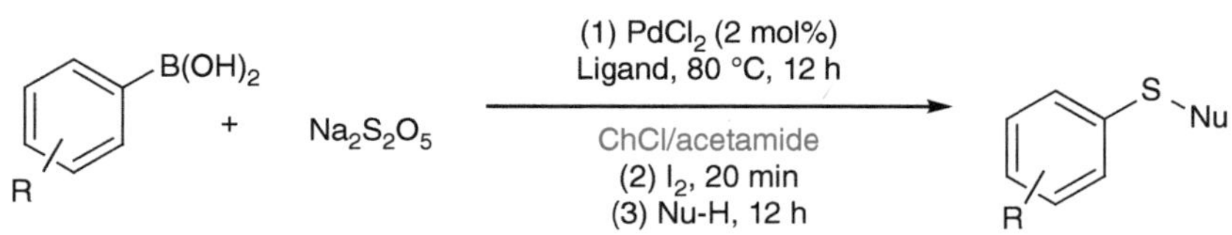

Scheme 6.35 C—S bond formation carried out in DES.

Shankarling and coworkers achieved an efficient nucleophilic aromatic substitution (S_NAr) reaction for C—S bond formation employing DES as both catalyst and green solvent. An array of thioethers are prepared in good yields. This protocol provides many advantages such as atom economy, mild reaction conditions, and reuse of DES up to four turns (Scheme 6.36) [37].

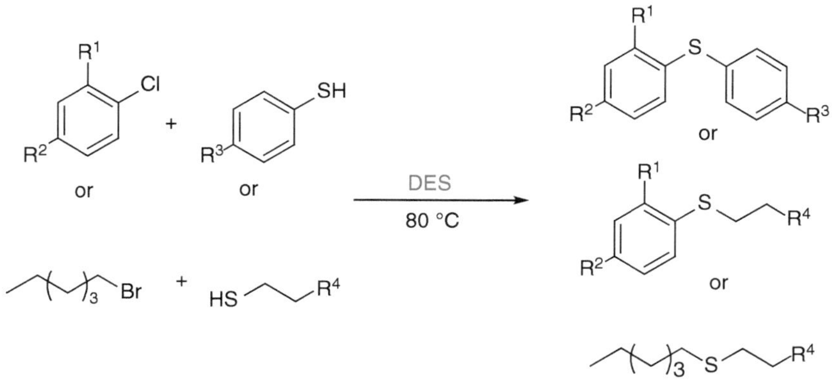

Scheme 6.36 DES mediated C—S coupling of halides with thiols.

6.3 Oxidation Reactions in Deep Eutectic Solvents

6.3.1 Metal-catalyzed Oxidation

Among a variety of oxidation reactions, the oxidation of toluene into benzaldehyde plays a significant role in chemical transformations due to toluene being the cheapest and primary source of aromatic compounds. Because of the easier tendency of the aldehyde to readily oxidize to benzoic acid, the selective partial oxidation of toluene still remains a big challenge. Guajardo and coworkers demonstrated an approach to convert toluene into benzaldehyde using $FeCl_3$ as a catalyst and hydrogen peroxide as an oxidant in DESs with a higher selectivity (Scheme 6.37) [38].

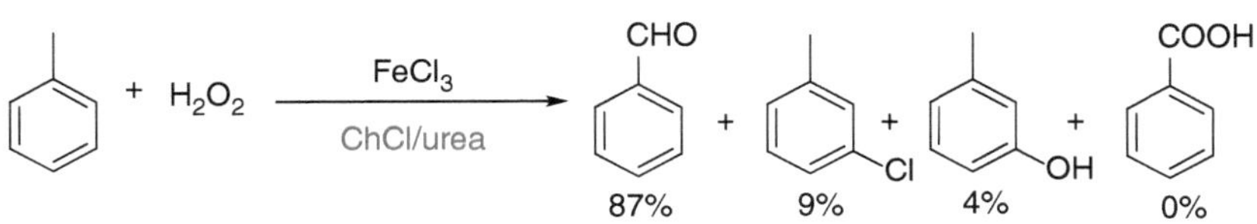

Scheme 6.37 Toluene oxidation by hydrogen peroxide in DES.

Tetrahydroisoquinoline compounds have attracted extensive attention from the academic community because of their excellent biological and pharmacological activities. In 2016, Ramón and coworkers reported a highly selective method to achieve cross-dehydrogenative coupling reaction of tetrahydroisoquinoline with a variety of nucleophilic reagents. They successfully prepared the expected products by using ChCl/ethylene glycol as a DES and copper (II) impregnated on magnetite as a catalyst, using bio-renewable air as the only oxidant. Therefore, the mixture of solvent and catalyst is highly recyclable, which makes the entire reaction system more environmentally friendly (Scheme 6.38) [39].

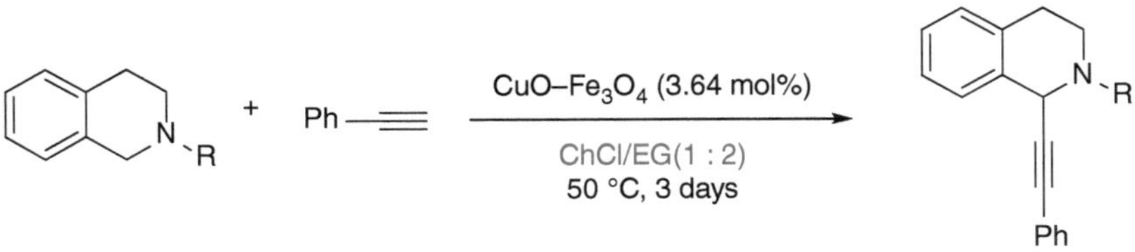

Scheme 6.38 Cross-dehydrogenation coupling of tetrahydroisoquinoline in DES.

6.3.2 Other Oxidative Processes

Pyrimidine compounds exhibit a wide range of biological and pharmacological activities. In addition, they also have many valuable applications in polymer and supramolecular chemistry. In 2015, Chaskar and coworkers introduced a more economical and environmentally friendly way to synthesize partially and completely substituted pyrimidines using α,β-unsaturated ketone and benzylamine hydrochloride as raw materials. Notably, choline hydroxide is regarded both as a catalyst and a reaction medium, which can be recycled to make the [3 + 3] tandem annulation–oxidation approach more environmentally friendly (Scheme 6.39) [40].

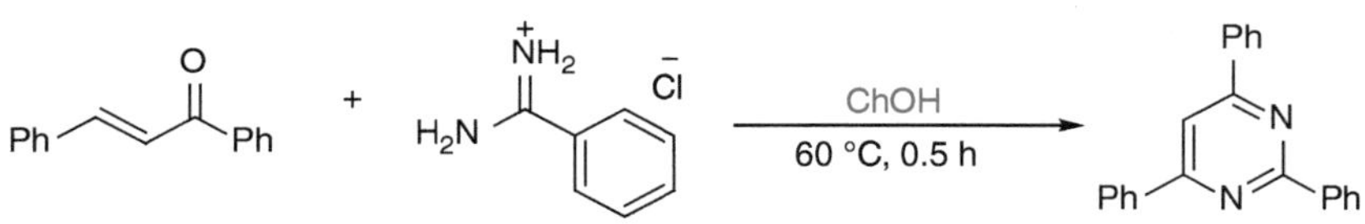

Scheme 6.39 The [3 + 3] tandem annulation–oxidation approach for the synthesis of substituted pyrimidines in DES.

Imines are many synthetic intermediates for organic conversion and are widely used in the synthesis of organic chemicals and synthetic drugs. Therefore, the oxidation of primary amines to secondary imines has attracted great interest from organic chemists. In 2017, Sharma and coworkers reported a novel 2-Iodoxybenzoic acid (IBX)-facilitated primary amine oxidative coupling reaction. With high-valent iodine as the oxidant and ChCl/urea as the natural DES, the direct synthesis of imines and bisamides from pseudo-four-component oxidative Ugi reaction was successfully achieved under mild and metal-free conditions. Noteworthily, the oxidant (IBX) and solvent could be reused up to five times without obvious loss in activity (Scheme 6.40) [41].

Carboxylic acid is widely used in chemical industry, medicine, natural products, and so on. Nowadays, many methods of carboxylic acid synthesis have been reported, among which oxidation to carboxylic acid with aldehydes is a very practical, simple, and direct reaction. In 2018, Nagarkar and coworkers developed a method for synthesizing carboxylic acid by oxidizing aldehydes using ChCl/urea as solvent and H_2O_2 as an oxidant, which can be performed at room temperature without a catalyst. This reaction yielded the products with a high conversion and

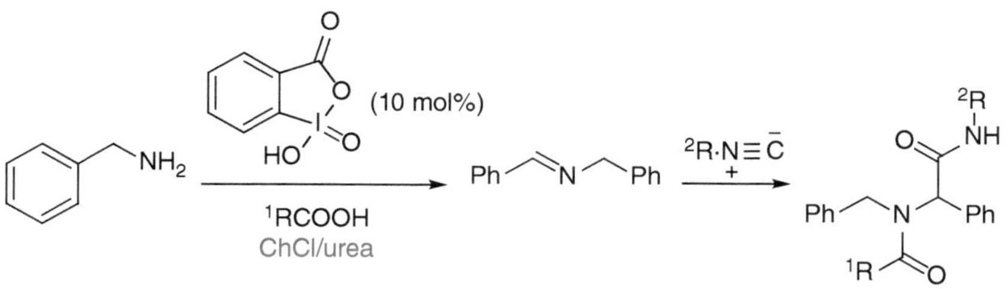

Scheme 6.40 IBX promoted primary amine oxidative coupling in DES.

good selectivity. Moreover, the utilization of ChCl/urea DESs can make this effective oxidation approach more environmentally friendly (Scheme 6.41) [42].

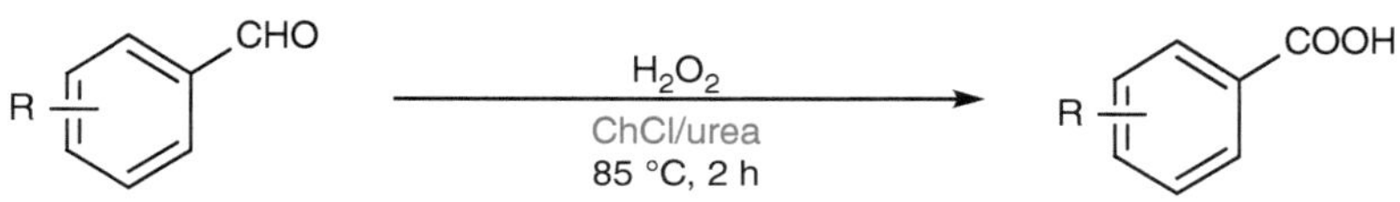

Scheme 6.41 The oxidation of benzaldehyde by hydrogen peroxide in DES.

Maleic acid and fumaric acid together with their anhydrides are significant chemical intermediates and have a wide range of applications in many fields of the chemical industry. In 2019, Yan and coworkers proposed a simple method to oxidize furfural to maleic acid and fumaric acid by using ChCl/oxalic acid as both solvent and acidic catalyst. The reaction was carried out at a mild reaction temperature using hydrogen peroxide as the oxidant with full conversion and high yield, which provides a new green and efficient route to convert biomass into valuable chemicals (Scheme 6.42) [43].

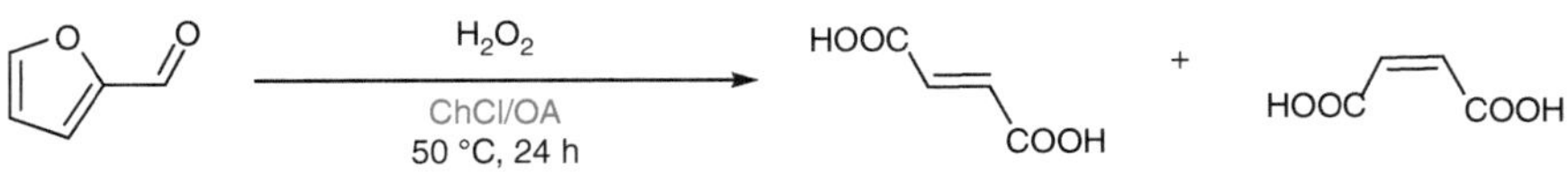

Scheme 6.42 The oxidation of furfural to fumaric acid and maleic acid in DES.

Pyrazine derivatives are an important class of heterocyclic compounds that are widely used in metal coordination chemistry and medicine. They can be isolated from nature or prepared by chemical synthesis. In 2019, Capriati and coworkers presented the synthesis of phenacyl azides and symmetrical 2,5-diarylpyrazines using a DES of ChCl/glycerol (1 : 2) as a sustainable reaction medium and Pd/C as a catalyst. This one-pot azidation/cyclization process minimizes the generation of chemical waste and allows for the direct synthesis of functionalized pyrazine derivatives without any intermediate separation/purification steps, as well as overall chemical yields higher than those of two single steps (Scheme 6.43) [44].

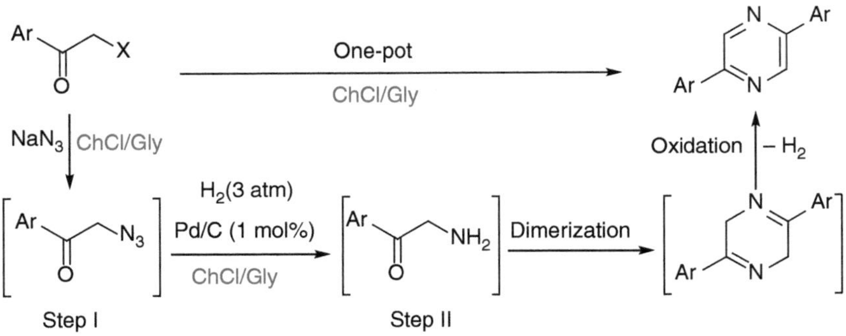

Scheme 6.43 One-pot (two steps) synthesis of 2,5-diarylpyrazines in DES.

In 2020, the same group described the first deprotonation of aliphatic and aromatic secondary phosphine (HPR_2) with ^{n}BuLi to synthesize highly polarized lithium phosphide ($LiPR_2$). Then, using eutectic mixture ChCl/Gly as the reaction medium, $LiPR_2$ generated in-situ was added to aldehydes or epoxides for rapid and chemo-selective reaction under air and room temperature conditions to obtain the corresponding α-hydroxyl or β-hydroxyphosphine oxides. This new advance makes it more possible to combine organometallic chemistry with aerobic conditions and protic bio-based solvents, making green chemistry more achievable (Scheme 6.44) [45].

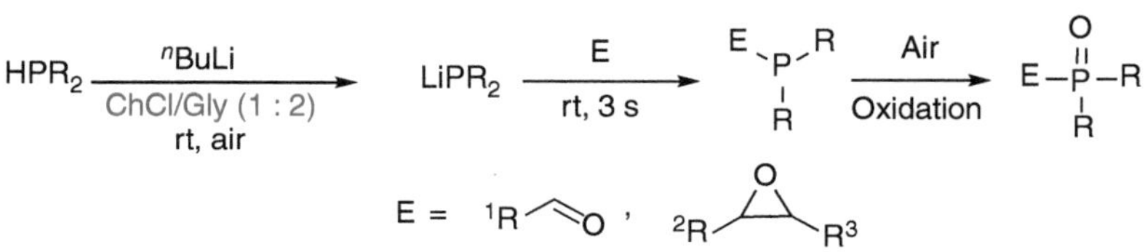

Scheme 6.44 Synthesis of α- and β-hydroxyphosphine oxides in a one-pot two-step process in DES.

Oxidative desulfurization is regarded as an efficient route in terms of the advantages of high desulfurization efficiency for aromatic sulfides under mild reaction conditions. These aromatic sulfides can be oxidized to polar sulfoxide or sulfones by employing some oxidants such as molecular oxygen, air, hydrogen peroxide, organic peroxyacid, and ozone, whereafter which are easily removed *via* adsorption or extraction. In 2019, Lü and coworkers discovered an efficient coupled redox biomimetic catalytic protocol to oxidize the sulfides in diesel into their corresponding sulfones selectively (Scheme 6.45) [46]. They used the Anderson-type polyoxometallate $(NH_4)_3Co(OH)_6Mo_6O_{18}$ and *p*-toluenesulfonic acid-based DESs as cocatalysts, which are regarded as electron donors and electron-transfer mediators (ETMs). The biomimetic approach provides outstanding desulfurization performance under mild conditions by using oxygen as an oxidant without sacrificial agents. In 2022, the same group described another biomimetic catalytic route for aerobic oxidative desulfurization by coupling the Mo_7 POM with carboxylic acid-based DES as cocatalysts [47].

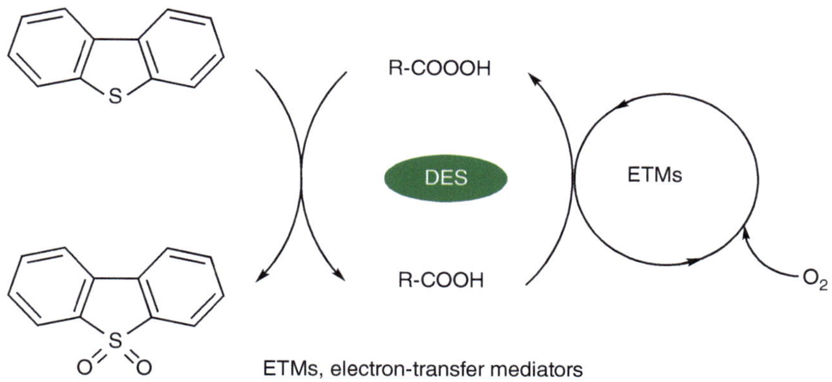

Scheme 6.45 The biomimetic aerobic oxidative desulfurization process in DES.

6.4 Reduction Reactions in Deep Eutectic Solvents

6.4.1 Metal-catalyzed Reduction

Metal-mediated hydrogenation reaction has gained great attention as it can allow the synthesis of a series of fine chemicals, commodities, and pharmaceutical substances in environmentally friendly routes. In 2006, König and coworkers disclosed that Wilkinson's complex [$RhCl(PPh_3)_3$] catalyzed hydrogenation of methyl α-cinnamide in the carboxylic acid-based eutectic mixture (Scheme 6.46) [10].

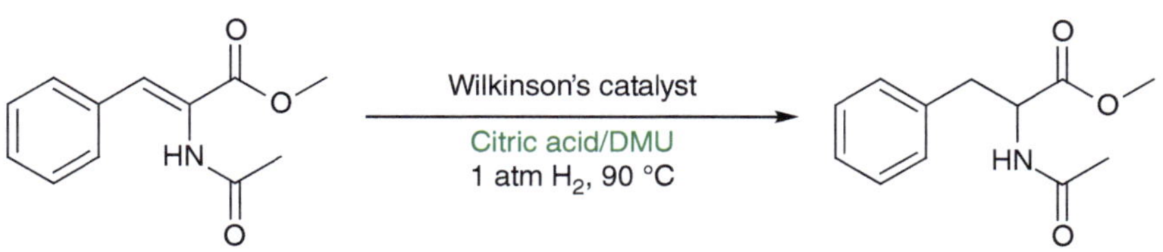

Scheme 6.46 Rh(I)-catalyzed hydrogenation of methyl α-cinnamide in DES.

The reduction of carbonyl compounds to their corresponding alcohols is a useful synthetic strategy. Azizi's team pioneered an efficient and green technology for the regio-selective and chemo-selective reduction of carbonyl compounds and epoxides in the presence of $NaBH_4$ using ChCl/urea as solvent. This reaction was carried out under mild conditions to give the corresponding alcohols in high yields (Scheme 6.47) [48].

Controlling the sulfur content is very important due to the environmental pollution caused by sulfur compounds in fuel oil. In 2021, Shu and coworkers employed DESs as solvents in the reductive desulfurization process in the presence of $NaBH_4$ and $NiCl_2$ for the first time. In this reaction, DES played a dual role of stabilizer and extractant. The results suggested that the desulfurization of aromatic sulfides underwent a stereospecific oxidative-addition/reductive-elimination process. Moreover,

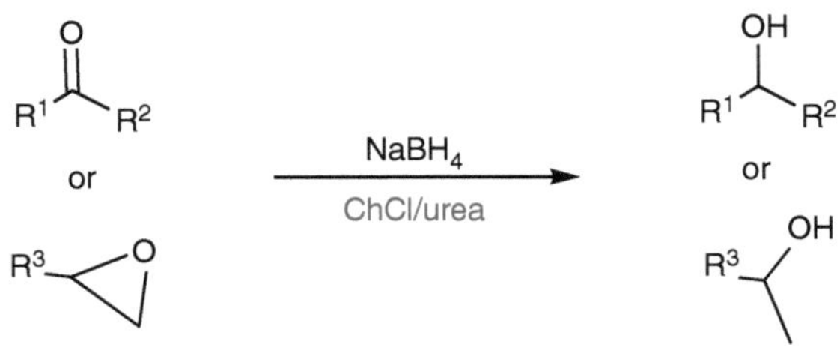

Scheme 6.47 Reduction of carbonyl compounds and epoxides in DES.

the structures of the DESs remained the same even after regeneration. This protocol may accelerate the industrialization of the desulfurization (Scheme 6.48) [49].

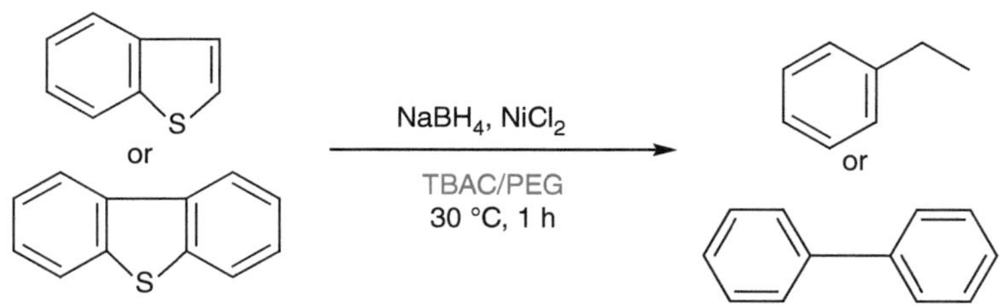

Scheme 6.48 Reductive desulfurization process in DES.

6.4.2 Other Catalytic Reduction

Biocatalysis such as oxidoreductases have become increasingly useful in redox processes due to their high activity and selectivity. Enzymes can promote reductive reactions of compounds containing the C=O bond and the C=C bond. Alcohol dehydrogenase (ADH) generally can be used as a sustainable and efficient catalyst to synthesize alcohols starting from carbonyl compounds as inexpensive substrates. In 2020, Santiago's group successfully reported new progress in the reduction of ketones and aldehydes using ADH from horse liver (HLADH) as a catalyst in DESs composed of ChCl and 1,4-butanediol. In this study, the combination of DES and smart cosubstrate systems may provide more chances for sustainable chemical transformations (Scheme 6.49) [50].

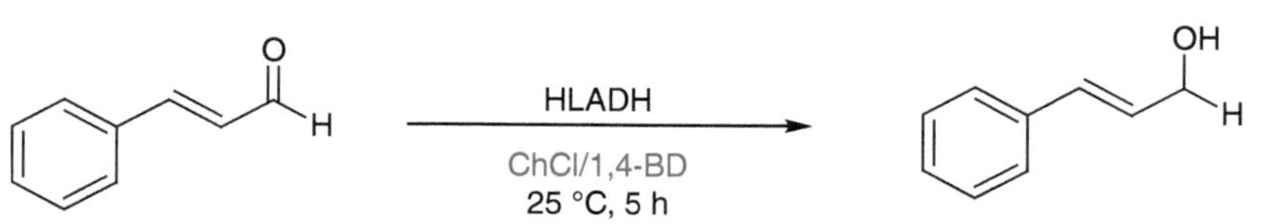

Scheme 6.49 HLADH-catalyzed cinnamaldehyde reduction in DES.

Nitroalkanes are valuable precursors of amines, which can be prepared through the reductive strategy. In 2021, Benaglia and coworkers developed a green synthetic organic protocol for the chemo-selective reduction of nitroalkene in the combination of an atom economically convenient and cheap reagent such as ammonia borane (AB) with bio-based and bio-renewable DESs (Scheme 6.50) [51].

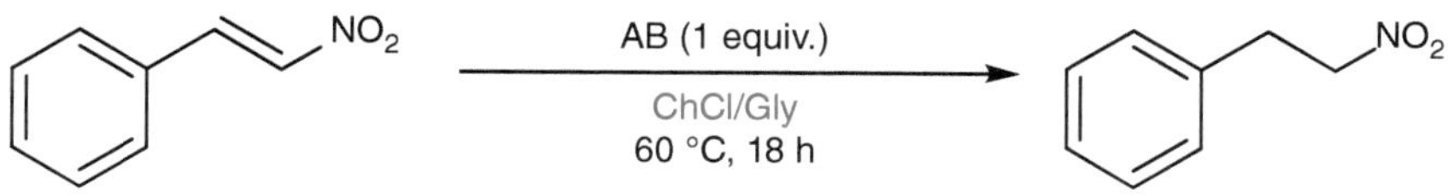

Scheme 6.50 Reduction reaction of nitrobenzene in DES.

(*R*)-1-phenyl-1,2-ethanol (PED) is the key synthon in the synthesis of enantiopure drugs such as β-adrenergic blocking agents for curing sympathetic nervous system disorder and cardiovascular disease. The biocatalytic route for synthesizing enantiomeric PED has attracted more attention. Lou's team disclosed biocatalysis (Kurthiagibsonii SC0312) promoted the reduction of 2-hydroxyacetophenone (HAP) using the ChCl/1,4-butanediol (ChCl/Bd) as a solvent to yield the corresponding (*R*)-1-phenyl-1,2-ethanediol with high selectivity, good yield and environmental friendliness (Scheme 6.51) [52].

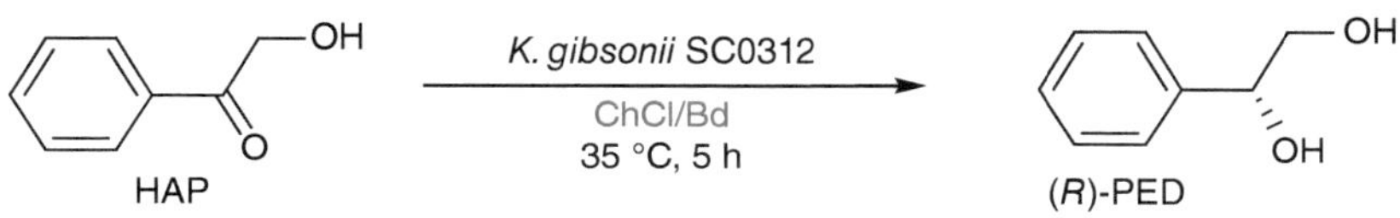

Scheme 6.51 The reduction of HAP to (*R*)-PED by K. gibsonii SC0312 in DES.

6.5 Cyclization Reactions in Deep Eutectic Solvents

6.5.1 Synthesis of Five-membered Ring

In 2021, Ali and coworkers established the Paal–Knorr synthetic route for the generation of various valuable substituted pyrroles in the presence of tartaric acid and *N,N′*-dimethylurea (TA/DMU) as a dual catalyst and solvent system. This particular protocol is simple, sustainable, and rapid (Scheme 6.52) [53].

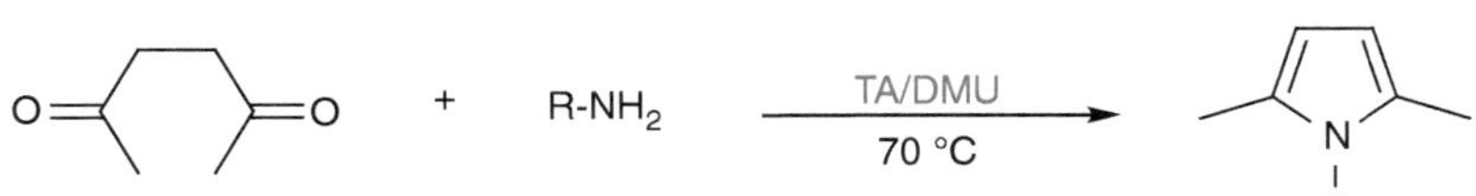

Scheme 6.52 Paal–Knorr reaction in DES.

Hameed, Malik, and coworkers demonstrated a novel route for the preparation of a variety of indoles *via* Fischer indole synthesis and 1*H*-tetrazoles *via* click reaction in *N*-alkylated 1,4-diazabicyclo [2.2.2]octane/polyethylene glycol (PEG) (Scheme 6.53). These ILs were prepared by the *N*-alkylation of DABCO using

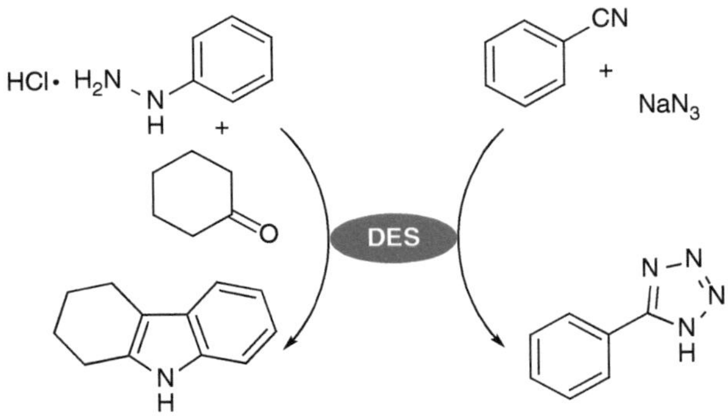

Scheme 6.53 Fischer indole synthesis and click reaction in DES.

different kinds of alkyl halides to vary chain lengths and mixed with PEGs of varying molar masses as HBDs to yield DESs. This methodology proved to be safe, environmentally friendly, and useful for the construction of *N*-heterocycles [54].

Shankarling and coworkers described a simple and efficient method for the utilization of ChCl/urea as a solvent to obtain oxadiazoles using carboxylic acid and acid hydrazide as substrates. Cetyltrimethylammoniumperoxodisulphate (CTAPS) *in situ* produced CTA-bisulfate which efficiently catalyzed cyclization. The protocol offers advantages such as mild reaction conditions, simplicity in workup and purification, high isolated yields, and recyclable solvents and catalysts (Scheme 6.54) [55].

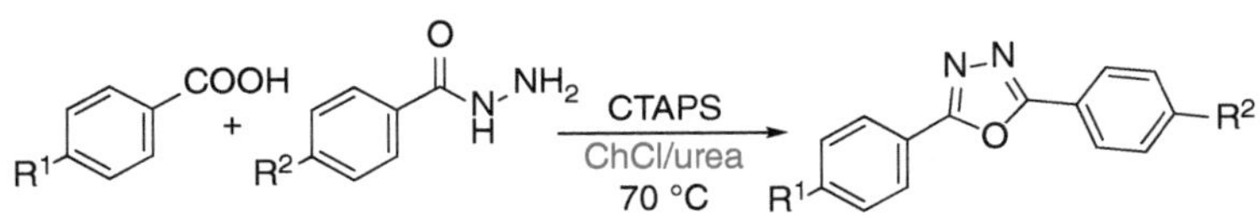

Scheme 6.54 Synthesis of oxadiazole derivatives in DES.

6.5.2 Synthesis of Six-membered Ring

In 2020, D'Anna and coworkers presented an efficient approach for the Diels–Alder reaction with *N*-ethylmaleimide and different dienes as substrates using DESs as solvent in the presence of both conventional heating and ultrasonic irradiation (Scheme 6.55). DES was indicated to be the optimal solvent for the Diels–Alder reaction, affording better yields and faster reactions in comparison with traditional organic solvents [56].

In 2022, Kouznetsov and coworkers discovered an efficient microwave-assisted irradiation for the preparation of new pharmacologically active 6-aryl-phenanthridines *via* Pictet–Spengler process using a zinc chloride/[Bmim]BF_4 mixture as both a practical catalytic system and solvent. This practical methodology achieves sustainable and straightforward access to novel phenanthridine-based molecules. Notably, the obtained compounds have potential biological activities such as antitumor and antibacterial activities (Scheme 6.56) [57].

In the same year, Singh, Devi, and coworkers established a simple, convenient, and wide-scope protocol for the synthesis of highly functionalized polysubstituted

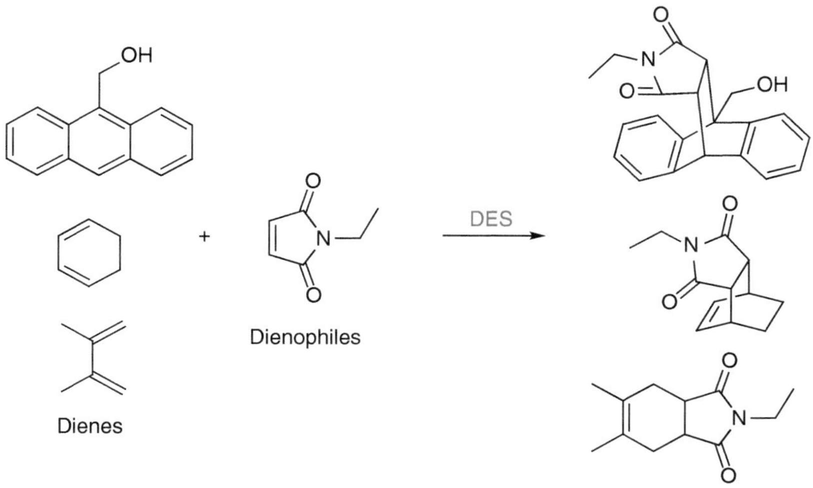

Scheme 6.55 Diels–Alder Reaction in DES.

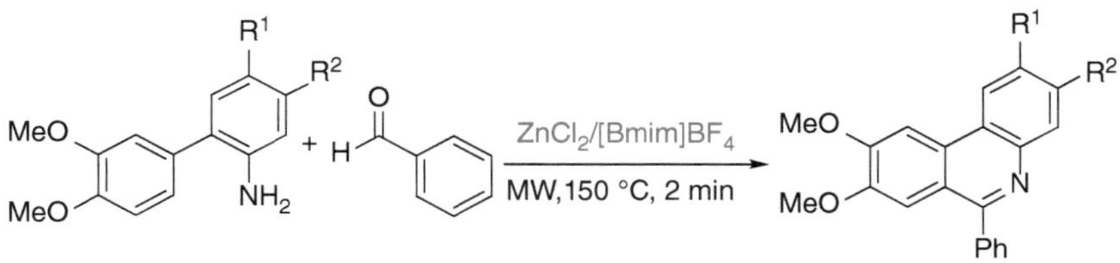

Scheme 6.56 Pictet–Spengler dehydrogenative cyclization in DES.

quinolines through Friedlander annulation reaction using choline-based DESs as both an efficient catalyst and reaction medium. This system could accomplish this reaction with excellent yields, sustainable choline-based DES catalytic systems, and easy work-up procedures (Scheme 6.57) [58].

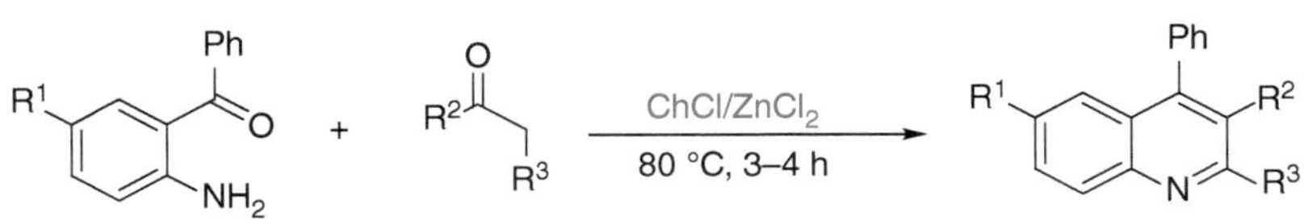

Scheme 6.57 Choline-based DES-mediated Friedlander annulation.

6.6 Condensation Reactions in Deep Eutectic Solvents

6.6.1 DES as the Catalyst/Solvent System for Condensation

Shankarling and coworkers reported a greener protocol for one-pot synthesis of coumarinstyryl dyes using a DES. This facile one-pot cyclocondensation reaction was carried out by adding 4-(diethylamino)-2-hydroxybenzaldehyde, ethyl acetoactetate, malononitrile, and aromatic aldehydes sequentially. The DES can serve the dual role of catalyst as well as reaction medium. The advantages of this procedure were short reaction times, mild conditions, and environmental protection (Scheme 6.58) [59].

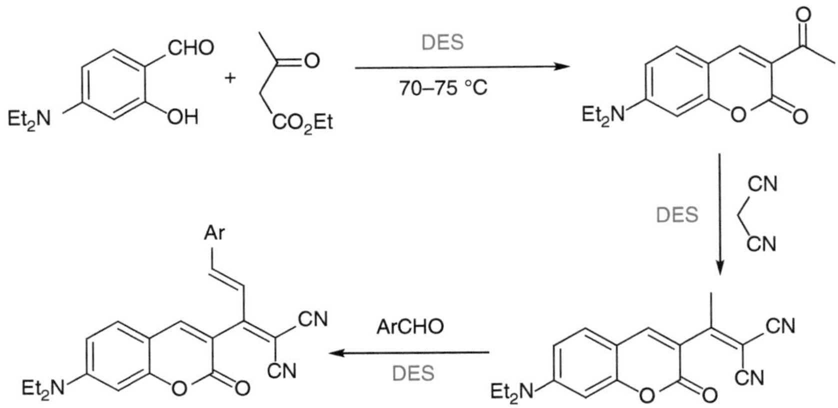

Scheme 6.58 One-pot cyclocondensation for the synthesis of coumarin styryl dyes in DES.

Qi and coworkers discovered an efficient Knoevenagel condensation catalyzed by imidazole-based halogen-free DES at room temperature. The DESs [3Im/PTSA] and [4Im/PTSA] were synthesized by imidazole (Im) and *p*-toluenesulfonic acid (PTSA), which were green and economical. This methodology yields unsaturated carbonyl compounds in good to excellent yields (Scheme 6.59) [60].

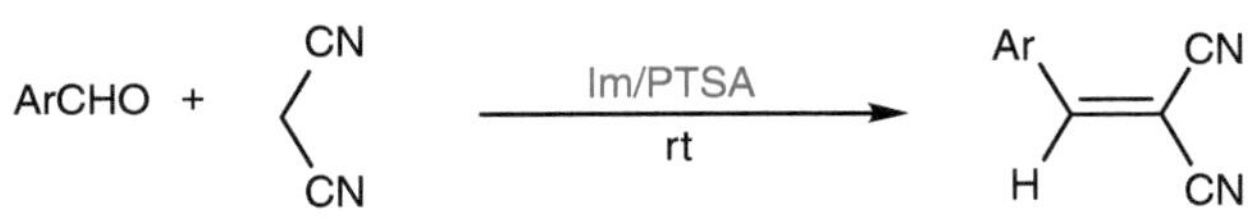

Scheme 6.59 Knoevenagel condensation in DES.

Bis-enol derivatives are widely employed due to their significant pharmaceutical properties. Srivastava demonstrated a facile synthesis of a library of bis-enol derivatives using ChCl/PEG-based DES served as both catalyst and solvent. This protocol underwent Knoevenegal condensation followed by Michael addition. Interestingly, the DES has many advantages such as nontoxicity, good biodegradability, easy preparation, and good recyclability (Scheme 6.60) [61].

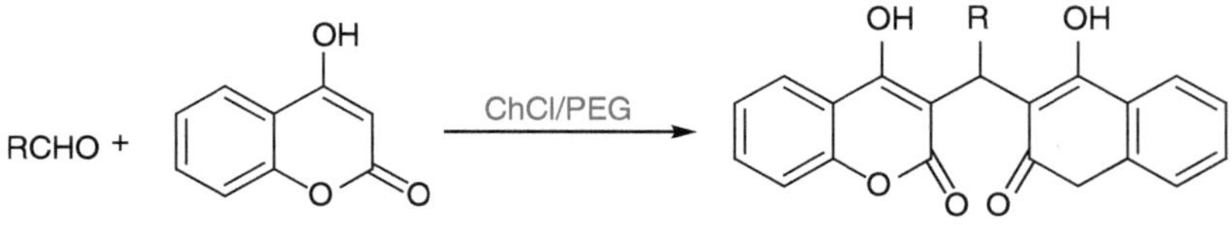

Scheme 6.60 Synthesis of a library of bis-enol derivatives in DES.

Molnar and coworkers found an expeditious and environmentally friendly synthesis of a series of rhodanine derivatives *via* Knoevenagel condensation using ChCl/urea (1 : 2) DES. Notably, some of the prepared benzylidene derivatives can exhibit better DPPH radical scavenging activity (Scheme 6.61) [62].

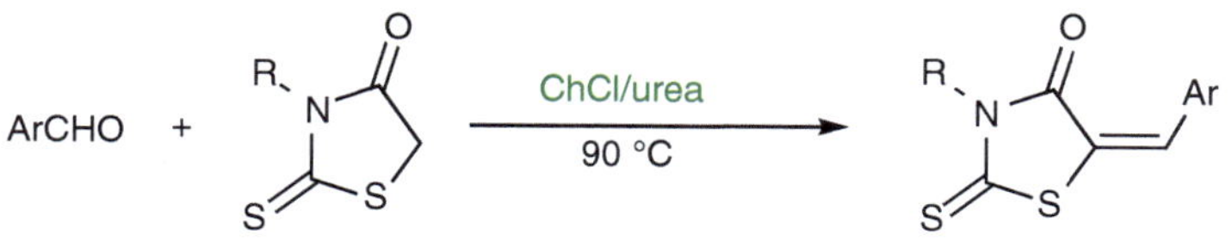

Scheme 6.61 Synthesis of rhodanine derivatives *via* Knoevenagel condensation in DES.

Wang and coworkers developed the synthetic method for the preparation of fructosazine (FZ) and deoxyfructosazine (DOF) through a self-condensation of *D*-glucosamine in ChCl-based DES. It indicated that the DESs played an important role in the reaction because they served not only as a solvent but also as a catalyst. The advantages of this protocol were cheap, recyclable, and sustainable (Scheme 6.62) [63].

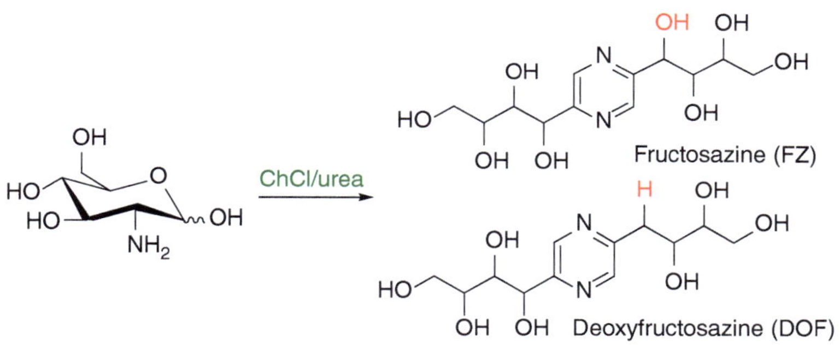

Scheme 6.62 Self-condensation of *D*-glucosamine to DOF and FZ in DES.

6.6.2 Other Catalytic System for Condensation

Capriati and coworkers disclosed a green and efficient condensation in the synthesis of 2-aminoimidazoles in ChCl-based DESs as nonconventional, safe, and environmentally friendly reaction medium. It proved to be practical in DES for shortening the reaction time compared to those usually conducted in toxic and volatile organic solvents. Moreover, the expected products could be isolated by filtration and crystallization in good yields (Scheme 6.63) [64].

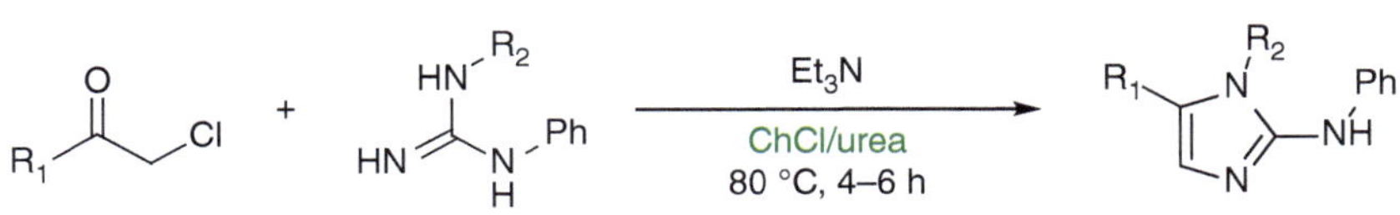

Scheme 6.63 Synthesis of 2-aminoimidazoles *via* condensation in DES.

The same group uncovered a highly tunable and bio-renewable reaction medium for distinct stereoselective Michael addition reaction mediated by a chiral primary amine to yield functionalized molecules in higher enantioselectivity. The DES can be recycled by simple extraction, which facilitates the recovery and the recyclability of the chiral catalyst in order to respond to the dwindling petroleum resources (Scheme 6.64) [65].

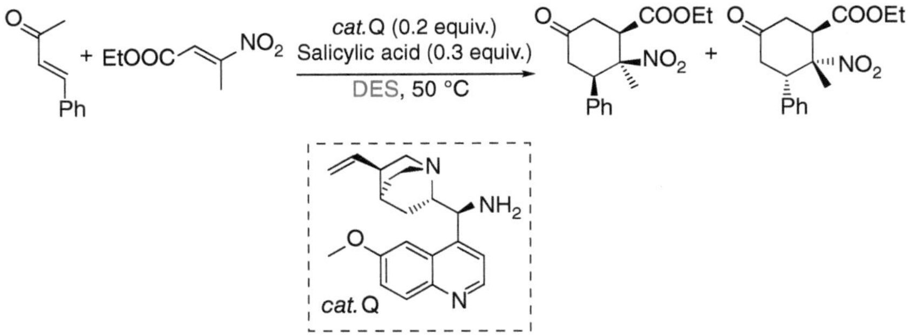

Scheme 6.64 Distinct stereoselective Michael addition reaction in DES.

Wang and coworkers elaborated an efficient *L*-proline-catalyzed Knoevenagel condensation of acetylacetone and aromatic aldehyde using ChCl-based DESs as reaction medium. In this reaction, the solvent and catalyst can be reused at least for four runs, providing a potential for industrial application (Scheme 6.65) [66].

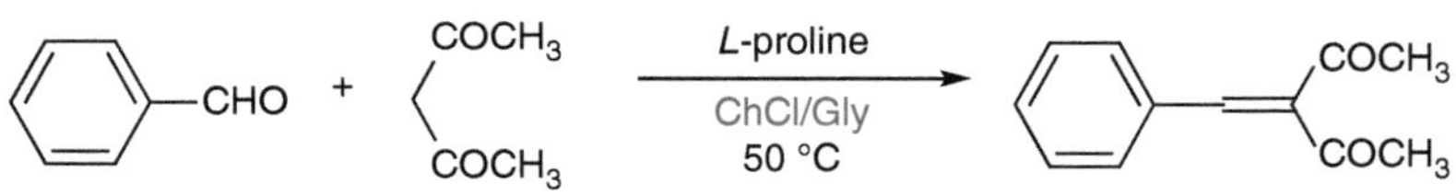

Scheme 6.65 *L*-proline-catalyzed Knoevenagel condensation in DES.

6.7 Multicomponent Reactions in Deep Eutectic Solvents

6.7.1 One-pot Three-component Reaction

Bhate's group reported a convenient one-pot three-component reaction to yield annulated pyrroles. Under acid-free and solvent-free conditions, the DES formed by ChCl and urea played a catalytic role in promoting the cyclization of enamine with free sugars. This synthetic methodology provides extra features, including easy workup and general applicability for the preparation of biologically active pyrrole derivatives (Scheme 6.66) [67].

Tavakol and coworkers developed a simple and efficient method for synthesizing quinoline derivatives through one-pot three-component reaction in the presence of ChCl/tin(II) chloride DES regarded both as catalyst and reaction medium.

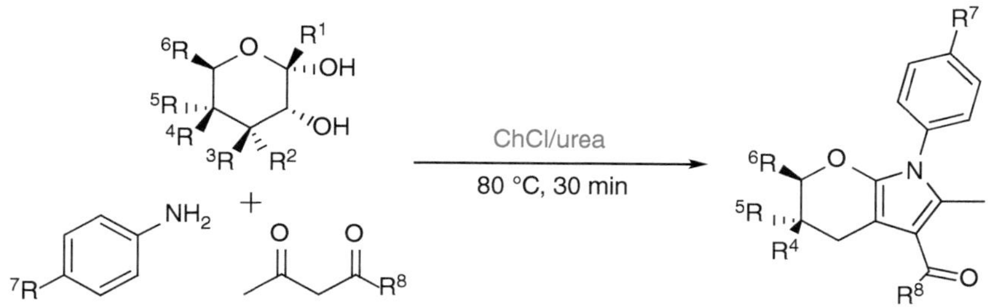

Scheme 6.66 Acid- and metal-free synthesis of annulated pyrroles in DES.

This reaction could perform well under mild conditions using aniline derivatives, aromatic aldehydes, and enolizable aldehydes as substrates. Furthermore, the method does not use toxic catalysts and solvents, in line with the concept of green chemistry (Scheme 6.67) [68].

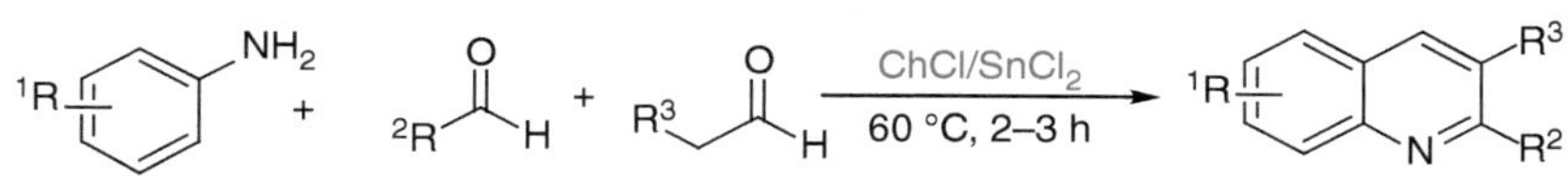

Scheme 6.67 Synthesis of quinoline derivatives in DES.

Ramón's group described a practical one-pot synthesis of sulfonamides from triarylbismuthines, $Na_2S_2O_5$, and nitro compounds in the presence of cuprous chloride without ligand. This process proved to be simple, chemo-selective, and air and moisture-compatible in a new acetyl ChCl/acetamide DES which was deemed the optimal reaction solvent (Scheme 6.68) [69].

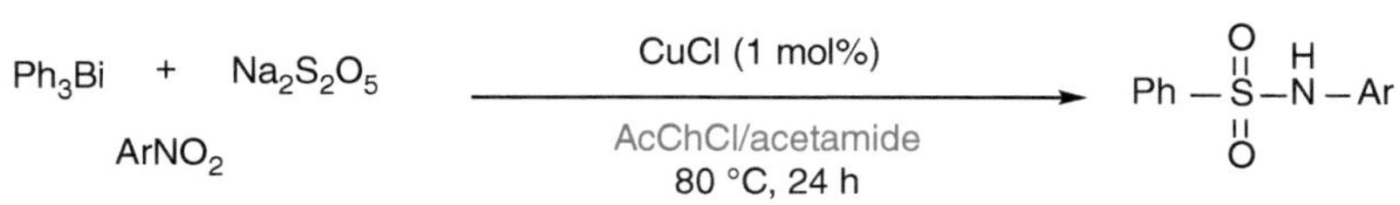

Scheme 6.68 Synthesis of sulfonamides in DES.

Karimi-Jaberi and coworkers synergistically promoted the three-component Strecker reaction of benzaldehydes, 2-aminobenzothiazole, and Me_3SiCN to yield α-amino nitrile derivatives using ChCl/urea DESs. The DES serves as solvent and co-catalyst. The methodology has features including shortened reaction times, simple separation, and mild reaction conditions (Scheme 6.69) [70].

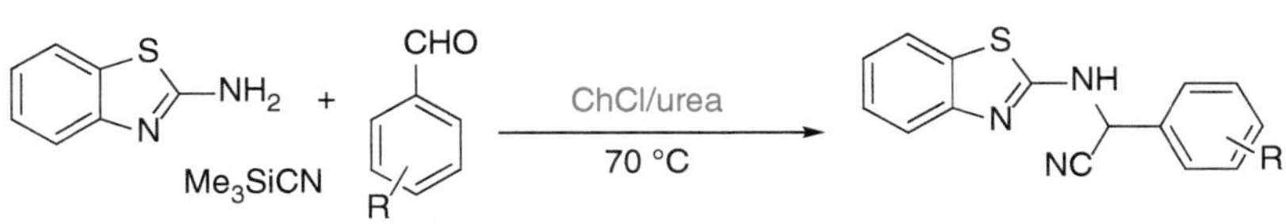

Scheme 6.69 One-pot synthesis of α-amino nitrile derivatives in DES.

Natural products containing pyrrole and imidazole heterocycles have anticancer, antibacterial, antiviral, or immunosuppressive properties, and the construction of drug molecules containing pyrrole and imidazole heterocycles has good development prospects. Kumar's group reported a highly efficient green synthesis of drug molecules containing pyrrole and imidazole in guanidine chloride/urea DESs.

The utilization of DES can promote the three-component domino reaction with special advantages including operational simplicity, high atom economy, high selectivity, and mild reaction conditions (Scheme 6.70) [71].

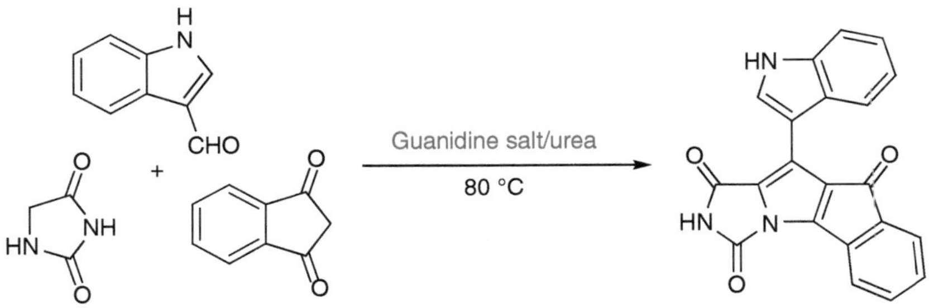

Scheme 6.70 Three-component domino reaction in DES.

1-Amidoalkyl-2-naphthols containing 1,3-amino oxygenated functional groups and polyhydroquinolines containing *N*-heterocyclics have many biological characteristics and pharmacological activities, which are widely used in the pharmaceutical field. Tran's group reported a method for the efficient three-component synthesis of 1-amidoalkyl naphthols and polyhydroquinolines. They used ChCl/zinc chloride DESs to promote straightforward three-component reactions at low temperatures with high selectivity, mild reaction conditions, and short reaction time. Interestingly, the process can be employed in industrial applications due to low costs and less environmental pollution (Scheme 6.71) [72].

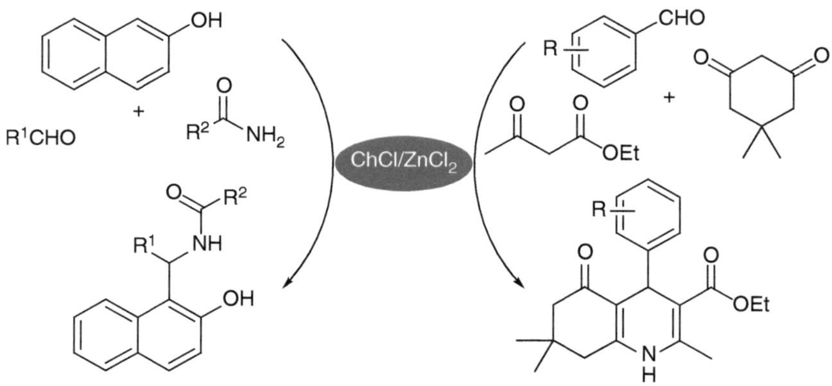

Scheme 6.71 Three-component synthesis of 1-amidoalkyl naphthols and polyhydroquinolines in DES.

Rhodanine derivatives have drug effects such as antiviral, anticancer, and antibacterial activities, and some rhodanine structures can regard as strong push–pull chromophores and fascinating structures to prepare dye-sensitized solar cells and optical materials. Azizi's group presented a practical route for the efficient three-component synthesis of 5-alkylidene-thiazolones in a ChCl/glycerol DESs as a green dual solvent/catalyst. This synthetic procedure provides valuable features with mild reaction conditions, operational simplicity, and large-scale synthesis (Scheme 6.72) [73].

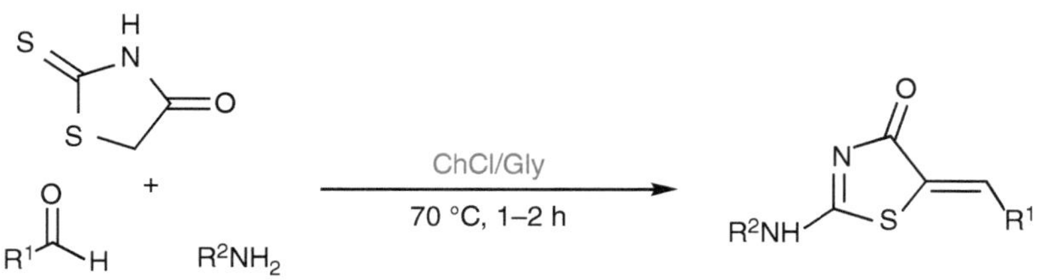

Scheme 6.72 Three-component synthesis of 5-alkylidene-thiazolones in DES.

The tetrazole ring is an important five-membered heterocyclic nucleus, which is present in many synthetic drugs and agrochemical products. Therefore, the efficient synthesis of tetrazoles has attracted great attention from organic chemists. In 2018, Xiong's team developed an effective copper acetate-catalyzed one-pot three-component synthesis of a series of 5-substituted-1*H*-tetrazoles by combining arylaldehydes, hydroxylamine hydrochloride with sodium azide in ChCl/urea DESs. The newly synthetic approach possesses special features including low cost, simple work-up, and sustainability (Scheme 6.73) [74].

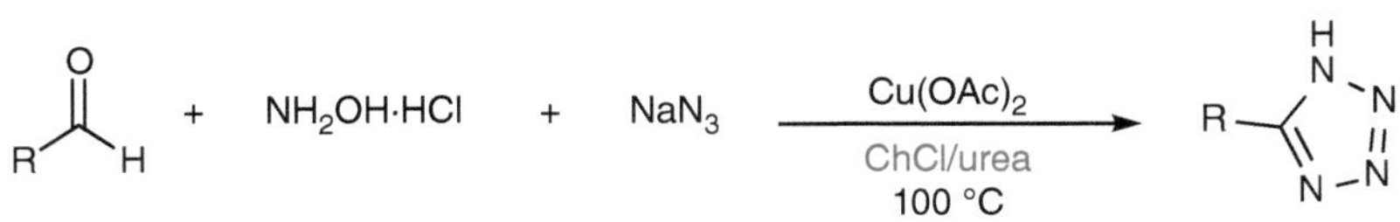

Scheme 6.73 Three-component synthesis of 5-substituted 1*H*-tetrazoles in DES.

6.7.2 One-pot Four-component Reaction

Pyrrole nucleus is one of the common structures in natural products, pesticides, drugs, and materials science. Therefore, the synthesis of pyrrole has received much attention. In 2015, Zhang and coworkers reported an efficient method of synthesizing different substituted pyrrole in ChCl/malonic acid DESs through a one-pot four-component coupling reaction of amines, aldehydes, 1,3-dicarbonyl compounds, and nitromethane. DES has good catalytic activity and excellent recyclability, which allows this methodology to prepare more powerful polysubstituted pyrroles on a large scale (Scheme 6.74) [75].

In 2017, Azizi and coworkers investigated an efficient one-pot four-component synthesis of a wide range of multisubstituted cyclohexa-1,3-dienamines using

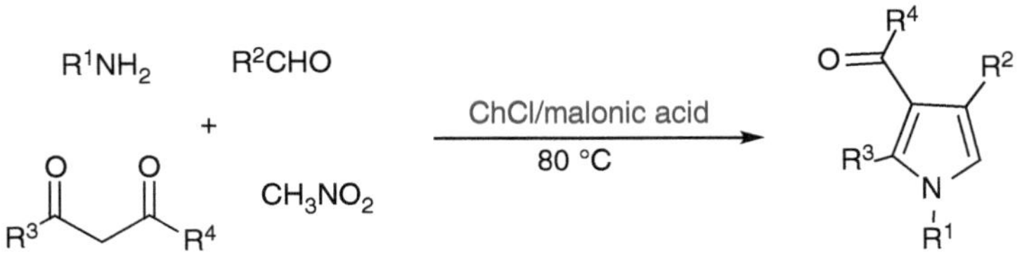

Scheme 6.74 Four-component synthesis of functionalized pyrroles in DES.

ChCl/urea DES as a solvent and catalyst to facilitate the multi-component domino reaction. The valuable synthetic approach provides advantages such as environmental sustainability, simple work-up procedures, and high isolated yields (Scheme 6.75) [76].

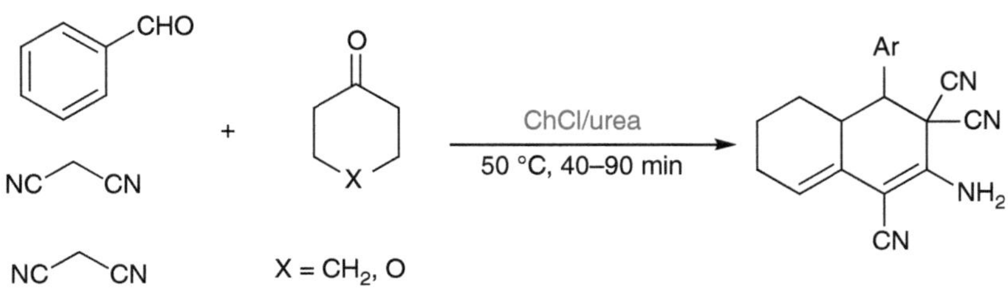

Scheme 6.75 Multicomponent domino reactions in DES.

6.8 Other Organic Reactions in Deep Eutectic Solvents

6.8.1 Isomerization Reaction

The Ru-catalyzed redox isomerization of allylic alcohols to yield saturated carbonyl compounds has some valuable features such as atom economy, mild reaction conditions, and practicality, which is very significant in organic synthesis. García-Álvarez and co-authors have extended the conversion of the redox isomerization of the commercially available allylic alcohol oct-1-en-3-ol into octan-3-one in ChCl/glycerol DESs with a highly efficient and selective ruthenium (IV) complex catalyst for the first time. The catalytic system is indicated to be competent with special advantages of recyclability of catalyst and solvent, simplicity of workup, and high yield of products (Scheme 6.76) [77].

In 2014, the same group disclosed an efficient synthetic route for the cycloisomerization of γ-alkynoic acids into cyclic enol-lactones using ChCl/urea DES as a solvent under standard bench experimental conditions by employing a new iminophosphorane–Au(I) complex as the catalyst for the first time. Notably, the protocol proved to be efficient and regioselective for the synthesis of cyclic enol-lactones in DES (Scheme 6.77) [78].

The *trans*-resveratrol served as a natural bioactive compound is well known for positive health effects. This compound can undergo a photochemically induced

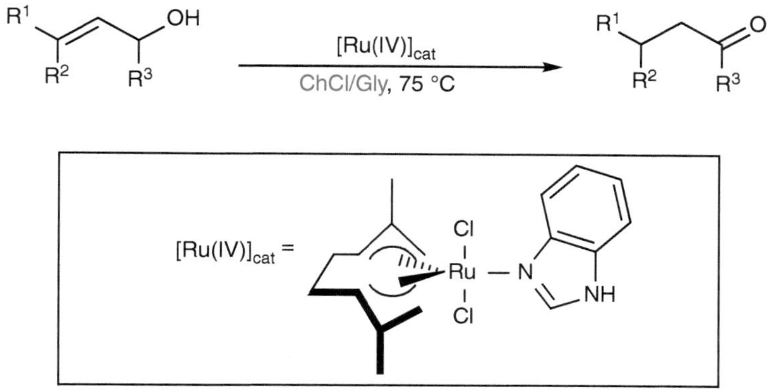

Scheme 6.76 Ru-catalyzed isomerization of allylic alcohols into saturated carbonyl compounds in DES.

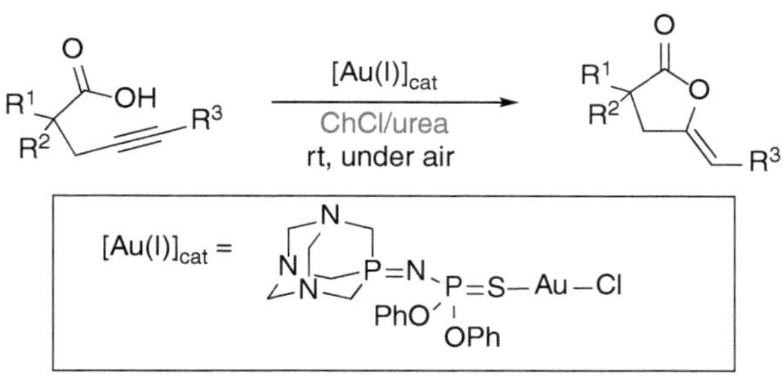

Scheme 6.77 Cycloisomerization of γ-alkynoic acids into cyclic enol-lactones in DES.

trans/cis isomerization and a 6π electrochemical cyclization subsequently for the generation of 2,4,6-trihydroxyphenanthrene (THP) irradiated by UV light. THP is potentially toxic as it can exert genotoxic effects. In 2022, Francioso, Mosca, and coworkers investigated the natural DESs as photo-protective agents to enhance resveratrol photostability and limit the formation of THP. The present method allows the improvement of green, eco-sustainable, and biocompatible resveratrol photo-stable formulations (Scheme 6.78) [79].

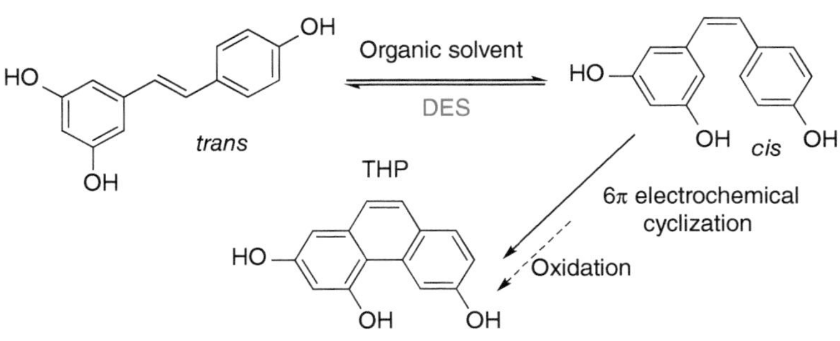

Scheme 6.78 DES-shifted the equilibrium to the left stabilizing the *trans*-isomeric form.

6.8.2 Ring-opening Reaction

Epoxides are very important raw materials and intermediates in organic transformations. 1,2-Bifunctional ring-opened compounds play a major role in the synthesis of novel therapeutic drugs and biologically active compounds. Azizi and coworkers reported an efficient approach in the synthesis of 1, 2-difunctional ring-opening products *via* the reaction of epoxides with various nucleophiles such as thiols, aromatic amines, azide, alcohols, and cyanide using choline DES as a catalyst for the first time (Scheme 6.79) [80].

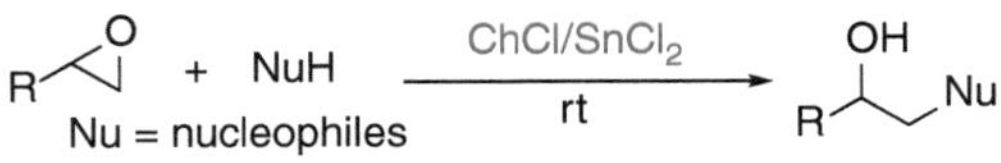

Scheme 6.79 Ring-opening of epoxides with different nucleophiles in DES.

Directed *ortho*-metallation (DoM) has received much attention in the field of organometallic chemistry. DoM provides an efficient synthetic strategy for direct functionalization of aromatic systems. Capriati and coworkers discovered an unexpected lateral-lithiation-promoted alkylative ring-opening reaction of *o*-tolyl tetrahydrofuran derivatives for the construction of new C—C bond in ChCl/glycerol DESs with a highly regioselectivity. This is a new method to synthesize functionalized primary alcohols (Scheme 6.80) [81].

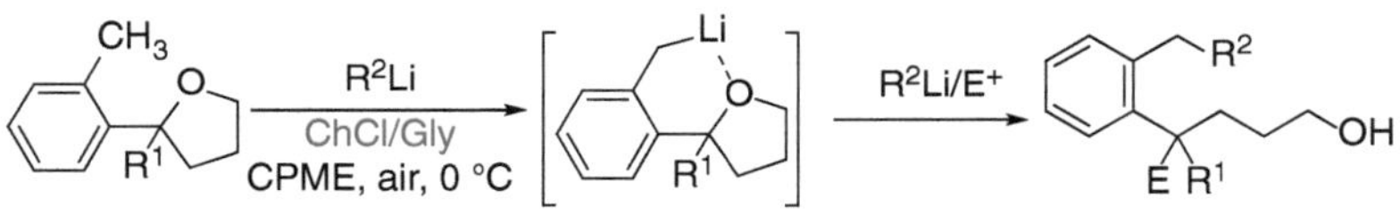

Scheme 6.80 Ring-opening of *o*-tolyl tetrahydrofuran derivatives in DES.

6.8.3 Esterification Reaction

Organic esters are widely used in the cosmetics industry and pharmaceutical fields, and esterification is also a fundamental reaction in organic synthetic chemistry. The reaction is often catalyzed by inorganic acids, which cause a lot of environmental pollution and are difficult to reuse. In 2016, Zhang's group reported an efficient and selective method for the esterification reaction of acetic acid with alcohols at room temperature using ChCl/chromium chloride hexahydrate DES served as both a catalyst and solvent. The DES possesses high catalytic activity and good recyclability which makes this method more practical (Scheme 6.81) [82].

$$ {}^{n}BuOH + CH_3COOH \xrightarrow{ChCl/CrCl_3 \cdot 6H_2O} CH_3COO^{n}Bu $$

Scheme 6.81 The esterification of acetic acid and alcohol in DES.

DESs can, not only be used as catalysts and solvents in esterification reactions, but also as reactants of alkylation reagents. Wang's team proposed a method for esterification of carboxylic acid with quaternary ammonium salts of DESs. The reaction performs smoothly in DES and does not require additional additives. The green methodology provides advantages such as good functional group compatibility, fast and simple reaction process, and high yield of desired products (Scheme 6.82) [83].

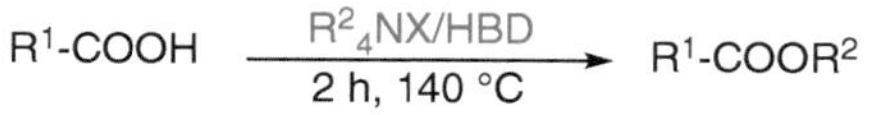

Scheme 6.82 Esterification of carboxylic acids in DES without any other additives.

6.9 Polymerization in DSEs

6.9.1 Anionic Polymerization of Alkenes

Anionic polymerization of olefins performs through organometallic end-chain propagation sites. Even though the anionic polymerization of styrene using ^{n}BuLi to serve as an initiator has been explored for a few decades, organolithiums still remain a more suitable choice for these transformations. However, the strict reaction conditions including extremely inert atmosphere conditions constitute a big challenge for industrial applications. In 2019, Soto and coworkers established a new air- and moisture-compatible organolithium-mediated method for the anionic polymerization of different olefins to obtain a range of synthetically relevant polyolefins in ChCl/glycerol DESs for the first time. This work has opened up new horizons for the application of organolithium reagents in DES and new avenues for the development of more sustainable polymer synthesis (Scheme 6.83) [84].

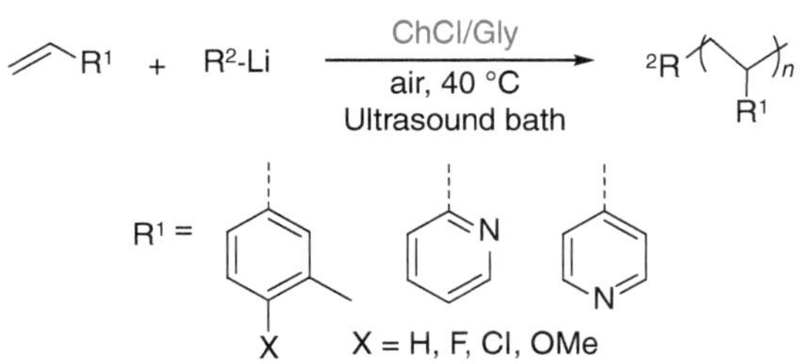

Scheme 6.83 Organolithium promoted anionic polymerization of alkenes in DES.

6.9.2 Glycolysis and Polyesterification

Poly(ethylene 2,5-furandicarboxylate) (PEF) is a polyester derived from C6 sugars, which has attracted great attention due to its similar chemical properties to PET. If PEF is not handled properly, it will inevitably accumulate in the environment. In 2022, Sousa and coworkers disclosed the potential of urea/zinc acetate DES system in continuous de-/re-polymerization for recycling PEF into *r*PEF. This methodology meets the requirements of green chemistry by designing a mild glycolytic reaction and favorably avoiding the usually complex and tedious procedure of recovering monomers, and is able to meet the high purity standards of the monomers required for the synthesis of polymers. Advantageously, it has great potential for other polyester applications (Scheme 6.84) [85].

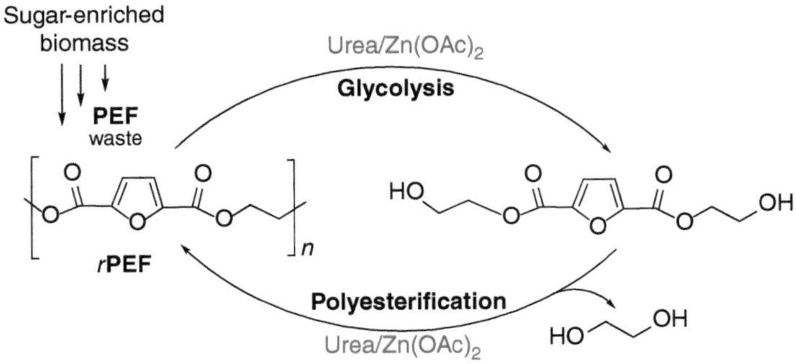

Scheme 6.84 Glycolysis and polyesterification reactions in DES.

6.9.3 Oxidative Polymerization

The suitable choice of solvents in the synthesis of conducting polymers is highly significant. Poly(3-alkylthiophene)s are conjugated polymers, which have gained increasing interest due to their intrinsic electrochemical and photoluminescence properties. To satisfy green chemistry, Lee and coworkers realized the $FeCl_3$-catalyzed simple one-step oxidative polymerization of 3-octylthiophene in ChCl/urea DES. The method has special features including fast polymerization speed, high yield, and green solvent (Scheme 6.85) [86].

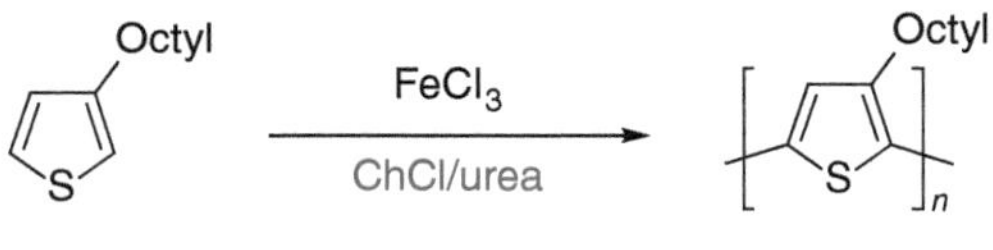

Scheme 6.85 $FeCl_3$-catalyzed oxidative polymerization in DES.

6.9.4 Visible-light-driven RAFT Polymerization

Reversible addition-fragmentation chain transfer (RAFT) polymerization receives the most attention due to the versatility of the chain transfer agents. However, the preparation of polymers is typically conducted in toxic organic solvents. It is really important to use greener solvents to eliminate the environmental effect of polymer generation and improve the efficiency of polymerization. In 2021, Yu's group disclosed an efficient visible-light-driven RAFT polymerization of various monomers containing methyl acrylate, methyl methacrylate, dimethylacrylamide, and styrene in tetrabutylammonium chloride/ethylene glycol DES. The practical protocol shows that DES is an effective and green medium in accelerating light-induced RAFT polymerization in the absence of additives or tedious degassing steps (Scheme 6.86) [87].

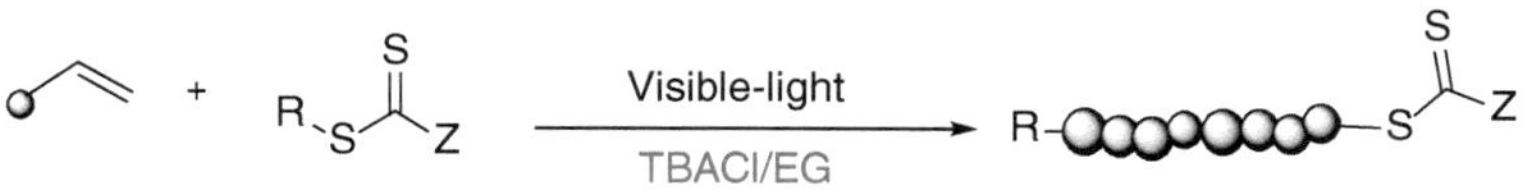

Scheme 6.86 RAFT polymerization driven by visible light in DES.

6.10 Conclusion

This chapter exemplifies that DESs are an environmentally benign sustainable alternative to conventional organic solvents in organic synthesis due to their properties (*e.g.*, high thermal stability, negligible vapor pressures, conductivity, non-flammability, and easy recycling). These eutectic mixtures have been used as green reaction media, catalysts, or even reagents in a variety of relevant organic transformations, such as cross-coupling reactions, oxidative reactions, hydrogenation reactions, cyclization reactions, condensation reactions, multicomponent reactions, other organic reactions, and polymerization. In most organic transformations, the satisfactory results have been obtained. With the development of green and sustainable chemistry, it is obvious that the use of DESs with their tailor-made properties will be able to develop environmentally friendly and economically viable synthetic protocols with their experimental simplicity and maximum synthetic efficiency. The organic transformations containing the utilization of DESs will be more attractive not only for academic research but also for industrial applications. We really hope that this chapter will provide some inspiration for future work in this promising research field, which is clearly open to many opportunities for new discoveries.

References

1 (a) Anastas, P. and Eghbali, N. (2010). *Chem. Soc. Rev.* 39: 301–312. (b) Zimmerman, J.B., Anastas, P.T., Erythropel, H.C., and Leitner, W. (2020). *Science* 367: 397–400.

2 (a) Zhang, W. and Cue, B.W. (2018). *Green Techniques for Organic Synthesis and Medicinal Chemistry. John Wiley & Sons.* (b) Ardila-Fierro, K.J. and Hernández, J.G. (2021). *ChemSusChem* 14: 2145–2162. (c) Chen, T.-L., Kim, H., Pan, S.-Y. et al. (2020). *Sci. Total Environ.* 716: 136998.

3 (a) Sheldon, R.A. (2019). *Curr. Opin. Green Sustain. Chem.* 18: 13–19. (b) Hessel, V., Tran, N.N., Asrami, M.R. et al. (2022). *Green Chem.* 24: 410–437.

4 Abbott, A.P., Capper, G., Davies, D.L. et al. (2003). *Chem. Comm.* 70–71.

5 (a) Dutta, S. and Nath, K. (2018). *J. Water Process. Eng.* 21: 163–176. (b) Smith, E.L., Abbott, A.P., and Ryder, K.S. (2014). *Chem. Rev.* 114: 11060–11082. (c) Abranches, D.O., Martins, M.A.R., Silva, L.P. et al. (2019). *Chem. Comm.* 55: 10253–10256.

6 (a) Chen, Y., Chen, W., Fu, L. et al. (2019). *Ind. Eng. Chem. Res.* 58: 12741–12750. (b) Stephens, N.M. and Smith, E.A. (2022). *Langmuir* 38: 14017–14024.

7 (a) Paiva, A., Craveiro, R., Aroso, I. et al. (2014). *Chem. Eng.* 2: 1063–1071. (b) Vanda, H., Dai, Y., Wilson, E.G. et al. (2018). *C. R. Chim.* 21: 628–638. (c) Płotka-Wasylka, J., de la Guardia, M., Andruch, V., and Vilková, M. (2020). *Microchem. J.* 159: 105539.

8 (a) Zhang, Q., De Oliveira Vigier, K., Royer, S., and Jérôme, F. (2012). *Chem. Soc. Rev.* 41: 7108–7146. (b) García-Álvarez, J. (2014). *Green Technologies for the Environment*, vol. 1186, 37–52. American Chemical Society. (c) García-Álvarez, J. (2015). *Eur. J. Inorg. Chem.* 2015: 5147–5157. (d) Liu, P., Hao, J.-W., Mo, L.-P., and Zhang, Z.-H. (2015). *RSC Adv.* 5: 48675–48704. (e) Alonso, D.A., Baeza, A., Chinchilla, R. et al. (2016). *Eur. J. Org. Chem.* 2016: 612–632. (f) Khandelwal, S., Tailor, Y.K., and Kumar, M. (2016). *J. Mol. Liq.* 215: 345–386. (g) Florindo, C., Lima, F., Ribeiro, B.D., and Marrucho, I.M. (2019). *Curr. Opin. Green Sustain. Chem.* 18: 31–36. (h) Hansen, B.B., Spittle, S., Chen, B. et al. (2021). *Chem. Rev.* 121: 1232–1285. (i) Perna, F.M., Vitale, P., and Capriati, V. (2020). *Curr. Opin. Green Sustain. Chem.* 21: 27–33. (j) Wu, J., Liang, Q., Yu, X. et al. (2021). *Adv. Funct. Mater.* 31: 2011102. (k) Yu, D., Xue, Z., and Mu, T. (2022). *Cell Rep. Phys. Sci.* 3: 100809.

9 Imperato, G., Vasold, R., and König, B. (2006). *Adv. Synth. Catal.* 348: 2243–2247.

10 Imperato, G., Höger, S., Lenoir, D., and König, B. (2006). *Green Chem.* 8: 1051–1055.

11 Zhao, X., Liu, X., and Lu, M. (2014). *Appl. Organomet. Chem.* 28: 635–640.

12 Marset, X., Khoshnood, A., Sotorríos, L. et al. (2017). *ChemCatChem* 9: 1269–1275.

13 Saavedra, B., González-Gallardo, N., Meli, A., and Ramón, D.J. (2019). *Adv. Synth. Catal.* 361: 3868–3879.

14 Delaye, P.-O., Pénichon, M., Boudesocque-Delaye, L. et al. (2018). *SynOpen.* 02: 0306–0311.

15 Dilauro, G., García, S.M., Tagarelli, D. et al. (2018). *ChemSusChem* 11: 3495–3501.

16 (a) Niakan, M. and Masteri-Farahani, M. (2022). *J. Mol. Liq.* 355: 118932. (b) Niakan, M., Masteri-Farahani, M., Karimi, S., and Shekaari, H. (2021). *J. Mol. Liq.* 324: 115078.

17 Ilgen, F. and König, B. (2009). *Green Chem.* 11: 848–854.

18 Dang, M.-H.D., Nguyen, T.T.T., Le, B.Q.G. et al. (2022). *J. Ind. Eng. Chem.* 111: 111–120.

19 Dilauro, G., Cicco, L., Vitale, P. et al. (2023). *Eur. J. Org. Chem.* 26: e202200814.

20 Marset, X., Saavedra, B., González-Gallardo, N. et al. (2019). *Front. Chem.* 7: 700.

21 Messa, F., Dilauro, G., Perna, F.M. et al. (2020). *ChemCatChem* 12: 1979–1984.

22 Niakan, M., Masteri-Farahani, M., Shekaari, H., and Karimi, S. (2021). *Carbohydr. Polym.* 251: 117109.

23 Marset, X., De Gea, S., Guillena, G., and Ramón, D.J. (2018). *ACS Sustain. Chem. Eng.* 6: 5743–5748.

24 Dilauro, G., Azzollini, C.S., Vitale, P. et al. (2021). *Angew. Chem. Int. Ed.* 60: 10632–10636.

25 Punzi, A., Coppi, D.I., Matera, S. et al. (2017). *Org. Lett.* 19: 4754–4757.

26 Teja, C. and Nawaz Khan, F.R. (2019). *ACS Omega* 4: 8046–8055.

27 Shariatipour, M., Salamatmanesh, A., Jadidi Nejad, M., and Heydari, A. (2020). *CatalCommun.* 135: 105890.

28 Marset, X. (2022). *Org. Biomol. Chem.* 20: 7071–7075.

29 Shaabani, A. and Afshari, R. (2018). *J. Colloid Interface Sci.* 510: 384–394.

30 Messa, F., Perrone, S., Capua, M. et al. (2018). *ChemComm* 54: 8100–8103.

31 Quivelli, A.F., Vitale, P., Perna, F.M., and Capriati, V. (2019). *Front. Chem.* 7: 723.

32 Cicco, L., Hernández-Fernández, J.A., Salomone, A. et al. (2021). *Org. Biomol. Chem.* 19: 1773–1779.

33 Pant, P.L. and Shankarling, G.S. (2017). *ChemistrySelect* 2: 4892–4898.

34 Quivelli, A.F., Marinò, M., Vitale, P. et al. (2022). *ChemSusChem* 15: e202102211.

35 Niakan, M., Karimi, S., Masteri-Farahani, M., and Shekaari, H. (2021). *Colloids Surf. A Physicochem. Eng. Asp.* 620: 126603.

36 Marset, X., Guillena, G., and Ramón, D.J. (2017). *Chem. Eur. J.* 23: 10522–10526.

37 Pant, P.L. and Shankarling, G.S. (2017). *ChemistrySelect.* 2: 7645–7650.

38 Guajardo, N., Carlesi, C., and Aracena, Á. (2015). *ChemCatChem* 7: 2451–2454.

39 Marset, X., Pérez, J.M., and Ramón, D.J. (2016). *Green Chem.* 18: 826–833.

40 Vadagaonkar, K.S., Kalmode, H.P., Prakash, S., and Chaskar, A.C. (2015). *New. J. Chem.* 39: 3639–3645.

41 Singh, K., Kaur, A., Mithu, V.S., and Sharma, S. (2017). *J. Org. Chem.* 82: 5285–5293.

42 Wagh, R., Nerurkar, G., and Nagarkar, J. (2018). *ChemistrySelect* 3: 9654–9657.

43 Ni, Y., Bi, Z., Su, H., and Yan, L. (2019). *Green Chem.* 21: 1075–1079.

44 Vitale, P., Cicco, L., Messa, F. et al. (2019). *Eur. J. Org. Chem.* 2019: 5557–5562.

45 Cicco, L., Fombona-Pascual, A., Sánchez-Condado, A. et al. (2020). *ChemSusChem* 13: 4967–4973.

46 Sun, L., Su, T., Xu, J. et al. (2019). *Green Chem.* 21: 2629–2634.

47 Zhu, Z., Lin, H., Chi, M. et al. (2022). *Fuel* 308: 122070.

48 Azizi, N., Batebi, E., Bagherpour, S., and Ghafuri, H. (2012). *RSC Adv.* 2: 2289–2293.

49 Shu, C., Liu, X., Xie, W. et al. (2021). *New J. Chem.* 45: 10502–10509.

50 Chanquia, S.N., Huang, L., García Liñares, G. et al. (2020). *Catalysis* 10: 1013.

51 Faverio, C., Boselli, M.F., Camarero Gonzalez, P. et al. (2021). *Beilstein J. Org. Chem.* 17: 1041–1047.

52 Peng, F., Chen, Q.-S., Li, F.-Z. et al. (2020). *Mol. Catal.* 484: 110773.

53 Alvi, S. and Ali, R. (2021). *Org. Biomol. Chem.* 19: 9732–9745.

54 Ali Ghumro, S., Alharthy, R.D., Al-Rashida, M. et al. (2017). *ACS omega* 2: 2891–2900.

55 More, P.A., Gadilohar, B.L., and Shankarling, G.S. (2014). *Catal. Lett.* 144: 1393–1398.

56 Marullo, S., Meli, A., and D'Anna, F. (2020). *ACS Sustain. Chem. Eng.* 8: 4889–4899.

57 Villamizar-Mogotocoro, A.-F., Bonilla-Castañeda, S.-M., and Kouznetsov, V.V. (2022). *Green Chem.* 24: 7996–8004.
58 Singh, R.R., Devi, T.J., Devi, T.J., and Singh, O.M. (2022). *Curr. Res. Green Sustain. Chem.* 5: 100272.
59 Phadtare, S.B., Jarag, K.J., and Shankarling, G.S. (2013). *Dyes Pigm.* 97: 105–112.
60 Qin, H., Zhou, Y., Zeng, Q. et al. (2020). *Green. Energy Environ.* 5: 124–129.
61 Srivastava, S. (2020). *ChemistrySelect* 5: 799–803.
62 Molnar, M., Brahmbhatt, H., Rastija, V. et al. (1897). *Molecules* 2018: 23.
63 Wu, M., Ma, H., Ma, Z. et al. (2018). *Chem. Eng.* 6: 9434–9441.
64 Capua, M., Perrone, S., Perna, F.M. et al. (2016). *Molecules* 21: 924.
65 Massolo, E., Palmieri, S., Benaglia, M. et al. (2016). *Green Chem.* 18: 792–797.
66 Wang, Y., Yao, Q.-X., He, J.-R. et al. (2022). *Biomass Convers. Biorefin.* 12: 87–93.
67 Rokade, S.M., Garande, A.M., Ahmad, N.A.A., and Bhate, P.M. (2015). *RSC Adv.* 5: 2281–2284.
68 Shahabi, D. and Tavakol, H. (2016). *J. Mol. Liq.* 220: 324–328.
69 Marset, X., Torregrosa-Crespo, J., Martínez-Espinosa, R.M. et al. (2019). *Green Chem.* 21: 4127–4132.
70 Bahrani, A. and Karimi-Jaberi, Z. (2019). *Monatsh. Chem.* 150: 303–307.
71 Rushell, E., Tailor, Y.K., Khandewal, S. et al. (2019). *New J. Chem.* 43: 12462–12467.
72 Nguyen, V.T., Nguyen, H.T., and Tran, P.H. (2021). *New J. Chem.* 45: 2053–2059.
73 Azizi, N., Qomi, M., Asghari, M., and Farhadi, E. (2021). *Sustain. Chem. Pharm.* 22: 100457.
74 Xiong, X., Yi, C., Liao, X., and Lai, S. (2019). *Tetrahedron Lett.* 60: 402–406.
75 Hu, H.-C., Liu, Y.-H., Li, B.-L. et al. (2015). *RSC Adv.* 5: 7720–7728.
76 Azizi, N., Ahooie, T.S., and Hashemi, M.M. (2017). *J. Mol. Liq.* 246: 221–224.
77 García-Álvarez, J. (2015). *Chem. Eur. J.* 2015: 5147–5157.
78 Rodríguez-Álvarez, M.J., Vidal, C., Díez, J., and García-Álvarez, J. (2014). *Chem. Commun.* 50: 12927–12929.
79 Mattioli, R., Di Risola, D., Federico, R. et al. (2022). *Molecules* 27: 2348.
80 Azizi, N. and Batebi, E. (2012). *Catal. Sci. Technol.* 2: 2445–2448.
81 Sassone, F.C., Perna, F.M., Salomone, A. et al. (2015). *Chem. Commun.* 51: 9459–9462.
82 Cao, J., Qi, B., Liu, J. et al. (2016). *RSC Adv.* 6: 21612–21616.
83 Yasmin, S., Sheng, W.-B., Peng, C.-Y. et al. (2018). *Synth. Commun.* 48: 68–75.
84 Sánchez-Condado, A., Carriedo, G.A., Presa Soto, A. et al. (2019). *ChemSusChem* 12: 3134–3143.
85 Agostinho, B., Silvestre, A.J.D., and Sousa, A.F. (2022). *Green Chem.* 24: 3115–3119.
86 Park, T.-J. and Lee, S.H. (2017). *Green Chem.* 19: 910–913.
87 Li, C.-Y. and Yu, S.-S. (2021). *Macromolecules* 54: 9825–9836.

7

Recent Achievements in Organic Reactions in Ionic Liquids

Jianxiao Li and Huanfeng Jiang

South China University of Technology, Key Laboratory of Functional Molecular Engineering of Guangdong Province, School of Chemistry and Chemical Engineering, Guangzhou 510640, P. R. China

7.1 Introduction

Ionic liquids (ILs) are defined as molten salts with melting points at or below 100 °C. In 1914, Walden synthesized the first IL ethylammonium nitrate (EAN) by neutralizing ethylamine with concentrated HNO_3 [1]. Particularly, IL EAN displays similar physical properties to water as it is colorless, odorless, and has reasonably high viscosity. Remarkably, the conductivity of EAN was consistent with composition of purely anions and cations [2]. After this, Hurley and Wier prepared the second generation of ILs by mixing alkylpyridinium chlorides with $AlCl_3$ in 1951 [3]. Unfortunately, most of the second-generation ILs are not stable in the presence of moisture, and their acidity/basicity is not easy to regulate [4]. In 1992, Wilkes and Zaworotko prepared moisture- and air-stable ILs based on imidazolium and tetrafluoroborate [5]. Most importantly, imidazolium, phosphonium, ammonium, and pyrrolidinium cations, with halides, bis(trifluoromethane sulfonyl) amide, and tetrafluoroborate anions are the most common choices in further studies (Fig. 7.1) [6]. The most fascinating nature of ILs is their structural diversity. The physicochemical properties of ILs, including viscosity, polarity, and other properties vary according to their ionic structures [6, 7]. As a result, their physicochemical properties can be tuned by exploiting the diversity of component organic cations and organic or inorganic anions. ILs have consequently been described as designer solvents.

Compared to molecular liquids, ILs have various advantages that include extremely low volatility, low flammability, and high thermal stability. Much of the interest in ILs has centered on their possible use as the "green" alternatives to volatile organic solvents [8–10]. This claim usually rests on the fact that ILs are generally nonvolatile under ambient conditions. Hence, exposure risk to ILs is much lower than it is for a volatile solvent. This nonvolatility also leads to most ILs being nonflammable under ambient conditions [11]. For this reason, ILs have attracted increasing interest recently in the context of green organic synthesis [7, 12–14]. Particularly, several excellent reviews have been published describing the

Green Solvents in Organic Synthesis, First Edition.
Edited by Xiao-Feng Wu, Zhiping Yin, Liang-nian He, and Feng Wang.

Cations

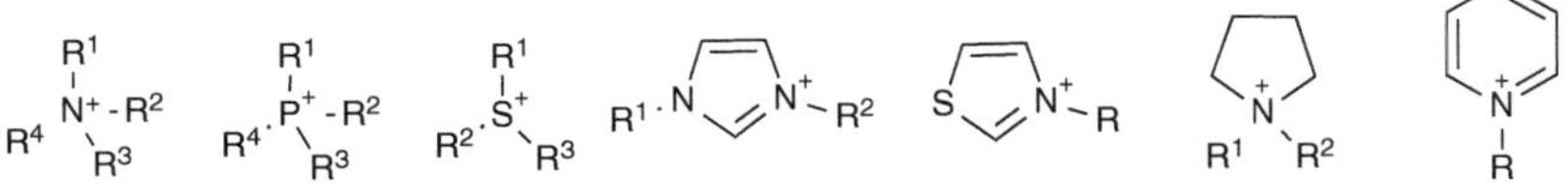

Anions

Cl^- Br^- I^- $AlCl_4^-$ $Al_2Cl_7^-$ BF_4^- PF_6^- CF_3COO^- $C_3F_7COO^-$ $CF_3SO_3^-$ SbF_6^-

NO_3^- $H_2PO_4^-$ HSO_4^- $(CF_3SO_2)_2N^-$ $(C_2F_5SO_2)N^-$ $(CF_3SO_2)_3N^-$ $CH_3(CH_2)_nCOO^-$

Figure 7.1 Some commonly used cations and anions for ionic liquids.

application of ILs in various fields, such as catalysis [15, 16], palladium-catalyzed reactions [17], heterocyclic chemistry [18], enzyme catalysis in ILs [19], electrochemical reactivity and voltammetry [20, 21], synthesis of nanomaterials [22], and inorganic complexes [23]. In addition, a 2016 review titled "Acidic ionic liquids" focused on specialized applications of acidic ILs [24]. In 2017, Ananikov and coworkers first summarized the biological activity of ILs and their application in pharmaceutics and medicine as well [25].

In the past 10 years, there have been several review articles published with the word "cross-coupling reactions in ionic liquids" in the title. For example, a distinguished review on palladium-catalyzed cross-coupling reactions in ILs has been dominated by Mastrorilli and coworkers in 2013 [26]. Our group also reviewed the application of the ILs in palladium-catalyzed unsaturated hydrocarbons coupling for the synthesis of structurally diverse heterocyclic scaffolds along with mechanistic studies and applications in 2016 [27]. Further, we also illustrated the most important advances in Pd-catalyzed cross-coupling reaction in ILs from 2013 to the beginning of 2018 [28]. In this chapter, we mainly attempt to comprehensively summarize the recent achievements in organic reactions, covering the recent literature up to early 2013. The scope of this review is wider than the previous reviews on this subdiscipline.

7.2 Transition Metal-catalyzed Reactions

Transition metal-catalyzed reactions have been extensively studied in ILs. Much of this has been descriptive in nature, and these studies have been the subject of several excellent recent reviews. Some studies have focused on the ways in which the ILs themselves affect the outcomes of catalytic reactions. Changing the solvent can affect a catalytic reaction in many ways; e.g., the solvents may interact with the reactants and/or interact directly with the catalyst. In biphasic systems, they can also affect the partitioning of solutes between the two phases. These interacting effects can lead to complex dependencies on changes in solvent that are difficult to interpret. In the discussion below, we only selected those studies from which reasonably robust inferences can be drawn.

7.2.1 Palladium-catalyzed Cascade Cyclization Reaction

Palladium-catalyzed cascade cyclization reactions have witnessed significant improvements in recent years, which can provide effective and straightforward methodologies for the construction of structurally diverse and complicated organic molecule architectures with excellent selectivity [29–31]. In this section, we mainly choose some representative methodologies using ILs as solvents in our lab.

For example, in 2016, we disclosed an attractive and concise approach for the synthesis of 3-sulfenyl benzofurans (**4**) and 3-sulfenyl indoles (**5**) through palladium- and copper- co-catalyzed cascade annulation/arylthiolation of 2-alkynylphenols (**1**) or 2-alkynylamines (**2**) with arylboronic acid (**3**) and S_8 in ILs [Bmim]Cl (Scheme 7.1) [32]. This strategy tolerated a broad range of electron-neutral, electron-withdrawing, and electron-donating substrates, thereby furnishing the corresponding sulfenylated heterocycles scaffolds in moderate-to-excellent yields. Additionally, various conventional solvents (such as DMF, DMSO, CH_3CN, and toluene) and different types ILs were examined, and ILs [Bmim]Cl was found to be the best solvent for this transformation. Other easily available sulfur sources such as Na_2S, S_8, and K_2S, were also screened, and S_8 was the most effective inorganic sulfur source. Considering the environmental impacts, ILs as the solvent make this transformation green and reusability.

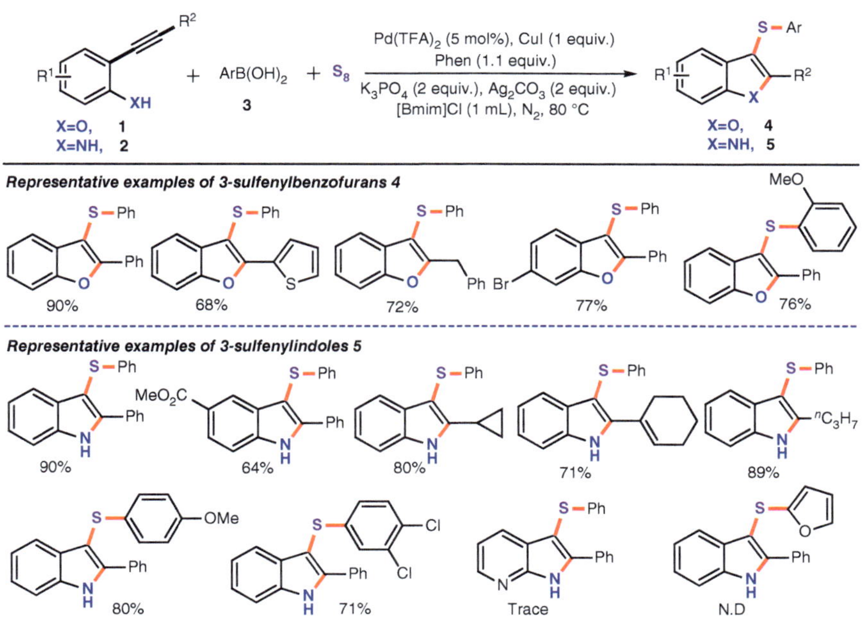

Scheme 7.1 Palladium-catalyzed cascade annulation/arylthiolation reaction.

Based on previous relevant reports [33], and the experimental results, a tentative pathway is proposed in Scheme 7.2. In this transformation, a Pd complex (**Int-1**) is formed *in situ* in ILs [34]. The nucleopalladation of 2-alkynylphenols or

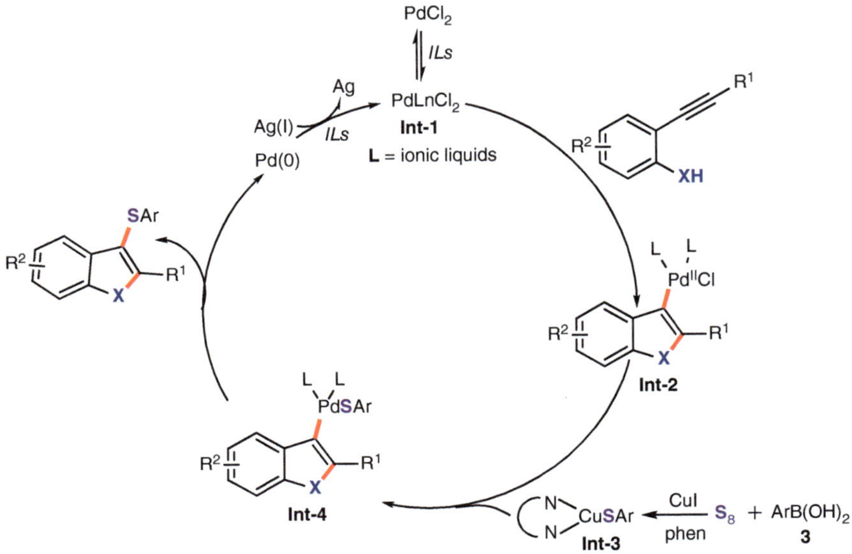

Scheme 7.2 Plausible mechanism for palladium-catalyzed cascade annulation/arylthiolation.

2-alkynylamines gives vinyl palladium species **Int-2**. Then, the transmetalation of organocopper thiolate complex **Int-3** with intermediate **Int-2** forms aryl palladium complex **Int-4**. Reductive elimination leads to the formation of the target products. Finally, the resulting palladium(0) is additionally oxidized by silver salt to palladium(II) to complete the catalytic cycle.

In 2019, we developed an effective and green *N*-heterocyclic carbene (NHC)-palladium-catalyzed three-component cascade cyclization/alkynylation of *O*-methyl oxime (**6**), secondary amines (**7**), and alkynes in ILs [Bmim]Cl (Scheme 7.3) [35]. A variety of *O*-methyloximes (**6**) and several secondary amines were well tolerated in this chemical transformation and furnished the corresponding 4-alkynylisoxazoles (**9**) in moderate-to-good yields. The ILs [Bmim]Cl as the green solvent played a profound effect on the efficiency of this reaction. This method complements existing procedures for the synthesis of 4-alkynylisoxazoles with full atom- and step-economical. However, the furanyl and pyridyl groups of *O*-methyloximes were not compatible with the developed reaction conditions.

Mechanistic studies suggested that *N*, *N*-diethylprop-2-yn-1-amine (**Int-6**) might be the possible intermediate in this chemical transformation. As shown in Scheme 7.4, oxypalladation of *O*-methyloximes (**6**) yields the vinyl palladium intermediate **Int-5**. Meanwhile, alkynes **Int-6** are produced *in situ* from secondary amines (**7**) with propargyl *p*-toluenesulfonate. And then, ligands exchange reaction of vinyl palladium species **Int-5** with alkynes **Int-6** gives intermediate **Int-7**. Subsequently, reductive elimination of intermediate **Int-7** gives the desired products (**9**). Finally, with the air or oxygen as the oxidant, palladium(0) is oxidized to palladium(II) (**8**) to complete the catalytic cycle.

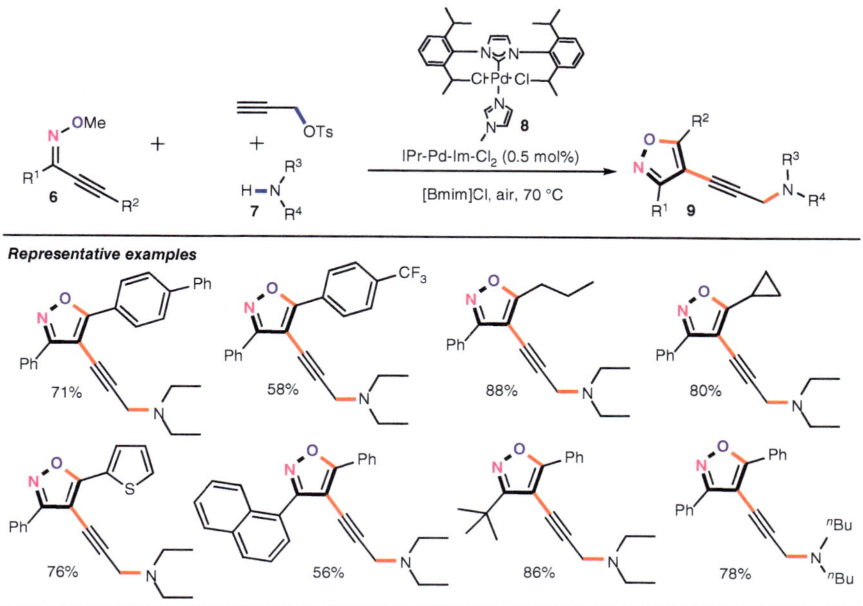

Scheme 7.3 Palladium-catalyzed three-component cascade cyclization/alkynylation.

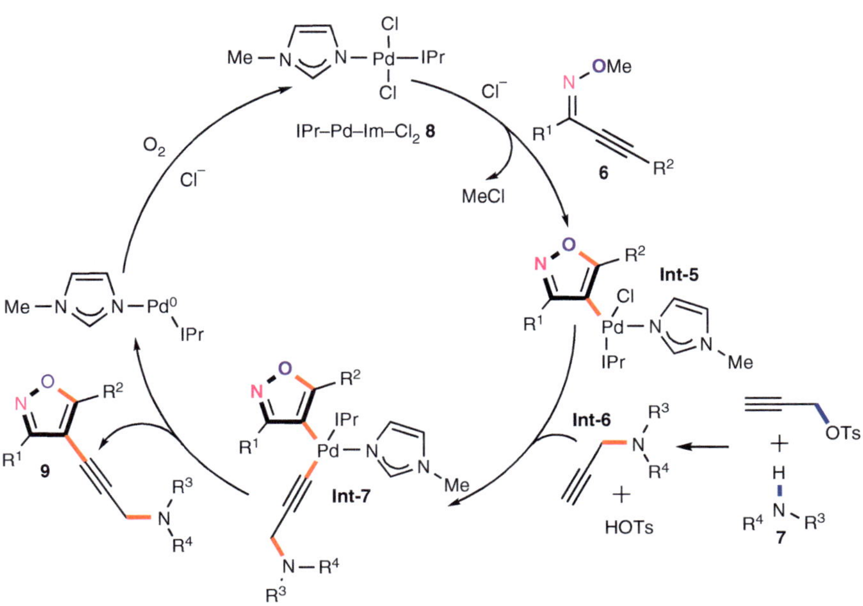

Scheme 7.4 General mechanism for three-component cascade cyclization/alkynylation.

After this, our group has exploited another green and expeditious synthetic approach for the assembly of various valuable 4-sulfenylisoxazole derivatives (**12**) by NHC–palladium-catalyzed three-component cascade *S*-transfer reaction in ILs [Cpmim]Cl (Scheme 7.5) [36]. Detailed investigations demonstrated that the best conditions for this chemical transformation were Ipr–Pd–Im–Cl_2 (**8**, 0.25 mol%), and [Cpmim]Cl (2 mL) under air at 80 °C for 12 h. Under the optimized reaction conditions, both aromatic and aliphatic acetylinic oximes (**10**) proceeded smoothly in this reaction system, and different functional groups worked well and produced the desired products **12** in satisfactory yield. However, when aliphatic halide (**11**) was explored as the substrate, no corresponding 4-sulfenylisoxazole product was obtained. Moreover, excellent stereoselectivity was observed for the formation of one C—O, and two C–S bonds in the present protocol. Particularly, by using ILs as both the green solvent and stabilizing agents, the palladium catalyst can be efficiently separated from the reaction mixture and recycled. Notably, $Na_2S_2O_3$ as an odorless sulfenylation reagent under aerobic conditions in ILs also makes this developed approach more eco-friendly.

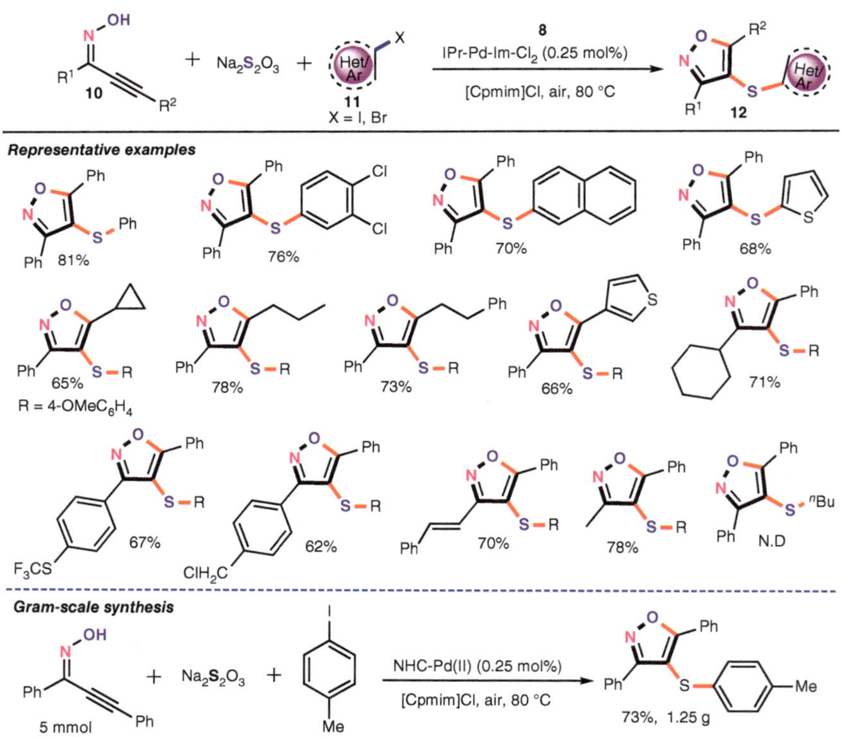

Scheme 7.5 Palladium-catalyzed three-component cascade *S*-transfer reaction.

A plausible mechanism is then described in Scheme 7.6. Initially, *trans*-oxypalladation of oximes (**10**) gives σ-isooxazolylpalladium species **Int-8** along with losing an HCl molecule. Meanwhile, aryl halides react with thiosulfate to

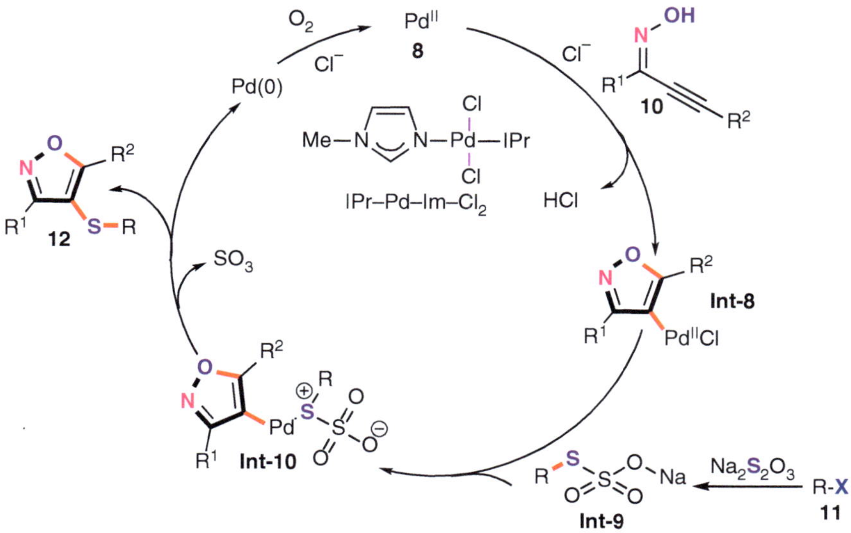

Scheme 7.6 Plausible mechanism for palladium-catalyzed 4-sulfenylisoxazoles synthesis.

yield thiosulfate salt **Int-9**. Subsequently, successive ligand exchange/reductive elimination generates the target products **12**. Finally, the Pd(0) is additionally oxidized by air to form the Pd(II) (**8**) species, completing the catalytic cycle.

Following this idea, our group described a palladium-catalyzed, IL-accelerated oxidative cascade annulation/functionalization of acetylenic oximes (**10**) with unactivated long-chain enols (**13**) under aerobic conditions for the assembly of synthetically versatile isoxazole building blocks. Under the optimized conditions using NHC–Pd complex IPr–Pd–Cin–Cl (**14**) in the presence of [C_2O_2mim]Cl in H_2O in an open air at 110 °C for 12 h, this cascade annulation/allylation was achieved in good-to-excellent yields with high atom- and step economy (Scheme 7.7) [37]. Notably, the requirement of IL [C_2O_2mim]Cl as the additive and H_2O as the green solvent was crucial for the success of this cascade process. With 0.05 mol% Ipr–Pd–Cin–Cl (**14**) complex as the catalyst, a diverse array of functional groups were well tolerated, producing the corresponding allylation products (**15**) in good yields. Additionally, even substrates containing sensitive halide groups could be coupled in excellent yields. Interestingly, a series of long-chain 1, *n*-enols and branched-chain enols also worked well, and delivered the desired products in satisfactory yields.

By changing the reaction conditions (using IL [Cpmim]BF_4 as the solvent), the NHC–Pd catalyzed cascade annulation/alkylation was successfully obtained for the straightforward synthesis of a library of functionalized 4-carbonyl isoxazole derivatives (**17**) (Scheme 7.8) [37]. Under an air atmosphere, this cascade annulation/alkylation reaction proceeded in the presence of 0.05 mol% of NHC–Pd complex (**14**) in [Cpmim]BF_4 conditions. The scope and limitations of the acetylenic oximes (**10**) and unactivated long-chain enols (**16**) were then investigated, and the results showed that both the alkyl and aryl-substituted enols performed well,

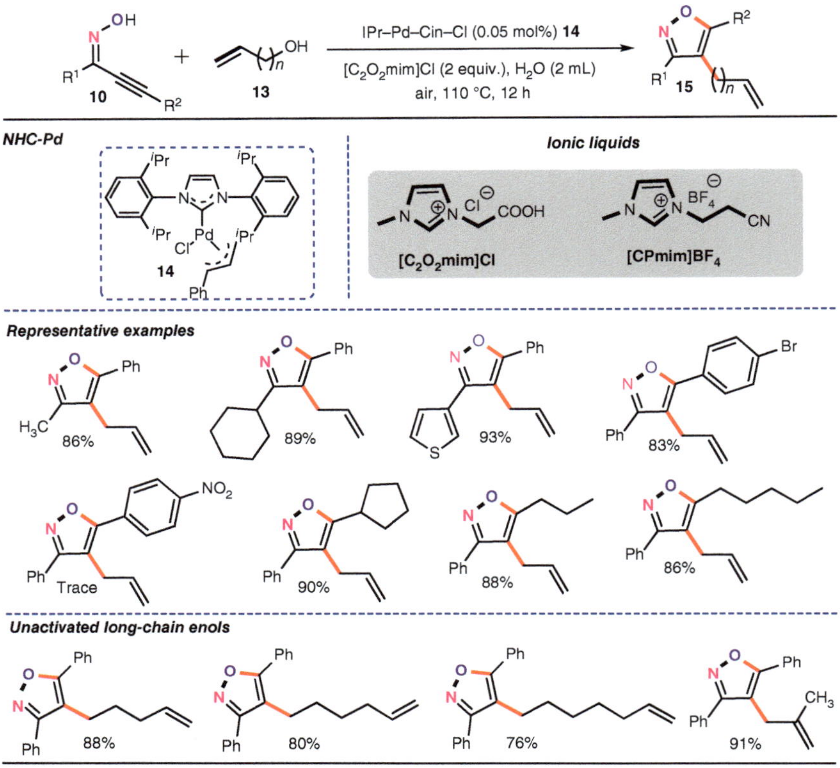

Scheme 7.7 NHC-Pd catalyzed cascade annulation/allylation of acetylenic oximes.

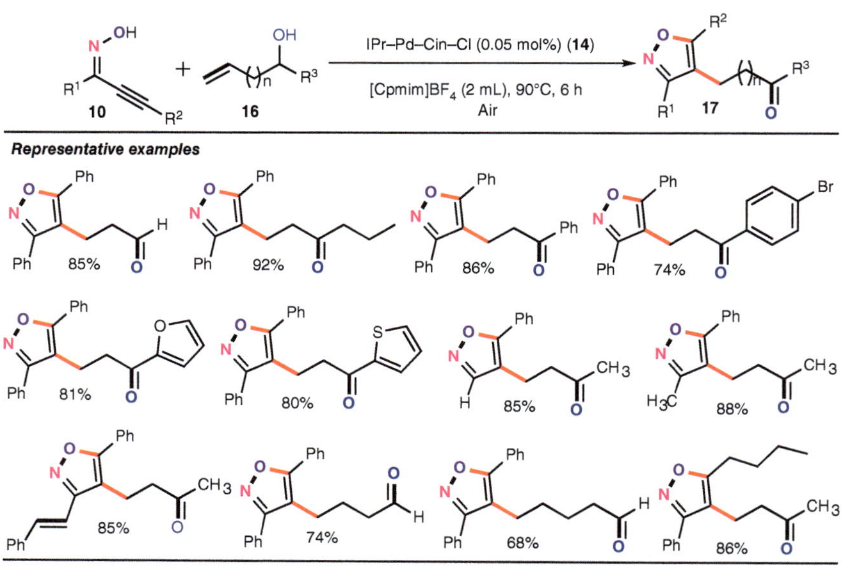

Scheme 7.8 Cascade annulation/alkylation of acetylenic oximes with long-chain 1, n-enols.

furnishing the corresponding products in excellent yields. Of note, the bromo group on the phenyl ring of enols provides handles for further derivatization.

Furthermore, by using a similar cascade cyclization/functionalization synthetic tactics, our group developed a palladium-catalyzed cascade cyclization/alkylation of oxime ethers (**6**) with unactivated alkenes (**18**) for the rapid construction of fully substituted isoxazole derivatives (**19**) (Scheme 7.9) [38]. Using 3 mol% of $Pd(Py)_2Cl_2$ as the catalyst and 1.2 equiv. NQ as the oxidant in [Aeim]Br at 90 °C for 10 h, both aromatic and aliphatic alkynone *O*-methyloximes (**6**) with different functional groups were smoothly transformed into the desired products in good-to-excellent yields with high regioselectivities. To our delight, this protocol is compatible with various halogenated aryl rings, such as fluorine, chlorine, and bromine atoms, rendering the desired products in fair yields. The IL [Aeim]Br as an environmentally friendly additive and solvent plays a crucial role in this reaction. Notably, but-3-en-1-ol also coupled well with *O*-methyloximes (**6**) to yield the expected products aldehyde in moderate yields. Detailed mechanistic studies revealed that this new protocol involved the "chain-walking" strategy through consecutive alkene migration insertion/β-hydride elimination process [39].

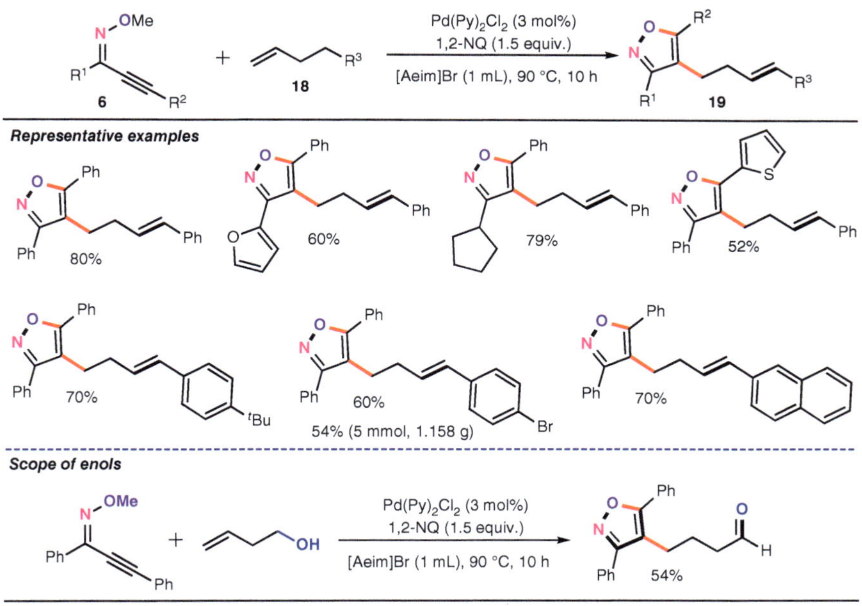

Scheme 7.9 Palladium-catalyzed cascade cyclization/alkylation of oxime ethers.

In a subsequent study, we also accomplished an NHC-palladium-catalyzed IL-accelerated regioselective oxyarylation of alkynes (**6**) with diaryl ethers (**20**) in a "task-specific" IL ([C_3NH_2mim]Br) *via* C(sp^3)—O and C(sp^2)—O bond cleavage of ethers for the assembly of structurally diverse 4-arylisoxazole derivatives (**21**) (Scheme 7.10) [40]. It is noteworthy that Ipr–Pd–Im–Cl_2 could efficiently

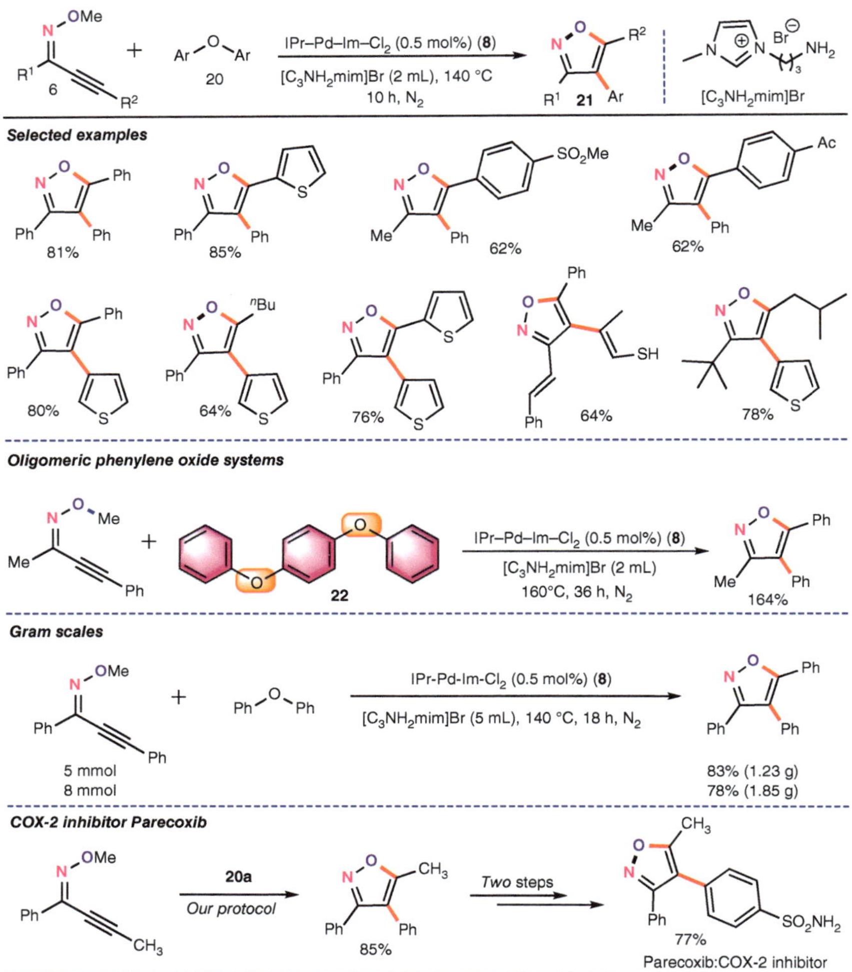

Scheme 7.10 Palladium-catalyzed ionic liquids-accelerated oxyarylation.

catalyze this cascade oxyarylation in the absence of phosphine ligands or other additional additives. Among a variety of solvents, IL [C_3NH_2mim]Br proved to be the suitable solvent for this chemical transformation. Under the optimized conditions, both alkynone *O*-methyloximes and a vast array of diaryl ethers were well accommodated in the transformation, giving the corresponding 4-arylisoxazoles in moderate-to-good yields. Notably, the oligomeric phenylene oxide systems, which bear two possible reaction sites, were well accommodated. Moreover, COX-2 inhibitor Parecoxib also has been successfully synthesized in 65% overall yield from the corresponding *O*-methyloximes. More importantly, this new approach provides a straightforward route to access a broad range of privileged 4-arylisoxazoles structures with good yields and excellent regioselectivities. Significantly, this catalytic system can be recycled up to eight times without significant loss of catalytic activity.

A plausible mechanism of this palladium-catalyzed ILs-accelerated oxyarylation reaction was then proposed, as shown in Scheme 7.11. Initially, oxidative addition of Pd(0) with diaryl ether gives Ar–Pd–OAr intermediate **Int-11** [41]. Then, the ILs *via* hydrogen bonding activate the –OAr motifs of the Pd–OAr species to form the palladium complexes **Int-12** [42]. Simultaneously, the exchange of ligands of the palladium complexes **Int-12** yields the Ar–Pd species **Int-13** through the loss of the phenol derivatives [43]. Subsequently, oxypalladation of the *O*-methyl oxime (**6**) generates oxonium palladium species **Int-14**, which can undergo the elimination of MeBr and gives the palladium intermediate **Int-15** in the presence of excessive bromine ion. Finally, a reductive elimination delivers the target products (**21**) and Pd(0) species to complete the catalytic cycle. The basic IL [C_3NH_2mim]Br is crucial for this catalytic transformation, and plays four roles: it acts as a green solvent and a bromine source, provides alkaline conditions, and helps in hydrogen bonding.

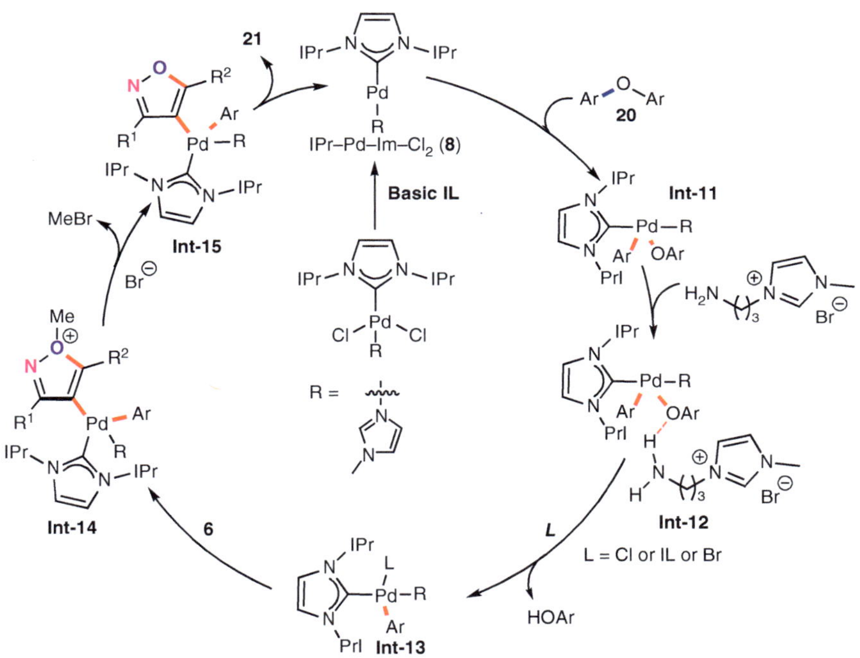

Scheme 7.11 Proposed mechanism for ionic liquids-accelerated oxyarylation.

Very recently, we also reported a palladium-catalyzed cascade annulation of alkynoic acids with unactivated alkenes under an oxygen atmosphere in ILs [Amim]Cl (Scheme 7.12) [44]. Satisfyingly, under the combination of 0.5 mol% NHC–palladacycle complex IPr–Pd–Py–Cl_2 (**24**) in [Amim]Cl at 30 °C for 8 h, this elegant cascade annulation approach can provide the corresponding alkyl-substituted α-methylene-γ-lactones (**25**) in moderate-to-good yields with excellent Z/E selectivity. Several NHC–palladium catalysts were tested, such as IPr–Pd–Allyl–Cl, IPr–Pd–Cin–Cl, IPr–Pd–Im–Cl_2, SIPr–Pd–Allyl–Cl, and

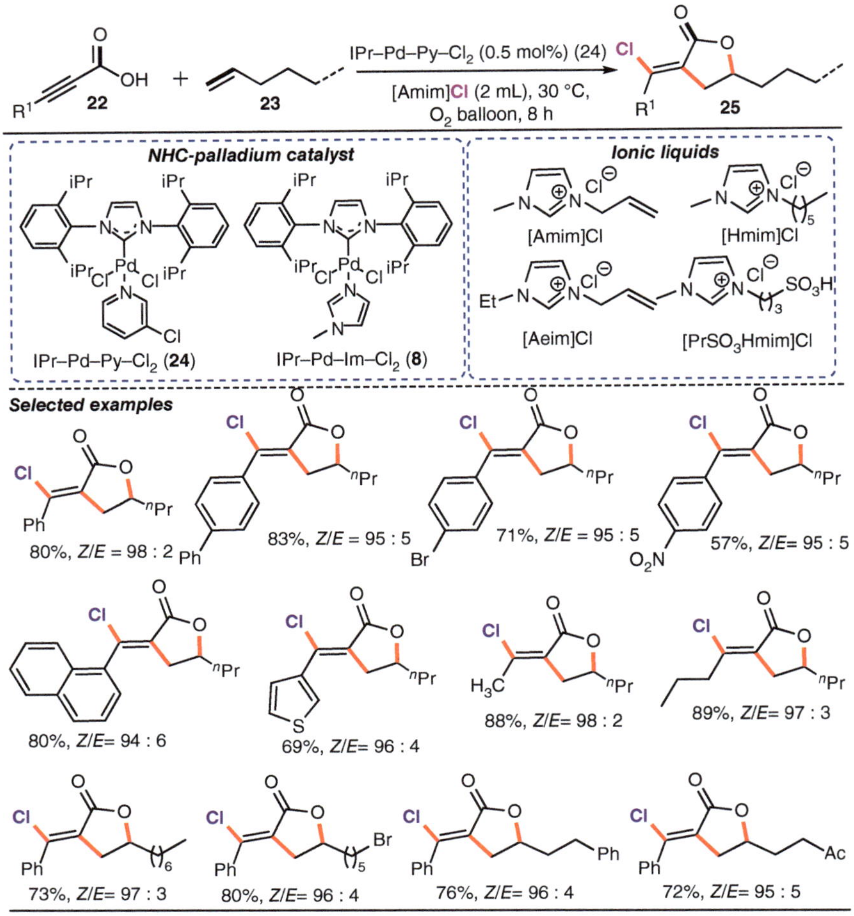

Scheme 7.12 Palladium-catalyzed cascade of alkynoic acids with unactivated alkenes.

IMes–Pd–Im–Cl_2, and the IPr–Pd–Py–Cl_2 and IPr–Pd–Im–Cl_2 catalyst provided the similar efficiency. Additionally, other chloride ILs such as [Bmim]Cl, [Aeim]Cl, [Hmim]Cl, and [$PrSO_3Hmim$]Cl, were also tested, while they gave inferior results compared with [Amim]Cl. It is particularly noteworthy that a library of steric surrounding internal alkynes as well as unactivated alkenes was also successfully tolerated under the optimized reaction conditions. It is noteworthy that O_2 as the final oxidant avoided the use of excess metal halides, such as cupric chloride. Notably, this novel methodology exhibits a new synthetic strategy for the assembly of synthetically versatile α-methylene-γ-lactones heterocyclic frameworks.

7.2.2 Carbonylation Reactions

Transition metal-catalyzed carbonylation reactions, making use of gaseous CO as the C1 source, provide an atom-economical route for the synthesis of aldehydes, ketones, carboxylic acid derivatives, and heterocyclic moieties. Aryl halides, alkenes, alkynes, and even alkanes are feedstocks for carbonylation [45]. This catalytic methodology, thus, paves the way for the synthesis of important bulk and fine chemicals. ILs as the green reaction medium have also been widely employed in the carbonylation process.

Our group reported an effective and straightforward palladium-catalyzed cascade carboxylative annulation of alkynoates (**26**) with homoallyl alcohol (**27**) in the presence of 1 atm CO/O_2 for the preparation of functionalized γ-lactones (**28**) in ILs $[C_2O_2mim]X$ with good atom- and step economies (Scheme 7.13) [46]. Under optimal conditions, the limitations and scope of alkynoates were then comprehensively investigated. Notably, hydroxyl, ester, chloro, and trifluoromethyl were amenable to this cascade heterocyclization reaction. Moreover, by changing the halogen ions in the ILs, the brominated saturated γ-lactones were obtained in moderate-to-good yields with high regioselectivities. Additionally, this novel and general methodology may open up a new viewpoint to the carbonylation of $C(sp^3)$-palladium bond.

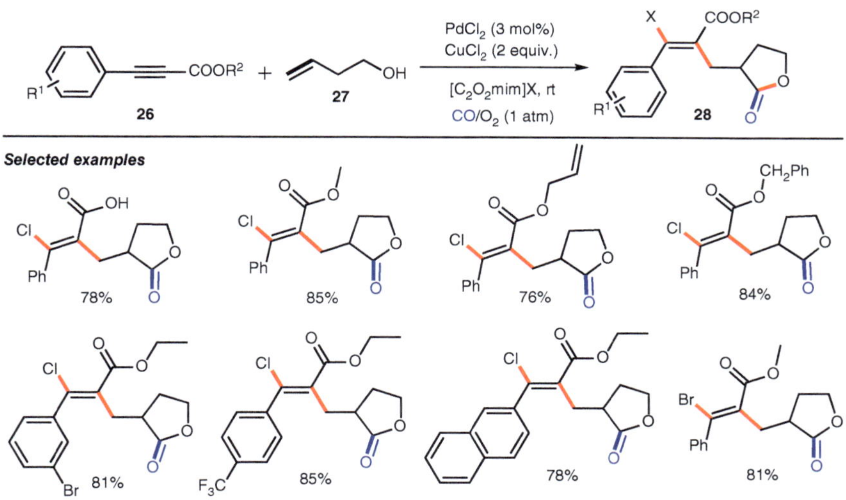

Scheme 7.13 Palladium-catalyzed cascade carboxylative to construct γ-lactones.

A possible catalytic mechanism is shown in Scheme 7.14. Pd complex is initially formed *in situ* in ILs, and the vinylpalladium intermediate **Int-16** is formed by *trans*-chloropalladation of the alkyne in a polar solvent system in the presence of excess chloride ions [47]. Then, intermediate **Int-16** could undergo alkene insertion. Simultaneously, the vinylpalladium species coordinates with both the oxygen atoms of OR^2 and the hydroxyl group to generate a Pd–alkyl intermediate **Int-17**. Subsequently, migratory insertion of CO into the palladium-carbon σ bond to produce intermediate **Int-18**. Finally, reductive elimination gives the target product and Pd(II) was regenerated in the presence of oxidant for the next cycle.

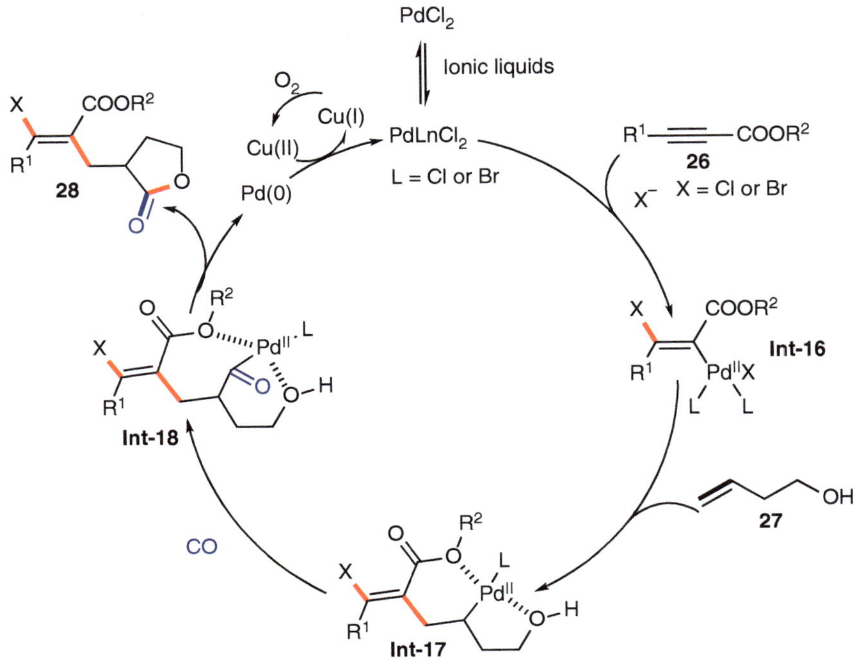

Scheme 7.14 The plausible pathway for cascade carboxylative.

In 2016, Mancuso and Gabriele group also accomplished a palladium-catalyzed oxidative carbonylation of propargylic amines (**29**) with CO (16 atm) and secondary amines (**30**) in [Emim]EtSO$_4$ for the preparation of a variety of 2-oxazolidinone derivatives (**31**) in good yields (Scheme 7.15) [48]. The reactions are carried out using a simple catalytic system consisting of PdI$_2$ in conjunction with an excess of KI, and the catalyst/solvent system could be recycled up to six times without appreciable loss of activity after extracting the organic product with Et$_2$O.

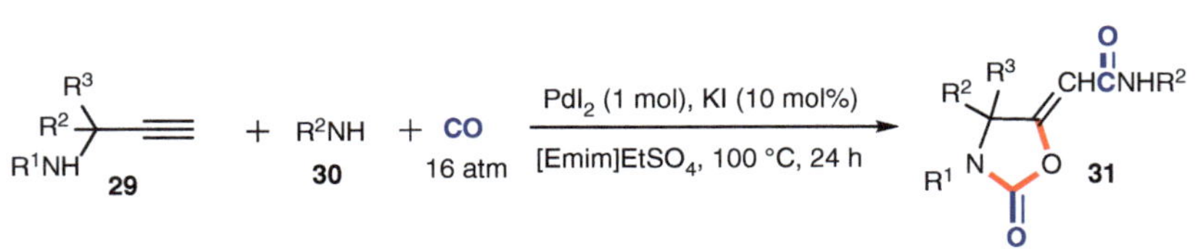

Scheme 7.15 Palladium-catalyzed oxidative carbonylation of propargylic amines.

In 2018, based on their work on oxidative carbonylation, the Godard and Riisager group developed a palladium-catalyzed selective oxidative carbonylation of aniline for the preparation of a library of diphenylurea compounds (**32**) in good yields with high regioselectivity (Scheme 7.16) [49]. The catalytic activity achieved with the benzoquinone oxidant is, to the best of our knowledge, the highest reported so

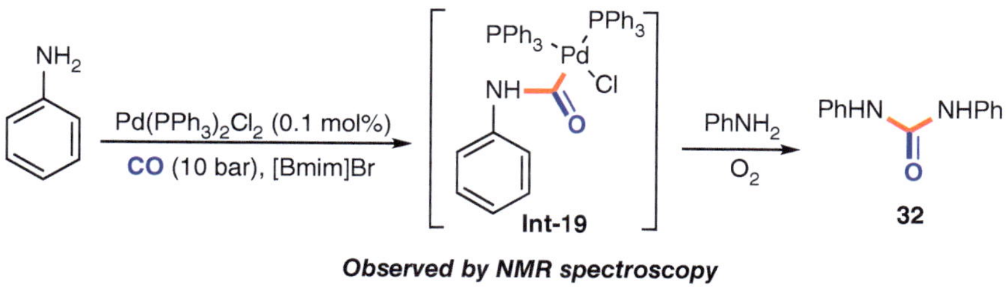

Scheme 7.16 Oxidative carbonylation of aniline to diphenylurea in [Bmim]Br.

far under comparatively mild reaction conditions (140 °C and 10 bar). In addition, they used high-pressure NMR spectroscopy to observe a suggested Pd–carbamoyl intermediate (**Int-19**) for the first time, which confirms the *trans* conformation of the complex [50]. To investigate the influence of the IL on the reaction, several IL ion compositions were tested, and the best results were achieved with [Bmim]Br, whereas the use of [Bmim][$(CF_3SO_2)_2N$] facilitated the spontaneous product separation. Recycling experiments with [Bmim][$(CF_3SO_2)_2N$] showed that the resulting catalytic system was reusable in three consecutive reaction runs, but with a significant loss of activity, which was attributed to the instability of the catalytically active Pd species.

In our subsequent study, we also accomplished a palladium-catalyzed aerobic oxidation aminocarbonylation of *O*-methyloximes (**6**) and amines (**33**) with an environmentally IL [Cpmim]Cl as the green additive in PEG-400 (Scheme 7.17) [51]. Particularly, this novel approach has opened up a new channel for the quick assembly of diverse and synthetically challenging 4-carboxamide isoxazole derivatives (**34**) with high atom- and step-economy. It is noteworthy that $Pd(Py)_2Cl_2$ could efficiently catalyze this cascade aminocarbonylation in the absence of phosphine ligands or other additional additives. Among a variety of solvents, PEG-400 proved to be the suitable solvent for this chemical transformation. Under the optimized conditions, both alkynone *O*-methyloximes and a vast array of primary amines were well accommodated in the transformation, yielding the corresponding 4-carboxamide isoxazoles in moderate-to-good yields. However, aromatic amine and secondary amine were not compatible with the current catalytic system. Notably, with 1 atm CO as the carbonyl source, O_2 or air as the terminal oxidant, IL [Cpmim]Cl as the green additive, and PEG-400 as the green solvent makes this cascade aminocarbonylation green and practical.

Mechanistic studies suggested that a Pd(0)/Pd(II) catalytic pathway is involved in this aminocarbonylation reaction. Based on the experimental results, we suggested a plausible mechanism as shown in Scheme 7.18. Similarly, migratory insertion of CO into the C–Pd bond of the vinylpalladium intermediate **Int-20** gives the acylpalladium species **Int-21**. Subsequently, ligand exchange of acylpalladium species **Int-21** with primary amines (**33**) leads carbamoylpalladium intermediate **Int-22** [52], which undergoes reductive elimination process to produce the aminocarbonylation products **34**.

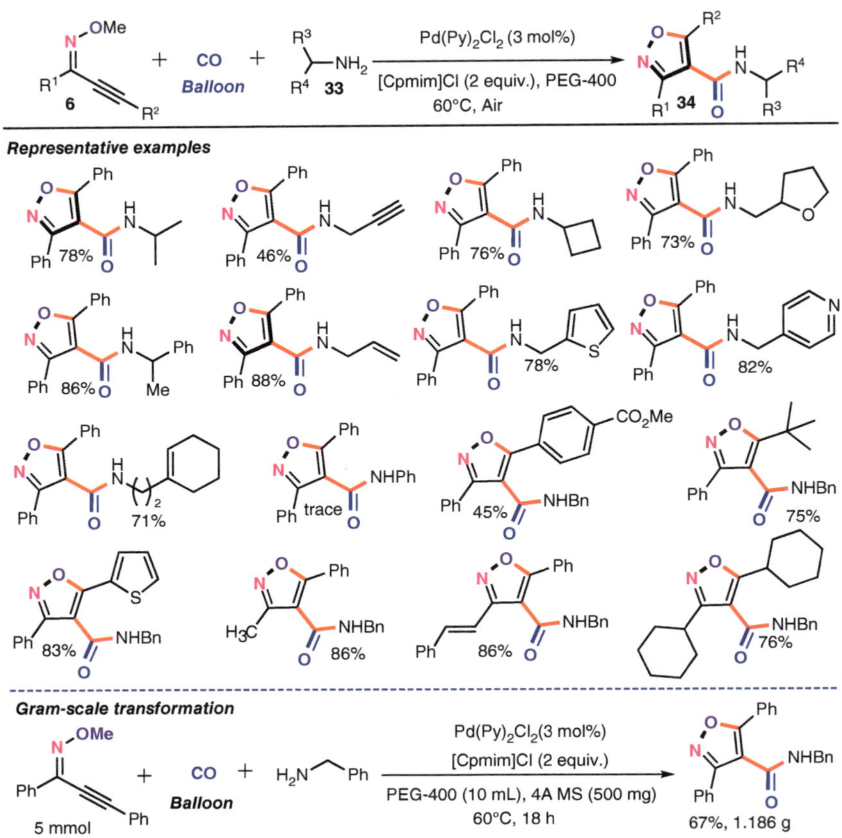

Scheme 7.17 Palladium-catalyzed cascade annulation/aminocarbonylation reaction.

7.2.3 Sonogashira Coupling Reactions

From the series of palladium-assisted C–C bond formation, the Sonogashira coupling reaction has been identified as a viable synthetic method for the preparation of various alkenyl- and arylacetylenes having great importance in the organic synthetic schemes of the pharmaceutical industry [53]. From the environmental point of view, Sonogashira reactions are usually performed in fossil-based common organic reaction mediums having high vapor pressure even at higher temperatures, toxicity, and flammability, which could result in several serious environmental concerns, especially when they are released into the atmosphere. To develop an environmentally benign alternative to this useful method, the reaction has been extended to green solvents such as ILs.

For example, Iranpoor and coworkers developed a copper and phosphane-free Sonagashira coupling reaction of aryl halides (**35**) with phenylacetylene in carboxylate-based ILs (**36**) (Scheme 7.19) [54]. The OH functionality of the ILs **36a** could be converted into $OPPh_2$ to generate a new diphenylphosphinite ILs **36c** from the reaction of **36a** (1 equiv.) with PPh_2Cl (1 equiv.) under argon gas. This class

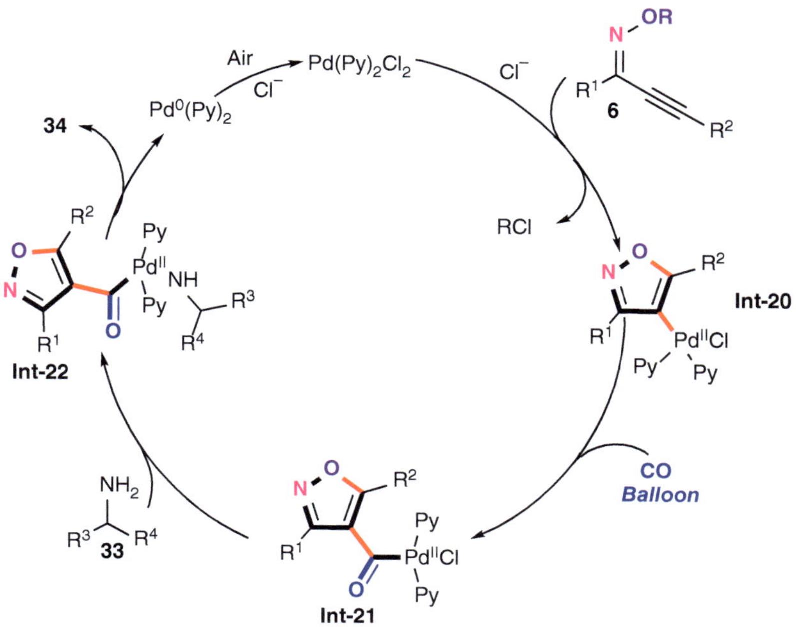

Scheme 7.18 Proposed mechanism for cascade annulation/aminocarbonylation reaction.

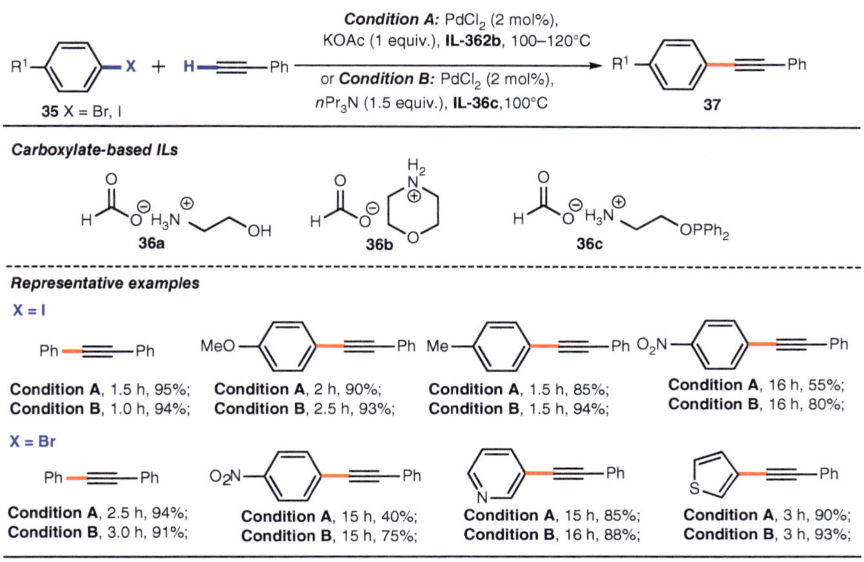

Scheme 7.19 Palladium-catalyzed Sonogashira reaction in two carboxylate-based Ils.

of ILs showed high efficiency and recyclability as a medium, and was also used as a ligand for the Sonogashira reactions under both phosphane- and copper-free conditions.

Substantial efforts were made to develop a new catalyst for this reaction. For instance, Liu and Wu synthesized a phosphine-ligated palladium complex **40**, which was used in the Sonogashira coupling reaction of aryl halides (**38**) with terminal acetylenes (**39**) in [Bmim]PF_6, under aerobic and CuI-free conditions (Scheme 7.20) [55]. This approach allowed the preparation of a variety of multi-substituted internal alkynes (**41**) in high GC yields (up to 100%) with 2 mol% Pd loading. Various functional groups, including alkyl, Ac, CF_3, CHO, and NO_2 groups, were compatible with the reaction conditions. There was no observed activity loss, Pd black precipitation, and Pd leaching even after being reused and recycled seven times. However, the reaction was less efficient when chlorobenzene derivatives were used as the coupling partners.

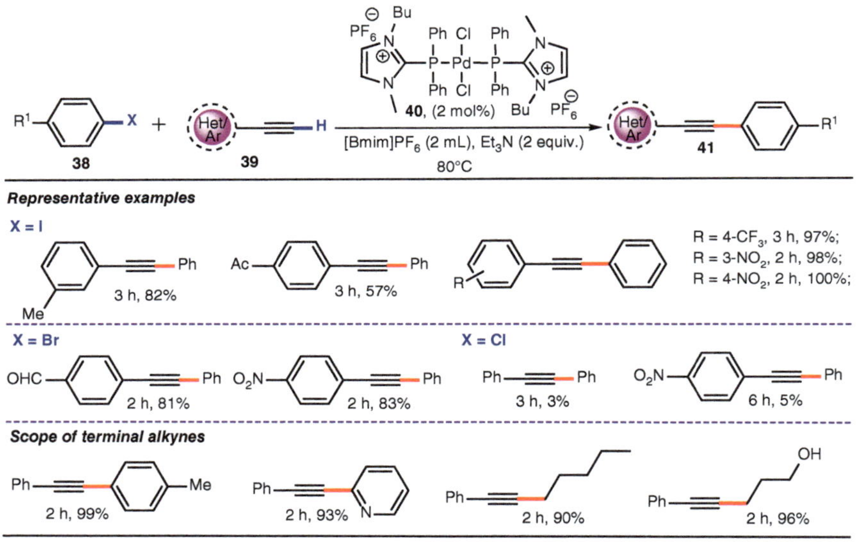

Scheme 7.20 Palladium-catalyzed Sonogashira coupling reaction in [Bmim]PF_6.

In 2019, Mika and coworkers described a novel approach for the synthesis of various acetylenes *via* copper- and auxiliary base-free Pd-catalyzed Sonogashira coupling reactions of aryl iodides (**42**) and functionalized acetylenes (**43**) in γ-valerolactone-based ILs [TBP][4EtOV] (Scheme 7.21) [56]. The γ-valerolactone-based IL [TBP][4EtOV] is a partially bio-based solvent, which can easily be synthesized from the renewable platform chemical γ-valerolactone and have negligible vapor pressures even at high temperatures. This reaction was performed by using 0.5 mol% catalyst loading and they also pointed out that both copper and external base could be eliminated from the reaction mixture without any decrease in catalytic activity. In addition, a vast array of value-added functionalized

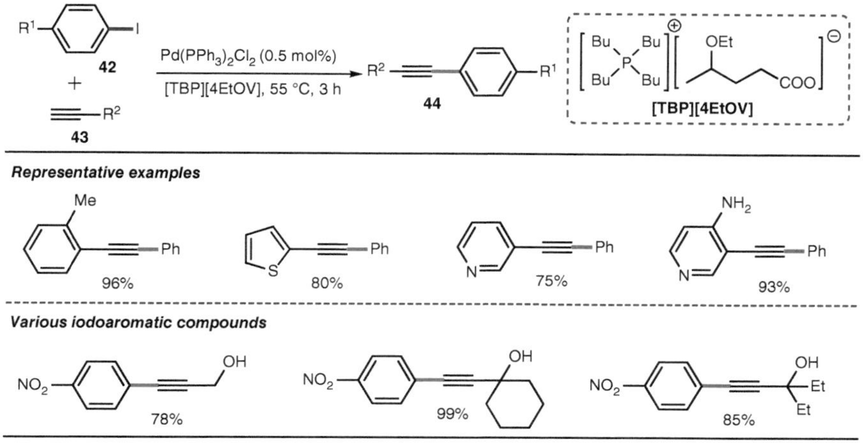

Scheme 7.21 Palladium-catalyzed Sonogashira coupling in γ-valerolactone-based ionic liquids.

internal acetylenes (**44**) has been prepared in good-to-excellent yields in an atom- and step-economical manner. Moreover, the Cu- and base-free reactions can be performed under air and are highly tolerant to moisture.

7.2.4 Suzuki Coupling Reactions

With the development of green chemistry and sustainable chemistry, the Suzuki coupling reaction in environmentally friendly reaction medium has been reported in recent years. This transformation used ILs and water as the media has been reviewed by Mastrorilli [26], Polshettiwar [57], and Ward [58], in 2015, 2010, and 2016, respectively. Thus, we will present recent improvements and new discoveries in the literature that have not been discussed before.

For example, Hu and coworkers reported a ligand-free and environmentally friendly $Pd_2(dba)_3$-catalyzed Suzuki coupling reaction of aryl halides (**45**) with arylboronic acids (**46**) for the synthesis of biphenyl derivatives (**47**) in ILs/supercritical CO_2 biphasic system (Scheme 7.22) [59]. With K_2CO_3 (1 equiv.) as the base in the presence of $scCO_2$ (supercritical carbon dioxide) and $[Hmim]NTf_2$ biphasic system, the corresponding products **47** were obtained in good-to-high yields. A broad range of functional groups, such as CH_3O, OH, NH_2, NO_2, CHO, F, and CN could be tolerated under these mild conditions. In addition, this catalytic system could be recycled up to six times and reused without significant loss of catalytic activity.

After this, Monopoli and coworkers developed a Pd nanoparticle-based catalyst supported on chitosan (Pd–NPs@Chitosan) for the Suzuki coupling of aryl halides (**48**) with arylboronic acids (**49**) in ILs TBAB. Using molten TBAB as the solvent, 0.1 mol% Pd–NPs@Chitosan as the catalyst, K_2CO_3 (1.5 equiv.) as the base, affording unsymmetrical biaryls (**50**) in good-to-excellent yields (Scheme 7.23) [60]. Additionally, Pd–NPs@Chitosan proved to be effective catalysts of up to 2400 ton. Recycling

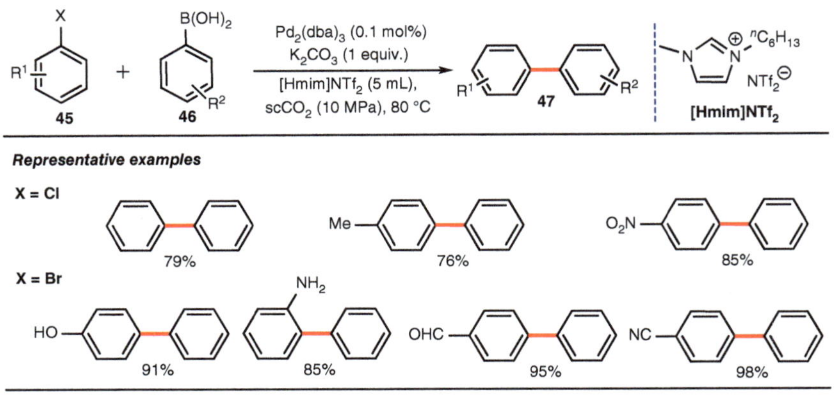

Scheme 7.22 Palladium-catalyzed Suzuki reaction in biphasic system.

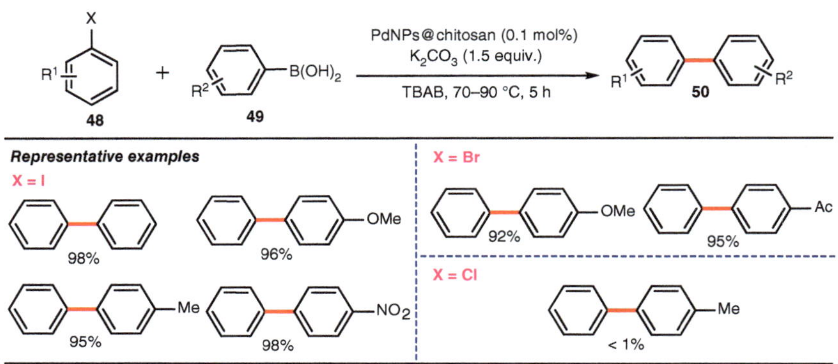

Scheme 7.23 Pd–NPs@Chitosan-catalyzed Suzuki coupling reaction.

experiments have shown that Pd–NPs@Chitosan underwent a loss of activity after the third cycle, probably due to the leaching of Pd during the reaction or the loss of material during the recycling procedure. Unfortunately, chlorobenzene proved to be unreactive under this condition.

In 2015, Remigy group reported palladium nanoparticles immobilized in a polymer membrane containing poly(ionic liquid) in Suzuki coupling reactions. ILs serve as stabilizers and immobilizers for the catalytic palladium nanoparticles. The active catalytic membrane showed remarkable catalytic performance on the coupling reaction of 1-iodo-4-nitrobenzene (**51**) with phenylboronic acid (**46a**) in [Mmpim]NTf_2 (Scheme 7.24) [61]. Notably, no leaching of palladium from the membrane was observed, neither during filtrations nor in the course of the catalytic reaction.

In 2017, Doddarevanna and coworkers reported the synthesis and characterization of a benzimidazole Schiff base palladium complex (**55**) and its application in the Suzuki coupling reaction in [Emim]PF_6 ILs (Scheme 7.25) [62]. The resulting Pd-complex **55** displays a robust air-stable system and very high catalytic activity

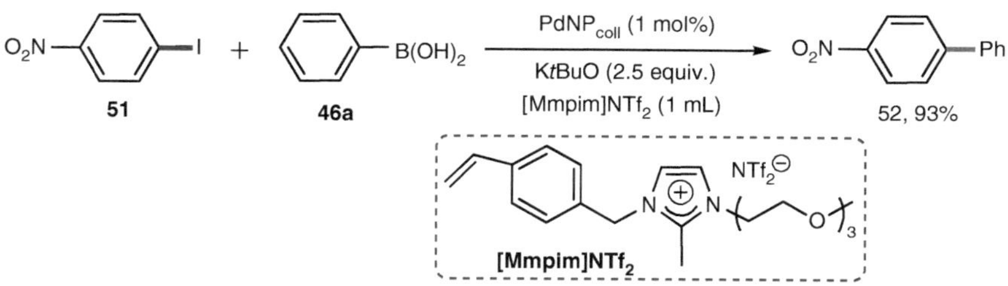

Scheme 7.24 Palladium nanoparticles catalyzed Suzuki coupling reaction.

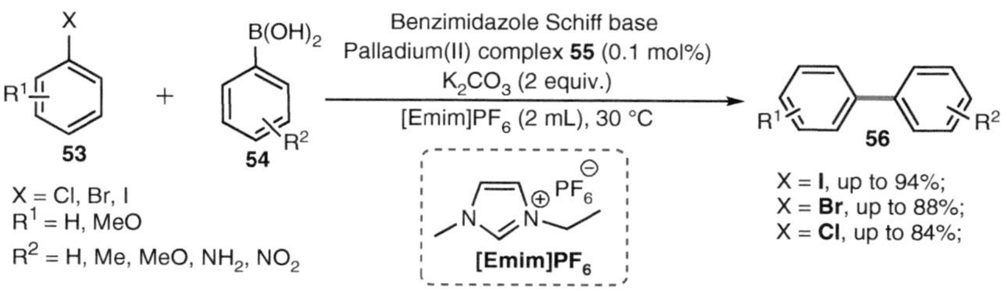

Scheme 7.25 Pd complex **55** catalyzed Suzuki coupling reaction.

for the formation of various functionalized biaryls (**56**) from aryl halides (**53**) and arylboronic acids (**54**) with a catalyst loading of 0.1 mol%, even at room temperature. Particularly, chlorobenzene could be tolerated in this protocol, albeit with somewhat reduced yields. In addition, the catalyst can be easily reused and recycled six times without much loss in activity.

7.2.5 Copper-catalyzed Coupling Reactions

Among the useful homogeneous catalytic transformations, the transition metal-catalyzed cross-coupling reactions are of utmost importance. Although several C–C, C–X coupling reactions were successfully performed, only a few studies have been published on the copper-catalyzed coupling reactions in ILs.

For instance, in 2018, Mika and coworkers discovered copper-catalyzed Ullmann-type coupling reactions between iodoaromatic compounds (**57**) and different amines (**58**) in bio-ILs [TBP][4EtOV] (Scheme 7.26) [63]. By replacement of the tetraalkylammonium cation with tetraalkyl phosphonium cations resulted in comparable densities and lower viscosities even in the presence of lower water content and higher thermal stability. Both ligands and bases could be eliminated from the reaction mixture without a significant decrease in activity of the catalyst system. The isolated yields of the desired products (**59**) were good to excellent, and no significant influence of the electronic effect of the amine substituents was detected. Notably, this base and ligand-free reactions can be carried out under air and are also highly tolerant to moisture.

In 2021, Kim and coworkers developed a convenient copper-catalyzed *N*-arylation of diverse primary amines (**60**) with aryltriazenes (**61**) for the preparation of

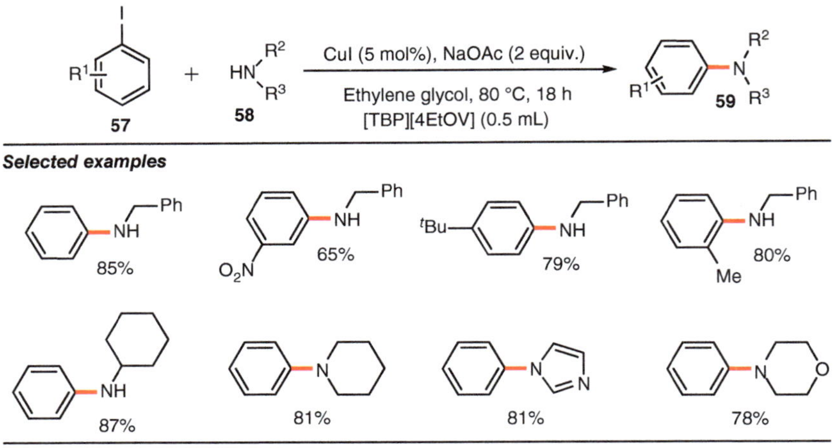

Scheme 7.26 Copper-catalyzed Ullmann-type coupling in ILs.

a library of *N*-arylated amines (**62**) in good yields (Scheme 7.27) [64]. In the presence of 20 mol% $Cu(OAc)_2$ as the catalyst, $[TMG][CH_3CO_2]$ as the solvent and promoter for this C–N cross-coupling reaction, both electron-donating and electron-withdrawing substituents on the phenyl ring of aromatic amine (**60**) were amenable for this transformation. Further screening of the reaction conditions revealed that $[TMG][CH_3CO_2]$ as the solvent and promoter was indispensable for this chemical transformation. Moreover, this protocol also displayed a wide tolerance for various functional groups on the aromatic amine and aryltriazene moiety, such as alkyl, cycloalkyl, aryl, nitro, and 2-pyridyl, and good yields of the corresponding products were obtained. This newly developed system involves ligand, additive, and base-free conditions in eco-friendly solvents with the prospects of improved yield and recycling/reuse of IL as opposed to the existing conventional volatile organic compounds (VOCs)-mediated methods.

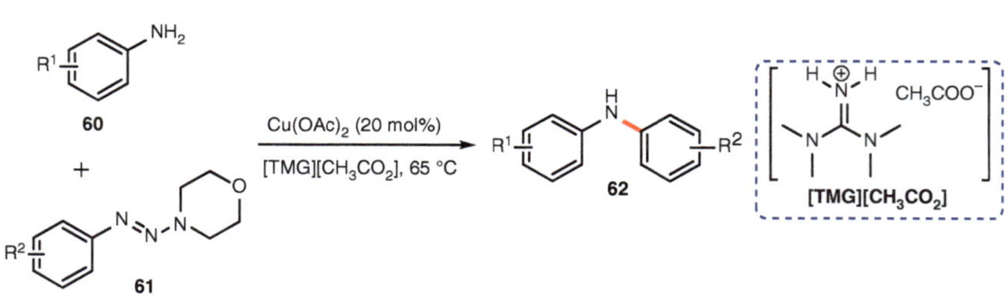

Scheme 7.27 Copper-catalyzed *N*-arylation of amines in ILs.

Additionally, ILs were also applied in other type of transition metal-catalyzed coupling reactions as the green and eco-friendly medium, such as C–H functionalization reaction [65], selective dimerization of unsaturated hydrocarbon [66],

Mizoroki–Heck reaction [67], olefin metathesis reaction [68], hydrogenation [69], difunctionalization of alkynes [70], Friedel–Crafts acylation [71], metal-catalyzed oxidation [72], metal-free oxidation [73], and other reactions.

7.3 Outlook

Nowadays, ILs are being applied in various fields as an interesting and promising medium, owing to their irreplaceably unique properties such as no volatility, large liquid range, nonflammability, high thermal stability, strong solubility capacity, and the tunability of structure and properties. There is no doubt that ILs chemistry has made great contributions to both academic and industrial applications. Transition metal-catalyzed coupling reactions in ILs, especially palladium-catalyzed coupling reactions, have established as a captivating and colorful branch of organic chemistry over the past several decades, since these reactions have provided structurally diverse and more elaborated organic molecules *via* an economical and environmentally benign synthetic manners. Notably, ILs were considered as nonparticipative or inert green solvents or co-solvents, and have been proven to enhance the overall reaction rates in these chemical transformations.

Nevertheless, there are several questionable issues that had been proposed by researchers. For instance, the chemical and thermal stability of ILs have been investigated [74]. Although a number of studies on the physicochemical, toxicological, and biodegradability of ILs have been reported, the toxicity and environmental impact of ILs are still uncertain. Besides the above critical issues, other properties of ILs, such as degradation and corrosion should also be addressed. We also hope that this review will play a stimulating and encouraging role in using ILs in transition metal-catalyzed organic reactions, thus emboldening the well-rounded acquaintance and comprehension of physicochemical characteristics of novel ILs.

List of Abbreviations

[Aeim]Br	1-allyl-3-ethylimidazolium bromide
[Aeim]Cl	1-allyl-3-ethylimidazolium chloride
[Amim]Cl	1-allyl-3-methylimidazolium chloride
[Bmim][$(CF_3SO_2)_2N$]	1-butyl-3-methylimidazolium trifluoromethanesulfonate
[Bmim]Br	1-butyl-3-methylimidazolium bromide
[Bmim]Cl	1-butyl-3-methylimidazolium chloride
[Bmim]PF_6	1-butyl-3-methylimidazolium hexafluorophosphate
[C_2O_2mim]Br	1-carboxymethyl-3-methylimidazolium bromide
[C_2O_2mim]Cl	1-carboxymethyl-3-methylimidazolium chloride
[C_3NH_2mim]Br	1-aminopropyl-3-methylimidazolium bromide
[Cpmim]BF_4	1-cyanopropyl-3-methylimidazolium tetrafluoroborate
[Cpmim]Cl	1-cyanopropyl-3-methyl imidazolium chloride
[Cpmim]Cl	1-cyanopropyl-3-methylimidazolium chloride.

[Emim]$EtSO_4$	1-ethyl-3-methylimidazolium ethyl sulfate
[Emim]PF_6	ethyl-methyl imidazolium hexafluorophosphate
[Hmim]NTf_2	1-hexyl-3-methylimidazolium bis(trifluoromethylsulfonyl)imide
[Mmpim]NTf_2	1,2-dimethyl-3-propylimidazolium bis(trifluoromethylsulfonyl)imide
[$PrSO_3$Hmim]Cl	1-propylsulfonate-3-methylimidazolium chloride
[TBP][4EtOV]	tetrabutylphosphonium 4-ethoxyvalerate
[TMG][CH_3COO]	tetramethylguanidinium acetate
EAN	ethylammonium nitrate
NHC	N-heterocyclic carbene
PEG-400	Polyethylene glycol 400
TBAB	tetrabutylammonium bromide

References

1 Walden, P. (1914). *Bull. Acad. Imp. Sci. St.-Petersbourg, Ser.* 6 (8): 405–422.
2 Hayes, R., Warr, G.G., and Atkin, R. (2015). *Chem. Rev.* 115: 6357–6426.
3 Hurley, F.H. and Wier, T.P. (1951). *J. Electrochem. Soc.* 98: 207–212.
4 Fedorov, M.V. and Kornyshev, A.A. (2014). *Chem. Rev.* 114: 2978–3036.
5 Wilkes, J.S. and Zaworotko, M. (1992). *J. Chem. Soc., Chem. Commun.* 13: 965–967.
6 Ananikov, V.P. (2011). *Chem. Rev.* 111: 418–454.
7 Welton, T. (1999). *Chem. Rev.* 99: 2071–2083.
8 Adams, D.J., Dyson, P.J., and Tavener, S.J. (ed.) (2004). *Chemistry in Alternative Reaction Media*. Chicester, U.K.: Wiley.
9 Kerton, F.M. (ed.) (2009). *Alternative Solvents for Green Chemistry*. Cambridge, U.K.: Royal Society of Chemistry.
10 Leitner, W., Jessop, P.G., Li, C.-J. et al. (2010). *Handbook of Green Chemistry-Green Solvents*, vol. 6. Weinheim, Germany: Wiley-VCH.
11 Fox, D.M., Awad, W.H., Gilman, J.W. et al. (2003). *Green Chem.* 5: 724–727.
12 Wasserscheid, P. and Welton, T. (ed.) (2003). *Ionic Liquids in Synthesis*. Weinheim, Germany: Wiley-VCH.
13 Rogers, R.D. and Seddon, K.R. (2002). *Ionic Liquids: Industrial Applications for Green Chemistry, ACS Symposium Series 818*. Oxford: Oxford University Press.
14 Hallett, J.P. and Welton, T. (2011). *Chem. Rev.* 111: 3508–3576.
15 Prvulescu, V.I. and Hardacre, C. (2007). *Chem. Rev.* 107: 2615–2665.
16 Dai, C., Zhang, J., Huang, C., and Lei, Z. (2017). *Chem. Rev.* 117: 6929–6983.
17 Welton, T. and Smith, P.J. (2004). *Adv. Organomet. Chem.* 51: 251–284.
18 Martins, M.A.P., Frizzo, C.P., Moreira, D.N. et al. (2008). *Chem. Rev.* 108: 2015–2050.
19 Itoh, T. (2017). *Chem. Rev.* 117: 10567–10607.
20 Hapiot, P. and Lagrost, C. (2008). *Chem. Rev.* 108: 2238–2264.
21 Zhang, G.-R. and Etzold, B.J.M. (2016). *J. Energy Chem.* 25: 199–207.

22 Kaur, P. and Chopra, H. (2019). *Curr. Org. Chem.* 23: 2815–2881.
23 Abbott, A.P., Frisch, G., and Ryder, K.S. (2008). *Annu. Rep. Prog. Chem., Sect. A.* 104: 21–45.
24 Amarasekara, A.S. (2016). *Chem. Rev.* 116: 6133–6183.
25 Egorova, K.S., Gordeev, E.G., and Ananikov, V.P. (2017). *Chem. Rev.* 117: 7132–7189.
26 Mastrorilli, P., Monopoli, A., Dell'Anna, M.M. et al. (2015). *Top Organomet. Chem.* 51: 237–285.
27 Li, J. and Jiang, H. (ed. X.F. Wu) *Noble Metal Noble Value: Ru-, Rh-, Pd-catalyzed Heterocycle Synthesis*, vol. Vol. 15, 351. Singapore: World Scientific Publishing Co Pte Ltd.
28 Li, J., Yang, S., and Jiang, H. (2018). *Eur. J. Org. Chem.* 1284–1306.
29 Wu, X.-F., Neumann, H., and Beller, M. (2013). *Chem. Rev.* 113: 1–35.
30 Wang, D., Weinstein, A.B., White, P.B., and Stahl, S.S. (2018). *Chem. Rev.* 118: 2636–2679.
31 Li, J., Yang, S., Wu, W., and Jiang, H. (2019). *Chem. Asian J.* 14: 4114–4128.
32 Li, J., Li, C., Yang, S. et al. (2016). *J. Org. Chem.* 81: 2875–2887.
33 Peng, P., Tang, B.-X., Pi, S.-F. et al. (2009). *J. Org. Chem.* 74: 3569–3572.
34 Li, J., Zhu, Z., Yang, S. et al. (2015). *J. Org. Chem.* 80: 3870–3879.
35 Li, J., Huang, R., Li, C. et al. (2019). *Chem. Asian J.* 14: 2309–2315.
36 Li, J., Wu, Y., Hu, M. et al. (2019). *Green Chem.* 21: 4084–4089.
37 Hu, M., Lin, Z., Li, J. et al. (2020). *Green Chem.* 22: 5584–5588.
38 Lai, Y.-L., Yan, S.-X., Zhang, S.-L. et al. (2022). *Chem. Asian J.* 17: e202200806.
39 Kochi, T., Hamasaki, T., Aoyama, Y. et al. (2012). *J. Am. Chem. Soc.* 134: 16544–16547.
40 Li, J., He, D., Lin, Z. et al. (2022). *Green Chem.* 24: 1983–1988.
41 Cao, D., Zeng, H., and Li, C.-J. (2018). *ACS Catal.* 8: 8873–8878.
42 Wang, H., Zhao, Y., Zhang, F. et al. (2020). *Angew. Chem. Int. Ed.* 59: 11850–11855.
43 Li, Y., Karlen, S.D., Demir, B. et al. (2020). *ChemSusChem* 13: 4660–4665.
44 Li, J., Luo, Y., Hu, L. et al. (2023). *Eur. J. Org. Chem.* 26: e202201387.
45 Peng, J.-B., Wu, F.-P., and Wu, X.-F. (2019). *Chem. Rev.* 119: 2090–2127.
46 Li, J., Yang, S., Wu, W., and Jiang, H. (2014). *Chem. Commun.* 50: 1381–1383.
47 Zhang, Q., Xu, W., and Lu, X. (2005). *J. Org. Chem.* 70: 1505–1507.
48 Mancuso, R., Maner, A., Ziccarelli, I. et al. (2016). *Molecules* 21: 897.
49 Zahrtmann, N., Claver, C., Godard, C. et al. (2018). *ChemCatChem* 10: 2450–2457.
50 Hiwatari, K., Kayaki, Y., Okita, K. et al. (2004). *Bull. Chem. Soc. Jpn.* 77: 2237–2250.
51 Li, J., Yu, J., Xiong, W. et al. (2020). *Green Chem.* 22: 465–470.
52 Liu, S., Wang, H., Dai, X., and Shi, F. (2018). *Green Chem.* 20: 3457–3462.
53 Sonogashira, K. (2002). *Palladium-catalyzed alkynylation: sonogashira alkyne synthesis*. In: *Handbook of Organopalladium Chemistry for Organic Synthesis* (ed. E. Negishi), 493–529. New York, NY, USA: John Wiley & Sons.
54 Iranpoor, N., Firouzabadi, H., and Ahmadi, Y. (2012). *Eur. J. Org. Chem.* 305–311.

55 Zhang, J., Đaković, M., Popović, Z. et al. (2012). *Catal. Commun.* 17: 160–163.

56 Orha, L., Tukacs, J.M., Kollár, L., and Mika, L.T. (2019). *Beilstein J. Org. Chem.* 15: 2907–2913.

57 Polshettiwar, V., Decottignies, A., Len, C., and Fihri, A. (2010). *ChemSusChem* 3: 502–522.

58 Chatterjee, A. and Ward, T.R. (2016). *Catal. Lett.* 146: 820–840.

59 Wang, H., Hua, Y.-L., and Li, D.-J. (2016). *J. Mol. Liq.* 218: 429–433.

60 Cotugno, P., Casiello, M., Nacci, A. et al. (2014). *J. Organomet. Chem.* 752: 1–5.

61 Gu, Y., Favier, I., Pradel, C. et al. (2015). *J. Membrane Sci.* 492: 331–339.

62 Chidananda, V.K.B., Ramakrishna, D., Kaur, M., and Doddarevanna, R.H. (2017). *J. Coord. Chem.* 70: 1573–1584.

63 Orha, L., Tukacs, J.M., Gyarmati, B. et al. (2018). *ACS Sustainable Chem. Eng.* 6: 5097–5104.

64 Kim, K.M. and Kim, K.M. (2021). *ChemistrySelect* 6: 2709–2715.

65 Aloia, A., Casiello, M., D'Accolti, L. et al. (2022). *Chem. Eur. J.* 28: e202202350.

66 Onuki, M., Ota, M., Otokozawa, S. et al. (2020). *Tetrahedron* 76: 130899.

67 Lerch, S., Fritsch, S., and Strassner, T. (2022). *Eur. J. Org. Chem.* e202200008.

68 Thomas, P.A. and Marvey, B.B. (2016). *Molecules* 21: 184.

69 Li, R., Wang, Y., Zhao, Y. et al. (2021). *ACS Sustainable Chem. Eng.* 9: 14216–14223.

70 Lai, Y.-L., Yan, S., He, D. et al. (2021). *RSC Adv.* 11: 28447–28451.

71 Lerch, S., Fritsch, S., and Strassner, T. (2023). *Beilstein J. Org. Chem.* 19: 212–216.

72 Betz, D., Altmann, P., Cokoja, M. et al. (2011). *Coordin. Chem. Rev.* 255: 1518–1540.

73 Hirashita, T., Nakanishi, M., Uchida, T. et al. (2016). *ChemCatChem* 8: 2704–2709.

74 Wang, B., Qin, L., Mu, T. et al. (2017). *Chem. Rev.* 117: 7113–7131.

8

Recent Achievements in Organic Reactions in Ketones and Esters

Fan-Lin Zeng and Bing Yu

Zhengzhou University, Green Catalysis Center, College of Chemistry, Zhengzhou 450001, China

8.1 Introduction

Solvents play a central role in defining green and sustainable processes. Most chemical processes of organic synthesis are conducted in organic solvents. After the reactions, the purification stage frequently involves the use of a sizable volume of organic solvent. Generally, conventional organic solvents including methanol, ethanol, benzene, toluene, and xylene are widely applied in chemical industries and could be partially recovered by distillation [1]. However, in those processes the release of volatile organic compounds (VOCs) into the environment is inevitable. As a result, around 20 million tons of VOCs are released into the environment each year [2]. In this context, the discovery and application of novel and environmentally friendly solvents for organic synthesis is one of the most important issues in the field of green chemistry. Although there isn't a perfect green solvent that works in every situation, using alternative solvents has drawn a lot of interest. In this chapter, we would like to summarize the recent progress of organic reactions in sustainable solvents like ketones and esters.

8.2 Organic Reactions in Ketones

Traditional ketone solvents such as acetone and cyclohexanone are widely used in organic synthesis reactions [3]. Over the past several years, some non-volatile ketones such as dihydrolevoglucosenone (Cyrene) and *N*-butylpyrrolidinone (NBP) have been developed as green solvents for organic synthesis (Scheme 8.1) [4]. The organic reactions conducted in Cyrene or NBP will be summarized in the following sections.

8.2.1 Organic Reactions in Cyrene

Cyrene is a biodegradable, non-mutagenic, and non-toxic solvent that can be produced from biomass in two steps, i.e., pyrolysis of cellulose and hydrogenation of

Green Solvents in Organic Synthesis, First Edition.
Edited by Xiao-Feng Wu, Zhiping Yin, Liang-nian He, and Feng Wang.

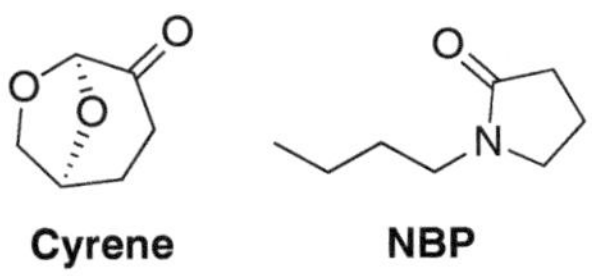

Scheme 8.1 The solvents of Cyrene and NBP.

alkene (Scheme 8.2) [5]. Cyrene has a very high boiling point (227 °C), which offers the possibility of reactions over a wide temperature range. Moreover, it has a similar π^* (dipolarizability) value to *N*-methylpyrrolidone (NMP), validating its definition as an aprotic solvent [6]. In addition, Cyrene shows a dispersion term (δ_D) similar to DMSO, a polar term (δ_P) similar to DMAc, and hydrogen bonding interactions (δ_H) similar to NMP in terms of Hansen solubility parameters (HSP)) [7]. Most notably, Cyrene has no known toxicity problems and is biodegradable compared to conventional solvents [8]. Its "green" profile, in combination with its physical properties, makes Cyrene a plausible replacement for some widely used toxic organic solvents. Thus, Cyrene as a bio-based solvent has been successfully used as a solvent in numerous research areas such as materials chemistry, peptide chemistry, and organic synthesis.

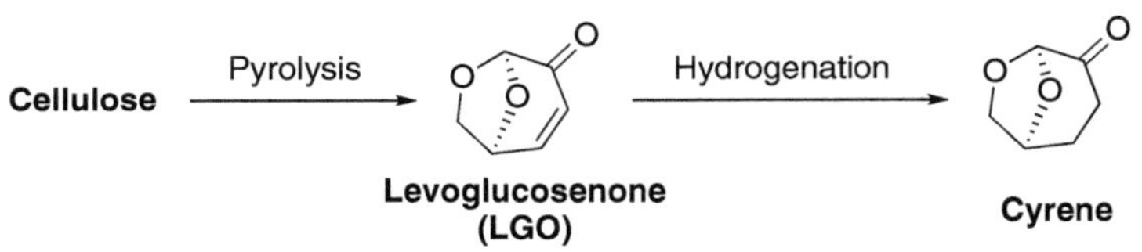

Scheme 8.2 Production of Cyrene from cellulose.

Sonogashira reaction generally takes place in a polar aprotic solvent such as DMF. In 2016, Watson's group made the first attempt to achieve a Pd-catalyzed Sonogashira coupling reaction using Cyrene as a solvent (Scheme 8.3) [9]. Notably, it is suggested that strong acids, strong oxidizing and reducing agents, and strong bases should be avoided when Cyrene is applied as a solvent. In this study, it was found that organic bases such as pyridine, DIPEA, and Et_3N were tolerated at 25 °C, with DIPEA and Et_3N also tolerated at 50 °C. While the stronger organic base DBU was not tolerated at any temperature. At room temperature, almost all inorganic bases resulted in an aldol reaction of Cyrene except KOAc.

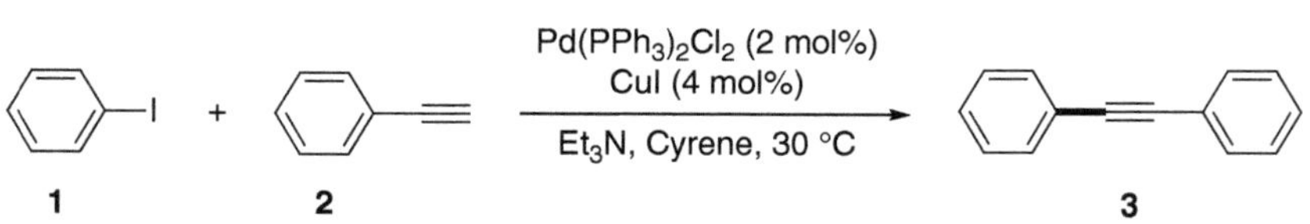

Scheme 8.3 Sonogashira reaction using Cyrene as the solvent.

Generally, DMF, THF, or 1,4-dioxane are usually employed as suitable solvents for Suzuki–Miyaura reactions. The application of Cyrene as a solvent in other

Pd-catalyzed cross-coupling reactions such as the Suzuki–Miyaura coupling reaction has also been explored with success (Scheme 8.4) [10]. The use of Cyrene as a solvent provides a greener pathway for the Suzuki-Miyaura coupling reaction.

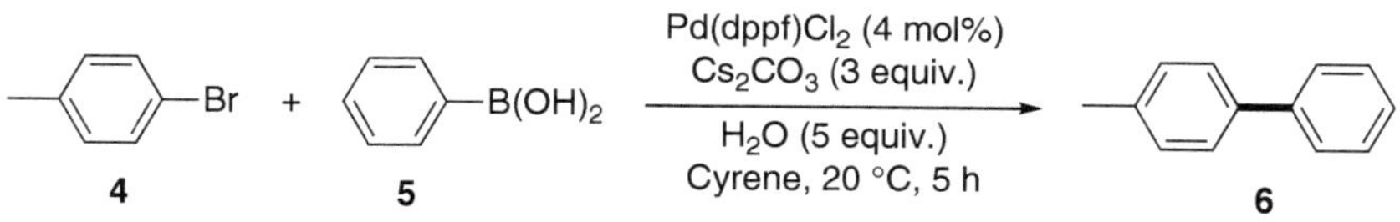

Scheme 8.4 Suzuki–Miyaura reaction using Cyrene as the solvent.

Considering the key role of urea in various fields, in 2017, Camp's group used Cyrene instead of DMF to achieve the synthesis of urea from secondary amines and isocyanates (Scheme 8.5) [11]. This environmentally friendly method eliminated the use of toxic DMF and established a column-chromatography-free water wash procedure to isolate the target products. Compared with the standard industrial aqueous work-up, this water wash procedure led to a 28-fold increase in molar efficiency.

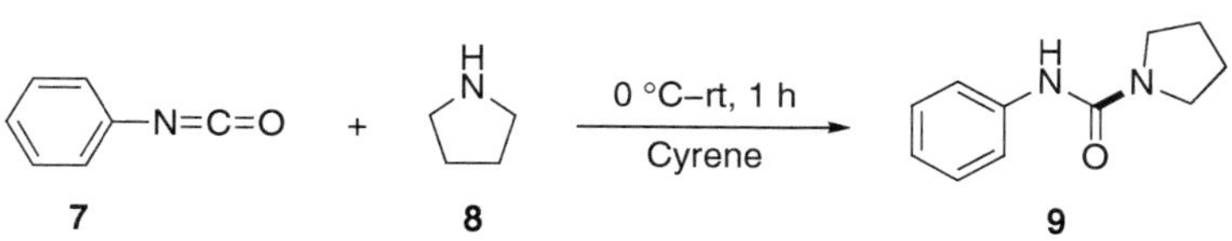

Scheme 8.5 Synthesis of urea in Cyrene.

In addition, Cyrene was used as a solvent to achieve the heterogeneous C—H difluoromethylation reactions of heteroarenes and terminal alkynes with (difluoromethyl)trimethylsilane ($TMSCF_2H$) as a difluoromethylating reagent and 9,10-phenanthraquinone (PQ) as an additive. In this procedure, the heterogeneous catalyst CuI/BNNS, i.e., copper iodide (CuI) nanoparticles/boron nitride nanosheets (BNNSs), were prepared *via* an ultrasonication method. It was found that Cyrene as a solvent can shorten the reaction time and improve the yield compared to conventional solvents such as NMP and DMF (Scheme 8.6) [12].

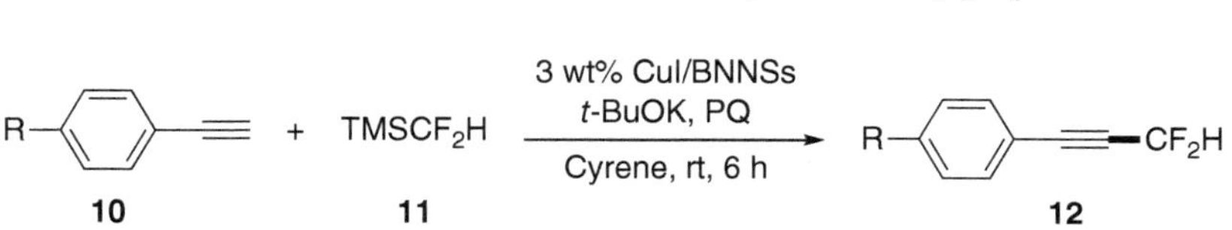

Scheme 8.6 Difluoromethylation of terminal alkynes in Cyrene.

Amides are widely found as important structures in drugs and peptides. Therefore, the development of novel methods for the synthesis of amides has drawn huge attention. Usually, dipolar aprotic solvents, such as DMF and NMP, are used for the synthesis of amides. In 2018, Watson's group reported an *N*-[(dimethylamino)-1

H-1,2,3-triazolo-[4,5-*b*]pyridin-1-ylmethylene]-*N*-methyl-methanaminium hexafluorophosphate *N*-oxide (HATU)-mediated amide bond formation procedure from various carboxylic acids and amines using Cyrene as a solvent (Scheme 8.7) [13]. In the condition optimization, it was found that a high stirring rate (800 rpm) delivered higher yields than that of low stirring rates (200 rpm or 400 rpm), suggesting the importance of stringent control of the stirring rate. However, the dependency of yield on stirring rate is decreased by further increasing the stoichiometry of DIPEA.

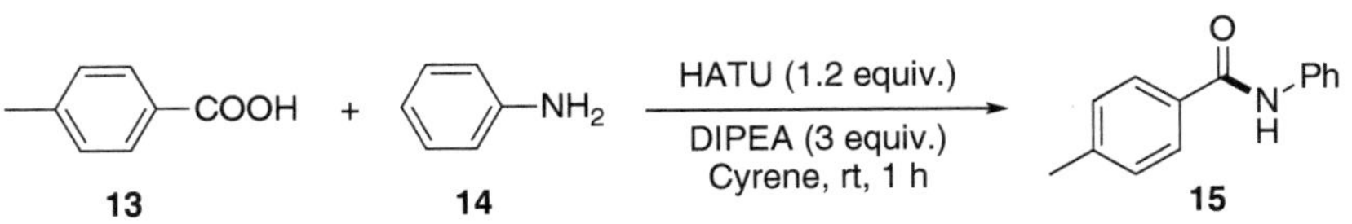

Scheme 8.7 Synthesis of amide in Cyrene.

Menschutkin reaction is an S_N2-type substitution reaction, which involves two neutral reactants, leading to a salt. It is widely used in the preparation of imidazole-containing ionic liquids [14]. Usually, the reaction is performed in a polar solvent such as DMF. Typically, in the reaction of 1,2-dimethylimidazole and 1-bromodecane to synthesize 1-decyl-2,3-dimethylimidazolium bromide, the reaction rate was found to be proportional to the dipolarity of the solvent, which was suppressed in protic solvents. In the established linear solvation energy relationship (LSER), Cyrene was demonstrated to be one of the best solvents most other dipolar aprotic solvents (e.g., dioxane, DMF, DMAc, and NMP) for this reaction (Scheme 8.8) [15].

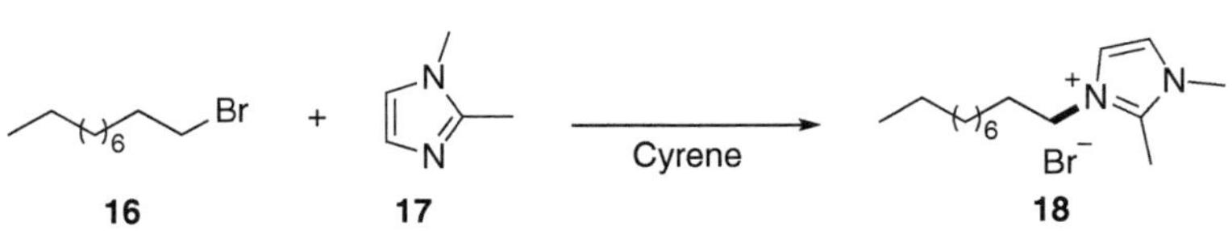

Scheme 8.8 Menschutkin reaction using Cyrene as a solvent.

8.2.2 Organic Reactions in NBP

Compared with the conventional dipolar aprotic solvents, such as *N*-methylpyrrolidone (NMP), *N*-butylpyrrolidone (NBP) is non-mutagenic (according to OECD 471 test method), non-reprotoxic (according to OECD 414 test method), and biodegradable (according to OECD 302B test method), which make NBP a promising candidate of non-toxic dipolar aprotic solvents [16].

Cross-coupling reactions are widely used in the pharmaceutical industry and are an important tool for the synthesis of drug molecules. Traditionally, highly dipolar aprotic solvents such as DMF, NMP, and DMAc have been preferred for cross-coupling reactions such as the Heck reaction and the Suzuki reaction [17].

NBP has been reported and applied as a green solvent to replace traditional highly dipolar aprotic solvents (Scheme 8.9) [18]. The results showed that the reaction rate and conversion to the product in NBP are comparable to that of NMP and DMSO.

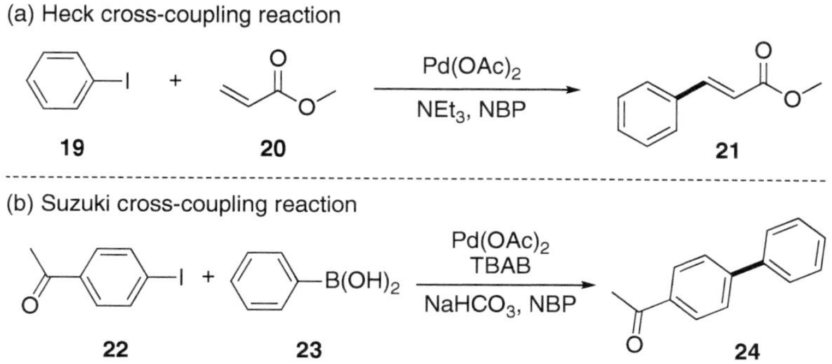

Scheme 8.9 (a) Heck reaction and (b) Suzuki reaction in NBP.

Heterocycles are widely found in a variety of drug molecules, therefore the synthesis of heterocyclics is of great importance. Typically, highly polar solvents are used to perform heterocyclic synthesis. NBP has also been used as a green solvent in the field of heterocyclic synthesis. For example, the Biginelli reaction, an important route for the synthesis of dihydropyrimidinones, usually uses ethanol or DMF as the reaction solvent. When the Biginelli reaction was conducted with NBP as the solvent, the yield of the product exceeded that obtained when ethanol or DMF was used as the solvent (Scheme 8.10) [19].

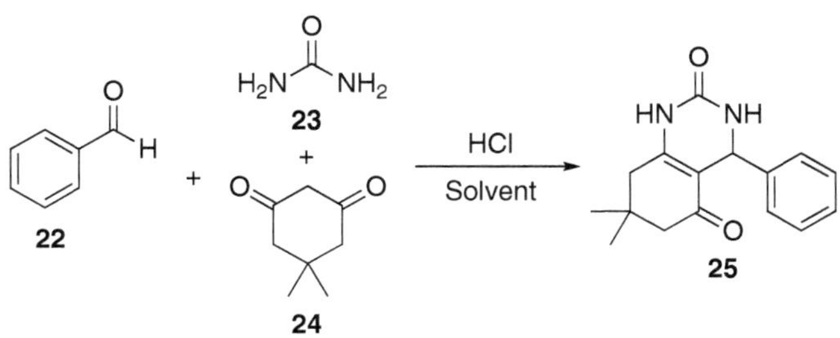

Solvent	Yield
Ethanol	29%
DMF	34%
NBP	46%

Scheme 8.10 Biginelli reaction using NBP as the solvent.

The Maitland-Japp reaction, an important route for the synthesis of highly functionalized piperidines, usually uses MeCN or DMF as the reaction solvent. When the Maitland–Japp reaction was performed with NBP as the solvent, the yield of the product exceeded that obtained when MeCN or DMF was used as the solvent (Scheme 8.11) [20].

The use of therapeutic peptides is of increasing interest to the pharmaceutical industry. Most of the marketed peptide drugs are produced chemically and mainly by solid-phase peptide synthesis (SPPS). Usually, a large number of toxic solvents

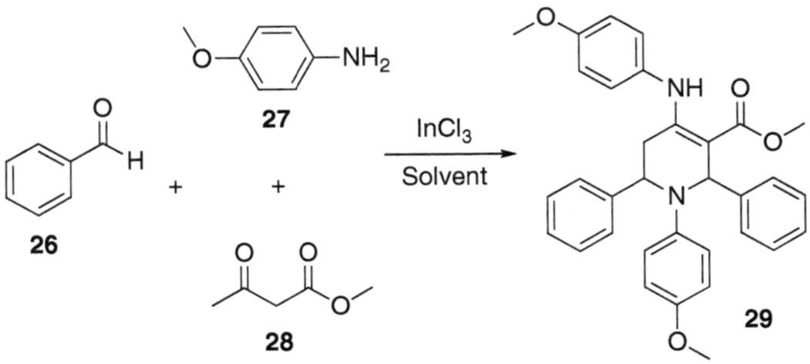

Solvent	Yield
MeCN	63%
DMF	63%
NBP	67%

Scheme 8.11 Maitland-Japp reaction using NBP as the solvent.

such as DMF, NMP, DMA, and DCM are used in the SPPS process. In 2018, Lopez's group tested the feasibility of several green solvents to replace the DMF as a solvent used in SPPS by a systematic approach. NBP could be compatible with the most popular polystyrene-based resins. The results showed that NBP is the best green solvent candidate to replace DMF in SPPS [21].

8.3 Organic Reactions in Esters

Esters are used extensively in organic reactions as an important class of solvent. In particular, some green ester solvents such as organic carbonates, γ-valerolactone (GVL), ethyl lactate (EL) have been applied as reaction mediums in organic synthesis and offer the possibility of replacing traditional toxic solvents.

8.3.1 Organic Reactions in Organic Carbonates

In general, organic carbonates can be roughly divided into linear carbonates and cyclic carbonates (Scheme 8.12). Linear carbonates, such as dimethyl carbonate (DMC), could be applied as environmentally friendly alkylation reagents and also good aprotic solvents to replace toxic solvents. In addition, cyclic carbonates, e.g., propylene carbonate, own a high boiling point and a low vapor pressure. In particular, it is inflammable and nontoxic. It can be therefore considered a green solvent also taking in mind its environmentally friendly synthesis from carbon dioxide and propylene oxide [22].

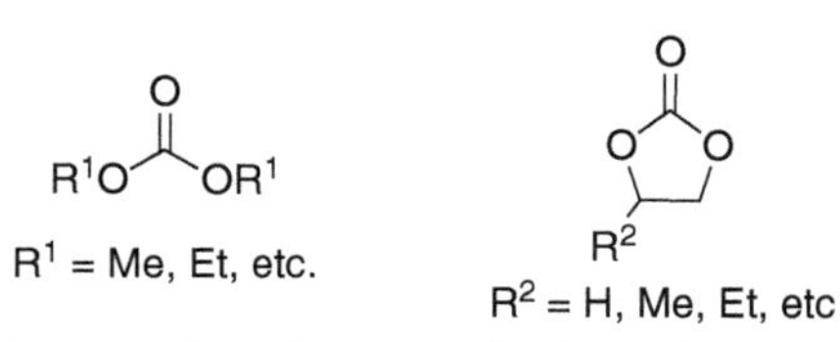

Scheme 8.12 Linear carbonates and cyclic carbonates.

In 2008, Börner's group reported their results on the use of organic carbonates as a reaction medium for Pd-catalyzed asymmetric allylic substitution reactions. For the first time, the suitability of the green solvents propylene carbonate (PC), butylene carbonate (BC), and diethyl carbonate (DEC) has been successfully demonstrated in the Pd-catalyzed asymmetric allylic alkylation reaction using bidentate diphosphanes as chiral ligands. Improved yields were observed compared to the conventional solvent CH_2Cl_2 (Scheme 8.13) [23]. Notably, in the reaction of **30** and dimethyl malonate **31**, the optimized enantioselectivity of up to 98% was observed by using $[Pd(\pi\text{-}C_3H_5)Cl]_2$ as a catalyst in BC.

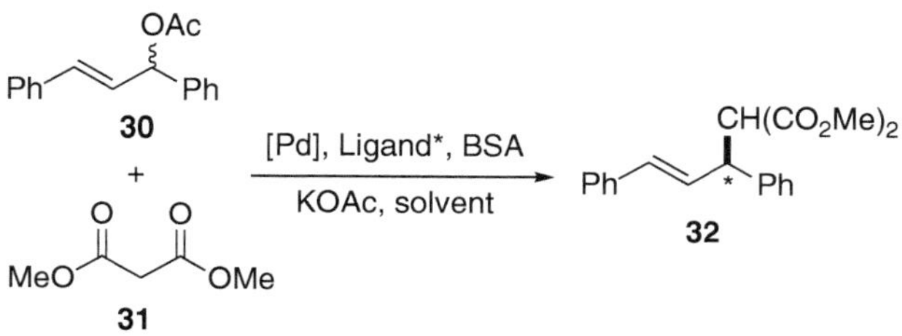

Solvent	Yield	ee
CH_2Cl_2	80%	44%
PC	91%	29%
BC	76%	36%
DEC	98%	35%

Scheme 8.13 Pd-catalyzed asymmetric allylic alkylation in different solvents.

In 2009, Börner's group also investigated the Rh-catalyzed asymmetric hydrogenation of olefins in organic carbonates (PC and BC) and other solvents such as DMF, DMSO, THF, CH_2Cl_2, and acetonitrile. Compared to DMSO, DMF, and acetonitrile, high conversions and excellent enantioselectivities were usually observed when using PC and BC (Scheme 8.14) [24].

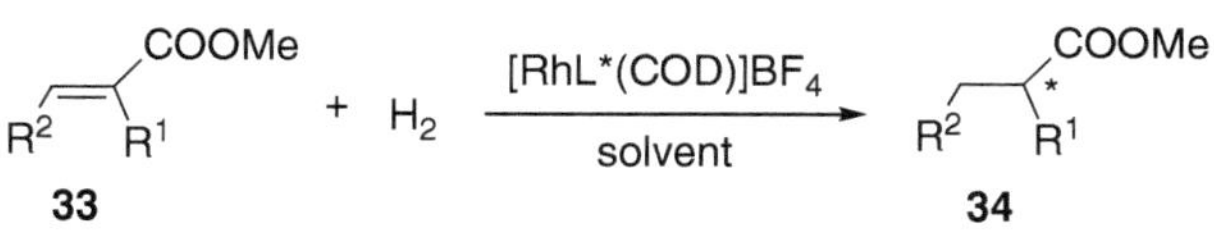

Solvent	Yield	ee
DMSO	53%	75%
PC	>99%	98%
BC	>99%	98%

Scheme 8.14 Rh-catalyzed asymmetric hydrogenation in different solvents.

In 2011, Willis et al applied propylene carbonate as a solvent for the chelation-assisted, rhodium-catalyzed intermolecular hydroacylation of S-chelating aldehydes with a variety of alkynes. The results showed that PC was an excellent solvent for rhodium-catalyzed intermolecular hydrogenation reactions. The use of acetone or DCE as a solvent is eliminated by the effective use of PC (Scheme 8.15) [25]. The commercially available rhodium complex $[Rh(nbd)_2]BF_4$ did not require pre-activation by hydrogenation.

In 2014, Clark's group explored organic carbonates such as ethylene carbonate and propylene carbonate as green alternative solvents for the Heck reaction. The results show that organic carbonate is a viable green solvent for the Heck reaction of aryl halide and methyl acrylate, providing an efficient alternative to conventional dipole aprotic solvents such as NMP, DMF, and DMAc. In addition, this work also

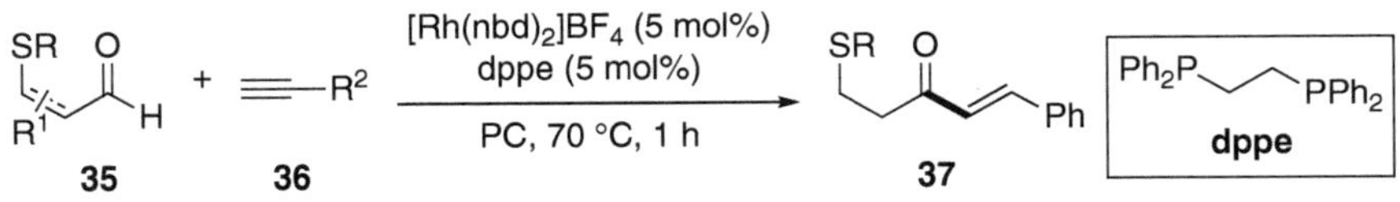

Scheme 8.15 PC as a solvent in Rh-catalyzed alkyne hydroacylation reactions.

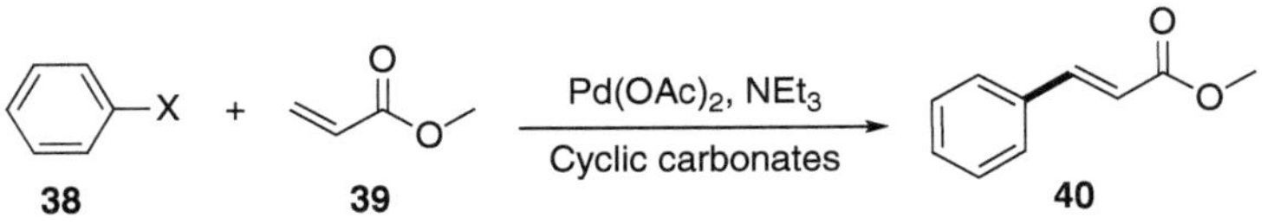

Scheme 8.16 Organic carbonates as a solvent in Heck reaction.

illustrated that the solvent polarity should be considered in the Heck reaction when developing greener reaction methods in the future (Scheme 8.16) [26].

In recent years, organic carbonates have also been applied as green solvents in the field of visible-light-induced radical chemistry. For example, Yu's group in 2020 reported a visible-light-promoted protocol for the construction of C—P bond to access various phosphorylated imidazo[1,2-*a*]pyridines with DEC as a green solvent (Scheme 8.17) [27]. This method avoids the use of toxic solvents compared to conventional methods.

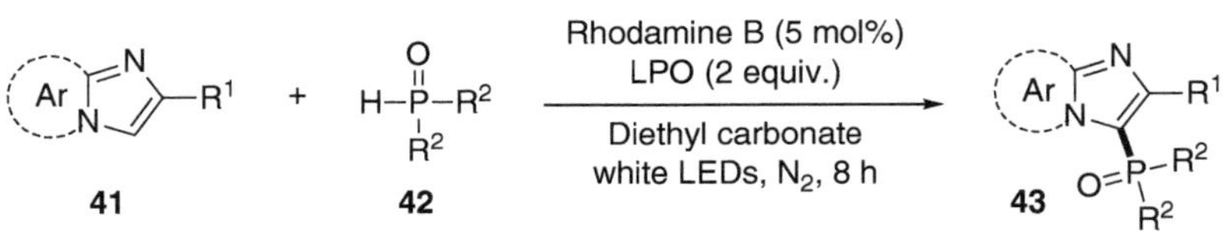

Scheme 8.17 Visible-light-promoted C—P bond formation in DEC.

Similarly, organic carbonates are also well-used as green solvents in visible-light-induced radical cascade cyclization reactions. In 2021, Yu's group also reported a general photocatalytic thiocyanation protocol using DMC as a green solvent (Scheme 8.18) [28]. It is assumed that the polar organic carbonate might play an important role in the stabilization of radical intermediates.

8.3.2 Organic Reactions in γ-Valerolactone

γ-Valerolactone (GVL) is readily derived from lignocellulosic biomass and has been considered as an ideal bio-based renewable, biodegradable, and non-toxic alternative to dipolar aprotic solvents. Significant progress has been achieved in the application of GVL as a green reaction medium in various organic reactions. In 2014, Dumesic's group disclosed that GVL significantly improved reaction rates and product selectivities in Brønsted-acid-catalysed biomass conversions, compared to other solvents such as 1,4-dioxane (Scheme 8.19) [29]. For the probe reaction, i.e., the dehydration of xylose to furfural, GVL changes the activation energies of the process.

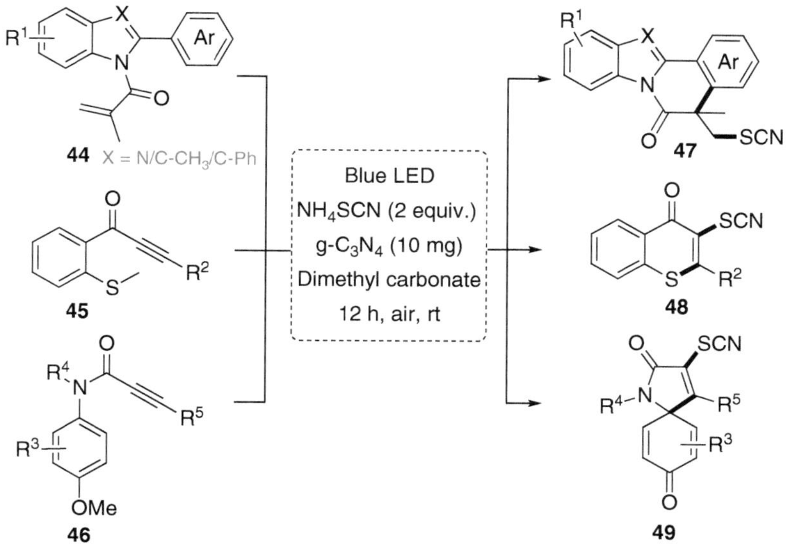

Scheme 8.18 Visible-light-promoted cascade cyclization reactions in DMC.

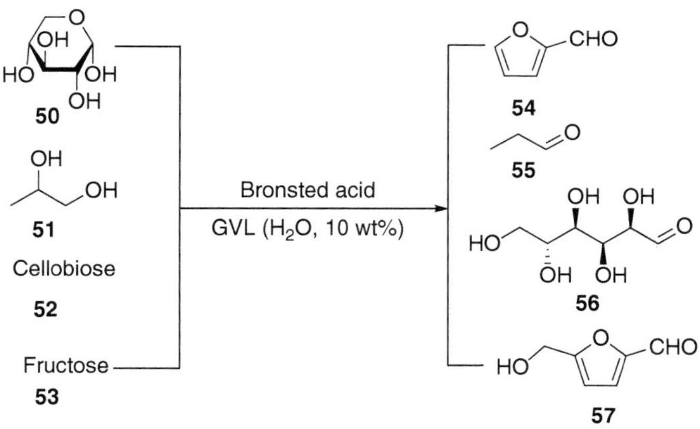

Scheme 8.19 Acid-catalyzed biomass conversions in GVL.

GVL is also claimed to be a valid substitute for traditional dipolar aprotic media in cross-coupling reactions. Vaccaro's group in 2014 employed GVL as the reaction medium for the Hiyama reaction. A sustainable protocol using a commercially available heterogeneous palladium catalyst (Pd/C) was developed for the representative Hiyama reaction between nitrobenzene and triethoxy(aryl)silane without the use of any ligand (Scheme 8.20a) [30]. In 2015, the same group also reported the Sonogashira cross-coupling reaction catalyzed by a similar heterogeneous Pd/C catalytic system with DABCO as the base using GVL as a green solvent. Compared to conventional solvents such as DMF, high yields of isolated products demonstrated that the biomass-derived solvent of GVL afforded very good performance (Scheme 8.20b) [31].

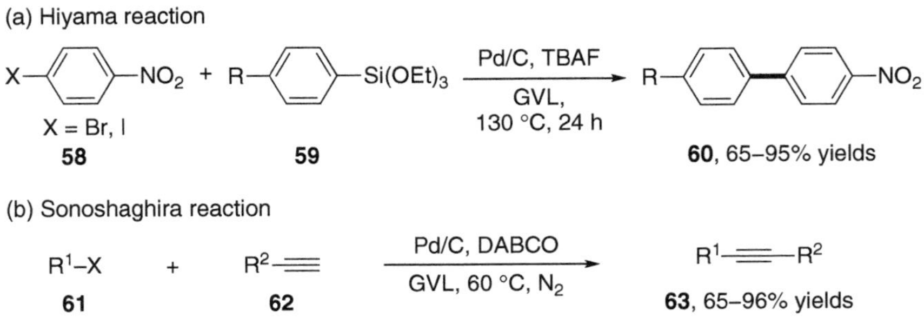

Scheme 8.20 Pd-Catalyzed cross-coupling reactions in GVL.

In 2017, Cravotto's group reported a microwave (MW)-assisted Pd-catalyzed coupling of aryl halides with thiophenes using GVL as the solvent. In this ligand-free protocol, the amount of palladium catalyst loading has been lowered to 0.2 mol%. Poly(3-hexyl)thiophene (P_3HT), a useful reference polymer for flexible solar cells, was successfully synthesized in three hours of microwave heating. Significantly, computational studies of this Pd-catalyzed C—H arylation of thiophenes with aryl halide revealed that GVL played the dual role of the reaction medium and ligand (Scheme 8.21) [32].

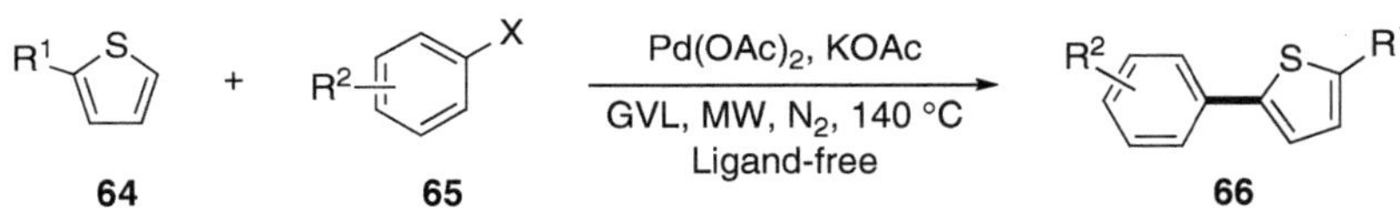

Scheme 8.21 Pd-Catalyzed C–H arylation of thiophenes with aryl halide in GVL.

In 2016, Vaccaro's group reported a palladium-catalyzed Catellani reaction using GVL as solvent. The encapsulated $Pd(OAc)_2$ (Pd EnCat™ 30) or Pd/Al_2O_3 proved to be an efficient catalyst for this transformation. However, the reaction performance of the Pd EnCat™ 30 is most likely contributed by leached homogeneous palladium species. While the heterogeneous Pd/Al_2O_3 catalyst can be recovered and reused for the consecutive reactions. The renewable biomass-derived GVL was identified as an effective reaction medium compared to conventional solvents such as DMF, NMP, and MeCN (Scheme 8.22) [33].

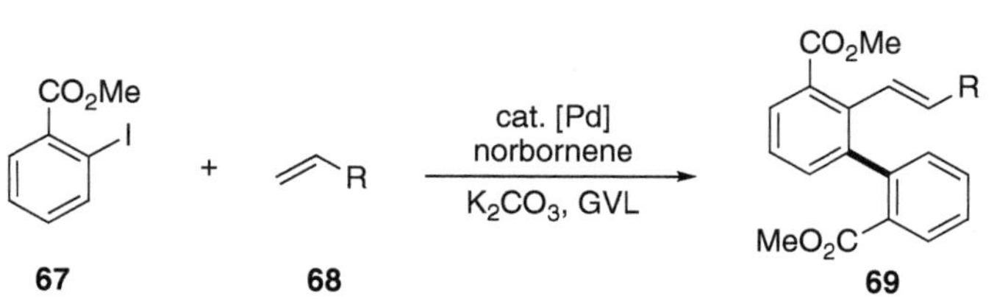

Scheme 8.22 Heterogeneous palladium-catalyzed Catellani reaction in GVL.

8.3.3 Organic Reactions in Ethyl Lactate

Biomass-derived ethyl lactate (EL) is attracting increasing attention as a green organic synthesis medium [34]. As a green reaction medium, EL has the unique advantages of non-toxicity, low cost, complete biodegradability, high boiling point, and excellent compatibility with water and most organic compounds [35]. Over the past years, EL has been successfully used as a medium in various types of organic reactions, such as Suzuki–Miyaura reactions, Glaser reactions, the synthesis of various heterocyclic compounds, and so on.

In 2012, Wan's group reported the first Suzuki coupling reaction in a green catalytic system of water-EL without the participation of any ligand [36]. It is worth mentioning that the use of traditional organic solvents such as DMF, DMSO, ethyl acetate, and dioxane as reaction solvents resulted in relatively low yields. It is suggested that EL probably acts not only as a solvent but also as a ligand to form a complex with palladium to catalyze the reaction (Scheme 8.23a). In 2014, the same group also developed a Glaser coupling reaction for the synthesis of 1,3-conjugated diynes using EL as the green medium [37]. The method does not require any ligands and can be efficiently achieved by heating at 50 °C with O_2 as the oxidant and CuI as the catalyst (Scheme 8.23b).

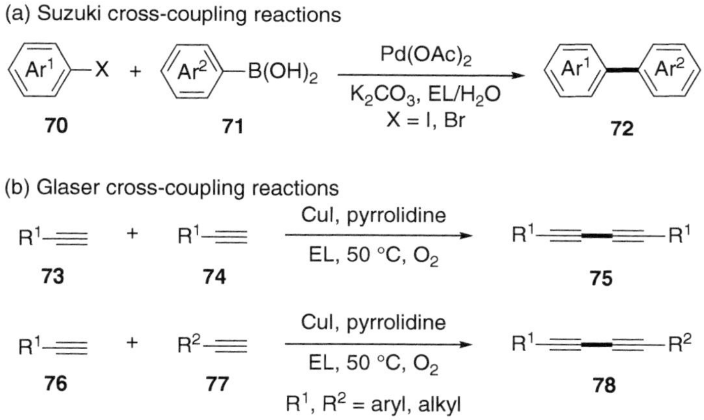

Scheme 8.23 (a) Suzuki reaction and (b) Glaser reactions in EL.

EL is miscible with water and common organic solvents. By varying the mixing ratio of the solvents, a range of reaction media with different polarities can be obtained, which will provide ideal reaction conditions and improve the efficiency of the reaction. Based on this concept, Bennett's group reported an efficient and green method for the synthesis of aryl aldimines which could be crystallized directly out of solution within minutes. The method uses EL as solvent and water as co-solvent to adjust the polarity of the reaction medium at room temperature resulting in the desired imines with excellent yields without further purification (Scheme 8.24) [38].

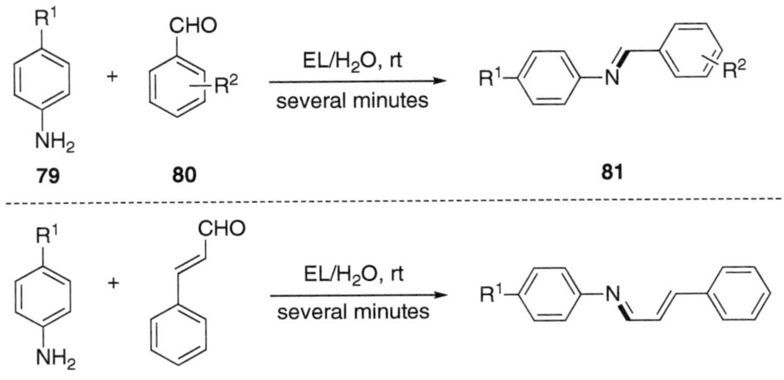

81 82 83

Scheme 8.24 Synthesis of aryl aldimines in EL or EL/H_2O system.

3,4-Dihydropyrimidinones and thiones have been extensively studied for biological activity. The Biginelli reaction is an important method for the synthesis of these compounds. In 2014, Wan's group reported the Biginelli reaction in the green reaction medium of EL [39]. This new synthetic system using EL as the green medium provides a new environmentally friendly method for the synthesis of 3,4-dihydropyrimidinothiones (Scheme 8.25).

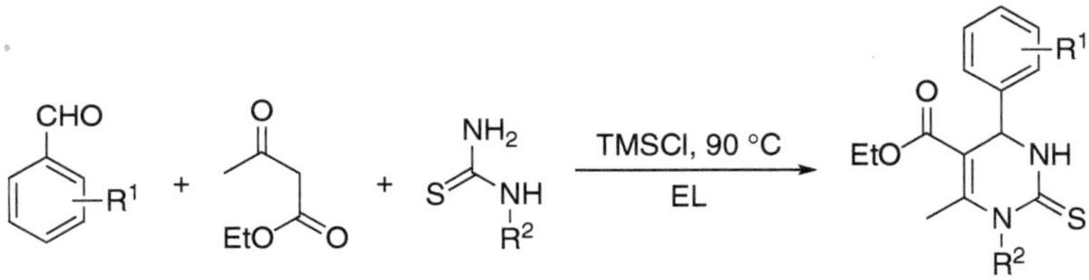

84 85 86 87

Scheme 8.25 Synthesis of 3,4-dihydropyrimidinothiones in EL.

In 2016, Yu's group reported a simple, efficient, and environmentally friendly method for the synthesis of C2-substituted benzothiazoles. This method enabled the efficient synthesis of a series of C2-substituted benzothiazoles using EL as the reaction solvent. EL is relatively more productive as a solvent than traditional organic solvents such as benzene, xylene, 1,4-dioxane, methanol, and ethanol (Scheme 8.26) [40].

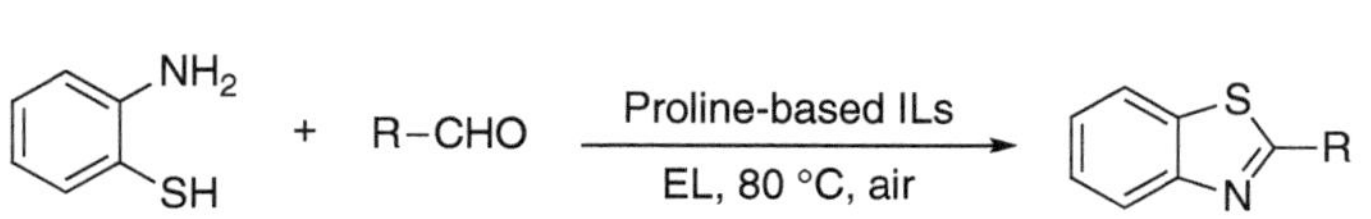

88 89 90

Scheme 8.26 Synthesis of two-substituted benzothiazoles in EL.

8.4 Conclusion

In summary, great achievements have been achieved in the organic transformations in the non-toxic, biodegradable, and environmentally friendly green ketone solvents (Cyrene, NBP) and ester solvents (organic carbonates, ethyl lactate) in the past decades. For example, Cyrene has been used as a solvent for coupling reactions, isothiocyanate synthesis, amide synthesis, Baylis–Hillman reaction, etc [41]. NBP is a promising solvent without reproductive toxicity and made good progress in replacing traditional dipolar aprotic solvents like NMP etc. in synthetic chemistry. Organic carbonates, due to their intrinsic and attractive properties such as low viscosity, high heat capacity, and good solubility, have been used as a new class of "green" sustainable alternatives to conventional dipolar aprotic solvents, which also showed good compatibilities in visible-light-induced radical reactions. Ethyl lactate has been successfully applied as a new green reaction medium in a variety of organic reactions and has demonstrated a unique catalytic effect in some reactions.

Overall, significant progress has been made in replacing traditional organic solvents with the aforementioned green solvents. However, these green solvents should not be simple alternatives. Further studies are still required to understand the unique advantages of the reaction in these mediums. On the other hand, some challenges have yet to be overcome, including how to reduce the costs of those sustainable solvents, and how to separate the products from the solution efficient in the industry.

References

1 Hessel, V., Tran, N.N., Asrami, M.R. et al. (2022). *Green Chem.* 24: 410–437.

2 Sheldon, R. (2001). *Chem. Commun.* 2399–2407.

3 Byrne, F.P., Jin, S., Paggiola, G. et al. (2016). *Sustain. Chem. Process* 4: 7.

4 (a) Stini, N.A., Gkizis, P.L., and Kokotos, C.G. (2022). *Green Chem.* 24: 6435–6449. (b) Stini, N.A., Gkizis, P.L., and Kokotos, C.G. (2023). *Org. Biomol. Chem.* 21: 351–358. (c) Sullivan, C., Zhang, Y., Xu, G. et al. (2022). *Green Chem.* 24: 7184–7193.

5 (a) Krishna, S.H., McClelland, D.J., Rashke, Q.A. et al. (2017). *Green Chem.* 19: 1278–1285. (b) Wang, Z., Lu, Q., Zhu, X.-F., and Zhang, Y. (2011). *ChemSusChem* 4: 79–84. (c) Mouterde, L.M.M., Allais, F., and Stewart, J.D. (2018). *Green Chem.* 20: 5528–5532.

6 Camp, J.E. (2018). *ChemSusChem* 11: 3048–3055.

7 Gao, F., Bai, R., Ferlin, F. et al. (2020). *Green Chem.* 22: 6240–6257.

8 Kar, S., Sanderson, H., Roy, K. et al. (2022). *Chem. Rev.* 122: 3637–3710.

9 Wilson, K.L., Kennedy, A.R., Murray, J. et al. (2016). *Beilstein J. Org. Chem.* 12: 2005–2011.

10 Wilson, K.L., Murray, J., Jamieson, C., and Watson, A.J.B. (2018). *Synlett* 29: 650–654.

11 Mistry, L., Mapesa, K., Bousfield, T.W., and Camp, J.E. (2017). *Green Chem.* 19: 2123–2128.
12 Veerabagu, U., Jaikumar, G., Lu, F., and Quero, F. (2021). *React. Chem. Eng.* 6: 1900–1910.
13 Wilson, K.L., Murray, J., Jamieson, C., and Watson, A.J.B. (2018). *Org. Biomol. Chem.* 16: 2851–2854.
14 Schleicher, J.C. and Scurto, A.M. (2009). *Green Chem.* 11: 694–703.
15 Sherwood, J., De bruyn, M., Constantinou, A. et al. (2014). *Chem. Commun.* 50: 9650–9652.
16 (a) Poliakoff, M., Fitzpatrick, J.M., Farren, T.R., and Anastas, P.T. (2002). *Science* 297: 807–810. (b) Clark, J.H., Farmer, T.J., Hunt, A.J., and Sherwood, J. (2015). *Int. J. Mol. Sci., Vol.* 16: 17101–17159.
17 Ramgren, S.D., Hie, L., Ye, Y., and Garg, N.K. (2013). *Org. Lett.* 15: 3950–3953.
18 Sherwood, J., Parker, H.L., Moonen, K. et al. (2016). *Green Chem.* 18: 3990–3996.
19 Clark, J.H., Macquarrie, D.J., and Sherwood, J. (2013). *Chem. Eur. J.* 19: 5174–5182.
20 Clarke, P.A., Zaytsev, A.V., and Whitwood, A.C. (2008). *Synthesis* 2008: 3530–3532.
21 Lopez, J., Pletscher, S., Aemissegger, A. et al. (2018). *Org. Process Res. Dev.* 22: 494–503.
22 Schäffner, B., Schäffner, F., Verevkin, S.P., and Börner, A. (2010). *Chem. Rev.* 110: 4554–4581.
23 Schäffner, B., Holz, J., Verevkin, S.P., and Börner, A. (2008). *ChemSusChem* 1: 249–253.
24 Schäffner, B., Andrushko, V., Bayardon, J. et al. (2009). *Chirality* 21: 857–861.
25 Lenden, P., Ylioja, P.M., González-Rodríguez, C. et al. (2011). *Green Chem.* 13: 1980–1982.
26 Parker, H.L., Sherwood, J., Hunt, A.J., and Clark, J.H. (2014). *ACS Sustain. Chem. Eng.* 2: 1739–1742.
27 Gao, F., Sun, K., Chen, X.-L. et al. (2020). *J. Org. Chem.* 85: 14744–14752.
28 Zeng, F.-L., Zhu, H.-L., Chen, X.-L. et al. (2021). *Green Chem.* 23: 3677–3682.
29 Mellmer, M.A., Sener, C., Gallo, J.M.R. et al. (2014). *Angew. Chem. Int. Ed.* 53: 11872–11875.
30 Ismalaj, E., Strappaveccia, G., Ballerini, E. et al. (2014). *ACS Sustain. Chem. Eng.* 2: 2461–2464.
31 Strappaveccia, G., Luciani, L., Bartollini, E. et al. (2015). *Green Chem.* 17: 1071–1076.
32 Tabasso, S., Calcio Gaudino, E., Rinaldi, L. et al. (2017). *New J. Chem.* 41: 9210–9215.
33 Rasina, D., Kahler-Quesada, A., Ziarelli, S. et al. (2016). *Green Chem.* 18: 5025–5030.
34 Dolzhenko, A.V. (2020). *Sustain. Chem. Pharm.* 18: 100322.
35 Wei, L., Chen, X., Liu, Y., and Wan, J. (2016). *Chin. J. Org. Chem.* 36: 954–961.
36 Wan, J.-P., Wang, C., Zhou, R., and Liu, Y. (2012). *RSC Adv.* 2: 8789–8792.
37 Wan, J.-P., Cao, S., and Jing, Y. (2014). *Appl. Organomet. Chem.* 28: 631–634.

38 Bennett, J.S., Charles, K.L., Miner, M.R. et al. (2009). *Green Chem.* 11: 166–168.

39 Xu, Z., Jiang, Y., Zou, S., and Liu, Y. (2014). *Phosphorus, Sulfur Silicon Relat. Elem.* 189: 791–795.

40 Yu, Z.-Y., Fang, Q.-S., Zhou, J., and Song, Z.-B. (2016). *Res. Chem. Intermed.* 42: 2035–2045.

41 (a) Nickisch, R., Conen, P., Gabrielsen, S.M., and Meier, M.A.R. (2021). *RSC Adv.* 11: 3134–3142. (b) Bousfield, T.W., Pearce, K.P.R., Nyamini, S.B. et al. (2019). *Green Chem.* 21: 3675–3681. (c) Sangon, S., Supanchaiyamat, N., Sherwood, J. et al. (2020). *React. Chem. Eng.* 5: 1798–1804.

9

Recent Achievements in Organic Reactions in Polyethylene Glycol

Zhiping Yin

Jiangsu University, School of Pharmacy, 301 Xuefu Road, Zhenjiang, 212013, P. R. China

9.1 Introduction

Solvent plays an essential role in the prosperity and rapid development of modern organic chemistry. A broad range of chemical processes, such as setting reactions, cleaning, chromatography, and crystallization depend on the excess use of organic solvent. Solvation effects also provide several major advantages, including selectivity control, reaction efficiency, and simple operating procedures [1]. Thus, accurately understanding the properties of various solvents is a key issue for achieving diverse chemical transformations. Along with this process, the large consumption of nonrenewable, unsustainable, and environmentally detrimental organic solvents is unavoidable. For example, solvents account for above 80% proportion of chemicals of concern used in the typical pharmaceutical or fine chemicals synthesis, which poses great risks to both human health and our environment [2].

Over the past century, the concept of “green and sustainable chemistry” has received considerable scholarly attention. There is an urgent need to address the safety problems caused by organic solvents. Replacing traditional organic solvents with recyclable green and safe solvents is one of the efficient strategies. Polyethylene glycol (PEG, $HOCH_2$-[CH_2-O-CH_2]$_n$-CH_2OH) is the polymeric form of ethylene glycol. With two hydroxyl groups at adjacent positions along a hydrocarbon chain, PEG possesses special properties including colorless, relatively nonvolatile, completely miscible with different polar solvents (alcohols, water, and acetone), and only slightly soluble in nonpolar solvents (toluene, benzene, and dichloroethane). According to their molecular weight, they can be divided into different grades and physical characteristics. For example, PEGs having molecular weights >2000 are hard crystalline solids, those between 1000 and 2000 are soft solids, and PEGs with molecular weights from 100–700 are liquids at room temperature [3].

As per the 2022 Global-PEG market research report analysis, PEG market size is estimated to be worth USD 3279.9 million in 2021 and is forecast to a readjusted size of USD 4817.7 million by 2028 with a compound annual growth rate (CAGR) of 5.6% during review period. These data clearly indicated that PEG is an important industrial commodity and has a broad scope of applications in the personal healthcare and

Green Solvents in Organic Synthesis, First Edition.
Edited by Xiao-Feng Wu, Zhiping Yin, Liang-nian He, and Feng Wang.

chemistry industry. For example, in healthcare applications, PEG was widely used as the polymer in drug delivery systems because of its adjustable and well-established safety characteristics [4]. PEGylated compounds involving small molecular drugs, nanoparticles, and enzymes represent an effective method for anti-cancer therapy by modifying the pharmacokinetics (PK) of various drugs [5]. With respect to organic synthesis, PEGs or PEGlayted compounds were generally used as a green catalyst as well as a solvent in different chemical reactions. The presence of two hydroxyl groups in PEG provides distinctive properties such as coordinating ability, stability at harsh conditions, and recyclability [6].

This brief chapter mainly focuses on employing PEGs as green organic solvents in various chemical transformations, those PEGylated compounds or using PEG as a catalyst have been summarized and will not be included [7]. The main contexts were organized based on the different types of organic reactions, including transition metal-catalyzed coupling, organocatalysis, multicomponent, and cyclization reactions (Scheme 9.1). A broad range of reaction examples and advantages of PEG will be summarized and discussed in detail. Besides, the mechanistic rationale for representative examples is also provided to better understand the function of PEG. We hope that this work will arouse increased interest and stimulate more and more chemists to develop diverse green and sustainable approaches for organic reactions.

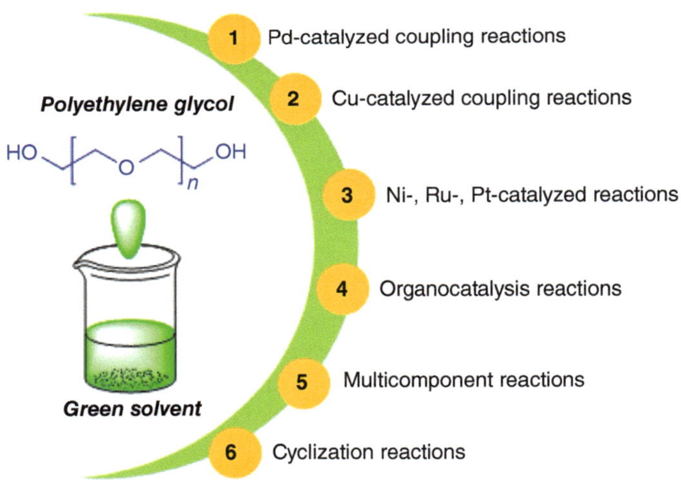

Scheme 9.1 Different organic reactions performed in PEG solvents.

9.2 PEG in Pd-catalyzed Coupling Reactions

9.2.1 Pd-catalyzed C—C, C—Si Bonds Formation in PEG

Transforming easily available, cost-effective, and simple organic molecules into value-added compounds is one of the long-term goals for modern organic chemists. Among the various strategies, the construction of C—C bonds in the presence of palladium such as Heck, Negishi, and Suzuki cross-coupling reactions, represents one of the most exciting protocols, which also have been recognized with the Noble Prize for Chemistry in 2010. PEG-2000 was first applied as a solvent to the regioselective Heck reactions by Chandrasekhar and coworkers [8]. Various aryl bromides substituted with electron-donating or electron-withdrawing groups proceeded smoothly with alkenes and formed corresponding products in good diastereoselection (>70/30, E/Z). Noteworthy, the conventional solvents (DMF, DMSO, and CH_3CN) also led to the formation of mixture products with different ratios. The author suggested that PEG acts as both an efficient solvent medium and a phase transfer catalyst (PTC) for C—C bond coupling (Scheme 9.2).

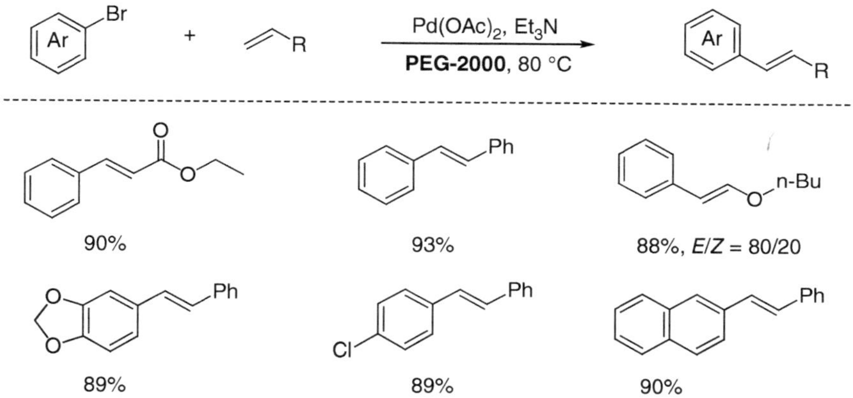

Scheme 9.2 Palladium-catalyzed Heck cross-coupling of aromatic bromides in PEG-2000.

Microwave (MW) accelerated reactions are generally faster and consequently cleaner than the traditional method and become a powerful technique in part of chemical synthesis [9]. In 2006, Lamaty and coworkers illustrated an MW-assisted intramolecular Heck reaction in PEG-3400 for the synthesis of benzazepines [10]. Solvent plays an essential role in producing cyclized products. When PEG 300 or a mixture of PEG 3400 and PEG 300 was used, the reaction efficiency was decreased dramatically and gave a complicated mixture of products, including Tsuji–Trost rearrangement byproducts (Scheme 9.3a). These results were consistent with related literature reported by Zhang and coworkers [11], that is Pd(II) is readily reduced into palladium nanoparticles by larger PEGs (with an average molecular weight equal to or higher than 2000) than the small molecular weights. The palladium nanoparticles were also verified by XRD and TEM investigations. In 2010, Liu and coworkers established that the *in situ*-generated palladium nanoparticles can also be found in PEG-400 [12]. According to their results, the average diameters of palladium nanoparticles generated *in situ* under the Heck conditions were approximately 2.5 nm, which displayed higher catalytic activity than the *ex situ*-generated palladium nanoparticles (approximately 3.6 nm) (Scheme 9.3b).

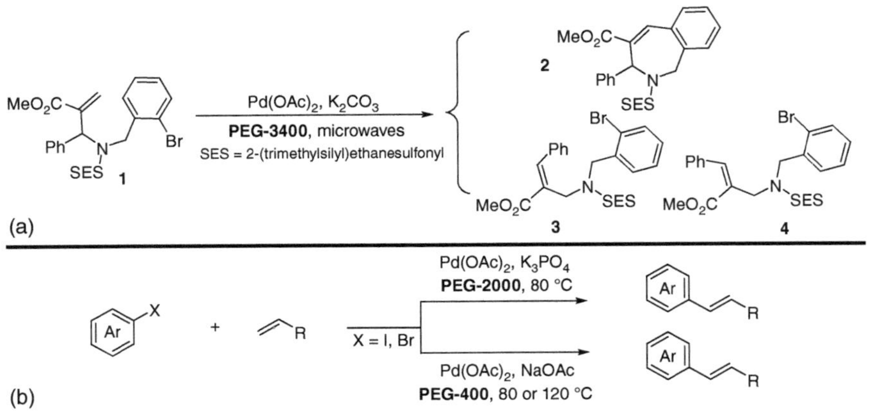

Scheme 9.3 Intramolecular and intermolecular Heck cross-coupling of aromatic halides in PEG.

In 2006, Larhed and coworkers achieved the first Pd-catalyzed selective terminal Heck arylation of vinyl ethers with aryl chlorides in the environmentally friendly PEG-200 [13]. This reaction was conducted at non-inert conditions and controlled MW heating to minimize handling and processing times. DFT calculations indicated a ligand-driven selectivity rationale, where the electronic and steric influence of bulky $P(t\text{-}Bu)_3$ ligand provides improved β-selectivity in the essential insertion step with aryl chlorides. In agreement with the DFT calculation results, aryl chlorides with electron-withdrawing substituents producing relatively strong Cl–Pd coordinations, correlate with a preference for β-selectivity. Conversely, electron-donating groups producing weaker Cl–Pd coordinations favor the formation of branched α-products. More importantly, this method was further demonstrated as the key step of a new synthetic route to Betaxolol, an approved well-known clinically used β-blocker (Scheme 9.4a). In 2009, Ackermann and coworkers developed the first unprecedented Pd(0)-catalyzed C—H bond arylations of 1,2,3-triazole in PEG-20000 under air [14]. With carboxylic acid $MesCO_2H$ as a co-catalyst, diverse *N*-benzylated triazoles could be generated from aryl bromides with 1,2,3-triazoles in 52–85% yields without phosphine ligands (Scheme 9.4b).

The selective Pd-catalyzed coupling of aryl halides offers a powerful strategy for preparing symmetric and unsymmetrical biaryls [15]. Typically, reducing agents such as hydrogen gas, alcohol, and zinc are indispensable in these reactions, which unavoidably cause problematic issues, for example, reductions of carbonyl or nitro groups as well as hydrodehalogenation. In 2006, Wang, Zhang, and coworkers discovered Pd-catalyzed coupling of aryl halides to give symmetrical and unsymmetrical biaryl products by employing PEG as a reductant and solvent medium [16]. Both homo-coupling of aryl iodides and cross-coupling of aryl iodides with aryl bromides delivered the corresponding biaryls in good yields and selectivities (Scheme 9.5). The substrates endowed with nitro or carbonyl groups can be tolerated well. Interestingly, the reaction selective was controlled by the amount of PEG. Excessive PEGs induce the increase of hydrodehalogenation product, and the

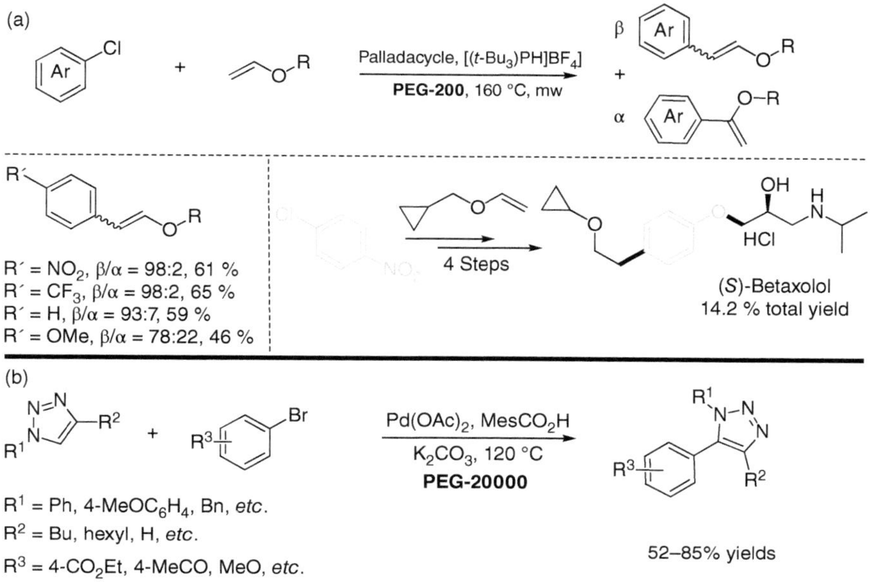

Scheme 9.4 Terminal β-arylation of alkyl vinyl ethers and arylation of 1,2,3-triazole in PEG.

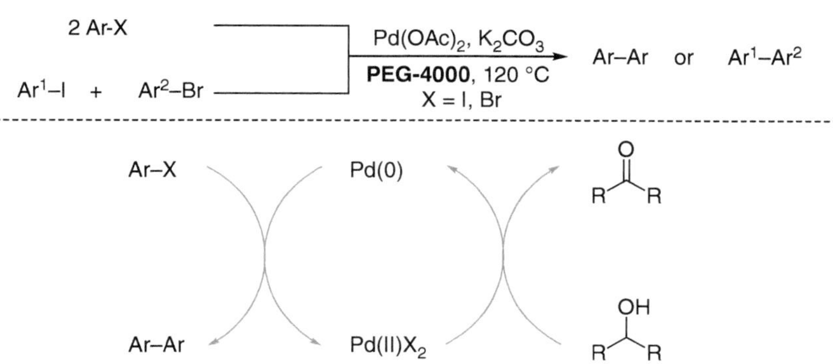

Scheme 9.5 Pd-catalyzed homo- and cross-coupling of aryl halides to synthesize symmetrical and unsymmetrical biaryl

best selectivity to biaryl is obtained when the concentration of the hydroxyl group in PEG achieves 100% relative to aryl halides. According to previous literature, the terminal hydroxyl group in PEG could act as the reductant [17]. A possible mechanism pathway was proposed in Scheme 9.5, the real reductant is the Pd(0) generated *in situ* through a reductive process of Pd(II) by the hydroxyl group in PEG. In addition, the product can be easily separated through diethyl ether extraction and the catalyst-containing PEG phase can be reused five times without loss of reaction activity.

Ullmann homocoupling reactions provide a direct route for the synthesis of symmetric biaryls. However, this reaction was generally catalyzed by copper catalyst under high temperatures (over 200 °C). Karimi and coworkers successfully used *N*-heterocyclic carbene (NHC) palladium polymers as precatalysts for Ullmann homocoupling of aryl halides with PEG-200 as solvent (Scheme 9.6) [18]. The reaction was conducted at 75 °C with high efficiency in the presence of 0.5 mol% catalysts, a wide range of aryl halides (I, Br, and Cl), and heteroaryl halides are converted to corresponding biaryls with excellent yields. A trace amount of palladium nanoparticles generated from the *in situ* decomposition of NHC–Pd polymer could catalyze this coupling reaction, which is consistent with their previously reported NHC–Pd-catalyzed Suzuki–Miyaura coupling reaction [19]. In addition, a Pd(0)–Pd(II)–Pd(IV) mechanism pathway was proposed to explain this reductive homocoupling.

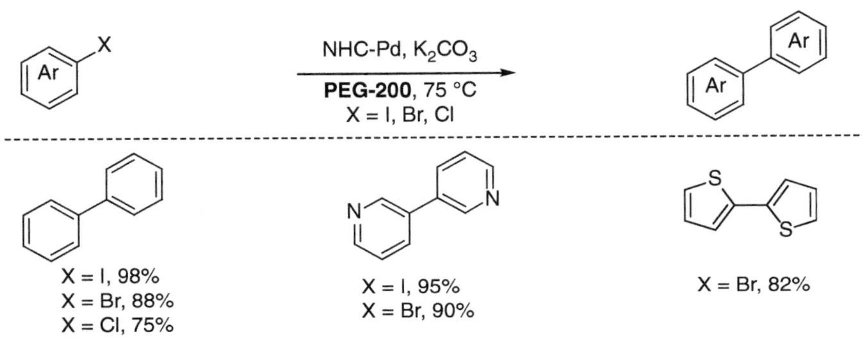

Scheme 9.6 NHC–Pd-catalyzed homocoupling in PEG-200.

Suzuki–Miyaura coupling reaction represents a practical synthetic method for the direct construction of new C—C bonds and has been widely applied both in the academic and industrial over more than 40 years of history [20]. In 2001, Varma and coworkers probably discovered the first MW-accelerated Suzuki cross-coupling reaction in PEG-400 [21]. Different from other methods, self-coupling of the reactants has been eliminated or reduced to provide the product with 66–92% yields (Scheme 9.7). The 2 g/mmol substrate was found to be the best amount of PEG. Aromatic iodides and bromides were more reactive than the chloride, but aliphatic

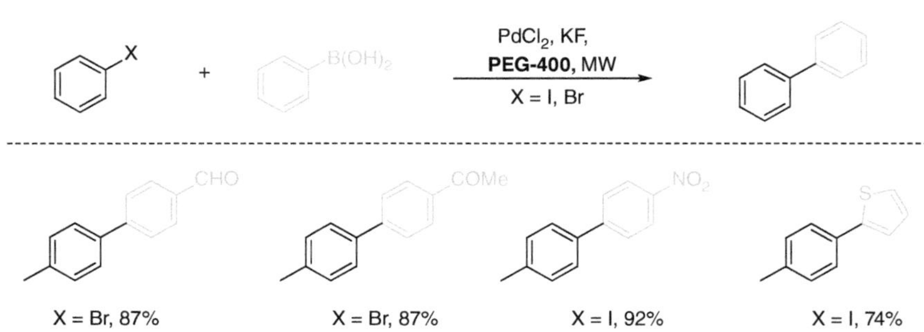

Scheme 9.7 Microwave-assisted Suzuki–Miyaura coupling in PEG-400.

halides were unsuitable reaction partners under standard conditions. Functional groups such as aldehyde, nitro, ketone, and ester were well tolerated.

Subsequently, Pd-catalyzed Suzuki–Miyaura coupling reactions were demonstrated by various other groups. For example, In 2005, Li and coworkers reported an efficient and reusable $Pd(OAc)_2$/DABCO/PEG system for the Suzuki–Miyaura coupling [22]. Different aryl iodides and bromides were transformed into biaryls in 32–100% yields and high turnover numbers (the maximal TONs were up to 960,000 for the reaction of 1-iodo-4-nitrobenzene with phenylboronic acid) by employing PEG-400 as the solvent (Scheme 9.8a). It is noteworthy that tetrabutylammonium bromide (TBAB) plays a pivotal role in improving the reaction efficiency of aryl chlorides. Without TBAB, only a 54% yield of product was produced after 15 hours from 1-(4-chlorophenyl)ethanone with phenylboronic acid. The same group also extends the $Pd(OAc)_2$/DABCO catalytic system to Stille, Heck, and Sonogashira cross-coupling with a catalytic amount of PEG-400 as the phase-transfer catalyst [23]. One year later, Zhang and coworkers developed a Pd-catalyzed coupling reaction of aryl boronic acid with carboxylic anhydride or acyl chloride in PEG-2000 to produce corresponding ketones in 9–98% yields [24]. The catalytic system was perfectly stable to air and could be conducted without phosphine ligands (Scheme 9.8b). Although the acyl chlorides are sensitive to moisture, various aryl or heteroaryl acyl chlorides could be converted to the corresponding ketones in good yields, and aliphatic hexanoyl chloride only furnished the product with 9% yield.

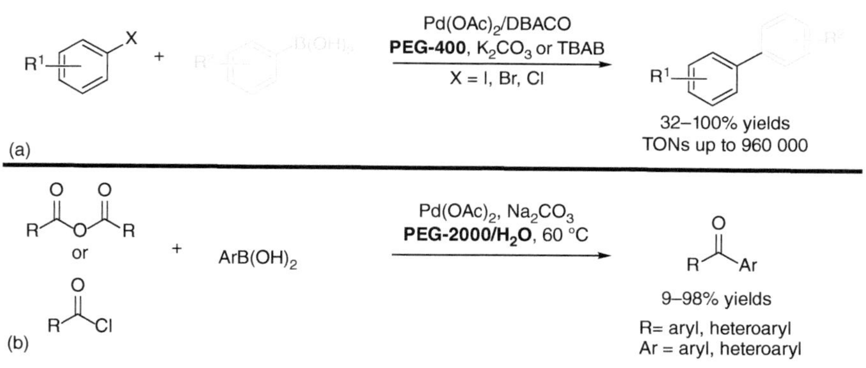

Scheme 9.8 Suzuki–Miyaura coupling of aromatic halides, carboxylic anhydride, and acyl chlorides in PEG.

The ratio of water and PEG is crucial to the Suzuki reaction. Zhang and coworkers found that the best Suzuki coupling condition of 4-bromotoluene (1 mmol) and phenylboronic acid (1.5 mmol) is 3.5 g PEG-2000 in water (3 mL) [25]. Further addition of PEG-2000 or using neat PEG-2000 will lead to a decrease in yields. In addition, the mild temperature (50 °C) and free use of phosphine ligands and MW make it a useful approach for the synthesis of biaryl products. TBAB has generally been used as an additive to promote Suzuki coupling of aryl chlorides. In 2005, Li and coworkers developed an efficient Suzuki coupling system in the presence of 0.01 to 3 mol% of $Pd(OAc)_2$, 10 mol% of TBAB, 2 equivalents of K_2CO_3, and 4 g of PEG-400, a wide range of aryl chlorides were transformed into the corresponding biaryl products in excellent yields under ligand-free conditions [26]. According to their proposed mechanism, TBAB plays a significant role in the reaction: (i) activation of the active Pd(0) species with the formation of anionic species; (ii) stabilization of the low-coordinate Pd(0) species; (iii) as phase-transfer catalyst for the inorganic base/solvent/substrate/product phases. One year later, Wang and coworkers also established PEG-300 as an effective medium for the Pd-catalyzed Suzuki coupling of aryl chlorides with arylboronic acids [27]. A broad range of functional groups such as ketone, aldehyde, and nitro were tolerated well in the presence of $PdCl_2/K_2CO_3$. Interestingly, both *p*-dichlorobenzene and *o*-dichlorobenzene afford the corresponding terphenyl in acceptable yields (Scheme 9.9). Utilizing 0.01% $Pd_2(dba)_3$, a ligand-free Suzuki coupling between aryl halides and boronic acids was achieved by Souza and coworkers [28]. In 1 hour, the reaction efficiency between 1-iodo-4-nitrobenzene and phenylboronic acid reached 98% of the yield and 9700 of TON.

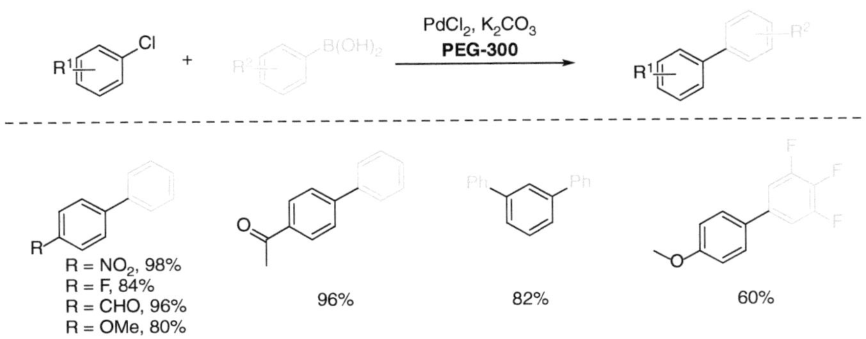

Scheme 9.9 Pd-catalyzed Suzuki–Miyaura coupling of aromatic chlorides in PEG-300.

In 2007, the oxygen-promoted ligand Suzuki–Miyaura cross-coupling between aryl chlorides and boronic acids was demonstrated by Liu and coworkers [29]. The substrates in open air or pure oxygen systems were consumed in only 50 minutes, whereas it took 6.5 hours to complete the coupling in N_2 (Scheme 9.10a). Different sizes of palladium nanoparticles generated *in situ* through the oxidation of hydroxyl groups in PEG-400 could clarify why the reaction took place much faster in air than in nitrogen. In addition, by employing UV-visible spectroscopy, they also found that phenylboronic acid can accelerate the formation of palladium nanoparticles [30]. In 2009, Mao and coworkers found that Suzuki coupling could be performed in metal-free conditions [31]. With iodine, potassium carbonate, and polyethylene glycol 400 (I_2/K_2CO_3/PEG-400) as an efficient catalytic system, various aryl halogens and phenyboronic acids were converted to biaryl products in the air (Scheme 9.10b). Moreover, this iodine-promoted catalytic system was also applied to the homocoupling of arylboronic acids in PEG-400/H_2O under air conditions [32]. Razler and coworkers also reported a practical ligand-free Suzuki–Miyaura coupling protocol by using 1 mol% of $Pd(OAc)_2$ and 10 mol% of

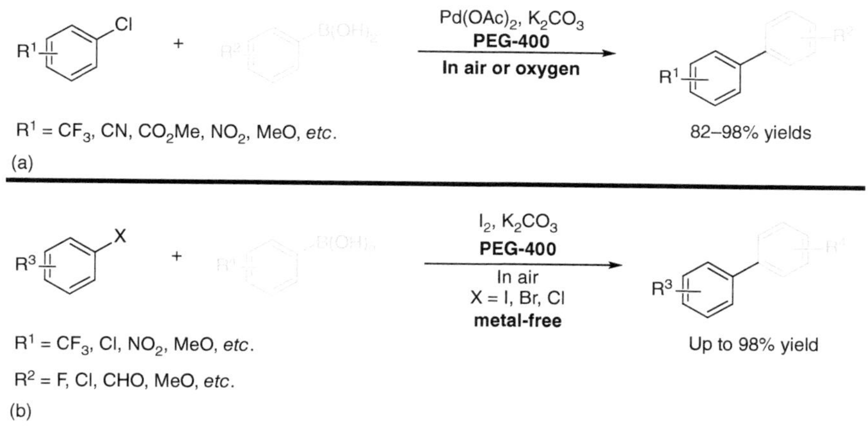

Scheme 9.10 Suzuki–Miyaura coupling of aromatic halides in PEG-400.

PEG 2000. Transmission electron microscopy (TEM) analysis of the Pd-PEG 2000 system indicated that palladium nanoparticles were formed in only one minute [33]. Both aryl and heteroaryl bromides are suitable reaction partners and gave the target product in good isolated yields.

The Sonogashira reactions represented one of the efficient methods for sp^2–sp coupling between aryl halides and terminal alkynes [34]. In its primary edition, the reaction was usually performed by using palladium species as catalysts under mild conditions. Recently, PEG was largely applied to this type of reaction due to its environmental and sustainable properties. In 2009, the first case of palladium-free, copper-free, and amine-free catalytic system for Sonogashira couplings was demonstrated by Mao and coworkers [35]. Samarium powder in PEG-600 was found to be the best catalytic system for the synthesis of diverse internal alkynes from aryl halides and terminal alkynes. Subsequently, Firouzabadi and coworkers exploited gelatin [36] or agarose [37] as a bioorganic ligand and support for the *in situ* generations of palladium nanoparticles. The generation of Pd(0) nanoparticles proceeded smoothly in the absence of reducing agents from the Pd(II) salt. Besides, the *in situ* generated palladium nanoparticles exhibited high catalytic activity for the Sonogashira coupling reaction of various aryl halides as well as heteroaryl halides and β-bromo styrene in PEG-400. Homocoupling of phenylacetylene was also achieved using similar conditions and isolated the *E*-enyne in 70% yield (Scheme 9.11).

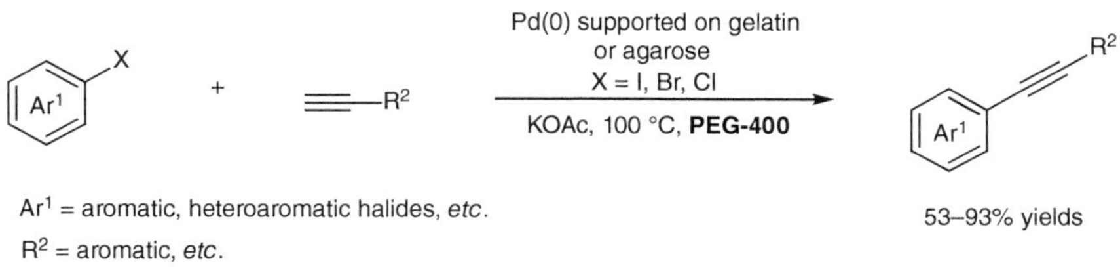

Scheme 9.11 Sonogashira reactions in PEG-400.

Over the past decades, palladium-catalyzed carbonylation reactions have emerged as a powerful strategy for the transformation of carbon monoxide, the simple and important C1 molecules, into value-added compounds, which are both interesting in academia and industry [38]. In 2013, Cai and coworkers found that $PdCl_2(PPh_3)_2$ displayed high catalytic activity for the carbonylative Sonogashira coupling reaction of aryl iodides with terminal alkynes in a mixture of water and PEG-2000 [39]. Under an atmospheric pressure of carbon monoxide, different aryl, heteroaryl iodides, and alkyl, aryl terminal alkynes proceeded smoothly in these conditions and provided alkynyl ketones in 57–95% yields (Scheme 9.12). This method was used for the synthesis of biologically active flavones from *o*-iodophenol and terminal alkynes. The recycling experiments indicated that this system can be reused for up to six runs without much change in yields and rates (93%, 91%, 94%, 92%, 93%, and 92%, respectively). Besides, ICP analysis of the extract suggested that palladium content was less than 0.65 ppm in the product.

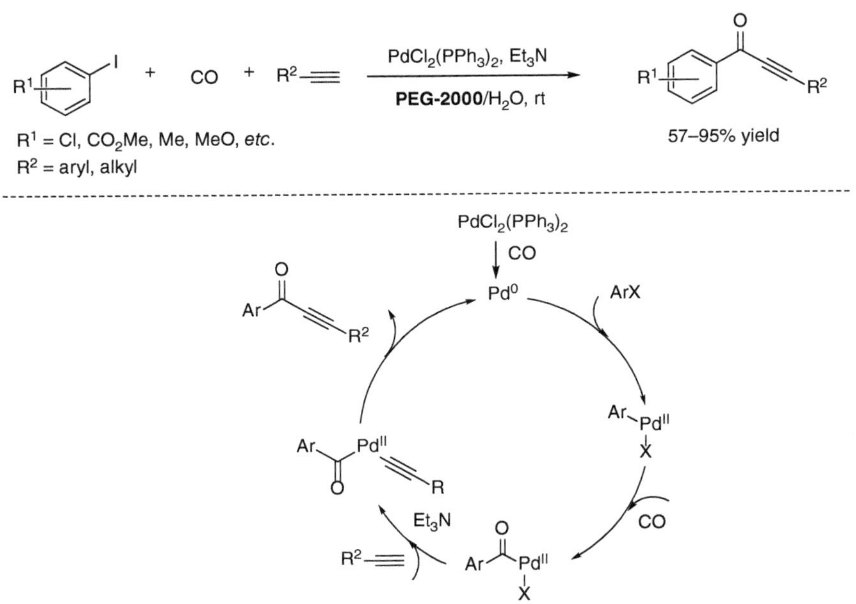

Scheme 9.12 Carbonylative Sonogashira reactions in PEG-2000.

Tetraphenylstannane is known to be an important organometallic reagent, it is a potentially useful candidate for Stille cross-coupling reactions [40]. Over the past decades, employing tetraphenylstannane as an atom-economical organometallic reagent to react with more than 1 equivalent of the electrophiles has received increased attention. In 2009, palladium-catalyzed Stille coupling of tetraphenylstannane with different aryl iodides or bromides in PEG-400 was established by Wang and coworkers [41]. Using 4 equivalents of aryl iodides or aryl bromides coupled effectively with 1 equivalent of tetraphenylstannane to furnish 4 equivalents of the corresponding functionalized biaryls in 41–97% yields (Scheme 9.13). With

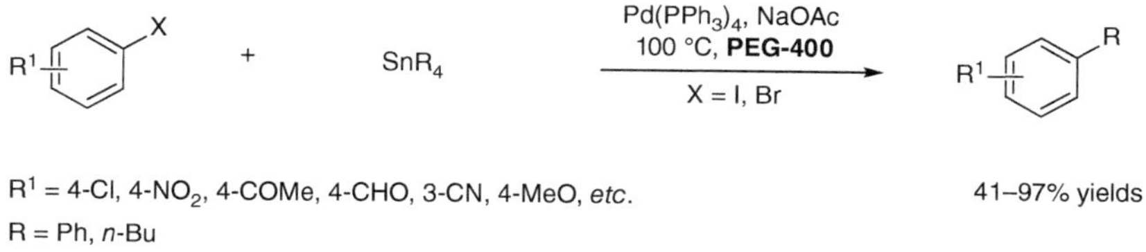

Scheme 9.13 Cross-coupling of tetraphenylstannane and aromatic halides in PEG-400.

NaOAc as the base, a wide range of aryl iodides or bromides was transformed into biphenyls with excellent yields [42]. This system presents significant features such as being atom-efficient under mild conditions, simple and rapid isolation of the product, and PEG can be reused several times. The same Stille coupling in the presence of DMAP under MW irradiation was also reported by Du and coworkers [43]. This method exhibits wide scope and good tolerance of functional groups for various aryl iodides and bromides, but no reaction occurred when chlorobenzene was used.

The Hiyama cross-coupling is characterized by the Pd-catalyzed C—C bond formation between aryl halides and organosilanes. The key to success of this reaction is the polarization of the C—Si bond, thus bases or fluoride ions (TASF, TBAF) are usually applied to activate the C—Si bond. In 2005, Kabalka and coworkers described a cross-coupling reaction between acetates of Baylis–Hillman adducts and organosilanes with PEG-600 as a green reaction media [44]. It is interesting to note that adducts derived from methyl acrylate, acrylonitrile, and vinyl ketones couple with organosilanes to give allylic rearranged (SN_2') products, whereas adducts derived from unsaturated cyclic ketones gave SN_2 products (Scheme 9.14a). In 2007, Zhang and coworkers discovered $Pd(OAc)_2$-catalyzed fluoride-free cross-coupling between the aryl bromides and arylsiloxanes in PEG-200/H_2O [45]. The amount of PEG 2000 had a strong influence on the activity of the reaction, the low yields or no reaction were obtained when an excess amount of PEG or neat PEG was presented in the catalytic system. With 3 equivalents of sodium hydroxide as the base, a wide range of biphenyls can be easily isolated by simple extraction with diethyl ether in 60–99% yields (Scheme 9.14b). Besides, the catalytic system could be reused eight times with only a small decrease in activity. Employing a Fischer carbene complex

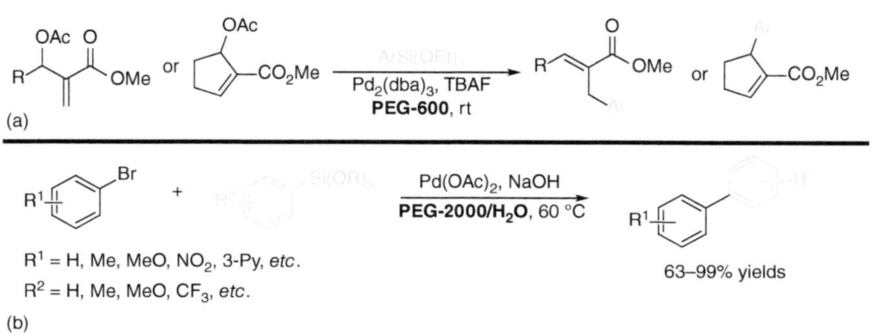

Scheme 9.14 Hiyama cross-coupling in PEG.

of tungsten as the reductant, Sarkar and coworkers also achieved the colloidal palladium-catalyzed Hiyama coupling in PEG under air [46]. A broad range of functional groups such as aldehyde and nitro were tolerated well and produced the desired products in good yields.

Li and coworkers reported Pd-catalyzed hydroalkylation cyclization of alkenyl β-keto esters and amides in PEG-400 [47]. No additives such as Me_3SiCl and $Ln(OTf)_3$ were required for the complete conversion, and 1 equivalent of $CuCl_2$ was used to accelerate the reaction rate sharply. Furthermore, the system could be recycled and reused five times without any loss of activity (Scheme 9.15).

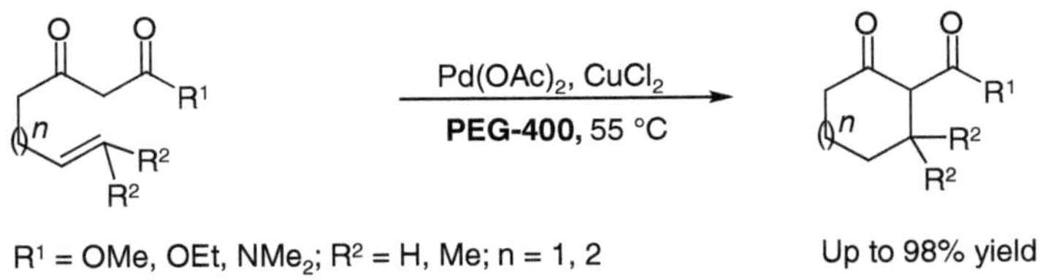

Scheme 9.15 Pd-catalyzed hydroalkylation cyclization of alkenyl β-keto esters.

9.2.2 Pd-catalyzed C—N, C—P Bond Formation in PEG

In 2004, Rao and coworkers reported a one-pot reaction for the synthesis of aryl-substituted pyrrole from but-2-ene-1,4-diones or but-2-yne-1,4-diones with ammonium formate through amination-cyclization [48]. Different ene-diones or yne-diones were transformed into pyrroles in 56–92% yields in the presence of 5% Pd/C in PEG-200 under MW irradiation for several minutes (Scheme 9.16). It is interesting to note that the steric and electronic effects on the amino group of the alkyl/aryl amines play a role in the rate-determining step during the pyrrole formation.

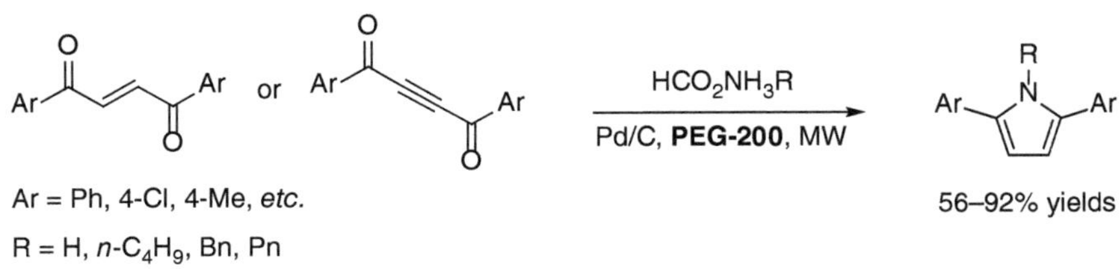

Scheme 9.16 Synthesis of aryl-substituted pyrrole in PEG-200.

In 2013, a general and greener protocol for the synthesis of aryl phosphonates through cross-coupling of aryl halides and diethyl phosphite by using $Pd(PPh_3)_4/Et_3N/PEG-600$ as an efficient catalytic system has been established by Zhang and coworkers [49]. Aryl halides having electron-withdrawing groups gave higher yields compared to those containing electron-donating groups (Scheme 9.17). Furthermore, aryl chlorides were successfully employed to afford the desired product with moderate yields. This procedure was also compatible with the presence of different functional groups such as ester, ketone, aldehyde, nitro, dialkylamine, amides, and hydroxy in the substrates.

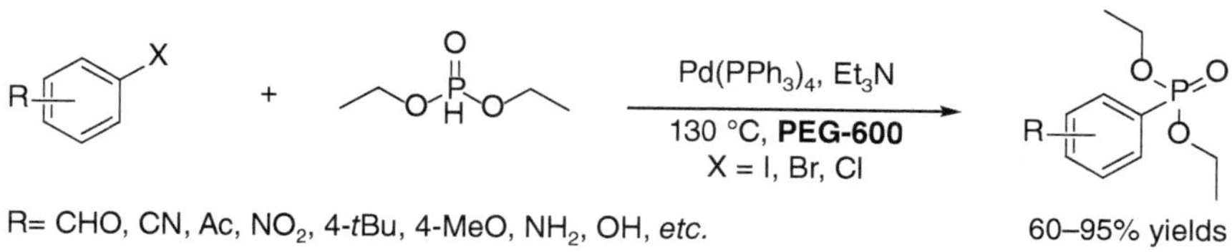

Scheme 9.17 Pd-catalyzed C–P bond formation in PEG-600.

9.2.3 Pd-catalyzed C—O Bond Formation in PEG

In 2007, He and coworkers reported Pd-catalyzed aerobic oxidation of styrene in the supercritical carbon dioxide and poly(ethylene glycol) biphasic system [50]. It is interesting to note that this reaction produced two possible products, benzaldehyde and acetophenone (the Wacker reaction product). The co-catalyst CuCl plays an essential role in controlling the selectivity, from 92% in favor of acetophenone in the presence of CuCl to 85% favoring benzaldehyde without any co-catalyst (Scheme 9.18a). The exact mechanism is not clear but the author suggested a Pd(II)–OOH species-promoted oxygen transfer pathway to terminal alkenes through a five-membered pseudocyclic peroxypalladation intermediate [51]. Later, Chandrasekhar and coworkers further demonstrated the Pd/Cu-catalyzed oxidation of alkynes in PEG-400/H_2O [52]. Different diaryl acetylenes and aryl alkyl acetylenes reacted smoothly and formed the corresponding diketone in 63–87% yields (Scheme 9.18b).

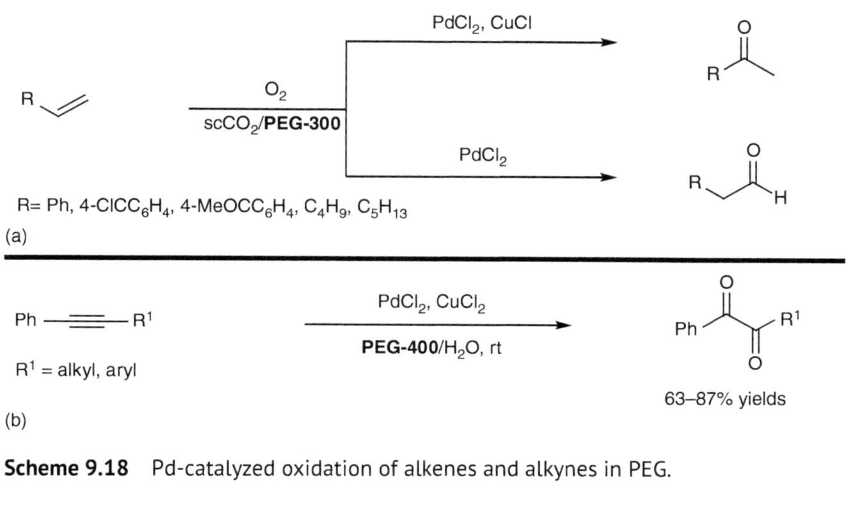

Scheme 9.18 Pd-catalyzed oxidation of alkenes and alkynes in PEG.

9.2.4 Pd-catalyzed C—B Bond Formation in PEG

Aryl boronic acids and their esters are important synthons owing to their wide application in organic reactions and pharmaceutical synthesis. In 2011, an efficient approach for the synthesis of arylboronates through a coupling of aryl halides with bis(pinacolato)diboron in PEG-600 was discovered by Zhang and coworkers [53].

With $Pd(PPh_3)_2Cl_2$/NaOAc/PEG 600 system, a diversity of aryl iodides and bromides proceeded smoothly and provided the corresponding arylboronates in 56–93% yields (Scheme 9.19). In addition, this protocol was also compatible with a wide range of aryl chlorides with electron-withdrawing or electron-donating groups, and the desired products were produced in 56–93% yields.

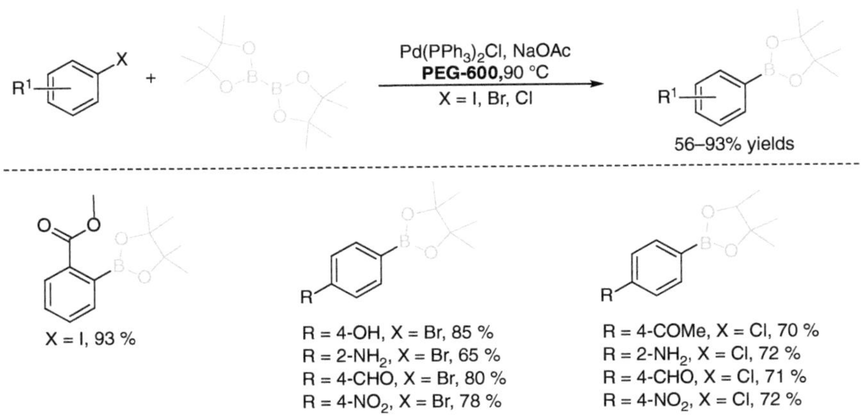

Scheme 9.19 Pd-catalyzed C—B bond formation in PEG-600.

9.3 PEG in Cu-catalyzed Reactions

9.3.1 Cu-catalyzed C—C Bond Formation in PEG

In 2008, copper-catalyzed Suzuki–Miyaura cross-coupling reactions in PEG-400 were reported by Ji, Mao, and coworkers [54]. PEGs acted not only as a recycling solvent but could also accelerate the coupling reactions. Molecular iodine plays a crucial role in the coupling reactions of aryl bromide or chloride with arylboronic acids. Other iodides such as potassium iodide or copper iodide gave poor reactivity. Aryl chlorides only afford the corresponding products in 23–51% yields because of the stronger C—Cl bonds (average energy of $C(SP^2)$–Cl bonds lie in the 370–390 KJ/mol). In addition, the way to recover the catalytic system was very simple, only by adding diethyl ether to the reaction and decanting the upper layer (Scheme 9.20).

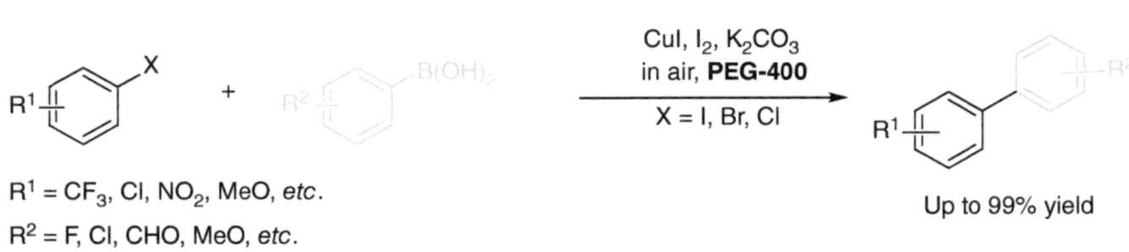

Scheme 9.20 Cu-catalyzed Suzuki–Miyaura cross-coupling in PEG-400.

Leadbeater and coworkers discovered the first examples of metal-free Sonogashira couplings [55]. By using PEG and water as co-solvents, sodium hydroxide as a base, various aryl bromides or iodides can be converted to the corresponding internal alkynes under 170 °C for 6 minutes. Inspired by this pioneering work, Lamaty and coworkers established a practical copper-catalyzed Sonogashira coupling in PEG (PEG; liquid or solid, various molecular weights: 300 < MW < 3400) under MW activation [56]. K_2CO_3 gave the best yield compared with another group 1 metal carbonates or acetates, this result can be rationalized and explained by the fact that PEG can generate a loose complex with large alkali metal ions such as K^+, leaving the basic anion as a free ion in the medium and increasing the base solubility and strength (Scheme 9.21). In addition, the author also investigated the effects of PEG molecular weight, chain end effects, and nature of the associated cations and anions. In 2011, a copper-catalyzed Sonogashira coupling of aryl iodides with terminal acetylenes in water–polyethylene glycol has been reported by Wan and Xie [57]. Both electron-rich and electron-deficient aryl iodides worked well under MW irradiation or conventional thermal heating conditions. Even the hindered *o*-iodotoluene could provide a good yield (90%) within 30 minutes. Besides, alkyl terminal alkyne hept-1-yne was also a suitable reaction partner and produced the corresponding product in good yields within 50 minutes.

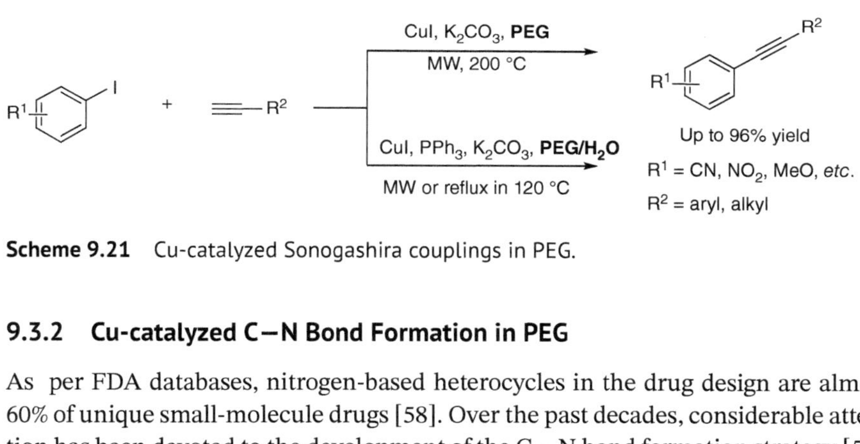

Scheme 9.21 Cu-catalyzed Sonogashira couplings in PEG.

9.3.2 Cu-catalyzed C—N Bond Formation in PEG

As per FDA databases, nitrogen-based heterocycles in the drug design are almost 60% of unique small-molecule drugs [58]. Over the past decades, considerable attention has been devoted to the development of the C—N bond formation strategy [59]. The Buckwald–Hartwig reaction has emerged as an efficient and practical method for the construction of C—N bonds both in academics and industry. The reaction is usually conducted in the presence of palladium or copper catalysts, ligands, and a stoichiometric of bases was added to achieve high reaction efficiency.

In 2010, a new protocol method for the *N*-arylation of aryl halides with anilines using Cu nanoparticles in PEG-400 as an efficient and reusable catalytic system was reported by Kidwai and coworkers [60]. The insoluble Cu NP/PEG-400 phase was reused for three cycles, and there was an approximately 5% weight loss of PEG from every cycle (Scheme 9.22a). Anilines endowed with electron-donating groups displayed greater reactivity in comparison with electron-withdrawing groups such as 4-chloro-, 4-bromo-, and 4-nitroaniline. Replace iodobenzene with bromo- or

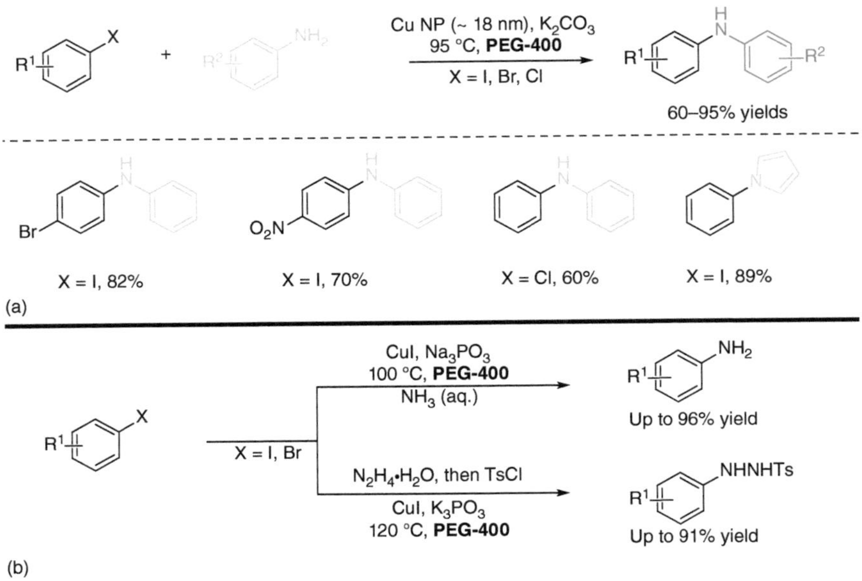

Scheme 9.22 Cu-catalyzed C—N bond formation in PEG-400.

chlorobenzene leads to lower yields. Besides, *NH* heterocycles also proved to be a suitable reaction partner with iodobenzene and provided the corresponding *N*-arylated products in 60–95% yields. In 2011, Chen and coworkers extended the cross-coupling reaction to aryl halides with aqueous ammonia in the CuI/PEG-400 system [61]. Various aryl iodides and bromides were transformed into primary arylamines in 50–96% yields. Aryl chlorides cannot afford the corresponding products under their standard conditions. Two years later, the same group further achieved copper-catalyzed cross-coupling of aryl halides with aqueous hydrazine in PEG-400 (Scheme 9.22b) [62]. Both electron-withdrawing and electron-donating groups substituted aryl iodides and bromides, and even sterically hindered substrates were converted to aryl hydrazines in good yields.

In 2010, Lamaty and coworkers found that heterocycles such as indole and benzimidazole could be *N*-arylated in the presence of cuprous oxide, cesium carbonate, and PEG-3400 under MW conditions [63]. Simple treatment by precipitation in Et_2O and filtration provided the expected product after evaporation and recovery of the catalytic system as a precipitate (Scheme 9.23a). The aryl chlorides showed lower reactivity compared with aryl bromides and iodides. In 2014, selective *N*-arylation was also achieved with CuO/PEG-400/140 °C catalyst system by using 1,3-benzothiazole derivatives [64]. Both aryl iodides and bromides are proven to be compatible reaction partners and give the desired product in excellent yields. It is interesting to note that the *S*-arylation product could also be obtained by CuI/PEG-400/50 °C catalyst system (Scheme 9.23b).

In 2009, Burley, Davies, and coworkers reported Cu-catalyzed *N*-alkynylation of imidazoles, benzimidazoles, indazoles, and pyrazoles using PEG-400 as a green

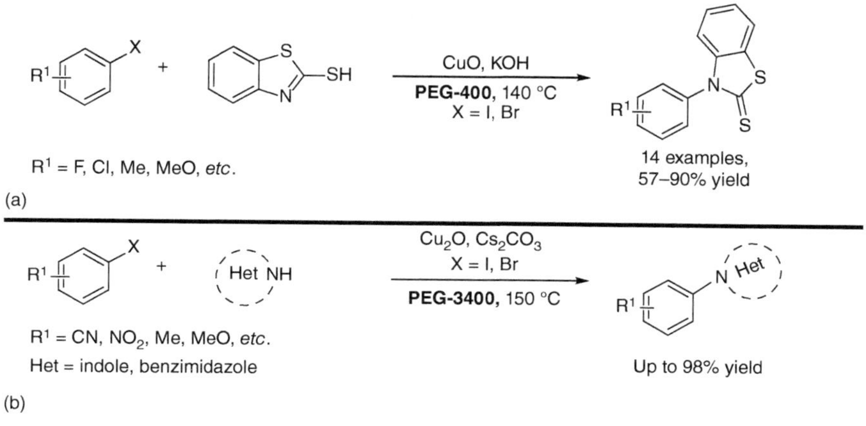

Scheme 9.23 Cu-catalyzed *N*-arylation in PEG.

and sustainable solvent [65]. The reaction utilizes inexpensive heteroaromatic and bromoalkynes to afford either *N*-alkynylated or *N*-bromoalkenylated products in good yield under MW conditions (Scheme 9.24a). Bulky bromoalkynes such as (bromoethynyl)triisopropylsilane exclusively formed *N*-alkynylheteroarenes, whereas (bromoethynyl)benzene formed predominantly *N*-alkynylheteroarene products with non-hindered imidazole and benzimidazole substrates. One year later, a CuI-catalyzed intramolecular *N*-arylation of 2-haloaryldiazonium salts in PEG-400/water has been demonstrated by the group of Mukhopadhyay [51d]. The protocol reacted well for different diazoaminobenzenes of 2-iodo, 2-bromo, and 2-chloro anilines, and gave the corresponding benzotriazoles in 75–92% yields. In 2012, a Cu-catalyzed *N*-arylation of *NH*-sulfoximines with diaryliodonium salts in PEG-400 was achieved by Varma and coworkers [66]. The application of high-intensity ultrasound dramatically reduced the reaction time to 5 minutes (Scheme 9.24b). Various *NH*-sulfoximines showed good reaction effects, while

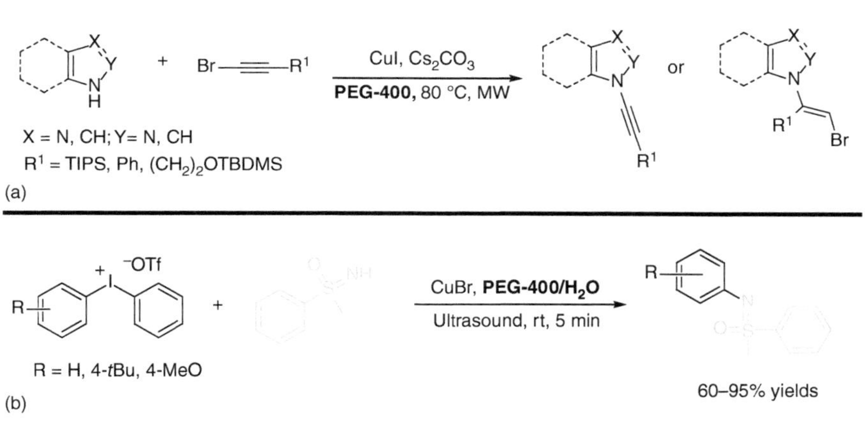

Scheme 9.24 Cu-catalyzed C-N bonds formation in PEG-400.

a significant amount of the *S,S*-diphenylsulfoximine remained unreacted in the standard conditions due to steric hindrance.

Click chemistry is one of the immensely powerful techniques and finds wide applications in drug discovery, bioorthogonal chemistry, nanoparticle modification, and so on, which has also been confirmed by the 2022 Nobel Prize in chemistry [67]. In 2007, Kumaraswamy and coworkers developed a CuI-catalyzed Click reaction of azide and alkyne to form β-hydroxy 1,2,3-triazole using environmentally PEG-400 as a reaction medium [68]. This one-pot epoxide-azide-Cu(I)-catalyzed [3+2] dipolar cycloaddition facilitates the rapid synthesis of small molecules possessing pharmacophores in high yields without racemization (Scheme 9.25a). Billault, Scherrmann, and coworkers also extended the copper-catalyzed azide-alkyne cycloaddition of lipophilic and hydrophilic substrates in molten PEG-2000 [69]. Based on the different solubility of products, the author provided two different purification methods that are, precipitation of the polymer-solvent for the compounds soluble in organic solvents, or ultrafiltration for the water-soluble compounds (Scheme 9.25b).

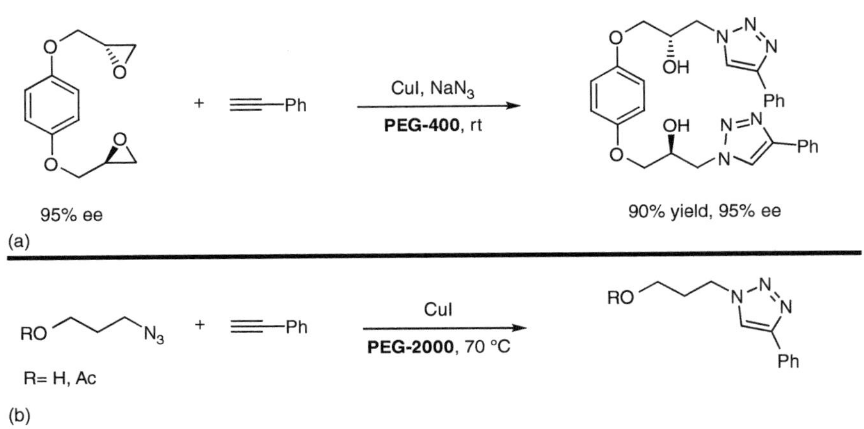

Scheme 9.25 CuI-catalyzed Click reaction of azides and alkynes in PEG.

9.3.3 Cu-catalyzed C—O, C—S, and C—Se Bond Formation in PEG

In 2009, Firouzabadi, Iranpoor, and coworkers developed CuI-catalyzed one-pot thioetherification of aryl halides with alky bromides in wet PEG-200 [70]. The best advantage of this method is located in the avoided use of foul-smelling thiols and the transforming of commercially available alkyl bromides into high-value arylthio ethers under mild conditions. The reaction showed good functional group tolerance, both aromatic and heteroaromatic halides or tosylates reacted well and formed the target products in up to 91% yields [71]. Respecting the possible mechanism, they proposed that the reaction begins with the *in situ* generations of the *S*-alkylisothiouronium salt, followed by hydrolyzation to produce the thiolate anion and urea. Then the generated thiolate anion reacts with Cu(I) to produce Cu(SR), which undergoes oxidative addition and reductive elimination with aryl halides giving the final products (Scheme 9.26). Except for thiourea, carbon disulfide is also used as a sulfur source in the same system and produced corresponding alky thioethers and diaryl disulfides in reasonably good-to-excellent yields by the same group [72].

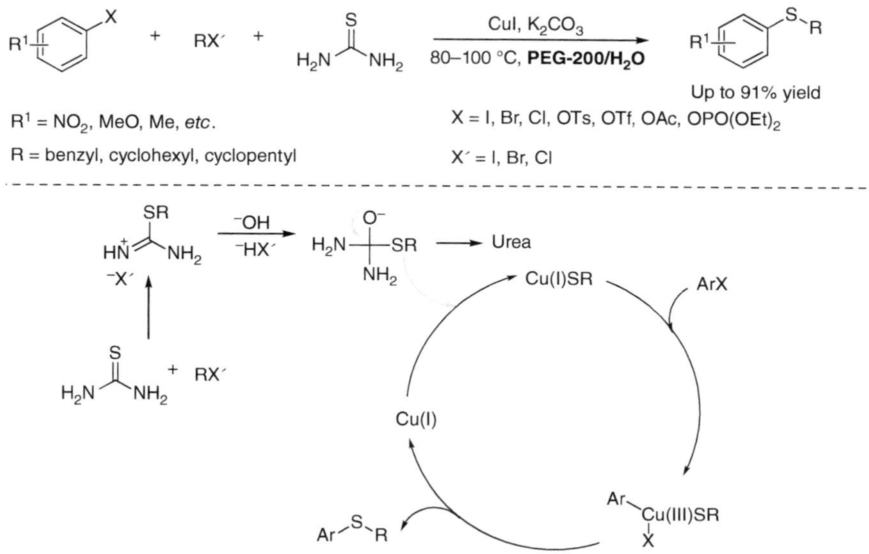

Scheme 9.26 CuI-catalyzed thioetherification of aryl halides with alky bromides.

Another Cu-catalyzed C—S bond formation in PEG was demonstrated by Han and coworkers [73]. They provided facile access to 2-(arylthio)anilines and 2-phenoxyanilines through a Cu-catalyzed ring-opening arylation of benzazoles with aryl iodides. TEM confirmed that the active catalyst results from the *in situ* generations of copper nanoparticles (Scheme 9.27). Besides, the system can be recycled and reused up to six times. The possible reaction mechanism may be initiated with the deprotonation of benzazole under the assistance of a base and subsequently hydrolyzed to a ring-opening intermediate. Then a traditional Cu-catalyzed cross-coupling with aryl iodides occurred to furnish the final product.

Diaryl selenides act as versatile reagents in organic synthesis and display various important biological activities. Traditional methods for the synthesis of C—Se bonds rely on the reaction of metal selenides or selenocyanates with aryl halides. The synthesis of diaryl/aryl–vinyl selenides by the Cu nanoparticle was widely investigated by the group of Ranu [74]. For example, an efficient protocol has been demonstrated for the construction of organochalcogenides through a coupling of organoboronic acids and dichalcogenides catalyzed by $CuFe_2O_4$ nanoparticles in PEG-400 [75]. In 2013, the same group also reported alumina-supported Cu-catalyzed selenylation of aryl iodides and aryl bromides with diselenides in the

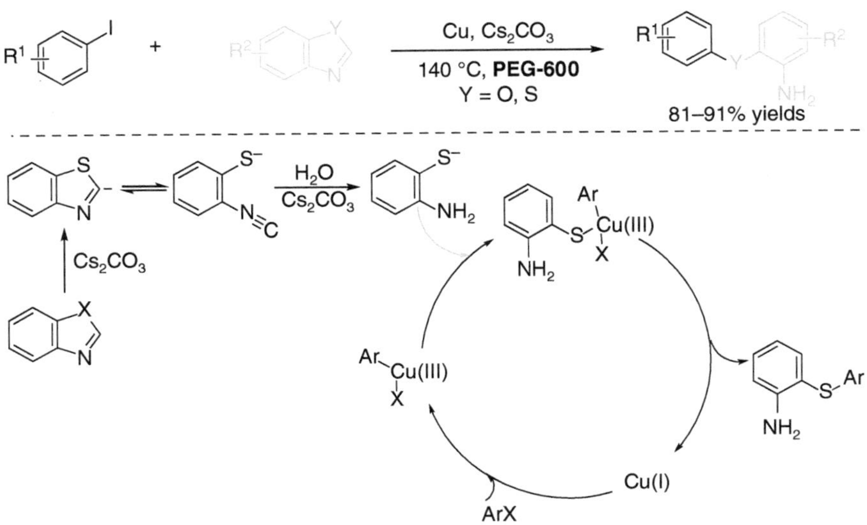

Scheme 9.27 Cu-catalyzed ring-opening arylation of benzazoles with aryl iodides.

green solvent PEG-600 (Scheme 9.28a) [76]. A wide range of substituted aryl- and heteroaryl halides proceeded smoothly in the standard conditions and furnished the corresponding product in 71–91% yields. In addition, the reaction is highly chemoselective as Br on the phenyl ring of aryl iodides remained inert during coupling with diphenyl diselenide. In 2012, Nageswar and coworkers developed an efficient $CuFe_2O_4$ nanoparticle-catalyzed reaction of organoboranes with PhSeBr in PEG-400 for the synthesis of diverse unsymmetrical diaryl selenides (Scheme 9.28b) [77]. The copper ferrite nanoparticles were magnetically separated, recycled, and reused for up to four cycles. It is interesting to note that halo-substituted aryl boronic acids exhibited good chemoselectivity and even active aromatic C–Br bonds can be retained in the standard conditions.

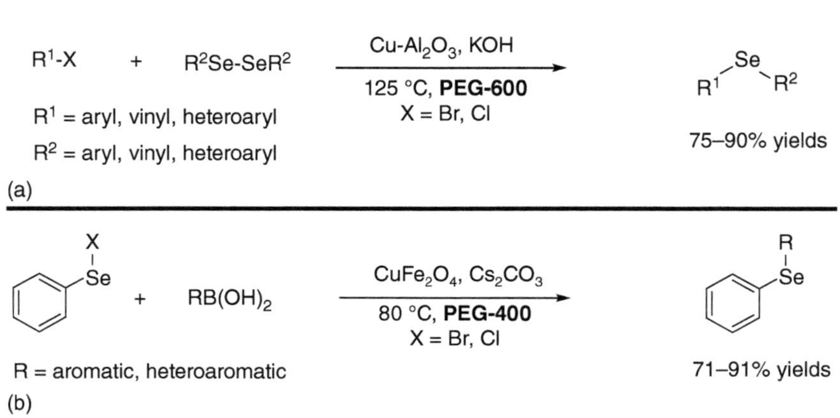

Scheme 9.28 Cu-catalyzed C–Se bonds formation in PEG.

9.4 PEG in Ni, Ru, and Pt-catalyzed Reactions

In 2018, Li and coworkers reported a one-pot approach for the reduction of alkenes using $NaBH_4$–$NiCl_2{\cdot}6H_2O$ in EtOH/PEG-400 under 30 °C [78]. A wide range of alkenes (including trisubstituted alkene α-pinene) were reduced to corresponding alkanes in 91–100% yields (Scheme 9.29a). Key to the success was the hydrogen source for the reduction can be sufficiently derived from the B–H of $NaBH_4$ and the O–H of ethanol. One year later, the same group also demonstrated the reduction of α-pinene by $NaBH_4$ using $NiCl_2{\cdot}6H_2O$ in the PEG-800/ethanol system under room temperature (Scheme 9.29b) [79]. α-Pinene could be transformed into *cis*-pinane with 97% conversion and 98% selectivity. Whereas the reaction proceeded sluggishly in higher molecular weight PEGs, including molecular weights of 1000–4000. Based on the TEM and poisoning experiments, the Ni nanoparticle generated *in situ* probably acts as a reservoir for active homogeneous catalyst species. In 2015, the Ni-catalyzed Sonogashira coupling of aryl iodides with alkynes in PEG-400 was established by the group of Cai and coworkers [80]. In the presence of $NiCl_2(PPh_3)_2$ and CuI, diverse arylacetylenes were formed in 24–95% yields, and the catalytic system could be recycled and reused six times without significant loss of activity (Scheme 9.29c).

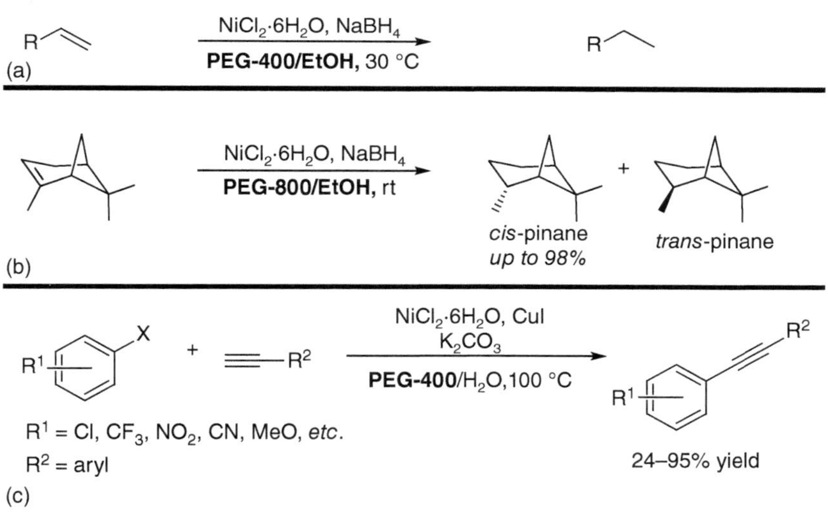

Scheme 9.29 Ni-catalyzed reduction and cross-coupling in PEG.

In 2015, Cai and coworkers established the Ru-catalyzed cross-dehydrogenative C—H bond alkenylation reaction between benzoic acids and alkenes in PEG-400/H_2O. The reaction proceeded smoothly at 80 °C using $Cu(OAc)_2{\cdot}H_2O$ as oxidant, producing a variety of phthalide derivatives in 42–93% (Scheme 9.30a) [51c]. The catalytic system could also be recycled and reused six times without any loss of activity. The possible reaction mechanism might be initiated with the *in situ* generations of ruthenium(II) diacetate complex, which coordinates with the

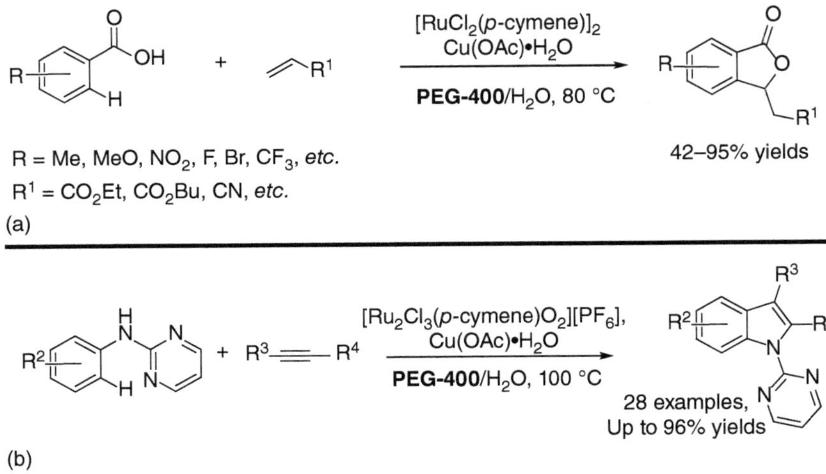

Scheme 9.30 Ru-catalyzed C—H bond activation in PEG-400.

carboxyl oxygen of benzoic acid. Subsequent *ortho*-C—H bond ruthenation to form a ruthenacycle intermediate, alkene insertion, and β-hydride elimination successively occur to produce an olefinated intermediate. The latter undergoes an intramolecular oxa-Michael addition reaction to afford the desired phthalide. Two years later, the same group further extended the Ru-catalyzed oxidative annulation of alkynes with *N*-2-pyrimidylanilines for the synthesis of substituted indoles [81]. A range of valuable functional groups, including methyl, methoxy, trifluoromethyl, hydroxy, ketone, and aldehyde were tolerated well. Both symmetric and non-symmetric internal alkynes furnished the desired indoles in highly regioselective (Scheme 9.30b).

In 2016, Bhanage and coworkers achieved the convergent synthesis of isoquinolinones, isocoumarins, *N*-methyl isoquinolinones, and olefination of the Weinreb amides through C—H bond activation by employing Ru(II)/PEG-400 as green and recyclable media [82]. Various C—C, C—N, and C—O bonds are constructed in high regioselective and stereoselective through the cleavage of C—H, N—H, O—H, and N—O bonds. The reaction tolerates a wide range of functional groups such as Br, Cl, NO_2, and pyridine, and produces the desired compounds with up to 94% yield (Scheme 9.31).

The asymmetric transfer hydrogenation of ketones provides an attractive alternative method to asymmetric hydrogenation for the preparation of chiral alcohols because of its operational simplicity and the easy availability of hydrogen sources [83]. In 2007, Fan, Gu, Chan, and coworkers discovered the asymmetric transfer hydrogenation of nonfunctionalized aromatic ketones catalyzed by Ru-TsDPEN catalyst in a mixture of PEG and water with HCOONa as the reductant [84]. The unmodified catalyst could be easily recovered after extraction of the reduced product with a less polar solvent such as hexane and was reused at least 14 times without obvious loss in enantioselectivity. Besides, the functional groups on the *ortho* position in the acetophenones influenced the enantioselectivity

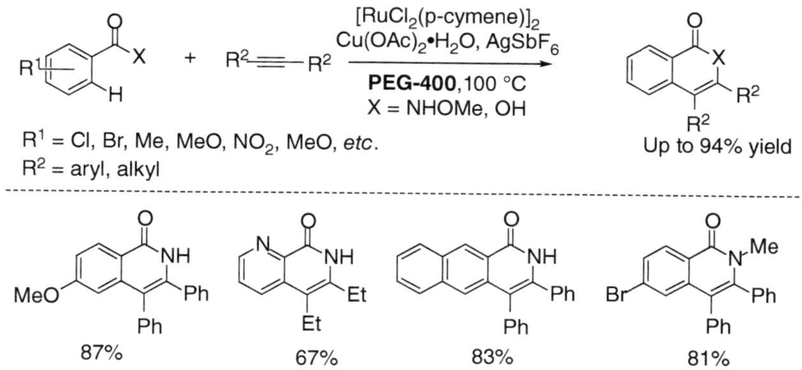

Scheme 9.31 Ru-catalyzed C–H activation/cyclization in PEG-400.

greatly. The electron-donating substrates gave low enantioselectivities for example *o*-methoxylacetophenone afforded the corresponding alcohols with only 81% ee (Scheme 9.32).

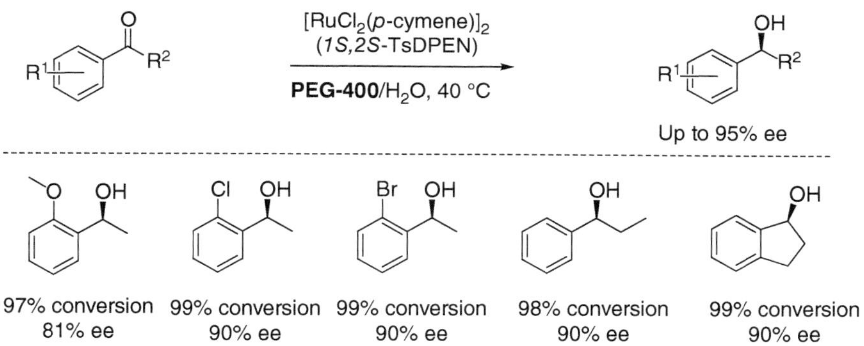

Scheme 9.32 Ru-catalyzed asymmetric hydrogenation of ketones in PEG-400.

Adams' catalyst (PtO_2) in PEG-400 has been found to be a superior catalytic system in promoting the hydrogenation of alkenes and nitro groups [85]. At atmospheric pressure H_2, a variety of alkenes and nitro groups can be reduced to the desired product in 60–97% yields (Scheme 9.33a). For example, the sugar derivative, having a hydrogenation-sensitive benzyl ether was effectively saturated in greater than 90% yields in PEG. The recyclability of both PEG as the solvent and PtO_2 as the catalyst can be reached to 10 runs and substrate cross-contamination was not observed despite a change in the substrate four times. Lamaty and coworkers developed the unprecedented Pt-catalyzed 5-endo-dig cycloisomerization of readily available α-amino ynones under MW irradiation in PEG-3400 [86]. A wide range of pyrrolin-4-ones was obtained in 80–98% yields directly through straightforward precipitation–filtration workup (Scheme 9.33b).

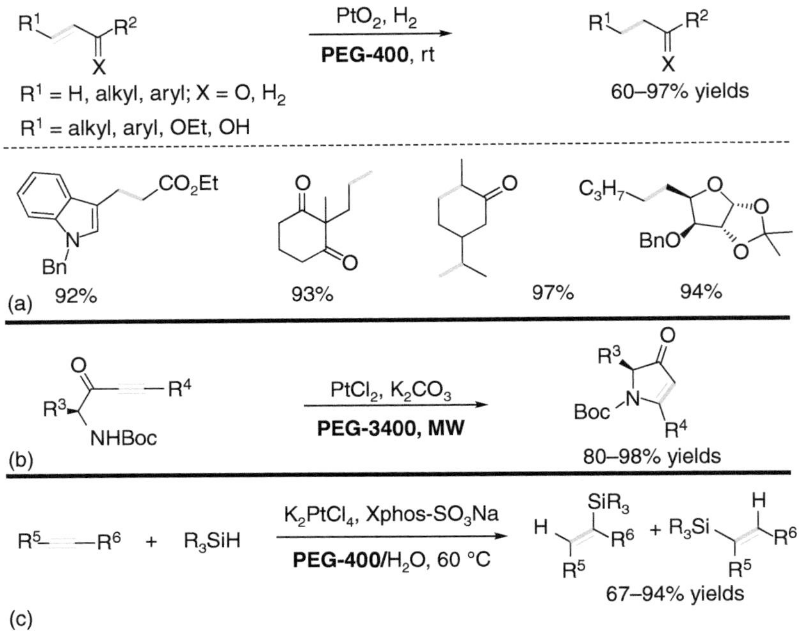

Scheme 9.33 Pt-catalyzed reduction and cyclization in PEG.

Vinylsilanes represent a type of special and useful intermediates that can undergo several valuable organic transformations such as Hiyama cross-couplings [87]. In 2017, Cai and coworkers demonstrated a highly regio- and stereoselective K_2PtCl_4-catalyzed hydrosilylation of terminal alkynes with hydrosilanes [88]. Different aromatic and alkyl terminal alkynes worked well and resulted in the corresponding vinylsilanes in 67–94% yields with a total β-(*E*)-selectivity (Scheme 9.33c). Furthermore, hydrosilylation of aromatic or aliphatic symmetrical internal alkynes could undergo the *cis*-hydrosilylation smoothly to give exclusively (*E*)-vinylsilanes in 86–96% yields. The isolation of the products is readily performed by the extraction with cyclohexane and K_2PtCl_4 and Xphos–SO_3Na in PEG-400/H_2O system and could be easily recycled and reused at least eight times without any loss of catalytic activity.

9.5 PEG in Organocatalysis Reactions

Over the past decades, organocatalysis, the use of small organic molecules as catalysts, has emerged as an important and powerful approach for the synthesis of various value-added molecules [89]. Fantin, Massi, and coworkers have demonstrated that 1,2-diketones may serve as acyl anion equivalents in organocatalyst NHC-promoted α-diketone homo-couplings [90]. PEG-400 was used as an effective and reusable reaction medium for this transformation (Scheme 9.34). When the

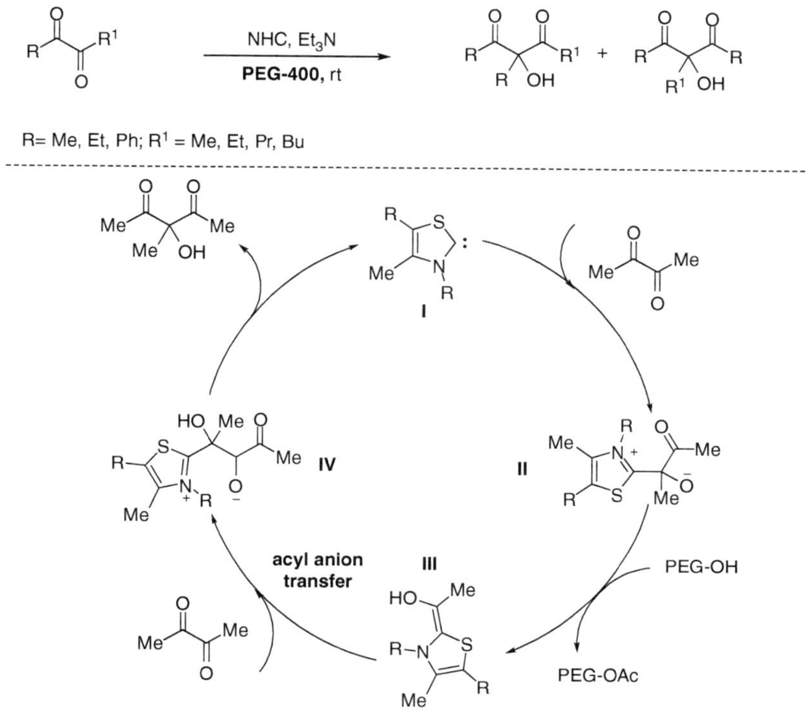

Scheme 9.34 NHC-catalyzed homocoupling of α-diketones in PEG-400.

protocol was extended to the unsymmetrically substituted dialkyl α-diketones, it was observed that acetyl anion transfer ($MeCO^-$) was predominant over the migration of its higher carbanion counterpart. The mechanism that accounts for all these findings may begin with the formation of the intermediate **II** resulting from the addition of the thiazolin-2-ylidene **I** to the α-diketone, and its evolution to the Breslow intermediate **III** by the attack of the PEG solvent to the carbonyl of **II**. Subsequent addition of the Breslow intermediate to the acceptor, that is, the second molecule of α-diketone, leads to the formation of the final product and regeneration of the catalyst **I**.

In 2005, the L-proline-catalyzed direct asymmetric aldol reaction of acetone with various aldehydes in PEG-400 was described by the group of Chandrasekhar and coworkers [91]. Diverse aromatic and alkyl aldehydes worked well in these conditions and produced chiral β-hydroxy ketones in good yields and enantioselectivities. The reusability of this solvent was studied over 10 runs without loss of activity of either the catalyst or the solvent. Shang and coworkers found that aromatic aldehydes with active methylene compounds proceeded Knoevenagel condensation efficiently in PEG-400 at room temperature with L-proline as the catalyst [92]. The presence of L-proline plays a vital role in the formation of *E*-isomer because of steric effects. Meanwhile, it requires a longer time and gives a lower yield without L-proline (Scheme 9.35a). A highly stereoselective organocatalytic Michael addition of aldehyde to *trans*-β-nitrostyrene using PEG-400 as a recyclable solvent

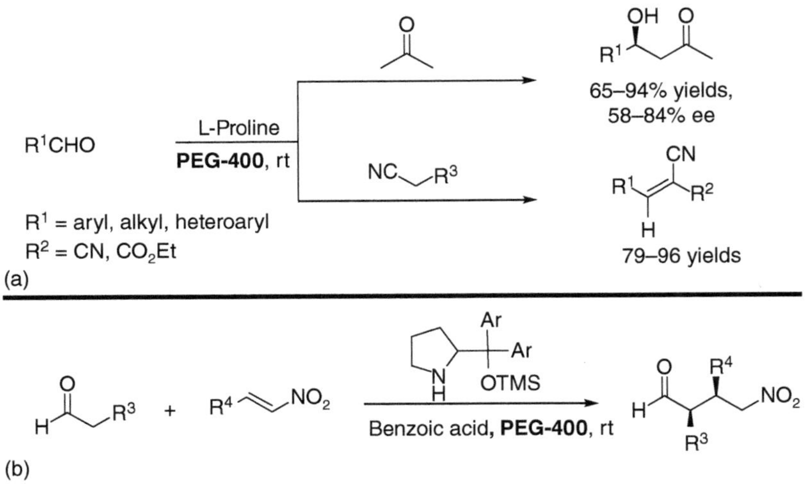

Scheme 9.35 Asymmetric organocatalytic Aldol condensation and Michael addition reactions.

medium was discovered by the group of Paixão and coworkers [93]. The key to the success of this reaction was the use of diaryl prolinol silyl ether organocatalyst. The author suggested that hydrophobic alkyl side chains greatly influence the reaction efficiency, with the better organocatalysts being the ones that possess ethyl and hexyl chains in their structure (Scheme 9.35b).

9.6 PEG in Multicomponent Reactions

Multicomponent reactions (MCRs) are attractive strategies for the selective construction of complex organic molecules from three or more readily accessible starting materials, where basically most of the atoms contribute to the newly formed product [94]. In 2007, Sain and coworkers established PEG-promoted solvent and catalyst-free synthesis of 3,4-dihydropyrimidinones at 100 °C [95]. A wide range of aldehydes, β-dicarbonyl compounds, and ureas were involved in PEG-400 under standard neutral conditions to give corresponding 3,4-dihydropyrimidinones in 86–98% yields. The exact role played by PEG in this reaction remains unclear, and no product was obtained when the reaction was conducted in the presence of acetonitrile or toluene without PEG-400 (Scheme 9.36).

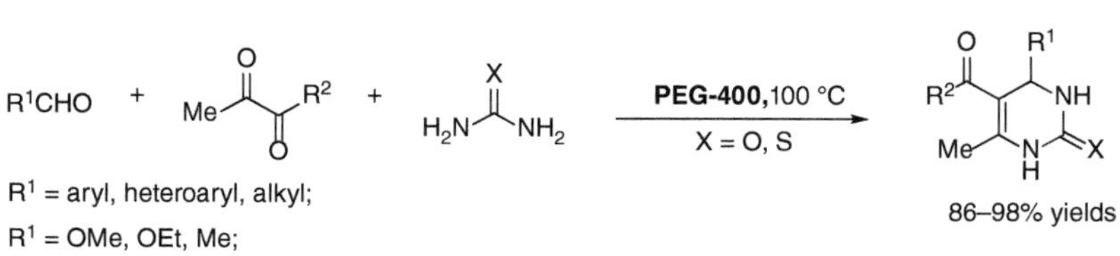

Scheme 9.36 Synthesis of 3,4-dihydropyrimidinones in PEG-400.

Coumarins and isatin widely exist in many natural products and exhibit various potent biological activities [96]. In 2012, Shafiee and coworkers achieved a straightforward synthesis of functionalized furo[3,2-*c*]coumarins via a one-pot oxidative pseudo three-component condensation of aldehydes and 4-hydroxycoumarin in PEG-300 [97]. With a mixture of I_2 and $K_2S_2O_8$ as an oxidative reagent, different aryl or alkyl aldehydes were transformed into the corresponding furocoumarins in 42–75% yields (Scheme 9.37a). Biscoumarin could act as the key intermediate, which undergoes sequential nucleophilic ring-opening, decarboxylation, iodination, cyclization, and oxidation to produce the target product. In 2017, Jeong and coworkers reported a method for the synthesis of functionalized benzo[*d*]imidazo[2,1-*b*]thiazol-3-yl)-2H-chromen-2-one derivatives from 4-hydroxycoumarin, 2-aminobenzothiazole, and arylglyoxal monohydrate in ethanol/PEG-600 [98]. This reaction provides a novel method for the efficient incorporation of two biologically important scaffolds. In 2017, the nickel-catalyzed one-pot three-component synthesis of pyrazolophthalazinyl spirooxindoles from isatin, malononitrile or cyanoacetic ester, and phthalhydrazide was demonstrated by Zhang and coworkers. A broad range of nitrogen heterocycles was obtained in 82–95% yields with PEG-600 as a green solvent (Scheme 9.37b) [99]. Besides, isatins were also employed as reaction partners for the synthesis of diverse heterocycles in PEG media by the group of Modugu [100] and Hasaninejad [101].

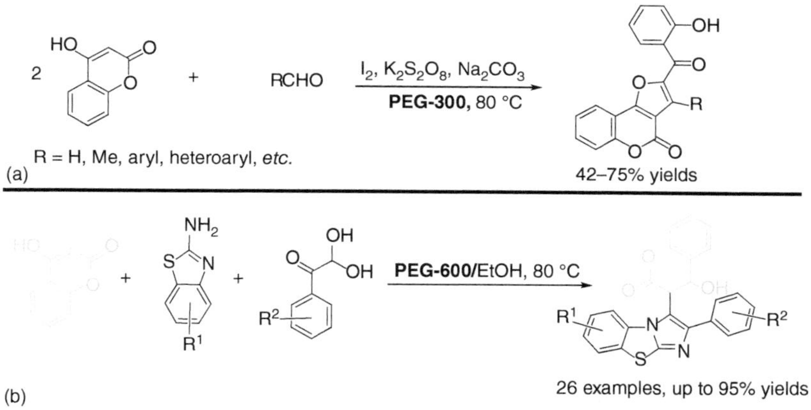

Scheme 9.37 Multicomponent reactions of hydroxycoumarin in PEG.

Cerium(IV) ammonium nitrate (CAN) has emerged as an important reagent for the synthesis of carbon–carbon, and carbon–heteroatom bonds based on its electron transfer capacity and Lewis acidic property. In 2008, CAN in PEG was used as an efficient and recyclable solvent system for a one-pot, three-component Mannich reaction of acetophenone with aromatic aldehydes and aromatic amines [102]. In the presence of 5 mol% CAN, diverse β-amino carbonyl compounds were formed in 98–98% yields (Scheme 9.38). Electron-donating group-substituted amines showed better reaction activity compared with the electron-withdrawing group-substituted

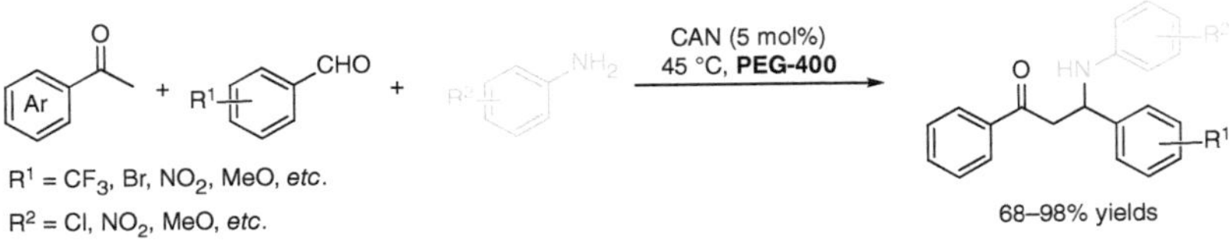

Scheme 9.38 CAN-catalyzed Mannich reaction in PEG-400.

ones. By replacing acetophenone with cyclohexadione, the same group also achieved the synthesis of *N*-substituted decahydroacridine-1,8-diones in PEG-400 [103].

In 2014, Jeong and coworkers developed an eco-friendly method for the synthesis of benzo[4,5]imidazo[1,2-*a*]-pyrimido[4,5-*d*]pyrimidin-4(1*H*)-ones under catalyst-free-conditions in PEG-400 [104]. According to their proposed mechanism, PEG-400 acts as an acidic catalyst and enhances the rate of reaction to complete the conversion of reactants at room temperature. Various aryl and heteroaryl aldehydes, 2-cyanoacetamide, and 1*H*-benzo[*d*]imidazol-2-amine reacted well and gave the corresponding products in 88–94% yields. When 2-cyanoacetamide was replaced with malononitrile, the synthesis of pyrimido[1,2-*a*]benzimidazole derivatives could be achieved by using PEG-400 as a promoting reaction medium in water under reflux conditions [105]. A diversity of aryl aldehydes, 2-aminobenzimidazole, and malononitrile proceeded efficiently under these conditions and yielded the corresponding pyrimido[1,2-*a*]benzimidazole in high selectivity (Scheme 9.39).

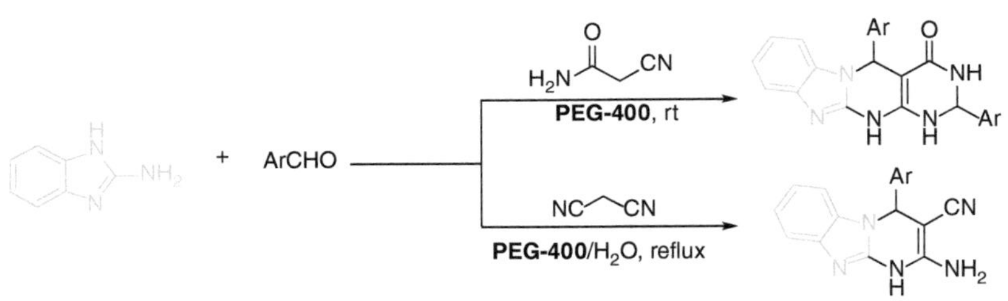

Scheme 9.39 Multicomponent reactions 1*H*-benzo[*d*]imidazol-2-amine in PEG-400.

9.7 PEG in Cyclization Reactions

9.7.1 Synthesis of Five-membered Ring Systems

Thiazoles, a unique heterocycle containing sulfur and nitrogen atoms, occupy an important place in the field of agriculture and medicinal chemistry [106]. In 2008, Chen, Wu, and coworkers reported a catalyst-free protocol for the synthesis of di- and trisubstituted thiazoles from α-haloketones and thioureas/thioamides in PEG-400 [107]. A wide range of α-bromoacetophenones bearing electron-donating or electron-withdrawing groups on the aromatic ring could be converted to thiazoles in 88–98% yields. Later, the group of Mane [108] and Chen [109] also demonstrated the construction of thiazole heterocycles in PEG-400 by employing acetophenones and α-tosyloxyketones as reaction partners. Both two methods tolerated a broad range of functional groups and provided corresponding products in good-to-excellent yields (Scheme 9.40).

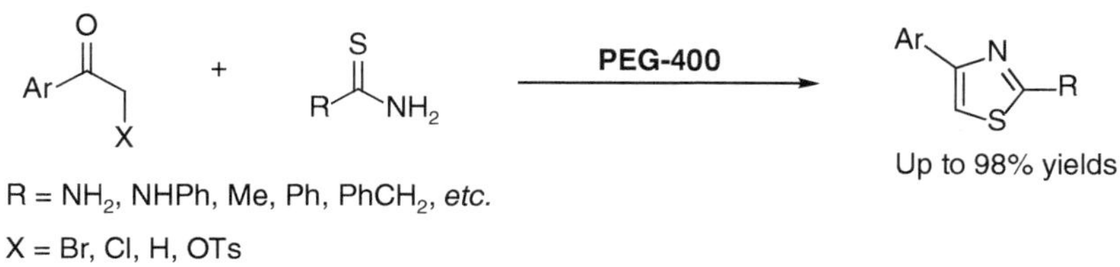

Scheme 9.40 Synthesis of thiazoles in PEG-400.

In 2009, Kaloyanova and coworkers reported an improved green protocol for the construction of 2-arylbenzothiazoles through condensation of equivalent amounts of 2,2′-diaminodiphenyldisulfides or 2-aminothiophenols with aromatic aldehydes in PEG 200/400 under MW irradiation [110]. Different aryl and heteroaryl aldehydes participated well under the standard conditions and produced corresponding benzothiazoles in 58–97% yields. In the case of the 2-hydroxy-1-naphthaldehyde, better yields were achieved by adding DMSO as an oxidant. Furthermore, the same benzothiazoles were synthesized through the condensation of bis(2-aminophenyl) disulfides with aryl aldehydes, catalyzed by NaSH in PEG-300 under low-energy MW irradiation (Scheme 9.41a) [111]. The PEG-400 also proved to be a very efficient green promoter for the construction of a wide variety of 2-substituted benzimidazoles and particularly bis-benzimidazoles under catalyst-free and solvent-less conditions at 110 °C (or 140 °C) [112]. A possible mechanism may include the formation of an imine intermediate, and cyclization followed by oxidation and dehydration leads to the desired product (Scheme 9.41b). Various aryl aldehydes and phenylenediamines are involved in the reaction smoothly, and benzimidazoles could be obtained by direct recrystallization from hot methanol.

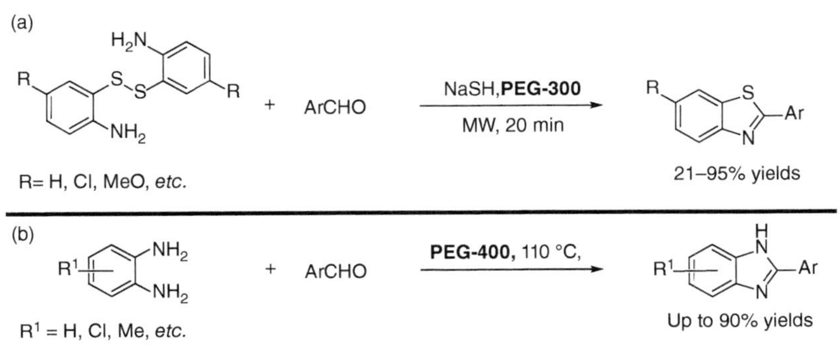

Scheme 9.41 Synthesis of benzothiazoles and benzimidazoles in PEG.

In 2014, the synthesis of 3,4,5-trisubstituted 1,2,4-triazoles and *N*-fused 1,2,4-triazoles via CAN-catalyzed oxidative cyclization of amidrazones and aldehydes in PEG-300 was discovered by the group of Vidavalur [113]. CAN acts as both a Lewis acid and an oxidant for this oxidative cyclization reaction, and both aryl and heteroaryl aldehydes participated well under these conditions. The easy recyclability of the reaction solvent makes the process economical and potentially viable for commercial applications (Scheme 9.42).

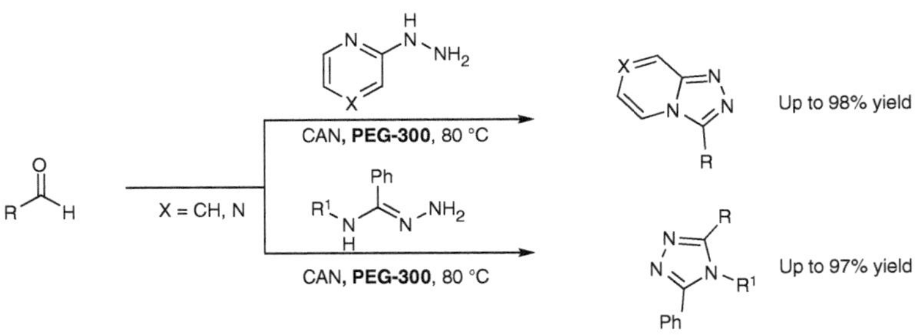

Scheme 9.42 CAN-catalyzed oxidative cyclization in PEG-300.

9.7.2 Synthesis of Six and Seven-membered Ring Systems

Quinoxalines are very important compounds because of their wide spectrum of biological activities [114]. PEG-400 was proved to be an efficient catalyst for the construction of six-membered quinoxaline derivatives through condensation of 1,2-diamines with 1,2-dicarbonyl compounds [115]. The substrates bearing electron-withdrawing groups gave lower yields than the electron-donating groups, and alicyclic 1,2-diamines could also proceed smoothly under these conditions (Scheme 9.43a). He and coworkers discovered a rapid procedure for the construction of triarylpyridine compounds [116]. Under MW irradiation, a wide range

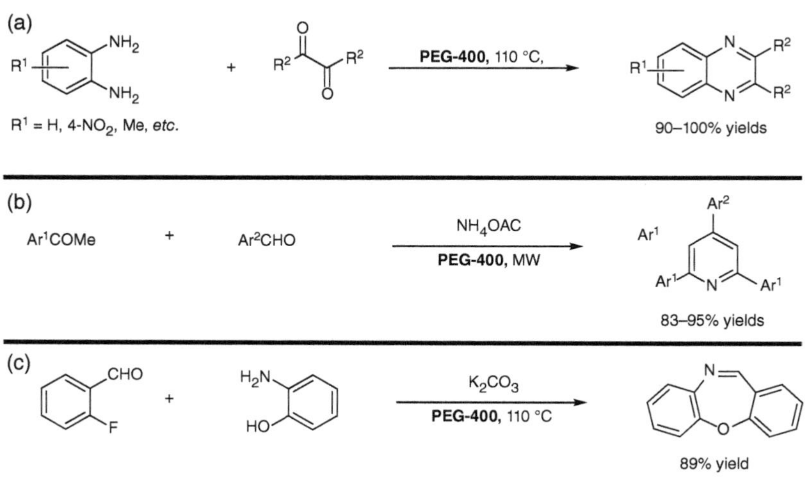

Scheme 9.43 Synthesis of six and seven-membered heterocycles in PEG-400.

of aryl aldehydes, aryl ketones, and ammonium acetate could be transformed into the corresponding triarylpyridine in 83–95% yields within 5–7 minutes (Scheme 9.43b). The electronic effect did not show any significant difference in the product yields, whereas the substituted acetophenones slightly affected product yields. Jorapur and coworkers achieved the construction of seven-membered dibenz[*b*,*f*]-1,4-oxazepine [117]. PEG-400 acts as both a phase transfer catalyst and a clean solvent by significantly enhancing intramolecular cyclization (Scheme 9.43c).

9.8 Conclusion

In summary, the recent applications of PEG in diverse organic transformations have been systematically covered and analyzed in this chapter. Owing to their nontoxicity, low cost, and recyclability, various molecular weights PEGs were largely employed as green and sustainable reaction solvents. For example, a wide range of transition metals (Pd, Cu, Ni, Pt, and Ru) catalyzed cross-coupling reactions, organocatalysis, multicomponent, and cyclization reactions proceeded smoothly in the presence of PEG. A broad range of chemical bonds, including C—C, C—Si, C—N, C—P, C—O, C—S, C—Se, and C—B bonds were constructed in high yields and selectivities. Besides, with two hydroxyl groups, PEG shows some reducibilities and could accelerate the *in situ* generations of nanoparticles, which generally allows reactions performed in open-air, water, and mild conditions. Moreover, heterogeneous catalysis with PEG as a reaction solvent has attracted considerable attention during the past two decades. In these reactions, the desired product usually can be separated via diethyl ether extraction, and the catalyst-containing PEG phase can be recycled and reused several times without significant loss of reactivities. Undoubtedly, the applications of PEG as a green solvent are increasingly recognized as a hot research topic, which also bridges the gap between academics and industries through heterogeneous catalysis.

Nevertheless, a systematic and deeper understanding of how PEG contributes to chemical synthesis is still essential for achieving further broader applications. First, the interactions and mechanistic details between PEG and the catalyst are unclear. Most of the studies proposed *in situ* generated nanoparticles as the catalytic center, but how PEG merges homogeneous and heterogeneous catalysis is less investigated. Second, considerable research focused on the use of transition metals palladium and copper, and other cost-effective and easily accessible transition metals such as iron, magnesium, and cobalt should be paid particular attention in the near future. Third, analyzing the phase behavior and solvation properties of various PEGs could give more inspiration on how to improve catalyst turnover and recovery. Lastly, there is relatively little literature on transition-metal catalyzed and organic small molecule-catalyzed asymmetric transformations in PEG. The results of these challenging problems will surely advance the field of PEG as a green solvent and motivate academic scientists to engage in this interesting research area.

Acknowledgments

We acknowledge the financial support from the Senior Talent Foundation of Jiangsu University (Grant No. 4111290008), the China Postdoctoral Science Foundation (Grant No. 2022M721379) and the Natural Science Foundation of the Jiangsu Higher Education Institutions of China (23KJB150007). We also appreciate the general support from Professor Yuan Wei and Professor Weidong Shi at Jiangsu University.

References

1 (a) E. Buncel, R. A. Stairs, H. J. C. Wilson (2003). *The Role of the Solvent in Chemical Reactions*. Oxford University Press; (b) Burrows, C.J., Harper, J.B., Sander, W., and Tantillo, D.J. (2022). *J. Org. Chem.* 87: 1599–1601.

2 (a) Constable, D.J.C., Jimenez-Gonzalez, C., and Henderson, R.K. (2007). *Org. Process Res. Dev.* 11: 133–137. (b) Sahoo, T., Panda, J., Sahu, J. et al. (2020). *Curr. Org. Synth.* 17: 426–439.

3 Soni, J., Sahiba, N., Sethiya, A., and Agarwal, S. (2020). *J. Mol. Liq.* 315.

4 (a) Veronese, F.M. and Pasut, G. (2005). *Drug Discovery Today* 10: 1451–1458. (b) Jain, A. and Jain, S.K. (2008). *Crit. Rev. Ther. Drug* 25: 403–447. (c) He, Z.-Y., Chu, B.-Y., Wei, X.-W. et al. (2014). *Int. J. Pharm.* 469: 168–178. (d) D'Souza, A.A. and Shegokar, R. (2016). *Expert Opin. Drug Delivery* 13: 1257–1275. (e) Turecek, P.L., Bossard, M.J., Schoetens, F., and Ivens, I.A. (2016). *J. Pharm. Sci.* 105: 460–475. (f) Armengol, E.S., Unterweger, A., and Laffleur, F. (2022). *Drug Dev. Ind. Pharm.* 48: 129–139.

5 (a) Milla, P., Dosio, F., and Cattel, L. (2012). *Curr. Drug Metab.* 13: 105–119. (b) Mishra, P., Nayak, B., and Dey, R.K. (2016). *Asian. J. Pharm. Sci.* 11: 337–348. (c) Fang, Y., Xue, J., Gao, S. et al. (2017). *Drug Delivery* 24: 22–32.

6 (a) Chen, J., Spear, S.K., Huddleston, J.G., and Rogers, R.D. (2005). *Green Chem.* 7: 64–82. (b) Zhou, H.-F., Fan, Q.-H., Tang, W.-J. et al. (2006). *Adv. Synth. Catal.* 348: 2172–2182. (c) Gu, Y. (2012). *Green Chem.* 14: 2091–2128. (d) Cecchini, M.M., Charnay, C., De Angelis, F. et al. (2014). *ChemSusChem* 7: 45–65. (e) Vafaeezadeh, M. and Hashemi, M.M. (2015). *J. Mol. Liq.* 207: 73–79. (f) Pires, M.J.D., Purificacao, S.I., Sofia Santos, A., and Marques, M.M.B. (2017). *Synthesis-Stuttgart* 49: 2337–2350. (g) Xiao, L., Dai, F., Li, Z. et al. (2019). *Chin. J. Org. Chem.* 39: 648–660. (h) Campos, J.F. and Berteina-Raboin, S. (2020). *Catalysts* 10. (i) Kardooni, R. and Kiasat, R.A. (2020). *Curr. Org. Chem.* 24: 1275–1314. (j) Hoffmann, M.M. (2022). *Curr. Opin. Colloid Interface Sci.* 57.

7 (a) Wu, C., Peng, J., Li, J. et al. (2008). *Catal. Commun.* 10: 248–250. (b) Zacchigna, M., Cateni, F., Drioli, S., and Bonora, G.M. (2011). *Polymers* 3: 1076–1090. (c) Puskas, J.E., Seo, K.S., and Sen, M.Y. (2011). *Eur. Polym. J.* 47: 524–534.

8 Chandrasekhar, S., Narsihmulu, C., Sultana, S.S., and Reddy, N.R. (2002). *Org. Lett.* 4: 4399–4401.

9 (a) Larhed, M., Moberg, C., and Hallberg, A. (2002). *Acc. Chem. Res.* 35: 717–727. (b) Nilsson, P., Ofsson, K., and Larhed, M. (2006). *Microwave Methods in Organic Synthesis*, vol. 266 (ed. M. Larhed and K. Olofsson), 103–144. (c) Xiao, S., Zhu, J., Mu, X., and Li, Z. (2013). *Chin. J. Org. Chem.* 33: 1668–1673.

10 Declerck, V., Ribière, P., Nédellec, Y. et al. (2007). *Eur. J. Org. Chem.* 2007: 201–208.

11 Luo, C., Zhang, Y., and Wang, Y. (2005). *J. Mol. Catal. A: Chem.* 229: 7–12.

12 Han, W., Liu, N., Liu, C., and Jin, Z.L. (2010). *Chin. Chem. Lett.* 21: 1411–1414.

13 Datta, G.K., von Schenck, H., Hallberg, A., and Larhed, M. (2006). *J. Org. Chem.* 71: 3896–3903.

14 Ackermann, L. and Vicente, R. (2009). *Org. Lett.* 11: 4922–4925.

15 (a) Zapf, A. and Beller, M. (2005). *Chem. Commun.* 431–440. (b) Wang, D.P., Zhang, X.D., Liang, Y., and Li, J.H. (2006). *Chin. J. Org. Chem.* 26: 19–26. (c) So, C.M. and Kwong, F.Y. (2011). *Chem. Soc. Rev.* 40: 4963–4972.

16 Wang, L., Zhang, Y., Liu, L., and Wang, Y. (2006). *J. Org. Chem.* 71: 1284–1287.

17 (a) Mukhopadhyay, S., Rothenberg, G., Gitis, D., and Sasson, Y. (2000). *Org. Lett.* 2: 211–214. (b) Mukhopadhyay, S., Joshi, A.V., Peleg, L., and Sasson, Y. (2003). *Org. Process Res. Dev.* 7: 44–46.

18 Karimi, B., Vafaeezadeh, M., and Akhavan, P.F. (2015). *ChemCatChem* 7: 2248–2254.

19 Karimi, B. and Fadavi Akhavan, P. (2011). *Inorg. Chem.* 50: 6063–6072.

20 (a) Lennox, A.J.J. and Lloyd-Jones, G.C. (2014). *Chem. Soc. Rev.* 43: 412–443. (b) Maluenda, I. and Navarro, O. (2015). *Molecules* 20: 7528–7557. (c) Beletskaya, I.P., Alonso, F., and Tyurin, V. (2019). *Coord. Chem. Rev.* 385: 137–173. (d) Kadu, B.S. (2021). *Catal. Sci. Technol.* 11: 1186–1221.

21 Namboodiri, V.V. and Varma, R.S. (2001). *Green Chem.* 3: 146–148.

22 Li, J.-H., Liu, W.-J., and Xie, Y.-X. (2005). *J. Org. Chem.* 70: 5409–5412.

23 Li, J.-H., Hu, X.-C., Liang, Y., and Xie, Y.-X. (2006). *Tetrahedron* 62: 31–38.

24 Xin, B., Zhang, Y., and Cheng, K. (2006). *J. Org. Chem.* 71: 5725–5731.

25 Liu, L., Zhang, Y., and Wang, Y. (2005). *J. Org. Chem.* 70: 6122–6125.

26 Liu, W.-J., Xie, Y.-X., Liang, Y., and Li, J.-H. (2006). *Synthesis* 2006: 860–864.

27 Yin, L., Zhang, Z.H., and Wang, Y.M. (2006). *Tetrahedron* 62: 9359–9364.

28 Silva, A.D.C., Senra, J.D., Aguiar, L.C.S. et al. (2010). *Tetrahedron Lett.* 51: 3883–3885.

29 (a) Han, W., Liu, C., and Jin, Z.-L. (2007). *Org. Lett.* 9: 4005–4007. (b) Liu, C. and Li, X. (2016). *Chem. Rec.* 16: 84–97.

30 Han, W., Liu, C., and Jin, Z. (2008). *Adv. Synth. Catal.* 350: 501–508.

31 Mao, J., Hua, Q., Xie, G. et al. (2009). *Adv. Synth. Catal.* 351: 635–641.

32 Mao, J., Hua, Q., Xie, G. et al. (2009). *Eur. J. Org. Chem.* 2009: 2262–2266.

33 Razler, T.M., Hsiao, Y., Qian, F. et al. (2009). *J. Org. Chem.* 74: 1381–1384.

34 (a) Wang, Y.F., Deng, W., Liu, L., and Guo, Q.X. (2005). *Chin. J. Org. Chem.* 25: 8–24. (b) Alonso, D.A., Baeza, A., Chinchilla, R. et al. (2018). *Catalysts* 8. (c) Heravi, M.M., Ghanbarian, M., Ghalavand, N., and Nazari, N. (2018). *Curr. Org. Chem.* 22: 1420–1457. (d) Akhtar, R., Zahoor, A.F., Parveen, B., and Suleman, M. (2019). *Synth. Commun.* 49: 167–192. (e) Murashkina, A.V., Mitrofanov, A.Y., and Beletskaya, I.P. (2019). *Russ. J. Org. Chem.* 55: 1445–1458. (f) Mohajer, F., Heravi, M.M., Zadsirjan, V., and Poormohammad, N. (2021). *RSC Adv.* 11: 6885–6925. (g) Mohjer, F., Mofatehnia, P., Rangraz, Y., and Heravi, M.M. (2021). *J. Organomet. Chem.* 936. (h) Nair, P.P., Philip, R.M., and Anilkumar, G. (2021). *Org. Biomol. Chem.* 19: 4228–4242. (i) Amrutha, S., Radhika, S., and Anilkumar, G. (2022). *Beilstein J. Org. Chem.* 18: 262–285.

35 Mao, J., Wu, M., Xie, G., and Ji, S. (2009). *Adv. Synth. Catal.* 351: 2101–2106.

36 Firouzabadi, H., Iranpoor, N., and Ghaderi, A. (2011). *Org. Biomol. Chem.* 9: 865–871.

37 Firouzabadi, H., Iranpoor, N., Kazemi, F., and Gholinejad, M. (2012). *J. Mol. Catal. A: Chem.* 357: 154–161.

38 (a) Cao, J., Zheng, Z.-J., Xu, Z., and Xu, L.-W. (2017). *Coord. Chem. Rev.* 336: 43–53. (b) Cheng, L.-J. and Mankad, N.P. (2019). *JACS*. (c) Gabriele, B., Mancuso, R., and Salerno, G. (2012). *Eur. J. Org. Chem.* 2012: 6825–6839. (d) Li, Y., Hu, Y., and Wu, X.-F. (2018). *Chem. Soc. Rev.* 47: 172–194. (e) Peng, J.B. and Wu, X.F. (2018). *Angew. Chem. Int. Ed.* 57: 1152–1160. (f) Peng, J.-B., Geng, H.-Q., and Wu, X.-F. (2019). *Chem* 5: 526–552. (g) Shen, C. and Wu, X.-F. (2017). *Chemistry (Weinheim an der Bergstrasse, Germany)*, vol. 23, 2973. (h) Wu, L., Fang, X., Liu, Q. et al. (2014). *ACS Catal.* 4: 2977–2989. (i) Wu, X.-F., Fang, X., Wu, L. et al. (2014). *Acc. Chem. Res.* 47: 1041–1053.

39 Zhao, H., Cheng, M., Zhang, J., and Cai, M. (2014). *Green Chem.* 16: 2515–2522.

40 (a) Valente, C., Calimsiz, S., Hoi, K.H. et al. (2012). *Angew. Chem. Int. Ed.* 51: 3314–3332. (b) Cordovilla, C., Bartolome, C., Martinez-Ilarduya, J.M., and Espinet, P. (2015). *ACS Catal.* 5: 3040–3053. (c) Devendar, P., Qu, R.-Y., Kang, W.-M. et al. (2018). *J. Agric. Food. Chem.* 66: 8914–8934.

41 Zhou, W.-J., Wang, K.-H., and Wang, J.-X. (2009). *Adv. Synth. Catal.* 351: 1378–1382.

42 Zhou, W.-J., Wang, K.-H., and Wang, J.-X. (2009). *J. Org. Chem.* 74: 5599–5602.

43 Xu, G., Zhang, Y., Wang, K. et al. (2015). *J. Chem. Res.* 39: 399–402.

44 Kabalka, G.W., Dong, G., Venkataiah, B., and Chen, C. (2005). *J. Org. Chem.* 70: 9207–9210.

45 Shi, S. and Zhang, Y. (2007). *J. Org. Chem.* 72: 5927–5930.

46 Srimani, D., Sawoo, S., and Sarkar, A. (2007). *Org. Lett.* 9: 3639–3642.

47 Li, J.-H., Zhu, Q.-M., Liang, Y., and Yang, D. (2005). *J. Org. Chem.* 70: 5347–5349.

48 Rao, H.S.P., Jothilingam, S., and Scheeren, H.W. (2004). *Tetrahedron* 60: 1625–1630.

49 Wang, P., Lu, J., and Zhang, Z.-H. (2013). *J. Chem. Res.* 37: 359–361.

50 Wang, J.-Q., Cai, F., Wang, E., and He, L.-N. (2007). *Green Chem.* 9: 882–887.

51 (a) Mimoun, H., Charpentier, R., Mitschler, A. et al. (1980). *JACS* 102: 1047–1054. (b) Mimoun, H. (1982). *Angew. Chem. Int. Ed.* 21: 734–750. (c) Zhao, H., Zhang, T., Yan, T., and Cai, M. (2015). *J. Org. Chem.* 80: 8849–8855. (d) Mukhopadhyay, C., Tapaswi, P.K., and Butcher, R.J. (2010). *Org. Biomol. Chem.* 8: 4720–4729.

52 Chandrasekhar, S., Reddy, N.K., and Kumar, V.P. (2010). *Tetrahedron Lett.* 51: 3623–3625.

53 Lu, J., Guan, Z.-Z., Gao, J.-W., and Zhang, Z.-H. (2011). *Appl. Organomet. Chem.* 25: 537–541.

54 Mao, J., Guo, J., Fang, F., and Ji, S.-J. (2008). *Tetrahedron* 64: 3905–3911.

55 Leadbeater, N.E., Marco, M., and Tominack, B.J. (2003). *Org. Lett.* 5: 3919–3922.

56 Colacino, E., Daïch, L., Martinez, J., and Lamaty, F. (2007). *Synletter* 2007: 1279–1283.

57 Chen, G., Xie, J., Weng, J. et al. (2011). *Synth. Commun.* 41: 3123–3133.

58 Lang, K.D., Kaur, R., Arora, R. et al. (2020). *Anti-Cancer Agents Med. Chem.* 20: 2150–2168.

59 (a) Louillat, M.-L. and Patureau, F.W. (2014). *Chem. Soc. Rev.* 43: 901–910. (b) Wang, Q., Su, Y., Li, L., and Huang, H. (2016). *Chem. Soc. Rev.* 45: 1257–1272. (c) Zhao, Y. and Xia, W. (2018). *Chem. Soc. Rev.* 47: 2591–2608.

60 Kidwai, M., Mishra, N.K., Bhardwaj, S. et al. (2010). *ChemCatChem* 2: 1312–1317.

61 Chen, J., Yuan, T., Hao, W., and Cai, M. (2011). *Tetrahedron Lett.* 52: 3710–3713.

62 Chen, J., Zhang, Y., Hao, W. et al. (2013). *Tetrahedron* 69: 613–617.

63 Colacino, E., Villebrun, L., Martinez, J., and Lamaty, F. (2010). *Tetrahedron* 66: 3730–3735.

64 Li, X., Yuan, T., Yang, Y., and Chen, J. (2014). *Tetrahedron* 70: 9652–9660.

65 Burley, G.A., Davies, D.L., Griffith, G.A. et al. (2010). *J. Org. Chem.* 75: 980–983.

66 Vaddula, B., Leazer, J., and Varma, R.S. (2012). *Adv. Synth. Catal.* 354: 986–990.

67 (a) Thirumurugan, P., Matosiuk, D., and Jozwiak, K. (2013). *Chem. Rev.* 113: 4905–4979. (b) Zou, Y., Zhang, L., Yang, L. et al. (2018). *J. Controlled Release* 273: 160–179. (c) Jiang, X., Hao, X., Jing, L. et al. (2019). *Expert Opin. Drug Discovery* 14: 779–789. (d) Kim, E. and Koo, H. (2019). *Chem. Sci.* 10: 7835–7851. (e) Fitzgerald, P.R. and Paegel, B.M. (2021). *Chem. Rev.* 121: 7155–7177. (f) Geng, Z., Shin, J.J., Xi, Y., and Hawker, C.J. (2021). *J. Polym Sci.* 59: 963–1042. (g) Taiariol, L., Chaix, C., Farre, C., and Moreau, E. (2022). *Chem. Rev.* 122: 340–384.

68 Kumaraswamy, G., Ankamma, K., and Pitchaiah, A. (2007). *J. Org. Chem.* 72: 9822–9825.

69 Billault, I., Pessel, F., Petit, A. et al. (2015). *New J. Chem.* 39: 1986–1995.

70 Firouzabadi, H., Iranpoor, N., and Gholinejad, M. (2010). *Adv. Synth. Catal.* 352: 119–124.

71 Firouzabadi, H., Iranpoor, N., Gholinejad, M., and Samadi, A. (2013). *J. Mol. Catal. A: Chem.* 377: 190–196.

72 Firouzabadi, H., Iranpoor, N., and Samadi, A. (2014). *Tetrahedron Lett.* 55: 1212–1217.

73 Yao, L., Zhou, Q., Han, W., and Wei, S. (2012). *Eur. J. Org. Chem.* 2012: 6856–6860.

74 (a) Bhadra, S., Saha, A., and Ranu, B.C. (2010). *J. Org. Chem.* 75: 4864–4867. (b) Bhadra, S., Sreedhar, B., and Ranu, B.C. (2009). *Adv. Synth. Catal.* 351: 2369–2378.

75 Kundu, D., Mukherjee, N., and Ranu, B.C. (2013). *RSC Adv.* 3: 117–125.

76 Chatterjee, T. and Ranu, B.C. (2013). *J. Org. Chem.* 78: 7145–7153.

77 Reddy, K.H.V., Satish, G., Ramesh, K. et al. (2012). *Chem. Lett.* 41: 585–587.

78 Li, K., Liu, C., Wang, K. et al. (2018). *RSC Adv.* 8: 7761–7764.

79 Li, K., Wang, K., Liu, C. et al. (2019). *Synth. Commun.* 49: 672–678.

80 Wei, T., Zhang, T., Huang, B. et al. (2015). *Appl. Organomet. Chem.* 29: 846–849.

81 Liao, Y., Wei, T., Yan, T., and Cai, M. (2017). *Tetrahedron* 73: 1238–1246.

82 Yedage, S.L. and Bhanage, B.M. (2016). *Green Chem.* 18: 5635–5642.

83 (a) Wang, D. and Astruc, D. (2015). *Chem. Rev.* 115: 6621–6686. (b) Werkmeister, S., Neumann, J., Junge, K., and Beller, M. (2015). *Chem. Eur. J.* 21: 12226–12250. (c) Ai, W., Zhong, R., Liu, X., and Liu, Q. (2019). *Chem. Rev.* 119: 2876–2953. (d) Seo, C.S.G. and Morris, R.H. (2019). *Organometallics* 38: 47–65. (e) Nie, R., Tao, Y., Nie, Y. et al. (2021). *ACS Catal.* 11: 1071–1095.

84 Zhou, H.-F., Fan, Q.-H., Huang, Y.-Y. et al. (2007). *J. Mol. Catal. A: Chem.* 275: 47–53.

85 Chandrasekhar, S., Prakash, S.J., and Rao, C.L. (2006). *J. Org. Chem.* 71: 2196–2199.

86 Spina, R., Colacino, E., Gabriele, B. et al. (2013). *J. Org. Chem.* 78: 2698–2702.

87 (a) Sore, H.F., Galloway, W.R.J.D., and Spring, D.R. (2012). *Chem. Soc. Rev.* 41: 1845–1866. (b) Monfared, A., Mohammadi, R., Ahmadi, S. et al. (2019). *RSC Adv.* 9: 3185–3202. (c) Noor, R., Zahoor, A.F., Irfan, M. et al. (2022). *Molecules* 27.

88 Xu, C., Huang, B., Yan, T., and Cai, M. (2018). *Green Chem.* 20: 391–397.

89 (a) Pellissier, H. (2012). *Adv. Synth. Catal.* 354: 237–294. (b) Aleman, J. and Cabrera, S. (2013). *Chem. Soc. Rev.* 42: 774–793. (c) Hong, L. and Wang, R. (2013). *Adv. Synth. Catal.* 355: 1023–1052. (d) Ryan, S.J., Candish, L., and Lupton, D.W. (2013). *Chem. Soc. Rev.* 42: 4906–4917. (e) Caruana, L., Fochi, M., and Bernardi, L. (2015). *Molecules* 20: 11733–11764. (f) Menon, R.S., Biju, A.T., and Nair, V. (2016). *Beilstein J. Org. Chem.* 12: 444–461. (g) Silvi, M. and Melchiorre, P. (2018). *Nature* 554: 40–48. (h) Chen, X., Wang, H., Jin, Z., and Chi, Y.R. (2020). *Chin. J. Chem.* 38: 1167–1202. (i) Dai, L. and Ye, S. (2021). *Chin. Chem. Lett.* 32: 660–667.

90 Bortolini, O., Fantin, G., Fogagnolo, M. et al. (2011). *Tetrahedron* 67: 8110–8115.

91 Chandrasekhar, S., Reddy, N.R., Sultana, S.S. et al. (2006). *Tetrahedron* 62: 338–345.

92 Liu, Y., Liang, J., Liu, X.H. et al. (2008). *Chin. Chem. Lett.* 19: 1043–1046.

93 Feu, K.S., de la Torre, A.F., Silva, S. et al. (2014). *Green Chem.* 16: 3169–3174.

94 (a) de Graaff, C., Ruijter, E., and Orru, R.V.A. (2012). *Chem. Soc. Rev.* 41: 3969–4009. (b) Climent, M.J., Corma, A., and Iborra, S. (2012). *RSC Adv.* 2: 16–58. (c) Singh, M.S. and Chowdhury, S. (2012). *RSC Adv.* 2: 4547–4592. (d) Slobbe, P., Ruijter, E., and Orru, R.V.A. (2012). *Medchemcomm* 3: 1189–1218. (e) Reguera, L. and Rivera, D.G. (2019). *Chem. Rev.* 119: 9836–9860.

95 Jain, S.L., Singhal, S., and Sain, B. (2007). *Green Chem.* 9: 740–741.

96 (a) Anand, P., Singh, B., and Singh, N. (2012). *Bioorg. Med. Chem.* 20: 1175–1180. (b) Bubols, G.B., Vianna, D.D.R., Medina-Remon, A. et al. (2013). *Mini Rev. Med. Chem.* 13: 318–334. (c) Srikrishna, D., Godugu, C., and Dubey, P.K. (2018). *Mini Rev. Med. Chem.* 18: 113–141. (d) Stefanachi, A., Leonetti, F., Pisani, L. et al. (2018). *Molecules* 23. (e) Cao, D., Liu, Z., Verwilst, P. et al. (2019). *Chem. Rev.* 119: 10403–10519. (f) Stringlis, I.A., de Jonge, R., and Pieterse, C.M.J. (2019). *Plant Cell Physiol.* 60: 1405–1419. (g) Akkol, E.K., Genc, Y., Karpuz, B. et al. (2020). *Cancers* 12. (h) Singh, G.S. and Desta, Z.Y. (2012). *Chem. Rev.* 112: 6104–6155. (i) Pavlovska, T.L., Redkin, R.G., Lipson, V.V., and Atamanuk, D.V. (2016). *Mol. Diversity* 20: 299–344. (j) Xu, Z., Zhang, S., Gao, C. et al. (2017). *Chin. Chem. Lett.* 28: 159–167. (k) Varun, Sonam, and Kakkar, R. (2019). *Medchemcomm.* 10: 351–368.

97 Zareai, Z., Khoobi, M., Ramazani, A. et al. (2012). *Tetrahedron* 68: 6721–6726.

98 Balwe, S.G., Lim, K.T., Cho, B.G., and Jeong, Y.T. (2017). *Tetrahedron* 73: 3564–3570.

99 Zhang, X.-N., Li, Y.-X., and Zhang, Z.-H. (2011). *Tetrahedron* 67: 7426–7430.

100 Modugu, N.R. and Pittala, P.K. (2017). *New J. Chem.* 41: 14062–14066.

101 Hasaninejad, A. and Beyrati, M. (2018). *RSC Adv.* 8: 1934–1939.

102 Kidwai, M., Bhatnagar, D., Mishra, N.K., and Bansal, V. (2008). *Catal. Commun.* 9: 2547–2549.

103 Kidwai, M. and Bhatnagar, D. (2010). *Tetrahedron Lett.* 51: 2700–2703.

104 Veeranarayana Reddy, M., Kim, J.S., Lim, K.T., and Tae Jeong, Y. (2014). *Tetrahedron Lett.* 55: 6459–6462.

105 Survase, D., Bandgar, B., and Helavi, V. (2017). *Synth. Commun.* 47: 680–687.

106 (a) Jain, A.K., Vaidya, A., Ravichandran, V. et al. (2012). *Bioorg. Med. Chem.* 20: 3378–3395. (b) Sharma, P.C., Bansal, K.K., Sharma, A. et al. (2020). *Eur. J. Med. Chem.* 188.

107 Zhu, D., Chen, J., Xiao, H. et al. (2009). *Synth. Commun.* 39: 2895–2906.

108 Jawale, D.V., Lingampalle, D.L., Pratap, U.R., and Mane, R.A. (2010). *Chin. Chem. Lett.* 21: 412–416.

109 Lin, P.-Y., Hou, R.-S., Wang, H.-M. et al. (2009). *J. Chin. Chem. Soc.* 56: 455–458.

110 Deligeorgiev, T.G., Kaloyanova, S., Vasilev, A., and Vaquero, J.J. (2010). *Phosphorus, Sulfur, and Silicon and the Related Elements*, vol. 185, 2292–2302.

111 Liu, L., Zhang, F., Wang, H. et al. (2017). *Phosphorus, Sulfur, and Silicon and the Related Elements*, vol. 192, 464–468.

112 Mukhopadhyay, C. and Tapaswi, P.K. (2008). *Tetrahedron Lett.* 49: 6237–6240.

113 Nakka, M., Tadikonda, R., Rayavarapu, S. et al. (2014). *Synthesis* 47: 517–525.

114 (a) Abdel-Wahab, B.F., Mohamed, H.A., and El-Mansy, M.F. (2013). *J. Sulfur Chem.* 34: 289–300. (b) Fleischer, S., Zhou, S., Werkmeister, S. et al. (2013). *Chem. Eur. J.* 19: 4997–5003. (c) Tristan-Manzano, M., Guirado, A., Martinez-Esparza, M. et al. (2015). *Curr. Med. Chem.* 22: 3075–3108. (d) Zhang, Z., Xie, C., Feng, L., and Ma, C. (2016). *Synth. Commun.* 46: 1507–1518. (e) Kalinin, A.A., Islamova, L.N., and Fazleeva, G.M. (2019). *Chem. Heterocycl.*

Compd. 55: 584–597. (f) Kaushal, T., Srivastava, G., Sharma, A., and Negi, A.S. (2019). *Bioorg. Med. Chem.* 27: 16–35. (g) Chauhan, D.S., Singh, P., and Quraishi, M.A. (2020). *J. Mol. Liq.* 320. (h) Biesen, L. and Mueller, T.J.J. (2021). *Adv. Synth. Catal.* 363: 980–1006. (i) Maikhuri, V.K., Prasad, A.K., Jha, A., and Srivastava, S. (2021). *New J. Chem.* 45: 13214–13246.

115 Zhang, X.Z., Wang, J.X., Sun, Y.J., and Zhan, H.W. (2010). *Chin. Chem. Lett.* 21: 395–398.

116 He, Q., Wang, J., and Feng, R. (2010). *Lett. Org. Chem.* 7: 172–177.

117 Jorapur, Y.R., Rajagopal, G., Saikia, P.J., and Pal, R.R. (2008). *Tetrahedron Lett.* 49: 1495–1497.

10

Recent Advances in Organic Reactions Using Water as Solvent

Chang-Sheng Wang[1,*], *Qiao Sun*[2,*], *Guowei Wang*[2], *Wei He*[1], *Zheng Fang*[1], *and Kai Guo*[1]

[1] *Nanjing Tech University, College of Biotechnology and Pharmaceutical Engineering, 30 Puzhu Rd S., Nanjing 211816, P. R. China*
[2] *Nanjing Tech University, College of Food Science and Light Industry, 30 Puzhu Rd S., Nanjing 211816, P. R. China*

10.1 Introduction

In the past two decades, water has become a green and sustainable alternative to traditional organic solvents which are generally hazardous and waste productive [1]. Due to its safe, nontoxic, nonflammable, cost-effective, and abundant nature, water has been increasingly used in various types of organic reactions ranging from conventional oxidation [2], reduction, addition [3], substitution, condensation [4], cyclization reactions to classical cross-coupling [5], C—H activation [6] reactions and more recent photoredox [7], electrochemical and chemoenzymatic reactions, etc. Apart from accelerating the reaction rate and increasing the reactivity/selectivity of a specific reaction, the use of water as a solvent could also simplify the reaction work-up by convenient filtration or extraction since the products are generally immiscible/insoluble in water, and the catalytic system could be easily recycled and reused.

Despite the outstanding advantages of water and the enormous progress made in aqueous organic reactions, scientists are still encountering challenges due to the poor water solubility and the water sensitivity and instability of certain reactants, reagents, and catalysts. Micellar catalysis has emerged as an attractive tool to address such shortcomings by adding surfactants in bulk water [6c, 8]. The addition of surfactants to water triggers spontaneous formation of micelles which act as nanoreactors to accommodate reactants and catalysts in the inner core of the micelles, thus increasing the water solubility and stability of these species. The high local concentration created by the micelles would also accelerate the reaction rate and allow the reaction to occur under much milder conditions with

*These authors contributed equally to this work.

Green Solvents in Organic Synthesis, First Edition.
Edited by Xiao-Feng Wu, Zhiping Yin, Liang-nian He, and Feng Wang.

less catalyst and ligand loadings at faster rates when compared with conventional reactions performed in organic solvents. Generally, the reactions performed using water as a reaction medium could be categorized into "in water" or "on water" reactions [9]. According to a recent review from Lipshutz's group, "in water" implies the substrate/catalyst solubility in this medium, and micellar reactions could also be categorized as "in water" reactions since the use of additives enables the solubilization of starting materials and catalysts. The reactions fall into "on water" category if no solvation of the reaction components takes place.

This chapter summarizes recent (2017–2022) representative examples of various types of reactions using water as the solvent and covers related reaction mechanisms and the roles that water plays in certain reactions. Micellar-enabled aqueous reactions facilitated by various types of surfactants will also be discussed. Most reactions discussed herein utilize neat water as the reaction medium and only several examples require the addition of cosolvents to improve the water solubility. Moreover, most reactions described have not been discussed in previous reviews or chapters.

10.2 Cross-Coupling Reactions

In the last few decades, transition metal-catalyzed cross-couplings have become one of the most efficient methods for construction of chemical bonds in organic synthesis. This section elaborates on different types of aqueous cross-coupling reactions according to the type of bond that is constructed (C—C, C—N, C—O, C—S, C—P). Examples of photocatalyzed cross-coupling reactions are also included in this part.

10.2.1 C—C Cross-Coupling

Suzuki–Miyaura (SM) Coupling

The Suzuki–Miyaura (SM) reactions are the most extensively studied C—C cross-coupling reactions in aqueous environments. In 2019, Lipshutz and coworkers accomplished ppm Pd-catalyzed (500–5000 ppm; i.e. 0.05-0.5 mol%) Suzuki cross-couplings under micellar catalysis of the surfactant TPGS-750-M using an EvanPhos-bearing *N,C*-diisopropyl biarylpalladacycle as precatalyst (Scheme 10.1) [10]. The highly lipophilic *N,C*-disubstituted palladacycle was designed to minimize the undesired catalyst deactivation caused by the carbazole by-product. After decantation or filtration of products, the aqueous reaction medium could be recycled several times with an E factor of nearly zero. A one-pot, four-step sequence involving SM coupling, reduction, benzylation, and acylation was also developed.

In the same year, the Lipshutz's group further reduced the Pd loading of Suzuki reaction to 25-500 ppm by using a new biaryldiisopropyl-substituted palladacycle containing a HandaPhos ligand (Scheme 10.2) [11]. In the cases of less reactive aryl chlorides, 500 ppm Pd precatalyst was required, and 10% THF cosolvent was added to increase the solubility of substrates and expand the interior micellar volume available for reaction.

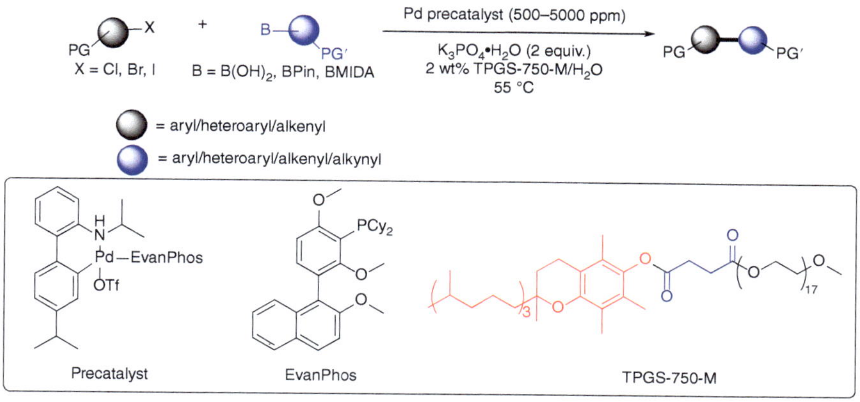

Scheme 10.1 Micellar-enhanced ppm Pd-catalyzed Suzuki coupling using EvanPhos-bearing *N,C*-diisopropyl biarylpalladacycle.

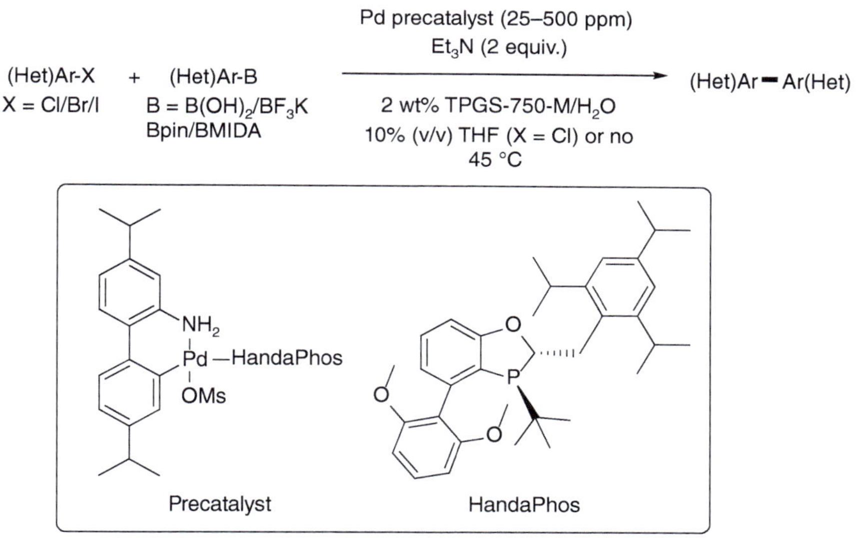

Scheme 10.2 Micellar-enhanced ppm Pd-catalyzed Suzuki coupling using HandaPhos-containing palladacycle.

Later, an organic solvent-free and surfactant-free ppm Pd-catalyzed Suzuki reaction of electron-poor aryl chlorides was disclosed by Schaub and coworkers using a simplified $Pd(OAc)_2/Cy_2P(t\text{-}Bu)$ catalytic system at a lower Pd loading of 50-100 ppm (Scheme 10.3) [12]. The melted aryl chloride substrate servers as an organic phase for this reaction, thus avoiding the use of additional organic solvents.

Gao and coworkers developed a ligand free and highly recyclable ppm Pd-catalyzed Suzuki reaction between (hetero)aryl halides (X=Br, I) and arylboronic acids under aqueous Triton X-100 micellar catalysis in air using water-soluble $Pd(NH_3)_4Cl_2$ as precatalyst and tetrabutylammonium hydroxide (TBAOH) as base

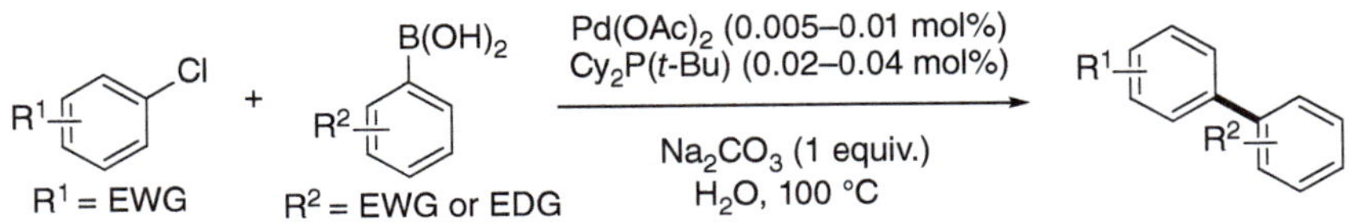

Scheme 10.3 Ppm Pd-catalyzed Suzuki reaction of electron-poor aryl chlorides.

(Scheme 10.4) [13]. TBAOH could prevent the reduction of the Pd(II) species and the formation of aggregated Pd black, thus stabilizing the active palladium species.

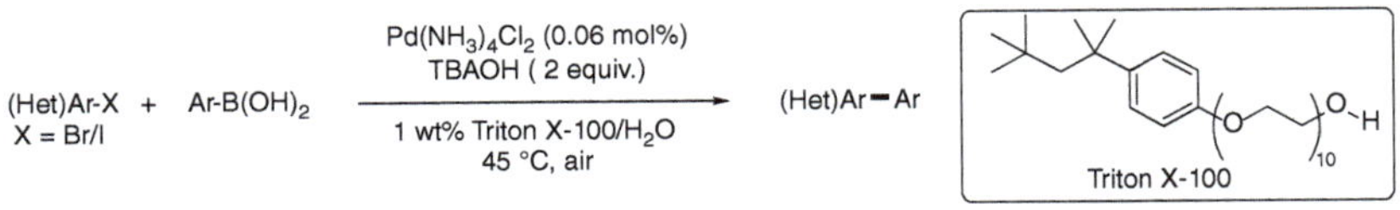

Scheme 10.4 Ligand-free Suzuki–Miyaura coupling between arylboronic acids and aryl halides.

Huang and coworkers developed a DNA-compatible Suzuki coupling between DNA-linked aryl halides and a wide variety of (hetero)arylboronic acids/esters in aqueous media using water-soluble sSPhos-Pd-G2 as the precatalyst and CsOH as the base (Scheme 10.5) [14]. The protocol preserves the structural integrity and ligation activity of the attached DNA oligomers.

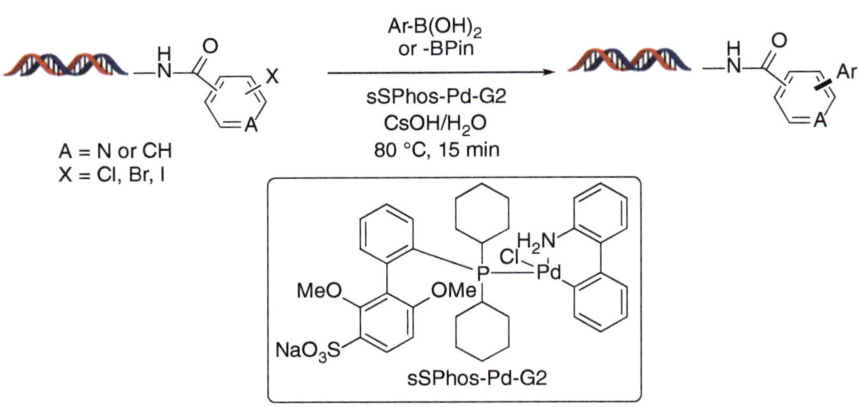

Scheme 10.5 DNA-compatible Suzuki–Miyaura coupling.

Pd/NHC-catalyzed aqueous Suzuki coupling reactions have also been reported. In 2018, Huynh and coworkers synthesized a series of substituted biphenyls via Suzuki coupling of various aryl/heteroaryl halides with phenyl boronic acid using a water-soluble ammonium-functionalized bis(1,2,4-triazolin-5-ylidene) palladium(II) complex as catalyst (0.001-0.1 mol%) in refluxed water (Scheme 10.6) [15]. The catalyst complex is not hydroscopic and thus could be stored and handled with ease.

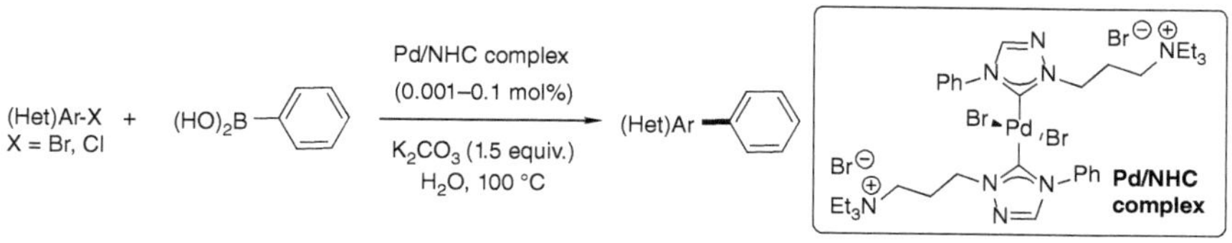

Scheme 10.6 Pd/NHC-catalyzed Suzuki–Miyaura coupling.

More recently, Tang and coworkers accomplished BaryPhos-facilitated asymmetric SM cross-couplings for the synthesis of various *ortho*-sulfur- or nitrogen-substituted axially chiral biaryls and the asymmetric synthesis of isoplagiochin D [16]. The tertiary alcohol moiety of the BaryPhos ligand participated in hydrogen bonding with the substrates and therefore mediated the asymmetric induction. By using a hydrophilic polyethylene glycol (PEG)-supported BaryPhos (PEG2000-BaryPhos) as the ligand, they achieved aqueous asymmetric syntheses of *ortho*-nitro-substituted axially chiral biaryls (Scheme 10.7). The supported palladium catalyst was reused multiple times by simply recycling the aqueous phase.

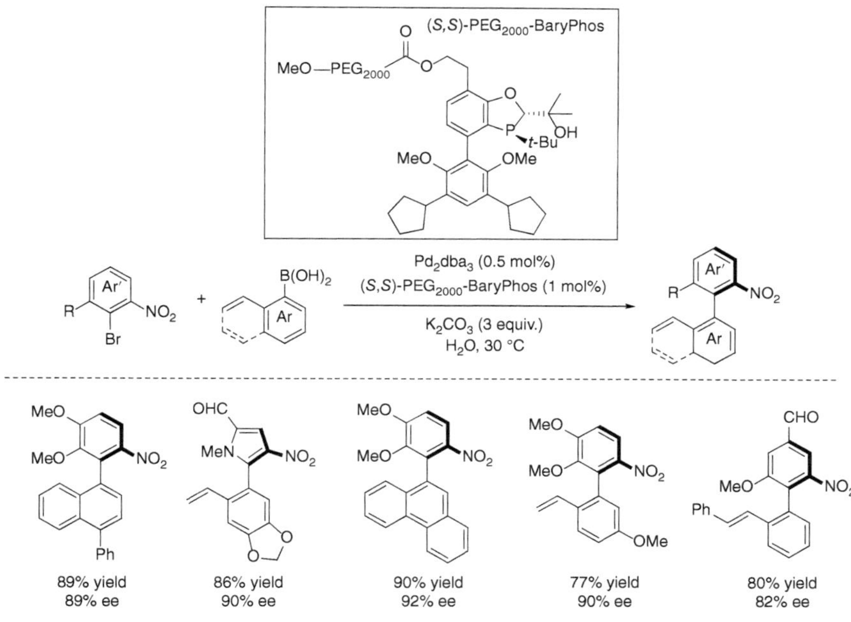

Scheme 10.7 BaryPhos-facilitated asymmetric Suzuki–Miyaura cross-coupling.

Apart from the above-mentioned Suzuki couplings via cleavage of carbon-halogen bonds, in 2020, Zeng and coworkers disclosed a C—N cleavage-mediated Suzuki-coupling of *N*-acylsuccinimides with potassium aryltrifluoroborates in water for the synthesis of aryl ketones at room temperature under ligand-free conditions (Scheme 10.8) [17]. The aqueous phase was recyclized four times with only a slight loss of yield. The reaction was proposed to occur through oxidative addition of the C—N bond onto Pd(0) to afford an acylpalladium(II) complex **A**, which undergoes ligand exchange with the hydroxide anion generated in basic water to afford an acyl-Pd(II)OH species **B**. Transmetalation of this species with potassium aryltrifluoroborates through a Pd—O—B complex **C** followed by reductive elimination produces the final product.

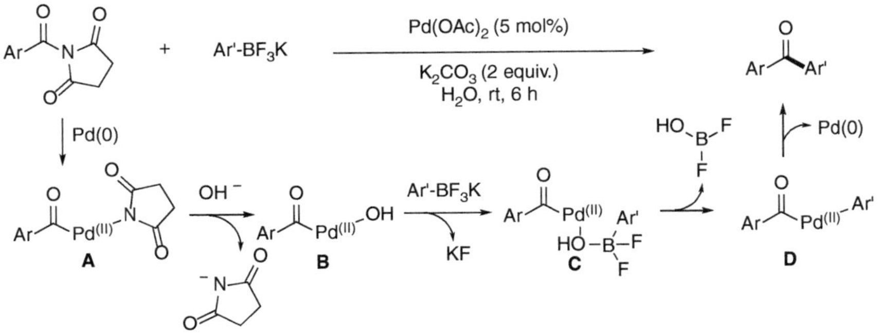

Scheme 10.8 Ligand-free Suzuki–Miyaura coupling of amides via C–N cleavage.

Liebeskind–Srogl Coupling

Liebeskind–Srogl cross-coupling is a well-known reaction for C—C bond construction via desulfurization. Recently, Pratap and coworkers reported a Pd-catalyzed CuTC-mediated Liebeskind–Srogl coupling of a series of thiomethylated 2*H*-pyran-2-ones, 2-oxobenzo[*h*]chromones and α-aroyl ketene-dithioacetals with arylboronic acids in water under base-free condition using cetyltrimethylammonium bromide (CTAB) as the surfactant (Scheme 10.9) [18]. They also accomplished the synthesis of 3-amino chalcone and diaryl azoles via one-pot arylation/amination of α-aroyl ketene dithioacetals. Unlike Suzuki–Miyaura coupling, the Liebeskind–Srogl reaction does not require a base since the boronic acid is activated

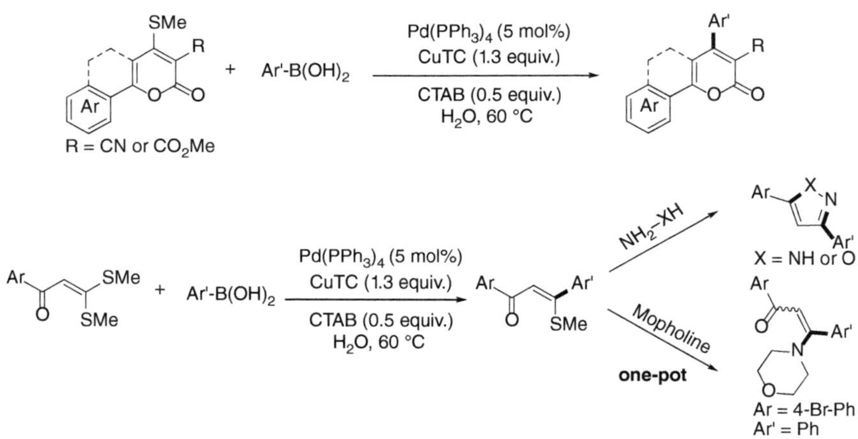

Scheme 10.9 Base-free desulfitative coupling of thioethers with arylboronic acids.

by CuTC. The reaction was proposed to occur through oxidative addition of CuTC-activated thioether onto the Pd(0) catalyst, transmetalation of the resulting Pd(II) complex with arylboronic acid via a six-membered ternary complex followed by reductive elimination of a newly generated Pd(II) complex to afford the final product.

Murahashi Coupling

The Murahashi cross-coupling involving moisture- and air-sensitive organolithium reagents could also be carried out under aqueous conditions. In 2019, Capriati and coworkers achieved ultrafast, recyclable Pd-catalyzed C(sp^2)—C(sp^3), C(sp^2)—C(sp^2), and C(sp^2)—C(sp) cross-couplings between (hetero)aryl halides and organolithiums at room temperature under air using water as the reaction medium and NaCl as the additive (Scheme 10.10) [19]. The organolithium reagent was rapidly added in one portion to the vigorously stirred suspension of the substrate, catalyst, and NaCl in water under the air. The reaction was completed within 20 s without the formation of undesired dehalogenation/homocoupling side products. The presence of NaCl additive, dissolved oxygen, and the addition rate of organolithium reagent are all critical to the success of this transformation.

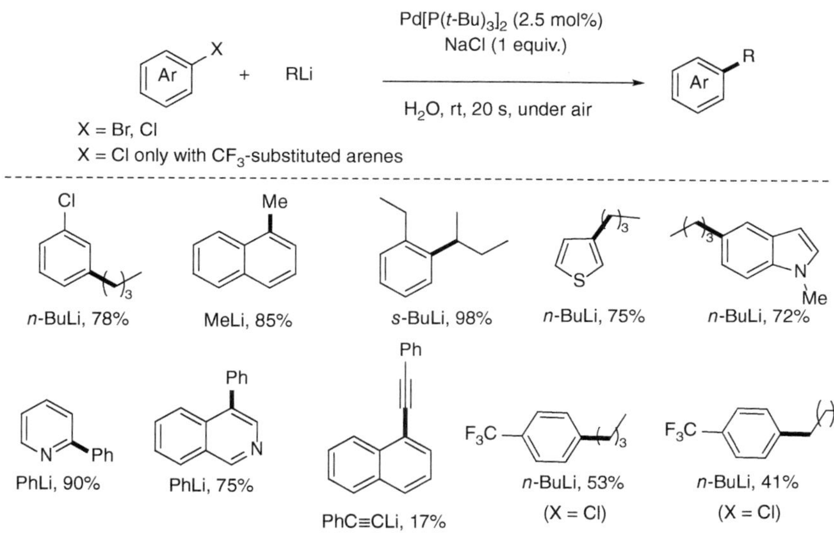

Scheme 10.10 Murahashi coupling between (hetero)aryl halides and organolithium reagents on water at rt under air.

Negishi Coupling

In 2021, the Capriati's group further disclosed an "on water" Negishi coupling between (hetero)aryl bromides and organozinc compounds by employing an analogous condition to the Murahashi coupling described above (Scheme 10.11) [20]. The organozinc reagent was prepared as a solution in 2-MeTHF and was rapidly canulated into the reaction mixture. The addition of NaCl caused a "halide effect" to facilitate the oxidative addition step and reduce the dehalogenation product. As a complementary to the water-based reactions, the author also performed reactions in deep eutectic solvent (ChCl/urea) at an increased temperature.

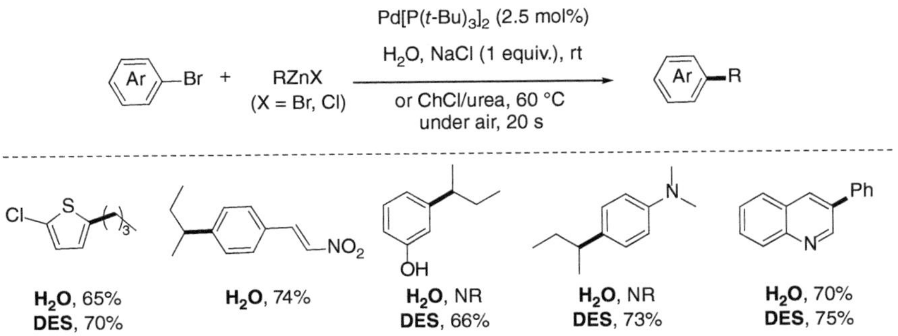

Scheme 10.11 Negishi cross-coupling "on water" or in deep eutectic solvent.

Considering the water-sensitivity and limited availability of organozinc reagents, Lipshutz and coworkers developed a Fe/ppm Pd nanoparticle-catalyzed direct reductive cross-coupling between (hetero)aryl bromides and alkyl bromides in the presence of Zn powder and TMEDA under aqueous TPGS-750-M micellar catalysis (Scheme 10.12) [21]. An organozinc species was generated *in situ* and participated in the Negishi coupling. The organozinc species could enter the hydrophobic inner

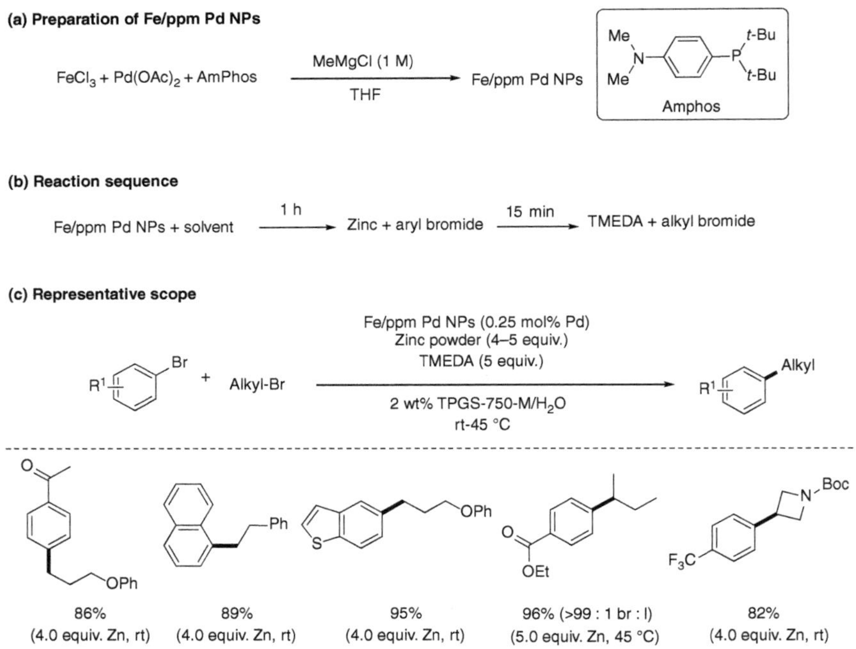

Scheme 10.12 Negishi coupling under micellar catalysis using Fe/ppm Pd NPs.

micellar core, thus escaping the hydrolysis by surrounding water. The superiority of the newly developed method was demonstrated when compared with representative recent examples of Negishi couplings. Furthermore, a tandem one-pot four-step sequence involving double allylation, ring-closing metathesis (RCM), Negishi coupling, and ketone α-arylation reaction was also developed.

Stille Coupling

Organotin reagents are generally more stable toward moisture compared to the corresponding boron and zinc reagents. In 2021, a triphenylphosphine-based palladacycle was designed by Lipshutz for ppm Pd-catalyzed Stille coupling of various (hetero)aryl/vinyl halides (Br, I) with (hetero)aryl/vinyl/alkynyl stannanes in water using inexpensive Brij-30 as surfactant (Scheme 10.13) [22]. An *N*-isopropyl group was introduced in the Pd precatalyst to reduce the toxification effect caused by competitive ligation of the carbazole by-product to palladium. A direct comparison of the Stille and Suzuki coupling of three heterocyclic stannanes or boronic acids indicated the superiority of the former. Furthermore, a one-pot two-step sequence involving stannylation followed by Stille coupling of the resulting stannane with a second (hetero)aryl bromide was developed and was found to be free of homocoupling byproduct.

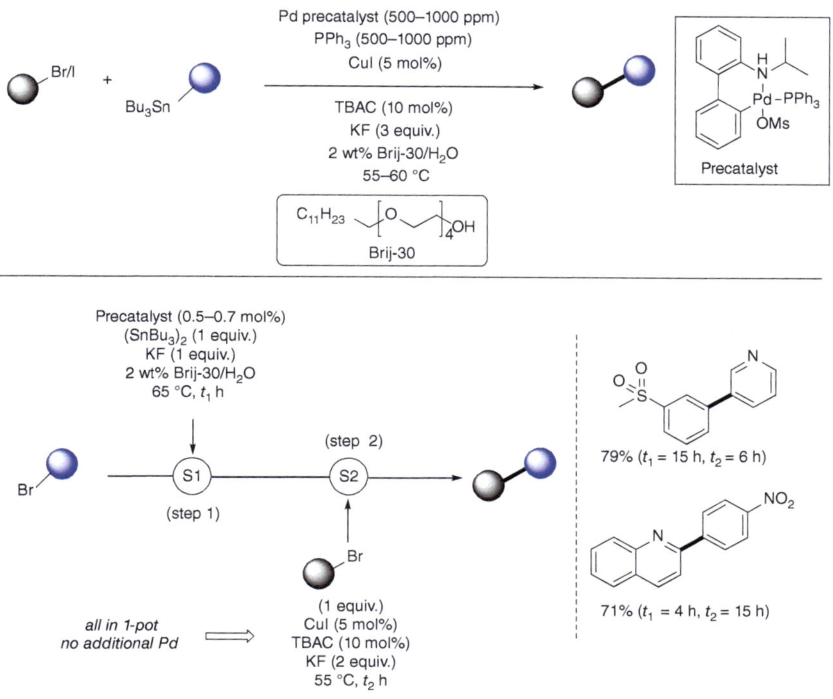

Scheme 10.13 Ppm-catalyzed Stille coupling in water using triphenylphosphine-based palladacycle.

In the same year, Tsai and coworkers described the synthesis of isoflavones via Stille cross-coupling between 3-bromochromones and aryltributylstannanes in water under a water-soluble and reusable $PdCl_2(NH_3)_2$/2,2'-cationic bipyridyl catalytic system using TBAF as an additive to accelerate the transmetalation step (Scheme 10.14) [23]. For substituted 3-bromochromones, a higher temperature (120 °C) and the addition of $NaHCO_3$ were necessary to give satisfactory yields. The product could be simply extracted by EtOAc and the aqueous phase was reused for three additional runs. Moreover, the simple synthetic protocol was employed in a one-pot synthesis of naturally occurring isoflavone daidzein.

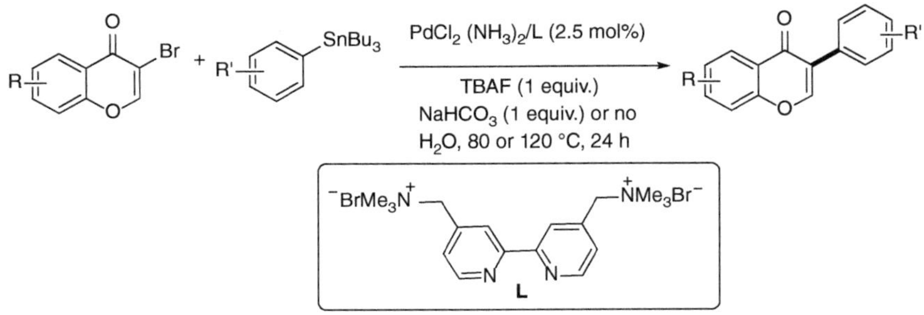

Scheme 10.14 Stille coupling for the synthesis of isoflavones in water.

Mizoroki–Heck Coupling and Related Reactions

The use of neat water in Pd-catalyzed Mizoroki–Heck cross-coupling has been recently explored. In 2017, Rode and coworkers described an additive-free Pd-catalyzed Mizoroki–Heck cross-coupling of aryl bromides/chlorides with acrylates/allyltrimethylsilane in water under aerobic conditions with the assistance of a $Pd(PPh_3)_4/Et_3N$ catalytic system (Scheme 10.15) [24]. Oxygen is important for this transformation since it could mediate continuous reoxidation of Pd(0) species to Pd^{2+}, thus preventing agglomeration of inactive Pd^0 black. Most reactions afforded products with complete *E* selectivity, with the exception being 1-naphthyl bromide or allyltrimethylsilane, which led to a pair of *E*/*Z* isomers with 1:1 ratio. A strategy

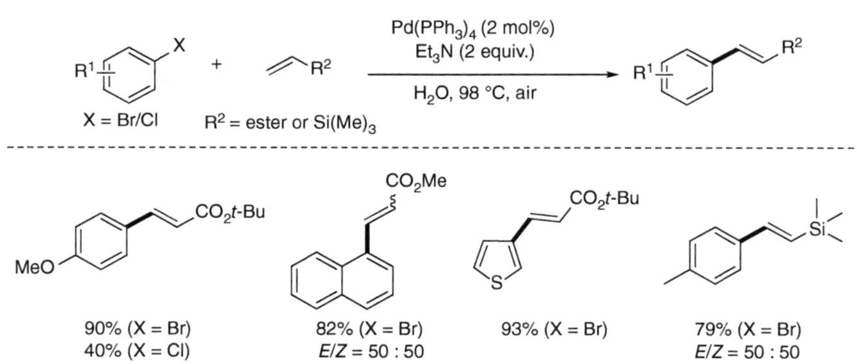

Scheme 10.15 Additive-free Pd-catalyzed Mizoroki–Heck cross-coupling of aryl bromides/chlorides with acrylates/allyltrimethylsilane.

was also evolved to recycle the Pd metal ten times without obvious reduction in its catalytic behavior.

In 2018, Frantz and coworkers reported the synthesis of substituted allenes via a cationic alkynyl Heck reaction between aryl triflates and alkynes in an aqueous DIPEA/H_2O biphasic system (Scheme 10.16) [25]. The development of a new hybrid Pd(0)-BobCat catalyst incorporating a water-soluble 4-CO_2Na dba ligand and a biaryl phosphine ligand is critical to this transformation since it could reduce the dba-derived Heck byproducts and provide reproducible results. A broad range of symmetrical alkynes and several unsymmetrical alkynes were tolerated and afforded the products in moderate to good yields. Under the basic biphasic condition, the water-soluble dba ligand dissociates to the aqueous layer and liberates a catalytically active mono-ligated Pd(0) complex in the organic layer. The reaction was proposed to follow a Heck-type mechanism via a Pd-mediated β-hydride elimination rather than base-assisted deprotonation/elimination.

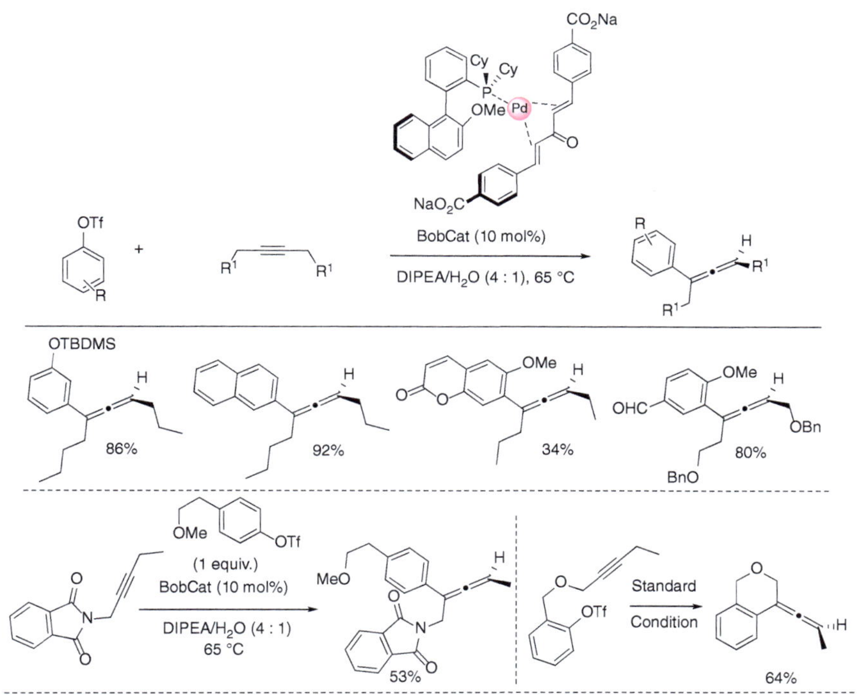

Scheme 10.16 Cationic alkynyl Heck reaction between aryl triflates and alkynes.

In the same year, diverse chalcone derivatives and 1,4-dicarbonyl esters were prepared by Xu and coworkers via Pd-catalyzed carbonylative Heck coupling of aryl tosylates with styrenes or acrylates in water with the assistance of a novel hydrophilic methoxy-PEG-modified *s*-triazinal N,P-ligand (Scheme 10.17) [26]. The hydrophilic catalytic system could be recycled several times with slightly diminished yields.

This reaction represents the first example of recyclable Pd-catalyzed carbonylative Heck reactions in water.

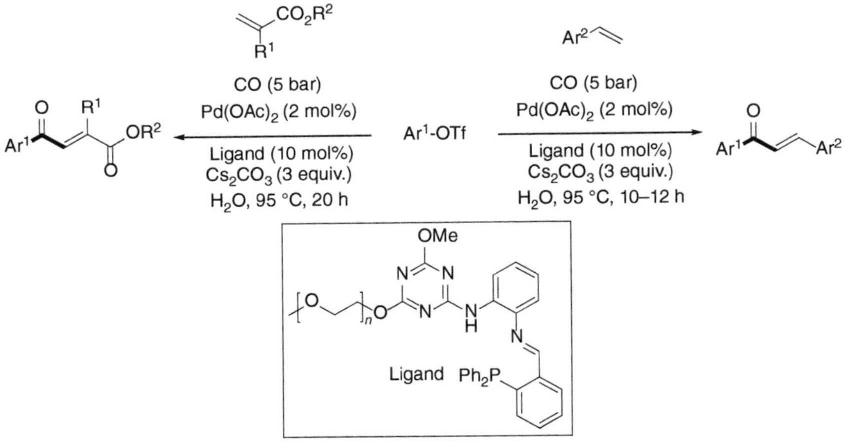

Scheme 10.17 Pd-catalyzed carbonylative Heck coupling of aryl tosylates with styrenes or acrylates.

Sonogashira Coupling

In 2019, Lipshutz and coworkers developed a ppm Pd-catalyzed, Cu-free Sonogashira coupling of (hetero)aryl bromides with alkynes under aqueous TPGS-750-M micellar catalysis using commercially available and cost-effective [(cinnamyl)PdCl]$_2$/cBRIDP as the catalytic system (Scheme 10.18) [27]. The increased lipophilicity and steric bulkiness of the cinnamyl ligand could inhibit undesired catalyst dimerization and increase the binding constant of this catalyst to the inner micellar core. In many cases, the products can be isolated by simple filtration and the aqueous reaction medium containing the surfactant can be easily recycled.

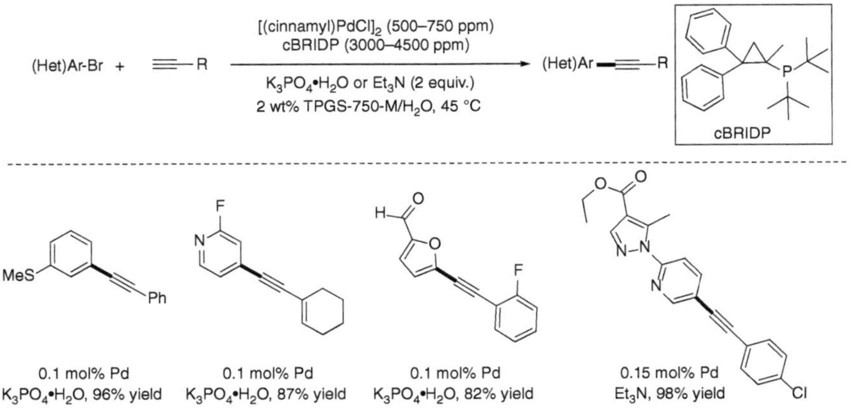

Scheme 10.18 Ppm Pd-catalyzed Cu-free Sonogashira coupling.

In 2020, Bian and coworkers reported the rapid synthesis of polysubstituted aromatic alkynes via microwave-assisted multiple Sonogashira couplings of poly-halogenated arenes with trimethylsilylacetylene using water as the solvent and 18-crown-6 as the phase-transfer catalyst (Scheme 10.19) [28]. This method is highly promising for the synthesis of porous aromatic frameworks without the need for strict exclusion of oxygen.

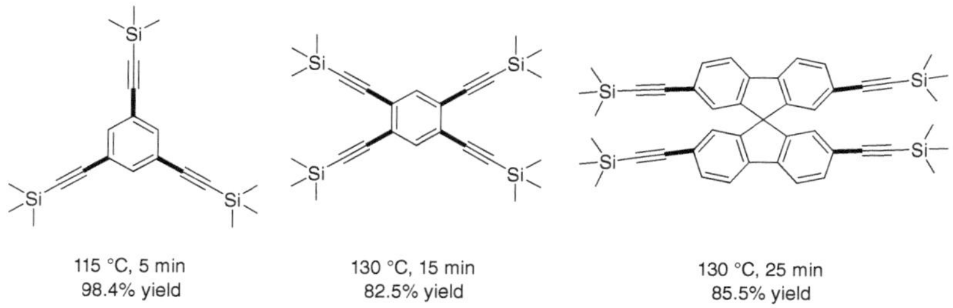

Reaction conditions: Ar-Br (1 equiv.), trimethylsilylacetylene (2.5 × *n* equiv.), Et_3N (2.5 × *n* equiv.), PPh_3 (0.33 equiv.), $Pd(PPh_3)_2Cl_2$ (0.16 equiv.), CuI (0.22 equiv.), 18-crown-6 (0.5 equiv.), H_2O, MW (270 W). The symbol "*n*" 3denotes the number of bromides in substrate

Scheme 10.19 Microwave-assisted multiple Sonogashira couplings of poly-halogenated arenes with trimethylsilylacetylene.

More recently, Trzeciak and coworkers described a copper- and ligand-free Sonogashira coupling between aryl iodides and aryl alkynes in an aqueous medium using $[bmpy]_2[Pd_2Cl_6]$, an anionic Pd(II) complex with pyridinium cations, as the catalyst and Et_3N as the base (Scheme 10.20) [29]. A one-pot, two-step Sonogashira-hydroarylation reaction was also developed using the same catalyst under microwave irradiation by using $NaBPh_4$ as the phenyl source and water as the proton donor, providing triarylethenes in moderate to good yields.

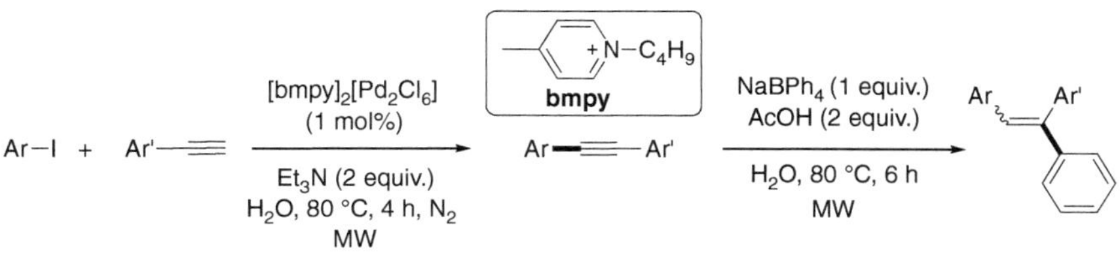

Scheme 10.20 Microwave-assisted one-pot, two-step Sonogashira-hydroarylation reaction.

Despite the extensive study of Sonogashira couplings using terminal alkynes as the coupling partners, homocoupling of terminal alkynes is a technical problem that needs to be solved. In 2017, a $PdCl_2$/XPhos-catalyzed deacetonative coupling of aryl propargylic alcohols with aryl chlorides in water was reported by Chang and coworkers using TBAB as the surfactant and K_2CO_3 as the base (Scheme 10.21) [30]. The aryl propargylic alcohols are cost-effective, stable, easily prepared, and only generate corresponding ketones as the by-products. Moreover, by using 2-methyl-3-butyn-2-ols and electron-neutral/electron-rich aryl chlorides as substrates, consecutive Sonogashira/deacetonative reactions also occurred to afford symmetrically alkynes in good yields.

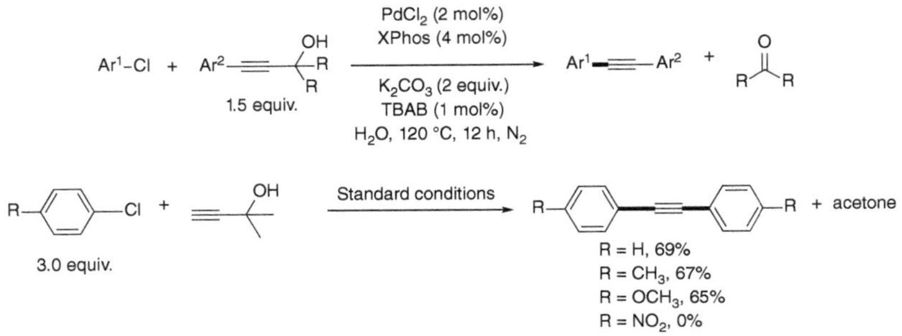

Scheme 10.21 Pd-catalyzed deacetonative coupling of aryl propargylic alcohols with aryl chlorides.

Tsuji–Trost Type Allylic Arylation

In 2018, Eppinger and coworkers developed a base-free Tsuji–Trost type allylic arylation of cinnamyl acetates with sodium tetraphenylborate in water under an air atmosphere using a low loading (0.002 mol%) of Bedford-type palladacycle catalyst (Scheme 10.22) [31]. A series of *para*-, *meta*-, and even *ortho*-substituted aryl cinnamyl acetates smoothly participated in the reactions to afford moderate to good yields of products within a short reaction time of 8 h. Water was found to facilitate the breakdown of the catalyst dimer by coordinating to the Pd(II) center.

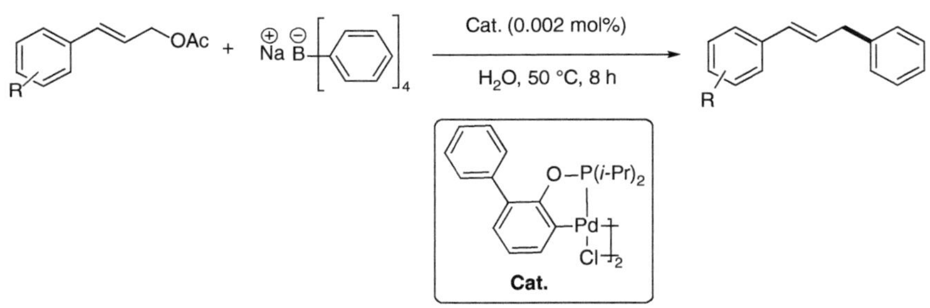

Scheme 10.22 Tsuji–Trost type allylic arylation of cinnamyl acetates with sodium tetraphenylborate.

Cyanation Reaction

In 2021, Lipshutz and coworkers accomplished mild and ppm Pd-catalyzed late-stage cyanations of highly complex molecules as well as a wide variety of (hetero)aryl iodides, bromides, and chlorides with $Zn(CN)_2$ in aqueous micellar media of 2 wt% Brij-30 using Xantphos Pd G3 as the catalyst and polymethylhydrosiloxane (PMHS) as the reductant (Scheme 10.23) [32]. A labeled source of CN (i.e. $^{13}C^{15}N$) was also introduced by using this protocol. Side-by-side comparisons of this ppm-catalyzed aqueous protocol with existing alternatives that rely on organic solvents indicated faster rates and higher yields of the current water-based protocol. Moreover, much lower levels of residual palladium were detected in the products based on ICP-MS analysis.

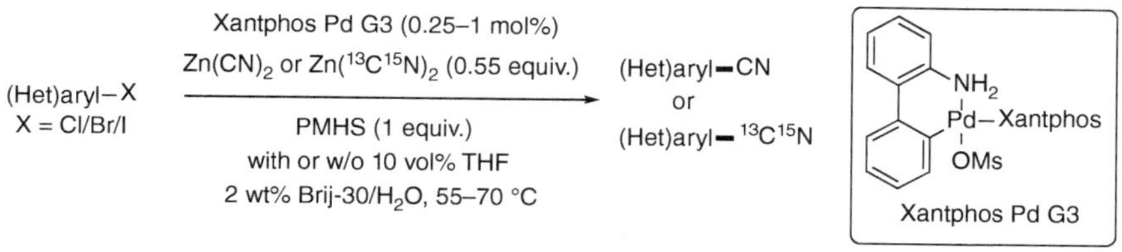

Scheme 10.23 Ppm Pd-catalyzed cyanation of aryl/heteroaryl halides.

Carbonyl Cross-Coupling

In 2017, Wu and coworkers reported the synthesis of *N*-(2-carbonylaryl)benzamides via recyclable Pd/C-catalyzed aminocarbonylation of aryl iodides with anthranils in water using $Mo(CO)_6$ as a safe solid CO source under air in the presence of TBAB as a phase-transfer catalyst (Scheme 10.24) [33]. No inert atmosphere protection was required for this transformation. Mechanistic studies indicated that $Mo(CO)_6$ serves the dual roles of promoting the generation of 2-aminobenzaldehyde as well as a CO source.

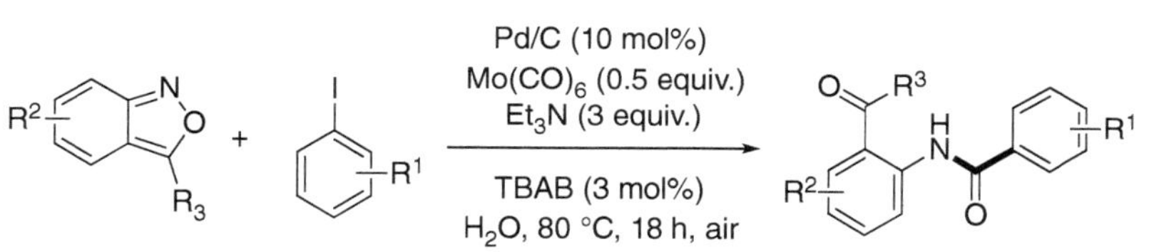

Scheme 10.24 Pd/C-catalyzed aminocarbonylation of aryl iodides with anthranils.

Later, Yin and coworkers reported a one-pot synthesis of 2,5-furandicarboxylic acid (FDCA) from 5-bromo-furoic acid via Pd-catalyzed carboxylation in a weakly basic $NaHCO_3$ solution under low CO pressure using water-soluble TPPTS (tris(3-sulfophenyl)phosphine trisodium salt) as the ligand (Scheme 10.25) [34]. The success of this reaction was heavily dependent on the pH values of the reaction media, with stronger bases favoring the formation of dehalogenation product and the weaker bases promoting the carboxylation process. Because 5-bromo-furoic acid has been successfully industrialized from the furfural-based platform molecule derived from bulk hemicellulose, this process offers an alternative route to original 5-hydroxymethylfurfural (HMF)-based FDCA synthesis considering the technical bottlenecks encountered in the transformation of bulk cellulose to HMF.

Reductive Coupling

In 2018, Li and coworkers described the synthesis of diarylethanes via a Cu-catalyzed, Zn-mediated radical reductive arylation of styrenes with aryl iodides

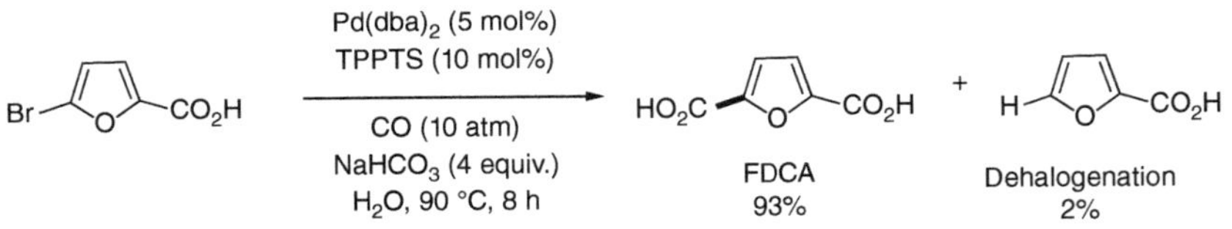

Scheme 10.25 Pd-catalyzed carboxylation of 5-bromo-furoic acid.

in water using $Cu(acac)_2$ as the catalyst and electron-rich 3,4-dimethoxyaniline as the ligand (Scheme 10.26) [35]. 3,4-Dimethoxyaniline not only participated in the formation of active ligands but also prevented the formation of diarylcopper species that accounts for the undesired Wurtz-type homocoupling of aryl halides. Mechanistically, the aryl iodide undergoes a single electron transfer (SET) with Zn or Zn/Cu to give an aryl radical anion **A** which is trapped by an active $Cu^{I}L$ catalyst to afford an aryl$Cu^{II}L$ species **B**. Coordination of the Cu^{II} species with the olefin followed by migratory insertion of the double bond generates a new Cu^{II} species **D** which provides the final product either by a hydrolysis/reduction or a reduction/hydrolysis pathway.

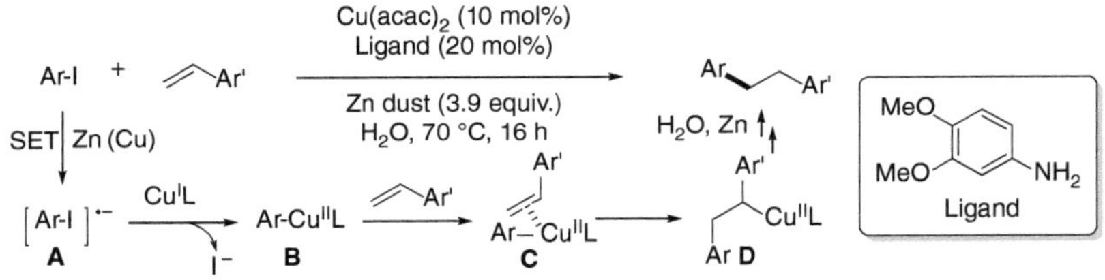

Scheme 10.26 Cu-catalyzed, Zn-mediated radical reductive arylation of styrenes.

In 2019, Lipshutz and coworkers reported a Fe-catalyzed reductive cross-coupling between terminal (hetero)aryl alkenes and alkyl bromides/iodides under TPGS-750-M micellar catalysis using CH_3CN as a cosolvent (Scheme 10.27) [36]. The reaction demonstrated broad substrate scope and various primary/secondary/tertiary alkyl halides and vinyl aromatic/heteroaromatic substrates were tolerated. The new C—C bond was regiospecifically formed at the β-site of the alkene. Mechanistically, the C—C bond is formed via the radical addition of an alkyl radical onto the vinyl group to give a benzyl radical species, which undergoes Zn-mediated reduction and subsequent protonation to produce the final product.

In 2020, König and coworkers reported an Ir-catalyzed, photoreductive cross-coupling between nonactivated alkyl chlorides and electron-deficient olefins in an aqueous sodium lauryl oligoethylene glycol sulfate (SLES) solution (Scheme 10.28) [37]. By using an assembly promoted SET (APSET) strategy, carbon-centered radicals were generated from nonactivated alkyl chlorides and were involved in subsequent radical addition reactions. The use of an aqueous environment allows the stabilization of an iridium complex and facilitates the cleavage of the C–Cl bond to form the alkyl radical species.

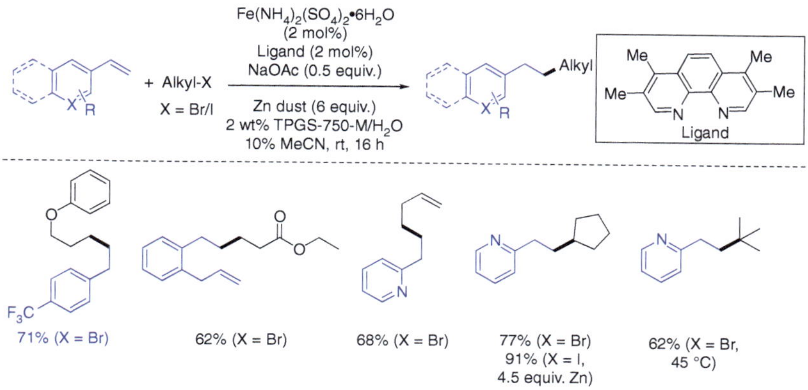

Scheme 10.27 Fe-catalyzed reductive coupling of terminal (hetero)aryl alkenes and alkyl halides.

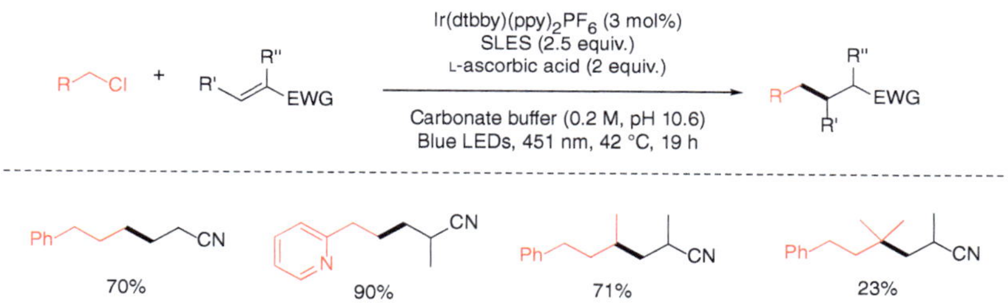

Scheme 10.28 Ir-catalyzed, photoreductive cross-coupling between nonactivated alkyl chlorides and electron-deficient olefins.

Cross-Dehydrogenative Coupling (CDC)

In recent years, cross-dehydrogenative coupling (CDC) has become a research hotspot in organic chemistry due to the avoidance of substrate pre-activation and generation of stoichiometric wastes. Dong and coworkers have summarized the lately progress of CDC reactions of C–H bonds in aqueous media in an elegant review [5c].

In 2018, Rao and coworkers developed a $K_2S_2O_8$-mediated, metal-free CDC coupling between 2-aryl pyridines and cyclic or acyclic ethers using water as the solvent (Scheme 10.29) [38]. The protocol regioselectively delivers the coupling products at

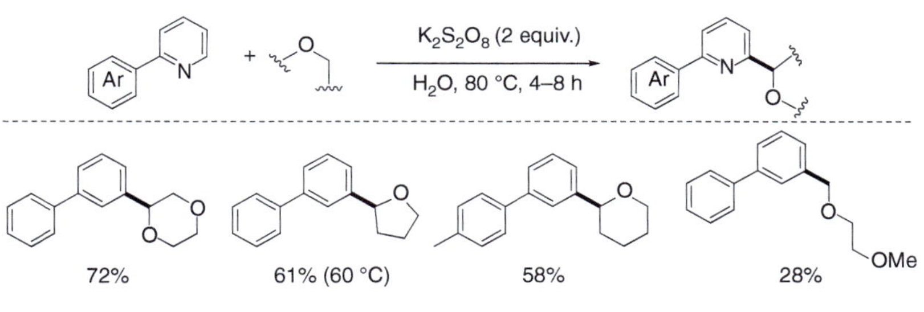

Scheme 10.29 Metal-free CDC between 2-aryl pyridines and cyclic or acyclic ethers.

C-2 position of the pyridine ring. The reaction was proposed to involve the radical addition of an α-oxylalkyl radical on the pyridine ring to produce a delocalized amine radical, which gets oxidized and loses a proton to afford the final product.

Deprotonative Cross-Coupling

Palladium-catalyzed α-arylation of ketones is an important method for the C(sp^2)—C(sp^3) bond formation. However, performing α-arylation entirely in water is quite difficult due to the moisture-sensitive nature of enolates. In 2021, Lipshutz and coworkers developed a micellar-enabled methodology for such transformations in water using TPGS-750-M as the surfactant, *t*-BuOK as the base, and [Pd(μ-Br)(*t*-Bu)$_3$P]$_2$ as the precatalyst (Scheme 10.30) [39]. The use of highly lipophilic *t*-BuOK as the base facilitates the reaction by easy access of the base into the lipophilic micellar inner cores to mediate the enolization. Arylation of aryl methyl ketones generally led to double-arylation products. By increasing the steric bulkiness of aryl halides, mono-arylation could also be achieved. Noteworthily, a one-pot α-arylation/deuteration was also realized by using D_2O as the bulk reaction medium.

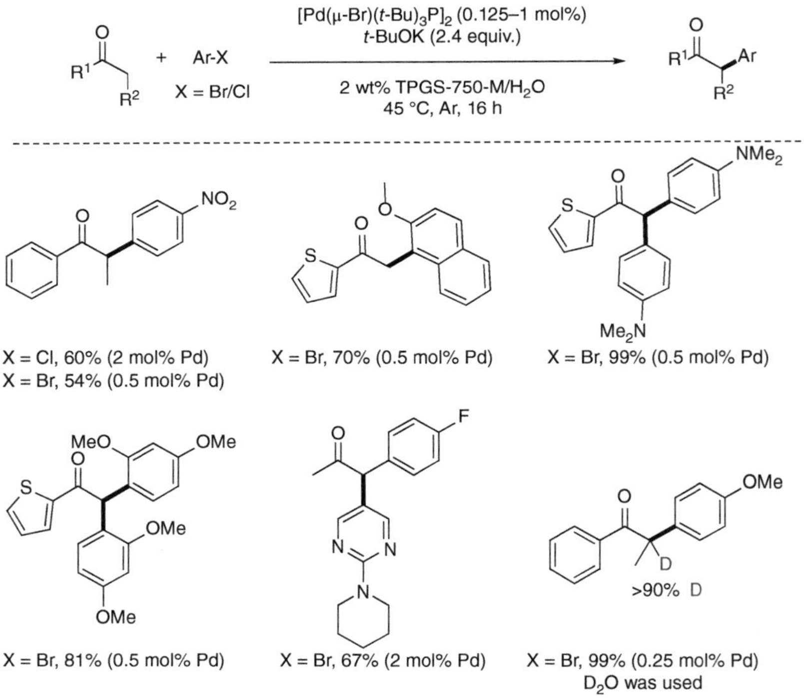

Scheme 10.30 Pd-catalyzed α-arylation of (hetero)aryl ketones.

Dehydrative Cross-Coupling

In 2019, Tang and coworkers reported an additive-free synthesis of terminal skipped dienes from ketone-stabilized phosphonium ylides and allylic alcohols via a one-pot dehydrative cross-coupling and Wittig olefination (Scheme 10.31) [40]. Exclusive γ-selectivity was observed for α-branched terminal allylic alcohols including tertiary allylic alcohols. Water plays an important role in the C–O cleavage step through hydrogen-bond-assisted activation of the allylic hydroxyl group to give a π-allylpalladium complex. This protocol was not suitable for water-sensitive phosphonium ylides.

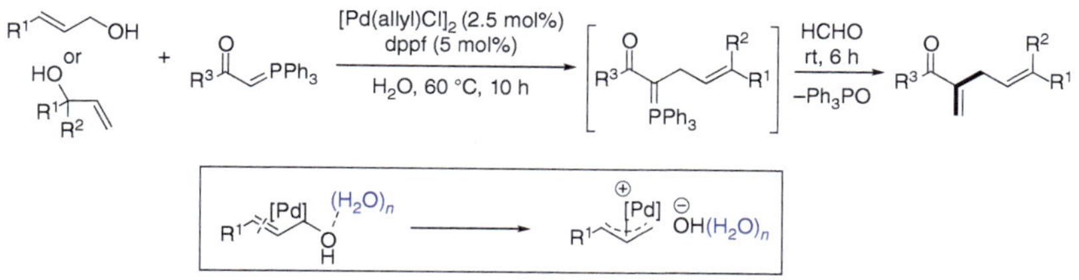

Scheme 10.31 Dehydrative cross-coupling of stabilized phosphonium ylides and allylic alcohols.

Decarboxylative Coupling

In 2020, a visible-light-induced, Ir-photoredox-catalyzed decarboxylative radical coupling of *N*-aryl amino acids with aldehydes and ketones was disclosed by Zhong and coworkers for the synthesis of 1,2-aminoacohols in water at ambient conditions (Scheme 10.32) [41]. The photoredox radical reaction proceeded smoothly without any additives even under the irradiation of sunlight. As to the reaction mechanism, an α-amino radical **A** is first generated via excited $Ir^{[III]}$*-mediated oxidative decarboxylation, which is then cross-coupled with a carbonyl radical anion **B** produced by Ir^{II}-mediated reduction to give the cross-coupled product upon protonation.

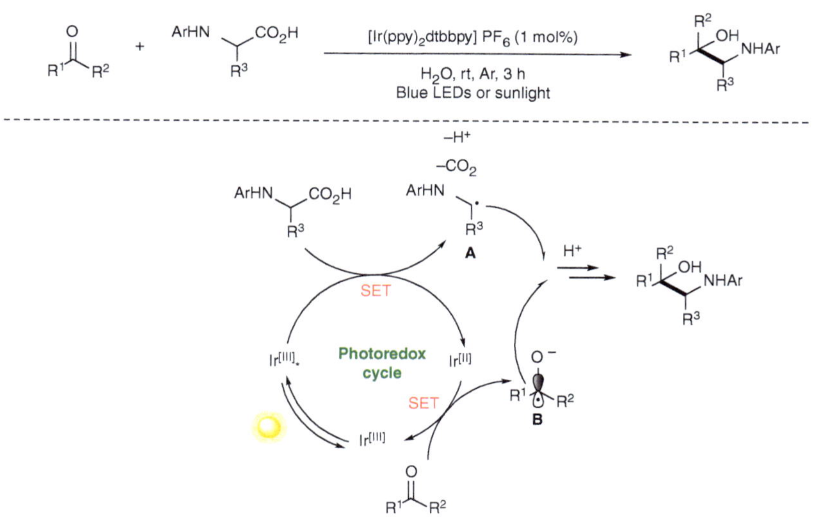

Scheme 10.32 Visible-light photoredox-catalyzed decarboxylative radical coupling reaction of *N*-aryl amino acids with aldehydes or ketones.

In the same year, a Ag_2CO_3-catalyzed, $K_2S_2O_8$-promoted oxidative decarboxylative coupling of α,α-difluoroarylacetic acids with allyl sulfones was described by Yao and coworkers for the synthesis of β,β-difluorinated alkenes in water (Scheme 10.33) [42]. The reaction was proposed to occur through oxidative decarboxylative of the α,α-difluoroarylacetic acid to afford a difluoroalkyl radical, radical addition of the difluoroalkyl radical onto allyl sulfone followed by β-elimination of the sulfone to give the allylated product.

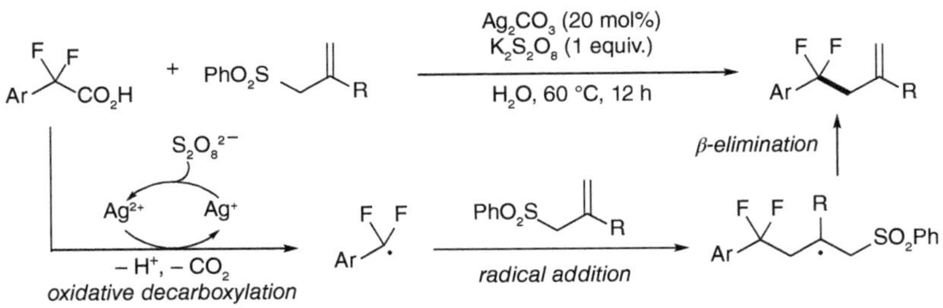

Scheme 10.33 Oxidative decarboxylative coupling of α,α-difluoroarylacetic acids with allyl sulfones.

10.2.2 C–N Cross-Coupling

Buchwald Reaction

In 2019, Lipshutz and coworkers reported a mild, sustainable, and recyclable aqueous micellar-enabled ppm Pd-catalyzed Buchwald reaction between aryl bromides and amines using mono-ligated [*t*-BuXPhosPd(cinnamyl)]OTf as the precatalyst at temperatures between rt and 45 °C (Scheme 10.34) [43]. The protocol compares very favorably with traditional amination conditions in organic media and could occur at faster rates at much lower catalyst loadings and temperatures.

In the same year, Salomé and coworkers disclosed the first *N*-arylation of oxetanyl amines under Buchwald–Hartwig amination conditions in water using TPGS-750-M as the surfactant and *t*-BuONa as the base (Scheme 10.35) [44]. With the assistance of either a $[Pd(cinnamyl)Cl]_2$/*t*-BuXPhos or $[Pd(cinnamyl)Cl]_2$/BippyPhos catalytic system (for difficult substrates), various alkyl, alkenyl, alkynyl, and (hetero)aryl substituted oxetanylamines were smoothly arylated with diverse (hetero)aryl bromides or chlorides at 50 °C. Intramolecular cyclization was also feasible under such conditions.

Recently, Yuan and coworkers reported an ultrafast micelle-enabled *N*-arylation of sulfoximines with (hetero)aryl bromides catalyzed by *t*-BuXPhos Pd G3 in water using TPGS-750-M as the surfactant (Scheme 10.36) [45]. Valuable features of this protocol include low catalyst loading, ultrafast reaction rate (generally 15 min in contrast to several hours required in organic solvents), and equimolar amount of coupling partners. The micellar-induced acceleration effect is crucial for the ultrafast and highly efficient reactions.

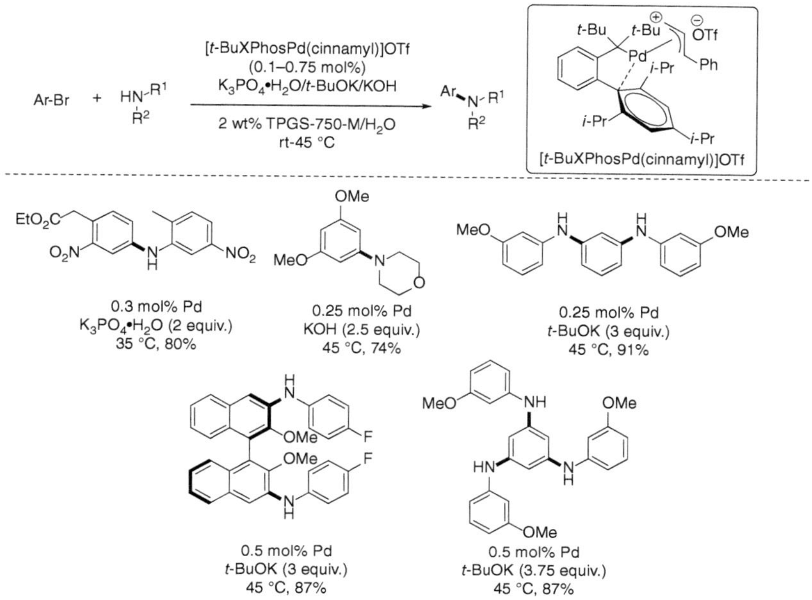

Scheme 10.34 Micellar-enabled ppm Pd-catalyzed Buchwald reaction.

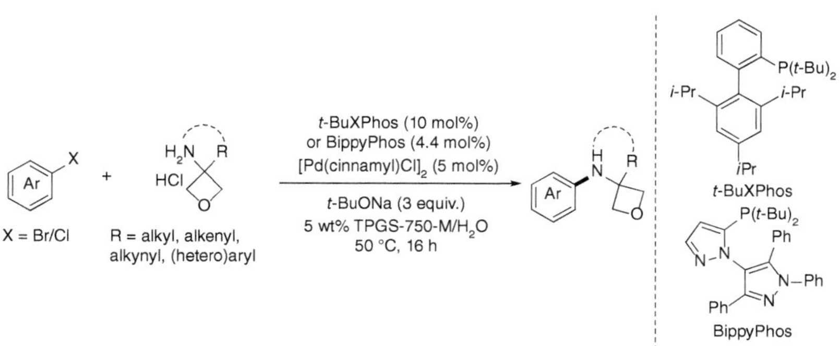

Scheme 10.35 *N*-arylation of oxetanylamines.

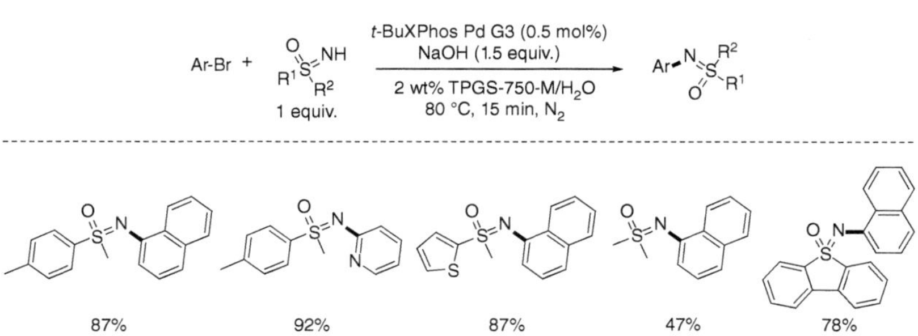

Scheme 10.36 Micelle-enabled *N*-arylation of sulfoximines.

Ullmann-Type Reactions

In 2017, Schmitt and coworkers developed an efficient and cost-effective Ullman-type *N*-arylation of various amides, carbamates, and azoles with (hetero) aryl iodides in aqueous TPGS-750-M solution at mild temperatures (25 or 50 °C) using $CuBr_2$ as a cheap catalyst, racemic *trans*-dimethyl 1,2-cyclohexyldiamine as the ligand and D-glucose as a Cu^{II} reductant (Scheme 10.37) [46]. TPGS-750-M is critical to this transformation since no reaction was observed in its absence. The aqueous catalytic system could be reused twice with a slight decrease in reactivity.

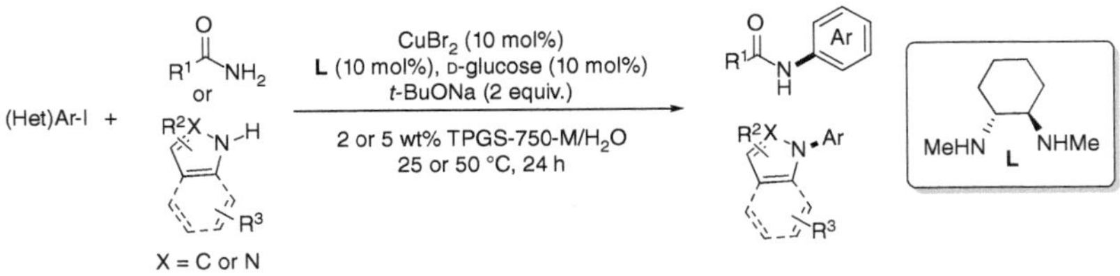

Scheme 10.37 Cu-catalyzed *N*-arylation of amides, carbamates, and azoles.

In the same year, Ren and coworkers reported the synthesis of 1-aryl-1*H*-indazoles via *N*-1 regioselective Ullmann reactions of indazoles with aryl iodides in water using Tween 20 as the surfactant (Scheme 10.38) [47]. The micelle-enabled protocol features mild temperature and short reaction time and was extended to other heterocycles such as indole and indolin-2-one. The aqueous micelles could improve the solubility of starting materials and increase the local concentration of reactants, thus accelerating the reaction rate.

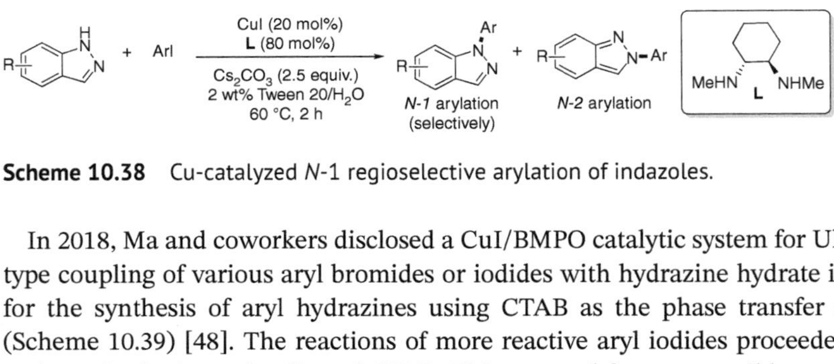

Scheme 10.38 Cu-catalyzed *N*-1 regioselective arylation of indazoles.

In 2018, Ma and coworkers disclosed a CuI/BMPO catalytic system for Ullmann-type coupling of various aryl bromides or iodides with hydrazine hydrate in water for the synthesis of aryl hydrazines using CTAB as the phase transfer catalyst (Scheme 10.39) [48]. The reactions of more reactive aryl iodides proceeded faster and required a lower loading of CTAB. This protocol features a mild temperature (80 °C), a low catalytic loading (2 mol%), and a broad substrate scope.

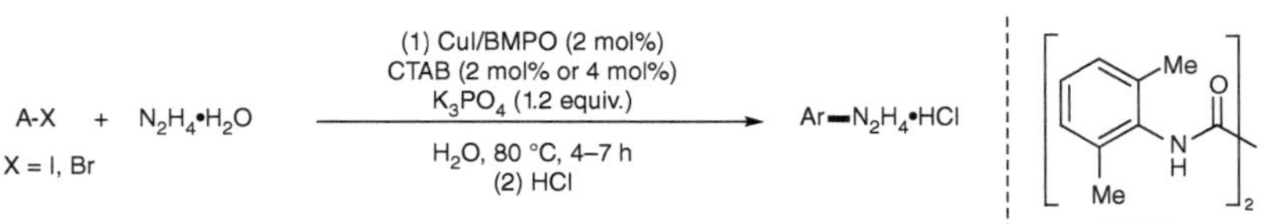

Scheme 10.39 CuI/BMPO-catalyzed coupling of aryl halides with hydrazine hydrate.

Xia and coworkers described a PTC-free, room-temperature Ullmann-type coupling of various aliphatic amines with a wide range of aryl and heteroaryl iodides in neat water under the catalysis of CuI/6,7-dihydroquinolin-8(5*H*)-one oxime in the presence of KOH (Scheme 10.40) [49]. The reaction of (hetero)aryl bromides required a moderate temperature of 65 °C and that of aryl chlorides showed poor reactivity even at 100 °C. A wide variety of functional groups including cyano, nitro, halides (Cl and Br), and even free amino, hydroxyl, and carboxylic acid groups were well-tolerated.

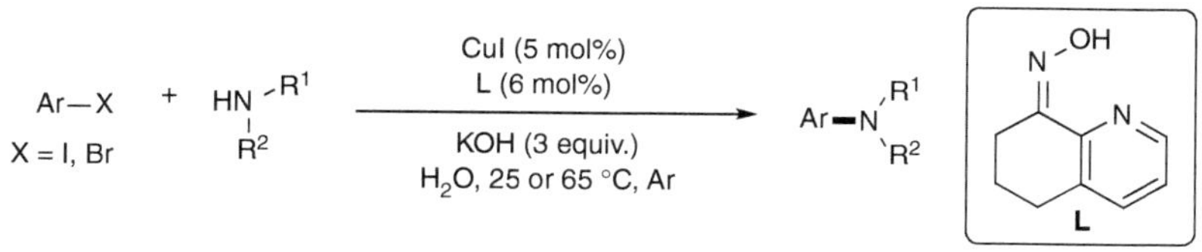

Scheme 10.40 Cu-catalyzed *N*-arylation of aliphatic amines.

In 2020, a CuI/NHC-catalyzed selective *N*-arylation of aminophenols was described by Gudimetla and coworkers in water using 1,3-bis-[2-hydroxyphenyl] imidazolium chloride (IHPHCl) as the ligand (Scheme 10.41a) [50]. The presence of chelating hydroxyl groups on the ligand enhances *N*-selectivity of this transformation. Dash and coworkers also achieved selective *N*-arylation of aminophenols using an easily accessible triazolyl prolinamide derivative as the ligand (Scheme 10.41b) [51]. This protocol was successfully utilized for step-economical syntheses of carbazole alkaloids under protecting group-free conditions.

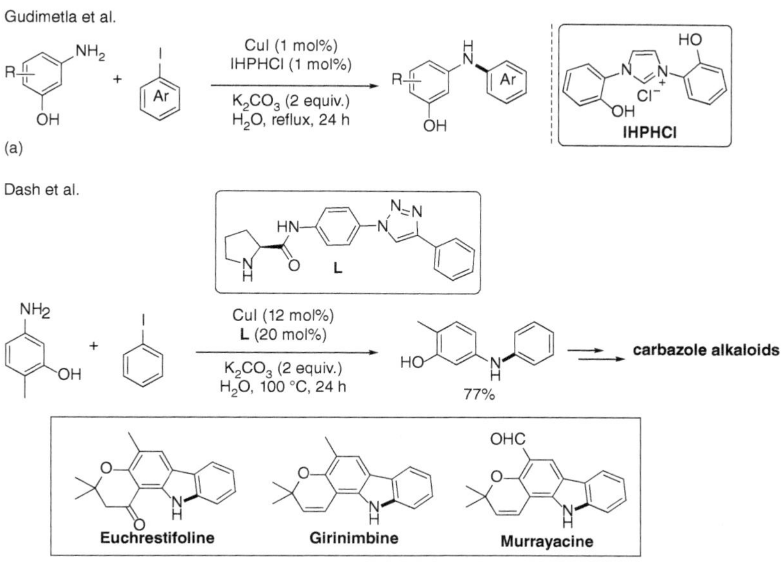

Scheme 10.41 Cu-catalyzed selective *N*-arylation of aminophenols.

In the same year, Chen and coworkers described Cu_2O-catalyzed amination of aryl bromides with aqueous ammonia and hydroxylation of water-soluble aryl iodides with NaOH in water using L-(-)-quebrachitol (L-Que), an optically active cyclitol obtained from wastewater of the rubber industry, as the ligand under PTC-free condition in the presence of air (Scheme 10.42) [52]. For water-insoluble aryl iodides, direct hydroxylation was performed using 95% ethanol as the solvent.

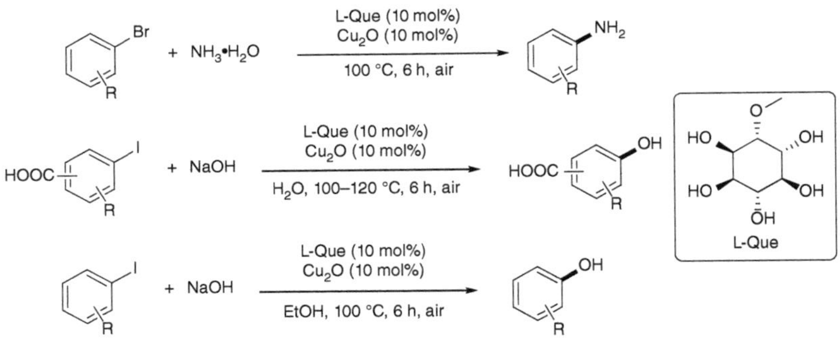

Scheme 10.42 Cu-catalyzed amination and hydroxylation of aryl halides.

Later, Kapdi, and coworkers achieved a room-temperature $Cu(OAc)_2$-catalyzed amination of chloroheteroarenes with a series of primary, secondary amines, and amino acid esters in water using water-soluble PTABS (phosphaadamantan-1-ium-1-yl butane-1-sulfonate or Kapdiphos) as the ligand (Scheme 10.43) [53]. Excellent reactivity of the Cu(II)/PTABS system was also observed even at a very low catalyst loading of 0.0001 mol% and the catalytic system can be recycled 12 times without loss of activity. Notably, the product can be isolated under column-free conditions.

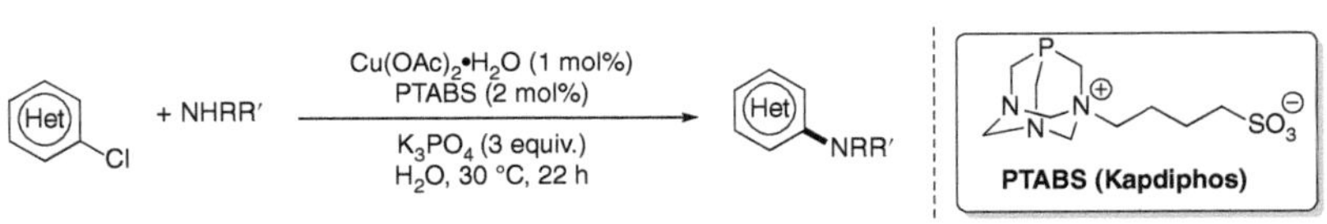

Scheme 10.43 Room-temperature amination of chloroheteroarenes.

Tsuji–Trost *N*-Allylation

In 2020, Lipshutz and coworkers established a versatile ppm Pd-catalyzed Tsuji–Trost reaction under aqueous micellar condition by coupling a variety of allylic acetates, carbonates, benzoates, and even vinylic epoxides with a vast range of nitrogen, carbon and oxygen nucleophiles [54]. An intramolecular version of this reaction was also developed and eventually led to the synthesis of various vinyl-substituted heterocycles such as piperazines, morpholines, and tetrahydroquinoxalines (Scheme 10.44). Various primary/secondary amines and *N*-heterocycles were compatible with the *N*-allylation reactions. Moreover, incorporation of ammonia equivalent was also realized by coupling with di-Boc-protected ammonia.

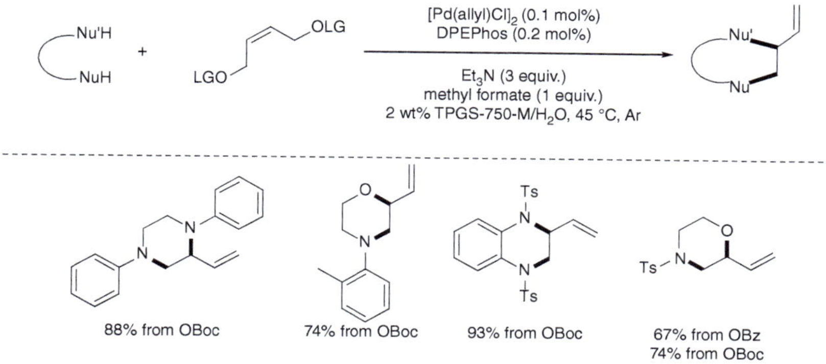

Scheme 10.44 Intramolecular cyclization involving Pd-catalyzed Tsuji–Trost *N*-allylation.

In 2022, a Ni(0)-catalyzed Tsuji–Trost type *N*-allylation of amines was reported by Kumar and coworker using nonactivated allylic alcohols as the allylation sources in aqueous TPGS-750-M solution in the absence of any additives, bases, and activators (Scheme 10.45) [55]. This protocol enables mild and highly selective preparation of allylic amines with excellent chemoselectivities (N vs O), regioselectivities (linear vs branched), and stereoselectivities (E vs Z). The synthetic utility of this protocol was demonstrated in "all-water" syntheses of three pharmaceuticals including flunarizine, cinnarizine, and naftifine. It should be noted that the nano-micelles formed in aqueous TPGS-750-M could protect the Ni(0) species from deactivation and no reaction occurred in neat water where no micelles were present. Moreover, the hydrogen-bonding network of water-assisted facile oxidative addition of allylic alcohol to Ni(0) species.

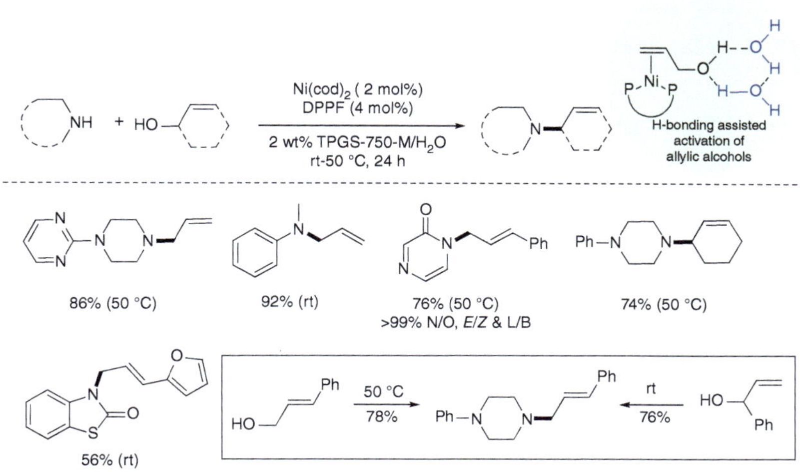

Scheme 10.45 Nickel-catalyzed allylic amination using allylic alcohols.

10.2.3 C–S Cross-Coupling

Babazadeh and coworkers have summarized the progress of metal-catalyzed cross-coupling of aryl halides with thiols during the period of 2007–2017 [5a]. The field of aqueous C–S cross-coupling is rapidly developing and several representative examples reported thereafter are described herein.

Migita Cross-Coupling and Related Reactions

Since the first Pd-catalyzed Migita C–S cross-coupling between aryl halides and thiols, diverse examples of C–S coupling have been reported using palladium catalysts and other transition metal catalysts (Co, Ni, Cu, etc.). In 2021, Stefan and coworkers reported α-diimine Pd(II) complex-catalyzed C–S cross-coupling of a series of (hetero)aryl bromides with alkyl, benzyl or heteroaryl thiols in water under aerobic conditions at a mild temperature using NaOH as the base (Scheme 10.46) [56]. The α-diimine Pd(II) complexes are air- and moisture-stable and were easily prepared from iminopyridine- and acenaphthene-based ligand cores.

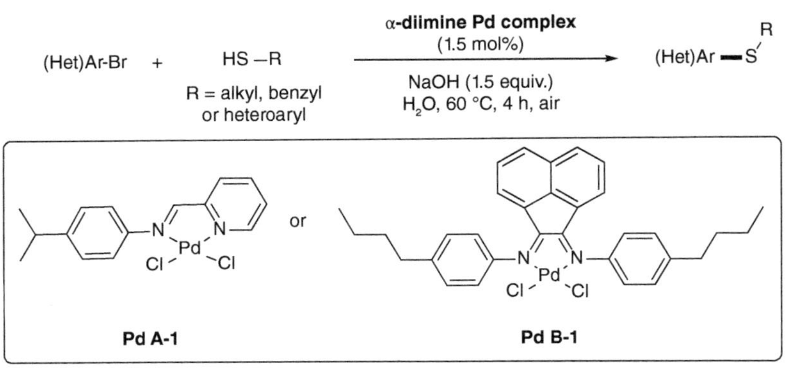

Scheme 10.46 α-Diimine Pd(II) complex-catalyzed C–S cross-coupling.

As a more impressive example, in 2018, Pentelute, Buchwald, and coworkers disclosed a protein–protein cross-coupling of two cysteine-containing proteins under aerobic aqueous condition through the formation of an air-stable Pd-protein oxidative addition complex (Pd-protein OAC) (Scheme 10.47) [57]. The reaction involves initial cross-coupling of the first cysteine-containing protein with a 1,4-diiodobenzene-derived sSPhos-Pd-I complex to provide a hypothetical π-complex intermediate which after an intramolecular oxidative addition affords a stable electrophilic *S*-aryl-Pd-I complex (i.e. Pd-protein OAC). The complex could be isolated and cross-coupled with a second thiol-containing protein to afford the final protein–protein conjugate.

In 2020, Wu and coworkers developed a Cu-catalyzed ligand-free methylthiolation of (hetero)aryl iodides or bromides with dimethyl disulfide in water under air using TBAB as the phase transfer catalyst (Scheme 10.48) [58]. The reaction of aryl bromides required a higher temperature (130 °C) than that of aryl iodides (100 °C).

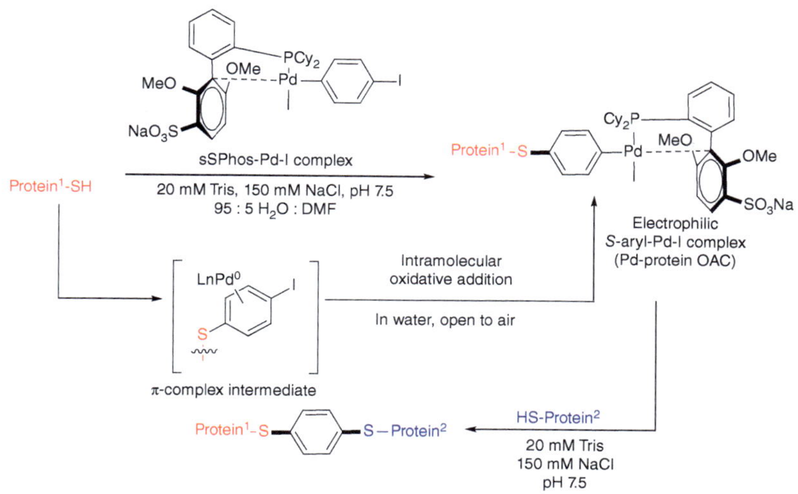

Scheme 10.47 Chemoselective Pd-protein OAC formation and protein–protein cross-coupling.

This protocol features a broad substrate scope and the tolerance of a variety of functional groups such as NO_2, CN, NH_2, SMe, and halogens (F, Cl, Br).

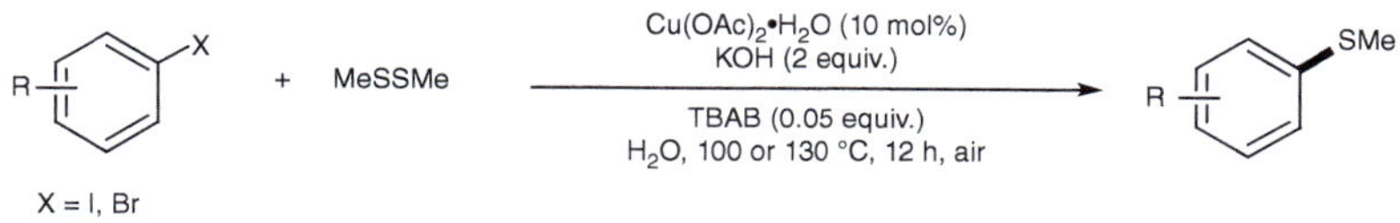

Scheme 10.48 Cu-catalyzed methylthiolation of aryl halides with dimethyl disulfide.

In 2021, a mild, inexpensive, and scalable nickel-catalyzed Migita-like C–S cross-coupling was reported by Lipshutz and coworkers for the synthesis of (hetero)aryl thioethers in recyclable water under the micellar catalysis of TPGS-750-M (Scheme 10.49) [59]. A low loading (0.7-3 mol%) of $Ni(Phen)_2Br_2$ pre-catalyst was

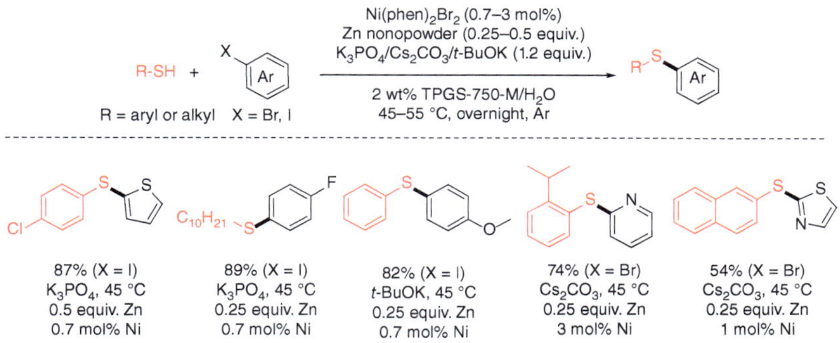

Scheme 10.49 Micellar-enabled nickel-catalyzed Migita-like C–S cross-coupling.

utilized, resulting in a low Ni residue in the final product. A tandem one-pot, two-step sequence involving photocatalysis and thioether formation was also developed in aqueous solution.

Oxidative C–S Coupling

In 2018, a metal-free organophotoredox-catalyzed oxidative C–S cross-coupling between (hetero)arylhydrazines and thiols was reported by Hajra and coworkers at room temperature in water using Rose Bengal as the photocatalyst and air as the oxidant (Scheme 10.50) [60]. Preliminary mechanistic studies indicated the presence of a disulfide intermediate which is formed by dimerization of a thiyl radical species. Moreover, an aryl radical is evolved via photocatalyzed denitrogenation of arylhydrazines and is cross-coupled with the disulfide to afford the coupling product. By using diphenyldiselenide as the substrate, a (diaryl)selane product was also obtained.

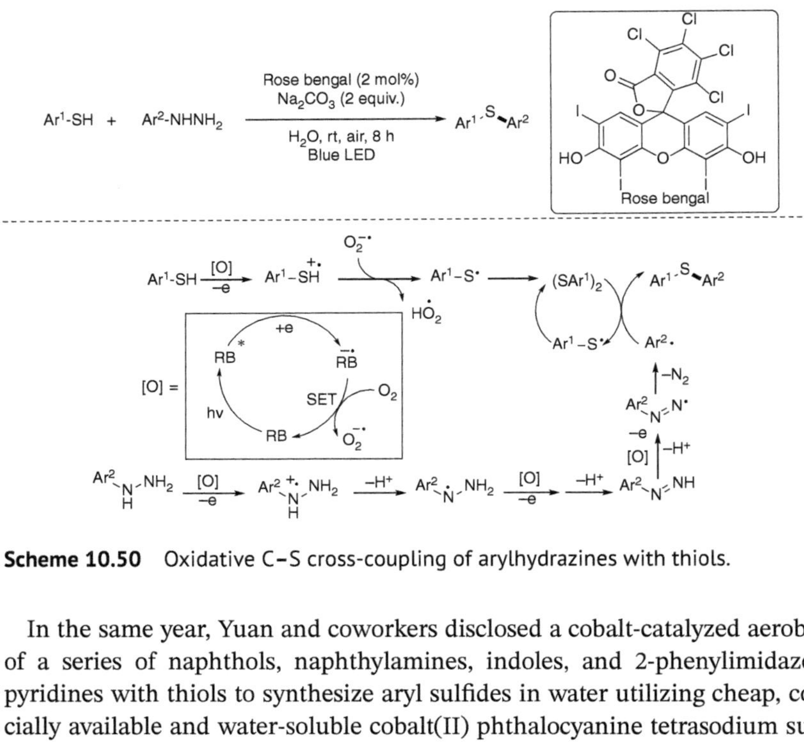

Scheme 10.50 Oxidative C–S cross-coupling of arylhydrazines with thiols.

In the same year, Yuan and coworkers disclosed a cobalt-catalyzed aerobic CDC of a series of naphthols, naphthylamines, indoles, and 2-phenylimidazo[1,2-*a*]pyridines with thiols to synthesize aryl sulfides in water utilizing cheap, commercially available and water-soluble cobalt(II) phthalocyanine tetrasodium sulfonate (CoPcS) as the catalyst and O_2 as the sole oxidant (Scheme 10.51) [61]. A variety of (hetero)aryl thiols and alkyl thiols were well-tolerated for this transformation. The proposed mechanism also involves a cross-coupling between an aryl radical and a disulfide intermediate which is formed by dimerization of a thiyl radical species.

Reductive C–S Coupling

In 2018, He and coworkers developed a step- and atom-economical protocol for the synthesis of sulfonylated *N*-heteroarenes from readily available

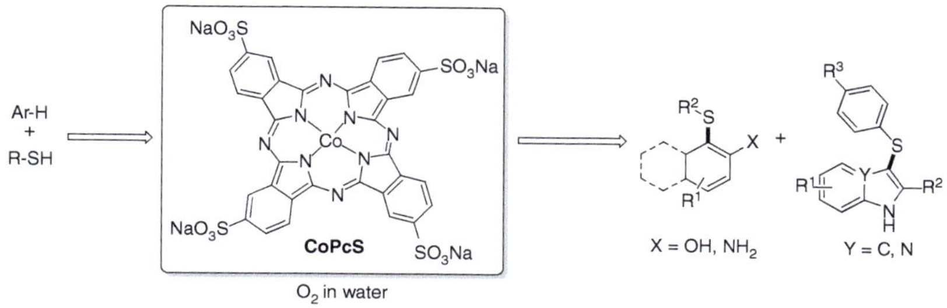

Scheme 10.51 Cobalt-catalyzed aerobic cross-dehydrogenative coupling of C–H and thiols.

(pseudo)halogenated *N*-heteroarenes and sulfonyl chlorides under metal- and base-free conditions in water using Na_2SO_3 as the reductant (Scheme 10.52) [62]. Water is essential for this transformation since no trace product was observed in organic solvents. Either by hydrolysis of a catalytic amount of sulfonyl chloride or by self-ionization, water provides a proton to activate the *N*-heteroarene toward nucleophilic attack by a sulfonyl anion generated *in situ* from reduction of the sulfonyl chloride. Notably, the pure products can be simply obtained by filtration and alcohol washing.

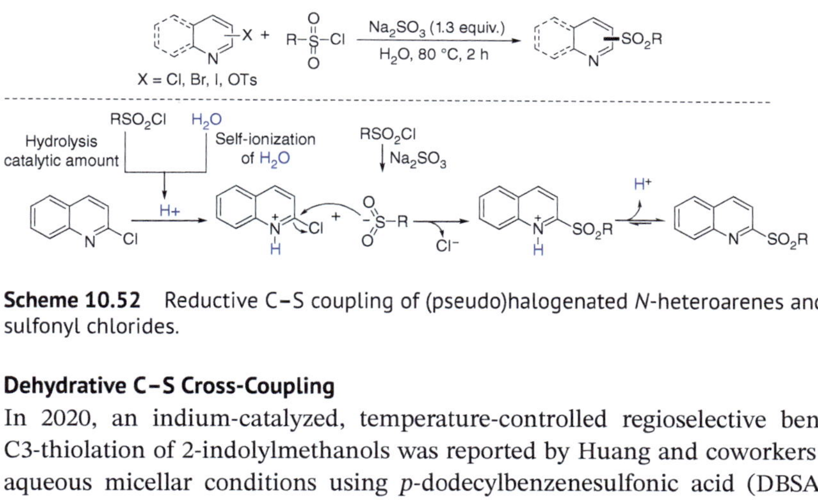

Scheme 10.52 Reductive C–S coupling of (pseudo)halogenated *N*-heteroarenes and sulfonyl chlorides.

Dehydrative C–S Cross-Coupling

In 2020, an indium-catalyzed, temperature-controlled regioselective benzyl or C3-thiolation of 2-indolylmethanols was reported by Huang and coworkers under aqueous micellar conditions using *p*-dodecylbenzenesulfonic acid (DBSA) as a Brønsted acidic surfactant-combined catalyst (Scheme 10.53) [63]. The dehydrative C–S cross-coupling regioselectively afforded benzylic sulfenylation products as kinetic products at 25 °C and C3-sulfenylation products as thermodynamic products at 90 °C. Notably, the benzylic sulfenylation products are heat sensitive and could be converted to C3-sulfenylation products at a higher temperature through a benzylic cation species.

Radical C–S Coupling

Recently, Huang and workers developed a metal-free, organophotocatalyzed coupling of acyl oxime acetates with thiophenols for the synthesis of nitrile-bearing

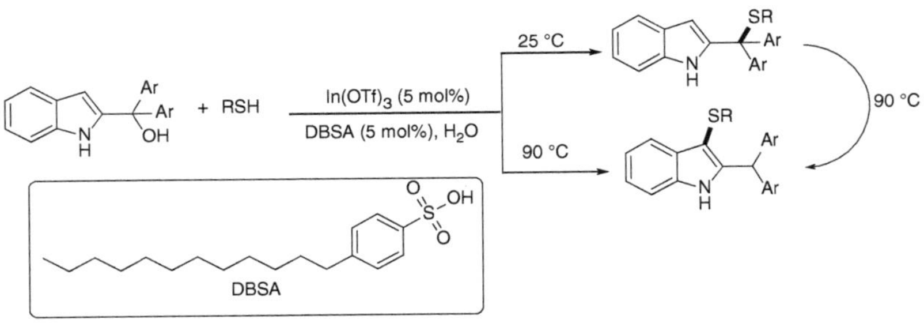

Scheme 10.53 In(OTf)$_3$-catalyzed, temperature-controlled regioselective benzyl or C3-thiolation of 2-indolylmethanols.

arylthioesters in water using Eosin Y as the photocatalyst and MPEG-550 [poly(ethylene glycol) methyl ether, average Mn 550] as the phase-transfer catalyst (Scheme 10.54) [64]. Preliminary mechanistic experiments suggested a radical–radical cross-coupling of an acyl radical and a thiyl radical species. The acyl radical is formed by ring-opening of an iminyl radical generated by photocatalyzed single electron reduction of acyl oxime acetate. Water is crucial for this transformation since no product was detected when MPEG-550 or THF was used as the sole solvent.

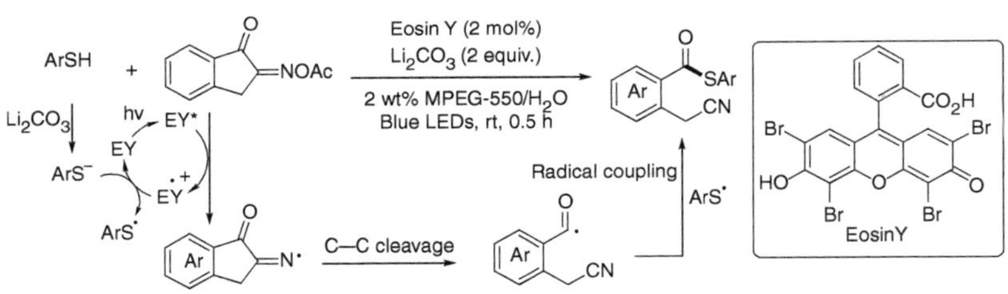

Scheme 10.54 Photoredox-catalyzed coupling of acyl oxime acetates with thiophenols.

10.2.4 C–P Cross-Coupling

Compared with other types of cross-coupling, the C–P cross-coupling is relatively fewer. In 2017, Uozumi and coworkers reported the synthesis of catalytically useful aryl(dicyclohexyl)phosphines through Pd-catalyzed C–P cross-coupling between aryl bromides and dicyclohexylphosphine under heterogeneous conditions in water using a palladium complex of an amphiphilic polystyrene–poly(ethylene glycol) (PS–PEG) resin-supported alkyl[di(*tert*-butyl)]phosphine ligand (Scheme 10.55) [65]. The alkyl[di(*tert*-butyl)]phosphine ligand coordinates stronger to palladium than the aryl(dicyclohexyl)phosphine product, thus resisting product inhibition in the catalytic cycle. The resin catalyst was easily recycled and was reused several times with constant high efficiency.

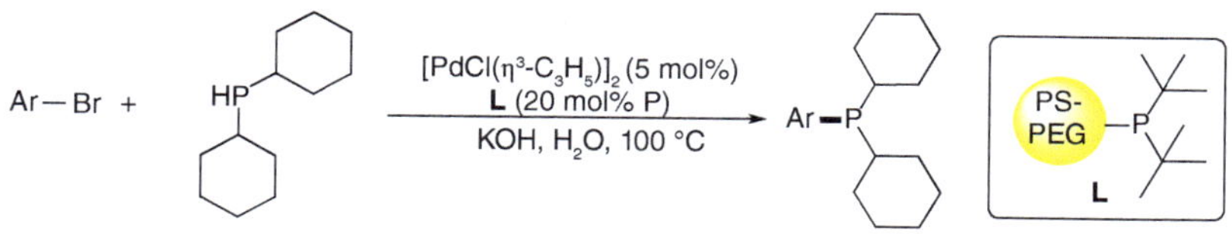

Scheme 10.55 Pd-catalyzed C–P cross-coupling for the synthesis of aryl(dicyclohexyl)phosphines.

10.3 C–H Functionalization

In the past two decades, transition metal-catalyzed C–H bond functionalization has become a suitable atom- and step-economical alternative to cross-coupling reactions since prefunctionalization is not required and no metal salts are generated as waste. This section will focus on different types of aqueous C–H functionalization/annulation reactions under metal-catalyzed, photo-catalyzed, or metal-free conditions and is organized by the reaction bond that is created (C–C, C–N, C–O, C–X).

10.3.1 C–C Bond Formation

C–H Arylation

In 2018, Požgan and coworkers developed a highly efficient and selective protocol for the synthesis of hexa(heteroaryl)benzenes via Ru(II)-catalyzed microwave-promoted multiple C–H heteroarylations of benzenes bearing one-, two-, or three-*N*-heteroaryl directing groups (Scheme 10.56) [66]. The multiple C–H bond activation proceeded smoothly in surfactant-free water in a propeller-like manner where a newly introduced pyridyl group severed as a chelation group for the neighboring C–H activation. Notably, C2 or C3 symmetric hexa(heteroaryl)benzenes were prepared by four- or three-fold C–H activation of 1,4-di(pyrimidin-2-yl)benzenes or 1,3,5-tris(pyridyl)benzenes. This C–H functionalization protocol enables the synthesis of multidentate ligands for selective complexation with metals to produce metal complexes for catalysis and photoredox systems. It should be noted that both water and microwave irradiation are crucial for the highly selective formation of hexa(heteroaryl)benzenes.

In 2019, Ackermann and coworkers reported a micellar-enabled Ru(II)-catalyzed weakly coordinating thioketone-directed C–H arylation of ferrocenes with (hetero)aryl bromides in water using TPGS-750-M as the surfactant (Scheme 10.57) [67]. The micelles formed in aqueous TPGS-750-M solution are key to this transformation since a trace product was formed when pure water was used as the reaction medium. It is notable that the micelles could be reused four times with unaffected catalytic performance.

Recently, Elias and coworkers synthesized a series of bis-arylated/methylated ferrocene derivatives via a ligand-free Pd-catalyzed, 8-aminoquinoline carboxamide or picolinamide-directed "on water" double C–H functionalization of ferrocene derivatives with aryl/methyl iodides using water as the sole solvent in an open

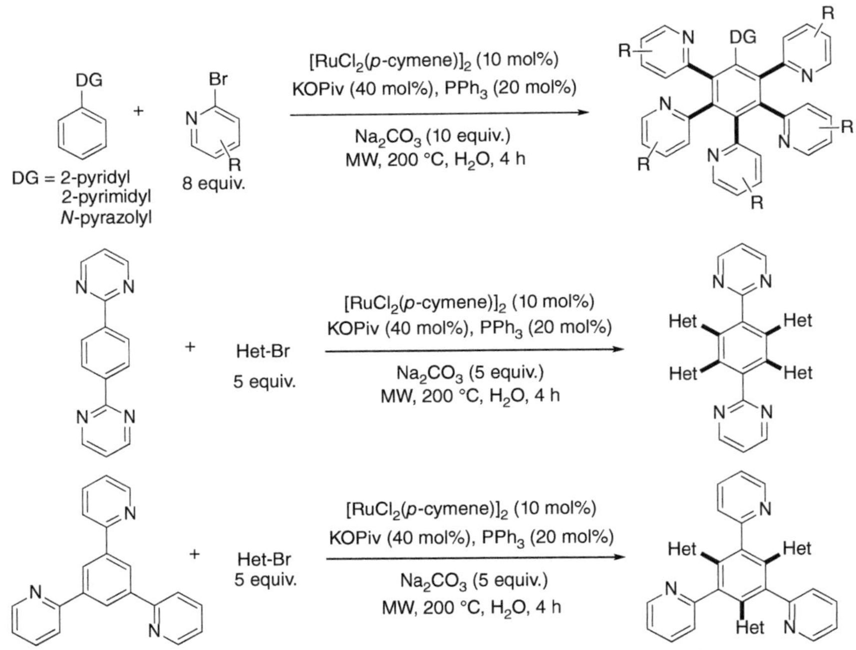

Scheme 10.56 Ru(II)-catalyzed microwave-promoted multiple C–H heteroarylations.

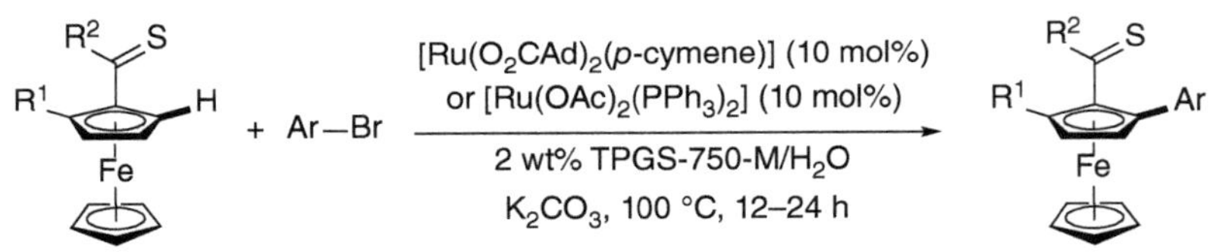

Scheme 10.57 Micellar-enabled Ru(II)-catalyzed thioketone-directed C–H arylation of ferrocenes.

atmosphere under surfactant-free conditions (Scheme 10.58) [68]. It is notable that the hydrophobic effect exerted by water accelerates the reaction by forming droplets surrounded by water molecules. The reusability of the solvent and catalyst was demonstrated and the *in situ* palladacycle formation was supported.

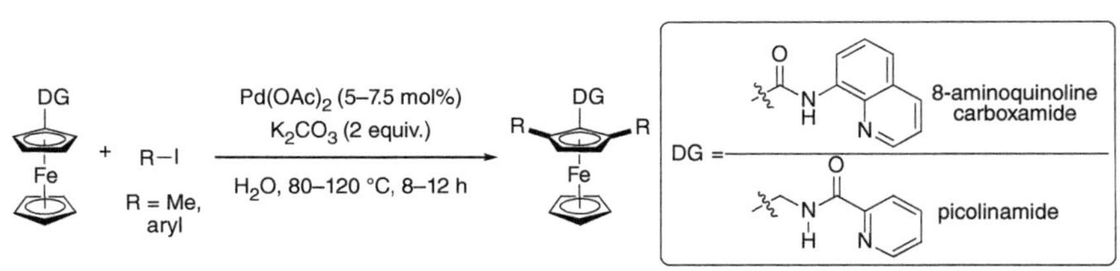

Scheme 10.58 "On water" double C–H functionalization of ferrocene derivatives.

Besides aryl halides, arylsilanes have also been utilized in aqueous C–H arylation reactions. In 2018, Szostak and coworkers reported a Ru(II)-catalyzed, *N*-pyrimidyl-directed C-2 oxidative C(sp^2)–H arylation of indoles with arylsilanes in water using $[RuCl_2(p\text{-cymene})]_2$ as the catalyst, $AgSbF_6$ as the co-oxidant, and CuF_2 as the silane activator and Ru(0) reoxidant (Scheme 10.59a) [69]. This protocol represents the first Ru(II)-catalyzed oxidative C–H arylation in water and it is also suitable for the direct C–H arylation of 2-phenylpyridines. More recently, a Rh(III)-catalyzed C–H Hiyama cross-coupling between benzimidates and arylsilanes was reported by Luo and coworkers for synthesizing biphenyl-2-carbonitriles in water (Scheme 10.59b) [70]. The benzimidate not only serves as a directing group but also as a nitrile source. The optimized condition utilized $[Cp^*RhCl_2]_2$ as the catalyst, Ag_2O as the oxidant, TBAF as the silane activator, and PivOH/$AgSbF_6$ as the additives.

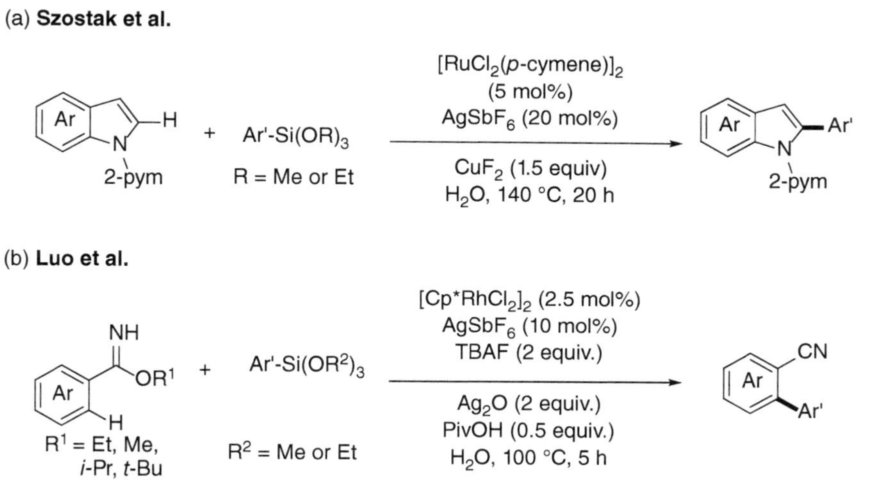

Scheme 10.59 Ru(II) or Rh(III)-catalyzed C–H arylation of indoles with arylsilanes.

In addition to the above aqueous C(sp^2)–H arylation reactions, examples of C(sp^3)–H arylation in water are also emerging. In 2018, Chandrasekharam and coworkers reported a copper-catalyzed α-(sp^3)–H indolation of tetrahydroquinolines in water under open-air conditions by using a CDC strategy (Scheme 10.60) [71]. Water and air were found to be beneficial for this transformation. A catalytic cycle involving nucleophilic attack of indole on an imine intermediate was proposed. Both the N–H protons of tetrahydroquinoline and indole are essential for this transformation.

In 2019, Chen and coworkers reported a Pd-catalyzed, picolinamide (PA)-directed synthesis of cyclophane-braced peptide macrocycles via intramolecular γ-C(sp^3)–H arylation of various aliphatic amino acid units at the *N*-terminus of peptides (Scheme 10.61) [72]. It is remarkable that the C(sp^3)–H cyclization reaction of unprotected peptide substrates bearing various free polar side chains proceeded well under aqueous condition. The authors proposed that the slightly acidic aqueous

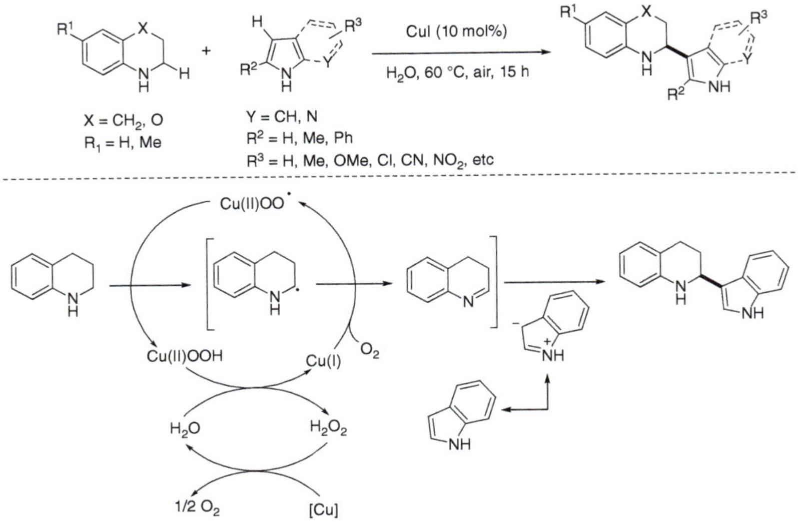

Scheme 10.60 Cu-catalyzed α-(sp^3)–H indolation of tetrahydroquinolines.

medium would weaken the undesired coordination between the polar groups and Pd(II), thus promoting the PA-directed C–H arylation.

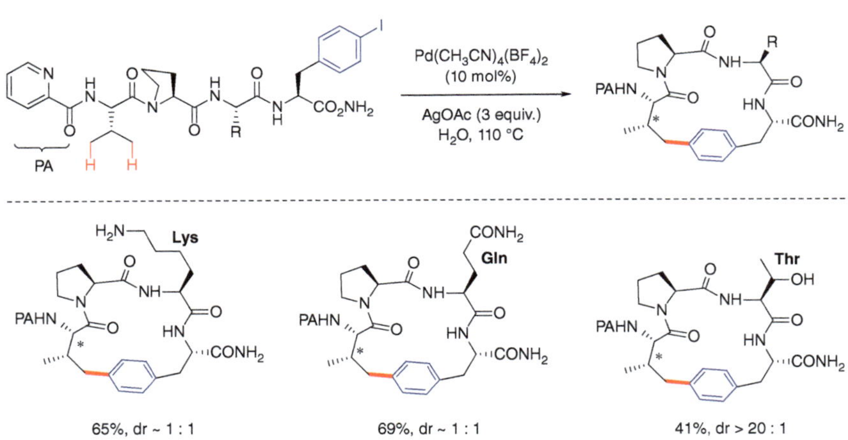

Scheme 10.61 Peptide macrocyclization via PA-directed intramolecular γ-C(sp^3)–H arylation.

In 2021, Hu and coworkers described a silver-catalyzed δ-C(sp^3)–H heteroarylation of free alcohols under gentle thermal condition in water via a 1,5-hydrogen atom transfer (HAT) process (Scheme 10.62) [73]. A variety of terminal secondary and tertiary alcohols were heteroarylated at the unactivated δ position with a series of electron-deficient heteroarenes using H_2SO_4 or $BF_3.Et_2O$ as the activator and

$K_2S_2O_8$ as the oxidant. The proposed mechanism involves 1,5-HAT of an alkoxy radical species to afford a δ-carbon radical, which then reacts with an acid-activated heteroarene followed by oxidation and deprotonation to afford the final product.

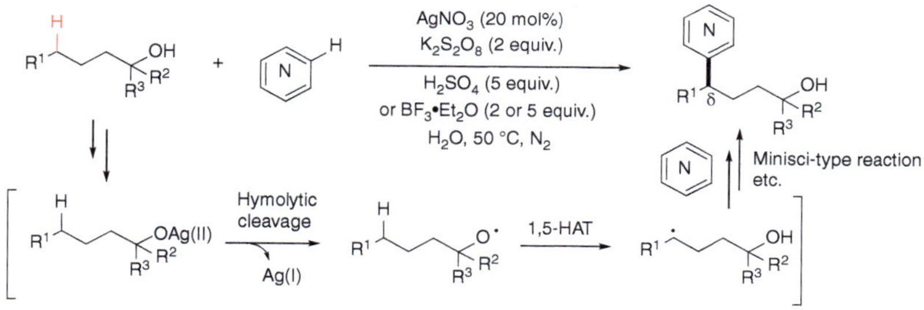

Scheme 10.62 Silver-catalyzed δ-C(sp^3)–H heteroarylation of free alcohols via 5-HAT.

More recently, Giedyk and coworkers developed a photocatalyzed chemo-divergent C(sp^3)–H arylation or *N*-dealkylation of *ortho*-halogenated benzamides under aqueous micellar catalysis using methylene blue (MB) as the photocatalyst and amine as the electron donor (Scheme 10.63) [74]. The choice of amine is critical to the reaction outcome of this transformation. While the combination of CTAB and TMEDA favors the formation of isoindolinones via intramolecular C(sp^3)–H arylation of an *N*-acyliminium radical intermediate, the use of SDS and *n*-$BuNH_2$ prefers the formation of *N*-dealkylation products through the intermediacy of a highly electrophilic *N*-acyliminium cation. The α-aminoalkyl radical is possibly generated by photo-induced reductive activation of an aromatic C–X bond followed by an intramolecular 1,5-HAT process.

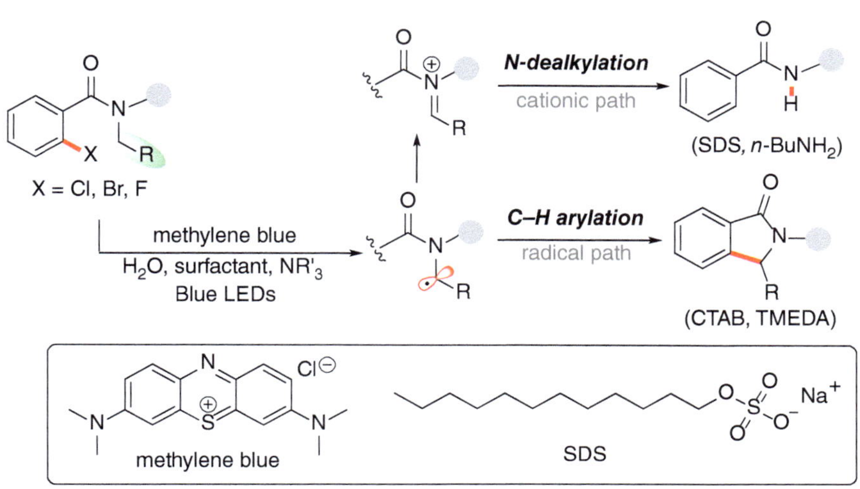

Scheme 10.63 Photocatalyzed chemo-divergent C(sp^3)–H arylation or *N*-dealkylation of *ortho*-halogenated benzamides.

C–H Alkenylation

In 2018, a transition-metal-controlled selective C(sp^2)–H mono- or di-olefination of arenes was reported by Wu and coworkers in water at a high temperature using [Ru(*p*-cymene)Cl_2]$_2$ or [Cp*RhCl_2]$_2$ as the catalyst, respectively (Scheme 10.64) [75]. One year later, by using [Cp*RhCl_2]$_2$ as the catalyst at room temperature in water, the same group further achieved mono-olefination of arenes using $AgSbF_6$ as the additive and Cu(OAc)$_2$ as the oxidant (Scheme 10.65) [76]. Solvent screening indicated that the reaction only occurred in hydrogen bond-forming solvents such as water, EtOH, and THF. Based on this observation, the C–H bond activation might occur through a water-assisted concerted metalation–deprotonation (CMD) process.

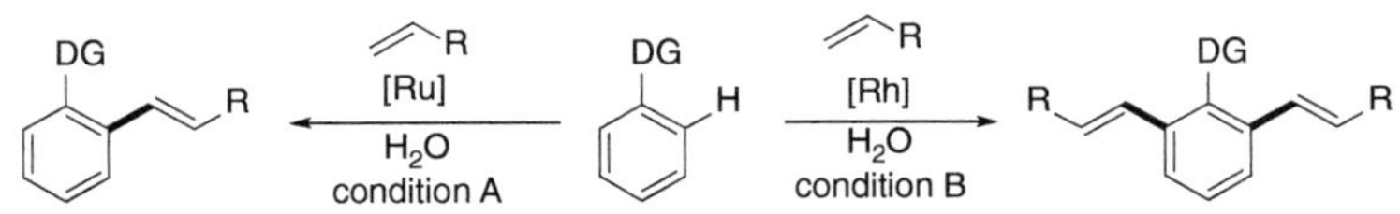

DG = 1,2,3-triazole, pyrazole, pyridine, pyrimidine, amide and oxime ether

Condition A: [Ru(*p*-cymene)Cl_2]$_2$ (5 mol%), $AgNTf_2$ (20 mol%), Cu(OAc)$_2$•H_2O (1 equiv), 120 °C, Ar, H_2O

Condition B: [Cp*RhCl_2]$_2$ (1 mol%), $AgNTf_2$ (20 mol%), Cu(OAc)$_2$•H_2O (4 equiv), 120 °C, Ar, H_2O

Scheme 10.64 Transition-metal-controlled selective C(sp^2)–H mono- or di-olefination of arenes.

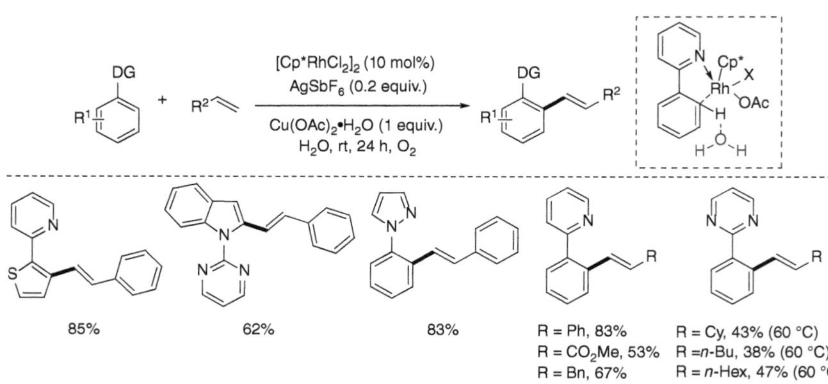

Scheme 10.65 [Cp*RhCl_2]$_2$-catalyzed monoolefination.

In 2019, Joshia and coworkers achieved a temperature-controlled selective mono- or di-vinylation of 1-phenylpyrazoles with styrenes or acylates in water using a low catalytic loading (0.5 mol%) of synthesized *N,O*-salicylaldimine Rh(III) catalyst (Scheme 10.66) [77]. The monovinylated product could undergo further vinylation with another styrene substrate, thus allowing the preparation of unsymmetrical di-alkenylated products.

Recent focus of C(sp^2)–H activation has been shifted from precious 4d and 5d transition metals to cost-effective, earth-abundant base metals. In 2020, Ruan and coworkers described an aqueous Mn-catalyzed C–H enaminylation of

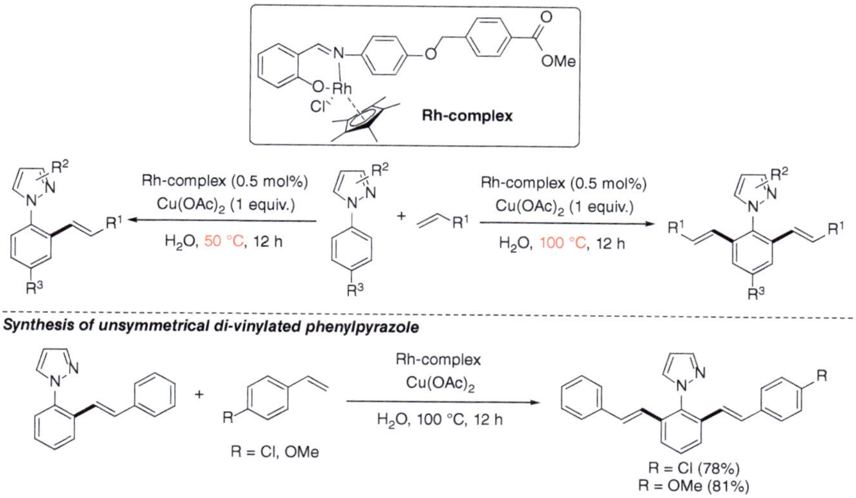

Scheme 10.66 Temperature-controlled selective mono- or di-vinylation of 1-phenylpyrazoles and synthesis of unsymmetrical di-vinylated phenylpyrazole.

6-(1*H*-indol-1-yl)-purines with various *N*-aryl or alkyl-substituted ketenimines to synthesize novel indole-purine hybrids with potent anti-tumor activities (Scheme 10.67) [78]. By using $MnBr(CO)_5$ as the catalyst and NaOAc as the additive, this transformation proceeded smoothly in water, delivering a series of C–H enaminylated products with high levels of chemo, site, and regioselectivities.

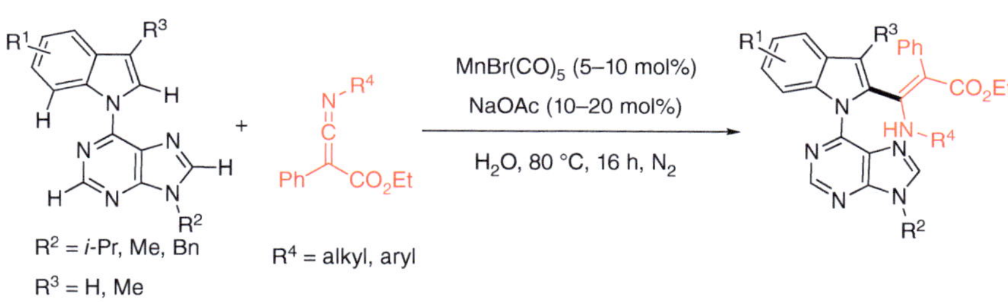

Scheme 10.67 Mn-catalyzed C–H enaminylation of 6-(1*H*-indol-1-yl)-purines with ketenimines.

In 2021, by using a water-soluble rhenium complex $[Re(CO)_3(H_2O)_3]Br$ as the catalyst, Yao and coworkers described an exclusive C2-*Z*-selective alkenylation of *N*-pyridyl- or *N*-pyrimidylindoles with ynamides in water under ligand and acid-free conditions using NaOPiv as the additive (Scheme 10.68) [79]. Unlike previous studies, this Re-catalyzed protocol delivered *N*-alkenylation products rather than expected C–H alkenylation products when *N*-pyrimidylanilines were used as substrates. A five-membered rhenacycle intermediate was involved in the tentative mechanism. After extraction, the aqueous layer can be reused for an additional catalytic cycle with only slight loss of activity.

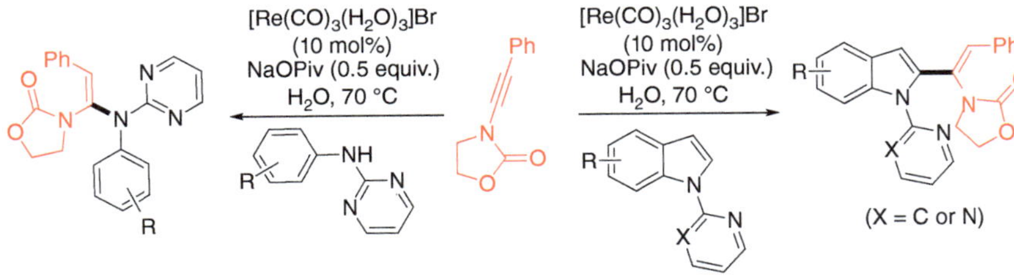

Scheme 10.68 C2-*Z*-selective alkenylation of *N*-pyridyl-/*N*-pyrimidylindoles and N–H alkenylation of *N*-pyrimidylaniline with ynamides.

C–H Acylation

In 2018, Singh and coworkers described a room-temperature Ag(I)-catalyzed, $K_2S_2O_8$-mediated C–H oxidative acylation of electron-deficient heteroarenes with alkynes using water as both the solvent as well as the oxygen source of the acyl group (Scheme 10.69) [80]. Various pyridine-*N*-oxides and fused heteroarenes *N*-oxides including quinoline *N*-oxide, quinoxaline *N*-oxide, and isoquinoline *N*-oxide were amenable for such transformation.

Scheme 10.69 Oxidative acylation of electron-deficient heteroarenes with alkynes.

In 2021, Laha and coworkers reported modified Minisci aroylation of isoquinolines, indoles, and quinolines with arylglyoxylic acids using environmentally benign choline persulfate or its pre-composition (choline acetate and $K_2S_2O_8$) in water at a mild temperature (Scheme 10.70) [81]. Unlike conventional Minisci acylation which produces toxic metal salts due to the use of metal persulfates, this protocol is more environmentally friendly and only produces biodegradable choline sulfate as the byproduct. A nucleophilic acyl radical is likely involved and reacts with the *N*-heterocycle to afford the aroylated product.

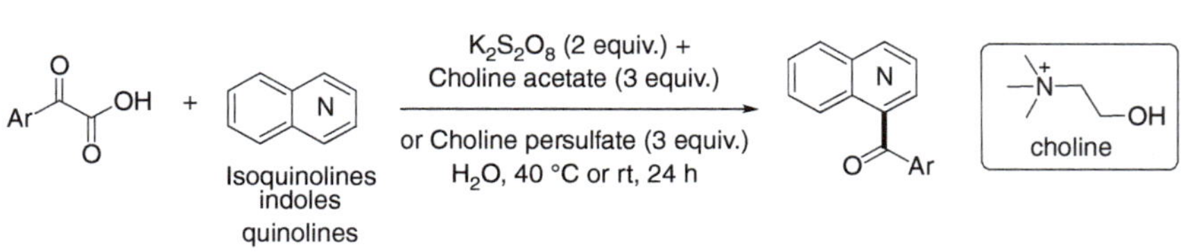

Scheme 10.70 Minisci aroylation of *N*-heterocycles using choline persulfate.

Xuan and coworkers described a photocatalyst-free, visible light- and PIDA-promoted C3–H decarboxylative acylation of quinoxaline-2(1*H*)-ones with α-oxo-carboxylic acids in water under a mild room-temperature condition (Scheme 10.71) [82]. This protocol tolerates a variety of *N*-unsubstituted and substituted quinoxalin-2(1*H*)-ones and allows late-stage modification of natural and pharmaceutically relevant molecules. Mechanistically, ligand exchange of α-oxo-carboxylic acids with PIDA affords a new hypervalent iodine(III) reagent. Visible light-promoted I–O bond homolysis of this reagent followed by fragmentation of the resulting *I*-centered and *O*-centered radical species provides the key acyl radical for subsequent C–H acylation.

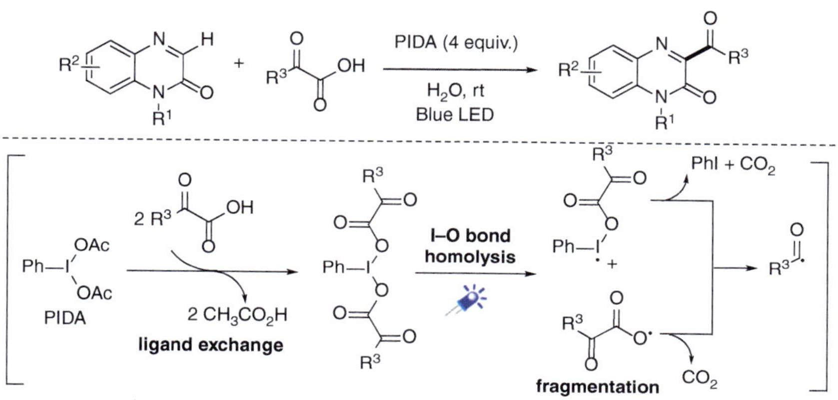

Scheme 10.71 C3–H decarboxylative acylation of quinoxaline-2(1*H*)-ones with α-oxo-carboxylic acids.

As a more interesting example, Correa and coworkers reported a Pd-catalyzed 2-pyridyl ether (OPy)-directed site-selective C–H acylation of Tyr-containing oligopeptides with aldehydes using water as the solvent and TBHP as the oxidant (Scheme 10.72) [83]. The reaction provided a mixture of mono- and di-acylated

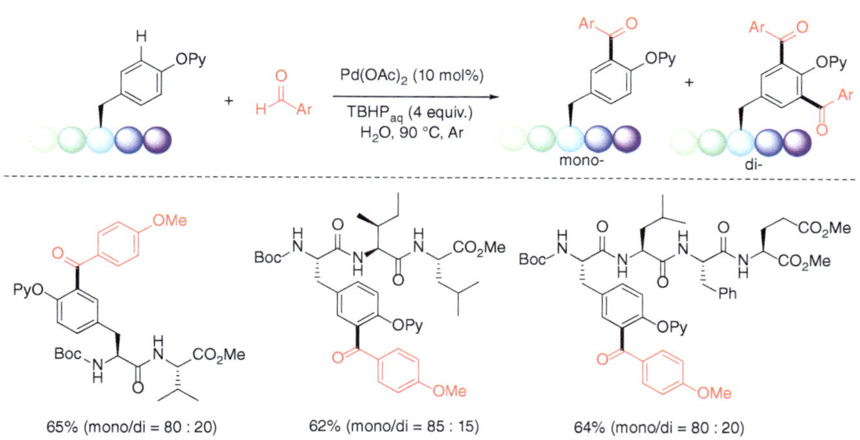

Scheme 10.72 Pd-catalyzed C(sp^2)–H aroylation of Tyr-containing oligopeptides with aldehydes.

products, with a preferential selectivity toward the monoacylated product. Consecutive C(sp^2)–H acylation was also realized by reacting with two distinct aldehydes, despite that the second acylation was carried out in toluene. The developed protocol allows selective late-stage labeling of the tyrosine residue within a peptide framework. A variety of Tyr-containing di-, tri-, and tetra-peptides were successfully acylated in moderate to good yields.

C–H Alkynylation

In 2019, a Pd-catalyzed, primary amine-directed C2-regioselective C(sp^2)–H alkynylation of indoles and pyrroles with bromoalkynes was developed by Jiang and coworkers using water as the solvent under air (Scheme 10.73) [84]. It is worthwhile to note that this transformation renders chemoselective construction of the C–C bond instead of the C–N bond. Moreover, water is beneficial to this reaction probably by mediating the protodemetalation step of the catalytic cycle.

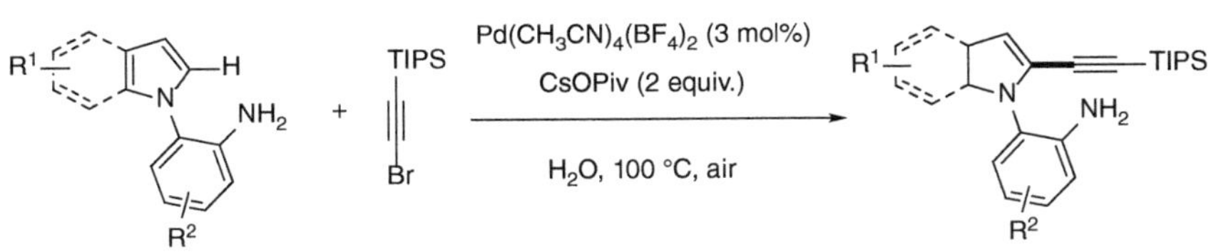

Scheme 10.73 Pd-catalyzed, primary amine-directed C2-regioselective C(sp^2)–H alkynylation of indoles and pyrroles with bromoalkynes.

C–H Cyanation

In 2019, Li and coworkers reported the synthesis of alkenyl nitriles via "on water" C(sp^2)–H cyanation of (*E*)-3-arylideneindolin-2-ones with TMSCN in the presence of KOH using $CeCl_3 \cdot 7H_2O$ as a mild water-tolerable Lewis acid catalyst and atmospheric air as the oxidant (Scheme 10.74) [85]. A brief solvent screening suggested water as the optimum solvent for this transformation. The reaction was proposed

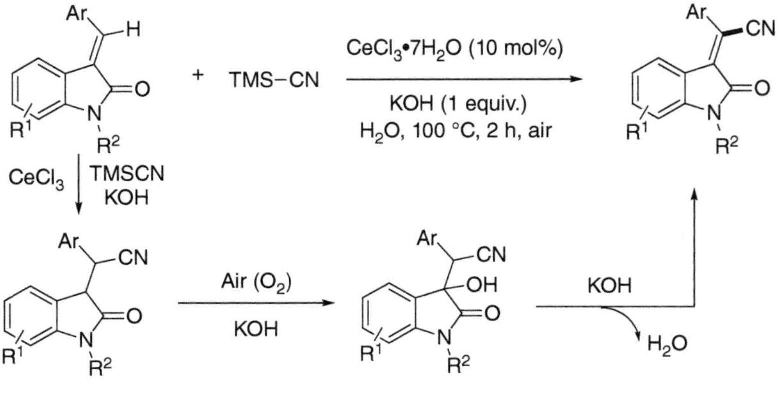

Scheme 10.74 $CeCl_3 \cdot 7H_2O$-catalyzed C(sp^2)–CN bond construction on water.

to occur through cerium-catalyzed Michael addition, KOH-promoted C–H hydroxylation of the Michael addition product with air, and dehydration to afford the final alkenyl nitrile product.

C–H Allylation

In 2017, Ackermann and coworkers disclosed a Mn(I)-catalyzed decarboxylative C(sp^2)–H allylation of indoles and aryl ketimines with substituted dioxolanones in water under air (Scheme 10.75) [86]. Water provided the best *E/Z* ratio of the newly formed double bond among all solvents screened.

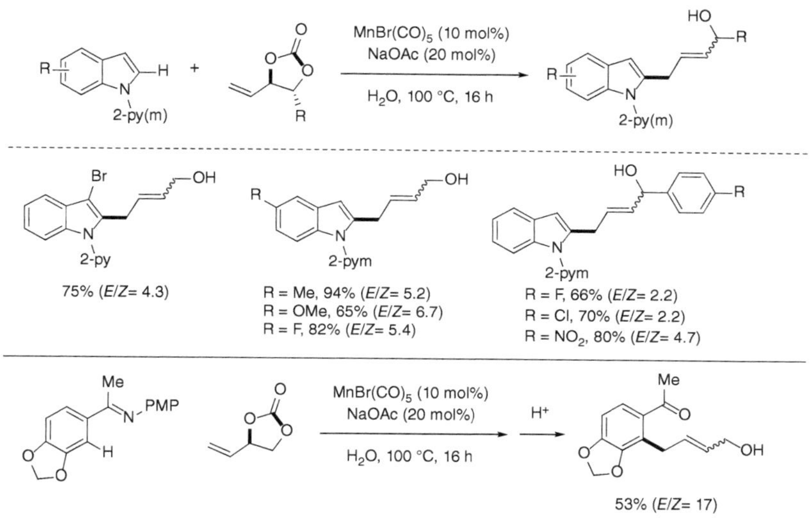

Scheme 10.75 Mn(I)-catalyzed decarboxylative C(sp^2)–H allylation.

In 2022, Kumar and coworkers described a water-enabled, Ni-catalyzed highly chemo-, regio-, and stereoselective C3-allylation of (NH)-indoles with allyl alcohols and allylamines under aqueous micellar catalysis of an anionic surfactant SDOSS (sodium dioctylsulfosuccinate) (Scheme 10.76) [87]. After extraction with ethyl acetate, the aqueous layer containing SDOSS could be reused for three subsequent fresh batches without compromising the reaction yield. A series of experiments were performed to probe the role of water and led to the conclusion that the hydrogen bond network of water plays a critical role in facilitating the oxidation of allyl alcohols/allylamines and promoting the intermolecular nitrogen to carbon allyl migration via π-allylNi complexation.

C–H Benzylation

In 2018, a Pd-catalyzed tandem benzylation of 2-morpholinoanilines with benzyl alcohols was reported by Hikawa and coworkers for the synthesis of *N*-(1,2-diphenylethyl)-2-morpholinoanilines on water under base- and additive-free conditions (Scheme 10.77) [88]. The reaction proceeds via a dehydrative Tsuji–Trost

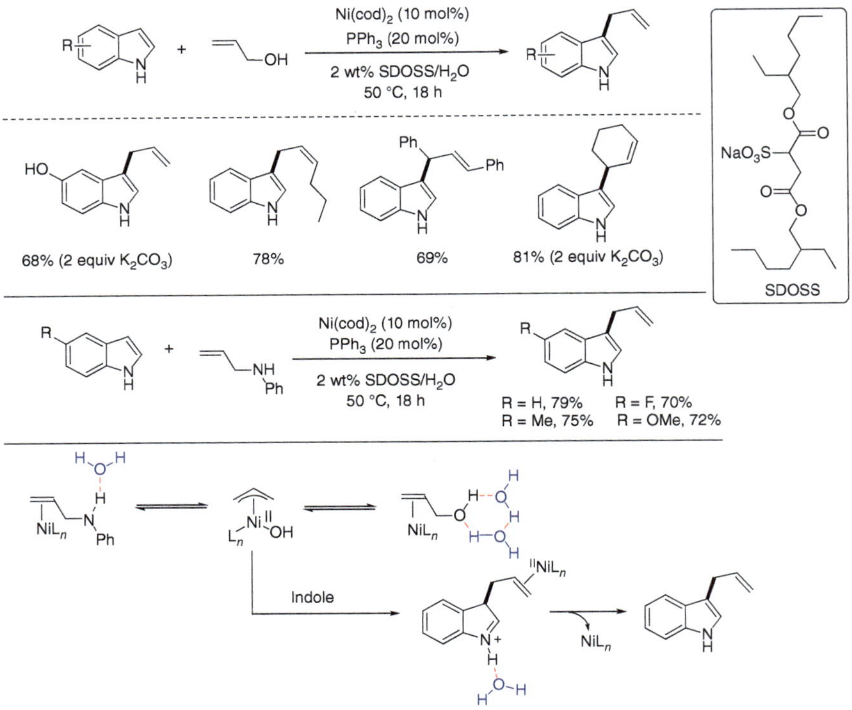

Scheme 10.76 Water-enabled, Ni-catalyzed C3-allylation of (NH)-indoles with allyl alcohols and allylamines.

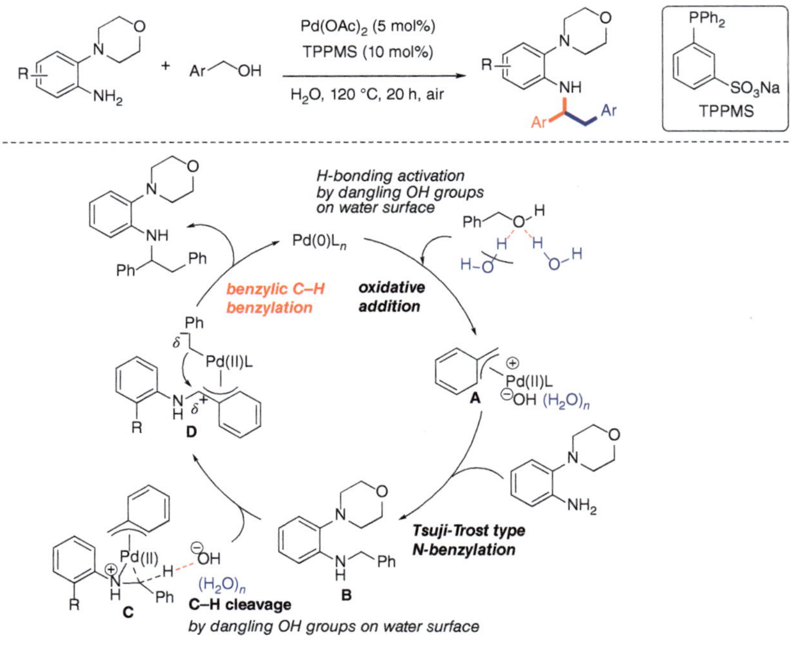

Scheme 10.77 *N*-benzylation/C–H benzylation of 2-morpholinoanilines.

type *N*-benzylation followed by dehydrative benzylic C(sp^3)–H benzylation of the *N*-monobenzylated intermediate, producing water as the sole co-product. Water is essential for this transformation and replacement of water with other organic solvents resulted in no reaction. Hydrogen bonding activation of the sp^3 C–O bond of the alcohol by dangling OH groups on the water surface facilitates oxidative addition of the benzyl alcohol to Pd(0) and the resulting cationic mono-π-benzylpalladium(II) complex **A** is also stabilized by water.

C–H Alkylation

In 2018, Gooßen and coworkers reported a Ru(II)-catalyzed carboxylate-directed *ortho*-alkylation of benzoic acids with α,β-unsaturated olefins in water using Li_3PO_4 or CsF as the base (Scheme 10.78) [89]. A series of α,β-unsaturated ketones/amides and even unprotected acrylic acids were smoothly hydroarylated. The widely available carboxylate directing group could be removed via proto-decarboxylation in a one-pot process.

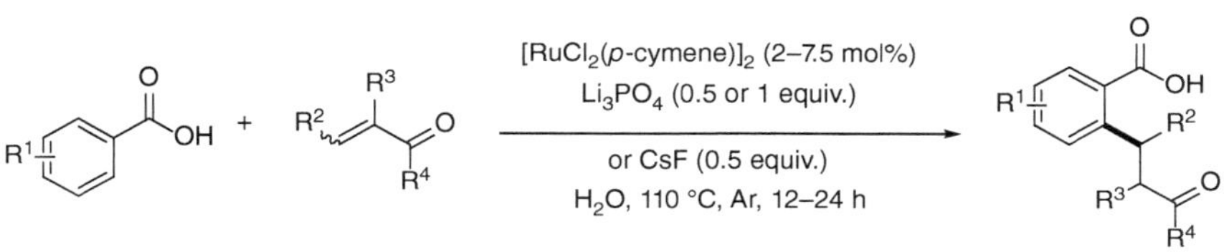

Scheme 10.78 Ru(II)-catalyzed *ortho*-alkylation of benzoic acids via carboxylate-directed hydroarylation.

Later, Kobayashi and coworkers developed a highly enantioselective C3-selective C(sp^2)–H alkylation of indoles with α,β-unsaturated ketones in water via a palladium triflate complex ligated with a hydroxy-bearing chiral 2,2'-bipyridine ligand (Scheme 10.79) [90]. The chiral 2,2'-bipyridine ligand reacts with the triflate anion as a hydrogen-bond donor to configure a chiral palladium aqueous complex in water that is effective for σ-indole palladation and subsequent asymmetric reaction. It should be noted that highly enantioselective reactions could only be achieved in water. The addition of even small amounts of organic solvents could cause deteriorated enantioselectivity. The water molecule located on the apical position of the aqueous complex could be easily replaced by an indole molecule to form an σ-indolylpalladium complex. By contrast, apical occupancy of a triflate anion in the anhydrous complex could block the access of indole molecules and thus impede the formation of indolylpalladium intermediate.

In 2021, an unprecedented C(sp^2)–H methylation of pyrazinones and quinoxalinones was developed by Kim and coworkers using trimethylsulfoxonium salt as the methyl source under basic aqueous conditions (Scheme 10.80) [91]. By using corresponding alkyl sulfoxonium salts, several C(sp^2)–H alkylation products were also synthesized. Impressively, the aqueous media accelerated the desired transformation, and no aziridine side-products arising from the Corey–Chaykovsky reaction were observed. The proposed reaction pathway involves nucleophilic

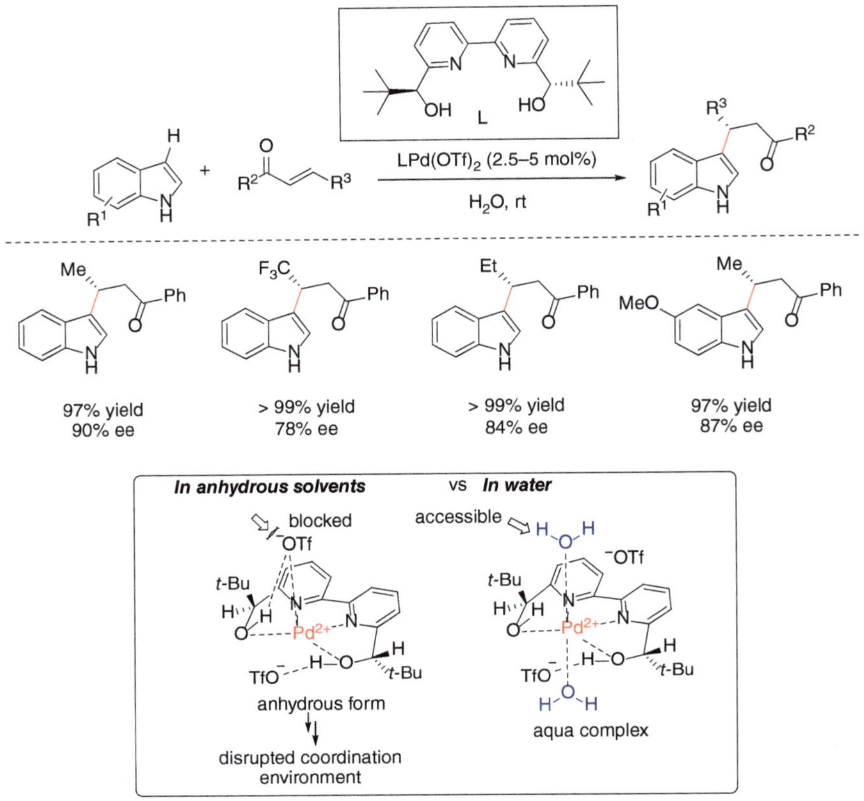

Scheme 10.79 Enantioselective C(sp^2)–H alkylation of indoles with α,β-unsaturated ketones.

addition of an *in situ* generated sulfoxonium ylide **A** to the imine moiety present in the heterocycle to form a nitrogen anion intermediate **B**. **B** undergoes protonation and base-mediated E2 elimination to deliver an exo-methylene intermediate **C** and DMSO. The final protonation of the exo-olefin in **C** affords the methylated product.

Giedyk and coworkers reported a micellar-enabled Minisci-type C–H alkylation of heteroarenes with nonactivated alkyl bromides under photoredox/bromide anion dual catalysis in the absence of any external radical-promoters, oxidants and acid additives (Scheme 10.81) [92]. The micellar condition allows spatial pre-aggregation of reacting species in the reaction mixture and is critical to the success of this transformation. The reaction was promoted either by a CBr_4/SDS or NaBr/CTAB system using [Ir(dtbbpy)(ppy)$_2$]PF_6 as the photocatalyst under blue LED irradiation. For the CBr_4/SDS system, photosensitized hydrolysis of CBr_4 produces a bromide anion necessary for the initiation of reaction. CTAB could act as the sole source of bromide anion without external NaBr added, affording slightly less product than that obtained from the NaBr/CTAB combination.

In 2022, Yu and coworkers reported a visible-light-promoted, rhodamine B-catalyzed C3-alkoxycarbonylmethylation or carbamylmethylation of 2-aryl-2*H*-indazoles/imidazo[1,2-*a*]pyridines using α-bromoesters or α-bromoamides as the

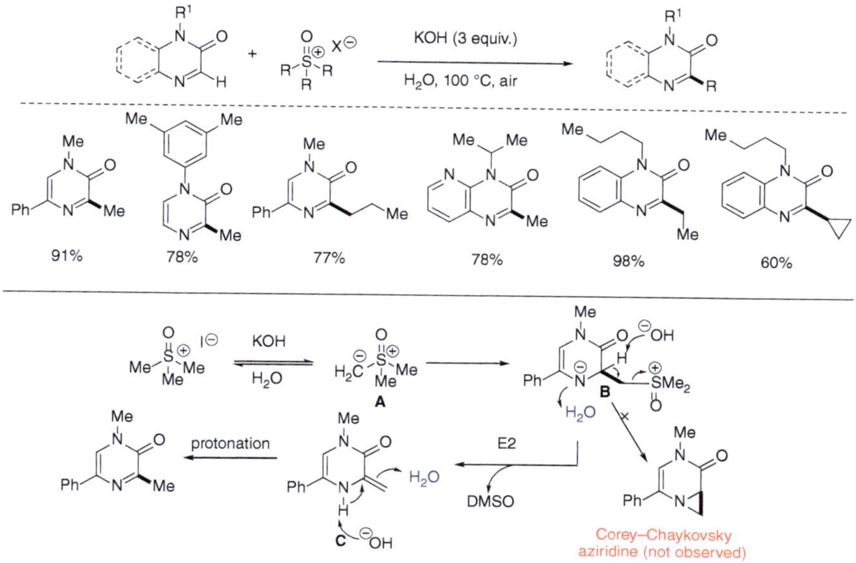

Scheme 10.80 C(sp^2)–H methylation and alkylation of pyrazinones and quinoxalinones with alkyl sulfoxonium salts.

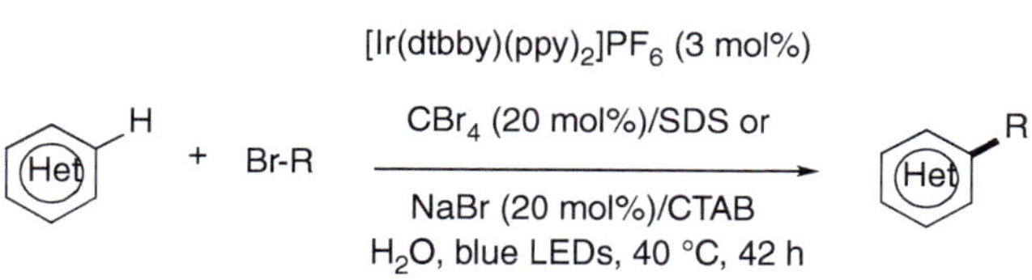

Scheme 10.81 Minisci C–H alkylation of heteroarenes enabled by dual photoredox/bromide catalysis in micellar solutions.

alkyl source in water under metal-free conditions (Scheme 10.82) [93]. Since direct activation of C–Br is difficult under visible light irradiation, the introduction of $EtOCS_2K$ as an additive is necessary to initiate the *in situ* formation of a xanthate ester which readily undergoes C–S homolytic cleavage to deliver an alkyl radical species. Water was found to be the best solvent after extensive solvent screening.

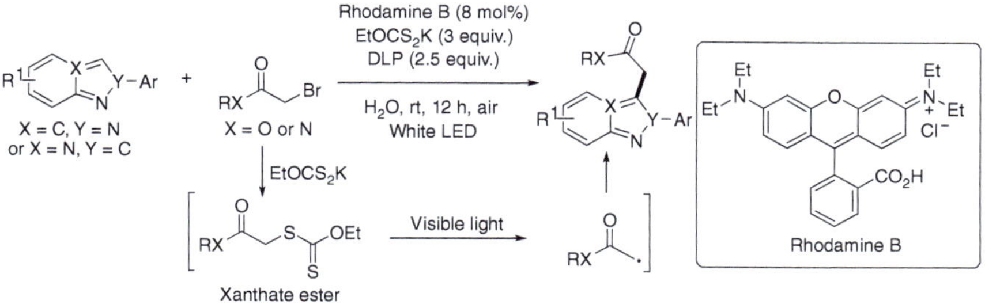

Scheme 10.82 Visible-light-induced direct C3-alkoxycarbonylmethylation and carbamylmethylation of 2-aryl-2*H*-indazoles/imidazo[1,2-*a*]pyridines.

Regarding aqueous C(sp^3)–H alkylation reactions, in 2019, Pérez and coworkers reported a micelle-enabled C(sp^3)–H alkylation of methane with ethyl diazoacetate (EDA) in water in the presence of a silver-based catalyst at room temperature (Scheme 10.83) [94]. Surfactant is critical to this reaction since no product was obtained in its absence. This might be explained by the negligible solubility of methane in net water. The addition of surfactant could increase the methane uptake by aggregated micelles and thus promote the desired reaction. Replacement of the water/surfactant mixtures with cyclohexane or chlorinated solvents led to undesired products arising from solvent C–H or C–Cl functionalization. Furthermore, C2-C4 gaseous alkanes were also functionalized and the alkylation of propane and butane delivered two regioisomers.

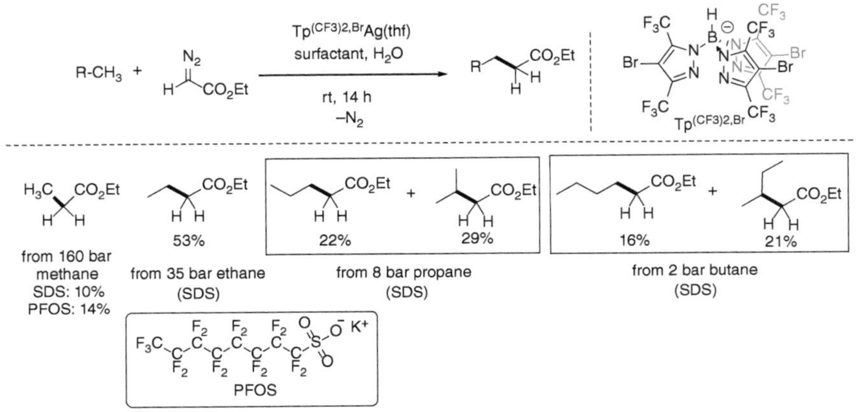

Scheme 10.83 Micelle-enabled C(sp^3)–H alkylation of methane and C2-C4 gaseous alkanes.

C–H Fluoroalkylation

In 2017, Zhang and coworkers described a Ni(II)-catalyzed, picolinamide (PA)-directed *ortho*-C(sp^2)–H trifluoromethylation of arylamines in water using Langlois' reagent (CF_3SO_2Na) as the CF_3 radical source and $K_2S_2O_8$ as the oxidant (Scheme 10.84) [95]. It is amazing that the aqueous catalytic system could be recycled eight times with a minimal loss of catalytic activity. The reaction was quite

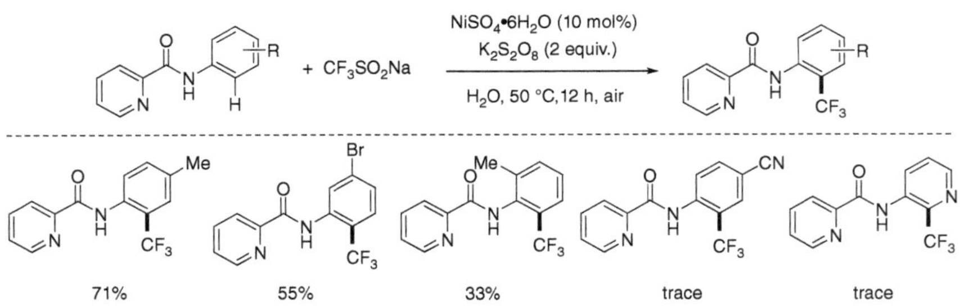

Scheme 10.84 Ni(II)-catalyzed, picolinamide (PA)-directed *ortho*-C(sp^2)–H trifluoromethylation of arylamines with CF_3SO_2Na.

sensitive to the steric and electronic nature of the arylamines and substrates bearing *ortho*-groups and electron-withdrawing (hetero)aryl amines were poor substrates for this transformation.

Later, by using a water-soluble Ir-complex as the photocatalyst, Conrad and coworkers achieved an oxidant-free photocatalytic C(sp^2)–H trifluoromethylation of an unprotected dipeptide in phosphate-buffered saline solvent using Langlois' reagent as the CF_3 radical source under blue LED irradiation (Scheme 10.85) [96]. The trifluoromethylation reaction requires low catalyst loading and few equivalents of Langlois reagent, thus is highly promising for further trifluoromethyl footprinting of proteins under physiological conditions.

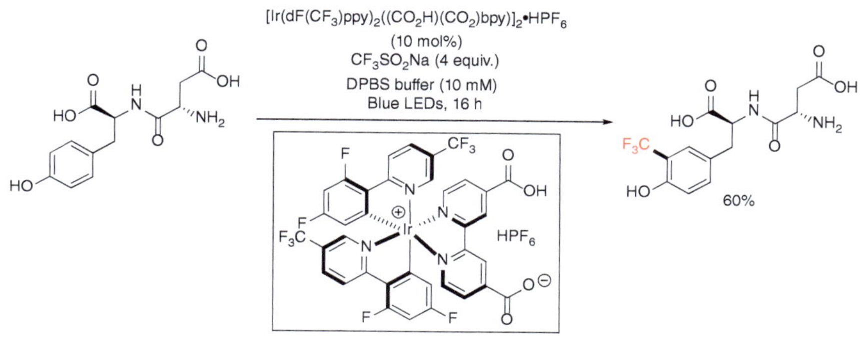

Scheme 10.85 Photocatalytic C(sp^2)–H trifluoromethylation of a dipeptide.

In 2021, Barata-Vallejo and coworkers developed a photocatalyzed radical C–H perfluoroalkylation of electron-rich anilines and alkoxyarenes with perfluoroalkyl bromides in water using Rose Bengal as the photocatalyst and unmodified vitamin B_{12} (cyano-cob(III)alamin) as the co-catalyst under green LED irradiation with TMEDA as a sacrificial electron donor (Scheme 10.86) [97]. In the cases of solid

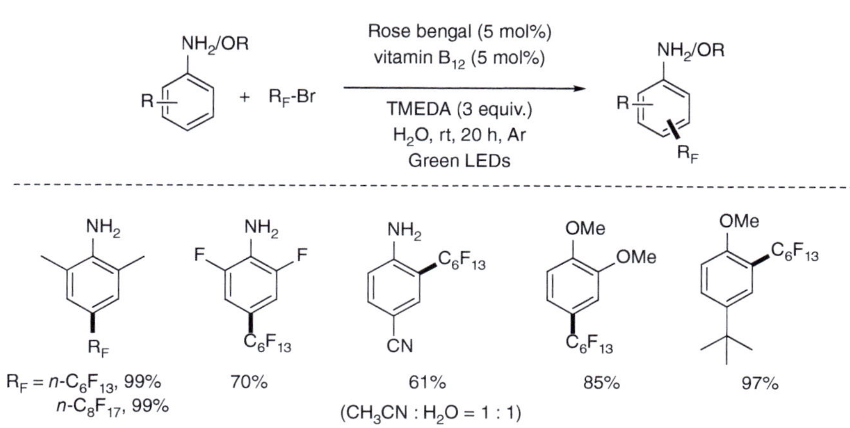

Scheme 10.86 Rose Bengal/vitamin B_{12} co-catalyzed radical C–H perfluoroalkylation of electron-rich anilines and alkoxyarenes.

substrates, a 1:1 mixture of water and acetonitrile was employed to solubilize the substrates and maximize the yields. The role of Rose Bengal is to mediate the photocatalyzed double reduction of cob(III)alamin to give a super-nucleophilic Co(I) species that is involved in subsequent perfluoroalkyl radical generation.

10.3.2 C–N Bond Formation

C–H Amination

In 2019, Jana and coworkers described a Cu_2O-catalyzed, 2-picolinamide (PA)-directed *ortho*-C–H amination of aryl amines with *O*-benzoylhydroxylamines in water in the absence of any external oxidants or additives (Scheme 10.87) [98]. A variety of anilines, bicyclic naphthyl, or heterocyclic amines were smoothly aminated with various five or six-membered cyclic amines and acyclic amines in moderate to good yields.

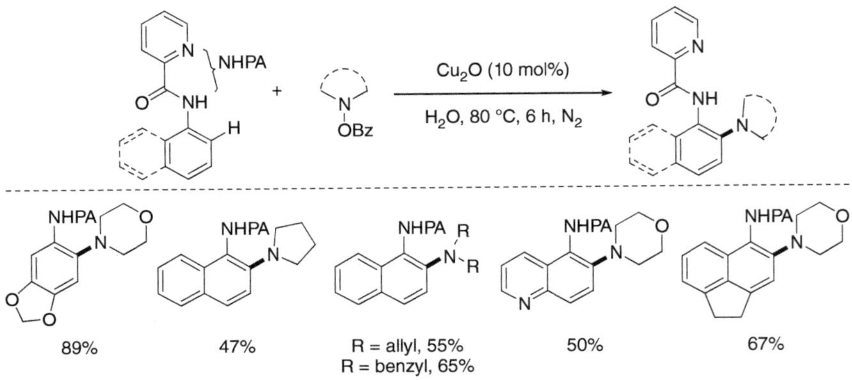

Scheme 10.87 *Ortho*-C–H amination of aryl amines with *O*-benzoylhydroxylamines.

In another elegant contribution, Xia and coworkers achieved picolinamide-directed *para*-selective C(sp^2)–H amination of arylamines with pyrazoles via nickelphotoredox catalysis using water as the solvent under ambient and aerobic conditions (Scheme 10.88) [99]. The reaction utilized cost-effective $NiSO_4.H_2O$ as catalyst and the organic dye Acr^+-$MesClO_4^-$ as photocatalyst in the presence of H_2O_2 as an oxidant. The catalyst-in-water system could be reused eight times without significant loss of activity. A single-electron-transfer mechanism was proposed for this C–N cross-coupling reaction.

In 2020, a directing-group-free *para*-selective C–H amination and aminative dearomatization of unprotected phenols was reported by Li and coworkers using azodicarboxylates as the amination source under silver catalysis in water (Scheme 10.89) [100]. Depending on the presence of a *para*-substituent or not, selective formation of amino cyclohexadieneones and *para*-aminophenols was achieved. Mechanistically, the silver catalyst interacts with the phenol-free hydroxyl group to activate the *para*-position toward attack by the azodicarboxylate. An amino

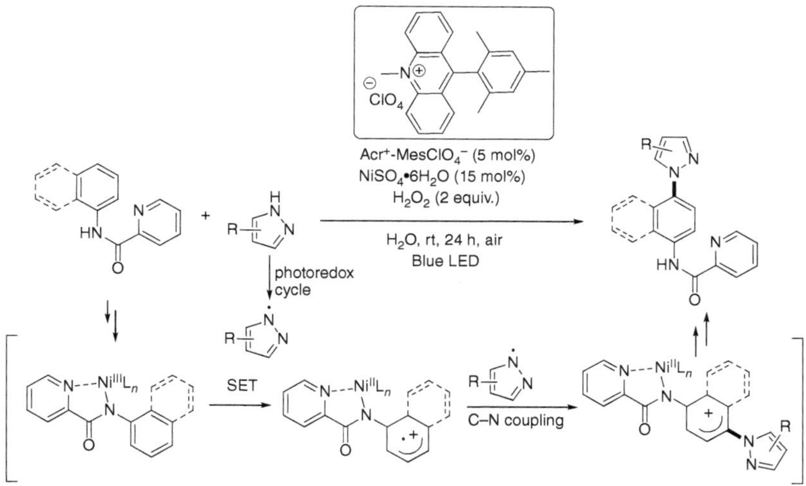

Scheme 10.88 Nickel/photoredox *para*-C–H amination of aryl amines with pyrazoles.

cyclohexadieneone is first formed and undergoes further aromatization to afford the C–H amination product if no *para*-substituent is present.

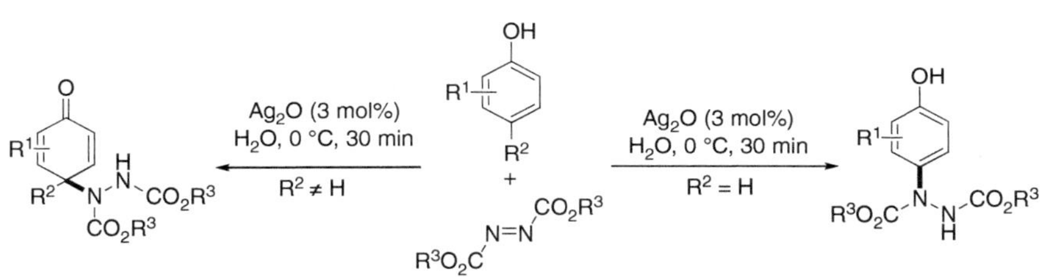

Scheme 10.89 Silver-catalyzed *para*-selective amination and aminative dearomatization of phenols with azodicarboxylates.

Recently, Favi and coworkers reported a Cu/Fe co-catalyzed intramolecular C(sp^2)–H amination of α-indolylhydrazones for the synthesis of pyrrolo[2,3-*b*] indoles in water using air as the terminal oxidant (Scheme 10.90) [101]. Notable features of this protocol include the use of cost-effective transition metals, the absence of external ligands/oxidants/additives, mild conditions, and the recyclability of the aqueous catalytic system. The acylhydrazone residue plays a dual role by acting as a bidentate directing group as well as an aminating agent.

C–H Sulfonamidation

Most reported C–H amidations were achieved using precious catalysts in the presence of chlorinated solvents. In 2018, by using $MnBr(CO)_5$ as a cheap and abundant base-metal catalyst, Xu and coworkers developed a Mn(I)-catalyzed directed C–H sulfonamidation of various 2-phenylpyridine and 2-pyrimidylindole derivatives using aromatic and aliphatic sulfonyl azides as amino sources in water with nitrogen

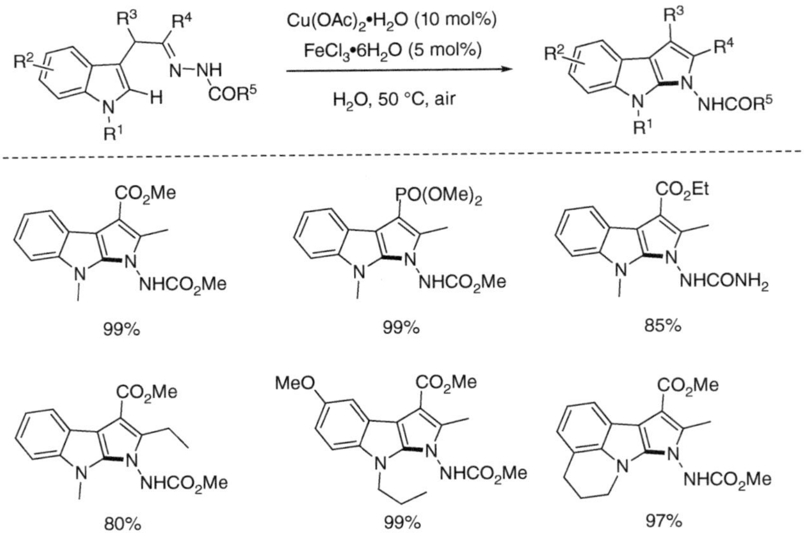

Scheme 10.90 Cu/Fe co-catalyzed intramolecular C(sp^2)–H amination.

produced as the only byproduct (Scheme 10.91) [102]. To enhance the mass transfer of the water-insoluble system, 10% TPAB [$(n\text{-}C_3H_7)_4NBr$] was added as a surfactant.

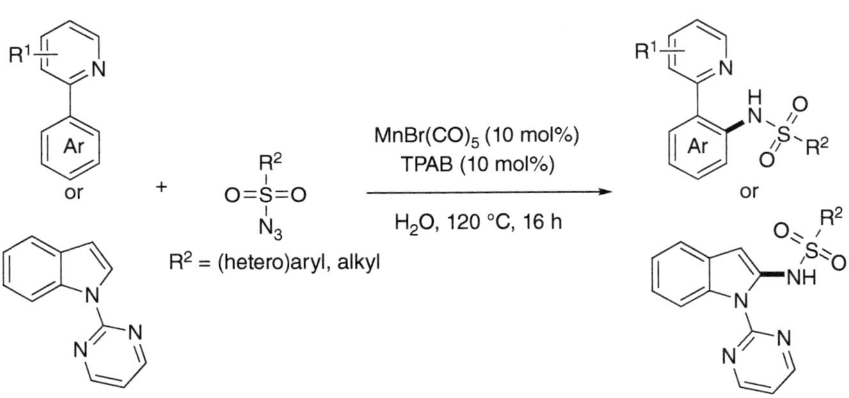

Scheme 10.91 Mn(I)-Catalyzed C–H amidation of 2-phenylpyridines and 2-pyrimidylindoles.

In 2021, Wang and coworkers reported an iridium-catalyzed regioselective C–H sulfonamidation of 1,2,4-thiadiazoles and *N*,3-diphenyl-1,2,4-thiadiazol-5-amines with sulfonyl azides in water using $AgSbF_6$ and $C_6F_5CO_2H/C_6F_5CO_2Na$ as the additives (Scheme 10.92) [103]. Water displayed the best performance among other solvents screened. This protocol regioselectively cleaves the C–H bond of the benzene ring located between two nitrogen atoms. Moreover, substrates bearing *meta*-substituted benzenes gave higher yields than those containing non-substituted or *para*-substituted benzenes due to the formation of less diaminylation products.

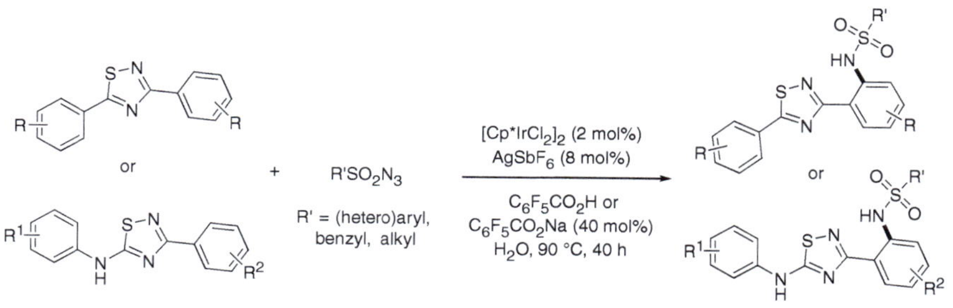

Scheme 10.92 Ir-catalyzed regioselective C–H sulfonamidation of 1,2,4-thiadiazoles and *N*,3-diphenyl-1,2,4-thiadiazol-5-amines.

10.3.3 C–O Bond Formation

C–H Oxygenation

In 2019, He and coworkers reported the synthesis of quinoxaline-2,3(1*H*,4*H*)-diones via metal-free direct C(sp^2)–hydroxylation of quinoxalin-2(1*H*)-ones in water using TPGS-750-M as the surfactant and $(NH_4)_2S_2O_8$ as both the oxidant and oxygen source (Scheme 10.93) [104]. The TPGS-750-M/H_2O solution can be recycled up to five times with slight loss in reaction efficiency. Mechanistically, radical addition of an *in situ* generated sulfate radical anion onto the imine bond produces a nitrogen-centered radical species **A**. Hydrolysis of the sulfonate followed by H atom abstraction of the resulting intermediate **B** by the sulfate radical anion affords the C3-hydroxylated product **C** which further tautomerizes to the ketone product.

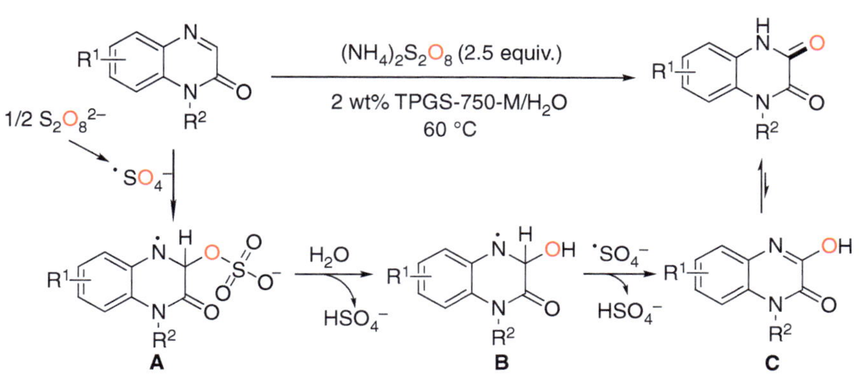

Scheme 10.93 Synthesis of quinoxaline-2,3(1*H*,4*H*)-diones via C3-hydroxylation of quinoxalin-2(1*H*)-ones.

In terms of aqueous C(sp^3)–H hydroxylation, in 2017, Sigman and coworkers reported a Ru-catalyzed tertiary- and benzylic-selective C(sp^3)–H hydroxylation of various amines and non-amine substrates in aqueous acetic acid using H_5IO_6 as the oxidant and CF_3SO_3H as the additive (Scheme 10.94) [105]. A series of oxidatively sensitive nitrogen functional groups including alkylamines, cyclic amines, and pyridines were well-tolerated. Cyclic imines were also synthesized from substrates bearing both a benzylic C(sp^3)–H bond and a primary amine group via ketone formation followed by intramolecular condensation. Notably, an optically active substrate bearing a chiral tertiary carbon center underwent complete stereoretentive hydroxylation, regardless of the presence or absence of triflic acid.

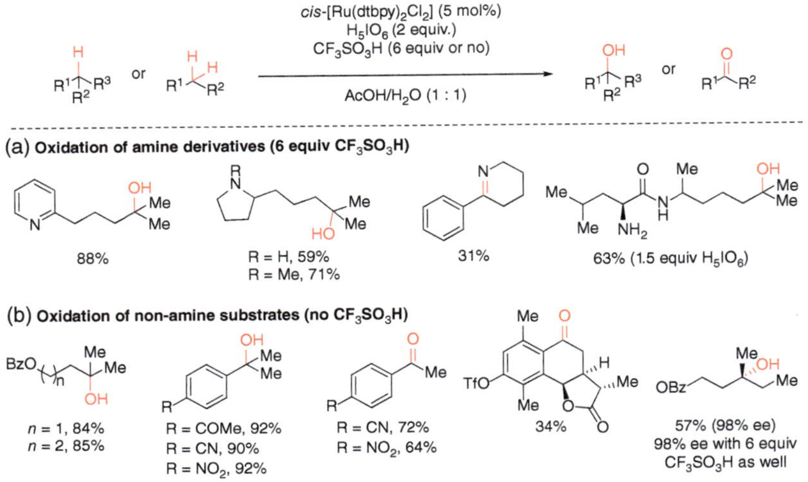

Scheme 10.94 Ru-catalyzed tertiary- and benzylic-selective C(sp^3)–H hydroxylation.

In the same year, Sanford and coworkers developed a transition metal-free remote C(sp^3)–H oxygenation of protonated aliphatic amines in aqueous potassium persulfate using H_2SO_4 as an additive (Scheme 10.95) [106]. Water was selected as the reaction medium to prevent competitive C–H oxidation of the solvent as well as to facilitate the dissolution of $K_2S_2O_8$ and protonated amines. Protonation renders the deactivation of proximal C(sp^3)–H bonds of the amine group and increases the solubility of amines in water. Various types of C(sp^3)–H bonds including tertiary, benzylic as well as secondary C–H bonds were oxygenated using this methodology. For the secondary C(sp^3)–H bond, a mixture of alcohol and ketone products was obtained and exclusive formation of the latter could be achieved by increasing the amount of oxidant.

In 2019, Wang and coworkers described α-C(sp^3)–H hydroxylation of β-keto esters and β-keto amides with cumene hydroperoxide (CHP) in water using TBAB as the surfactant under transition-metal, organic solvent, and base-free conditions (Scheme 10.96) [107]. The reaction was proposed to occur through heat-promoted homolysis of the peroxide to give an alkoxide radical and a hydroxy radical. The alkoxide radical abstracts α-hydrogen of the β-dicarbonyl compound to afford an alkyl radical species, which cross-couples with the hydroxyl radical to give the desired α-hydroxylation product.

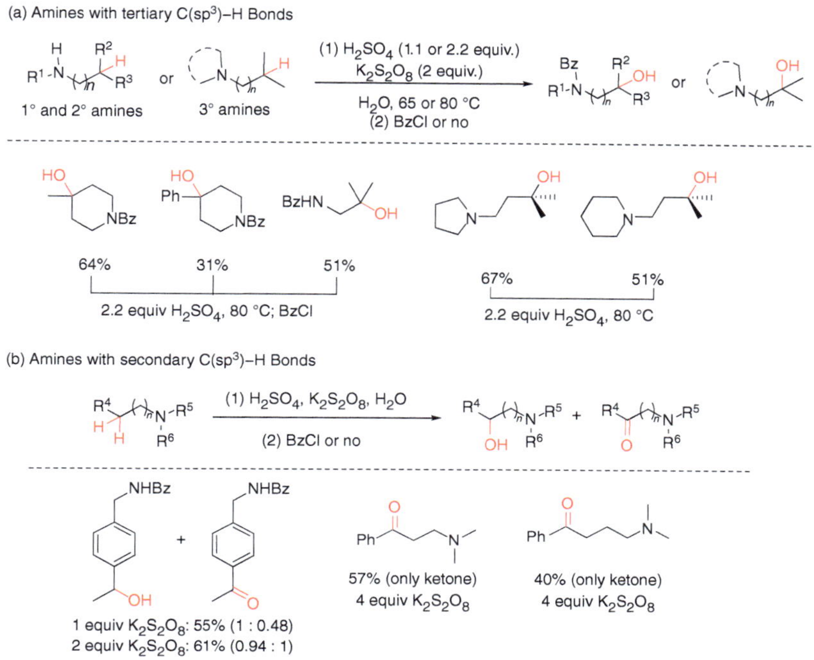

Scheme 10.95 Remote C(sp^3)–H oxygenation of protonated aliphatic amines.

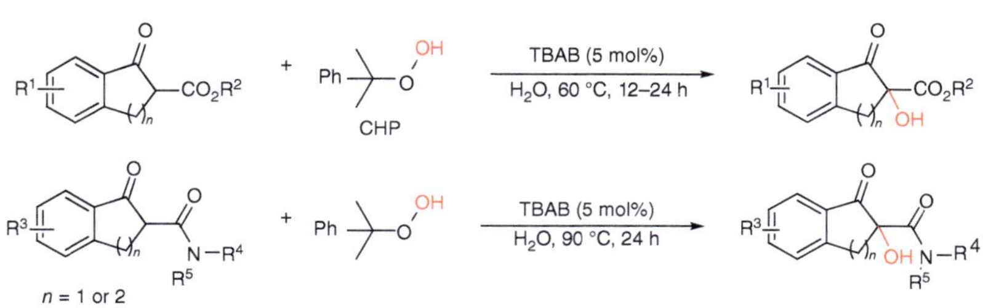

Scheme 10.96 α-Hydroxylation of β-keto esters and β-keto amides with CHP.

10.3.4 C–X Bond Formation

C–H Halogenation

In 2017, Wu and coworkers reported a metallaphotoredox-catalyzed *para*-selective bromination or iodination of 8-aminoquinoline amides with KBr/KI in water at room temperature using cost-effective $FeCl_3$ as catalyst, alizarin red S as photocatalyst and $K_2S_2O_8$ as oxidant under household light irradiation (Scheme 10.97) [108]. Mechanistically, reductive quenching of the excited photocatalyst with bromo or iodide anion produces a bromo or iodine radical. The halide radical adds to the C5-position of 8-aminoquinoline to form the C–Br or C–I bond.

Later, Handa and coworkers achieved a highly selective C(sp^2)–H monofluorination of indoles and arenes with *N*-fluorobenzenesulfonimide (NFSI) in a

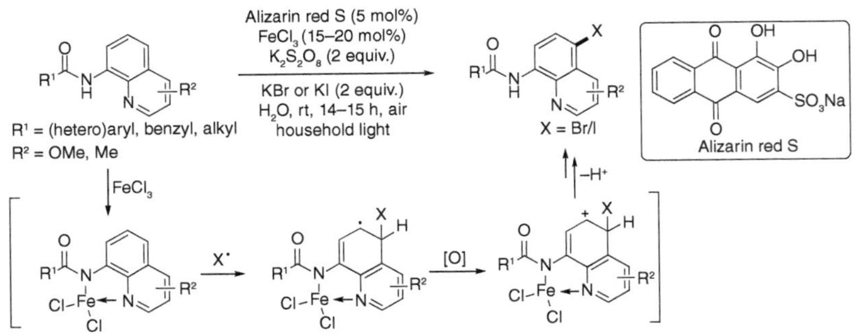

Scheme 10.97 C5–H bromination and iodination of 8-aminoquinoline amides under metallaphotocatalysis.

recyclable aqueous PS-750-M/H_2O medium by taking advantage of the shielding effect of micelles (Scheme 10.98) [109]. Site-specific distribution of the substrate and fluorine source in the micelle together with the minimal oxygen solubility in the inner core allow highly selective formation of monofluorinated products over overoxidized and difluorinated products. Poor selectivities were observed for the reactions carried out in organic solvents or neat water where no shielding effect was present. A radical pathway was proposed for the reaction based on control experiments.

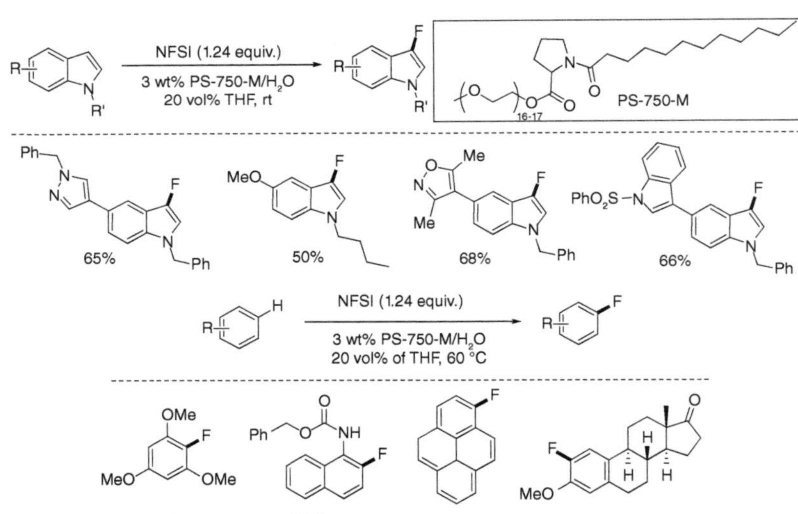

Scheme 10.98 C(sp^2)–H monofluorination of indoles and arenes with NFSI.

10.3.5 C–H Annulation/Cyclization

In 2017, Cheng and coworkers reported a Rh(III)-catalyzed, regioselective C–H annulation of benzamides with alkynes for the synthesis of isoquinolones in water using air as the sole oxidant (Scheme 10.99) [110]. The choice of solvents is critical and water outperformed all other solvents examined.

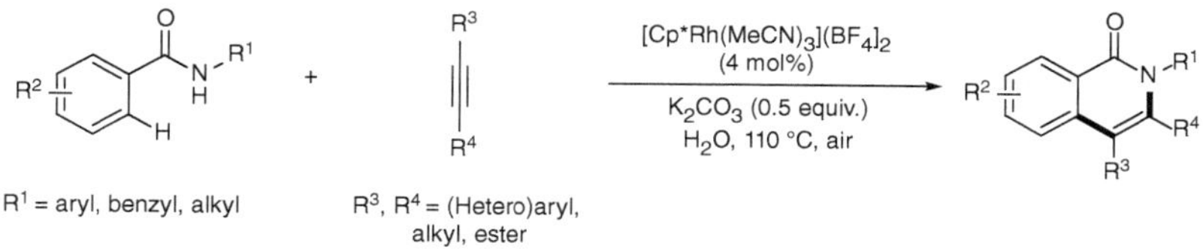

Scheme 10.99 Rh(III)-catalyzed C–H annulation of benzamides with alkynes.

Later, a Ru(II)-catalyzed C–H annulation of aromatic acids with alkynes was reported by Shang and coworkers for the synthesis of isocoumarins and α-pyrones using air the sole oxidant in water in the presence of PivOH as an additive (Scheme 10.100) [111]. The same catalytic system was also applied for the oxidative cyclization of benzoic acid with acrylate to construct the isobenzofuran-1(3*H*)-one scaffold.

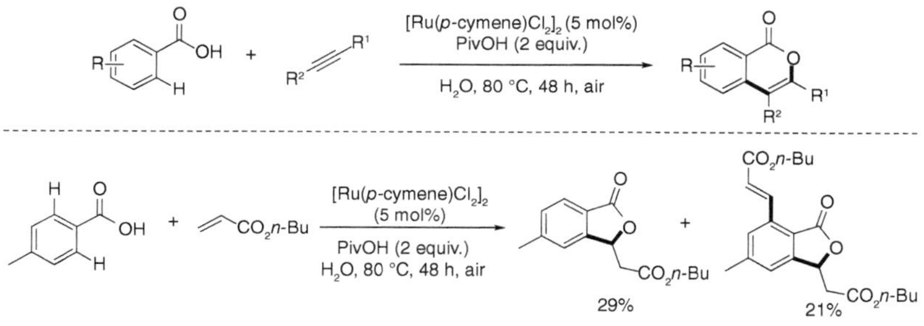

Scheme 10.100 Ru(II)-catalyzed oxidative annulation of aromatic acids and alkynes/acrylate.

In 2018, Ma and coworkers reported a Rh-catalyzed synthesis of isoindolin-1-ones from *N*-OMe substituted benzamides and propargylic acetates via a sequential C–H allenylation and cyclization process using water as the solvent under the assistance of a $[Cp^*RhCl_2]_2$/NaOAc/HOAc catalytic system (Scheme 10.101) [112]. External oxidants were not required for this transformation. The proposed mechanism involves a cascade of cyclometallation, alkyne insertion, β-oxygen elimination, ring-closing azametalation, and protonation.

In the same year, Sahoo and coworkers described the synthesis of dihydrobenzofurans via Ru(II)-catalyzed methyl phenyl sulfoximine (MPS)- or amide-directed intramolecular hydroarylation of *O*-tethered olefin-bearing arenes in water (Scheme 10.102) [113]. Compared to MPS-directed reactions, the amide-directed reaction was achieved at an elevated temperature (110 °C) in the presence of $AgSbF_6$. The MPS group could be easily removed by base-mediated hydrolysis.

In 2019, a variety of tetrahydro-β-carbolines with a C1 quaternary carbon center were synthesized by Li and coworkers through Pd-catalyzed C–H functionalization

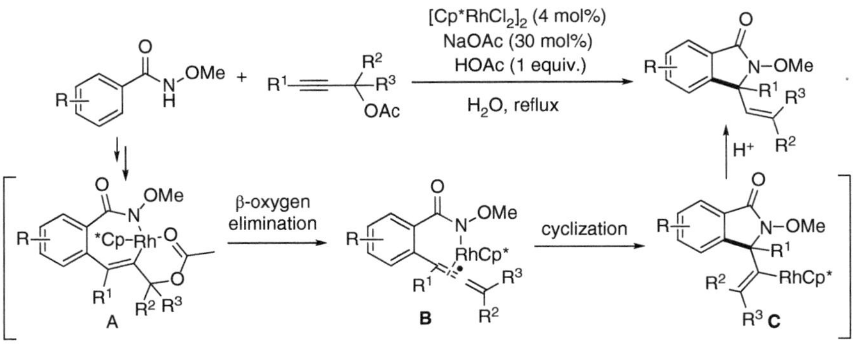

Scheme 10.101 Rh(III)-catalyzed C–H annulation for isoindolin-1-one synthesis.

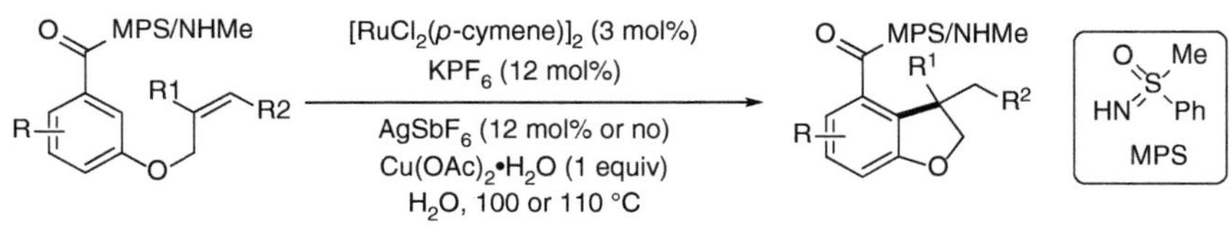

Scheme 10.102 Ru(II)-catalyzed intramolecular hydroarylation of *O*-tethered olefin-bearing arenes.

of tryptamine derivatives with ketones in water (Scheme 10.103) [114]. The reaction proceeded via a sequence of ketimine formation followed by C–H activation of indole and cyclization with the imine intermediate. The imine group serves a dual role of transient directing group (TDG) and reagent. Interestingly, bifunctional

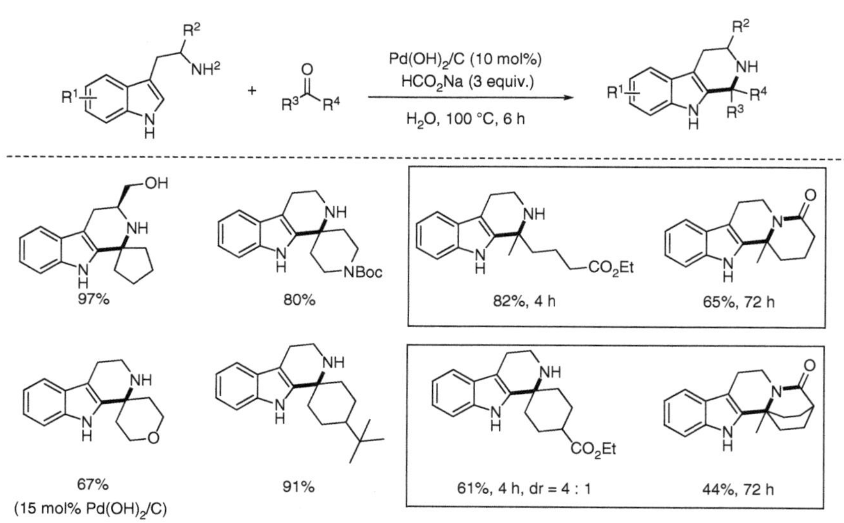

Scheme 10.103 Pd-catalyzed C–H functionalization of tryptamine derivatives with ketones.

δ-ketoester substrates underwent further intramolecular amidation to afford corresponding fused and bridged products when the reaction time was prolonged from 4 h to 72 h.

In the same year, Wu and coworkers synthesized a variety of *ortho*-C(sp²)–H alkylated arenes via water-mediated, additive-free C(sp²)–H activation of arenes with sulfoxonium ylides under the assistance of diverse directing groups such as pyridine, pyrimidine, pyrazole and isoquinoline (Scheme 10.104) [115]. A series of 1,2-benzothiazines and isoquinolines were also prepared via C–H annulation of NH-sulfoximine or benzylamine derivatives with sulfoxonium ylides. The proposed mechanism involves a reactive rhodium α-oxo carbene species which undergoes migratory insertion and protonation to give an alkylated product. Upon heating, the alkylated product participates in an intramolecular annulation of the carbonyl and amino/imino group to produce the cyclized product.

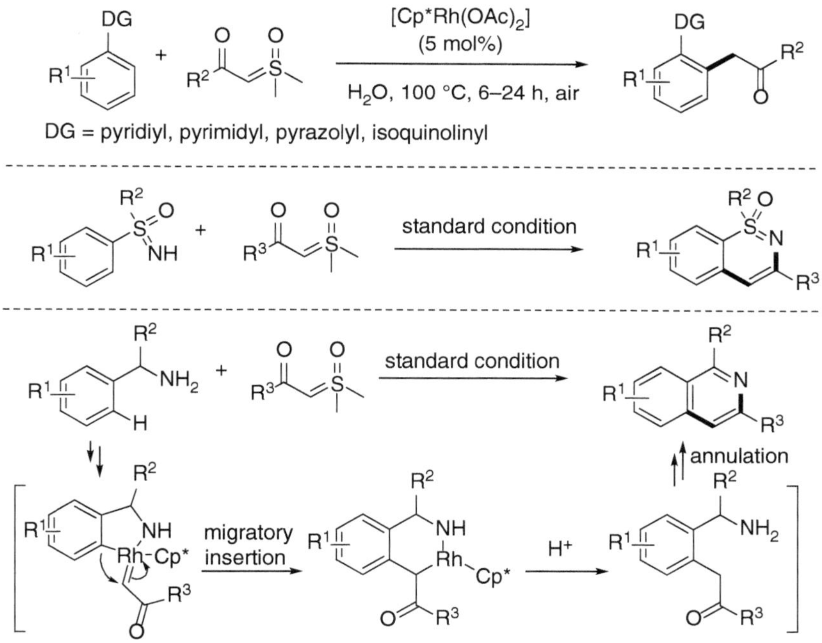

Scheme 10.104 C(sp²)–H activation/annulation of arenes with sulfoxonium ylides.

Maji and coworkers reported a Cp*Ir(III)-catalyzed C–H annulation of salicylaldehydes with α-diazo-ketones for the synthesis of chromones using water as the solvent and PivOH as the additive (Scheme 10.105) [116]. The novel protocol features low catalyst loading, no requirement of any silver salt/co-solvent/phase-transfer catalyst/surfactant and facile C3 or C5–H diversification of the chromone product. By using *tert*-butyl diazoester as the substrate, a one-pot annulation-hydrolysis-decarboxylation process occurred to provide C3-unsubstituted chromones as products. The reaction was proposed to involve a cyclo-iridium intermediate **A** which reacts with the diazo-compound to give an iridium-carbenoid species **B**. Migratory insertion of the carbenoid species followed by protonation liberates an *ortho*-hydroxyl bearing diketone compound. Subsequent intramolecular nucleophilic addition of this diketone followed by dehydration provides the final chromone product.

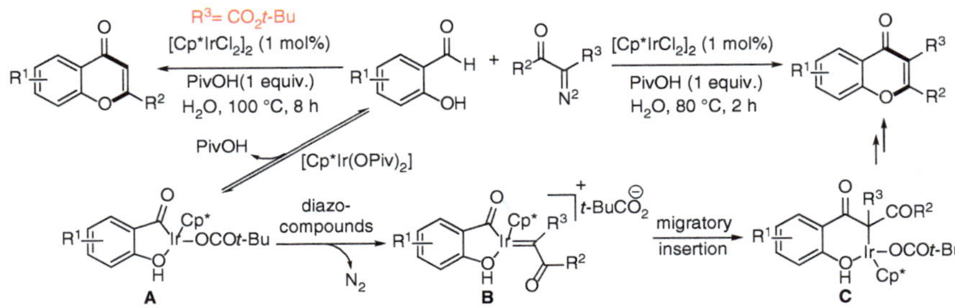

Scheme 10.105 Cp*Ir(III)-catalyzed C–H annulation of salicylaldehydes with α-diazo-ketones.

In 2020, Guo and coworkers described the synthesis of phenanthridinones via Pd-catalyzed, Ag_2O-mediated cyclization of *N*-aryl-2-aminopyridines with 2-iodobenzoic acids in water under air using a very low loading (0.1 mol%) of $Pd(OAc)_2$ catalyst (Scheme 10.106) [117]. Water was proved to be the optimum solvent for this transformation. The reaction was proposed to occur through pyridine-assisted *ortho* C–H arylation followed by intramolecular acylation to afford the final product. The pyridine directing group could be removed by a two-step quaternization/hydride-reduction process to furnish phenanthridine-6-one.

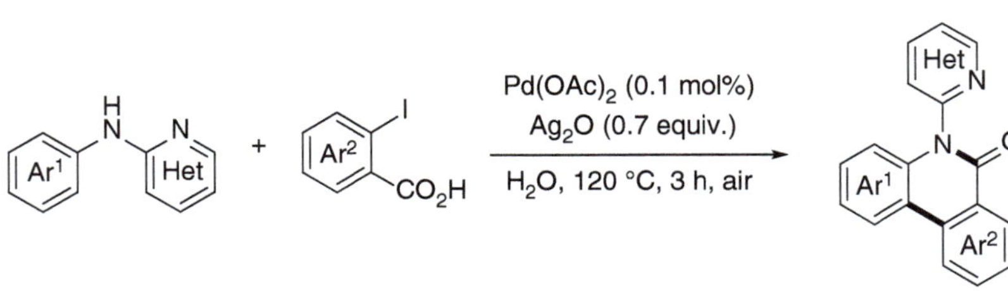

Scheme 10.106 Pd-catalyzed, Ag_2O-mediated cyclization of *N*-aryl-2-aminopyridines with 2-iodobenzoic acids.

10.4 C–C Activation

Compared with the well-developed C–H activation mode, the field of organometallic C–C activation is underdeveloped and most reactions are performed with precious and toxic metal catalysts using organic solvents as the reaction media. Organometallic C–C functionalizations in water are scarce because of the sensitive nature of the C–C activation transition metal catalysts.

In 2018, versatile C–C allylation, C–C alkenylation, and C–C alkylation reactions were developed by Ackermann and coworkers through $MnBr(CO)_5$-catalyzed chelation-assisted C–C activation of secondary/tertiary alcohol-bearing arenes with vinyl dioxolanones, alkynes and alkenes in water (Scheme 10.107) [118]. Mn(I) catalyst showed unique catalytic performance and water tolerance than commonly used precious metal complexes. Moreover, water gave better results than other organic solvents in terms of both yield and *E/Z* selectivity. The Mn-catalyzed system allowed faster C–C cleavages than C–H activations, thus enabling regioselective syntheses of 1,2,3-tri-substituted arenes. This 1,2,3-substitution pattern is totally distinct from the 1,2,5-substitution pattern accessed by sterically controlled C–H activation of *meta*-substituted arenes. In addition, the pyrazole directing group could be efficiently removed to afford synthetically useful anilines. Based on a series of mechanistic studies, the reaction was believed to occur through a

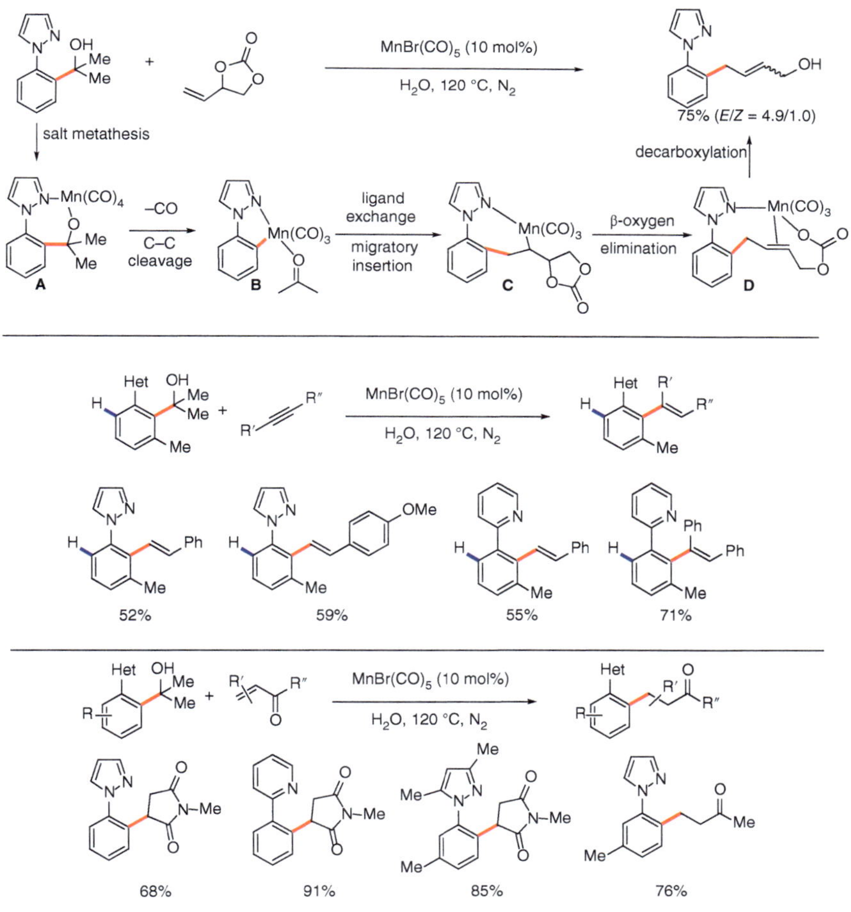

Scheme 10.107 C–C allylation, C–C alkenylation, and C–C alkylation by chelation-assisted manganese catalysis.

C–C activation mode via the formation of a catalytically active seven-membered manganese (I) alkoxide species **A**. C–C cleavage of the seven-membered species affords a five-membered intermediate **B** coordinated with acetone. In the case of C–C allylation with vinyl dioxolanone, exchange of acetone with the double bond followed by a sequence of migratory insertion, β-oxygen elimination, and decarboxylation provides the desired C–C allylation product.

10.5 C–O Cleavage Reactions

In 2019, Wald and coworkers developed several RCM-triggered C–O uncaging reactions of diolefinic-tethered esters, ethers, carbamates, and carbonates under physiological conditions through a "close-to-release" strategy (Scheme 10.108) [119]. Cargo molecules including carboxylic acids, alcohols, and amines were uncaged through spontaneous 1,4-elimination following RCM, paving the way for diverse applications such as prodrug activation, *in vivo* imaging, and metabolic engineering. The uncaging reactions were believed to be driven by the formation of an aromatic molecule.

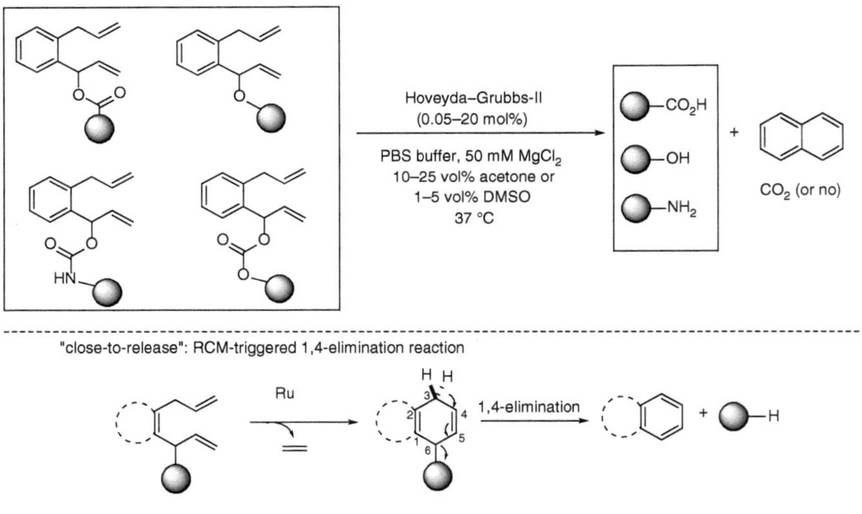

Scheme 10.108 RCM-triggered C–O uncaging reactions.

Considering the energy shortage and environmental protection issues, there is an urgent need to establish renewable approaches to produce aromatic chemicals from biomass instead of fossil resources. In 2020, Wang and coworkers reported a rhodium-terpyridine catalyzed selective C–O cleavage of lignin model substrates to produce aromatic ketones and phenols under redox-neutral conditions in water (Scheme 10.109) [120]. The C–O cleavage occurs through a Rh-catalyzed hydrogen autotransfer process where the C–O bond is reduced by a Rh hydride species generated from Rh-catalyzed alcohol dehydrogenation reaction.

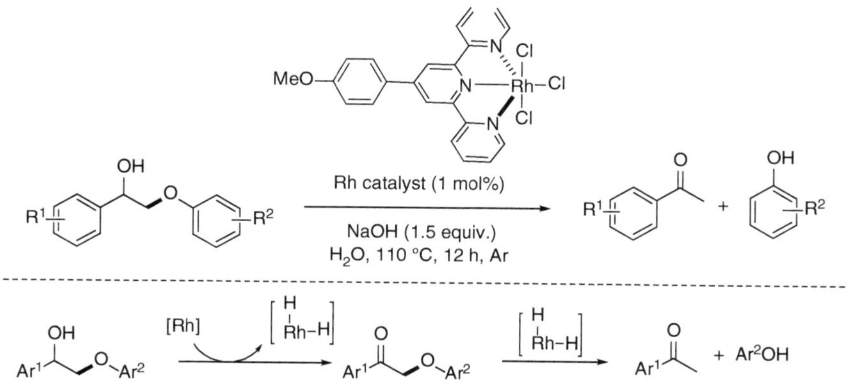

Scheme 10.109 Rh-catalyzed redox-neutral C–O cleavage of lignin model substrates.

10.6 Oxidative and Reductive Reactions

Oxidation and reductive reactions are among the most fundamental transformations in organic chemistry. Examples of aqueous oxidative and reductive reactions are abundant.

10.6.1 Electrochemical Oxidation

In recent years, electrochemical oxidation has become a sustainable and environmentally friendly alternative to traditional oxidation reactions due to the avoidance of toxic and contaminative stoichiometric oxidants. In 2019, Lei and coworkers reported a continuous-flow electrochemical oxidation of primary and secondary alcohols to aldehydes and ketones with concomitant hydrogen evolution in the absence of any external oxidants or mediators under neutral aqueous conditions (Scheme 10.110) [121]. Water was used as the solvent with surfactant *N,N,N*-trimethylhexadecan-1-ammonium sulfate serving as both the supporting electrolyte and solubility enhancer. A variety of benzylic, allylic alcohols were smoothly oxidized under such conditions. In the case of primary alcohols, the reaction stopped at the aldehyde stage without over-oxidation.

Zhang and coworkers successively achieved the synthesis of various nitriles and dihydroisoquinolines via anionic electro-oxidation of primary amines and selective semi-dehydrogenation of tetrahydroisoquinolines in aqueous KOH solution over a NiSe nanorod electrode or Ni_2P nanosheet electrode, respectively (Scheme 10.111) [122]. Mechanistic studies suggested that the *in situ* formed Ni^{II}/Ni^{III} serves as the redox active species for such transformations. In the case of electro-oxidation of primary amines, the nitrile products became more hydrophobic and could escape from the aqueous electrolyte/electrode interface, thus avoiding the deactivation of the catalyst and promoting continuous gram-scale synthesis.

Ke and coworkers described a catalyst-free and room-temperature electrochemical hydroxylation of aryl halides in water using ambient air as the oxygen source and

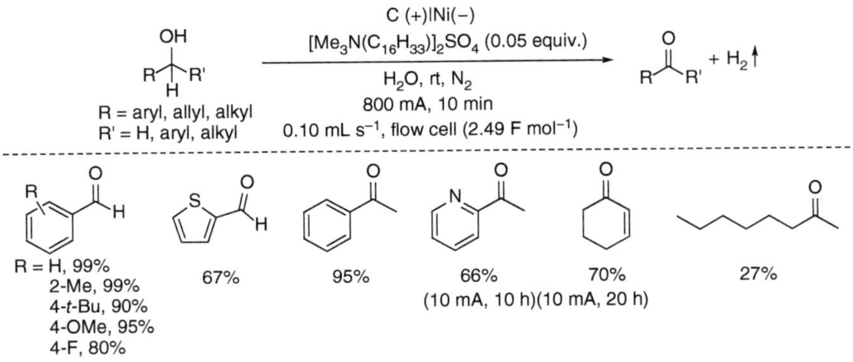

Scheme 10.110 Electrochemical oxidation of alcohols in water.

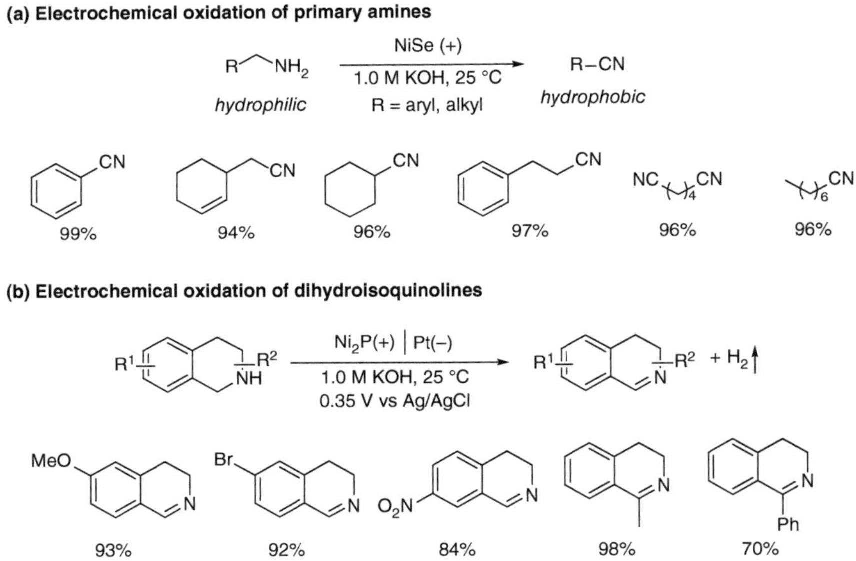

Scheme 10.111 Electro-oxidation of primary amines and selective semi-dehydrogenation of tetrahydroisoquinolines.

Et_3N as the mediator (Scheme 10.112) [123]. The reactions occurred in an undivided cell with a Pt anode and a Pt cathode using Bu_4NPF_6 as the electrolyte. A series of aryl iodides, bromides, and even aryl chlorides were amenable for such transformation to afford phenols in good yields. Mechanistically, the reaction was supposed to undergo a cathodic reduction of the aryl halide to afford an aryl radical **B** and capture of the aryl radical with oxygen produces a peroxyl radical **C**. Reduction of the peroxyl radical by Et_3N delivers the phenol product and an oxidized Et_3N radical **D** which undergoes SET to afford an amide by-product **E**.

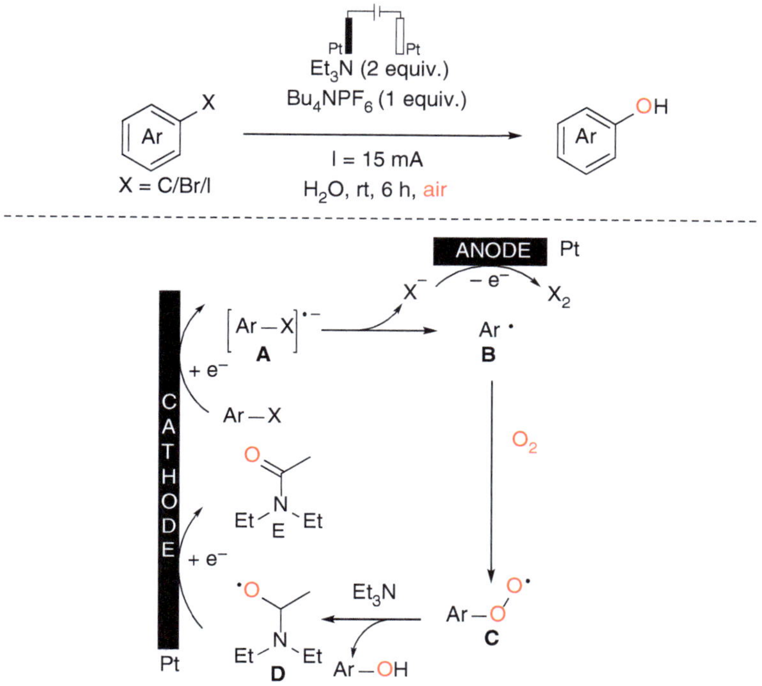

Scheme 10.112 Electrochemical-induced hydroxylation of aryl halides using air as the oxygen source.

10.6.2 Reduction and Related Reactions

In 2018, Albrecht and coworkers discovered that a single triazolylidene-based iridium complex could catalyze both the hydrogenation of quinoline and dehydrogenation of tetrahydroquinoline in water under mild conditions in the absence of external acid additives and hydrogen scavengers (Scheme 10.113) [124]. The reversible hydrogenation–dehydrogenation process catalyzed by a single catalyst holds great promise for reversible hydrogen storge in liquid organic hydrogen carriers.

Aqueous reductions performed with reductants other than hydrogen have also been reported. In 2021, Song and coworkers reported a $CuSO_4$-catalyzed reduction of *N*-heteroarenes in water using $NH_3{\cdot}BH_3$ as the hydrogen source (Scheme 10.114) [125]. A variety of *N*-heteroarenes containing sensitive OH-, NH_2-, Cl-, Br- groups and *N*-heteroarenes including quinolines, quinoxalines, 1,5-naphthyridines, and quinazolines were well-tolerated. According to the proposed mechanism, a [Cu–H] species generated from $NH_3{\cdot}BH_3$ and $H_2N{=}BH_2$ in the $CuSO_4/H_2O$ system was believed to be the active hydrogenating species for such transformation.

More recently, Szőllősi and coworkers achieved a water-assisted mechanochemical asymmetric transfer hydrogenation (ATH) of a series of acetophenone derivatives, carbo- and heterocyclic ketones using HCOONa as the hydrogen source

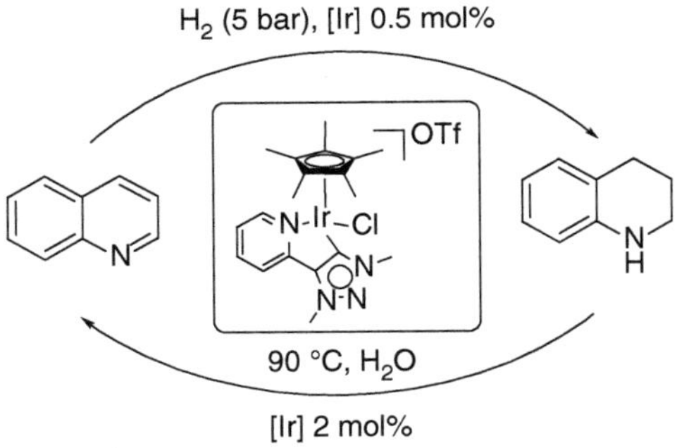

Scheme 10.113 NHC-based iridium catalysts for hydrogenation and dehydrogenation of *N*-heteroarenes.

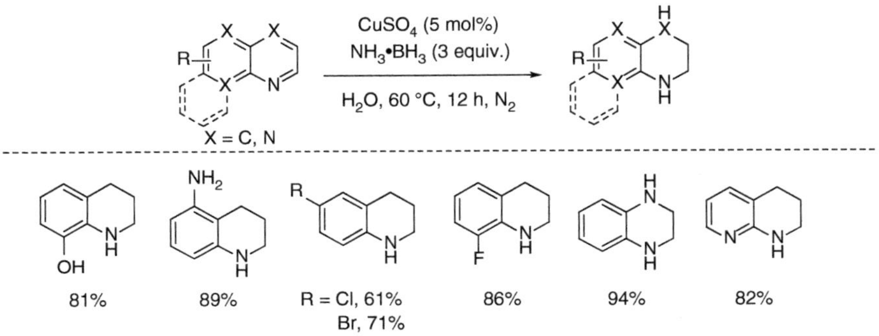

Scheme 10.114 Cu-catalyzed chemoselective reduction of *N*-heteroaromatics with $NH_3{\cdot}BH_3$.

under the assistance of a $[Ru(p\text{-cymene})Cl]_2Cl_2/(R,R)$-TsDPEN catalytic system (Scheme 10.115) [126]. The water additive could mediate the energy transmission process, solubilize the formate, promote the formation and stabilization of a catalytically active species as well as accelerate the reaction via an "on water" effect. In comparison to the traditional batch reactions, the mechanochemically activated protocol could achieve high conversions and high enantioselectivities within shorter reaction times using smaller amounts of hydrogen donor and water.

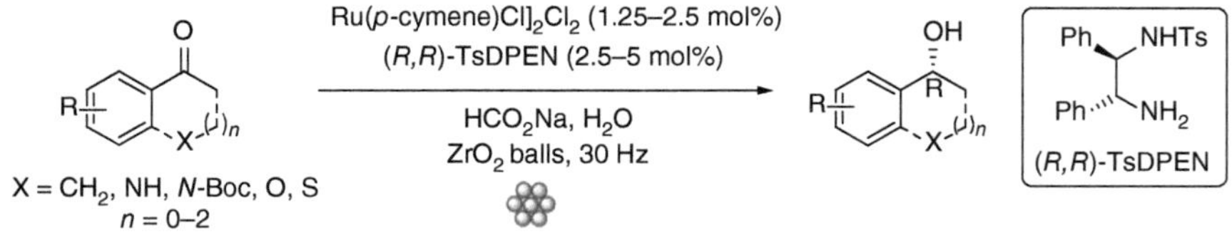

Scheme 10.115 Mechanochemical, water-assisted ATH of ketones.

Li and coworkers described a recyclable Pd/C-catalyzed *N*-cyclohexylation of a series of α-amino acids and small peptides with phenols in water using HCO_2Na as the reductant under biocompatible conditions (Scheme 10.116) [127]. The reaction was proposed to proceed via phenol reduction to ketone followed by reductive amination between the ketone and amine. The cyclohexylation of α-amino acids proceeded under protecting group-free conditions and various functional groups such as alcohols, carboxylic acids, and amides were tolerated. Moreover, the reaction of enantioenriched α-amino acids occurred smoothly without racemization.

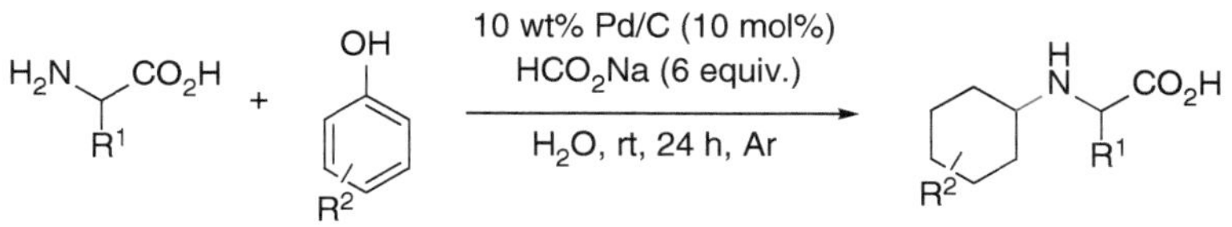

Scheme 10.116 Pd-catalytic *N*-cyclohexylation of α-amino acids/small peptides with phenols.

10.7 Substitution Reactions

10.7.1 Nucleophilic Substitution

In 2017, Yang and coworkers disclosed a catalyst-free synthesis of triaryl phosphates from phenols and $POCl_3$ in water (Scheme 10.117) [128]. To obviate quick hydrolysis of $POCl_3$, phenol was first transformed into more nucleophilic phenoxide in aqueous NaOH solution and was then exposed to $POCl_3$. The desired reaction took place almost immediately after mixing and the product gradually precipitated out as a solid. Notably, DCM and toluene gave inferior results than water due to the poor solubility of PhONa and the nonprecipitation nature of the product in these solvents. The presence of the appropriate amount of water is critical to the reaction efficiency.

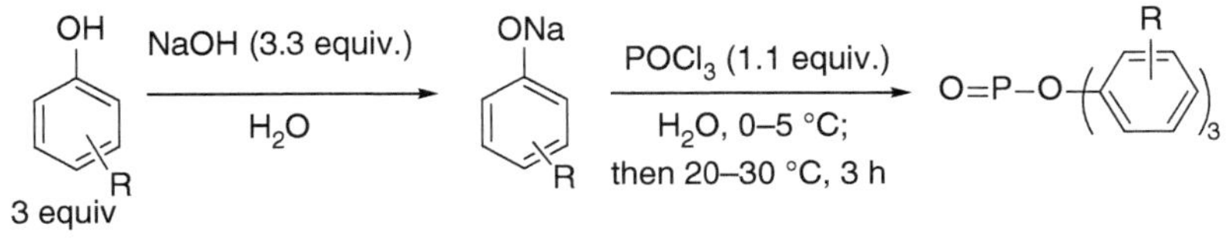

Scheme 10.117 Green synthesis of triaryl phosphates with $POCl_3$ in water.

In 2018, Miller and coworkers developed a rapid protecting group-free phenolic *O*-glycosylation of small molecules and complex unprotected peptides in water using water-stable glycosyl fluorides as glycosyl donors and $Ca(OH)_2$ as the promoter (Scheme 10.118) [129]. For fully unprotected peptides, the glycosylation occurred selectively at the phenol group of tyrosine residues with nearly all side chains of the proteinogenic amino acids remaining untouched. A singular exception belongs to cysteine where the highly nucleophilic thiol group reacted competitively with the phenol group of tyrosine.

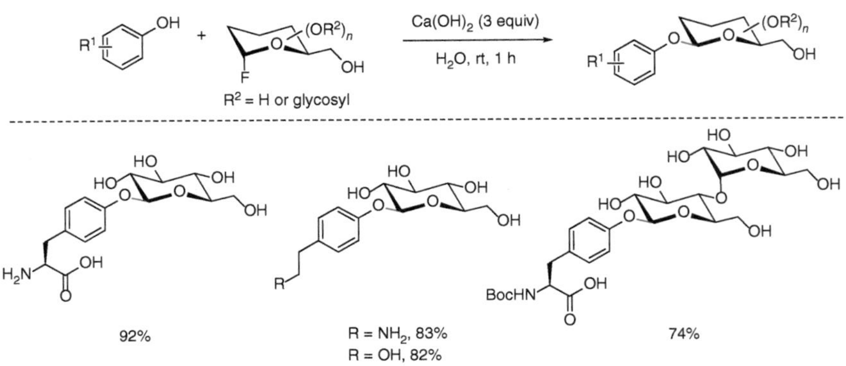

Scheme 10.118 Phenolic *O*-glycosylation of small molecules.

In 2019, Yao and coworkers described a metal-free *t*-BuONa-promoted selective formation of 2-aminopyridine derivatives through amination of various 2-fluoro-, 2-bromo-, 2-iodo-, and even 2-chloro-substitued polyhalogenated pyridines with a series of 1°, 2° alkyl amines and 3° alkyl or aryl dimethylamides using water as the solvent (Scheme 10.119) [130]. For the reaction of amide derivatives, a reaction pathway involving base-mediated formation of dimethylamine from *N,N*-dimethylamide was proposed.

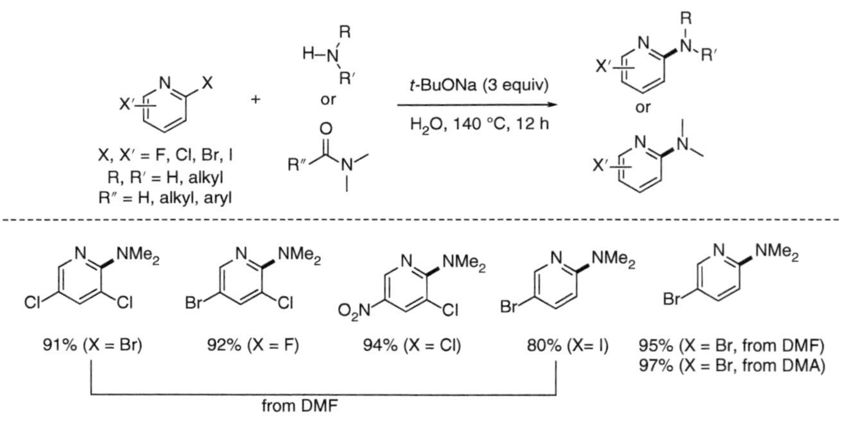

Scheme 10.119 Base-promoted selective amination of polyhalogenated pyridines.

In 2021, Miao and coworkers developed a direct allylic substitution of pyrazole and its derivatives with allylic alcohols in water under metal-free conditions using Bu_4NHSO_4 as a Brønsted acid and a phase transfer catalyst (Scheme 10.120) [131]. Water outperformed all organic solvents except 1,2-dichloroethane. The hydrogen bonding between water and two substrates facilitates the formation of a carbocation and subsequent nucleophilic attack of the carbocation by nitrogen nucleophiles. Furthermore, the hydrogen bonding effect between pyrazole and HSO_4^- also promotes the nucleophilicity of pyrazoles.

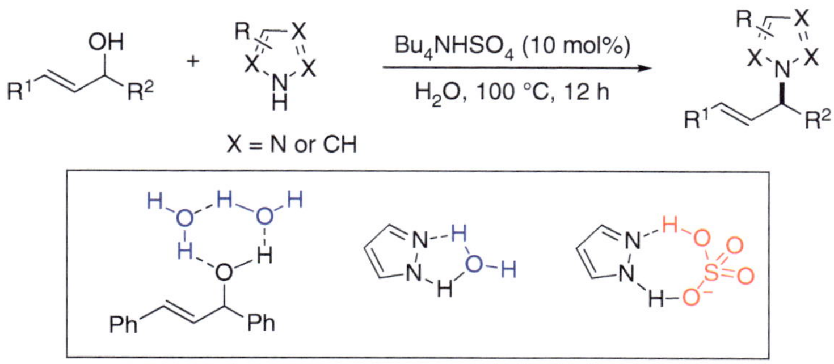

Scheme 10.120 Bu_4NHSO_4-catalyzed direct *N*-allylation of pyrazole and its derivatives with allylic alcohols.

10.7.2 Electrophilic Substitution

In 2020, Rao and coworkers reported a highly regioselective dehydrative Friedel–Crafts arylation of CF_3-containing-3-indolyl(2-thiophenyl)methanols with two-substituted indoles or electron-rich arenes via an unique 1,8-addition process using DBSA, an air-stable surfactant-type Brønsted acid, as catalyst in water (Scheme 10.121) [132]. Replacement of DBSA with other non-surfactant type protic acids led to trace amounts of products. It is worth noting that the steric hindrance at C2 position of indoles is critical for the regioselective 1,8-addition whereas less sterically hindered indoles afforded the 1,4-addition products as the major or sole products. Furthermore, the presence of a trifluoromethyl and indole moiety as well as non-substitution at the C4 position of thiophene group are crucial for the regioselective 1,8-addition process.

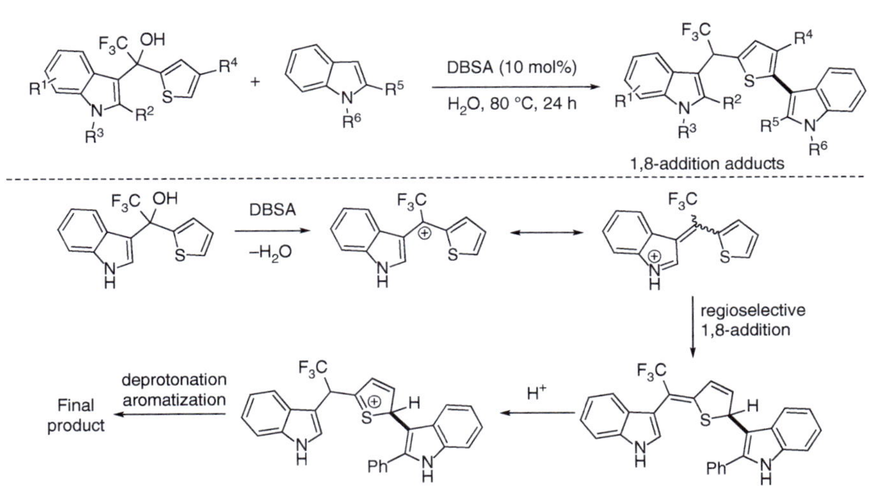

Scheme 10.121 DBSA-catalyzed regioselective dehydrative Friedel–Crafts arylation of CF_3-containing 3-indolyl(2-thiophenyl)methanols with two-substituted indoles.

In 2022, Ke and coworkers developed an efficient electrophilic aromatic bromination and hydroxybromination of olefins with NBS in water utilizing an amphiphilic indole phase-transfer catalyst bearing a hydrophilic side-chain (Scheme 10.122) [133]. The indole catalyst would react with NBS to form a brominated intermediate. This intermediate could assemble into micelles with the hydrophilic side chains stretching into water and the hydrophobic brominated indole heads aggregating in the center. The micelles could house the substrates inside and execute the bromination in the micellar microenvironment away from water, thereby avoiding side products arising from HOBr which is generated by the reaction of Br^+ and water.

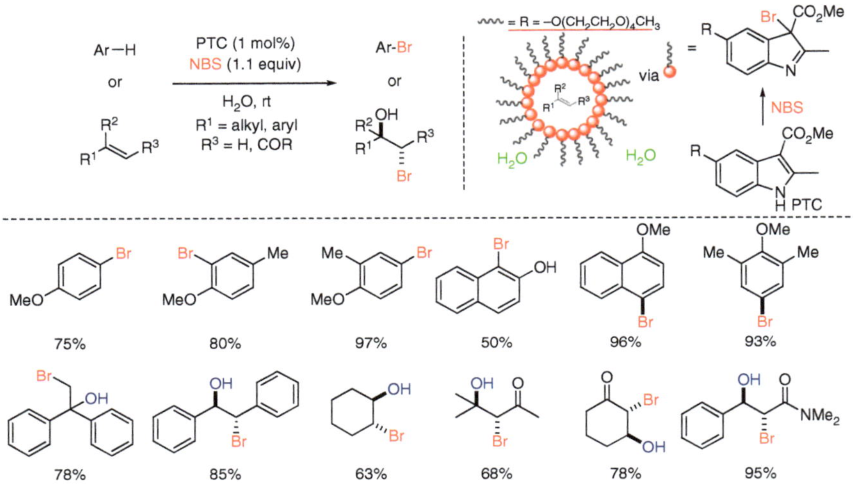

Scheme 10.122 Amphiphilic indoles as efficient phase-transfer catalysts for bromination in water.

10.7.3 Radical Substitution

In 2020, Wan and coworker reported an unprecedented radical deiodination-deuteration reaction of alkyl and aryl iodides in weakly basic aqueous solution using water-compatible and soluble deuterated calcium/sodium hypophosphites as deuterium sources (Scheme 10.123) [134]. The water-compatible deuteration protocol seems unusual when compared with conventional deuteration approaches which necessitate strict moisture exclusion. The unique water compatibility of this protocol is caused by the inertness of these deuterated hypophosphites toward H/D exchange. By using these cheap, nontoxic, shelf-stable, and easily handled deuteration reagents, a variety of substrates were smoothly deuterated in good yields with excellent D-incorporation levels, with the bromo/chloro substituents and H–D exchangeable functional groups remaining intact. The deuteration of water-insoluble alkyl iodides was also feasible by adding *t*-BuOH as the cosolvent.

More recently, Kobayashi and coworkers reported an "on water" chemoselective *C*-nitration of tyrosines and phenols using 2-methoxyethyl nitrite as an organic nitrite (Scheme 10.124) [135]. Water is essential to this transformation since

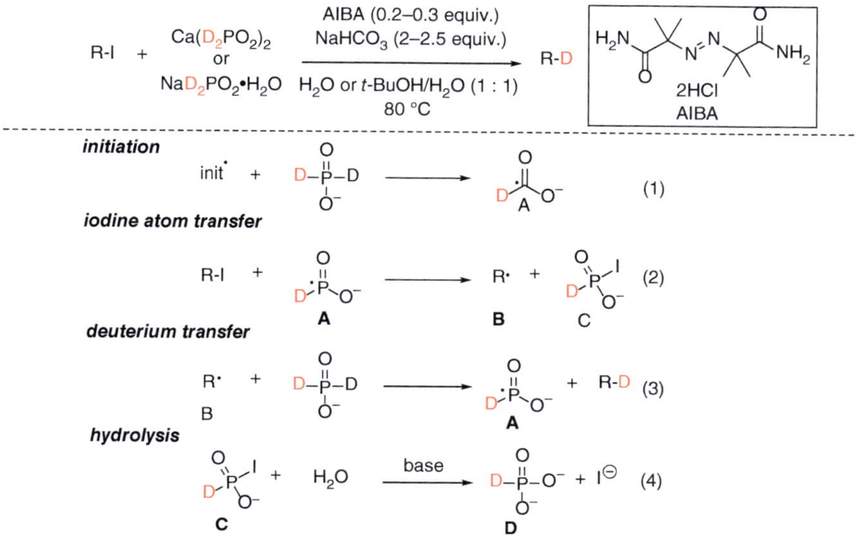

Scheme 10.123 Radical deiodination-deuteration reaction of alkyl and aryl iodides.

replacing it with other organic solvents led to decreased conversions and *C*/*N* selectivities. The proposed mechanism involves the radical coupling of a tyrosine radical and a nitro radical at theoil–water interface followed by water-assisted rearomatization to afford the aryl nitro compound. 2-Methoxyethyl nitrite possesses an ancillary unit hydrogen bonded to interfacial water molecules, thus inhibiting undesired, non-interfacial "in-oil" pathways such as *N*-nitrosation. Moreover, the hydrogen-bonding-facilitated interfacial tyrosyl radical-water interaction would also boost the desired *C*-nitration.

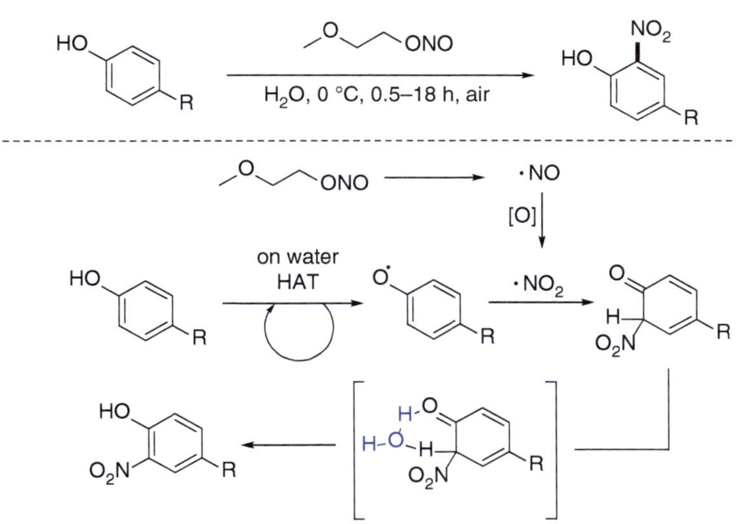

Scheme 10.124 "On water" nitration of tyrosines and phenols.

10.8 Addition Reactions

10.8.1 Nucleophilic Addition

Addition Reactions with Ketone/Imine/Nitrile

In 2017, Capriati and coworkers described an unprecedented "on-water" nucleophilic addition of Grignard and organolithium reagents to nonactivated imines and nitriles at room temperature under air (Scheme 10.125a) [136], which stands in contrast to classical methods performed under inert atmosphere in dry volatile organic solvents at low temperatures. The intermolecular hydrogen bonds between water molecules and those between water and the substrate may play a key role in disfavoring protonolysis of the organometal species and activating the substrate toward nucleophilic addition. Various secondary amines were prepared within seconds by rapidly spreading organolithium reagents into the suspension of imines. Furthermore, a one-pot sequential addition of organolithium and allyl Grignard reagents to nitriles gave easy access to a variety of tertiary carbinamines (Scheme 10.125b).

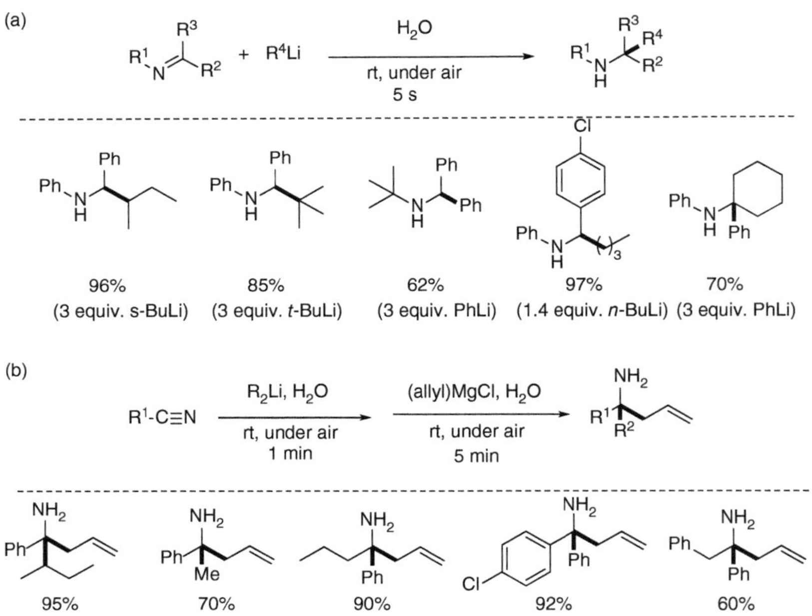

Scheme 10.125 Nucleophilic additions of organometallic compounds to imines and nitriles.

Later, a one-pot tandem protocol comprising organocatalyzed oxidation (AZADO/NaClO) of secondary alcohols and subsequent nucleophilic addition of organolithium reagents to the resulting ketones was reported by García-Álvarez and coworkers in aqueous media at room temperature in the presence of air, affording various tertiary alcohols in moderate to good yields (Scheme 10.126) [137]. The "on water" addition reaction was accomplished in an instantaneous manner and no Li–X exchange products were observed for halogen-bearing secondary alcohols.

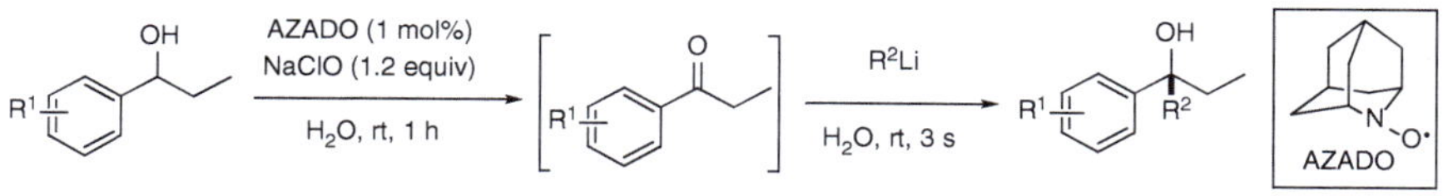

Scheme 10.126 One-pot tandem alcohol oxidation and nucleophilic addition.

In 2018, Han and coworkers developed a catalyst-free nucleophilic addition of thiols to silyl glyoxylates to access a variety of α-hydroxysilanes in water (Scheme 10.127) [138]. With the assistance of hydrogen bonding from water, the carbonyl group of silyl glyoxylate was activated toward nucleophilic addition, and the resulting alkoxide was protonated faster than the [1,2]-anionic migration responsible for the formation of undesired Brook rearrangement products. Water delivered the best result among all solvents examined in terms of both efficiency and selectivity.

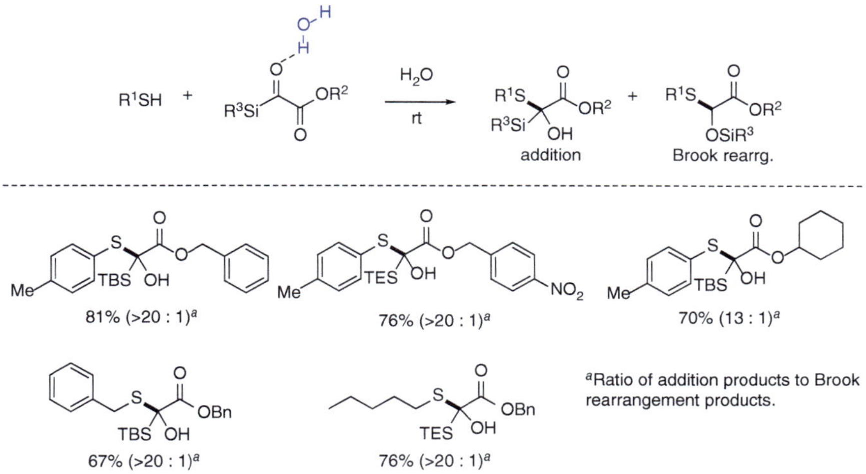

Scheme 10.127 Catalyst-free nucleophilic addition of thiols to silyl glyoxylates.

In 2019, Weng and coworkers described the synthesis of 3-aroyl coumarins via $Pd(OAc)_2$-catalyzed carbopalladation of 3-cyanocoumarins with arylboronic acids in water using 2,2'-bipyridine as the ligand and $TsOH.H_2O$ as the additive (Scheme 10.128) [139]. Notably, by using bromo- or iodo-substituted 3-cyanocoumarins as substrates, highly selective carbopalladation was achieved in preference to potential Suzuki cross-coupling. Water was proved to be superior for this transformation since a hydrolysis process was involved as the key step.

Cai and coworkers reported an additive-free "all-water" synthesis of α,α-difluoro-β-aminoketones through nucleophilic addition of fluorinated enol silyl ethers

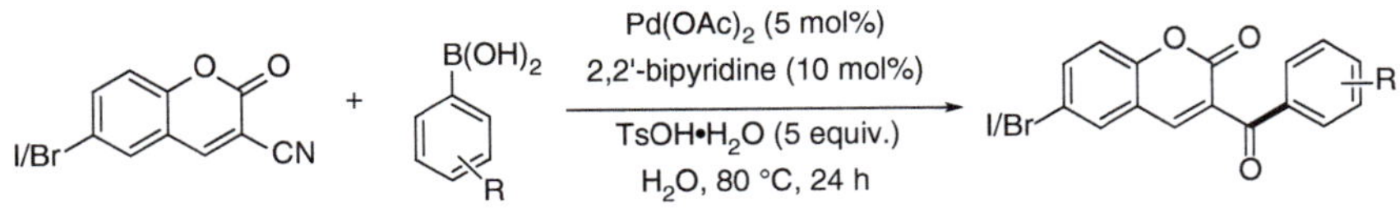

Scheme 10.128 Palladium-catalyzed addition for the synthesis of 3-aroyl coumarins.

with *in situ* generated aldimines (Scheme 10.129) [140]. Water plays a crucial and indispensable role in the nucleophilic addition process by forming hydrogen bonds with the fluorinated enol silyl ether. Later, they achieved the synthesis of gem-difluoromethylene azo compounds from arenediazonium salts and fluorinated enol silyl ethers by using the same strategy [141].

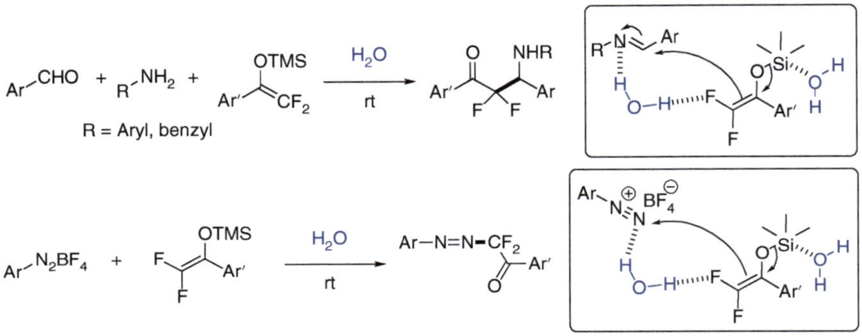

Scheme 10.129 "All-water" synthesis of α,α-difluoro-β-aminoketones, and gem-difluoromethylene azo compounds.

Recently, Wu and coworkers disclosed the synthesis of α,α-difluoro-β-hydroxyketones in water via indium-mediated Reformatsky reaction of iododifluoroketones with aldehydes or isatins (Scheme 10.130) [142]. The reaction was proposed to proceed via difluoromethyl radical formation via SET from indium, further reduction of the radical by indium or indium iodide to produce a difluoromethyl-indium reagent, isomerization, and nucleophilic addition of the resulting α,α-difluoroenol to the carbonyl group to give the final product upon water quench.

Regarding aqueous asymmetric nucleophilic addition, Wang and coworkers disclosed a highly enantioselective Henry reaction of β,γ-unsaturated α-ketoesters with nitromethane in water using $CuBr_2$ as the Lewis acid catalyst, a proline-derived tertiary alcohol as the chiral ligand and Et_4NCl as the phase transfer catalyst (Scheme 10.131) [143]. 2-Fluorophenol was added as a weak acid additive to enhance the stereoselectivity of this transformation. Water plays a key role in promoting the enantioselectivity and replacement of water with $CHCl_3$ and $MeNO_2$ led to significant drop of ee values. A diversity of unsaturated β-nitro-α-hydroxy esters bearing tetrasubstituted carbon stereocenters was prepared with excellent enantioselectivities at 0 °C.

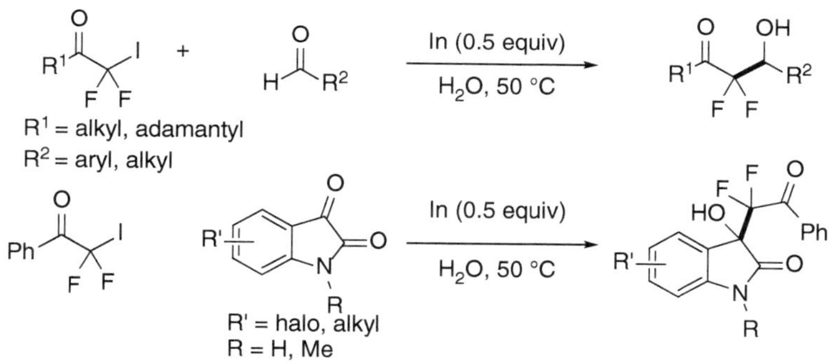

Scheme 10.130 Indium-mediated Reformatsky reaction of iododifluoroketones with aldehydes or isatins.

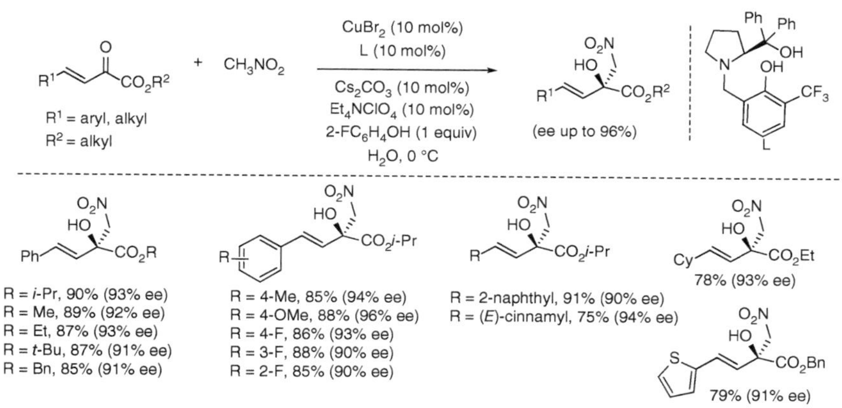

Scheme 10.131 Cu-catalyzed enantioselective Henry reaction of β,γ-unsaturated α-ketoesters with nitromethane.

Conjugate Addition

As an alternative to conventional air- and water-sensitive Cu-catalyzed/mediated conjugate addition with organometallic reagents, in 2018, Li and coworkers developed an unprecedented Cu-catalyzed, Zn-mediated direct conjugate addition of aryl/alkyl halides to electron-deficient alkenes in air and water without prior stoichiometric formation of organometallic reagents (Scheme 10.132) [144]. A variety of α,β-unsaturated esters, acrylonitriles, and α,β-unsaturated phosphonates were diversified under the catalytic system of $Cu(acac)_2$/3,4-dimethoxylaniline in the presence of the surfactant cetyltrimethyl ammonium hydrogensulfate (CTAS) or no. Surprisingly, the same reaction performed in an organic solvent was much inferior to the water-based reaction. A tentative mechanism involving an *in situ* formed organocopper species was proposed.

In the same year, Macmillan and coworkers described a photocatalyzed site- and chemo-selective *C*-terminal decarboxylative conjugation addition of peptides with α,β-unsaturated carbonyl compounds under mild and biocompatible aqueous conditions using water-compatible flavins as photocatalysts (Scheme 10.133) [145].

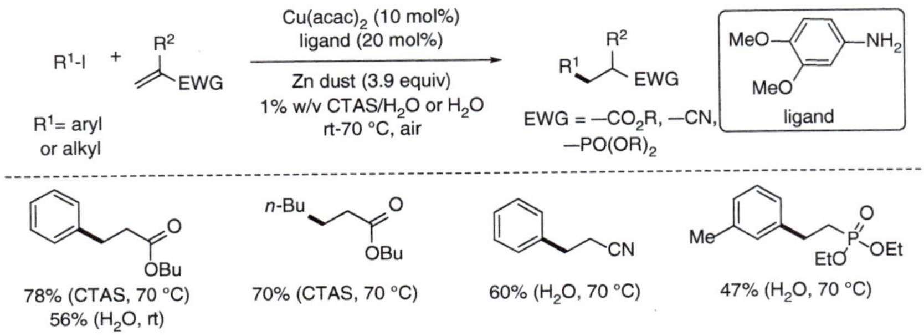

Scheme 10.132 Direct conjugate additions using aryl and alkyl halides in air and water.

The site-selectivity arises from innate differences in the oxidation potential between side-chain alkyl carboxylates and *C*-terminal α-amino carboxylates, with the more readily oxidized *C*-terminal carboxylates selectively undergoing the decarboxylative alkylation. The reaction was proposed to proceed via a single-electron oxidation decarboxylation of the *C*-terminal carboxylate by the excited photocatalyst to give an α-amino radical. Addition of the radical into the α,β-unsaturated carbonyl compound provides an α-acyl radical, which is reduced by the reduced photocatalyst and protonated to afford the *C*-terminal-functionalized product. The reaction in cesium formate buffer (pH 3.5, 10 mM) showed enhanced reactivity than that performed in phosphate buffer (pH 7, 10 mM).

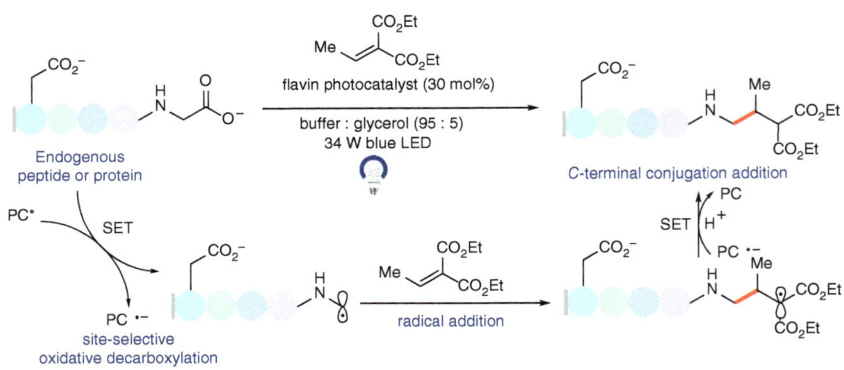

Scheme 10.133 Photocatalyzed *C*-terminal decarboxylative conjugation addition.

In 2022, Bloom and coworkers developed a biocompatible and synergistic metallaphotoredox-catalyzed conjugation addition of aromatic boronic acids to dehydroalanine (Dha)-containing peptides in water using lumiflavin as photocatalyst, $Pd(MeCN)_4(BF_4)_2$ as metal catalyst and bathocuproine (BC) as ligand (Scheme 10.134) [146]. The reaction was carried out in phosphonate buffer with the addition of 5% vol. TFE cosolvent to inhibit the undesired Heck byproduct. Apart from synthesizing proteinogenic and nonproteinogenic peptides via C–C conjugate addition, the developed protocol was also amenable to late-stage diversification of small organic molecules via C–H arylation and C–P bond formation. Mechanistically, flavin-catalyzed oxidation of hydrated arylboronic acid gives an aryl radical, which is captured by (BC)Pd^{II} to give an arylPd^{III}(BC) species. The arylPd^{III}(BC) species undergoes conjugate addition to the Dha peptide to forge the C–C bond.

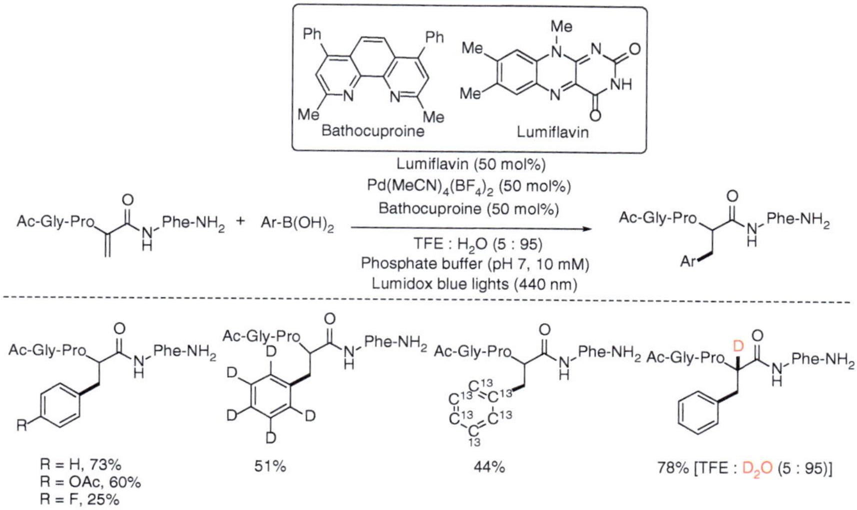

Scheme 10.134 Metallaphotoredox-catalyzed conjugation addition of aromatic boronic acids to dehydroalanine (Dha)-containing peptides.

More recently, Bae and coworkers reported the synthesis of β-aminosulfonyl fluorides from β-arylethenesulfonyl fluorides and anilines via water-accelerated, NHC-catalyzed "on water" aza-Michael addition mediated by a combination of SIMes.HCl and DABCO (Scheme 10.135) [147]. The reaction occurred smoothly in a saturated NaCl/H_2O solution with 3 equiv. of toluene added as a hydrophobic additive. Compared with the commonly used organic solvents, water significantly

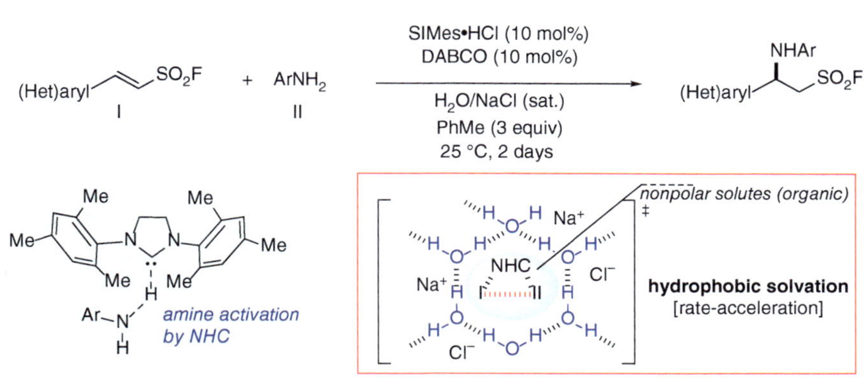

Scheme 10.135 Synthesis of β-aminosulfonyl fluorides via water-accelerated NHC catalysis.

accelerated the reaction rate with excellent chemo- and site-selectivities (up to >99:1). Such rate acceleration might be caused by hydrophobic amplification derived by biphasic hydrophobic solvation in a water-embraced organic cage. In this transformation, NHC activates the amine into a nucleophilic species for further C–N bond construction.

Besides 1,4-conjugation addition, there are also examples of 1,6-conjugate addition reactions in water. In 2019, Pawar and coworkers disclosed a catalyst-, reagent-, and additive-free protocol for fully atom-economical synthesis of diaryl methyl phosphine oxides via 1,6-hydrophosphonylation of *p*-quinone methides with diaryl phosphine oxides in water under an air atmosphere (Scheme 10.136) [148]. This protocol was amenable for gram-scale synthesis, and the product was isolated by simple filtration without resorting to column-chromatographic purification. The reaction was proposed to follow a non-free radical mechanism based on radical scavenging experiments.

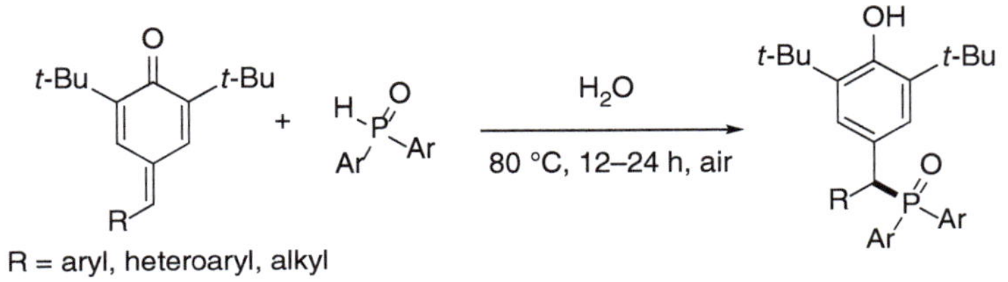

Scheme 10.136 Catalyst- and reagent-free 1,6-hydrophosphonylation of *p*-quinone methides.

10.8.2 Alkene/Alkyne Functionalization via Radical Addition

Functionalization of Alkynes

In 2018, Lei and coworkers reported a highly *Z*-selective radical addition of diaryl phosphine oxides to various terminal and internal alkynes via visible light-induced eosin Y-catalyzed hydrophosphonylation in water at room temperature (Scheme 10.137) [149]. Water was superior than other organic solvents in

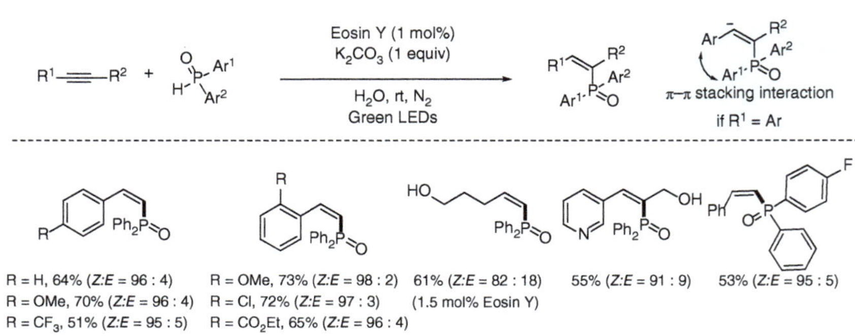

Scheme 10.137 Photocatalyzed *Z*-selective radical addition of diaryl phosphine oxides to alkynes.

terms of both efficiency and Z/E selectivity. Improved Z-selectivity was observed for aromatic alkynes due to the presence of an additional π–π stacking interaction between the aryl group and phosphorus substituents. A tentative mechanism involving the formation of a phosphinoyl radical via proton-coupled electron transfer (PCET) oxidation was proposed.

In the same year, Lu and workers reported a Mn(III)-mediated synthesis of β-ketosulfones through decarboxylative hydroxysulfonylation of arylpropiolic acids with sodium sulfinates using water as both the solvent and β-oxygen source (Scheme 10.138) [150]. Water and toluene afforded comparable yields for this transformation. However, the use of water as a solvent allowed the reaction to proceed faster and facilitated an easier separation of the product by simple filtration. The proposed mechanism involves the addition of a sulfonyl radical generated by Mn(III)-mediated single electron oxidation to alkyne to afford a reactive vinyl radical. The vinyl radical is oxidized to a carbocation species, which is further attacked by water and undergoes ketone-enol tautomerism and decarboxylation to afford the β-ketosulfone.

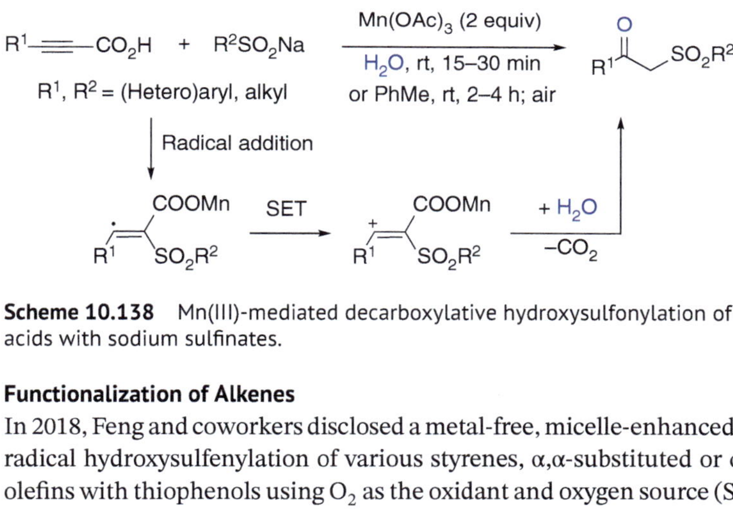

Scheme 10.138 Mn(III)-mediated decarboxylative hydroxysulfonylation of arylpropiolic acids with sodium sulfinates.

Functionalization of Alkenes

In 2018, Feng and coworkers disclosed a metal-free, micelle-enhanced auto-oxidative radical hydroxysulfenylation of various styrenes, α,α-substituted or α,β-substituted olefins with thiophenols using O_2 as the oxidant and oxygen source (Scheme 10.139) [151]. The auto-oxidation process enables the direct initiation of a thiyl radial by sustainable O_2 instead of commonly used transition metals or peroxides. It should be noted that a hydrothiolation product was exclusively formed in pure water. The addition of Brij L4 as a surfactant changed the chemoselectivity by forming a hydroxysulfenylated product instead. This altered chemoselectivity could be explained by increased O_2 concentration in the micelles which may accelerate the auto-oxidation process.

In the same year, Bihel and coworkers reported a metal-free TBHP-mediated radical dioxygenation of styrenes with molecular oxygen and N-hydroxyphthalimide (NHPI) in water at room temperature using Triton X-100 as the surfactant (Scheme 10.140) [152]. While terminal and 1,1-disubstituted styrenes provided good yields of products, the reaction of 1,2-disubstituted styrenes afforded lower yields due to the presence of steric hindrance.

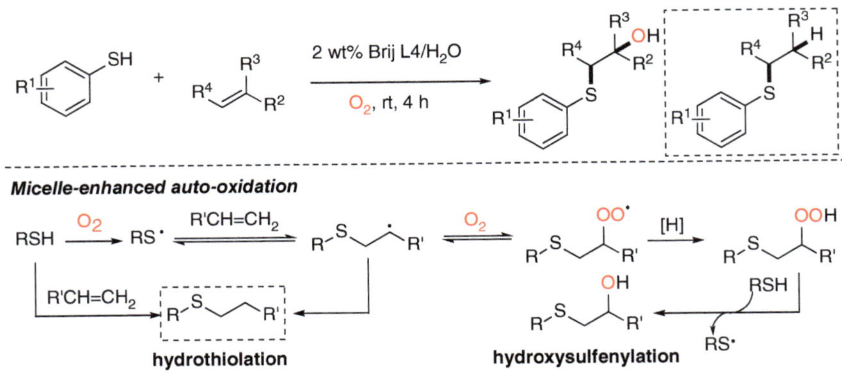

Scheme 10.139 Micelle-enhanced auto-oxidative radical hydroxysulfenylation of alkenes.

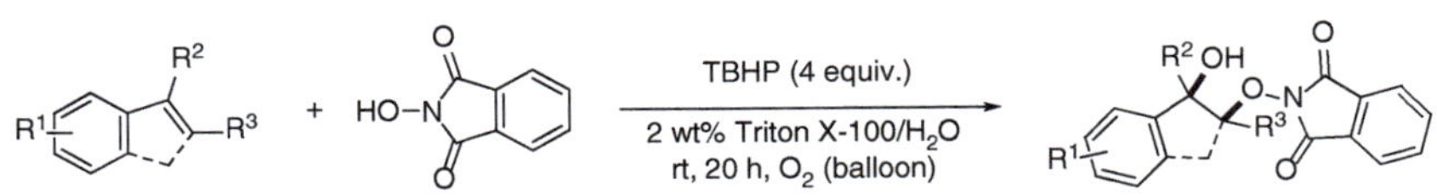

Scheme 10.140 Radical dioxygenation of styrenes with molecular oxygen and NHPI.

In 2020, Kokotos and coworkers developed a phenylglyoxylic acid-initiated photochemical hydroacylation of electron-rich olefins with various aromatic and aliphatic aldehydes in water using household lamps under directing group-, additive-, and metal-free conditions (Scheme 10.141) [153]. Water outperformed all organic solvents examined and was believed to promote the smooth propagation of acyl radicals from aldehydes. The reactions of branched α,α-disubstituted aldehydes proceeded well with suppressed formation of the decarbonyl radical addition side-product. Mechanistically, a benzoyl radical is generated by photo-excitation of $PhCOCO_2H$, and it abstracts hydrogen from the aldehyde to produce an acyl radical **A**. The acyl radical reacts with the olefin to give an alkyl radical intermediate **B**, which abstracts the aldehyde hydrogen to deliver the final product and initiate the chain propagation. Water could also promote the propagation by reacting with intermediate **B** to give a hydroxyl radical. The hydroxyl radical undergoes H abstraction from the aldehyde to initiate the propagation.

In 2021, Handa and coworkers reported several catalyst-free NBS-mediated one-pot oxyfunctionalizations of styrenes in water under the micelle effect induced by the amphiphile PS-750-M surfactant structurally mimicking the dipolar aprotic solvent (Scheme 10.142) [154]. A series of oxyazidation, oxyacyloxylation, oxylthiocyanation, and oxyamination reactions were achieved by using azides, carboxylic acids, thiocyanates, and amines as the nucleophiles, respectively. The reaction was proposed to occur through NBS-mediated bromination of the double bond to afford a stabilized benzyl radical **A**, which reacts with the water molecule to give an α-hydroxy bromide intermediate **B**. After reacting with NBS, the intermediate **B** is transformed to a bromo-ketone intermediate **D** via the intermediacy of a

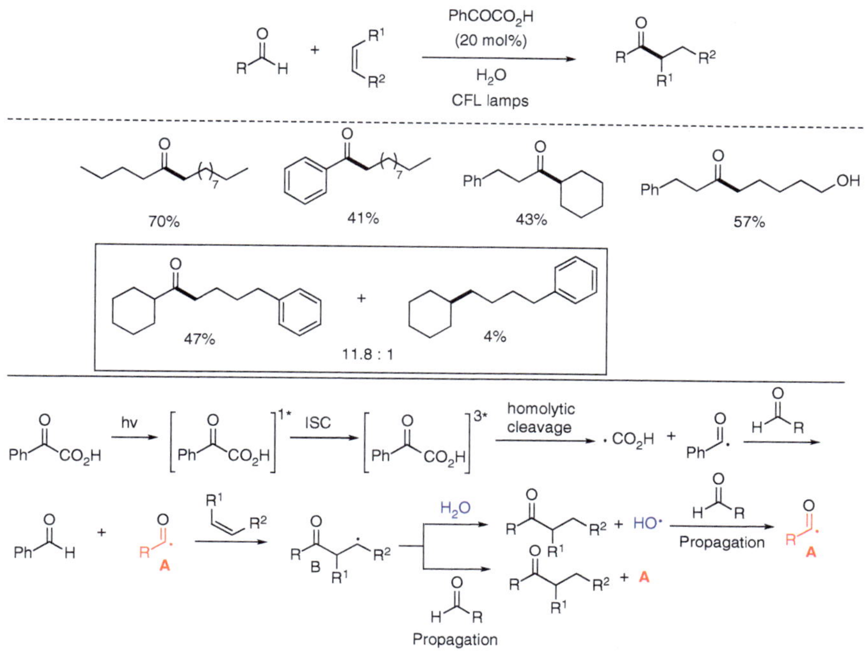

Scheme 10.141 Phenylglyoxylic acid-initiated photochemical hydroacylation of electron-rich olefins with various aldehydes.

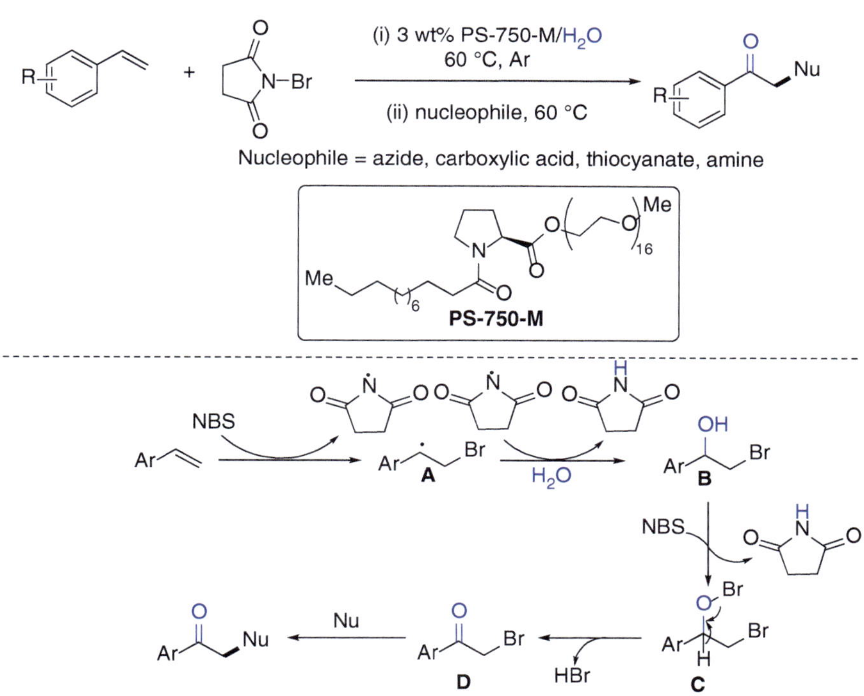

Scheme 10.142 NBS-mediated one-pot oxyfunctionalization of styrenes.

brominated species **C**. Nucleophilic substitution of the bromo-ketone with various nucleophiles finally affords the corresponding oxyfunctionalized products.

Besides the above 1,2-radical addition reactions, Xu and coworkers developed an eosin Y-catalyzed photo-mediated 1,5-trifluoromethylthio-sulfonylation of vinylcyclopropanes with $ArSO_2SCF_3$ in water using SDS as the surfactant (Scheme 10.143) [155]. A variety of aryl/alkyl vinylcyclopropanes were compatible with this transformation and afforded products with excellent regioselectivities, moderate to good yields, and *Z/E* stereoselectivities. The remote 1,5-difunctionalization protocol is facilitated by the initial addition of a photo-generated sulfonyl radical to the double bond followed by ring opening of the cyclopropane to afford a remote radical ready for trifluoromethylthiolation.

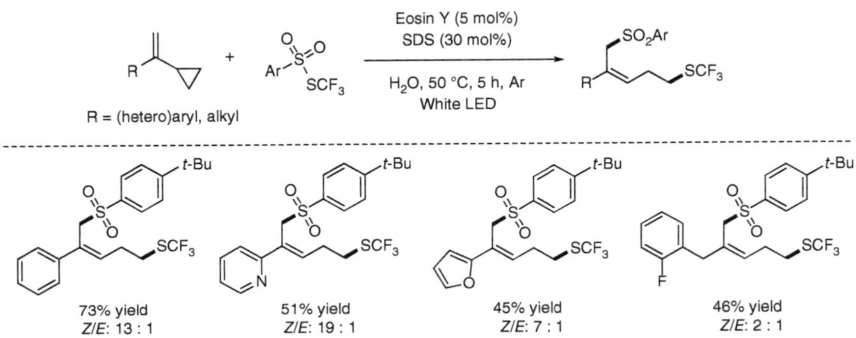

Scheme 10.143 Eosin Y-catalyzed remote 1,5-trifluoromethylthio-sulfonylation of vinylcyclopropane with $ArSO_2SCF_3$.

10.8.3 Alkene or Alkyne Functionalization via Radical-Free Addition

Functionalization of Alkynes

In 2017, Song and coworkers reported the synthesis of vinylboranes via $Cu(OTf)_2$/spiro(phosphoamidite)(SIPHOS)-catalyzed hydroboration of unactivated internal alkynes with B_2Pin_2 using water as both the solvent and proton source under air at room temperature (Scheme 10.144) [156]. Exclusive β-regio- and *Z*-stereoselectivities were obtained by using this protocol. Mechanistically, $Cu(OTf)_2$ is converted to a more activated $LCu(OH)_2$ species with the assistance of Cs_2CO_3 in water. B–B bond activation of B_2Pin_2 with $LCu(OH)_2$ gives a catalytically active Cu(II)–B species **A**, which mediates borylcupration of unactivated internal alkynes and the resulting vinyl copper species **C** undergoes hydrolysis to afford vinylboranes and regenerate the $LCu(OH)_2$ species.

In 2021, Jia and coworkers reported the synthesis of α,β-unsaturated carboxylic acids via water-soluble Pd-catalyzed hydroxycarbonylation of alkynes with CO and H_2O using water as both a reactant and solvent (Scheme 10.145) [157]. A sulfoxantphos ligand containing a water-soluble $-SO_3H$ group was employed for this transformation to facilitate the separation of product, catalyst recovery, and recycling. A variety of symmetrical and unsymmetrical diaryl alkynes, dialkyl alkynes, aryl

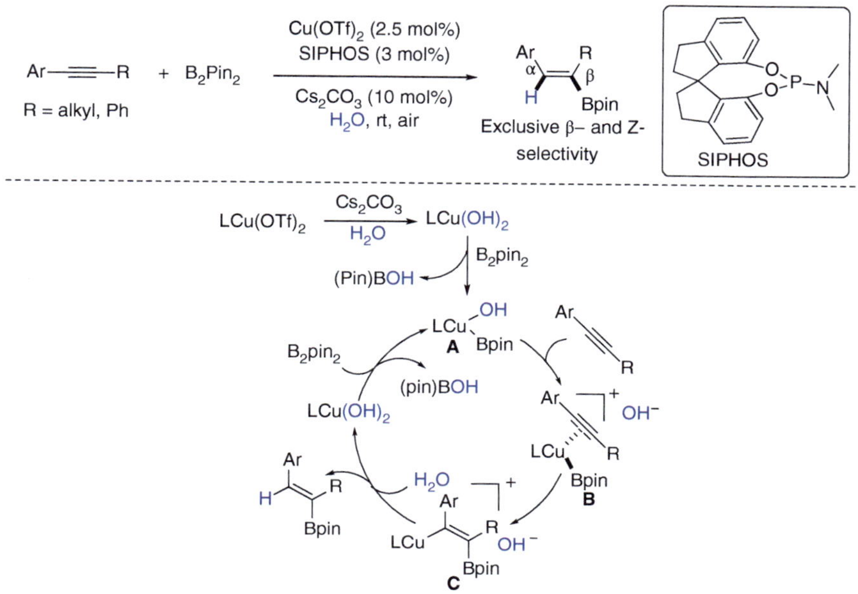

Scheme 10.144 Cu(OTf)$_2$/SIPHOS-catalyzed hydroboration of internal alkynes.

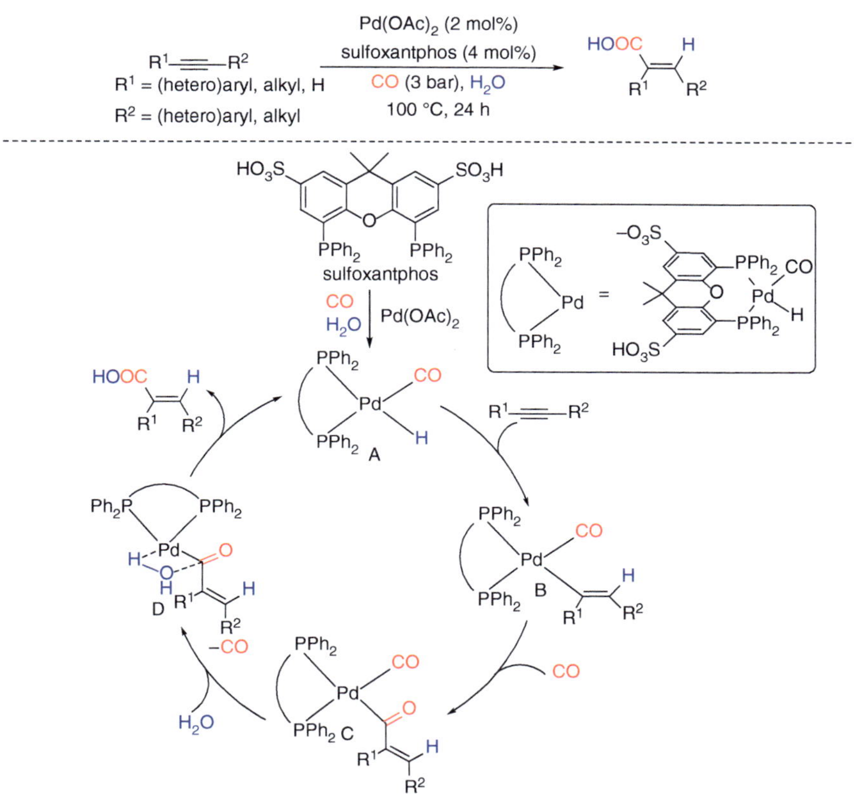

Scheme 10.145 Pd-catalyzed hydroxycarbonylation of alkynes with CO and H_2O.

alkyl alkynes, and terminal alkyne were hydroxycarbonylated with good yields and regioselectivities. The reaction was proposed to proceed via alkyne insertion into a catalytically active palladium hydride species **A** (generated by reacting Pd/L with CO/H_2O) to form an alkenyl-Pd intermediate **B**. CO insertion into the Pd–C bond affords an acylpalladium intermediate **C** which exchanges with water and undergoes σ-bond metathesis to afford the final product and regenerate the Pd–H species.

Functionalization of Alkenes

In 2021, Ostaszewski and coworkers reported a one-pot access to 1,2-diarylethanols via ligandless and base-free Pd-catalyzed α,β-homodiarylation of vinyl esters with arylboronic acids in water at room temperature using benzoquinone (BQ) as the oxidant (Scheme 10.146) [158]. As suggested by the possible catalytic cycle, water plays triple roles in this transformation: (1) promoting the formation of a catalytically active HO–Pd–OH species from $Pd(OAc)_2$; (2) stabilizing the six-membered migratory insertion intermediate **D** by coordinating to the palladium center; and (3) transforming BQ-Pd(0) to active OH–Pd–OH species and serving as proton donors in reduced hydroquinone.

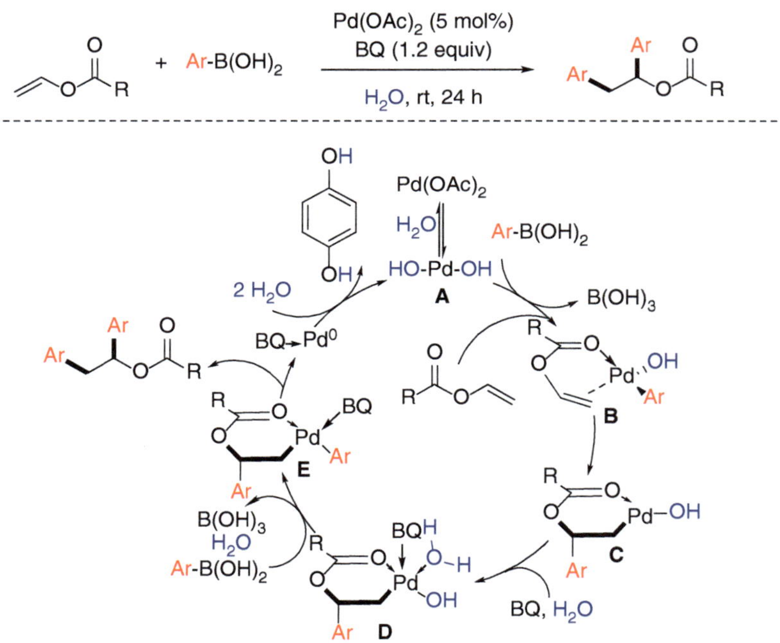

Scheme 10.146 Ligandless Pd-catalyzed direct α,β-homodiarylation of vinyl esters.

In the same year, Petricci and coworkers reported the micellar- and microwave-facilitated regioselective preparation of linear aldehydes via hydroformylation of terminal alkenes with syngas (a mixture of H_2 and CO) in water (Scheme 10.147) [159]. The protocol features mild conditions (70 °C and 9 bar) and short reaction times (40–60 min) using cost-effective $Rh(CO)H(PPh_3)_3$ as catalyst and Xantphos as ligand. This methodology was also used in tandem with intramolecular hemiacetalization to selectively synthesize six-membered cyclic hemiacetals from hydroxy group-bearing starting alkenes. By using $NaHSO_3$ as an additive, the product aldehyde was precipitated as corresponding Bertagnini's salt and was filtered, thus obviating organic solvent-mediated purification steps. Moreover, the micellar-catalyst system could be re-utilized at least five times without any impact on reaction yields.

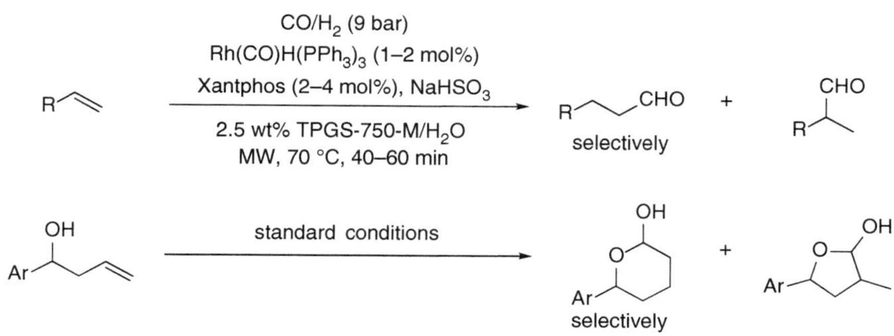

Scheme 10.147 Micellar- and microwave-facilitated hydroformylation of terminal alkenes.

10.8.4 Cycloaddition Reactions

Azide–Alkyne Cycloaddition

The azide–alkyne cycloaddition is a major synthetic tool for triazole synthesis. Since the first reports by the groups of Shapless and Meldal, copper-catalyzed azide–alkyne cycloaddition (CuAAC) has become an efficient method for the regioselective synthesis of 1,4-disubstituted 1,2,3-triazoles. This transformation not only occurs in organic solvents but also in aqueous media even at room temperature. Compared with the routine synthesis of 1,4-disubstituted triazoles, there are only limited reports regarding the synthesis of other regioisomeric triazoles under aqueous conditions.

In 2017, Hong and coworkers achieved selective syntheses of 1,5-disubstituted 1,2,3-triazoles via a nickel-catalyzed azide–alkyne cycloaddition (NiAAC) in air and water at room temperature with the assistance of a Cp_2Ni/Xantphos catalytic system (Scheme 10.148) [160]. The Ni-catalyzed methodology is complementary to the classical 1,4-seletive CuAAC reaction which typically affords 1,4-triazoles. The protocol demonstrates a broad substrate scope and a series of aromatic/aliphatic terminal alkynes and diaryl/arylalkyl internal alkynes including unsymmetrical alkynes were tolerated. Moreover, triazoles decorated with carbohydrates or amino acid moieties were also prepared using this method.

In 2021, Dash and coworkers reported the regioselective formation of N^2-carboxyalkylated triazoles from azidoformates and alkynes in water via a CuI/prolinamide-catalyzed domino [3 + 2] cycloaddition-1,2-acyl migration process (Scheme 10.149) [161]. Prolinamide could further catalyze the hydrolysis of

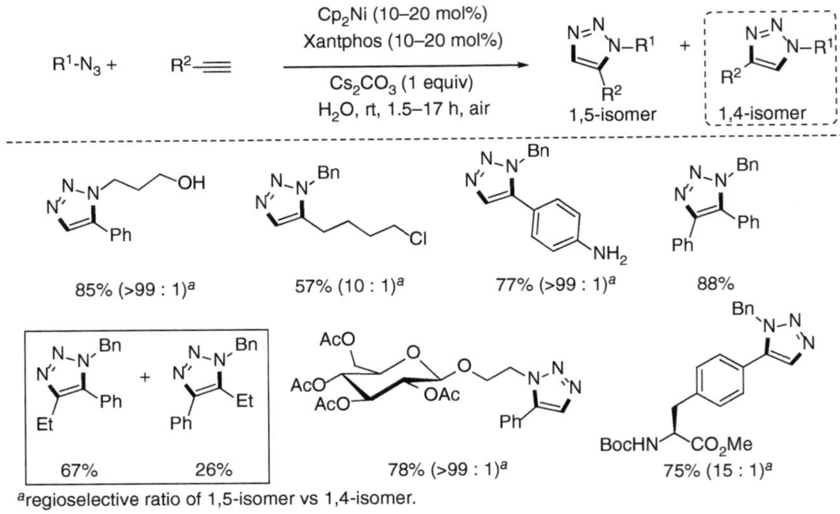

Scheme 10.148 NiAAC for the synthesis of 1,5-disubstituted 1,2,3-triazoles.

N^2-carboxyalkylated triazoles to produce free $2H$-1,2,3-triazoles. Mechanistically, the [3+2] cycloaddition initially affords an N^1-carboxyalkylated triazole which undergoes a rapid CuI- or prolinamide-catalyzed 1,2-acyl transfer to give the N^2-substituted product.

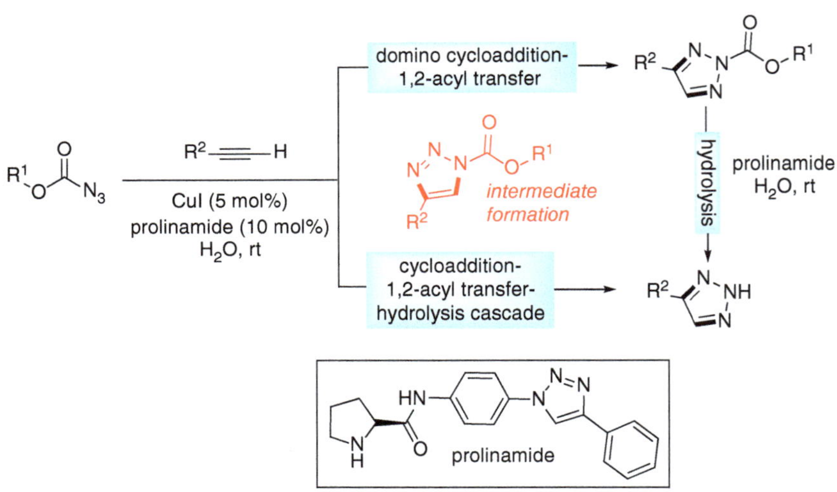

Scheme 10.149 Regioselective synthesis of N^2-carboxyalkylated- and free $2H$-1,2,3-triazoles.

[4 + 2] Cycloaddition

In the 1980s, Breslow and coworkers observed the hydrophobic acceleration of Diels–Alder (D–A) reactions in water. This pioneering work ignited a renewed interest in water-based reactions and many aqueous [4 + 2] cycloadditions have been developed since then [162].

In 2017, Ananikov and coworkers described a one-pot three-step sustainable synthesis of a nonplanar tricyclic compound from biobased 5-HMF in water featuring a Diels–Alder cycloaddition as the key step (Scheme 10.150) [163]. Due to the reluctance of HMF to undergo D–A reaction directly, it was first reduced to more reactive 2,5-bis(hydroxymethyl)furan (BHMF) by $NaBH_4$ in water. BHMF readily participated in the D–A reaction with HMF-derived maleimide to provide an oxabicyclic derivative, which was hydrogenated to give the double-bond reduced product.

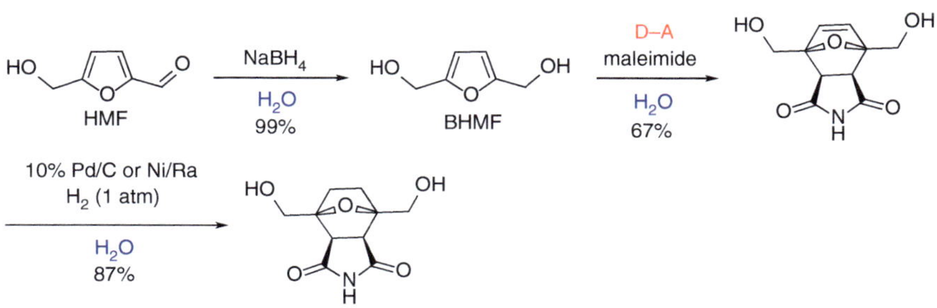

Scheme 10.150 One-pot three-step "all-water" synthesis of bis(hydroxymethyl) norcantharimide from HMF.

In 2021, a highly efficient and selective protocol was developed by Han and coworkers for the novel synthesis of spiro(indoline-2,3'-hydropyridazine) scaffold via "on water" [4 + 2] annulation between 2-methyl-3*H*-indolium salts (2-methylene indole precursors) and α-bromo *N*-acyl hydrazones (azoalkene precursors) using Cs_2CO_3 as the base (Scheme 10.151) [164]. Both substrates are insoluble in water and the "on water" nature is crucial for achieving high efficiency and selectivity. Theoretical calculations suggested that the reaction proceeds via a stepwise ionic annulation process rather than a concerted pathway. Water, as a strong polar and

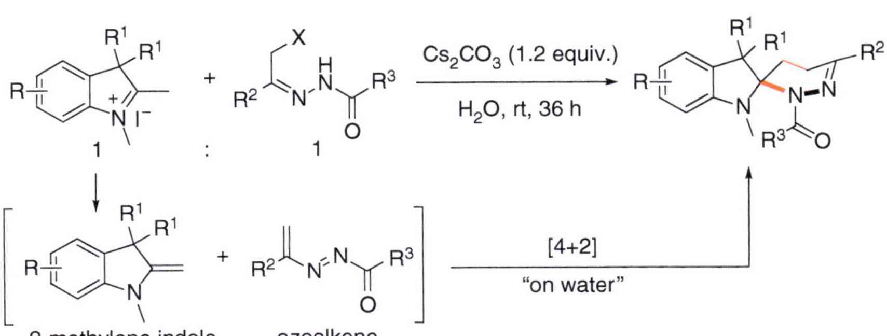

Scheme 10.151 Synthesis of spiro(indoline-2,3'-hydropyridazine) via "on-water" [4 + 2] annulation.

H-bond donor solvent, could increase the reaction rate by lowering the energy barrier of the rate-determining step and also stabilize the product by lower energy.

Very recently, Perekalin and coworkers reported a visible-light-induced generation of cyclopentadiene from cyclopentadienyl iron complex and subsequent D–A reactions with a series of dienophiles in water at room temperature using malonic acid as a scavenger for the released iron ion (Scheme 10.152) [165]. The use of water may prevent the precipitation of iron salts, facilitate the isolation of products, and probably accelerate the cycloaddition process. Various dienophiles such as substituted maleimides, dimethyl fumarate, dimethyl acetylenedicarboxylate, and quinone were amenable for such transformation.

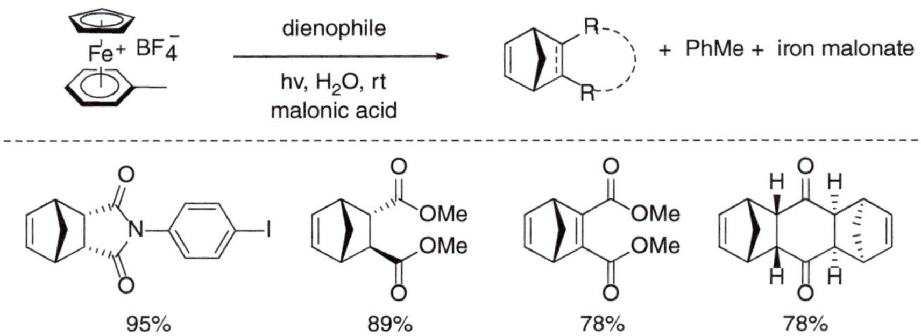

Scheme 10.152 Photochemical generation of cyclopentadiene and subsequent D–A reaction.

With regard to asymmetric aqueous D–A reactions, in 2019, Che and coworkers reported a silver-mediated preparation of axially chiral vinylallenes from optically active propargylamines and subsequent (hetero)-D–A reactions with azodicarboxylates or maleimides on water under mild conditions (Scheme 10.153) [166]. Excellent axial-to-point chirality transfer was observed in the D–A step to give the cycloadducts with up to >99% ee. Solvent screening of the hetero-D–A reaction revealed water as the best solvent in terms of both conversion and yield. The beneficial effect of water might be explained by the polarity effect pushing the organic molecules together for

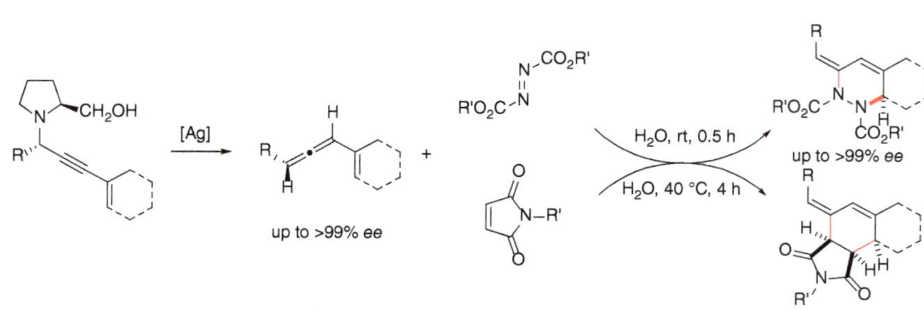

Scheme 10.153 Silver-mediated synthesis of axially chiral vinylallenes and subsequent (Hetero)-Diels–Alder reactions.

effective intermolecular reaction. The enantioselectivity of the reaction was hardly affected by the polarity of solvents and remained the same for all solvents screened.

Later, Wang and coworkers prepared various optically active 3,4-dihydro-2*H*-pyran derivatives via highly diastereoselective and enantioselective [4 + 2] cycloaddition between β,γ-unsaturated α-keto esters and terminal 2-vinylpyrroles in water under the catalysis of $Cu(OTf)_2$/prolinol/*t*-BuOK at a low loading (Scheme 10.154) [167]. The enantioselectivity of this transformation was dictated by a hydrogen-bonding interaction between the N–H of 2-vinylpyrrole and the oxygen atom of the free prolinol hydroxyl in the ligand. Attachment of a methyl group to the prolinol or phenol hydroxyl group of the ligand and the N–H of 2-vinylpyrrole all led to racemic products. The 2-vinylpyrrole attacks from the *Re* face of the β,γ-unsaturated α-keto ester due to the action of hydrogen bonding.

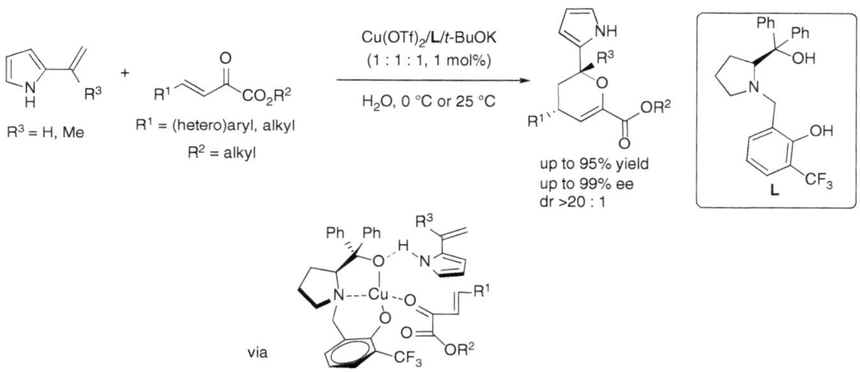

Scheme 10.154 Cu-catalyzed stereoselective [4 + 2] cycloaddition of β,γ-unsaturated α-keto esters and 2-vinylpyrroles.

10.9 Cyclization or Annulation Reactions

10.9.1 Radical-Free Cyclization/Annulation

In 2018, Satyanarayana and coworkers reported the synthesis of indenones via $Pd(OAc)_2$/L-proline catalyzed annulation of *ortho*-halobenzaldehydes with diarylacetylenes or unsymmetrical alkylarylacetylenes in water using TBAI as the phase-transfer catalyst and K_2CO_3 as the base (Scheme 10.155) [168]. The reaction of unsymmetrical alkylarylacetylenes displayed excellent regioselectivity, providing exclusively the regioisomer with the alkyl group located at the indenone C-2 position. The synthetic strategy was also amenable for the synthesis of a neolignan natural product.

In 2019, Mo and coworkers described a cost-effective and recyclable Ni(II)-catalyzed [5 + 1] annulation of 2-carbonyl-1-propargylindoles with hydroxylamine to synthesize pyrazino[1,2-*a*]indole-2-oxides in water using TBAI as the phase transfer catalyst and NaOAc as the base (Scheme 10.156) [169]. They successfully

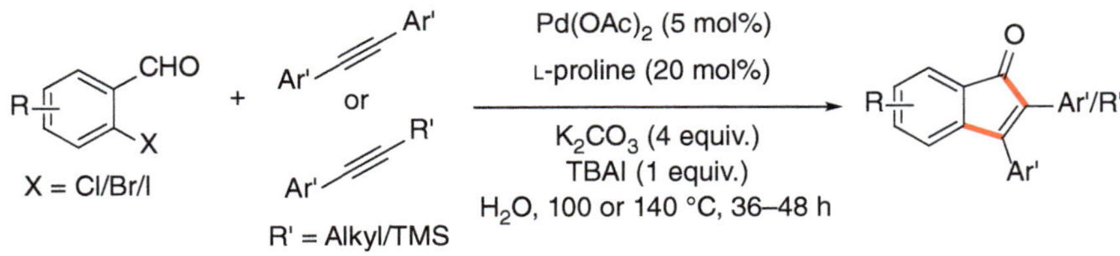

Scheme 10.155 Pd-catalyzed annulation of *ortho*-halobenzaldehydes with internal alkynes.

extended the scope to alkynyl pyrroles and benzoimidazoles, affording pyrrolo [1,2-*a*]pyrazine 2-oxide and benzo[4,5]imidazo[1,2-*a*]-pyrazine 2-oxide in useful yields. The reaction involves an initial condensation of ketone with hydroxylamine to afford an oxime intermediate **A**, which undergoes nickel(II)-catalyzed *6-exo-dig* cyclization, protonolysis, and isomerization to give the final product. It is noteworthy that the gram-scale pure products can be simply isolated by filtration and ethanol washing.

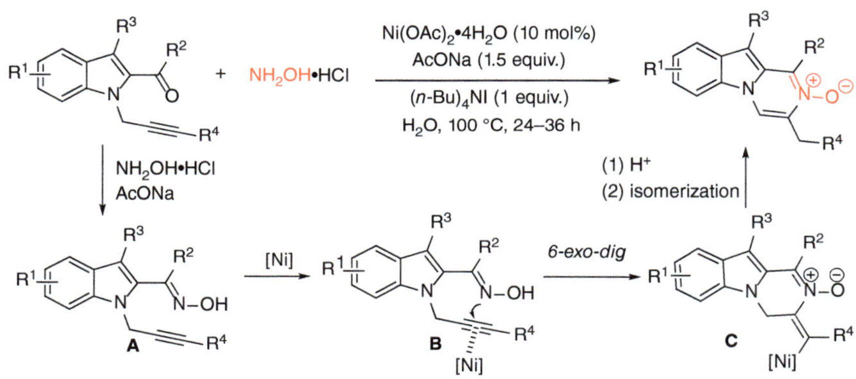

Scheme 10.156 Ni(II)-catalyzed [5 + 1] annulation of 2-carbonyl-1-propargylindoles with hydroxylamine.

In 2020, an Ir-catalyzed hydrogen borrowing annulation strategy was developed by Donohoe and coworkers for the synthesis of saturated azaheterocycles from amines and diols (Scheme 10.157) [170]. β-substituted diols were susceptible to racemization in the hydrogen borrowing alkylation processes. But remarkably, by using water as the solvent, the reaction of enantiopure β-substituted diols proceeded with minimal racemization to provide highly enantioenriched C3-substituted saturated aza-heterocycles. Moreover, base could be totally omitted from the reaction mixture under aqueous conditions without causing any detrimental effect.

In the same year, Wu and coworkers disclosed the cyclization of 2-aminobenzonitriles with CO_2 for the synthesis of quinazoline-2,4-(1*H*,3*H*)-diones in water using CO_2-philic and cost-effective diethanolamine (DEA) as an organocatalyst (Scheme 10.158) [171]. Water was proved to be the best solvent for this transformation, and it plays a dual role of solvent as well as co-catalyst in this catalytic process.

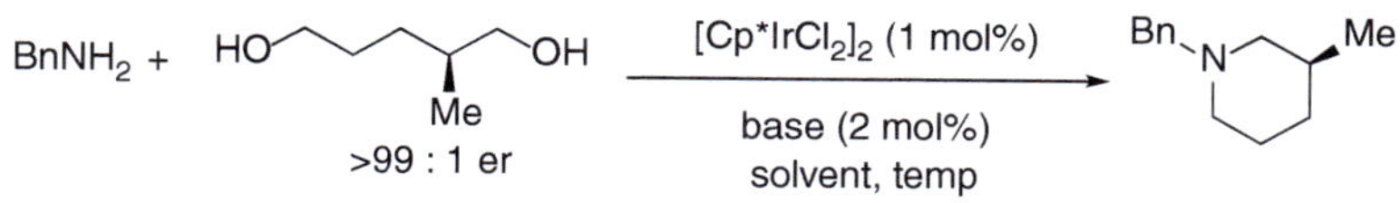

Entry	Base	Solvent	T/°C	Yield	er
1	$CsHCO_3$	PhMe	110	98	74 : 26
2	$CsHCO_3$	CPME	110	81	80 : 20
3	$CsHCO_3$	H_2O	110	79	84 : 16
4	–	H_2O	110	79	83 : 17
5	–	H_2O	80	72	90 : 10

Scheme 10.157 Ir-catalyzed enantiospecific hydrogen borrowing annulation.

DFT studies suggested that water could participate in the formation of a novel six-membered ring transition state and reduce the energy barrier for the formation of carbamate. The hydroxyl group of DEA is indispensable for the activation of both substrates. After filtration of the water-insoluble product, the catalyst was easily recovered and reused for several cycles with similar high yields.

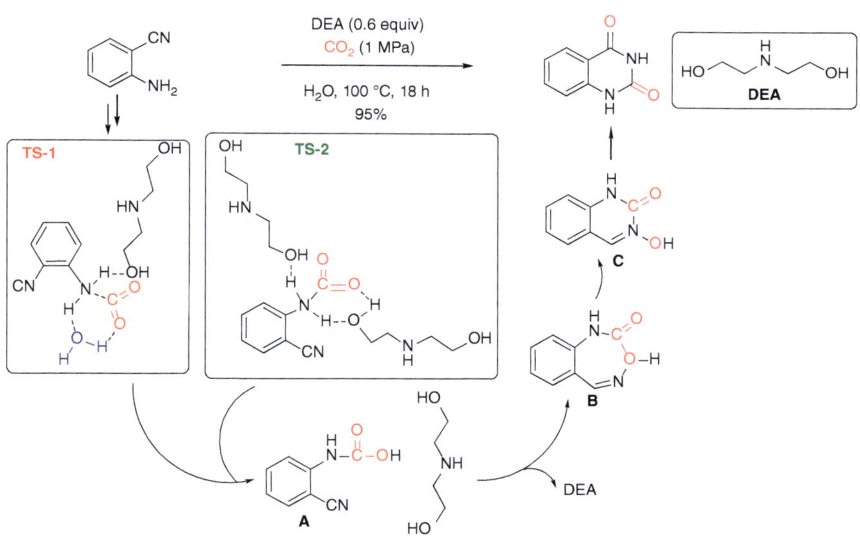

Scheme 10.158 DEA-catalyzed cyclization of 2-aminobenzonitrile and CO_2.

Ji and coworkers reported a catalyst- and additive-free synthesis of monobromo- and multibromobenzoxazines by treatment of olefinic amides with NBS in water (Scheme 10.159) [172]. Monobromobenzoxazines were mainly afforded at room temperature, while multibromobenzoxazines were preferentially formed at 80 °C with an increased amount of NBS. By employing Selectfluor and *N*-chlorosuccinimide as the electrophiles, monofluorobenzoxazine and dichlorobenzoxazine were also obtained. According to the proposed mechanism, the active bromo source is HOBr which is derived from the reaction between NBS/Br_2 (*in situ* formed) and water. For the formation of multibromobenzoxazines, a cascade process of electrophilic addition, cyclization, and electrophilic substitution was proposed.

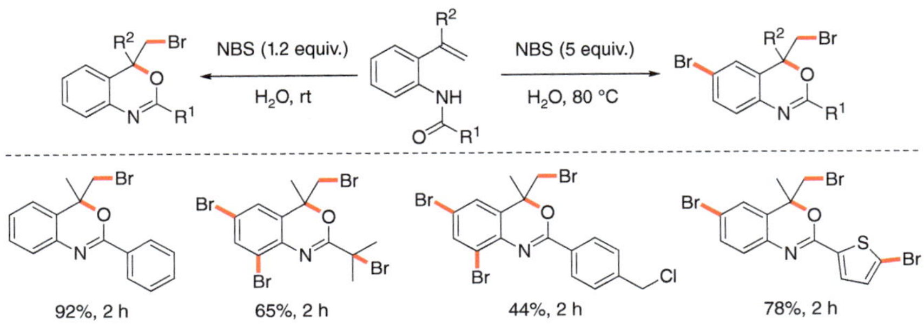

Scheme 10.159 Synthesis of monobromo- and multibromobenzoxazines via cyclization.

10.9.2 Radical Cyclization

Persulfate/Peroxide-Mediated Radical Cyclization

In 2020, Wei and coworkers reported a catalyst-free radical cyclization of enynes/dienes with alcohols in water using $K_2S_2O_8$ as a green oxidant (Scheme 10.160) [173]. Water was chosen as the solvent not only because of its green and safe nature but also because of the high solubility of $K_2S_2O_8$ in water and the lack of C–H bond to undergo competitive oxidation. While the reactions of enynes with propan-2-ol, ethanol, and methanol at 80 °C afforded the corresponding tertiary, secondary, and primary alcohol products, the reactions of dienes

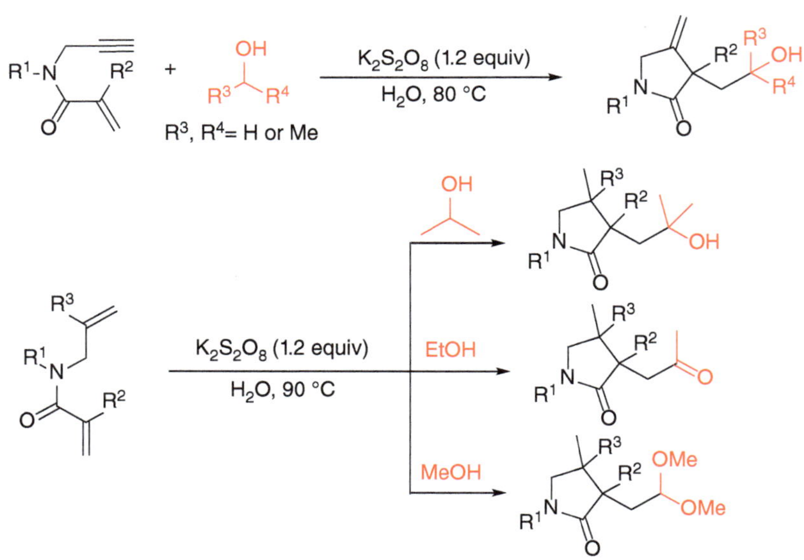

Scheme 10.160 Radical cyclization of enynes/dienes with alcohols.

with these alcohols at a higher temperature (90 °C) produced the tertiary alcohol, acetone, and dimethyl acetal products, respectively. The reaction was proposed to occur via thermal homolysis of the O–O bond of $S_2O_8^{2-}$ to afford a sulfate radical anion (SO4$^{\cdot-}$), which undergoes HAT with the alcohol to produce an alkyl radical. A sequence of radical addition, intramolecular cyclization, and H-abstraction then proceeds to afford the alcohol product.

Wei's group further developed a highly selective and controllable synthesis of three types of lactams from the same substrate via metal-free aryl sulfonyl radical-triggered 5-*exo-dig* cyclization, 5-*exo-dig* cyclization/iodination, or 6-*endo-dig* cyclization of 1,6-enynes with aryl or alkyl sulfonyl hydrazides in the presence of different phase transfer agents and halide additives using water as the solvent (Scheme 10.161) [174]. Various five-membered cyclic lactams, five-membered cyclic lactams containing C–I bond, and six-membered cyclic lactams were synthesized with moderate to good yields. It was found that the replacement of the halide additives with 18-crown-6 changed the selectivity from 5-*exo-dig* to 6-*endo-dig* cyclization. For the cyclization/iodination process, a radical cross-coupling of an iodine radical with an alkenyl radical was proposed.

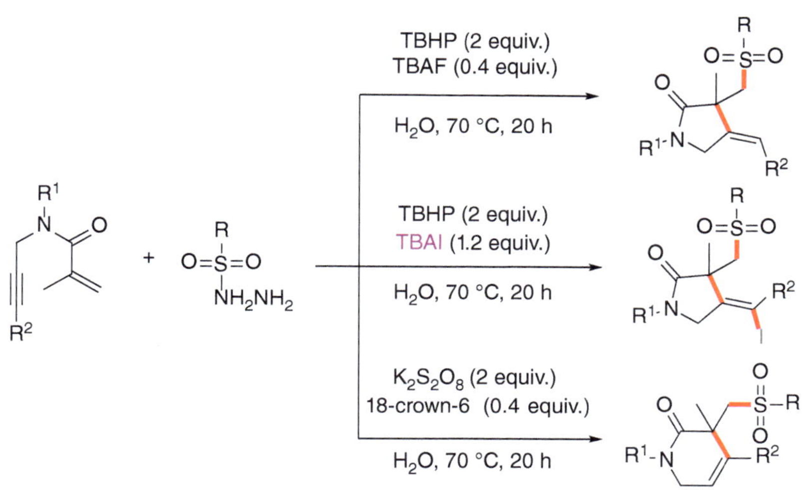

Scheme 10.161 Radical cyclization of 1,6-enynes for the selective and switchable synthesis of lactams.

In another recent contribution, a transition-metal-free radical cyclization of 2-arylbenzoimidazoles with unactivated alkanes was disclosed by the same group via C(sp^3)–H functionalization under aqueous micellar condition using di-*tert*-butyl peroxide (DTBP) as promoter and sodium dodecyl sulfate (SDS) as surfactant (Scheme 10.162) [175]. A series of unactivated alkanes including cyclic alkanes, straight-chain alkanes, and toluene derivatives were smoothly functionalized under the action of *tert*-butoxy radical. Notably, the volume of unactivated alkanes is critical and the best results were achieved with an alkane/H_2O ratio of 1:9.

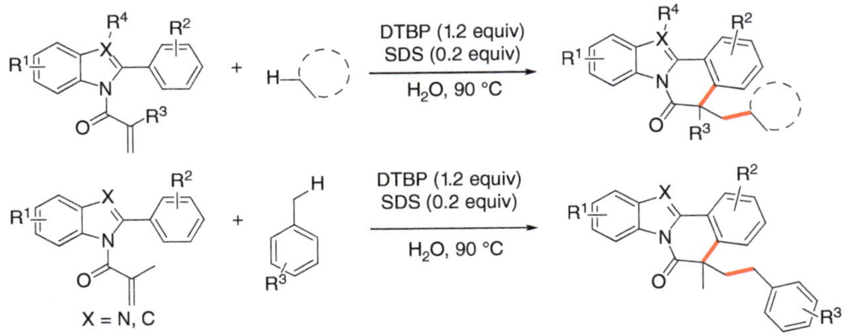

Scheme 10.162 Radical cyclization of 2-arylbenzoimidazoles with unactivated alkanes via C(sp^3)–H functionalization.

Photo-Induced Radical Cyclization

Recently, visible-light-induced reactions have emerged as versatile tools for chemical transformations. In 2020, Yu and coworkers successively developed two metal-free visible light-induced, 4-CzlPN-catalyzed phosphorylation/cyclization of methylthiolated alkynones and methylthiolated alkynes in water using diarylphosphine oxides as phosphoryl radical precursors (Scheme 10.163) [176]. A variety of 2-phosphorylated thioflavones and 3-phosphorylated benzothiophenes were synthesized at room temperature using dilauroyl peroxide (LPO) as an oxidant. The aqueous media can be recycled several times without significant reduction in reactivity. The second protocol was also amenable for the synthesis of benzoselenophene when the corresponding methylseleno substrate was used.

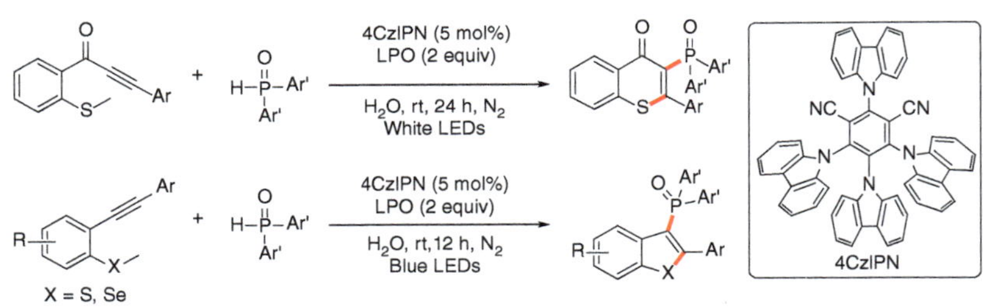

Scheme 10.163 Visible light-induced phosphorylation/cyclization of methylthiolated alkynones and methylthiolated alkynes.

More recently, Xia and coworkers reported a photo-induced radical addition–translocation–cyclization (RATC) reaction of terminal alkynes and α-oxocarboxylic acids in the absence of external metal catalysts and oxidants using water as the optimum solvent at room temperature (Scheme 10.164) [177]. Mechanistically, fragmentation of the triplet excited ArC(O)CO_2H affords an acyl radical. Direct addition of the acyl radical onto alkynes followed by a 1,5-HAT/radical cyclization cascade provides cyclopentane-containing ketones including various rigid bicyclic and spirocyclic products. Mechanistic studies indicated that the α-hydrogen of the carbonyl group might originate from water.

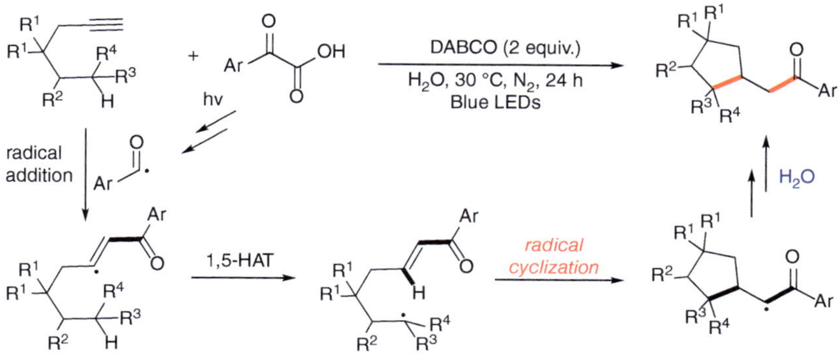

Scheme 10.164 Photo-induced radical addition–translocation–cyclization reaction.

Yu and coworkers established two sets of reaction conditions for water-promoted visible light-induced iodine atom transfer cyclization of 2-allyloxy(or prop-2-yn-1-yloxy)-3-iodo tetrahydropyrans and tetrahydrofurans using phenanthroline–diphosphine Cu(I) complexes as photocatalysts in pure water or a mixed CH_3CN/H_2O solution (v/v: 1 : 4) (Scheme 10.165) [178]. While the reaction in mixed solution relies on the use of a Xantphos-based complex using VC (ascorbic acid) as

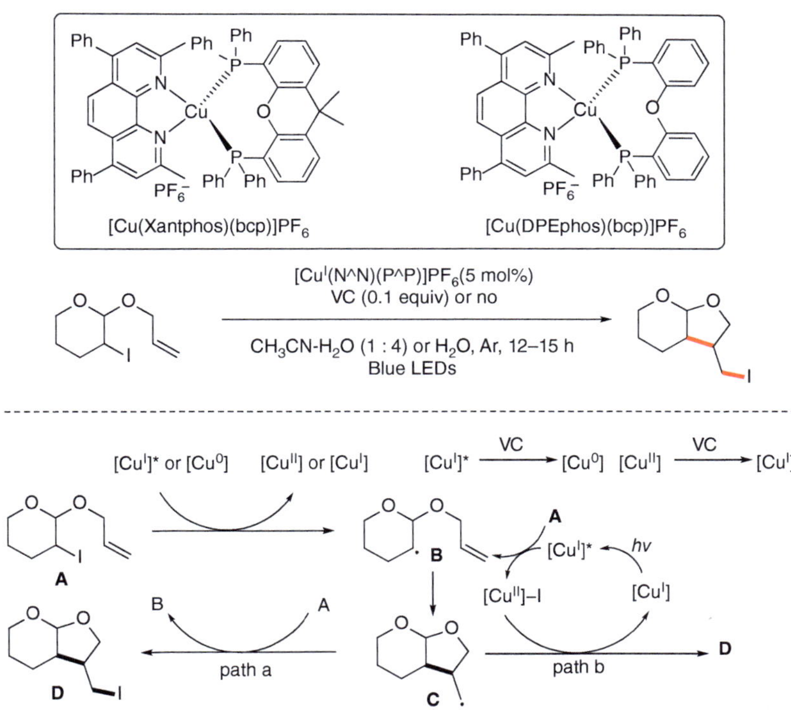

Scheme 10.165 Water-promoted visible light-induced iodine atom transfer cyclization.

the reductant, the reaction in pure water could occur with a DPEphos-based Cu complex in the absence of VC. Water is beneficial to this transformation since increasing the amount of water could considerably increase the reaction yields. The beneficial role of water could be explained by two aspects: (1) increasing the reducing capacity of $[Cu^I(N^\wedge N)(P^\wedge P)]$ complexes by shifting the redox equilibrium of $[Cu^I]/[Cu^{II}]$ toward $[Cu^{II}]$ by preferably coordinating to $[Cu^{II}]$; (2) increasing the heterogeneity and radical-initiating capacity of $[Cu^I]$ complexes by reducing its solubility in water.

10.10 Multicomponent Reaction (MCR)

Multicomponent reactions (MCRs) have attracted much attention from the synthetic community due to the capacity to generate several bonds in a single one-pot operation and the ease of constructing diverse and complex molecular skeletons efficiently. Compared with traditional multistep reactions, MCRs are highly atom- and step-economical methods that could reduce the time and energy required for unnecessary intermediate isolation and purification. Recent interests of scientists have been directed toward developing MCRs in water for sustainable and green synthesis [179].

Carbon dioxide and carbon disulfide are two important C1 synthons in organic synthesis. The efficient transformation of CO_2 (an earth-abundant greenhouse gas) into value-added chemicals has become a hot topic in recent years. In 2021, Huang and coworkers described the synthesis of octahydroacridines and dihydropyridines via [2 + 2 + 1 + 1] cycloaddition of ketones or β-ketone esters with NH_4Cl under a CO_2 atmosphere in an aqueous solution (Scheme 10.166) [180]. Since the solubility of carbon dioxide decreases with an increased temperature, a monoethanolamine (MEA) additive was added to increase the solubility in boiling aqueous solution.

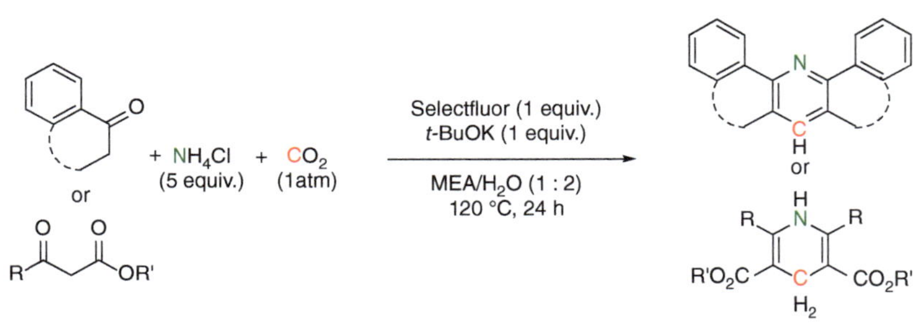

Scheme 10.166 [2 + 2 + 1 + 1] cycloaddition of ketones with NH_4Cl under CO_2 atmosphere.

In 2017, Halimehjani and coworkers reported solvent-controlled selective formation of substituted thiazole-2(3*H*)-thiones or thiazolidine-2-thiones via a one-pot three-component reaction between primary amines, CS_2, and nitroepoxides using THF and H_2O as the solvent, respectively (Scheme 10.167) [181]. The synthesis of thiazole-2(3*H*)-thiones was proposed to proceed via epoxide ring opening of nitroepoxides by an *in situ* generated dithiocarbamic acid followed by intramolecular hemiaminalization and dehydration. When water was used as the solvent, the dehydration step was inhibited to afford thiazolidine-2-thiones as products.

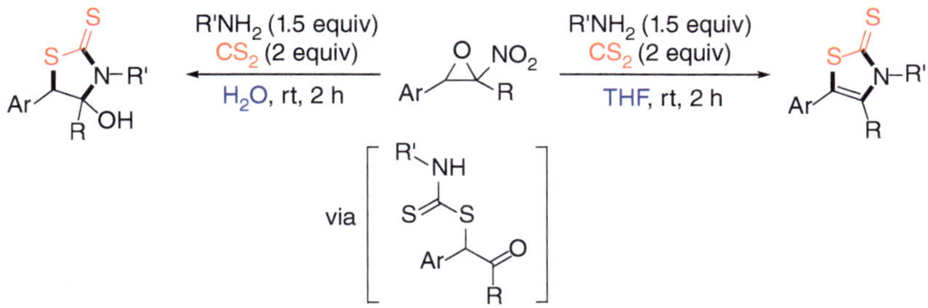

Scheme 10.167 Solvent-controlled selective formation of substituted thiazole-2(3*H*)-thiones or thiazolidine-2-thiones.

In 2020, the Halimehjani group further reported the one-pot synthesis of dithiocarbamates via catalyst-free and room-temperature four-component condensation between secondary amines, carbon disulfide, formaldehyde, and naphthols in water (Scheme 10.168) [182]. A one-pot pseudo-seven-component reaction was also developed for the synthesis of bis(dithiocarbamates) employing naphthalene-2,3-diol as the substrate. Notably, the crude products can be simply isolated by filtration.

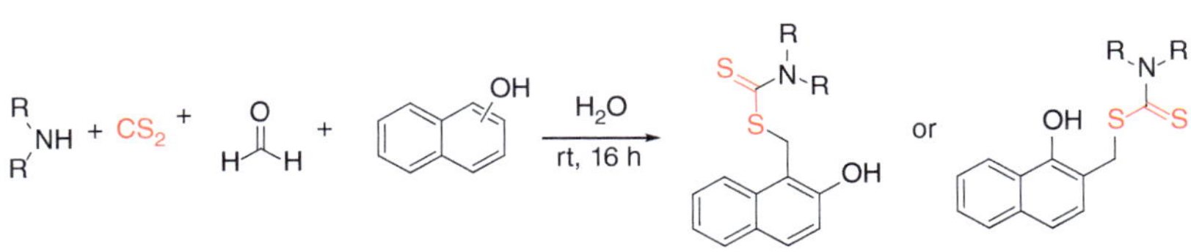

Scheme 10.168 Four-component condensation between secondary amines, CS_2, formaldehyde, and naphthols.

More recently, a visible light-promoted three-component reaction between Katritzky salts, CS_2, and secondary amines was reported by Yang and coworkers for the synthesis of *S*-alkyl dithiocarbamates in water without any external transition metals or photocatalysts (Scheme 10.169) [183]. Preliminary mechanistic studies indicated that the reaction involves *in situ* formation of an EDA complex **A** and photoexcited intermolecular charge transfer of this complex affords a thiyl radical **C** and a Katritzky salt radical **D.** Fragmentation of the latter produces an alkyl radical **E** which is then cross-coupled with the thiyl radical to give the desired product.

In 2020, Herrera and coworkers developed an ultrasound-promoted three-component synthesis of 2-amino-3-cyanopyrans from β-ketone ester/β-dicarbonyl compounds, aryl formaldehydes, and malononitrile in water using 20 mol% Et_3N as the catalyst (Scheme 10.170) [184]. The ultrasound-assisted multicomponent approach proceeded efficiently within shorter reaction times than another

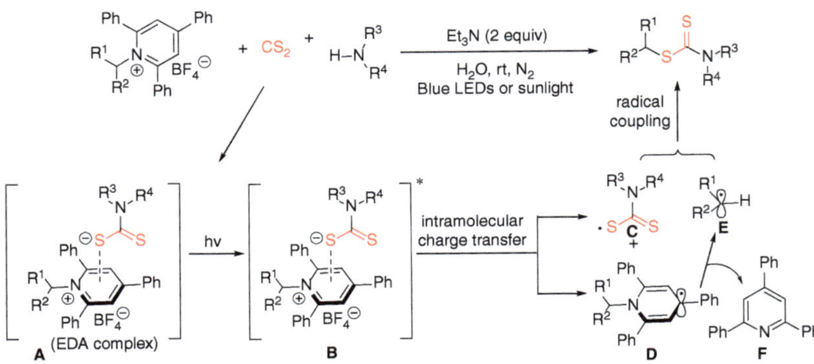

Scheme 10.169 Visible light-promoted three-component reaction between Katritzky salts, CS_2, and secondary amines.

non-ultrasound protocol employing preformed alkylidene malononitrile as substrate. The products were easily isolated by simple extraction and fast column chromatography on silica gel.

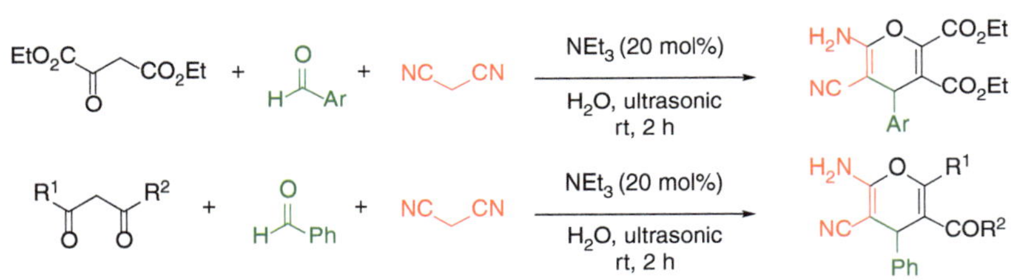

Scheme 10.170 Ultrasound-assisted multicomponent synthesis of 4*H*-pyrans in water.

One year later, an ultrasonic-promoted one-pot four-component synthesis of 3,5-disubstituted isoxazoles-sulfonates and sulfonamides from sulfonyl chlorides, propargyl alcohols/or propargylamines, aromatic aldehydes and hydroxylamine hydrochloride was reported by Bougrin and coworkers in water using NaDCC as both the base and oxidant (Scheme 10.171) [185]. The ultrasound irradiation protocol provided better yields of products in shorter reaction times compared with the classical magnetic stirring method. The reaction was proposed to proceed via three tandem steps including K_2CO_3-promoted sulfonylation, NaDCC-promoted oximation-oxidation, and 1,3-dipolar cycloaddition between the resulting nitrile oxides and alkynes.

In 2020, Olyaei and coworkers described a rapid, catalyst-free synthesis of 2*H*-furo[3,2-*c*]chromene-2,4(3*H*)-diones via a one-pot three-component condensation between equal molar of 4-hydroxycoumarin, glyoxalic acid monohydrate and various aryl or heteroaryl amines in water under reflux condition (Scheme 10.172) [186]. By replacing the above amines with 2-aminopyrimidine derivatives or piperidine, arylamino-bis(coumarin)methane derivatives were obtained instead.

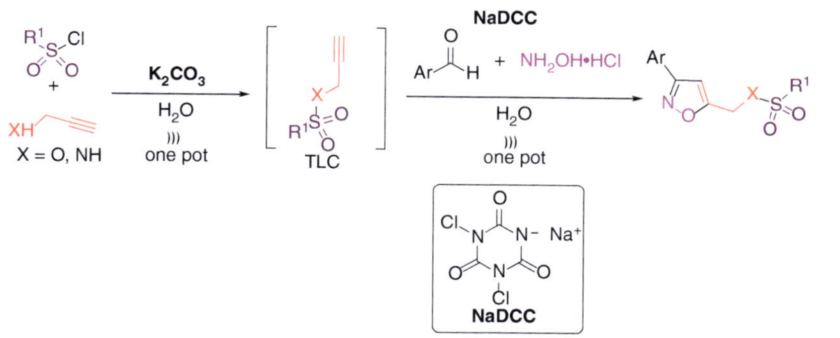

Scheme 10.171 Ultrasound-promoted one-pot four-component synthesis of 3,5-disubstituted isoxazoles sulfonamides and sulfonates catalyzed by NaDCC in water.

Water plays a key role in this transformation by activating the glyoxalic acid and an imine intermediate via hydrogen bonding.

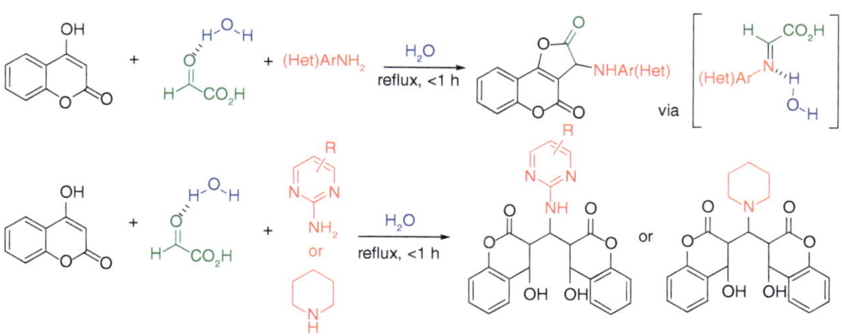

Scheme 10.172 Catalyst-free one-pot synthesis of 2*H*-furo[3,2-*c*]chromene-2,4(3*H*)-dione and arylamino-bis(coumarin)methane derivatives.

In 2021, Ábrányi-Balogh and coworkers disclosed a chromatography-free multicomponent synthesis of thioureas from isocyanides, amines/amidines, and elemental sulfur using aqueous polysulfide solution prepared from elemental sulfur and a pool of organic bases (Scheme 10.173) [187]. The products were easily isolated by simple filtration. Noteworthily, cyclic amidines underwent ring-opening in the presence of water leading to thioureas substituted with aminopropyl lactams or tetrahidropyrimidin-2-ones. The oxygen atom of the lactam or tetrahidropyrimidin-2-one moiety came from water.

Lately, Giustiniano and coworkers reported a photomicellar-catalyzed three-component synthesis of α-amino amides from *N*-methyl-*N*-alkyl aromatic amines, aliphatic, or aromatic isocyanides in 2 wt% SDS aqueous solution under the irradiation of either blue light or sunlight (Scheme 10.174) [188]. Water was used as both the solvent and reactant for this transformation. The micellar solvent system could be recyclized up to four cycles with no or slight change in yield. It is notable that negatively charged surfactant such as SDS could localize a positively charged

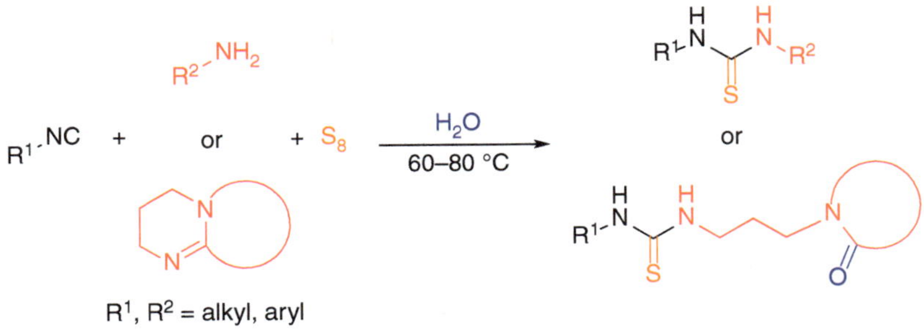

Scheme 10.173 Elemental sulfur-based multicomponent synthesis of thioureas.

photocatalyst on the micelles' surface, thus rendering an improved efficiency than that achieved in pure water.

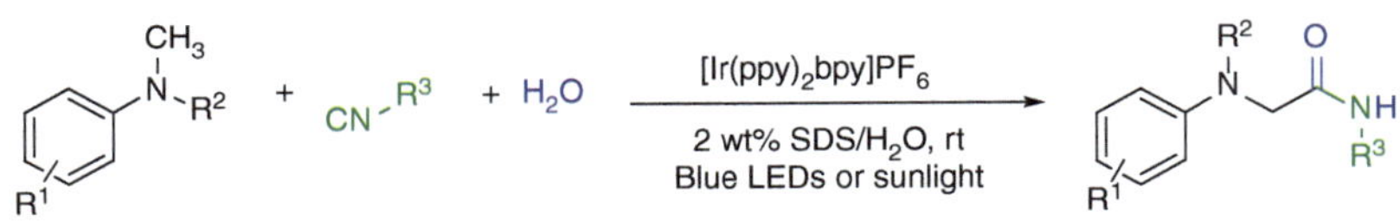

Scheme 10.174 Photomicellar-catalyzed synthesis of amides from isocyanides.

In 2019, Wang and coworkers reported a Cu-catalyzed three-component synthesis of 3-cyano, nitro, sulfones, and benzoyl 4-quinolone derivatives from substituted 3-(2-halophenyl)-3-oxopropane, aldehydes and aq. ammonium using water as a benign reaction media (Scheme 10.175) [189]. Various aldehydes including formaldehyde, aromatic, or heteroaromatic aldehydes were suitable reaction partners for this transformation. However, the reaction failed to work with aliphatic/benzyl aldehyde derivatives and simple acetophenone derivatives. The reaction was proposed to involve an aldol condensation, a Cu-catalyzed Ullmann amination, an intramolecular Michael addition, and an aerobic oxidation.

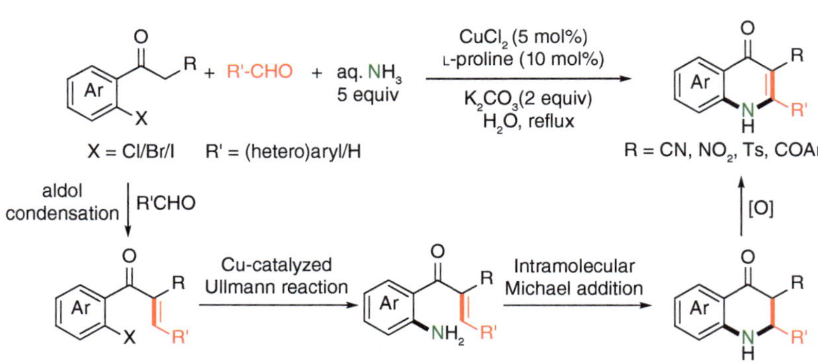

Scheme 10.175 Cu-catalyzed three-component synthesis of substituted 4-quinolones.

In the same year, Zhou and coworkers reported the synthesis of *N*-substituted 1-napthyamines via CuI-catalyzed 6-*endo-dig* benzannulation of terminal alkynes, 2-bromoaryl ketones, and amides in water under Pd- and ligand-free conditions (Scheme 10.176) [190]. Amides serve as the amine sources for this transformation through NaOH-mediated cleavage of the amide bonds. It is interesting to note that the Sonogashira coupling product was not observed in this transformation. Mechanistically, coordination of Cu(I) acetylide **A** to the carbonyl group of 2-bromoaryl ketone gives an intermediate **C**. Oxidative addition of the Cu(I) atom into the *ortho*-C–Br bond of **C** in the presence of amine anion **B** delivers a Cu(III) acetylide species **D.** The Cu(III) moiety activates the α-carbon atom of alkyne toward nucleophilic attack by the β-vinyl carbon atom, resulting in exclusive 6-*endo-dig* carbocyclization instead of the competitive 5-*exo-dig* annulation and the Sonogashira coupling.

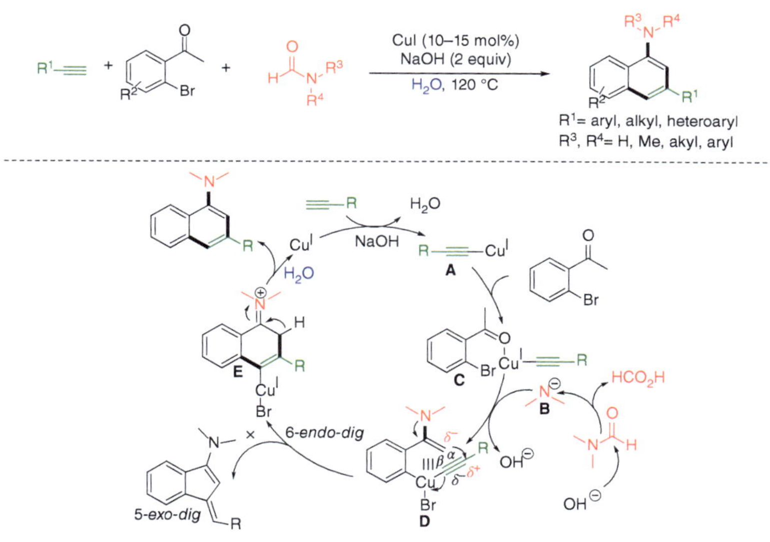

Scheme 10.176 Cu(I)-catalyzed 6-*endo-dig* cyclization of terminal alkynes, 2-bromoaryl ketones, and amides toward 1-naphthylamines.

Two years later, Zhang and coworkers described a multicomponent bifunctionalization of methyl ketones with quinoxalinones and *t*-BuOCl under a combined heterogeneous acid catalysis and solar photocatalysis in water at ambient temperature using air as the oxidant (Scheme 10.177) [191]. The solid acid Amberlyst 15 was used as the heterogeneous catalyst and it demonstrated certain recyclability. The reaction was proposed to proceed via acid-catalyzed formation of a Mannich-type product **A** from the methyl ketone and quinoxalinone followed by singlet oxygen-mediated oxidative dehydrogenation of the Mannich-type product to afford an α,β-unsaturated ketone intermediate **B**. The intermediate **B** is attacked by a chlorine (Cl•) radical generated by the homolysis of *t*-BuOCl to give a chlorinated carbon radical **C** which is oxidized by the *tert*-butoxy (*t*-BuO•) radical to provide the final product.

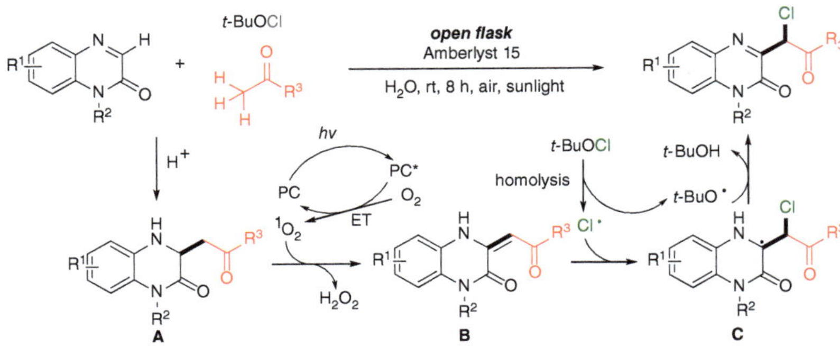

Scheme 10.177 Multicomponent bifunctionalization of methyl ketones enabled by heterogeneous catalysis and solar photocatalysis.

More recently, a water-accelerated organocatalytic three-component reaction between ketones/aldehydes, benzhydrazide, and allylboronic acid pinacol ester was realized by Bae and coworkers for the synthesis of α-tertiary homoallylic amines via allylation of an *in situ* generated benzoyl hydrazone (Scheme 10.178) [192]. A variety of carbonyl compounds including β-ketoester, conjugated ketone, and 2,2,2-trifluoromethyl ketone were tolerated to afford the products with high chemo-/site- selectivities. The reaction occurs via a "on water" mechanism under water-accelerated synergistic acid catalysis by taking advantage of a hydrophobic Brønsted acid (DBSA) in combination with a squaramide hydrogen-bonding donor (SQA-III). This synergistic catalysis occurs in a hydrophobic hydration shell

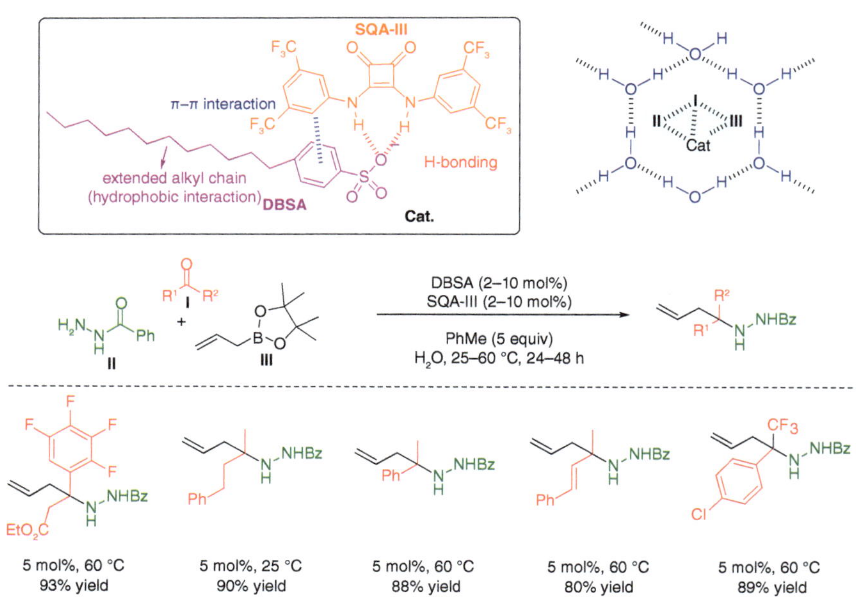

Scheme 10.178 Multicomponent organocatalytic allylation.

formed by the hydrogen-bonding networks of surrounding water molecules and a pseudo-high-pressure effect is possibly in operation to facilitate the reaction.

10.11 Domino/Tandem/Cascade Reactions

Similar to the MCRs, the one-pot domino/tandem/cascade processes are also gaining increasing attention from the synthetic community due to the ability to perform multistep reactions in a single vessel. In this section, both chemo- and chemoenzymatic domino/tandem/cascade reactions will be discussed.

10.11.1 Chemo-Domino/Tandem/Cascade Reactions

Transition Metal-Catalyzed

In 2018, Wang and coworkers reported a micelle-enabled, Pd/C-catalyzed domino synthesis of symmetrical and unsymmetrical ureas from aryl iodides, NaN_3, $CHCl_3$, and aryl/alkyl amines in aqueous TPGS-750-M solution using chloroform as a safe source of carbon monoxide in the presence of KOH (Scheme 10.179) [193]. A domino process of carbonylation, azidation, Curtius rearrangement, and nucleophilic addition was involved in this transformation. Due to the easy recyclability, the same batch of Pd/C catalyst could be used for six consecutive runs with only a slight decrease in catalytic efficiency.

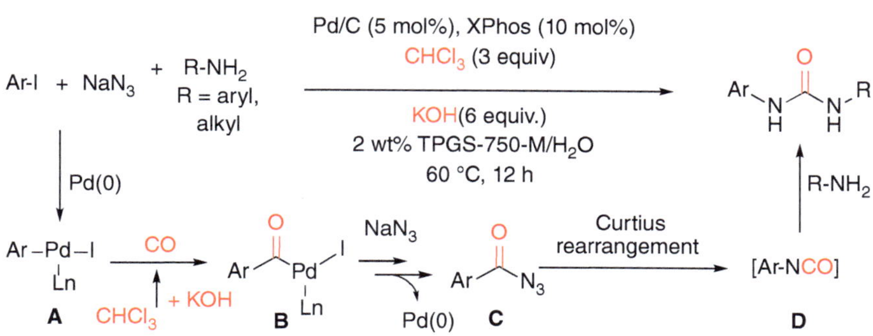

Scheme 10.179 Pd/C-catalyzed domino synthesis of ureas using chloroform as the carbon monoxide source.

Recently, Zhang and coworkers reported the synthesis of aromatic esters and diesters through Pd(II)-catalyzed four-component domino reaction between aryl iodides, alkyl chlorides, carbon monoxide, and water using water as both the solvent and reagent (Scheme 10.180) [194]. The reaction was proposed to occur through Pd-catalyzed carboxylation followed by base-mediated substitution of the alkyl halide with carboxylate to afford the ester product. By replacing monohalides with CH_2Cl_2, double esterification occurred to produce aromatic diesters in moderate to good yields.

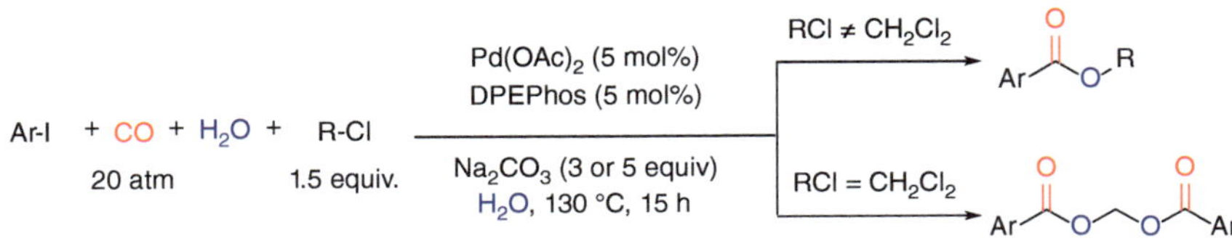

Scheme 10.180 Pd(II)-catalyzed four-component domino synthesis of aromatic esters and diesters.

In 2020, a Pd-catalyzed tandem addition/ring opening/cyclization reaction of epoxynitriles with arylboronic acids was reported by Chen and coworkers for divergent synthesis of 2,5-diaryl furans and pyrroles in aqueous media (Scheme 10.181) [195]. The chemoselectivity of this reaction was governed by the steric hindrance of arylboronic acids. Selective formations of furans and pyrroles were achieved by using sterically less-hindered and sterically bulkier arylboronic acids, respectively. Mechanistically, there is an additional hydrolysis step for the formation of furans, the sterically more bulkier aryl imine intermediate is reluctant to undergo hydrolysis, thus rendering more facile intramolecular cyclization of the ketimine intermediate to afford the pyrrole product. Water not only serves as the oxygen source of furan but also facilitates the solubilization of boronic acids, the activation of the C–B bond as well as promoting the hydrolysis process.

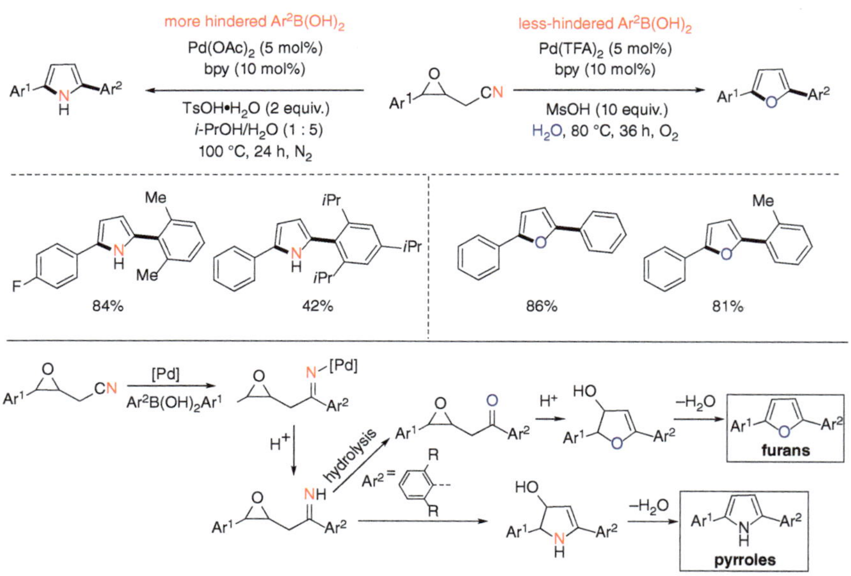

Scheme 10.181 Pd-catalyzed divergent synthesis of furans and pyrroles via tandem reaction of epoxynitriles with arylboronic acids.

More recently, Zhou and coworkers disclosed a water-promoted acid- and base-free Pd-catalyzed cascade reaction between *N*-cyanomethyl-substituted

2-acylimidazoles and arylboronic acids for the synthesis of multi-substituted imidazo[1,2-*a*]pyridazines as switchable dual-state emission (DSE), aggregation-induced emissions (AIE), and aggregation-caused quenching (ACQ) luminophores (Scheme 10.182) [196]. Water outperformed many other protic and aprotic solvents and was believed to promote the transmetalation between arylboronic acid and the catalytically active Pd(II) species **A** to give an aryl palladium intermediate **B**. The aryl palladium species then coordinates with the cyano group and undergoes intramolecular migratory insertion of nitrile to afford an imine-Pd(II) species **D**. Further intramolecular nucleophilic addition and dehydration affords the final product and regenerates the active Pd(II) species.

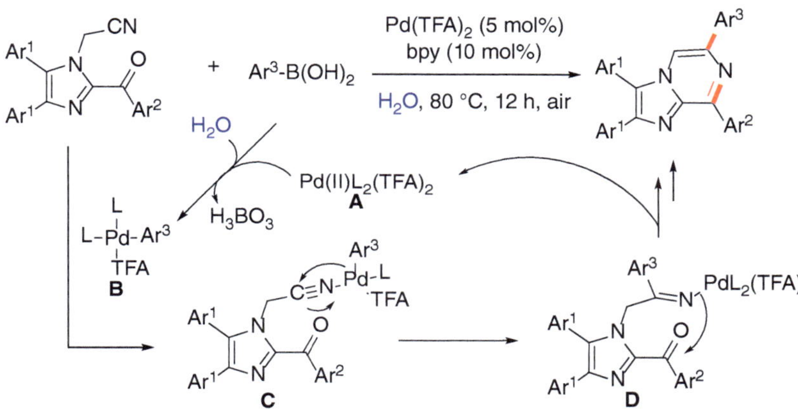

Scheme 10.182 Pd-catalyzed cascade reaction to imidazo[1,2-*a*]pyridazines.

Aqueous domino/tandem/cascade reactions catalyzed by non-palladium transition metals have also been reported. In 2022, Chen and coworkers developed a Rh-catalyzed highly branch-selective hydroaminomethylation (HAM) of vinyl arenes with anilines and syngas (CO/H_2) in water using a water-soluble diphosphine ligand derived from bis(2-(diphenylphosphaneyl)ethyl)amide (Scheme 10.183) [197]. Since the reaction consists of a cascade process of hydroformylation, imine condensation, and imine hydrogenation, the use of water as a reaction medium is quite impressive in light of the potential imine hydrolysis. For the reactions with less-soluble substrates, the addition of sodium dodecylbenzene sulfonate (SDBS) as a surfactant was beneficial.

In the same year, Zhou and coworkers described a one-pot synthesis of polyfunctional naphthyl aryl ethers from 2-bromoaryl ketones and terminal alkynes via CuI-catalyzed cascade 6-*endo-dig* cyclization and C(sp^2)–O coupling in water using 5-bromopyrimidin-2-amine as a critical additive (Scheme 10.184) [198]. Water is essential for this transformation since the replacement of water with other organic solvents did not afford the desired product. The unique role of water might be ascribed to the favored formation of 1-naphthol intermediate **E** for subsequent C(sp^2)–O coupling. 5-Bromopyrimidin-2-amine is critical for this reaction by acting

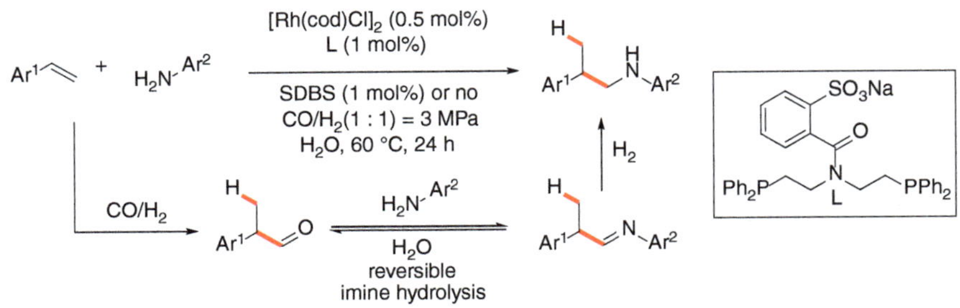

Scheme 10.183 Rh-catalyzed highly branch-selective hydroaminomethylation (HAM) of vinyl arenes with anilines and syngas.

as a ligand for intermediate **F** in the C(sp^2)–O coupling step as well as a hydrogen bond receptor in intermediate **C** to stabilize the enol tautomer for subsequent 6-*endo-dig* cyclization.

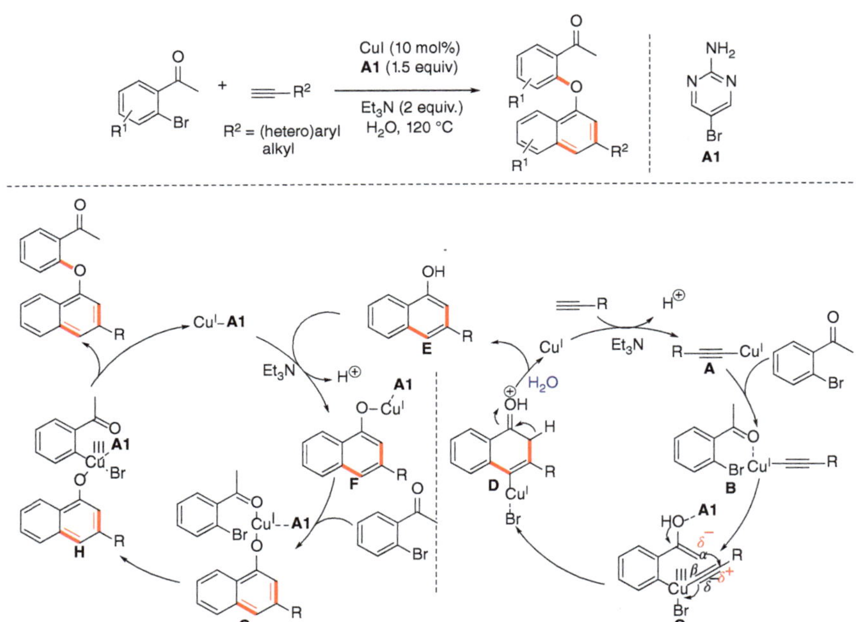

Scheme 10.184 Cu-catalyzed 6-*endo-dig* cyclization and C(sp^2)–O coupling of 2-bromoaryl ketones and terminal alkynes.

Organo-Catalyzed

In 2019, a water-enabled, highly chemoselective and enantioselective cross-reaction of α-keto acids with α-keto esters was developed by Li and coworkers through a cascade of asymmetric cross-aldol/lactonization/enolization using a naphthyl-containing chiral proline-imidazole as enamine catalyst (Scheme 10.185) [199]. The reaction chemoselectively afforded the cross-cyclization product with α-keto acid as the nucleophile and α-keto ester as the only effective aldol acceptor. Water is beneficial to this reaction since no or low reactivity was detected by using other organic solvents. DFT studies indicated that the high polarity of water facilitates the generation of water-stabilized zwitterionic enamine intermediates/transition states with a lower activation energy and higher face selectivity, thus rendering high activity and enantioselectivity. Water could also facilitate the differentiation of the nucleophilicities and electrophilicities of two substrates by collaborating with the new enamine catalyst, thus promoting the chemo-selective cross-reaction.

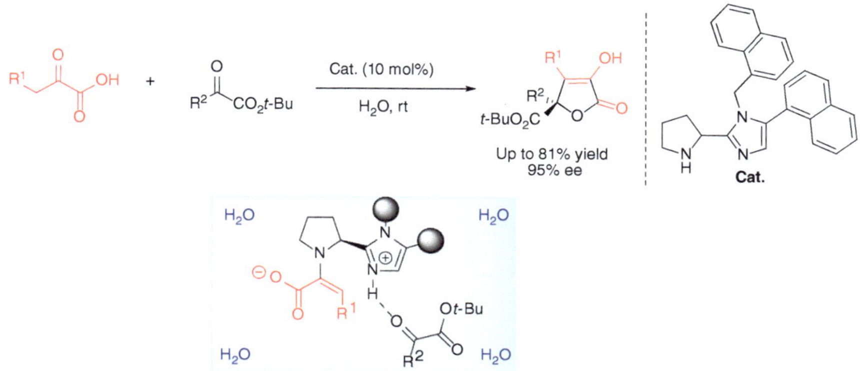

Scheme 10.185 Water-enabled asymmetric cross-reaction of α-keto acids with α-keto esters.

Catalyst Free

In 2017, Zhou and coworkers reported a $B_2(OH)_4$-mediated one-pot cascade cyclization/transfer hydrogenation process for synthesizing tetrahydroquinoxalines from 2-aminoanilines and 1,2-dicarbonyl compounds using water as both a solvent and a hydrogen donor (Scheme 10.186) [200]. 2-Nitroanilines and 1,2-dinitrobenzene were

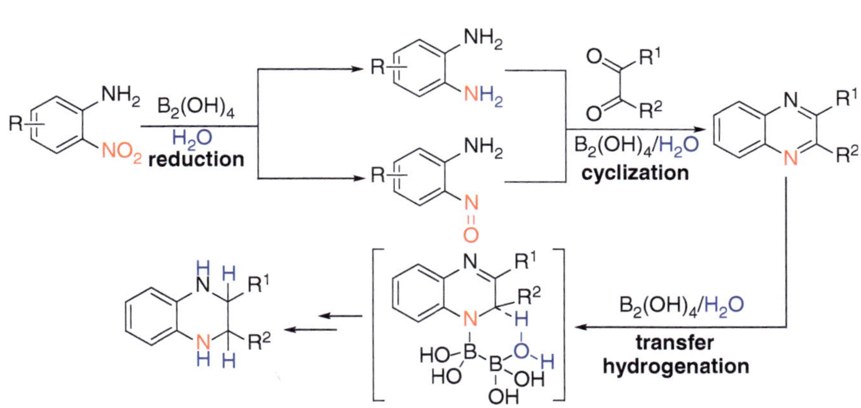

Scheme 10.186 $B_2(OH)_4$-mediated one-pot synthesis of tetrahydroquinoxalines from 2-amino(nitro)anilines and 1,2-dicarbonyl compounds.

also suitable partners for this cascade transformation through additional nitro reduction steps. The deuterium labeling experiments with D_2O confirmed that water is indeed the hydrogen donor.

In 2019, Zhao and coworkers described the first catalyst- and additive-free synthesis of 3-hydroxyisoindolinones on water via a highly atom-economical one-step cascade reaction between 2-alkynylbenzoic acids and a variety of nitrogen-containing nucleophiles such as alkyl/aryl amines, hydroxylamines, hydrazides, and hydrazines (Scheme 10.187) [201]. A one-pot two-step tandem construction of indole/pyrrole/thiophene-fused polycyclic compounds was also achieved, and the ketone moiety of the products was further reduced to afford tetrahydroberberine analogs. After extensive solvent screening, water was proved to be the best solvent. Deuterium-labeling experiments suggested that water not only serves as the solvent but also participates in H–H exchange with 2-alkynylbenzoic acids and corresponding reaction intermediates. Moreover, water could activate the carbonyl group of the final intermediate **E** through hydrogen bonding to accelerate the annulation process.

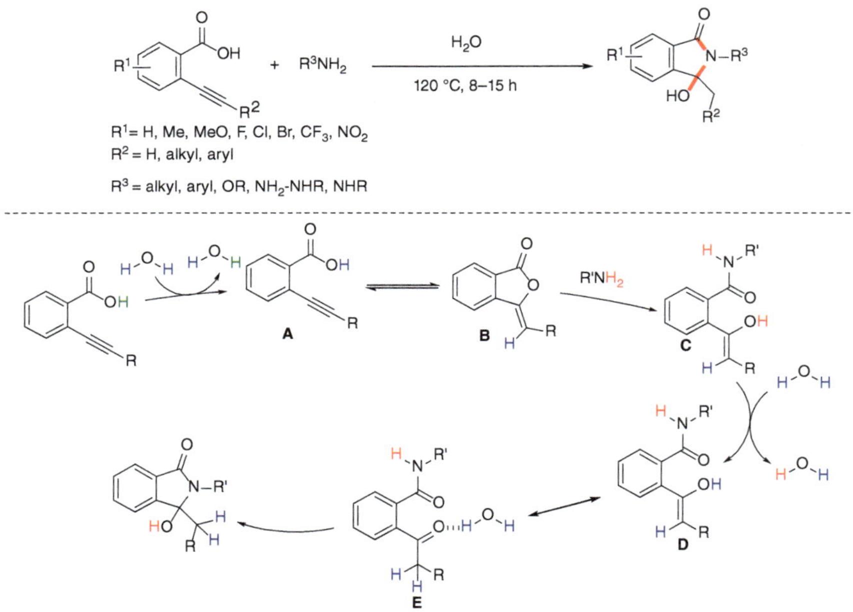

Scheme 10.187 Catalyst-free cascade synthesis of 3-hydroxyisoindolinones from 2-alkynylbenzoic acids and nitrogen-containing nucleophiles.

10.11.2 Chemoenzymatic Reactions

In recent years, chemoenzymatic reactions have gained increased popularity since the high efficiency of the chemical reaction and the high selectivity of the enzymatic reaction could be integrated simultaneously. Carrying out chemoenzymatic reactions in water is quite challenging due to the poor solubility of substrates and the potential mutual inactivation of the chemocatalysts and enzymes. In spite of these challenges, there has been significant progress in this field during these years.

In 2018, a cooperative chemoenzymatic reaction that integrates photocatalyzed alkene isomerization with ene-reductase mediated reduction of C–C double bonds was developed by Hartwig and coworkers for the synthesis of various enantioenriched products (Scheme 10.188) [202]. The highly compatible chemoenzymatic reaction occurred at room temperature in Tris buffer using DMSO as a cosolvent. This protocol enables stereoconvergent reduction of *E/Z* mixtures or unreactive stereoisomers of alkenes in comparable yields and enantiomeric excesses (ee) to those obtained from pure, more reactive isomers. Furthermore, the cooperative isomerization and reduction process afforded products with higher yields and ee values than those obtained through a sequential process.

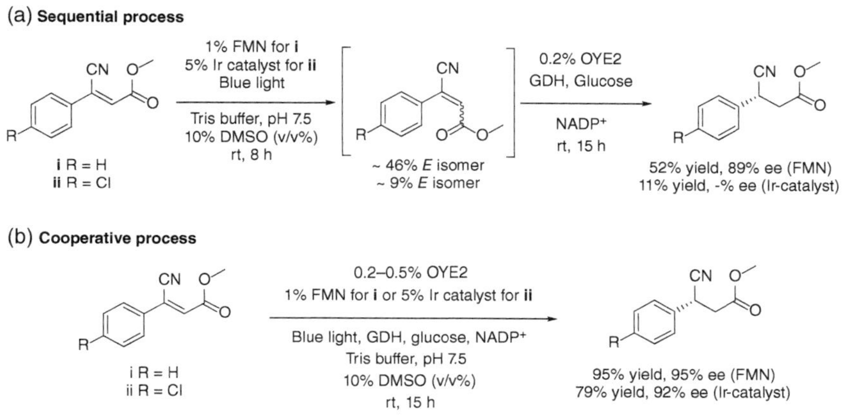

Scheme 10.188 Comparison of sequential and cooperative (simultaneous) isomerization and reduction of cyanoacrylates.

In 2019, Lipshutz and coworkers developed several aqueous micellar-catalyzed one-pot, two-step chemoenzymatic reactions integrating transition metal-catalyzed Sonogashira/Heck/1,4-addition reactions or hydration reactions with alcohol dehydrogenase (ADH)-catalyzed ketone reduction (Scheme 10.189) [203]. Enzymes were highly compatible with the metal catalysts and TPGS-750-M surfactant, with the latter even enhancing the enzymatic activity. Nanomicelles present in the TPGS-750-M/H_2O solution not only function as a reaction medium but also as a reservoir for the substrates, products, and catalysts, thereby moderating the degree of enzyme saturation and decreasing noncompetitive enzyme inhibition. A one-pot, three-step sequence consisting Rh-catalyzed 1,4-addition, nitro reduction, and enzymatic ketone reduction was also developed.

In the same year, Garg and coworkers achieved a one-pot sequential chemoenzymatic conversion of amides to enantioenriched alcohols in aqueous medium by coupling Ni-catalyzed SM coupling of amides with ketoreductase (KRED)-mediated asymmetric reduction of the ketone intermediate (Scheme 10.190) [204]. They also accomplished formal syntheses of both enantiomers of the pharmaceutical orphenadrine taking advantage of ketoreductases displaying complementary stereoselectivities. This example represents the first enantioselective conversion of amides through catalytic C–N bond activation.

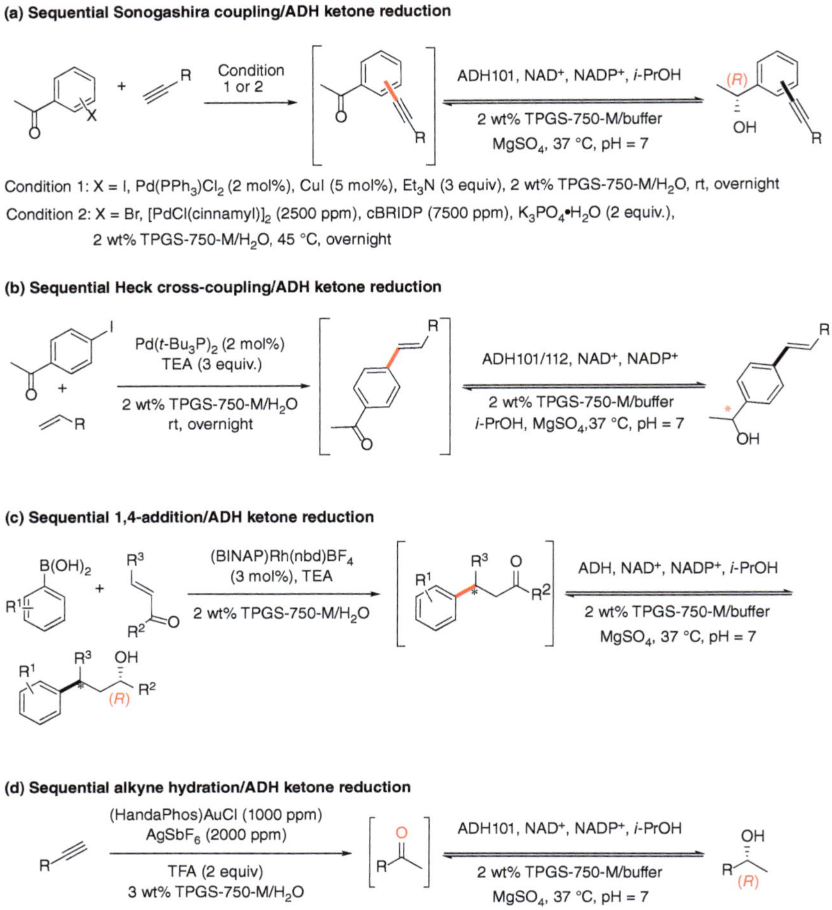

Scheme 10.189 Micellar-catalyzed tandem Sonogashira/Heck/1,4-addition/hydration reaction and enzymatic reduction.

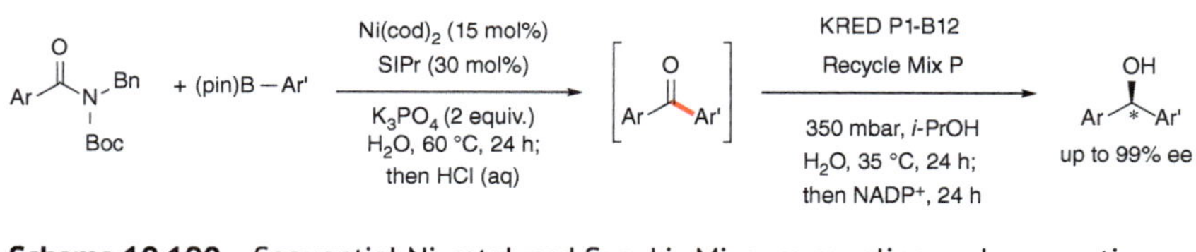

Scheme 10.190 Sequential Ni-catalyzed Suzuki–Miyaura coupling and enzymatic reduction.

Enabled by the micellar catalyst TPGS-750-M, Hastings and coworkers later reported three one-pot chemoenzymatic reactions in water by coupling Heck coupling, RCM, or cross-metathesis with enzymatic hydrolysis at room temperature (Scheme 10.191) [205]. The enzymatic reactions were compatible with many nonionic surfactants, and the transition metal-catalyzed reactions were almost unaffected by the addition of enzymes. Moreover, the sequential chemoenzymatic reactions generally gave higher yields than those carried out concurrently by adding metal catalysts and enzymes together.

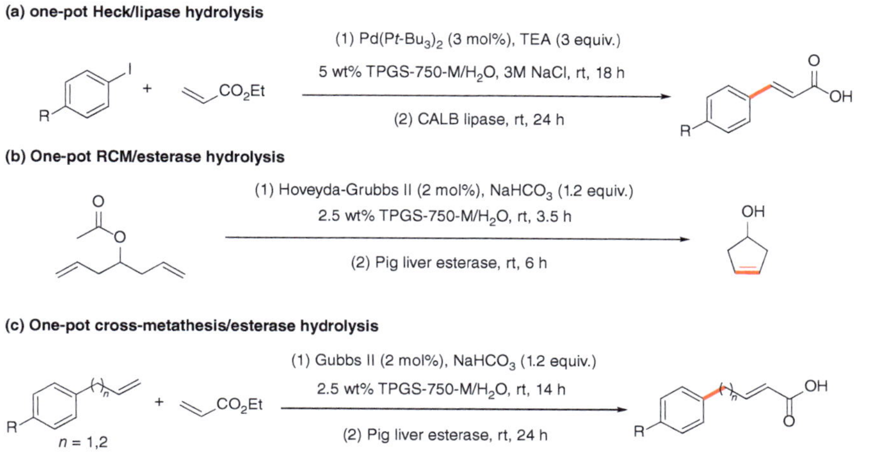

Scheme 10.191 Micellar-enabled one-pot Heck/lipase hydrolysis, RCM/esterase hydrolysis, or cross-metathesis/esterase hydrolysis.

As described above, most chemoenzymatic reactions rely on the use of metal catalysts. Recently, Zhu and coworkers reported a cooperative bio- and organocatalyzed system for chemoenzymatic synthesis of *N*-heterocycles from alcohols and amines in aqueous buffer with O_2 as the terminal oxidant (Scheme 10.192) [206]. The chemoenzymatic reaction is composed of horse liver alcohol dehydrogenase (HLADH)-catalyzed alcohol oxidation and flavin analog (F1)-catalyzed oxidative condensation between aldehyde and amine. The flavin organocatalyst (F1) plays dual roles in both bio- and chemo-process in terms of NAD^+ regeneration and oxidative cyclodehydrogenation, respectively. The yields of chemoenzymatic method were generally higher than those achieved by using an individual catalyst, which embodies the advantage of cooperative chemoenzymatic catalysis.

10.12 Rearrangement or Insertion Reactions

10.12.1 Rearrangement Reactions

In 2019, Malapaka and coworkers reported the facile synthesis of 7-amino-8-aroylated quinolines from 8-amidoquinolines and aroyl peroxides via oxygenative cleavage of the inert C_{Ar}–N bond with concomitant 8,7-amino migration on an aryl moiety (Scheme 10.193) [207]. The reaction proceeded well in an atom- and step-economical manner in water at room temperature in open air without any additives. Better yield was obtained with water and this might be caused by a hydrogen bonding-mediated acceleration effect. Mechanistically, a *N*-benzoyloxy

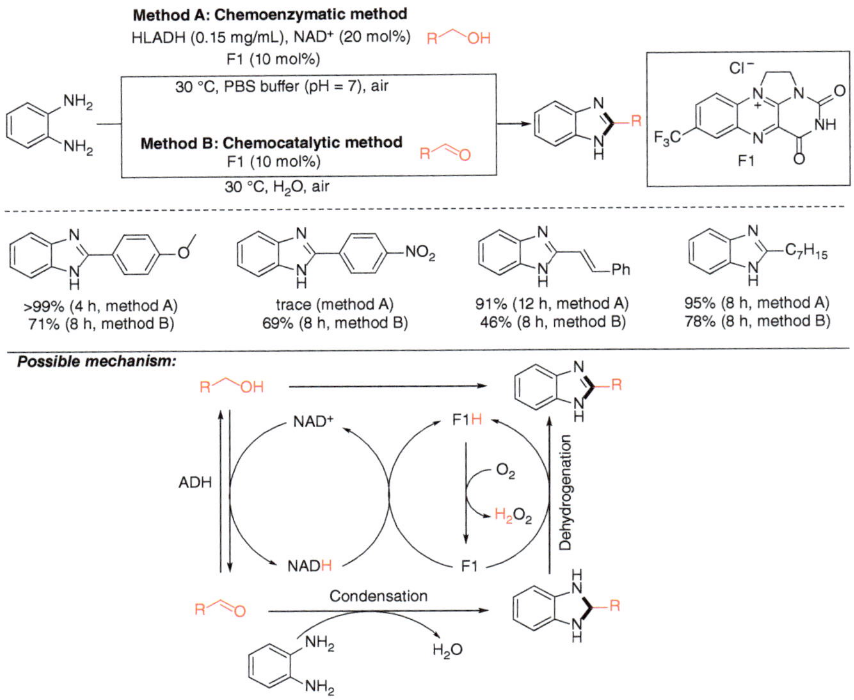

Scheme 10.192 Cooperative chemoenzymatic synthesis of *N*-heterocycles via synergizing bio- and organocatalysis.

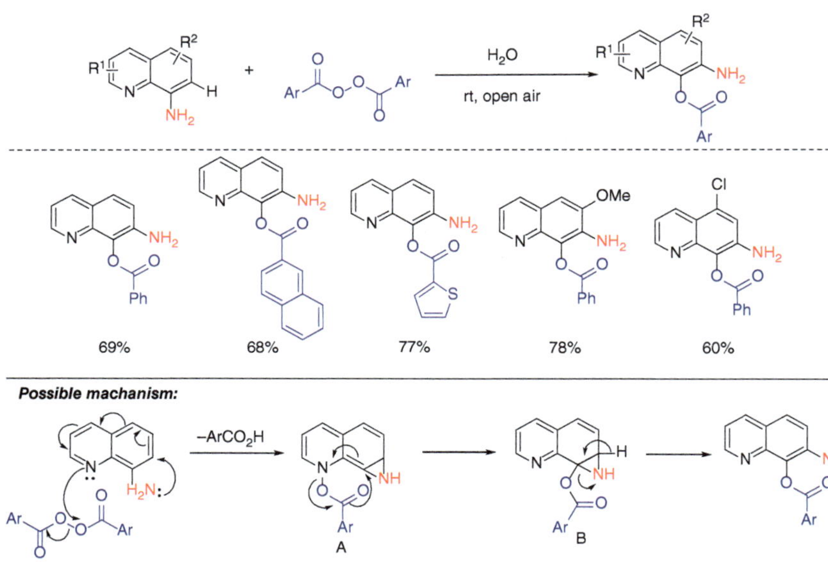

Scheme 10.193 Uncatalyzed oxygenative cleavage of inert C–N bond with concomitant 8,7-amino shift in 8-aminoquinoline derivatives.

tricyclic aziridine intermediate is formed which undergoes two successive migration processes (i.e. acyl migration from nitrogen to C8 and amine migration from C8 to C7) to give the final product.

In the same year, a fully atom-economical protocol was developed by He and coworkers for the clean preparation of quinolin-2-yl substituted ureas from quinoline *N*-oxides and carbodiimides under base-free aqueous conditions using $CuSO_4.5H_2O$ as a Lewis acid catalyst (Scheme 10.194) [208]. An intramolecular oxygen and HAT process was involved in this transformation with the oxygen atom of the urea product originating from the *N*-oxide. The reaction involves an initial [3 + 2] cycloaddition to produce an oxadiazolidinimine intermediate **A**. Coordination of Cu ion to intermediate **A** initiates an intramolecular rearrangement to produce **B** and is followed by an enol-ketone tautomerization to deliver the ureation product. Notably, the product could be easily isolated by simple filtration and alcoholic washing.

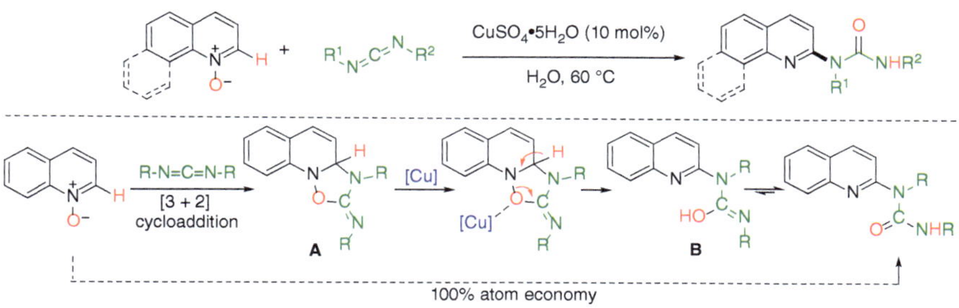

Scheme 10.194 Clean preparation of quinolin-2-yl substituted ureas from quinoline *N*-oxides and carbodiimides.

Zhang and coworkers successfully synthesized a β-bromo ketone compound via metal- and oxidant-free electrochemical bromination/semi-pinacol rearrangement of an allylic alcohol in water using $MgBr_2{\cdot}6H_2O$ as both the bromide source and electrolyte (Scheme 10.195) [209]. The reaction occurred at room temperature under air in an undivided cell using a carbon felt as the anode and a Pt plate as the cathode. To overcome the poor solubility of allylic alcohols in water, a CH_3CN/H_2O system was developed for the synthesis of a variety of β-halocarbonyls from allylic alcohols. The proposed mechanism involves anionic oxidation of bromide anion to afford molecular bromine, which participates in subsequent electrophilic bromination/semi-pinacol rearrangement to afford the final product.

In 2021, Sun and coworkers reported electrocatalytic syntheses of hydropyranones from biomass-derived furfuryl alcohols via an Achmatowicz reaction with concomitant H_2 evolution using a heterogenized Ni-DVBP electrocatalyst prepared by photo-induced polymerization of 5,5'-divinyl-2,2'-bipyridine (dvbpy) (Scheme 10.196) [210]. The reaction proceeded under ambient conditions without stoichiometric chemical oxidants using water as the solvent as well as the oxygen source. Hydropyranones were selectively formed with superior selectivity over furfurals and over-oxidized furanoic acids. The electrocatalytic Achmatowicz reaction afforded comparable or even better yields than those obtained from *m*-CPBA-mediated reactions.

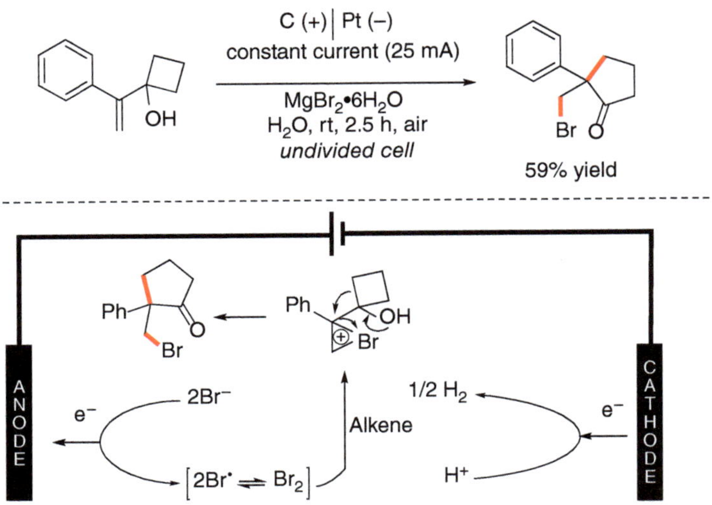

Scheme 10.195 Electrochemical bromination/semi-pinacol rearrangement of allylic alcohols.

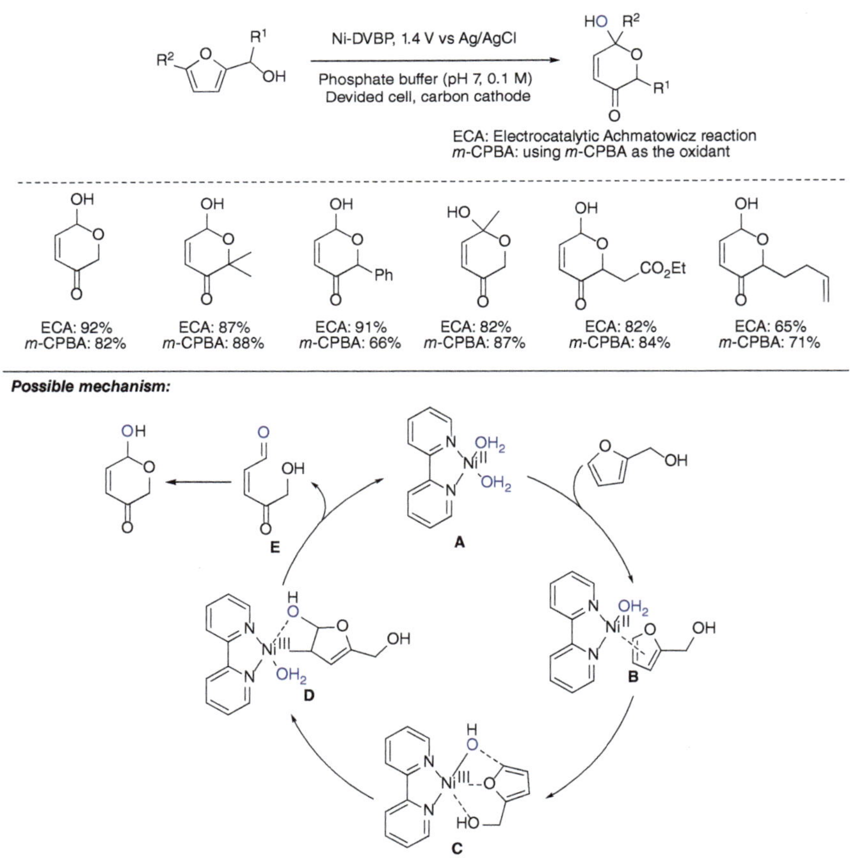

Scheme 10.196 Electrocatalytic Achmatowicz reaction for the synthesis of hydropyranones from furfuryl alcohols.

10.12.2 Carbene Insertion/Transfer Reactions

Carbene transfer reactions are among the most versatile transformations in organic chemistry and recent years have witnessed a remarkable development of such reactions in aqueous and even biological media [211]. The most common carbene transformations include cyclopropanations, C–H insertion, X–H insertion (X=O, N, S, Si), and ylide formation.

In 2019, a Rh-catalyzed carbene transfer reaction of diazoesters with diselenides in water was reported by Koenigs and coworkers leading to the synthesis of seleno-ketals via formal insertion of a carbene fragment into the Se–Se bond (Scheme 10.197) [212]. When the reaction was conducted in organic solvents, the diazoester should be added via a syringe pump over 2 h to avoid decomposition. Delightfully, the transformation can be conducted in a one-pot manner without the slow addition of the reactive diazoester when water is used as the solvent. An ionic mechanism was proposed based on preliminary control experiments.

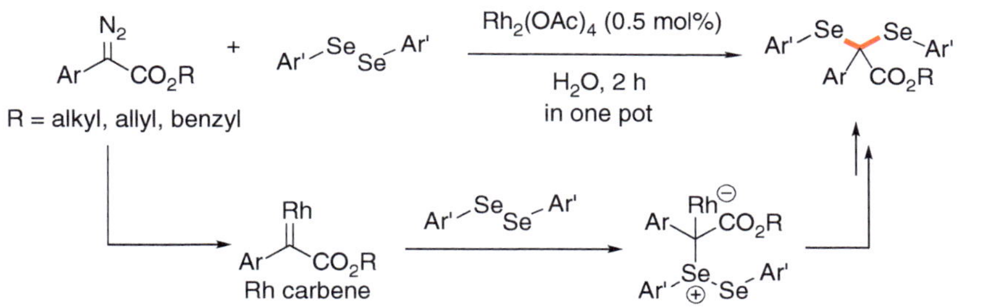

Scheme 10.197 Rh-catalyzed synthesis of seleno-ketals via carbene transfer reactions of diazoesters.

In the same year, Rastogi and coworkers reported the synthesis of 2,2-disubstituted indolin-3-ones via Rh-carbenoid mediated reaction between *N*-*o*-alkylamino benzoylbenzotriazoles and diazoacetates at reflux in water (Scheme 10.198) [213]. The reaction is quite clean since the only byproducts are nitrogen gas and benzotriazole and the latter can be reused for starting material preparation. The protocol features a broad substrate scope and takes advantages of the leaving group ability of the benzotriazole moiety to install the carbonyl group of the product.

One year later, Hu and coworkers disclosed the synthesis of difluoromethyl ethers via difluoromethylation of a series of primary, secondary, and tertiary alcohols in water using $TMSCF_2Br$ as a unique difluorocarbene reagent and KHF_2 as a mild activator (Scheme 10.199) [214]. While the difluoromethylation of liquid alcohols occurred at room temperature using water as the sole solvent, the reaction of solid alcohols either proceeded at a higher temperature to melt the alcohol or at ambient temperature with the addition of a minimum volume of CH_2Cl_2 to dissolve the alcohol. The reaction was proposed to occur in a biphasic oil–water system with the liquid alcohol and $TMSCF_2Br$ serving as the organic phase. Mechanistically, $TMSCF_2Br$ is activated by KHF_2 to form a pentacoordinate silicate intermediate and release the difluorocarbene species via fragmentation. The difluorocarbene species

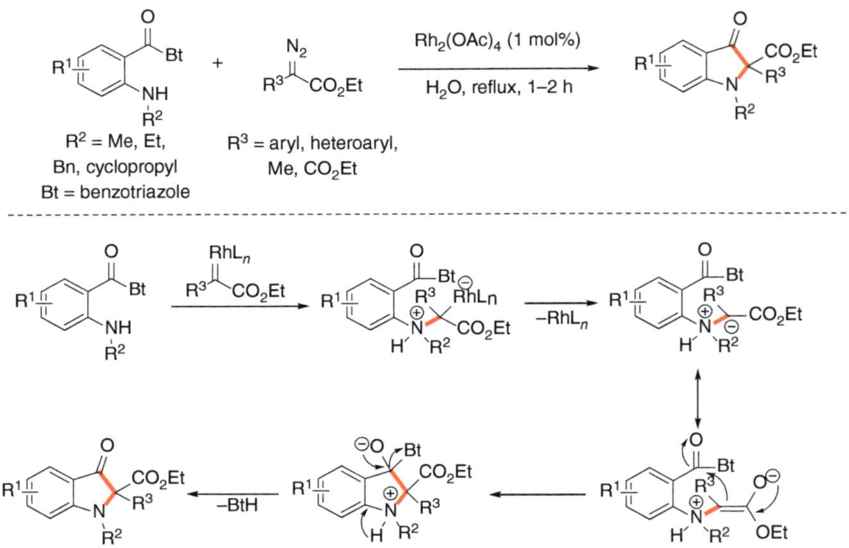

Scheme 10.198 Synthesis of 2,2-disubstituted indolin-3-ones via Rh-carbenoid mediated reaction.

interacts with two alcohol molecules to form a five-membered complex via hydrogen bonding and vacant p-orbital-lone pair interaction. Subsequent double proton transfer delivers the difluoromethyl ether and regenerates one alcohol molecule.

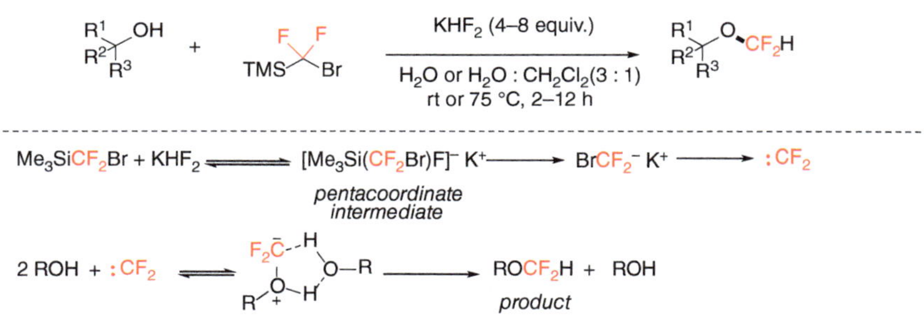

Scheme 10.199 Difluoromethylation of alcohols with $TMSCF_2Br$ via a difluorocarbene species.

In 2021, Wu and coworkers described several blue light-induced, metal-free cyclopropenizations, cyclopropanation, and X–H insertion reactions (X=N, S) of *N*-tosylhydrazones with alkynes, alkenes, or X–H bonds in water at room temperature using Et_3N as the base (Scheme 10.200) [215]. The proposed mechanism involves base-mediated decomposition of the hydrazone to form a diazo compound followed by blue light-induced release of nitrogen to give a carbene intermediate. Subsequent carbene migration insertion of the carbene intermediate with alkynes, alkenes, or other X–H bonds gives rise to the final products.

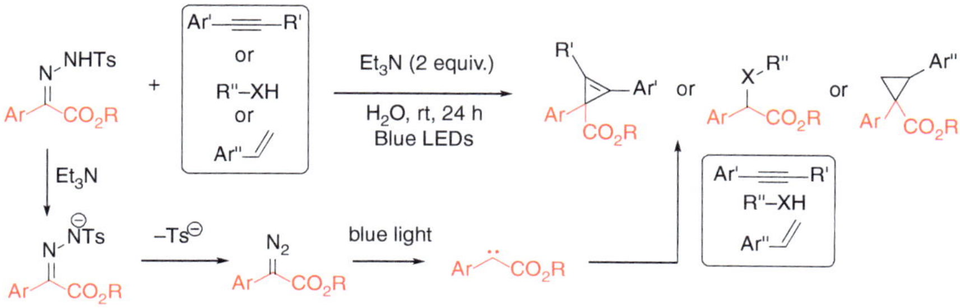

Scheme 10.200 Blue light-induced cyclopropenizations, cyclopropanation, and X–H insertion of *N*-tosylhydrazones with alkynes, alkenes or X–H bonds.

10.13 Amide Condensation Reactions

Amide bond is a structural unit prevalent in many natural products, pharmaceuticals, and biologically active molecules. Construction of the amide bond is one of the most encountered transformations in the pharmaceutical and chemical industries. Handa and coworkers have recently summarized the progress of amide condensation in water under micellar catalysis or polymer hydrophobic effects [4].

The amide condensation reactions are generally achieved by base-mediated reactions of acyl chlorides with amines or by activator-mediated couplings between carboxylic acids and amines in the presence of condensation reagents such as HATU, HOBt, EDCI, and DCC. In 2020, Wu and coworkers reported a simple and efficient amidation protocol by condensing labile acyl chlorides with amines in 2 wt% aqueous TPGS-750-M solution using DIPEA as the base (Scheme 10.201) [216]. The micellar effect of TPGS-750-M surfactant could protect acyl chlorides from surrounding water and mitigate the formation of undesired hydrolysis products. This protocol features a chromatography-free separation process and a good recyclability of the TPGS-750-M solution. It is notable that exclusive di-acetylation took place when 2-aminopyridine was used as the amine substrate.

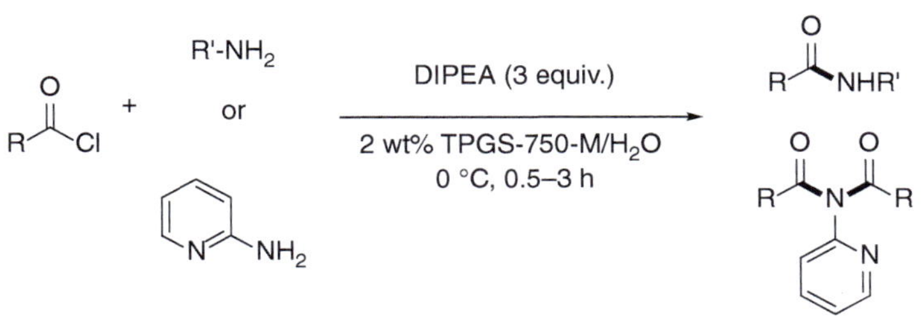

Scheme 10.201 Synthesis of amides via their acid chlorides in aqueous TPGS-750-M.

Later, Zhao and coworkers developed an improved aqueous amide condensation protocol for the synthesis of a cinchona alkaloid amide catalyst for enantioselective allylsilylation of aldehydes (Scheme 10.202) [217]. The improved amide condensation was performed with COMU/2,6-lutidine in TPGS-750-M solution using carboxylic acid as the starting material. This protocol avoids the use of corrosive acid chlorides and solvent-intensive column chromatography, thus greatly reducing the step PMI (process mass intensity) of the original protocol based on acid chloride. COMU is a water-soluble uranium-type coupling reagent showing a less hazardous safety profile than benzotriazole-based regents such as HATU and HBTU.

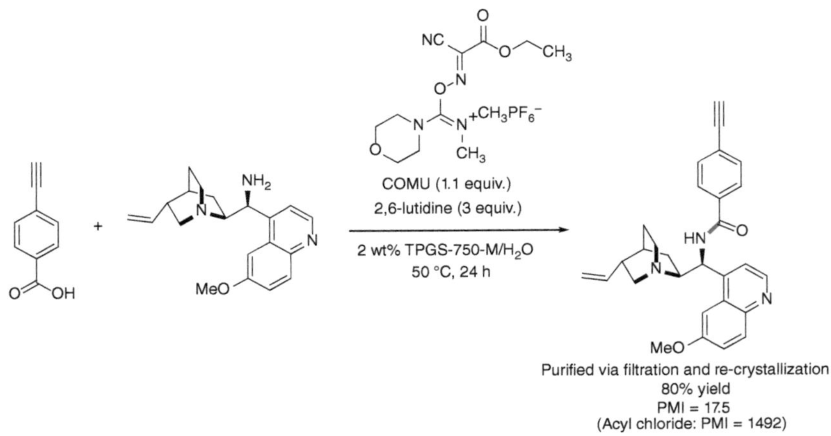

Scheme 10.202 COMU-mediated amide condensation in aqueous micellar solution.

Do and coworkers recently developed a surfactant-, base-, and cosolvent-free amine condensation of various carboxylic acids with amines in neat water under air using a water soluble and stable *N*-methylimidazole (NMI)-based pyridinium reagent (AMP_{im1}) as the coupling reagent (Scheme 10.203) [218]. AMP_{im1} serves multiple roles of a leaving group, activator, and base in this transformation. It was easily prepared from 2-chloro-1-methylpyridinium iodide (CMPI) by reacting with NMI in acetonitrile and showed improved water solubility and better coupling capacity than the parent CMPI reagent. Furthermore, AMP_{im1} is a green amide condensation reagent with a similar or lower PMI metric than that of conventional coupling reagents. It can also be used in tandem with one-pot oxidation-amidation or reduction-amidation reactions.

The traditional amide condensation reactions employing acyl chlorides or condensation reagents generally suffer from poor atom economy and generate large amounts of waste. In recent years, electro- or photo-chemical amide condensations have been developed to increase the atom economy and environmental sustainability of the amide coupling process. For example, Ke and coworkers developed an electrochemical *N*-acylation of amines with carboxylic acids for the synthesis of amides in water at room temperature using Pt-wire as the electrode, Cs_2CO_3 as the base, and TBAB as both the electrolyte and catalyst (Scheme 10.204) [219]. A variety of amines including primary/secondary aliphatic amines, (hetero)aryl amines, ammonium hydroxide, and hydrazine were smoothly cross-coupled with various acetic or aromatic acids.

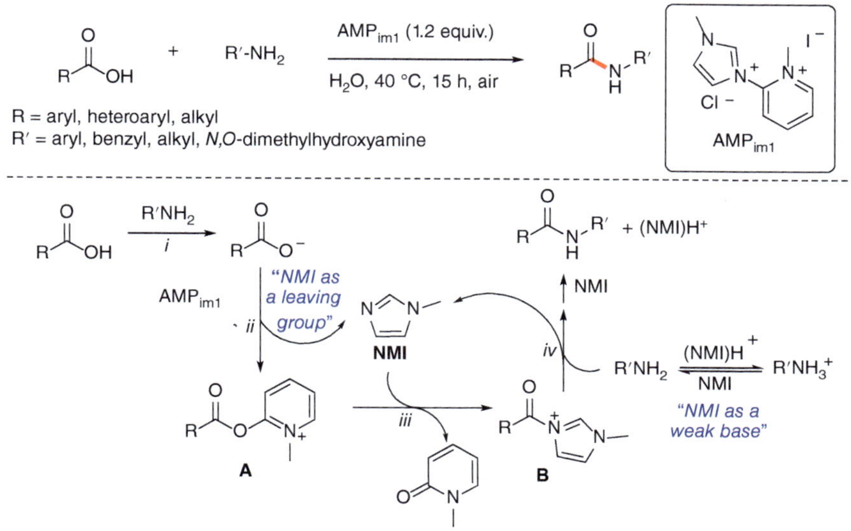

Scheme 10.203 AMP$_{im1}$-mediated amine condensation under surfactant- and base-free conditions.

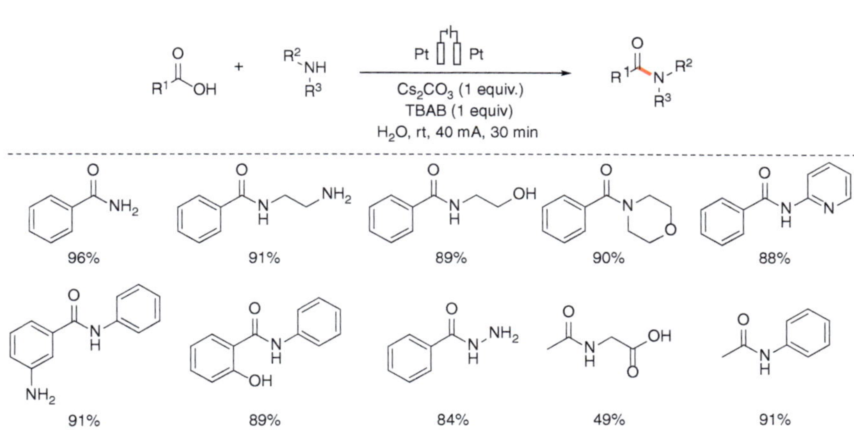

Scheme 10.204 Electrochemical *N*-acylation of amines with carboxylic acids.

The same group further achieved a metal-free photocatalytic *N*-acylation of amines with carboxylic acids in water using eosin Y as the photocatalyst in the presence of TBAB as the phase transfer catalyst and KOH as the base under ambient air condition (Scheme 10.205a) [220]. A tentative mechanism involving radical coupling of an amine radical with a benzoyl radical was proposed. Later, Xu and coworkers achieved iodine-catalyzed sunlight-promoted amide condensation in aqueous solution using the combination of KOH and TBAB (Scheme 10.205b) [221].

More recently, Liu and coworkers reported a solvent-controlled selective preparation of amides or symmetrical and unsymmetrical disubstituted thioureas from isothiocyanates by using the nitrile/KOH or H_2O/DMAP system, respectively (Scheme 10.206) [222]. For the nitrile/KOH system, the nitrile undergoes alkaline

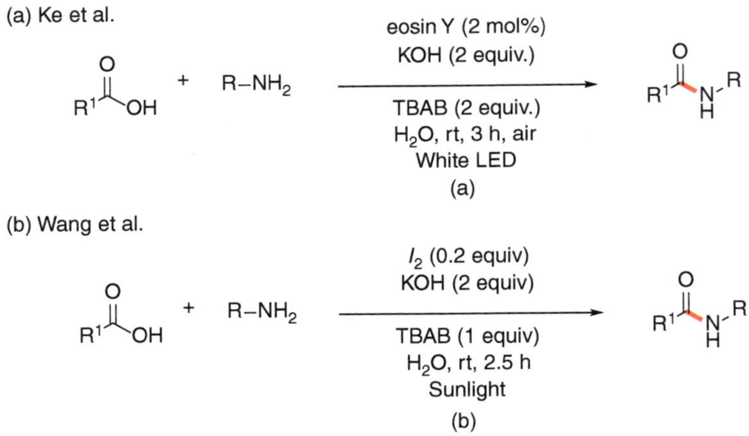

Scheme 10.205 Photochemical *N*-acylation of amines with carboxylic acids.

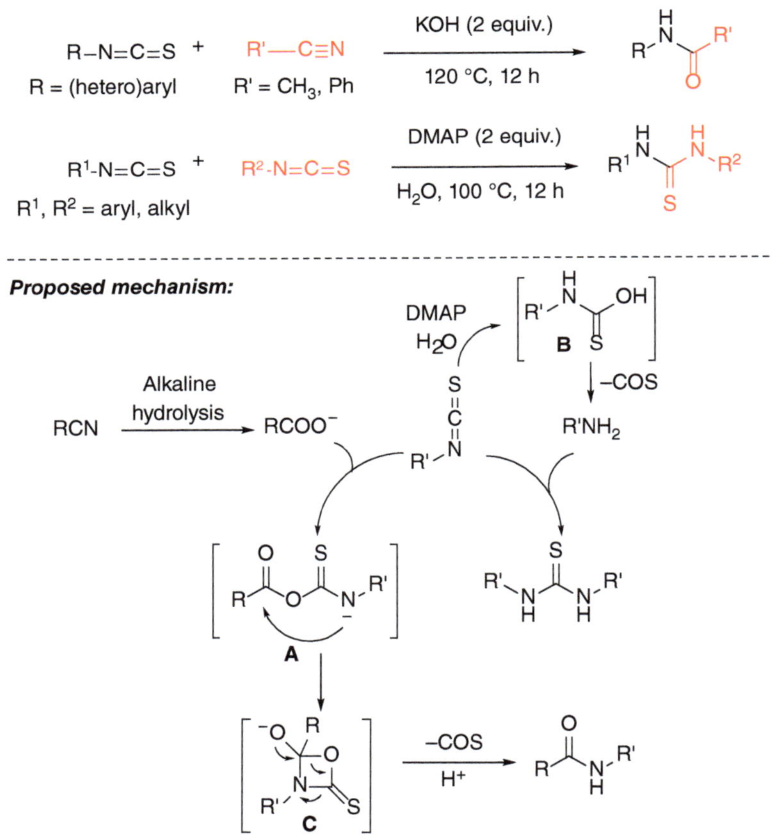

Scheme 10.206 Solvent-controlled selective synthesis of amides and thioureas from isothiocyanates.

hydrolysis to afford a carboxylic acid, which reacts with the isothiocyanate to produce a carbonic anhydride intermediate. Cyclization of this species followed by carbonyl sulfide (COS) release gives the amide product. For the H_2O/DMAP system, addition of H_2O to the isothiocyanate followed by COS release affords an amine which adds to the isothiocyanate to form the disubstituted thiourea.

10.14 Summary and Conclusions

In summary, water has become a powerful solvent in organic chemistry because of its unique physical and chemical properties. Many recent examples of aqueous reactions including metal catalysis, photocatalysis, biocatalysis, electrochemical reactions, and metal-free reactions are presented in this chapter. Despite the significant progress made in this field, many examples still require the addition of harmful metal catalysts, phase transfer catalysts, surfactants, or co-solvents to facilitate the reactions. Further efforts should be directed toward developing greener, more sustainable, and environmentally friendly aqueous reactions under metal-free and additive-free conditions. The development of aqueous reactions will continue and we hope that the chemistry explored in this chapter will be beneficial to scientists interested in this field and inspire future innovation in creating more structurally diverse organic compounds and different reaction types.

Acknowledgments

We gratefully acknowledge the Jiangsu Provincial Natural Science Foundation of China (Grant BK20220326 to C-S.W.) Start-Up Funding of Nanjing Tech University (Grant 39808137 to C-S.W.), Province Distinguished Professor Program (Grant No. 54908022 to C.-S. Wang), Discipline Fund of School of Food and Light Industry, Nanjing Tech University (Grant to QS), Start-Up Funding of Nanjing Tech University (Grant to QS) for financial support.

References

1 (a) Kitanosono, T., Masuda, K., Xu, P., and Kobayashi, S. (2018). *Chem. Rev.* 118: 679–746. (b) Romney, D.K., Arnold, F.H., Lipshutz, B.H., and Li, C.-J. (2018). *J. Org. Chem.* 83: 7319–7322. (c) Song, H.-X., Han, Q.-Y., Zhao, C.-L., and Zhang, C.-P. (2018). *Green Chem.* 20: 1662–1731. (d) Satpute, D.P., Vaidya, G.N., Lokhande, S.K. et al. (2021). *Green Chem.* 23: 6273–6300. (e) Shilpa, T., Neetha, M., and Anilkumar, G. (2021). *Adv. Synth. Catal.* 363: 1559–1582.

2 Hazra, S., Malik, E., Nair, A. et al. (2020). *Chem. Asian. J.* 15: 1916–1936.

3 (a) Mondal, A., Bhowmick, S., Ghosh, A. et al. (2017). *Tetrahedron* 28: 849–875. (b) Zhou, F. and Li, C.-J. (2019). *Chem. Sci.* 10: 34–46. (c) Goswami, P., Park, J.H., and Bae, H.Y. (2020). *Tetrahedron* 76: 131254–131274.

4 Hazra, S., Gallou, F., and Handa, S. (2022). *ACS Sustainable Chem. Eng.* 10: 5299–5306.

5 (a) Vessally, E., Didehban, K., Mohammadi, R. et al. (2018). *J. Sulfur Chem.* 39: 332–349. (b) Cicco, L., Dilauro, G., Perna, F.M. et al. (2021). *Org. Biomol. Chem.* 19: 2558–2577. (c) Peng, K. and Dong, Z.-B. (2021). *Adv. Synth. Catal.* 363: 1185–1201.
6 (a) Laskar, R., Pal, T., Bhattacharya, T. et al. (2022). *Green Chem.* 24: 2296–2320. (b) Albano, G., Punzi, A., Capozzi, M.A.M., and Farinola, G.M. (2022). *Green Chem.* 24: 1809–1894. (c) Borrego, E., Caballero, A., and Pérez, P.J. (2022). *Organometallics* 41: 3084–3098.
7 (a) Barata-Vallejo, S., Yerien, D.E., and Postigo, A. (2021). *ACS Sustainable Chem. Eng.* 9: 10016–10047. (b) Sun, K., Lv, Q.-Y., Chen, X.-L. et al. (2021). *Green Chem.* 23: 232–248. (c) Russo, C., Brunelli, F., Tron, G.C., and Giustiniano, M. (2022). *J. Org. Chem.* https://doi.org/10.1021/acs.joc.2c00805.
8 (a) Lipshutz, B.H., Ghorai, S., and Cortes-Clerget, M. (2018). *Chem. Eur. J.* 24: 6672–6695. (b) Lipshutz, B.H. (2018). *Curr. Opin. Green Sustain. Chem.* 11: 1–8. (c) Steven, A. (2019). *Synthesis* 51: 2632–2647. (d) Lorenzetto, T., Berton, G., Fabris, F., and Scarso, A. (2020). *Catal. Sci. Technol.* 10: 4492–4502.
9 (a) Kitanosono, T. and Kobayashi, S. (2020). *Chem. Eur. J.* 26: 9408–9429. (b) Cortes-Clerget, M., Yu, J., Kincaid, J.R.A. et al. (2021). *Chem. Sci.* 12: 4237–4266.
10 Thakore, R.R., Takale, B.S., Gallou, F. et al. (2019). *ACS Catal.* 9: 11647–11657.
11 Takale, B.S., Thakore, R.R., Handa, S. et al. (2019). *Chem. Sci.* 10: 8825–8831.
12 Orecchia, P., Petkova, D.S., Goetz, R. et al. (2021). *Green Chem.* 23: 8169–8180.
13 Wang, Y., Liu, Y., Zhang, W.-q. et al. (2019). *ChemSusChem* 12: 5265–5273.
14 Li, J.-Y. and Huang, H. (2018). *Bioconjugate Chem.* 29: 3841–3846.
15 Nguyen, V.H., El Ali, B.M., and Huynh, H.V. (2018). *Organometallics* 37: 2358–2367.
16 Yang, H. and Tang, W. (2022). *Nat. Commun.* 13: 4577.
17 Zhang, Y., Wang, Z., Tang, Z. et al. (2020). *Eur. J. Org. Chem.* 2020: 1620–1628.
18 Shaw, R. and Pratap, R. (2022). *Asian J. Org. Chem.* 11: e202200078.
19 Dilauro, G., Francesca Quivelli, A., Vitale, P. et al. (2019). *Angew. Chem. Int. Ed.* 58: 1799–1802.
20 Dilauro, G., Azzollini, C.S., Vitale, P. et al. (2021). *Angew. Chem. Int. Ed.* 60: 10632–10636.
21 Hu, Y., Wong, M.J., and Lipshutz, B.H. (2022). *Angew. Chem. Int. Ed.* 61: e202209784.
22 Takale, B.S., Thakore, R.R., Casotti, G. et al. (2021). *Angew. Chem. Int. Ed.* 60: 4158–4163.
23 Chang, Y.-T., Liu, L.-J., Peng, W.-S. et al. (2021). *J. Chin. Chem. Soc.* 68: 469–475.
24 Jadhav, S.N. and Rode, C.V. (2017). *Green Chem.* 19: 5958–5970.
25 Neff, R.K. and Frantz, D.E. (2018). *J. Am. Chem. Soc.* 140: 17428–17432.
26 Gao, P.-S., Zhang, K., Yang, M.-M. et al. (2018). *Chem. Commun.* 54: 5074–5077.
27 Jin, B., Gallou, F., Reilly, J., and Lipshutz, B.H. (2019). *Chem. Sci.* 10: 3481–3485.

28 Tian, Y., Wang, J., Cheng, X. et al. (2020). *Green Chem.* 22: 1338–1344.

29 Wójcik, E., Dziadas, M., and Trzeciak, A.M. (2022). *J. Organomet. Chem.* 962: 122269.

30 Chang, F. and Liu, Y. (2017). *Synth. Commun.* 47: 961–967.

31 Ghorpade, S.A., Sawant, D.N., Renn, D. et al. (2018). *New J. Chem.* 42: 6210–6214.

32 Thakore, R.R., Takale, B.S., Singhania, V. et al. (2021). *ChemCatChem* 13: 212–216.

33 Wang, Z., Yin, Z., and Wu, X.-F. (2017). *Chem. Eur. J.* 23: 15026–15029.

34 Zhang, S., Shen, G., Deng, Y. et al. (2018). *ACS Sustainable Chem. Eng.* 6: 13192–13198.

35 Zhou, F., Hu, X., Zhang, W., and Li, C.-J. (2018). *J. Org. Chem.* 83: 7416–7422.

36 Pang, H., Wang, Y., Gallou, F., and Lipshutz, B.H. (2019). *J. Am. Chem. Soc.* 141: 17117–17124.

37 Giedyk, M., Narobe, R., Weiß, S. et al. (2020). *Nat. Catal.* 3: 40–47.

38 Surendra Babu, G.V., Sai Prathima, P., Perumgani, P.C. et al. (2018). *Heteroatom Chem.* 29: e21423.

39 Wood, A.B., Roa, D.E., Gallou, F., and Lipshutz, B.H. (2021). *Green Chem.* 23: 4858–4865.

40 Ma, X., Yu, J., Han, C. et al. (2019). *Adv. Synth. Catal.* 361: 1023–1027.

41 Pan, S., Jiang, M., Hu, J. et al. (2020). *Green Chem.* 22: 336–341.

42 Li, X., Zhang, R., Zhang, X. et al. (2020). *Chem. Asian. J.* 15: 1175–1179.

43 Zhang, Y., Takale, B.S., Gallou, F. et al. (2019). *Chem. Sci.* 10: 10556–10561.

44 Bollenbach, M., Lecroq, W., Wagner, P. et al. (2019). *Chem. Commun.* 55: 1623–1626.

45 Song, M., Zhang, L., Wei, D. et al. (2022). *Green Chem.* 24: 6119–6124.

46 Bollenbach, M., Aquino, P.G.V., de Araújo-Júnior, J.X. et al. (2017). *Chem. Eur. J.* 23: 13676–13683.

47 Ding, X., Bai, J., Wang, H. et al. (2017). *Tetrahedron* 73: 172–178.

48 Kumar, S.V. and Ma, D. (2018). *Chin. J. Chem.* 36: 1003–1006.

49 Wang, D., Zheng, Y., Yang, M. et al. (2017). *Org. Biomol. Chem.* 15: 8009–8012.

50 Paul, S., Joy, B.P., Sasikala, G. et al. (2020). *ChemistrySelect* 5: 15004–15009.

51 Chakraborti, G., Paladhi, S., Mandal, T., and Dash, J. (2018). *J. Org. Chem.* 83: 7347–7359.

52 Liang, X., Li, H., Du, F. et al. (2020). *Tetrahedron Lett.* 61: 152222.

53 Parmar, U., Somvanshi, D., Kori, S. et al. (2021). *J. Org. Chem.* 86: 8900–8925.

54 Lee, N.R., Moghadam, F.A., Braga, F.C. et al. (2020). *Org. Lett.* 22: 4949–4954.

55 Vaidya, G.N., Choudhary, R.H., Nagpure, M. et al. (2022). *Green Chem.* 24: 3977–3984.

56 Talukder, M.M., Miller, J.T., Cue, J.M.O. et al. (2021). *Organometallics* 40: 83–94.

57 Dhanjee, H.H., Saebi, A., Buslov, I. et al. (2020). *J. Am. Chem. Soc.* 142: 9124–9129.

58 Wang, Y.-Y., Wu, X.-M., and Yang, M.-H. (2020). *Synlett* 31: 1226–1230.

59 Yu, T.-Y., Pang, H., Cao, Y. et al. (2021). *Angew. Chem. Int. Ed.* 60: 3708–3713.

60 Kibriya, G., Mondal, S., and Hajra, A. (2018). *Org. Lett.* 20: 7740–7743.
61 Huang, X., Chen, Y., Zhen, S. et al. (2018). *J. Org. Chem.* 83: 7331–7340.
62 Xie, L.-Y., Peng, S., Tan, J.-X. et al. (2018). *ACS Sustainable Chem. Eng.* 6: 16976–16981.
63 Zhu, S., Zhang, Y., Luo, J. et al. (2020). *Green Chem.* 22: 657–661.
64 Fu, Y., Zhu, S., Zhao, X., and Huang, S. (2022). *Green Chem.* 24: 6849–6853.
65 Hirai, Y. and Uozumi, Y. (2017). *Synlett* 28: 2966–2970.
66 Drev, M., Grošelj, U., Ledinek, B. et al. (2018). *Org. Lett.* 20: 5268–5273.
67 Yetra, S.R., Rogge, T., Warratz, S. et al. (2019). *Angew. Chem. Int. Ed.* 58: 7490–7494.
68 Verma, A., Dolui, P., Hazra, S., and Elias, A.J. (2022). *J. Organomet. Chem.* 964: 122303.
69 Nareddy, P., Jordan, F., and Szostak, M. (2018). *Org. Lett.* 20: 341–344.
70 Zhang, X., Zhang, F., Li, X. et al. (2022). *Org. Lett.* 24: 5029–5033.
71 Ramana, D.V. and Chandrasekharam, M. (2018). *Adv. Synth. Catal.* 360: 4080–4083.
72 Li, B., Li, X., Han, B. et al. (2019). *J. Am. Chem. Soc.* 141: 9401–9407.
73 Wang, M., Yin, C., and Hu, P. (2021). *Org. Lett.* 23: 722–726.
74 Cybularczyk-Cecotka, M., Predygier, J., Crespi, S. et al. (2022). *ACS Catal.* 12: 3543–3549.
75 Zhang, H., Yang, Z., Ma, Q. et al. (2018). *Green Chem.* 20: 3140–3146.
76 Zhang, H., Yang, Z., Liu, J. et al. (2019). *Org. Chem. Front.* 6: 967–971.
77 Satrawala, N., Williams, C., Srivastava, A.K. et al. (2019). *Catal. Commun.* 129: 105727.
78 Xu, Z., Wang, Y., Zheng, Y. et al. (2020). *Org. Chem. Front.* 7: 3709–3714.
79 Sarathkumar, S., Kavala, V., and Yao, C.-F. (2021). *Org. Lett.* 23: 1960–1965.
80 Sharma, S., Kumar, M., Vishwakarma, R.A. et al. (2018). *J. Org. Chem.* 83: 12420–12431.
81 Laha, J.K., Tinwala, U., and Hunjan, M.K. (2021). *New J. Chem.* 45: 22853–22859.
82 Lu, J., He, X.-K., Cheng, X. et al. (2020). *Adv. Synth. Catal.* 362: 2178–2182.
83 San Segundo, M. and Correa, A. (2020). *Chem. Sci.* 11: 11531–11538.
84 Wu, W., Fang, S., Jiang, G. et al. (2019). *Org. Chem. Front.* 6: 2200–2204.
85 Du, Y. and Li, Z. (2019). *Synth. Commun.* 49: 65–72.
86 Wang, H., Lorion, M.M., and Ackermann, L. (2017). *Angew. Chem. Int. Ed.* 56: 6339–6342.
87 Vaidya, G.N., Lokhande, S.K., Shinde, S.D. et al. (2022). *Green Chem.* 24: 4921–4927.
88 Hikawa, H., Ichinose, R., Kikkawa, S., and Azumaya, I. (2018). *Green Chem.* 20: 1297–1305.
89 Zhang, G., Jia, F., and Gooßen, L.J. (2018). *Chem. Eur. J.* 24: 4537–4541.
90 Kitanosono, T., Hisada, T., Yamashita, Y., and Kobayashi, S. (2021). *Angew. Chem. Int. Ed.* 60: 3407–3411.
91 Ghosh, P., Kwon, N.Y., Kim, S. et al. (2021). *Angew. Chem. Int. Ed.* 60: 191–196.

92 Santos, M.S., Cybularczyk-Cecotka, M., König, B., and Giedyk, M. (2020). *Chem. Eur. J.* 26: 15323–15329.

93 Ma, C.-H., Zhao, L., He, X. et al. (2022). *Org. Chem. Front.* 9: 1445–1450.

94 Gava, R., Ballestín, P., Prieto, A. et al. (2019). *Chem. Commun.* 55: 11243–11246.

95 Xu, J., Qiao, L., Shen, J. et al. (2017). *Org. Lett.* 19: 5661–5664.

96 Nguyen, T.-T.H., O'Brien, C.J., Tran, M.L.N. et al. (2021). *Org. Lett.* 23: 3823–3827.

97 Yerien, D.E., Postigo, A., Baroncini, M., and Barata-Vallejo, S. (2021). *Green Chem.* 23: 8147–8153.

98 Begam, H.M., Choudhury, R., Behera, A., and Jana, R. (2019). *Org. Lett.* 21: 4651–4656.

99 You, G., Wang, K., Wang, X. et al. (2018). *Org. Lett.* 20: 4005–4009.

100 Zhao, R., Zhou, Z., Liu, J. et al. (2020). *Org. Lett.* 22: 8144–8149.

101 Corrieri, M., Crescentini, L.D., Mantellini, F. et al. (2022). *Green Chem.* 24: 7340–7345.

102 Kong, X., Lin, L., and Xu, B. (2018). *Adv. Synth. Catal.* 360: 2801–2805.

103 Cao, X.-T., Wei, S.-N., Sun, H.-T. et al. (2021). *RSC Adv.* 11: 22000–22004.

104 Peng, S., Hu, D., Hu, J.-L. et al. (2019). *Adv. Synth. Catal.* 361: 5721–5726.

105 Mack, J.B.C., Gipson, J.D., Du Bois, J., and Sigman, M.S. (2017). *J. Am. Chem. Soc.* 139: 9503–9506.

106 Lee, M. and Sanford, M.S. (2017). *Org. Lett.* 19: 572–575.

107 Wang, Y., Wang, S., Gao, Q. et al. (2019). *Tetrahedron* 75: 3856–3863.

108 Qiao, H., Sun, S., Yang, F. et al. (2017). *Adv. Synth. Catal.* 359: 1976–1980.

109 Bora, P.P., Bihani, M., Plummer, S. et al. (2019). *ChemSusChem* 12: 3037–3042.

110 Upadhyay, N.S., Thorat, V.H., Sato, R. et al. (2017). *Green Chem.* 19: 3219–3224.

111 Wu, J., Qian, B., Liu, Y., and Shang, Y. (2020). *ChemistrySelect* 5: 10269–10275.

112 Wu, S., Wu, X., Fu, C., and Ma, S. (2018). *Org. Lett.* 20: 2831–2834.

113 Mukherjee, K., Ramesh, E., Ghosh, K., and Sahoo, A.K. (2018). *Asian J. Org. Chem.* 7: 1380–1384.

114 Zeng, H., Wang, Z., and Li, C.-J. (2019). *Angew. Chem. Int. Ed.* 58: 2859–2863.

115 Nie, R., Lai, R., Lv, S. et al. (2019). *Chem. Commun.* 55: 11418–11421.

116 Debbarma, S., Sk, M.R., Modak, B., and Maji, M.S. (2019). *J. Org. Chem.* 84: 6207–6216.

117 Ding, X., Zhang, L., Mao, Y. et al. (2020). *Synlett* 31: 280–284.

118 Wang, H., Choi, I., Rogge, T. et al. (2018). *Nat. Catal.* 1: 993–1001.

119 Sabatino, V., Rebelein, J.G., and Ward, T.R. (2019). *J. Am. Chem. Soc.* 141: 17048–17052.

120 Liu, Y., Li, C., Miao, W. et al. (2020). *Green Chem.* 22: 33–38.

121 Wang, D., Wang, P., Wang, S. et al. (2019). *Nat. Commun.* 10: 2796.

122 (a) Huang, Y., Chong, X., Liu, C. et al. (2018). *Angew. Chem. Int. Ed.* 57: 13163–13166. (b) Huang, C., Huang, Y., Liu, C. et al. (2019). *Angew. Chem. Int. Ed.* 58: 12014–12017.

123 Yang, L., Zhuang, Q., Wu, M. et al. (2021). *Org. Biomol. Chem.* 19: 6417–6421.

124 Vivancos, Á., Beller, M., and Albrecht, M. (2018). *ACS Catal.* 8: 17–21.

125 Gao, C., Xuan, Q., and Song, Q. (2021). *Chin. J. Chem.* 39: 2504–2508.

126 Kolcsár, V.J. and Szőllősi, G. (2022). *ChemCatChem* 14: e202101501.

127 Dominguez-Huerta, A., Perepichka, I., and Li, C.-J. (2018). *Commun. Chem.* 1: 45.

128 Huang, X., Zhao, X., Zhang, M. et al. (2017). *ChemistrySelect* 2: 11007–11011.

129 Wadzinski, T.J., Steinauer, A., Hie, L. et al. (2018). *Nat. Chem.* 10: 644–652.

130 Sun, C., Yao, W., Chen, X. et al. (2019). *ACS Omega* 4: 10534–10538.

131 Zhuang, H., Lu, N., Ji, N. et al. (2021). *Adv. Synth. Catal.* 363: 5461–5472.

132 An, D., Miao, X., Ling, X. et al. (2020). *Adv. Synth. Catal.* 362: 1514–1524.

133 Zhang, Q., Xu, Y., Liang, X., and Ke, Z. (2022). *ChemSusChem* 15: e202200574.

134 Song, Z., Zeng, J., Li, T. et al. (2020). *Org. Lett.* 22: 1736–1741.

135 Kitanosono, T., Hashidoko, A., Yamashita, Y., and Kobayashi, S. (2022). *Chem. Asian. J.* 17: e202200457.

136 Dilauro, G., Dell'Aera, M., Vitale, P. et al. (2017). *Angew. Chem. Int. Ed.* 56: 10200–10203.

137 Elorriaga, D., Rodríguez-Álvarez, M.J., Ríos-Lombardía, N. et al. (2020). *Chem. Commun.* 56: 8932–8935.

138 Han, M.-Y., Lin, J., Li, W. et al. (2018). *Green Chem.* 20: 1228–1232.

139 Weng, Y., Yang, T., Chen, H. et al. (2019). *ChemistrySelect* 4: 14233–14236.

140 Wu, Y.-B., Wan, L., Lu, G.-P., and Cai, C. (2017). *Eur. J. Org. Chem.* 2017: 3438–3441.

141 Wu, Y.-B., Wan, L., Lu, G.-P., and Cai, C. (2018). *J. Fluor. Chem.* 206: 125–127.

142 Li, Z., Huang, J., Ni, Z. et al. (2022). *SynOpen* 6: 19–30.

143 Li, Y., Huang, Y., Gui, Y. et al. (2017). *Org. Lett.* 19: 6416–6419.

144 Zhou, F., Hu, X., Zhang, W., and Li, C.-J. (2018). *Org. Chem. Front.* 5: 3579–3584.

145 Bloom, S., Liu, C., Kölmel, D.K. et al. (2018). *Nat. Chem.* 10: 205–211.

146 Chilamari, M., Immel, J.R., Chen, P.-H. et al. (2022). *ACS Catal.* 12: 4175–4181.

147 Park, J.H., Lee, S.B., Koo, B.J., and Bae, H.Y. (2022). *ChemSusChem* 15: e202201000.

148 Aher, Y.N. and Pawar, A.B. (2019). *Org. Biomol. Chem.* 17: 7536–7546.

149 Wang, H., Li, Y., Tang, Z. et al. (2018). *ACS Catal.* 8: 10599–10605.

150 Xiong, Y.-S., Weng, J., and Lu, G. (2018). *Adv. Synth. Catal.* 360: 1611–1616.

151 Zhang, B., Liu, T., Bian, Y. et al. (2018). *ACS Sustainable Chem. Eng.* 6: 2651–2655.

152 Tang, S.-Q., Wang, A.-P., Schmitt, M., and Bihel, F. (2018). *Tetrahedron Lett.* 59: 1465–1468.

153 Voutyritsa, E. and Kokotos, C.G. (2020). *Angew. Chem. Int. Ed.* 59: 1735–1741.

154 Hazra, S., Kaur, G., and Handa, S. (2021). *ACS Sustainable Chem. Eng.* 9: 5513–5518.

155 Liu, J., Yao, H., Li, X. et al. (2020). *Org. Chem. Front.* 7: 1314–1320.

156 Xuan, Q.-Q., Wei, Y.-H., and Song, Q.-L. (2017). *Chin. Chem. Lett.* 28: 1163–1166.

157 Lv, J., Zong, L., Zhang, J. et al. (2021). *Catal. Sci. Technol.* 11: 4708–4713.

158 Trzepizur, D., Brodzka, A., Koszelewski, D. et al. (2021). *Eur. J. Org. Chem.* 2021: 6028–6036.

159 Migliorini, F., Dei, F., Calamante, M. et al. (2021). *ChemCatChem* 13: 2794–2806.
160 Kim, W.G., Kang, M.E., Lee, J.B. et al. (2017). *J. Am. Chem. Soc.* 139: 12121–12124.
161 Chakraborti, G., Mandal, T., Roy, C.P., and Dash, J. (2021). *Chem. Commun.* 57: 7970–7973.
162 Rideout, D.C. and Breslow, R. (1980). *J. Am. Chem. Soc.* 102: 7816–7817.
163 Kucherov, F.A., Galkin, K.I., Gordeev, E.G., and Ananikov, V.P. (2017). *Green Chem.* 19: 4858–4864.
164 Zuo, W.-F., Zhou, J., Wu, Y.-L. et al. (2021). *Org. Chem. Front.* 8: 922–927.
165 Sokolov, A.S., Polezhaev, A.V., Nelyubina, Y.V., and Perekalin, D.S. (2022). *Helv. Chim. Acta.* 105: e202100246.
166 Lo, V.K.-Y., Chan, Y.-M., Zhou, D. et al. (2019). *Org. Lett.* 21: 7717–7721.
167 Zhao, S., Li, K., Sun, X. et al. (2022). *Org. Lett.* 24: 4224–4228.
168 Ramesh, K. and Satyanarayana, G. (2018). *Eur. J. Org. Chem.* 2018: 4135–4146.
169 Bi, H.-Y., Du, M., Pan, C.-X. et al. (2019). *J. Org. Chem.* 84: 9859–9868.
170 Chamberlain, A.E.R., Paterson, K.J., Armstrong, R.J. et al. (2020). *Chem. Commun.* 56: 3563–3566.
171 Sheng, Z.-Z., Huang, M.-M., Xue, T. et al. (2020). *RSC Adv.* 10: 34910–34915.
172 Zhang, X., Cao, W.-B., Xu, X.-P., and Ji, S.-J. (2019). *Synthesis* 51: 3805–3814.
173 Wang, D.-K., Fang, Y.-L., Zhang, J. et al. (2020). *Org. Biomol. Chem.* 18: 8491–8495.
174 Yu, X.-C., Zheng, Y.-N., Zhang, J.-H. et al. (2022). *ACS Sustainable Chem. Eng.* 10: 6057–6062.
175 Li, J.-Z., Mei, L., Cai, X.-E. et al. (2022). *Adv. Synth. Catal.* 364: 2080–2085.
176 (a) Liu, X.-C., Chen, X.-L., Liu, Y. et al. (2020). *ChemSusChem* 13: 298–303. b) Yuan, X.-Y., Zeng, F.-L., Zhu, H.-L. et al. (2020). *Org. Chem. Front.* 7: 1884–1889.
177 Shi, W., Yang, C., Guo, L., and Xia, W. (2022). *Org. Chem. Front.* 9: 6513–6519.
178 Yang, X. and Yu, W. (2022). *Chem. Commun.* 58: 11693–11696.
179 (a) Vachan, B.S., Karuppasamy, M., Vinoth, P. et al. (2020). *Adv. Synth. Catal.* 362: 87–110. (b) Nazeri, M.T., Nasiriani, T., Farhid, H. et al. (2022). *ACS Sustain. Chem. Eng.* 10: 8115–8134.
180 Xiang, S., Fan, W., Zhang, W. et al. (2021). *Green Chem.* 23: 7950–7955.
181 Ziyaei Halimehjani, A. and Lotfi Nosood, Y. (2017). *Org. Lett.* 19: 6748–6751.
182 Khalili Foumeshi, M., Haghi, R., Beier, P., and Ziyaei Halimehjani, A. (2020). *J. Sulfur Chem.* 41: 581–592.
183 Li, X., Cui, W., Deng, Q. et al. (2022). *Green Chem.* 24: 1302–1307.
184 Auria-Luna, F., Fernández-Moreira, V., Marqués-López, E. et al. (2020). *Sci. Rep.* 10: 11594.
185 Talha, A., Tachallait, H., Benhida, R., and Bougrin, K. (2021). *Tetrahedron Lett.* 81: 153366.
186 Olyaei, A., Moghadam, H.R., and Sadeghpour, M. (2020). *J. Heterocyclic Chem.* 57: 3029–3036.
187 Németh, A.G., Szabó, R., Domján, A. et al. (2021). *ChemistryOpen* 10: 16–27.

188 Cannalire, R., Santoro, F., Russo, C. et al. (2022). *Inorg. Au* 2: 66–74.

189 Gore, B.S., Lee, C.C., Lee, J., and Wang, J.-J. (2019). *Adv. Synth. Catal.* 361: 3373–3386.

190 Su, L., Ren, T., Dong, J. et al. (2019). *J. Am. Chem. Soc.* 141: 2535–2544.

191 Xu, J., He, L., Liang, C. et al. (2021). *ACS Sustainable Chem. Eng.* 9: 13663–13671.

192 Goswami, P., Cho, S.Y., Park, J.H. et al. (2022). *Nat. Commun.* 13: 2702.

193 Wang, L., Wang, H., Li, G. et al. (2018). *Adv. Synth. Catal.* 360: 4585–4593.

194 Zhu, F., Yin, P., and Zhang, P. (2022). *J. Org. Chem.* 88: 5153–5160.

195 Yu, S., Dai, L., Shao, Y. et al. (2020). *Org. Chem. Front.* 7: 3439–3445.

196 Lao, J., Li, J., and Zou, H. (2022). *Chem. Eur. J.* 28: e202202179.

197 Zhang, L., Ning, Y., Ye, B. et al. (2022). *Green Chem.* 24: 4420–4424.

198 Su, L., Xie, S., Dong, J. et al. (2022). *Org. Lett.* 24: 4569–4574.

199 Chen, P., Wang, K., Zhang, B. et al. (2019). *Chem. Commun.* 55: 12813–12816.

200 Liu, S., Zhou, Y., Sui, Y. et al. (2017). *Org. Chem. Front.* 4: 2175–2178.

201 Li, P., Jia, X., Ma, X. et al. (2019). *Asian J. Org. Chem.* 8: 2073–2091.

202 Litman, Z.C., Wang, Y., Zhao, H., and Hartwig, J.F. (2018). *Nature* 560: 355–359.

203 Cortes-Clerget, M., Akporji, N., Zhou, J. et al. (2019). *Nat. Commun.* 10: 2169.

204 Dander, J.E., Giroud, M., Racine, S. et al. (2019). *Commun. Chem.* 2: 82.

205 Hastings, C.J., Adams, N.P., Bushi, J., and Kolb, S.J. (2020). *Green Chem.* 22: 6187–6193.

206 Tan, Z., Zhang, X., Xu, M. et al. (2022). *Sci. Adv.* 8: eadd1912.

207 Botla, V., Pilli, N., and Malapaka, C. (2019). *Green Chem.* 21: 1735–1742.

208 Xie, L.-Y., Peng, S., Liu, F. et al. (2019). *ACS Sustain. Chem. Eng.* 7: 7193–7199.

209 Chen, C., Kang, J.-C., Mao, C. et al. (2019). *Green Chem.* 21: 4014–4019.

210 Liu, X., Li, B., Han, G. et al. (1868). *Nat. Commun.* 2021: 12.

211 Gutiérrez, S., Tomás-Gamasa, M., and Mascareñas, J.L. (2022). *Chem. Sci.* 13: 6478–6495.

212 Jana, S., Aseeva, P., and Koenigs, R.M. (2019). *Chem. Commun.* 55: 12825–12828.

213 Devi, L., Shukla, R., and Rastogi, N. (2019). *Org. Biomol. Chem.* 17: 135–139.

214 Zhang, R., Ni, C., Xie, Q., and Hu, J. (2020). *Tetrahedron* 76: 131676.

215 Yan, K., He, H., Li, J. et al. (2021). *Chin. Chem. Lett.* 32: 3984–3987.

216 Shi, M., Ye, N., Chen, W. et al. (2020). *Org. Process Res. Dev.* 24: 1543–1548.

217 Ng, X.Q., Kang, M.H., Toh, R.W. et al. (2022). *Green Synth. Catal.* 3: 272–277.

218 Nguyen, H.D., Tran, T.V., Taylor, C.R. et al. (2022). *ACS Sustain. Chem. Eng.* 10: 12968–12974.

219 Ke, F., Xu, Y., Zhu, S. et al. (2019). *Green Chem.* 21: 4329–4333.

220 Xu, Y., Xu, X., Wu, B. et al. (2020). *Asian J. Org. Chem.* 9: 1032–1035.

221 Wang, J., Hou, H., Hu, Y. et al. (2021). *Tetrahedron Lett.* 65: 152801.

222 Zhong, P., Wu, J., Wu, J. et al. (2022). *Tetrahedron Lett.* 107: 154099.

Index

a

acetamido-pyridine *N*-oxide 64
acetonitrile, reactions 51
N-acetylbenzoxazine 83
acid-catalyzed biomass conversions in GVL 271
acidic ionic liquids 123, 238
activation of DMSO enabled by single-electron transformation 151–162
β-acyl allylic methylsulfones 148, 154
S-acyl-and *N*-acylcysteines 88
N-acylbenzoxazides 83
N-(acyloxy)phthalimides 165
addition reaction 11–13, 28
  alkene/alkyne functionalization via radical addition 392–396
  alkene/alkyne functionalization via radical-free addition 396–399
  cycloaddition reactions 399–403
  nucleophilic addition 386–392
β-adrenergic blocking agents 219
alcohol dehydrogenase (ADH) 218, 423–425
alcohols
  dihydric 1
  fluorinated 1
  green solvents
    addition reaction 11–13
    condensation/ring condensation reaction 21–28
    coupling reaction 18–21
    cyclization reactions 13–18
    hydrogenation/reduction reaction 6–8
    oxidation reactions 8–10
    substitution reactions 10–11
  green solvents and catalysts
    addition reaction 28
    condensation reaction 30–35
    coupling reactions 30
    cyclization reactions 28–30
    green solvents and hydrogen donor 35–39
    metathesis reaction 35
  guidelines for solvents 3
  monohydric 1
  perfluoro *tert*-butyl 1
  physical and chemical properties of 2
  structure of 2
  tertiary 1
alicyclic carbonates and cyclic carbonates 268
alkene or alkyne functionalization via radical-free addition 396–399
alkenyl boronic esters 21
alkenyl C–H/aromatic C–H cross-coupling reaction 163
α-alkyl-acetophenone product 165
7-[4-alkyl-and (hetero)aryl-1*H*-1,2,3-triazol-1-yl]thio[3,2-b]pyridines 11
alkyl arylacetates 195
N-alkylated 1,4-diazabicyclo[2.2.2]octane/polyethylene glycol 219
2-alkylbenzoxazoles 152
alkylidene-cyclopropyl esters 80
5-alkylidene-thiazolones 227
N-alkyloxaziridines 195
γ-alkynoic acids 228–229
o-alkynylanilines 159
γ,*δ*-alkynyl oximes 15
"all-water" synthesis of *α*,*α*-difluoro-*β*-aminoketones 387–388
allylic alcohols 38, 175, 196, 228, 229, 334–335, 341, 377, 382–383, 427, 428
amide condensation reactions 431–435
1-amidoalkyl naphthols 226
1-amidoalkyl-2-naphthols 226

Green Solvents in Organic Synthesis, First Edition.
Edited by Xiao-Feng Wu, Zhiping Yin, Liang-nian He, and Feng Wang.

amino acid dithiocarbamate 29
5-amino-2,3-dihydro-7*H*-thiazolo[3,2-*a*] pyrimidin-7-one 25, 27
amino esters/amino acids 79
2-aminoimidazoles 223
β-amino ketones 138
α-amino nitrile derivatives 225
aminophenoxazinones 118
2-aminophenyl 155
8-aminoquinoline amides 209–210, 369–370
2-aminothiazoles 21–22
Anderson-type polyoxometallate 216
anionic polymerization of alkenes 231
annulated pyrroles 224–225
aromatic esters 19, 417–418
aromatic ketones 13, 300, 376
3-aroylimidazo[1,2-*a*]-*N*-heterocycles 144
N-aryl-*γ*-aminonitriles 85
2-aryl and 2-alkyl quinazolines 155
2-aryl aziridines 80
aryl boronic acids 19–20, 71, 212, 285, 291, 298
1-aryl-2,2-difluoroalkenes 191
N-arylethynylsulfonamides 176
N-arylhydroxylamine synthesis 8
4-aryl-NH-1,2,3-triazoles 14
4-aryl-NH-2,3,4-triazoles 14
2-aryloxy phenylacetylenes 95–96
aryloxy propanolamines 109
2-aryl pyridines and cyclic or acyclic ethers 333
N-arylpyrroles 22
4-aryl quinolines 140
2-aryl terephthalates 82
  via [3+3] cyclization 82
aryltrifluoroborates 203, 321, 322
arythioethers 73
assembly-promoted single electron transfer (APSET) 332
asymmetric copper-catalyzed Si–H insertion of 1-aryl-2,2,2-trifluoro-1-diazoethanes 189
asymmetric transfer hydrogenation (ATH) 37, 300, 379–380
azide-alkyne cycloaddition 296, 399–400
azomethine ylides 81

b
benzimidazole 24–25, 145, 256–257, 294–295, 306–307
benzo[g]-and dihydropyrano[2,3-g] chromene derivatives 85, 86
benzodiazepine 24, 25
benzodihydrofurans 74
benzofurans 76–77
benzoic acids 60–61, 213, 299–300, 359, 371
benzophosphole oxides 94
benzothiazoles 24–25, 76, 145, 152, 274, 307
benzotriazoles 82, 151, 295, 429, 432
benzoxazines 83, 96–97
benzoxazole 24, 25
N-benzyl-4-methylaniline 38
benzyne 1,2,4-trisubstitution and dearomatization 68
biaryl coupling of aryl iodides and organoboron reagents 186
bifunctional *N*-doped Co@C catalysts 37
Biginelli reaction using NBP 267
bimetallic Pd/Cu catalyst system 52
bio-based solvents
  cyclopentyl methyl ether (CPME) 117–120
  ethyl lactate (EL) 128–130
  eucalyptol (Euc) 130
  γ-valerolactone (GVL) 125–128
  glycerol 108–112
  2-methyltetrahydrofuran (2-MeTHF) 114–117
  organic carbonates 120–125
  polyethylene glycols (PEGs) 112–113
biomass-derived ethyl lactate (EL) 273
biomimetic aerobic oxidative desulfurization process 217
3,3′-bipyrazolo[1,5-*a*]pyridine derivatives 172
bis-enol derivatives 222
bis(4*H*-chromene)-and 4*H*-benzo[g] chromene-3,4-dicarboxylate derivatives 86
N-Boc-(prop-2-yn-1-yl)pyridine-2-amines 75
Buchwald reaction 336, 337

c
carbene insertion/transfer reactions 429–431
carbonylation reactions 114, 248–252, 253, 288
carbonylative Sonogashira reactions in PEG-2000 288
carbonyl cross-coupling 331
carboxylic acid synthesis 214
cascade aminooxygenation reaction 62

cascade annulation/alkylation of acetylenic oximes 243, 244
cascade annulation/aminocarbonylation reaction 252, 253
C–C activation 374–376
C–C bond couplings 201–210
C–C bond formation 292–293
  C–H acylation 354–356
  C–H alkenylation 352–354
  C–H alkylation 359–362
  C–H alkynylation 356
  C–H allylation 357, 358
  C–H arylation 347–351
  C–H benzylation 357–359
  C–H cyanation 356–357
  C–H fluoroalkylation 362–364
C–C cross-coupling
  carbonyl cross-coupling 331, 332
  cross-dehydrogenative coupling (CDC) 333–334
  cyanation reaction 330–331
  decarboxylative coupling 335
  dehydrative cross-coupling 334
  deprotonative cross-coupling 334
  Liebeskind–Srogl cross-coupling 322–323
  Mizoroki–Heck cross-coupling 326–328
  Murahashi coupling 323
  Negishi coupling 323–325
  reductive coupling 331–333
  Sonogashira coupling 328–330
  Stille coupling 325–326
  Suzuki–Miyaura reactions 318–322
  Tsuji–Trost type allylic arylation 330
C–C, C–Si bonds formation 90
cellulose-modified magnetite-graphene oxide nanocomposite 207
cetyltrimethyl ammonium hydrogensulfate (CTAS) 389
cetyltrimethylammonium bromide (CTAB) 322, 338, 351
cetyltrimethylammonium peroxodisulphate (CTAPS) 220
C–H acylation 354–356
C–H alkenylation 125–126, 352–354
C–H alkylation 171, 359–362
C–H alkynylation 356
C–H allylation 357
C-halogen bonds formation 98–102
C–H amination 83, 364–365
C–H annulation/cyclization 370–374
C–H arylation 126, 209, 272, 347–351, 374, 391
C–H benzylation 357–359
C–H cyanation 356
C–H difluoromethylation reactions 265
chemo-domino/tandem/cascade reactions
  catalyst-free 421–422
  organo-catalyzed 420–421
  transition metal-catalyzed 417–420
chemoenzymatic reactions 317, 422–425
C–H fluoroalkylation 362–364
C–H functionalization
  C–C bond formation 347–364
  C–H annulation/cyclization 370–374
  C–N bond formation 364–367
  C–O bond formation 367–369
C–H halogenation 369–370
1-chloromethyl-4-fluoro1,4-diazoniabicyclo[2.2.2]octane bis(tetrafluoroborate) 148
choline based-DES mediated Friedlander annulation 221
choline chloride/1,4-butanediol (ChCl/Bd) 219
C–H oxygenation 367–369
C–H sulfonamidation 365–367
C–N bond couplings 210–211
C–N bond formation 293–296
  C–H amination 364–365
  C–H sulfonamidation 365–367
C–N, C–P bond formation 93–96
C–N cross-coupling
  Buchwald reaction 336–337
  Tsuji–Trost N-Allylation 340–341
  Ullmann-type reactions 338–340
cobalt(II) bromide catalyzed allylation of alkyl halides 68
C–O bond couplings 211–212
C–O bond formation 211, 291
  C–H oxygenation 367–369
C–O cleavage reactions 376–377
C–O, C–S bond formation 96–98
condensation reaction 30–35
  catalyst/solvent system 221–223
  other catalytic system 223–224
condensation/ring condensation reaction 6, 21–28
conjugate addition 389–392
conventional organic solvents 4, 41, 123, 233, 263
copper-bisoxazoline-catalyzed allylic oxidation 56
copper-catalyzed azide–alkyne cycloaddition (CuAAC) 12, 399
copper-catalyzed carboamination of alkenes 94

copper-catalyzed Chan–Lam cross-coupling 73
copper-catalyzed coupling reactions 19, 257–259
copper-catalyzed oxidative coupling reaction of carboxylic acids 186
copper-mediated assisted tandem catalyzed double functionalization process 54
copper(I)-catalyzed asymmetric allylic oxidation 56
coumarinstyryl dyes 221
coupling reactions
- copper-catalyzed 19, 257–259
- green solvents and catalysts 18–21, 30

C–P cross-coupling 346–347
C–S bond couplings 212–213
C–S cross-coupling
- dehydrative 345
- Migita C–S cross-coupling 342–344
- oxidative C–S coupling 344
- radical 345–346
- reductive 344–345

cross-coupling of tetraphenylstannane and aromatic halides in PEG-400 289
cross-coupling reactions
- C–C bond couplings 201–210
- C–N bond couplings 210–211
- C–O bond couplings 211–212
- C–S bond couplings 212–213
- C–C cross-coupling
  - carbonyl cross-coupling 331
  - cross-dehydrogenative coupling (CDC) 333–334
  - cyanation reaction 330–331
  - decarboxylative coupling 335–336
  - dehydrative cross-coupling 334–335
  - deprotonative cross-coupling 334
  - Liebeskind–Srogl cross-coupling 322–323
  - Mizoroki–Heck cross-coupling 326–328
  - Murahashi coupling 323
  - Negishi coupling 323–325
  - reductive coupling 331–333
  - Sonogashira coupling 328–330
  - Stille coupling 325–326
  - Suzuki–Miyaura reactions 318–322
  - Tsuji–Trost type allylic arylation 330
- C–N cross-coupling 336–341
- C–P cross-coupling 346–347
- C–S cross-coupling 342–346
- in ionic liquids 238

cross-dehydrogenative coupling (CDC) 172, 333–334, 344, 345, 349
Csp3-H functionalization of ketones with alcohols 209
Cu catalyst/TBHP system 57
Cu-catalyzed diamination of styrenes 55
Cu-catalyzed intermolecular cycloaddition 76
Cu-catalyzed reactions
- C–C bond formation 292–293
- C–N bond formation 293–296
- C–O, C–S, C–Se bond formation 296–298

Cu/Fe catalyzed intermolecular oxidative amination 57
cyanation reaction 330–331
cyano-containing ketones 170
1-cyanoethylated 3,4-dihydronaphthalenes 90, 92
3-cyanoindoles 70, 71
cyanomethane 51
cyanomethylated benzoxazines 96
cyclic carbonates 41, 124, 191–192, 268
cyclic enol-lactones 228–229
cyclization/annulation reactions
- radical cyclization 406–410
- radical-free 403–406

cyclization reactions 13–18, 28–30
- five-membered ring synthesis 219–220
- six-membered ring synthesis 220–221

cycloaddition reactions
- azide-alkyne cycloaddition 399–400
- [4 + 2] cycloaddition 400–403

cyclodextrin 202
cyclopentadiene 190, 402
cyclopentyl methyl ether (CPME) 108, 117–121
Cyrene 263–266, 275

d

decarboxylative coupling 335–336
deep eutectic solvents (DESs)
- condensation reactions
  - catalyst/solvent system 221–223
  - other catalytic system 223–224
- cross-coupling reactions
  - C–C bond couplings 201–210
  - C–N bond couplings 210–211
  - C–O bond couplings 211–212
  - C–S bond couplings 212–213
- cyclization reactions
  - five-membered ring synthesis 219–220
  - six-membered ring synthesis 220–221
- multicomponent reactions

one-pot four-component reaction 227–228
one-pot three component reaction 224–227
organic reactions
esterification reaction 230–231
isomerization reaction 228–229
ring-opening reaction 230
oxidation reactions
metal-catalyzed oxidation 213–214
other oxidative processes 214–217
polymerization
anionic polymerization of alkenes 231
glycolysis and polyesterification 231
oxidative polymerization 232
visible-light-driven RAFT polymerization 232
reduction reactions
metal-catalyzed reduction 217–218
other catalytic reduction 218–219
dehydrative cross-coupling 334–335
dehydrative C–S cross-coupling 345
dehydroalanine (Dha)-containing peptides 390–391
deoxyfructosazine 223
deprotonative cross-coupling 334
2,5-diacylthiophenes 148
diamines 7, 54–55, 81
diamino dihydroquinazolines
derivatives 77
synthesis 78
β,*β*-diarylation of acrylates derivatives 19
diaryl ether motifs 211
1,1-diaryl-2-iodoethane 163
2,5-diarylpyrazines 215–216
diarylsulfides 73
1,4-diazabicyclo [2.2.2] octane (DABCO) 120, 170, 193, 205, 219, 271, 391
α-diazocarbonyls 189
1,3-dicarbonyl compounds 110–111, 119, 151, 227
di-μ-chlorobis[5-hydroxy-2-[1-(hydroxyimino-κN)ethyl]phenyl-κC]palladium(II) dimer 19
2,2-dichloroethenamines 84–85
Diels–Alder (D–A) reactions 25, 190, 220–221, 400–402
diethyl carbonate (DEC) 120–121, 269–270
N,*N*-diethylprop-2-yn-1-amine 240
1,2-difunctional ring-opening products 230
β,*β*-di(hetero)aryl-*α*,*α*-difluorinated ketones 174
dihydric alcohols 1
1,4-dihydropyridines 63
2,5-dihydropyridine derivatives 27
2,5-dihydropyridine skeleton 25
3,4-dihydropyrimidinothiones in EL 274
di(indolyl)methanes 32
2,5-dimethoxytetrahydrofuran 22
dimethyl carbonate (DMC) 270
alkyl arylacetates 195
N-alkyloxaziridines 195
1-aryl-2,2-difluoroalkenes 191
asymmetric copper-catalyzed Si–H insertion of 1-aryl-2,2,2-trifluoro-1-diazoethanes 189
biaryl coupling of aryl iodides and organoboron reagents 186
borrowing carbonate-enabled allylic amination reactions 196
copper-catalyzed oxidative coupling reaction of carboxylic acids 186
cyclic carbonates 192
cyclization between aliphatic and aromatic 1,4-bifunctional compounds 192
cyclopentadiene 190
α-diazocarbonyls 189
5-hydroxymethylfurfural 188
intramolecular hydroalkoxylation of double bonds 187
5-iodo-1,2,3-triazoles 190
iron-catalyzed hydrosilylation of dicarboxylic acids 191
iron-catalyzed methylation of secondary amines and imines 194
iron(II) caffeine-derived ionic salt catalyst 190
levulinic acid (LA) 193
lignin *β*-*O*-4 model compounds 194
malonate derivatives 192
$MnCO_3$-300 catalyzed transesterification of alcohols 196
$NaAlO_2$-catalyzed carboxymethylation 187
non-symmetrical alkyl carbonates 195
palladium-catalyzed carbonylative transformation of benzyl amines 187
palladium-catalyzed coupling of benzyl/allyl alcohols with malonates 194
silica-supported $FeCl_3$ catalyzed aminal synthesis 186
N-substituted cyclic amines 190
α-substituted homoallylamines 189

dimethyl carbonate (DMC) (*contd.*)
  sulfonimidates and sulfonimidamides 188
  1,5,7-triazabicyclo[4.4.0]dec-5-ene (TBD) 193
  1,4,5-trisubstituted 1,2,3-triazole-5-carboxamides 190
  Tsuji–Trost type reaction 194
  α,β-unsaturated N-acyl-oxazolidinones 190
dimethyl sulfoxide (DMSO)
  activation enabled by single-electron transformation 151–162
  acts as the metal ligand 171–174
  electrocatalytic synthesis enabled activation 163–164
  photocatalytic reaction enabled activation 164–171
  Pummerer-type activation 138–148
  Selectfluor-enabled activation 148–151
  special activation or usage of 174–181
1,8-dioxo-decahydroacridines 109
1,1-diphenylpropane-1,3-diol 179
dipolar aprotic solvents 107, 125, 265–267, 270, 275, 394
direct C-5 arylation of imidazole derivatives 209
directed *ortho*-metallation (DoM) 230
N,N'-disubstituted oxalamides 69
1,4-disubstituted pyrazoles 148

e
electrocatalytic synthesis enabled activation of DMSO 163–164
electrochemical oxidation 163, 377–379
electron-deficient polyfluoroarenes 76
electrophilic substitution 383–384
enol silyl ethers 93–94, 387, 388
(Z)-enynols 15
esterification reaction 88, 230–231
esters
  ethyl lactate 273–274
  γ-valerolactone (GVL) 270–272
  organic carbonates 268–270
2-(1-ethoxyethylideneamino) malonic acid dimethyl ester 81
ethylammonium nitrate (EAN) 237
ethyl 7-iodo-2-heptynoate 80
ethyl lactate (EL) 128–130, 268, 273–275
ethylene glycol 1–4, 6, 10, 13, 15–16, 25, 27, 112, 279
eucalyptol (Euc) 130

f
fac-Ir(ppy)$_3$ 165, 170
Fe-catalyzed reductive cross-coupling 332–333
FeI_2-catalyzed reductive Strecker reaction 53
Fe_3O_4@SiO_2-BT-Cu catalyzed O-arylation of phenols 212
Fischer indole synthesis 219–220
five-membered ring synthesis 219–220, 306–308
florfenicol phosphodiester 90
fluorinated alcohols 1, 5–6, 33, 35, 43, 45–46
fluoroalkylated (Z)-alkenes 169
α-fluoro-and α,α-difluoroimines 89–90
α-fluoromethyl selenoethers 116
fructosazine 223
functionalized pyrazine derivatives 215
functionalized pyrroles 228
functionalized quinolines 141

g
γ-valerolactone (GVL) 108, 111, 125–128, 254, 268, 270–272
G2/Co for Suzuki coupling reaction 204
D-glucosamine 223
glycerol 1, 4, 6, 9–11, 13–20, 24–26, 28–39, 45–46, 57, 108–112, 207, 209
glycolysis and polyesterification 231–232
gold(I) N-heterocyclic carbene complex 11
green solvents
  addition reaction 11–13
  and catalysts
    addition reaction 28
    condensation reaction 30–35
    coupling reactions 30
    cyclization reactions 28–30
    metathesis reaction 35
  condensation/ring condensation reaction 21–28
  coupling reaction 18–21
  cyclization reactions 13–18
  hydrogenation/reduction reaction 6
  and hydrogen donor 35–39
  oxidation reactions 8–10
  substitution reactions 10–11

h
Hansen solubility parameters (HSP) 264
2H-azirines 98, 117–118
4H-3,1-benzoxazin-4-one skeleton 114

Heck reaction 79, 266–267, 269–270, 281, 327–328
2-(hetero)aryl THF and THP derivatives 205
2,5-hexanedione 22
highly functionalized benzo[g]-and dihydropyrano[2,3-g]chromene derivatives 85
highly functionalized polysubstituted quinolines 220–221
2*H*-indazoles 154–155
Hiyama cross-coupling in PEG 289
Hiyama reaction 18, 126, 207–208, 271–272
HLADH-catalyzed cinnamaldehyde reduction 218
4*H*-pyrans 32–33, 412
1*H*-tetrazoles 30, 219, 227
hydrodehalogenation of aryl chlorides and aryl bromides 66
hydrogen bond acceptors (HBAs) 200–201
hydrogenation/reduction reaction 6
1,5-hydrogen atom transfer process 167, 169, 350
hydrophilic palladium(II) *cis*-[$PdCl_2$ $(DAPTA)_2$] 14
α-hydroxyacetophenone 16, 17, 110, 111
2-hydroxyacetophenone (HAP) 219
3-hydroxylpyridines 15
5-hydroxymethylfurfural 188, 331
4-hydroxy-1-methyl-2-quinolone 25
α-and *β*-hydroxyphosphine oxides 216

i
IBX promoted primary amine oxidative coupling 215
IBX-facilitated primary amine oxidative coupling reaction 214
imidazopyridines 97–98, 151
2-(imidazo[1,2-*α*]pyridine-3-yl)-*N*,*N*-dialkylacetamides 74
indium-mediated Reformatsky reaction of iododifluoroketones with aldehydes or isatins 388, 389
intramolecular hydroalkoxylation of double bonds 186, 187
β-iodoalkoxysulfonium ion 163
o-iodoanilines 145
5-iodo-1,2,3-triazoles 190
ionic liquids (ILs) 200
  molecular liquids 237
  physicochemical properties of 237
  transition-metal-catalyzed reactions
    carbonylation reactions 248–252
    copper-catalyzed coupling reactions 257–259
    palladium-catalyzed cascade cyclization reactions 239–248
    Sonogashira coupling reactions 252–255
    Suzuki coupling reactions 255–257
ionic liquids-accelerated oxyarylation 246–247
iron-catalyzed hydrosilylation of dicarboxylic acids 190–191
iron-catalyzed methylation of secondary amines and imines 194
iron-catalyzed reductive Strecker reaction 54
iron(II) caffeine-derived ionic salt catalyst 190
iron(II) cyclam complex 59
isomerization reaction 228–229
2,3-isopropylideneglycerol 61
isoquinolino[1,2-*b*]quinazolines 127–128

k
ketones
  Cyrene 263–265
  in NBP 266–268
α-keto-imides 176–177
γ-ketonitriles 90–91
3-ketoquinoline 141
Knoevenagel condensation 59, 222–224, 303
Kornblum oxidation reaction 166
$K_2S_2O_8$-promoted Pummerer-type reaction 138

l
levulinic acid (LA) 114, 125, 193
Lewis acid catalyzed [3+2] cycloaddition 81
Liebeskind–Srogl cross-coupling 322
ligand-free Cu(I) or Cu(II)-catalyzed C–O coupling reactions 212
lignin *β*-O-4 model compounds 194
liquid PEG 4
lithium phosphide 216
low molecular weight PEGs 4, 5, 41

m
magnetic Pd-nanocomposite mediated Sonogashira reaction 207
magnetic reduced graphene oxide-supported palladium catalyst 209

Maitland–Japp reaction using NBP 267–268
malonate derivatives 192
Menschutkin reaction using Cyrene 266
mesoionic *N*-heterocyclic carbene (NHC)-Pd complex 207
metal-catalyzed oxidation 56–64, 213–214
metal-catalyzed reduction 217–218
metathesis reaction 35, 259
methane carbonitrile (MeCN)
- C–X bonds formation with radicals involved
  - C–C, C–Si bonds formation 90–93
  - C-halogen bonds formation 98–102
  - C–N, C–P bonds formation 93–96
  - C–O, C–S bonds formation 96–98
- transition-metal catalyzed reactions without radicals involved
  - addition reactions 52–55
  - oxidation reactions 56–64
  - reduction reactions 64–66
  - substitution reactions 66–73
  - transition-metal catalyzed cyclization reactions 74–80
- transition-metal-free catalyzed reactions without radicals involved
  - C–X bonds formation 87–90
  - cyclization reactions 80–84
  - multi-component reactions 84–87

N-methylation/*N*-formylation 67
2-methylbenzoic acid 57–58
methyl α-cinnamide 217
N-methylindole 110–111
O-methyloximes 240, 245–246, 251
N-methylpyrrole 110
2-methyltetrahydrofuran (2-MeTHF) 108, 114–117
Micellar-enabled aqueous reactions 318
microwave-assisted Suzuki–Miyaura coupling in PEG-400 284
Minisci-type reaction 159
Mizoroki–Heck cross-coupling 204, 205, 326–328
$MnCO_3$-300 catalyzed transesterification of alcohols 196
mono-and difluoroimines 89
monobromination of phenols and anisoles 87
monohydric alcohols 1, 3
multicomponent domino reactions 228
multicomponent reactions (MCRs) 304–306, 410–417
- one-pot four-component reaction 227–228
- one-pot three component reaction 224–227

Murahashi coupling 323

n

$NaAlO_2$-catalyzed carboxymethylation 187
2-naphthol 4-hydroxy-6-methyl-2-pyrone, 25
1,8-naphthyridine derivatives 159
NCN-Pd pincer catalyst for Hiyama reaction 207
Negishi coupling 323–325
Negishi cross-coupling reactions 208
NHC-Pd catalyzed cascade annulation/allylation of acetylenic oximes 244
N-heterocyclic carbene (NHC) palladium polymers 284
Ni, Ru, Pt-catalyzed reactions 299–302
Ni-catalyzed C3-cyanation of *N*-aryl indoles 71
Ni-catalyzed cross-electrophile-coupling 70
nickel(II)-catalyzed *6-exo-dig* cyclization 404
nickel-catalyzed azide-alkyne cycloaddition (NiAAC) 399, 400
$NiCl_2$/2,2′-bipyridine-catalyzed synthesis 73
nitrile, defined 51
nitroarenes, reduction of 65
N-methyl-pyrrolidinone (NMP) 107, 114, 125, 264–269, 272, 275
non-symmetrical alkyl carbonates 195–196
n-substituted-4-aryl quinolines 140
nucleophilic addition
- addition reactions with ketone/imine/nitrile 386–389
- conjugate addition 389–392

nucleophilic substitution 175, 381–383
NZSM-5 zeolite catalyzed synthesis 195–196

o

oligoethylene glycols (OEGs) 7
one-pot four-component reaction 227–228, 412–413
one-pot Mizoroki-Heck cross-coupling/reduction 205
one-pot O_2-oxidation-HWE (Horner–Wadsworth–Emmons) reaction 61

one-pot three component reaction 27, 28, 33, 224–227, 410
organic carbonates 108, 119, 120–125, 268–270, 275
organic esters 230
organic reactions
  esterification reaction 230–231
  in esters
    γ-valerolactone (GVL) 270–272
    ethyl lactate 273–274
    organic carbonates 268–270
  isomerization reaction 228–229
  in ketones
    Cyrene 263–266
    in NBP 266–268
  ring-opening reaction 230
organolithium promoted anionic polymerization of alkenes 231
organosilicon reagents 207
2-organylselanyl pyridines 11
3-oxabicyclo[3.1.0]hexan-2-ones 80–81
oxidation of benzaldehyde 215
oxidation of ethylbenzene 42, 58
oxidation of toluene 57, 213
oxidation reactions 8–10
  metal-catalyzed oxidation 213–214
  other oxidative processes 214–217
oxidative polymerization 232
1-oxo-hexahydroxanthenes 32

p

Paal–Knorr reaction 219
palladium-catalyzed reactions
  allylative dearomatisation of naphthyl halides 114–115
  α-arylation of (hetero)aryl ketones 334
  carbonylative transformation of benzyl amines 187
  cascade cyclization reactions 239–248
    alkylation of oxime ethers 245
  cascade of alkynoic acids with unactivated alkenes 248
  C–C bond formation 207
  C–H arylation of thiophenes 272
  coupling of benzyl/allyl alcohols with malonates 194
  C–P bonds formation in PEG-600 291–292
  deoxygenation reaction 64
  formal [4+2] cycloaddition 172
  Heck cross-coupling of aromatic bromides 281
  ionic liquids-accelerated oxyarylation 246
  oxidative cross-trimerization 63
  Sonogashira reaction 206
  4-sulfenylisoxazoles synthesis 243
  Suzuki–Miyaura coupling of aromatic chlorides in PEG-300 286
  synthesis of tetrahydrofuroindoles 55
  tandem dehydrogenative[4 + 2] annulation 172
Pd-catalyzed coupling reactions
  C–B bonds formation 291–292
  C–C, C–Si bonds formation 280–290
  C–N, C–P bonds formation 290–291
  C–O bonds formation 291
Pd-catalyzed cross-coupling reaction 264
  in GVL 272
  in ILs 238
Pd/β-cyclodextrin cross-linked catalyst (Pd/CβCAT) 127
Pd/Cu-catalyzed diboration of propargylic alcohols 53
PdI_2-catalyzed carbonylation 75
Pd(II)/Cu(II) catalyzed synthesis 159
penta-2,4-dien-1-yl moiety 66
perfluoro *tert*-butyl alcohol 1
persulfate/peroxide-mediated radical cyclization 406
phenacyl azides 215
phenol, synthesis of 60
(*R*)-1-phenyl-1,2-ethanediol 219
(*R*)-1-phenyl-1,2-ethanol (PED) 219
2-phenylimidazo[1,2-*a*] pyridine 138, 344
phosphorylated pyrrolines 95
photo-induced decarboxylation of silacarboxylic acids 93
photo-induced radical cyclization 408–410
photocatalytic reaction enabled activation of DMSO 164–171
Pictet–Spengler dehydrogenative cyclization 221
poly(ethylene 2,5-furandicarboxylate) (PEF) 231–232
polycyclic spiroindolines 84
polyethylene glycol (PEG) 112–113
  in Cu-catalyzed reactions
    C–C bond formation 292–293
    C–N bond formation 293–296
    C–O, C–S, C–Se bond formation 296–298
  in cyclization reactions
    Five-Membered Ring Systems 306–308

polyethylene glycol (PEG) (*contd.*)
six and seven-membered ring systems 308–309
multicomponent reactions (MCRs) 304–306
in Ni, Ru, Pt-catalyzed reactions 299–302
organic reactions in 280
in organocatalysis reactions 302–304
in Pd-catalyzed coupling reactions
C–B bonds formation 291–292
C–C, C–Si bonds formation 280–290
C–N, C–P bonds formation 290–291
C–O bonds formation 291
radical-initiated oxidation reactions 41–43
as solvent for CO_2 capture and conversion 39–41
polyfunctional furans 145
polyhydroquinolines 226
polymerization
anionic polymerization of alkenes 231
glycolysis and polyesterification 231–232
oxidative polymerization 232
visible-light-driven RAFT polymerization 232–233
polypyrrole (PPy)/Cu(II) nanocomposite 14
polystyrene-stabilized iridium (Ir@PS) nanoparticles 7
L-proline-catalyzed Knoevenagel condensation 224
propargylamines 86–87, 402, 412
(*E*)-prop-1-en-1-ylbenzene 160
2-propylquinolines 17–18
Pummerer-type activation of DMSO 138–148
Pummerer-type mechanism 145
pyranoid heterocyclic derivatives 32
pyrazine derivatives 215
pyrazolo[1,5-*a*]pyridines 172
pyridines, synthesis of 75
pyrido[2,3-*d*]pyrimidine 111
pyrrolidinedione-fused hexahydropyrrolo [2,1-*a*]isoquinolines 79
pyrrolo[1,2-*a*]quinolines 121
pyrrolo[3,2-*c*]pyridines 22

q
quinazoline derivatives 155
quinoline derivatives 7, 224–225
quinoxalines 24–25, 308, 379

r
radical C–S coupling 345–346
radical cyclization
persulfate/peroxide-mediated 406–408
photo-induced 408–410
radical-relayed cyclization process 166
radical substitution 384–385
rearrangement reactions 15, 425–428
reduction reactions
metal-catalyzed reduction 217–218
other catalytic reduction 218–219
reductive C–S coupling 344–345
reductive coupling 331–333
reductive desulfurization process 217–218
reversible addition-fragmentation chain transfer (RAFT) polymerization 232–233
Rh-catalyzed asymmetric hydrogenation 269
Rh-catalyzed hydroformylation and hydrogenation 65
rhodanine derivatives 223, 227
ring-closing metathesis (RCM) 325, 376, 424, 425
ring-opening reaction 5–6, 43–45, 141, 170, 230

s
$Sc(OTf)_3$-catalyzed [4+2] annulation reaction 16, 17
1-selanylindolizines 16, 17
Selectfluor-enabled activation of DMSO 148–151
Selectfluor-mediated Pummerer-type reaction 151
short-chain aliphatic alcohols 1
silica-supported $FeCl_3$ catalyzed aminal synthesis 186
silicon heterocycle synthesis 92
six-membered ring synthesis 220–221
sodium lauryl oligoethylene glycol sulfate (SLES) 332
solid palladium nanoparticle 19
solid-phase peptide synthesis (SPPS) 267–268
Sonogashira coupling reactions 206–207, 252–255, 264, 287–288, 328–330
Sonogashira reactions in PEG-400 287
stereoselective Michael addition reaction 224
Stille coupling 288–289, 325–326
Stille cross-coupling reaction 208, 288, 326

3-substituted-1-aryl-1*H*-pyrazolo-[3,4-b] quinoline 141
substituted benzo[*d*]imidazo[5,1-*b*] thiazol-1-amine derivatives 75
N-substituted cyclic amines 190
α-substituted homoallylamines 189–190
2-substituted indoles 83–84
4-substituted, 3-substituted, or 3,4-disubstituted quinolines 138
5-substituted 1*H*-tetrazoles 30, 227
substitution reactions 10–11
  electrophilic substitution 383–384
  nucleophilic substitution 381–383
  radical substitution 384–385
1-sulfanyl-and 1-selanylindolizines 17
1-sulfanylindolizines 16
N-sulfonyl-2-aryloxoacetamides 176
3-sulfonyl benzofurans and indoles 79
(*E*)-*β*-sulfonyl enamines 98–99
sulfonamides 163–164, 225, 412–413
sulfonimidates and sulfonimidamides 188
Suzuki coupling reactions 204, 255–257, 273, 320
Suzuki–Miyaura (SM) coupling
  amides via C–N cleavage 322
  of aromatic halides
    carboxylic anhydride, and acyl chlorides in PEG 285
    in PEG-400 287
  arylboronic acids and aryl halides 320
  BaryPhos-facilitated asymmetric 321
  cross-coupling reactions in aqueous EL 273
  DNA-compatible 320
  electron-poor aryl chlorides 320
  EvanPhos-bearing *N,C*-diisopropyl biarylpalladacycle 318
  HandaPhos-containing palladacycle 319
  Pd/NHC-catalyzed 321
  reaction 202, 284
  reaction using Cyrene 265
symmetrical 2,5-diarylpyrazines 215
symmetrical methylene-bridged imidazoheterocycles 138

t

[3 + 3] tandem annulation–oxidation approach 214
t-BuOK/DMSO/air reaction system 161
TCCA induced chlorine radical cascade chlorination 101
C-terminal decarboxylative conjugation addition 389, 390
tertiary alcohols 1, 92, 321, 350, 375, 386, 388, 407, 429
tertiary alkyl halides 69, 99, 332
tetrabutylammonium bromide (TBAB) 63, 192, 255, 285–286, 342, 368, 432–433
tetrahydrofuroindoles 55
2,3,6,7-tetrahydro-1*H*-pyrrolo[3,2-c] pyridin-4(5*H*)-ones 22
tetrahydroisoquinoline compounds 113, 172, 213–214, 377, 378
2,2,6,6-tetramethylpiper-idine-1-oxy 43, 61
tetrasubstituted *β*-sulfonyl *N*-unprotected enamines 97
2-tetrazolyl-substituted 3-acylpyrroles 61
thiazolopyrimidine derivatives 25
thioamides 153
thiophene-aryl coupling 208, 209
thiophosphates 96–97
Z-thiovinyl sulfone 88–89
three-component domino reaction 226
toluene oxidation by hydrogen peroxide 213
p-toluene sulfonic acid 216
o-tolyl tetrahydrofuran derivatives 230
toxic nonrenewable fossil-based solvents 107
transition-metal-catalyzed reactions
  addition reactions 52–55
  carbonylation reactions 114, 248–252
  copper-catalyzed coupling reactions 257–259
  cyclization reactions 74–79
  oxidation reactions 56–64
  palladium-catalyzed cascade cyclization reactions 239–248
  reduction reactions 64–66
  Sonogashira coupling reactions 252–255
  substitution reactions 66–73
  Suzuki coupling reactions 255–257
transition-metal-free catalyzed reactions
  C–X bonds formation 87–90
  cyclization reactions 80–84
  multi-component reactions 84–87
trans-resveratrol 228
4,5-trans-1,5-scis-3-oxabicyclo[3.1.0]hex-2-ones 81
2,4,5-triaryl and 1,2,4,5-tetraaryl imidazole derivatives 33, 34
1,5,7-triazabicyclo[4.4.0]dec-5-ene (TBD) 193
1,2,3-triazole in PEG 282, 283
2,2,2-trichloroethanamines 84–85
trichloroisocyanuric acid (TCCA) 101
triethylene glycol (3-OEG) 7

α-trifluoromethylated ketones 166
trifluoromethyl tertiary alcohols 92
2,4,6-trihydroxyphenanthrene 229
2-(trimethylsilyl) aryltriflates 82
2,3,4-trisubstituted imidazole derivatives 23
trisubstituted nicotinonitriles 76, 77
1,4,5-trisubstituted 1,2,3-triazole-5-carboxamides 190
L-tryptophan methyl ester and aldehydes 34
Tsuji–Trost type reaction 194
  allylic arylation 330

u

Ullmann amine synthesis 211
Ullmann homocoupling reactions 283–284
Ullmann-type reactions 338–340
Z-α,β-unsaturated esters 61
α,β-unsaturated ketones 81, 141, 214, 359–360, 415
α,β-unsaturated *N*-acyl-oxazolidinones 190
α,β-unsaturated nitriles 59, 60
2-unsubstituted benzothiazoles 145
unsymmetrical thioethers 71, 72

v

visible-light-driven RAFT polymerization 232–233

w

water as solvent
  addition reactions
    alkene/alkyne functionalization via radical addition 392–396
    alkene/alkyne functionalization via radical-free addition 396–399
    cycloaddition reactions 399–403
    nucleophilic addition 386–392
  amide condensation reactions 431–435
  carbene insertion/transfer reactions 429–431
  C–C activation 374–376
  C–H functionalization
    C–C bond formation 347–364
    C–H annulation/cyclization 370–374
    C–N bond formation 364–367
    C–O bond formation 367–369
    C–X bond formation 369–370
  C–O cleavage reactions 376–377
  cross-coupling reactions
    C–C cross-coupling 318–336
    C–N cross-coupling 336–341
    C–P cross-coupling 346–347
    C–S cross-coupling 342–346
  cyclization/annulation reactions
    radical cyclization 406–409
    radical-free 403–406
  domino/tandem/cascade reactions
    chemo 417–422
    chemzoenzymatic reactions 422–425
  electrochemical oxidation 377–379
  multicomponent reactions (MCRs) 410–417
  rearrangement reactions 425–428
  reduction and related reactions 379–381
  substitution reactions
    electrophilic substitution 383–384
    nucleophilic substitution 381–383
    radical substitution 384–385
water-soluble hypervalent iodine(III) reagent (phenyliodonio)sulfamate 83
water-soluble PEG 4

x

xanthene-1,8(*2H*)-diones 32